V.9 Grzimek, Bernhard.
Grzimek's Animal
life encyclopedia.

WITHDRAWN

Grzimek's ANIMAL LIFE ENCYCLOPEDIA

Volume 1
LOWER ANIMALS

Volume 2
INSECTS

Volume 3
MOLLUSKS AND ECHINODERMS

Volume 4
FISHES I

Volume 5
FISHES II AND AMPHIBIA

Volume 6
REPTILES

Volume 7
BIRDS I

Volume 8
BIRDS II

Volume 9
BIRDS III

Volume 10
MAMMALS I

Volume 11
MAMMALS II

Volume 12
MAMMALS III

Volume 13
MAMMALS IV

Grzimek's ANIMAL LIFE ENCYCLOPEDIA

Editor-in-Chief

Dr. Dr. h.c. Bernhard Grzimek

Professor, Justus Liebig University of Giessen
Director, Frankfurt Zoological Garden, Germany
Trustee, Tanzania National Parks, Tanzania

 VAN NOSTRAND REINHOLD COMPANY

New York   Cincinnati   Toronto   London   Melbourne

First published in paperback in 1984

Copyright © 1968 Kindler Verlag A.G. Zurich

Library of Congress Catalog Card Number 79-183178

ISBN 0-442-23043-5

All rights reserved. No part of this work covered by
the copyright hereon may be reproduced or used in any
form or by any means—graphic, electronic, or mechanical,
including photocopying, recording, taping, or information
storage and retrieval systems—without written
permission of the publisher.

Printed in Federal Republic of Germany

Van Nostrand Reinhold Company Inc.
135 West 50th Street
New York, New York 10020

Van Nostrand Reinhold Company Limited
Molly Millars Lane
Wokingham, Berkshire RG11 2PY, England

Van Nostrand Reinhold
480 Latrobe Street
Melbourne, Victoria 3000, Australia

Macmillan of Canada
Division of Gage Publishing Limited
164 Commander Boulevard
Agincourt, Ontario M1S 3C7 Canada

16 15 14 13 12 11 10 9 8 7 6 5 4 3 2 1

EDITORS AND CONTRIBUTORS

Editor-in-Chief
DR. DR. H.C. BERNHARD GRZIMEK
Professor, Justus Liebig University of Giessen, Germany
Director, Frankfurt Zoological Garden, Germany
Trustee, Tanzania and Uganda National Parks, East Africa

DR. MICHAEL ABS Curator, Ruhr University	BOCHUM, GERMANY
DR. SALIM ALI Bombay Natural History Society	BOMBAY, INDIA
DR. RUDOLF ALTEVOGT Professor, Zoological Institute, University of Münster	MÜNSTER, GERMANY
DR. RENATE ANGERMANN Curator, Institute of Zoology, Humboldt University	BERLIN, GERMANY
EDWARD A. ARMSTRONG Cambridge University	CAMBRIDGE, ENGLAND
DR. PETER AX Professor, Second Zoological Institute and Museum, University of Göttingen	GÖTTINGEN, GERMANY
DR. FRANZ BACHMAIER Zoological Collection of the State of Bavaria	MUNICH, GERMANY
DR. PEDRU BANARESCU Academy of the Roumanian Socialist Republic, Trajan Savulescu Institute of Biology	BUCHAREST, RUMANIA
DR. A. G. BANNIKOW Professor, Institute of Veterinary Medicine	MOSCOW, U.S.S.R.
DR. HILDE BAUMGÄRTNER Zoological Collection of the State of Bavaria	MUNICH, GERMANY
C. W. BENSON Department of Zoology, Cambridge University	CAMBRIDGE, ENGLAND
DR. ANDREW BERGER Chairman, Department of Zoology, University of Hawaii	HONOLULU, HAWAII, U.S.A.
DR. J. BERLIOZ National Museum of Natural History	PARIS, FRANCE
DR. RUDOLF BERNDT Director, Institute for Population Ecology, Heligoland Ornithological Station	BRAUNSCHWEIG, GERMANY
DIETER BLUME Instructor of Biology, Freiherr-vom-Stein School	GLADENBACH, GERMANY
DR. MAXIMILIAN BOECKER Zoological Research Institute and A. Koenig Museum	BONN, GERMANY
DR. CARL-HEINZ BRANDES Curator and Director, The Aquarium, Overseas Museum	BREMEN, GERMANY
DR. DONALD G. BROADLEY Curator, Umtali Museum	UMTALI, RHODESIA
DR. HEINZ BRÜLL Director; Game, Forest, and Fields Research Station	HARTENHOLM, GERMANY
DR. HERBERT BRUNS Director, Institute of Zoology and the Protection of Life	SCHLANGENBAD, GERMANY
HANS BUB Heligoland Ornithological Station	WILHELMSHAVEN, GERMANY
A. H. CHRISHOLM	SYDNEY, AUSTRALIA
HERBERT THOMAS CONDON Curator of Birds, South Australian Museum	ADELAIDE, AUSTRALIA
DR. EBERHARD CURIO Director, Laboratory of Ethology, Ruhr University	BOCHUM, GERMANY

DR. SERGE DAAN Laboratory of Animal Physiology, University of Amsterdam	AMSTERDAM, THE NETHERLANDS
DR. HEINRICH DATHE Professor and Director, Animal Park and Zoological Research Station, German Academy of Sciences	BERLIN, GERMANY
DR. WOLFGANG DIERL Zoological Collection of the State of Bavaria	MUNICH, GERMANY
DR. FRITZ DIETERLEN Zoological Research Institute, A. Koenig Museum	BONN, GERMANY
DR. ROLF DIRCKSEN Professor, Pedagogical Institute	BIELEFELD, GERMANY
JOSEF DONNER Instructor of Biology	KATZELSDORF, AUSTRIA
DR. JEAN DORST Professor, National Museum of Natural History	PARIS, FRANCE
DR. GERTI DÜCKER Professor and Chief Curator, Zoological Institute, University of Münster	MÜNSTER, GERMANY
DR. MICHAEL DZWILLO Zoological Institute and Museum, University of Hamburg	HAMBURG, GERMANY
DR. IRENÄUS EIBL-EIBESFELDT Professor and Director, Institute of Human Ethology, Max Planck Institute for Behavioral Physiology	PERCHA/STARNBERG, GERMANY
DR. MARTIN EISENTRAUT Professor and Director, Zoological Research Institute and A. Koenig Museum	BONN, GERMANY
DR. EBERHARD ERNST Swiss Tropical Institute	BASEL, SWITZERLAND
R. D. ETCHECOPAR Director, National Museum of Natural History	PARIS, FRANCE
DR. R. A. FALLA Director, Dominion Museum	WELLINGTON, NEW ZEALAND
DR. HUBERT FECHTER Curator, Lower Animals, Zoological Collection of the State of Bavaria	MUNICH, GERMANY
DR. WALTER FIEDLER Docent, University of Vienna, and Director, Schönbrunn Zoo	VIENNA, AUSTRIA
WOLFGANG FISCHER Inspector of Animals, Animal Park	BERLIN, GERMANY
DR. C. A. FLEMING Geological Survey Department of Scientific and Industrial Research	LOWER HUTT, NEW ZEALAND
DR. HANS FRÄDRICH Zoological Garden	BERLIN, GERMANY
DR. HANS-ALBRECHT FREYE Professor and Director, Biological Institute of the Medical School	HALLE A.D.S., GERMANY
GÜNTHER E. FREYTAG Former Director, Reptile and Amphibian Collection, Museum of Cultural History in Magdeburg	BERLIN, GERMANY
DR. HERBERT FRIEDMANN Director, Los Angeles County Museum of Natural History	LOS ANGELES, CALIFORNIA, U.S.A.
DR. H. FRIEDRICH Professor, Overseas Museum	BREMEN, GERMANY
DR. JAN FRIJLINK Zoological Laboratory, University of Amsterdam	AMSTERDAM, THE NETHERLANDS
DR. DR. H.C. KARL VON FRISCH Professor Emeritus and former Director, Zoological Institute, University of Munich	MUNICH, GERMANY
DR. H. J. FRITH C.S.I.R.O. Research Institute	CANBERRA, AUSTRALIA
DR. ION E. FUHN Academy of the Roumanian Socialist Republic, Trajan Savulescu Institute of Biology	BUCHAREST, RUMANIA
DR. CARL GANS Professor, Department of Biology, State University of New York at Buffalo	BUFFALO, NEW YORK, U.S.A.
DR. RUDOLF GEIGY Professor and Director, Swiss Tropical Institute	BASEL, SWITZERLAND

DR. JACQUES GERY		ST. GENIES, FRANCE
DR. WOLFGANG GEWALT Director, Animal Park		DUISBURG, GERMANY
DR. DR. H.C. DR. H.C. VIKTOR GOERTTLER Professor Emeritus, University of Jena		JENA, GERMANY
DR. FRIEDRICH GOETHE Director, Institute of Ornithology, Heligoland Ornithological Station		WILHELMSHAVEN, GERMANY
DR. ULRICH F. GRUBER Herpetological Section, Zoological Research Institute and A. Koenig Museum		BONN, GERMANY
DR. H. R. HAEFELFINGER Museum of Natural History		BASEL, SWITZERLAND
DR. THEODOR HALTENORTH Director, Mammalology, Zoological Collection of the State of Bavaria		MUNICH, GERMANY
BARBARA HARRISSON Sarawak Museum, Kuching, Borneo		ITHACA, NEW YORK, U.S.A.
DR. FRANCOIS HAVERSCHMIDT President, High Court (retired)		PARAMARIBO, SURINAM
DR. HEINZ HECK Director, Catskill Game Farm		CATSKILL, NEW YORK, U.S.A.
DR. LUTZ HECK Professor (retired), and Director, Zoological Garden, Berlin		WIESBADEN, GERMANY
DR. DR. H.C. HEINI HEDIGER Director, Zoological Garden		ZURICH, SWITZERLAND
DR. DIETRICH HEINEMANN Director, Zoological Garden, Münster		DÖRNIGHEIM, GERMANY
DR. HELMUT HEMMER Institute for Physiological Zoology, University of Mainz		MAINZ, GERMANY
DR. W. G. HEPTNER Professor, Zoological Museum, University of Moscow		MOSCOW, U.S.S.R.
DR. KONRAD HERTER Professor Emeritus and Director (retired), Zoological Institute, Free University of Berlin		BERLIN, GERMANY
DR. HANS RUDOLF HEUSSER Zoological Museum, University of Zurich		ZURICH, SWITZERLAND
DR. EMIL OTTO HÖHN Associate Professor of Physiology, University of Alberta		EDMONTON, CANADA
DR. W. HOHORST Professor and Director, Parasitological Institute, Farbwerke Hoechst A.G.		FRANKFURT-HÖCHST, GERMANY
DR. FOLKHART HÜCKINGHAUS Director, Senckenbergische Anatomy, University of Frankfurt a.M.		FRANKFURT A.M., GERMANY
FRANCOIS HÜE National Museum of Natural History		PARIS, FRANCE
DR. K. IMMELMANN Professor, Zoological Institute, Technical University of Braunschweig		BRAUNSCHWEIG, GERMANY
DR. JUNICHIRO ITANI Kyoto University		KYOTO, JAPAN
DR. RICHARD F. JOHNSTON Professor of Zoology, University of Kansas		LAWRENCE, KANSAS, U.S.A.
OTTO JOST Oberstudienrat, Freiherr-vom-Stein Gymnasium		FULDA, GERMANY
DR. PAUL KÄHSBAUER Curator, Fishes, Museum of Natural History		VIENNA, AUSTRIA
DR. LUDWIG KARBE Zoological State Institute and Museum		HAMBURG, GERMANY
DR. N. N. KARTASCHEW Docent, Department of Biology, Lomonossow State University		MOSCOW, U.S.S.R.
DR. WERNER KÄSTLE Oberstudienrat, Gisela Gymnasium		MUNICH, GERMANY
DR. REINHARD KAUFMANN Field Station of the Tropical Institute, Justus Liebig University, Giessen, Germany		SANTA MARTA, COLOMBIA

DR. MASAO KAWAI
Primate Research Institute, Kyoto University — KYOTO, JAPAN

DR. ERNST F. KILIAN
Professor, Giessen University and Catedratico Universidad Austral, Valdivia-Chile — GIESSEN, GERMANY

DR. RAGNAR KINZELBACH
Institute for General Zoology, University of Mainz — MAINZ, GERMANY

DR. HEINRICH KIRCHNER
Landwirtschaftsrat (retired) — BAD OLDESLOE, GERMANY

DR. ROSL KIRCHSHOFER
Zoological Garden, University of Frankfurt a.M. — FRANKFURT A.M., GERMANY

DR. WOLFGANG KLAUSEWITZ
Curator, Senckenberg Nature Museum and Research Institute — FRANKFURT A.M., GERMANY

DR. KONRAD KLEMMER
Curator, Senckenberg Nature Museum and Research Institute — FRANKFURT A.M., GERMANY

DR. ERICH KLINGHAMMER
Laboratory of Ethology, Purdue University — LAFAYETTE, INDIANA, U.S.A.

DR. HEINZ-GEORG KLÖS
Professor and Director, Zoological Garden — BERLIN, GERMANY

URSULA KLÖS
Zoological Garden — BERLIN, GERMANY

DR. OTTO KOEHLER
Professor Emeritus, Zoological Institute, University of Freiburg — FREIBURG I. BR., GERMANY

DR. KURT KOLAR
Institute of Ethology, Austrian Academy of Sciences — VIENNA, AUSTRIA

DR. CLAUS KÖNIG
State Ornithological Station of Baden-Württemberg — LUDWIGSBURG, GERMANY

DR. ADRIAAN KORTLANDT
Zoological Laboratory, University of Amsterdam — AMSTERDAM, THE NETHERLANDS

DR. HELMUT KRAFT
Professor and Scientific Councillor, Medical Animal Clinic, University of Munich — MUNICH, GERMANY

DR. HELMUT KRAMER
Zoological Research Institute and A. Koenig Museum — BONN, GERMANY

DR. FRANZ KRAPP
Zoological Institute, University of Freiburg — FREIBURG, SWITZERLAND

DR. OTTO KRAUS
Professor, University of Hamburg, and Director, Zoological Institute and Museum — HAMBURG, GERMANY

DR. DR. HANS KRIEG
Professor and First Director (retired), Scientific Collections of the State of Bavaria — MUNICH, GERMANY

DR. HEINRICH KÜHL
Federal Research Institute for Fisheries, Cuxhaven Laboratory — CUXHAVEN, GERMANY

DR. OSKAR KUHN
Professor, formerly University Halle/Saale — MUNICH, GERMANY

DR. HANS KUMERLOEVE
First Director (retired), State Scientific Museum, Vienna — MUNICH, GERMANY

DR. NAGAMICHI KURODA
Yamashina Ornithological Institute, Shibuya-Ku — TOKYO, JAPAN

DR. FRED KURT
Zoological Museum of Zurich University, Smithsonian Elephant Survey — COLOMBO, CEYLON

DR. WERNER LADIGES
Professor and Chief Curator, Zoological Institute and Museum, University of Hamburg — HAMBURG, GERMANY

LESLIE LAIDLAW
Department of Animal Sciences, Purdue University — LAFAYETTE, INDIANA, U.S.A.

DR. ERNST M. LANG
Director, Zoological Garden — BASEL, SWITZERLAND

DR. ALFREDO LANGGUTH
Department of Zoology, Faculty of Humanities and Sciences, University of the Republic — MONTEVIDEO, URUGUAY

LEO LEHTONEN
Science Writer — HELSINKI, FINLAND

BERND LEISLER
Second Zoological Institute, University of Vienna — VIENNA, AUSTRIA

DR. KURT LILLELUND Professor and Director, Institute for Hydrobiology and Fishery Sciences, University of Hamburg	HAMBURG, GERMANY
R. LIVERSIDGE Alexander MacGregor Memorial Museum	KIMBERLEY, SOUTH AFRICA
DR. DR. KONRAD LORENZ Professor and Director, Max Planck Institute for Behavioral Physiology	SEEWIESEN/OBB., GERMANY
DR. DR. MARTIN LÜHMANN Federal Research Institute for the Breeding of Small Animals	CELLE, GERMANY
DR. JOHANNES LÜTTSCHWAGER Oberstudienrat (retired)	HEIDELBERG, GERMANY
DR. WOLFGANG MAKATSCH	BAUTZEN, GERMANY
DR. HUBERT MARKL Professor and Director, Zoological Institute, Technical University of Darmstadt	DARMSTADT, GERMANY
BASIL J. MARLOW, B.SC. (HONS) Curator, Australian Museum	SYDNEY, AUSTRALIA
DR. THEODOR MEBS Instructor of Biology	WEISSENHAUS/OSTSEE, GERMANY
DR. GERLOF FOKKO MEES Curator of Birds, Rijks Museum of Natural History	LEIDEN, THE NETHERLANDS
HERMANN MEINKEN Director, Fish Identification Institute, V.D.A.	BREMEN, GERMANY
DR. WILHELM MEISE Chief Curator, Zoological Institute and Museum, University of Hamburg	HAMBURG, GERMANY
DR. JOACHIM MESSTORFF Field Station of the Federal Fisheries Research Institute	BREMERHAVEN, GERMANY
DR. MARIAN MLYNARSKI Professor, Polish Academy of Sciences, Institute for Systematic and Experimental Zoology	CRACOW, POLAND
DR. WALBURGA MOELLER Nature Museum	HAMBURG, GERMANY
DR. H.C. ERNA MOHR Curator (retired), Zoological State Institute and Museum	HAMBURG, GERMANY
DR. KARL-HEINZ MOLL	WAREN/MÜRITZ, GERMANY
DR. DETLEV MÜLLER-USING Professor, Institute for Game Management, University of Göttingen	HANNOVERSCH-MÜNDEN, GERMANY
WERNER MÜNSTER Instructor of Biology	EBERSBACH, GERMANY
DR. JOACHIM MÜNZING Altona Museum	HAMBURG, GERMANY
DR. WILBERT NEUGEBAUER Wilhelma Zoo	STUTTGART-BAD CANNSTATT, GERMANY
DR. IAN NEWTON Senior Scientific Officer, The Nature Conservancy	EDINBURGH, SCOTLAND
DR. JÜRGEN NICOLAI Max Planck Institute for Behavioral Physiology	SEEWIESEN/OBB., GERMANY
DR. GÜNTHER NIETHAMMER Professor, Zoological Research Institute and A. Koenig Museum	BONN, GERMANY
DR. BERNHARD NIEVERGELT Zoological Museum, University of Zurich	ZURICH, SWITZERLAND
DR. C. C. OLROG Institut Miguel Lillo San Miguel de Tucuman	TUCUMAN, ARGENTINA
ALWIN PEDERSEN Mammal Research and Arctic Explorer	HOLTE, DENMARK
DR. DIETER STEFAN PETERS Nature Museum and Senckenberg Research Institute	FRANKFURT A.M., GERMANY
DR. NICOLAUS PETERS Scientific Councillor and Docent, Institute of Hydrobiology and Fisheries, University of Hamburg	HAMBURG, GERMANY
DR. HANS-GÜNTER PETZOLD Assistant Director, Zoological Garden	BERLIN, GERMANY

DR. RUDOLF PIECHOCKI		
Docent, Zoological Institute, University of Halle	HALLE A.D.S., GERMANY	
DR. IVO POGLAYEN-NEUWALL		
Director, Zoological Garden	LOUISVILLE, KENTUCKY, U.S.A.	
DR. EGON POPP		
Zoological Collection of the State of Bavaria	MUNICH, GERMANY	
DR. DR. H.C. ADOLF PORTMANN		
Professor Emeritus, Zoological Institute, University of Basel	BASEL, SWITZERLAND	
HANS PSENNER		
Professor and Director, Alpine Zoo	INNSBRUCK, AUSTRIA	
DR. HEINZ-SIBURD RAETHEL		
Oberveterinärrat	BERLIN, GERMANY	
DR. URS H. RAHM		
Professor, Museum of Natural History	BASEL, SWITZERLAND	
DR. WERNER RATHMAYER		
Biology Institute, University of Konstanz	KONSTANZ, GERMANY	
WALTER REINHARD		
Biologist	BADEN-BADEN, GERMANY	
DR. H. H. REINSCH		
Federal Fisheries Research Institute	BREMERHAVEN, GERMANY	
DR. BERNHARD RENSCH		
Professor Emeritus, Zoological Institute, University of Münster	MÜNSTER, GERMANY	
DR. VERNON REYNOLDS		
Docent, Department of Sociology, University of Bristol	BRISTOL, ENGLAND	
DR. RUPERT RIEDL		
Professor, Department of Zoology, University of North Carolina	CHAPEL HILL, NORTH CAROLINA, U.S.A.	
DR. PETER RIETSCHEL		
Professor (retired), Zoological Institute, University of Frankfurt a.M.	FRANKFURT A.M., GERMANY	
DR. SIEGFRIED RIETSCHEL		
Docent, University of Frankfurt; Curator, Nature Museum and Research Institute Senckenberg	FRANKFURT A.M., GERMANY	
HERBERT RINGLEBEN		
Institute of Ornithology, Heligoland Ornithological Station	WILHELMSHAVEN, GERMANY	
DR. K. ROHDE		
Institute for General Zoology, Ruhr University	BOCHUM, GERMANY	
DR. PETER RÖBEN		
Academic Councillor, Zoological Institute, Heidelberg University	HEIDELBERG, GERMANY	
DR. ANTON E. M. DE ROO		
Royal Museum of Central Africa	TERVUREN, SOUTH AFRICA	
DR. HUBERT SAINT GIRONS		
Research Director, Center for National Scientific Research	BRUNOY (ESSONNE), FRANCE	
DR. LUITFRIED VON SALVINI-PLAWEN		
First Zoological Institute, University of Vienna	VIENNA, AUSTRIA	
DR. KURT SANFT		
Oberstudienrat, Diesterweg-Gymnasium	BERLIN, GERMANY	
DR. E. G. FRANZ SAUER		
Professor, Zoological Research Institute and A. Koenig Museum, University of Bonn	BONN, GERMANY	
DR. ELEONORE M. SAUER		
Zoological Research Institute and A. Koenig Museum, University of Bonn	BONN, GERMANY	
DR. ERNST SCHÄFER		
Curator, State Museum of Lower Saxony	HANNOVER, GERMANY	
DR. FRIEDRICH SCHALLER		
Professor and Chairman, First Zoological Institute, University of Vienna	VIENNA, AUSTRIA	
DR. GEORGE B. SCHALLER		
Serengeti Research Institute, Michael Grzimek Laboratory	SERONERA, TANZANIA	
DR. GEORG SCHEER		
Chief Curator and Director, Zoological Institute, State Museum of Hesse	DARMSTADT, GERMANY	
DR. CHRISTOPH SCHERPNER		
Zoological Garden	FRANKFURT A.M., GERMANY	

DR. HERBERT SCHIFTER
Bird Collection, Museum of Natural History — VIENNA, AUSTRIA

DR. MARCO SCHNITTER
Zoological Museum, Zurich University — ZURICH, SWITZERLAND

DR. KURT SCHUBERT
Federal Fisheries Research Institute — HAMBURG, GERMANY

EUGEN SCHUHMACHER
Director, Animals Films, I.U.C.N. — MUNICH, GERMANY

DR. THOMAS SCHULTZE-WESTRUM
Zoological Institute, University of Munich — MUNICH, GERMANY

DR. ERNST SCHÜT
Professor and Director (retired), State Museum of Natural History — STUTTGART, GERMANY

DR. LESTER L. SHORT, JR.
Associate Curator, American Museum of Natural History — NEW YORK, NEW YORK, U.S.A.

DR. HELMUT SICK
National Museum — RIO DE JANEIRO, BRAZIL

DR. ALEXANDER F. SKUTCH
Professor of Ornithology, University of Costa Rica — SAN ISIDRO DEL GENERAL, COSTA RICA

DR. EVERHARD J. SLIJPER
Professor, Zoological Laboratory, University of Amsterdam — AMSTERDAM, THE NETHERLANDS

BERTRAM E. SMYTHIES
Curator (retired), Division of Forestry Management, Sarawak-Malaysia — ESTEPONA, SPAIN

DR. KENNETH E. STAGER
Chief Curator, Los Angeles County Museum of Natural History — LOS ANGELES, CALIFORNIA, U.S.A.

DR. H.C. GEORG H. W. STEIN
Professor, Curator of Mammals, Institute of Zoology and Zoological Museum, Humboldt University — BERLIN, GERMANY

DR. JOACHIM STEINBACHER
Curator, Nature Museum and Senckenberg Research Institute — FRANKFURT A.M., GERMANY

DR. BERNARD STONEHOUSE
Canterbury University — CHRISTCHURCH, NEW ZEALAND

DR. RICHARD ZUR STRASSEN
Curator, Nature Museum and Senckenberg Research Institute — FRANKFURT A.M., GERMANY

DR. ADELHEID STUDER-THIERSCH
Zoological Garden — BASEL, SWITZERLAND

DR. ERNST SUTTER
Museum of Natural History — BASEL, SWITZERLAND

DR. FRITZ TEROFAL
Director, Fish Collection, Zoological Collection of the State of Bavaria — MUNICH, GERMANY

DR. G. F. VAN TETS
Wildlife Research — CANBERRA, AUSTRALIA

ELLEN THALER-KOTTEK
Institute of Zoology, University of Innsbruck — INNSBRUCK, AUSTRIA

DR. ERICH THENIUS
Professor and Director, Institute of Paleontology, University of Vienna — VIENNA, AUSTRIA

DR. NIKO TINBERGEN
Professor of Animal Behavior, Department of Zoology, Oxford University — OXFORD, ENGLAND

ALEXANDER TSURIKOV
Lecturer, University of Munich — MUNICH, GERMANY

DR. WOLFGANG VILLWOCK
Zoological Institute and Museum, University of Hamburg — HAMBURG, GERMANY

ZDENEK VOGEL
Director, Suchdol Herpetological Station — PRAGUE, CZECHOSLOVAKIA

DIETER VOGT — SCHORNDORF, GERMANY

DR. JIRI VOLF
Zoological Garden — PRAGUE, CZECHOSLOVAKIA

OTTO WADEWITZ — LEIPZIG, GERMANY

DR. HELMUT O. WAGNER Director (retired), Overseas Museum, Bremen	MEXICO CITY, MEXICO
DR. FRITZ WALTHER Professor, Texas A & M University	COLLEGE STATION, TEXAS, U.S.A.
JOHN WARHAM Zoology Department, Canterbury University	CHRISTCHURCH, NEW ZEALAND
DR. SHERWOOD L. WASHBURN University of California at Berkeley	BERKELEY, CALIFORNIA, U.S.A.
EBERHARD WAWRA First Zoological Institute, University of Vienna	VIENNA, AUSTRIA
DR. INGRID WEIGEL Zoological Collection of the State of Bavaria	MUNICH, GERMANY
DR. B. WEISCHER Institute of Nematode Research, Federal Biological Institute	MÜNSTER/WESTFALEN, GERMANY
HERBERT WENDT Author, Natural History	BADEN-BADEN, GERMANY
DR. HEINZ WERMUTH Chief Curator, State Nature Museum, Stuttgart	LUDWIGSBURG, GERMANY
DR. WOLFGANG VON WESTERNHAGEN	PREETZ/HOLSTEIN, GERMANY
DR. ALEXANDER WETMORE United States National Museum, Smithsonian Institution	WASHINGTON, D.C., U.S.A.
DR. DIETRICH E. WILCKE	RÖTTGEN, GERMANY
DR. HELMUT WILKENS Professor and Director, Institute of Anatomy, School of Veterinary Medicine	HANNOVER, GERMANY
DR. MICHAEL L. WOLFE Utah State University	UTAH, U.S.A.
HANS EDMUND WOLTERS Zoological Research Institute and A. Koenig Museum	BONN, GERMANY
DR. ARNFRID WÜNSCHMANN Research Associate, Zoological Garden	BERLIN, GERMANY
DR. WALTER WÜST Instructor, Wilhelms Gymnasium	MUNICH, GERMANY
DR. HEINZ WUNDT Zoological Collection of the State of Bavaria	MUNICH, GERMANY
DR. CLAUS-DIETER ZANDER Zoological Institute and Museum, University of Hamburg	HAMBURG, GERMANY
DR. DR. FRITZ ZUMPT Director, Entomology and Parasitology, South African Institute for Medical Research	JOHANNESBURG, SOUTH AFRICA
DR. RICHARD L. ZUSI Curator of Birds, United States National Museum, Smithsonian Institution	WASHINGTON, D.C., U.S.A.

Volume 9

BIRDS III

Edited by:

BERNHARD GRZIMEK
WILHELM MEISE
GÜNTHER NIETHAMMER
JOACHIM STEINBACHER

ENGLISH EDITION

GENERAL EDITOR:
George M. Narita

SCIENTIFIC EDITOR:
Erich Klinghammer

TRANSLATORS:
Emil O. Hohn
Erich Klinghammer

ASSISTANT EDITORS:
Jeanine Grau
Peter W. Mehren
Sharon Rinkoff

PRODUCTION DIRECTOR:
James V. Leone

EDITORIAL ASSISTANTS:
Karen Boikess
John B. Brown

ART DIRECTOR:
Lorraine K. Hohman

INDEX:
Suzanne C. Klinghammer

CONTENTS

For a more complete listing
of animal names, see systematic classification or the index.

1. **THE CORACIIFORMES**

Introduction by Joachim Steinbacher	21
Kingfishers by Bernhard Grzimek	22
Todies by Joachim Steinbacher	34
Motmots by Alexander F. Skutch	35
Bee-eaters by Claus König	36
Coraciidae by Herbert Schifter	42
Cuckoo-rollers by Jean Dorst	42
Ground rollers by Jean Dorst	42
True rollers by Herbert Schifter	44
Hoopoes by Ernst Sutter	45
Hornbills by Kurt Sanft	49

2. **WOODPECKERS, BARBETS, TOUCANS, AND RELATED FORMS**

Woodpeckers by Günther Niethammer	62
Jacamars by Joachim Steinbacher	63
Puffbirds by Joachim Steinbacher	64
Barbets by Herbert Schifter	65
Honey guides by Herbert Friedmann	77
Toucans by Alexander F. Skutch	79
Woodpeckers by Herbert Wendt	87
Wrynecks by Dieter Blume	88

3. **PASSERINE BIRDS**

Passerine birds by Wilhelm Meise	118

4. **BROADBILLS, NOISEMAKERS, AND LYREBIRDS**

Desmodactylae by Bertram E. Smythies	123
Broadbills by Bertram E. Smythies	123
Clamatores by Wilhelm Meise	127
Woodcreepers by Alexander F. Skutch	128
Horneros and relatives by Wilhelm Meise	129
Antbirds by Helmut Sick	132
Pittas by Nagamichi Kuroda	141
Asitys by Jean Dorst	147
New Zealand wrens by C. A. Heming	147
Tyrant flycatchers by Alexander F. Skutch	148
Sharpbills by Helmut Sick	157
Manakins by Helmut Sick	157
Cotingas by Ernst Schäfer	160
Plantcutters by Ernst Schäfer	163
Primary songbirds by Wilhelm Meise	164
Lyrebirds by A. H. Chisholm	164
Scrub birds by A. H. Chisholm	165

5. **SONGBIRDS**

Songbirds by Wilhelm Meise	167

6. **THE LARKS**

Larks by Michael Abs	171

7. **THE SWALLOWS**

Swallows by Dieter Stefan Peters	182

8. WAGTAILS, CUCKOO-SHRIKES, BULBULS, FAIRY BLUEBIRDS, AND LEAFBIRDS

Wagtails by Wilhelm Meise	192
Cuckoo-shrikes by Bertram E. Smythies	198
Bulbuls by Bertram E. Smythies	200
Leafbirds by Salim Ali	204
Fairy bluebirds by Bertram E. Smythies	206

9. SHRIKES AND WAXWINGS

Shrikes and Waxwings by Günther Niethammer	207
Helmeted shrikes by C. W. Benson	207
Bush shrikes by C. W. Benson	208
True shrikes by Werner Münster	209
Wood Shrikes by Bertram E. Smythies	213
Vangas by Jean Dorst	213
Waxwings by Hans Bub	215
Phainopepla by Alexander F. Skutch	216
Hypocoliinae by Wilhelm Meise	217
Palmchats by Alexander Wetmore	217

10. WRENS AND RELATED FORMS AND ACCENTORS

Dippers by Otto Jost	219
Wrens by Edward A. Armstrong	223
Mockingbirds and thrashers by Emil Otto Höhn	227
Accentors by Herbert Schifter	229

11. BABBLING THRUSHES OR BABBLERS AND OLD WORLD WARBLERS

Flycatcher-like birds by Joachim Steinbacher	233
Babbling thrushes by Bertram E. Smythies	233
Rail-babblers by Herbert Thomas Condon	248
Old World warblers by Bernd Leisler	249
Wren warblers by Herbert Thomas Condon	265
Kinglets by Ellen Thaler-Kottek	267
Hyliinae by Wilhelm Meise	268

12. FLYCATCHERS AND THRUSHES

Flycatchers and thrushes by Wilhelm Meise	272
True flycatchers by Rudolf Berndt	272
Fantails by Wilhelm Meise	278
Paradise flycatchers by Wilhelm Meise	281
Thickheads by Wilhelm Meise	283
Thrushes by Jean Dorst	284

13. TITS (CHICKADEES), NUTHATCHES, AND TREE CREEPERS

Tits, nuthatches, and tree creepers by Joachim Steinbacher	310
Long-tailed tits by Rudolf Berndt	310
Penduline tits by Ragnar Kinzelbach	311
True tits by Rudolf Berndt	314
True nuthatches by Rudolf Berndt	321
Wall creepers by Hans Psenner	323
Tree runner-like birds by Herbert Thomas Condon	324
Philippine creepers by Joachim Steinbacher	325
Spotted tree creeper by C. W. Benson	326
True tree creepers by Rudolf Berndt	326

14. THE SUNBIRDS, FLOWER PECKERS, HONEY EATERS, AND WHITE-EYES

The honey eaters and relatives by Joachim Steinbacher	329
Flower peckers (mistletoe birds) by Bertram E. Smythies	329
Sunbirds by J. Berlioz	332
White-eyes by Gerlof Fokke Mees	339
Honey eaters by D. L. Serventy	344
Sugarbirds by Joachim Steinbacher	347

15. BUNTINGS AND THEIR RELATIVES

Buntings and their relatives by Wilhelm Meise	348
Buntings by Emil Otto Höhn	349
Cardinals by Herbert Schifter	366
Tanagers by Alexander F. Skutch	370
Wood warblers by Lester L. Short Jr.	380
Bananaquits-conebills by Wilhelm Meise	385
Wren thrushes by Alexander F. Skutch	386
Hawaiian honey creepers by A. J. Berger and Wilhelm Meise	386
Vireos by Herbert Schifter	394
Icterids by Ernst Schäfer	396

16. THE FINCH FAMILY

The finch family by Ian Newton	402
The domestic canary by Herbert Wendt	408

17. WEAVERS AND WEAVERFINCHES

Weavers and weaverfinches by Herbert Wendt	421
Weavers by Hans Edmund Wolters	421
Sparrows by Hans Edmund Wolters	422
Parasitic weavers by Herbert Friedmann	427
Viduinae by Jürgen Nicolai	427
Whydahs by Hans Edmund Wolters	433
True weavers by Hans Edmund Wolters	437
Thick-billed weavers by Hans Edmund Wolters	442
Buffalo weavers by Hans Edmund Wolters	442
Sparrow weavers by Hans Edmund Wolters	443
Scaly weavers by Hans Edmund Wolters	443
Weaver finches by Hans Edmund Wolters and K. Immelmann	444

18. STARLINGS, OLD WORLD ORIOLES, AND DRONGOS

Starlings by Herbert Bruns	463
Ox-peckers by C. W. Benson	475
Orioles by H. H. Reinsch	476
Drongos by Bertram E. Smythies	477

19. CROWS AND RELATED BIRDS

Crows by Joachim Steinbacher	482
New Zealand wattlebirds by C. A. Fleming	482
Magpie-larks by D. L. Serventy	483
Wood swallows by Bertram E. Smythies	484
Song shrikes by D. L. Serventy	485
Birds of paradise by Bernhard Grzimek and Thomas Schultze-Westrum	487
Crow-like birds by Walter Wüst	505

Appendix

Systematic Classification	522
On the Zoological Classification and Names	563
Animal dictionary:	565
English-German-French-Russian	565
German-English-French-Russian	576
French-German-English-Russian	596
Russian-German-English-French	606
Conversion Tables of Metric to U.S. and British Systems	614
Supplementary Readings	619
Picture Credits	624
Index	625
Abbreviations and Symbols	648

1 The Coraciiformes

Order: coraciiformes, by J. Steinbacher

The order Coraciiformes, consisting of the superfamilies kingfishers, todies, motmots, with the families of bee-eaters, cuckoo-rollers, rollers, hoopoes, wood hoopoes, and hornbills, is named after one of its component families, the Coraciidae (cuckoo-rollers to wood hoopoes in the list above); it forms a group of very diverse birds. Only a few common structural characteristics and habits unite the families of this order; as a result, this order was previously subject to many different decisions concerning its relationship and position in the zoological system. Today, these birds have been combined into one order on the basis of their three partially fused anterior toes (syndactylism), because of their desmognathous palatal structure, their leg muscles (no ambiens muscle), and because of particular plumage developments and arrangement. The feet are generally noticeably small. There are two notches in the rear edge on each side of the sternum (only one in hoopoes and hornbills), ten primaries (often with a vestigial eleventh one), and twelve tail feathers (only ten in motmots, hoopoes and hornbills). The L is 9–105 cm. These are largely colorful tropical or subtropical land birds. The beak is often large and of a peculiar shape. These birds are primarily meat, fish, and insect-eaters, although some also take fruit and berries. The young are naked and blind at hatching (with the exception of the hoopoes). The sexes are similar except in many of the hornbills and some kingfishers.

Most families are confined to the east of the Old World, but kingfishers are also found in the New World. Todies and motmots occur only in the New World. Bee-eaters, hoopoes, and hornbills probably originated in Africa, while the kingfishers originated in southeastern Asia. Geologically this order is very old and it was able to evolve distinctive features which still clearly distinguish it today. We recognize seven families: Kingfishers (Alcedinidae), todies (Todidae), motmots (Momotidae), bee-eaters (Meropidae), rollers (Coraciidae), hoopoes (Upupidae), and hornbills (Bucerotidae); there are 53 genera with 190 species.

KINGFISHERS (family Alcedinidae) are characterized by a compact, often very strong body, a short neck, and a large head with a long, pointed, or slightly compressed beak. The L is 10–47 cm. The wings are round, short, and may be up to medium length. The tail is either short and square (as in the genus *Alcedo*), or relatively long and used as a rudder (in the genera *Ceryle* and *Halcyon*). The small short-winged species have a linear, whirring flight. The legs are short, and the feet are small and weak. The three front toes are fused for more than one-third of their length, thereby enlarging the sole of the foot (this is important in ground-dwelling hole nesters which scoop out soil loosened with the beak by pushing it backwards with their feet). A few species, like the green and long-tailed kingfishers, have an erectile crest. The plumage is generally iridescent green or blue, but many birds also have bright white or red spots and bands (these decorative features further enhance the effect of the plumage which, as a whole, is often green, blue, or rust-colored). Other species have a black plumage with gray spots, or they have bands on a white background. There are two subfamilies: 1. Water kingfishers (Alcedininae), and 2. Tree kingfishers (Daceloninae). There are a total of 15 genera with 84 species.

Like the diurnal raptors, parrots, and swifts, the kingfishers have a lateral fovea in the eye, in addition to the central fovea. An image perceived sharply by both eyes simultaneously is formed on the lateral fovea. Since these birds pursue live prey, this second fovea is useful to them. The eggs are white (as in all hole nesters) and almost round. Because these birds are poor walkers, they merely jump from stump to stump or stones, or they move in stiff hops. Their center of distribution is in the Old World. About sixty species live from Central and East Asia to the Pacific Islands as far south as Australia. Africa has about fifteen species, while America has only six. Kingfishers, with very few exceptions, are tropical or subtropical birds. There are only two species north of Texas; the belted kingfisher and the green kingfisher. The pied kingfisher is found in northeastern Asia as far as northern Japan, eventually meeting the Eurasian kingfisher in Eurasia.

WATER KINGFISHERS (subfamily Alcedininae) feed mainly on fish and water insects. They have narrow, pointed beaks, and dig their nests out of slopes or banks. Their distribution is worldwide. The long-tailed and hovering kingfishers are distinguished from the short-tailed species, which plunge after their prey from a perch, by their method of hunting. However, they too hover over the surface and dive beneath the water surface to some extent.

The EURASIAN KINGFISHER (*Alcedo atthis*; Color plate, p. 37) has an L of up to 17 cm and weighs about 35 g. This bird is the most colorful member of its group. The ♂ is distinguished from the ♀ by the red or rusty base of its lower mandible and, more rarely, by its totally red

Family: kingfisher, by B. Grzimek

Subfamily: water kingfishers

beak. There is normally only one brood a year; once in a while there are as many as three. The incubation period lasts 19-21 days. Both sexes incubate and feed the young, which are fledged after 22-27 days. There are seven subspecies which are spread over Europe into Asia (Ceylon and the Solomons) near mountain streams at altitudes of up to 1800 meters. The other six *Alcedo* species are very similar to the Eurasian kingfisher.

This unusual and strikingly colored bird has attracted human attention and stimulated the imagination for a long time. According to old myths, it was originally an inconspicuous gray bird which left Noah's ark in such a hurry that its lower parts were burnt brown by the sun and its upper parts took on the steel blue color of the sky. The ancient Greeks believed it nested on the open sea; hence its Greek name, halcyon, means "the one that conceives at sea." The gods so favored the halcyons that, during their supposed breeding season, around Christmas, they established a calm period with smooth seas for fourteen days. Another myth tells the story of one of the Pleiades, Alcyone (in Latin the name is Alcedo, hence the scientific name of the genus), who married Keyx, a son of the evening star Hesperos. When Alcyone's husband was drowned, she threw herself into the sea. The gods, in their mercy, changed the two into a pair of kingfishers. Later Alcyone became the brightest star of the Pleiades, along with her six sisters who became stars also.

Kingfishers are popular in many fish hatcheries, because they take small fish from seven to nine centimeters (rarely ten) in length. Sticklebacks form thirty percent of their diet. They always take the smallest members from a swarm of fish, usually the sick or ailing individuals. Sticklebacks and other undesired species, because of their numerous offspring, often outnumber the other species in a mixed population, and kingfishers thus weed out the overstock. A kingfisher regularly visited a pool in the Rhineland which had an excessive population of 600 goldfish; in one year the kingfisher took about sixty. There are still fishermen and fish hatchery operators, however, who do not realize how useful this bird is to them. If kingfishers are to be kept away from the smallest fish in a hatchery, all possible perches along the shores should be removed, and nylon threads should be stretched about thirty centimeters above and over the water surface. The best protection of all, practiced in China and Vietnam, is to cover the pools completely with wire netting.

Eurasian kingfishers are not sociable; on the contrary, they expel conspecifics from their individual territories. They claim an area which extends 120 meters above and below their nests along a stream. As these birds never occur in numbers, their damage to the local fish population is only slight. Only during very hard winters, when all the rivers are frozen, does one occasionally see several birds together on the few open patches in rivers or along the sea. Incidentally, it is useless to shoot these birds, because the depopulated area is usually very rapidly reoccupied by others. Unfortunately, ninety kingfishers were shot and killed within a year in a Swiss fish hatchery.

Christopher Scherpner observed the activities of kingfishers kept on an artificial stream in the Frankfurt Zoo. One bird took seven to eleven small fish daily (an average of ten); in order to do so, it had to make 87 to 114 plunges into the water (an average of 101, or ten per fish). Oskar Heinroth fed an adult kingfisher that he reared twenty grams of fish and a few cockroaches each day. The kingfisher's prey is generally beaten to death on a branch and swallowed head first. Sticklebacks are usually whacked several times until the spines break off and remain stuck in the wood.

When the members of a kingfisher pair meet in the spring, the male offers the female a fish, holding it with the head toward her (the young are also fed in this manner). The pair builds the nest together, using steep earth banks for this purpose. The birds fly toward a particular spot and repeatedly knock out small clumps of dirt with their beaks. After about a day, the female, with her very small feet, is able to cling to the edge of the hole, and then the real tunneling begins. The male joins her and shovels out earth with his feet. The passage has a slight upward slope, and it is anywhere from forty centimeters to one meter in length. The nest chamber is between eight and ten centimeters high and slightly wider.

The kingfisher hole may be anywhere from sixty centimeters to thirty-six meters above the water, depending upon a favorable site. Two nests are never found together. Holes which are only slightly above the ground are easily dug out by foxes. The nest is not lined; pellets accumulate in it gradually. These consist of regurgitated bones and fish scales; they are white and are as easily squashed as cigar ash. The female lays between six and eight eggs (usually seven) from April to June or July. A second brood, when it occurs, helps to build up the reduced population after a hard winter.

The young require very small fish on their first day of life. The male helps with the feeding, but often ceases to do so at an early stage. Heyn once observed a male which was mated to two females. Each female had her own nest and brood, but often also fed the young of her "neighbor" as well. When an adult enters the passage of the nest, it naturally darkens it; this causes the young nearest the inner opening of the passage to open its bill. The same behavior will occur when a hand is placed over the outer opening. After the first young is fed, the other young move around so that the next in line reaches the opening. Soon the young turn about and shoot their liquid feces into the passage. If one looks into the nest hole, one may well get an eyeful. The adults must climb over the offensive fecal material and, as a result, they bathe very often during the twenty-three to twenty-six days that young are in the nest. When bathing, kingfishers do not plunge head first into the water as they do when hunting fish. Instead they splash about on the surface, thoroughly wetting their plumage. The young eat about six tiny fish a day, which they gulp down whole.

The young separate only a few days after fledging, and they are as intolerant of one another as the adults are. The parents separate soon thereafter as well. Although the catching of fish is inborn, it takes much practice. A young kingfisher must make many more dives than an adult before it captures a fish. Many young perish due to the repeated wetting of their plumage in cold or wet weather. Once they have established their own territories, kingfishers remain very much in the same place. Horst Reetz once netted kingfishers in the Black Forest and liberated them thirty kilometers away; two months later he accidently recovered one in its original territory.

In winter kingfishers move to lower areas—to the coast, and probably to the south as well. However, they are replaced by other kingfishers from further north. Hard winters, such as those of 1928/29, 1939/40, and 1962/63, in Europe are the worst times for these birds and eighty to ninety-five percent of them may perish. On the other hand, kingfishers are more easily seen at holes in the ice during these times. Their shining colors show up particularly well against the white. At night, their wings or tails may freeze to the branches. August Theil found a bird in such a predicament. It had tried (in vain) to fly away and had nearly fallen into the water as a result; its beak had evidently become frozen to the feathers below its wing. After Theil had warmed it for a quarter of an hour in his breast pocket, the bird took off. Kingfishers disappeared from Siberia in the spring of 1929 as a result of the hard winter in that area. They did not reappear there until 1938. The winter of 1962/63 also caused the disappearance of these birds from large areas east of the Elbe River and from the mountainous area around the Weser. Unfortunately, these "flying gems" have become increasingly rare in Germany. In 1962, 292 to 369 breeding pairs were estimated to inhabit West Germany, but in 1963 there were only 41 to 92 pairs. Therefore, this bird should definitely be protected.

Fig. 1-1. Malachite kingfisher (Corythornis cristatas).

The smaller kingfisher species include the MALACHITE KINGFISHER (Corythornis cristatas), with a L of 13.5 cm, found from Senegal to Madagascar and the Comoro Islands, and to the Cape Province; the NATAL KINGFISHER (Ispidina picta) with a L of 11.5 cm, found in tropical and southeastern Africa; and the DWARF KINGFISHER (Myioceyx lecontei) with a L of 10 cm, found in the West African jungles except one race, from Uganda.

Fig. 1-2. Natal kingfisher (Ispidina picta).

The MALACHITE KINGFISHER is one of the most common African birds. It flies extremely swiftly and favors wide waters and swamps, particularly reed-grown lake shores. During the breeding season it prefers to frequent narrower rivers where it can find earth banks in which to make its breeding holes. In Uganda, these birds breed from February to April, in June and July, or from September to December. They feed on fish, water insects, and dragonflies. The NATAL KINGFISHER is very small, colorful bird. This land-loving species is common in many areas, especially in bush districts. It feeds on arthropods, including water insects, and makes plunges into the grass, just as the malachite kingfisher does into water. The rare DWARF KINGFISHER has a flat, rounded beak, with which it catches mainly arthropods and young lizards.

E. M. Boehm of the United States was the first person successful in rearing NATAL KINGFISHERS in captivity. The first zoo to be successful in rearing these birds was the Frankfurt Zoo in 1969. The Frankfurt malachites live together with sunbirds *(Anthrapates collaris)* in a flight cage containing natural vegetation, a pool with an earth bank, and running water. In the winter of 1969 the pair flew about the bank searching for a nest, and then accepted a pre-bored hole above the pool. They deepened it with their beaks and feet. In nine days the hole was large enough to enable the birds to turn around in it and emerge head first. After a month, the breeding hole was ready. The male then displayed vigorously for two days, flew around the female and tried to feed her. Both birds incubated in alternation; however the female did so for longer periods and generally took the night shift.

The parent kingfishers were very restless in the period before the young hatched, and they often called. They even chased the sunbirds of which they had previously taken no notice. After the young hatched, the parents fed them with live arthropods and cut-up hearts; they refused fish, however. The prey was beaten to death more thoroughly, and was chewed more in the parents' beaks than was the parents' own food. Food was always placed head first into the beaks of the young. After eleven days, the parents entered the nest hole for only one or two minutes at a time. Four broods were reared between March and August, and four young were fledged. Unlike the adults, the fledged young have short beaks which, like the legs, are brownish at first instead of being red. The tip of the beak is white; the belly, which is red in adults, is pale in the young, as are the sides of the head, which are purple in adults.

The THREE-TOED KINGFISHERS (genus *Ceyx*) have a L of about 12.2 cm. This genus includes species which are primarily insect eaters, as well as species which hunt fish in woodland waters. These birds are found from India to the Solomons and Tasmania. There are three species, but we will mention only the CELEBES THREE-TOED KINGFISHER *(Ceyx fallax)*, which inhabits dense forests away from water.

The group of LONG-TAILED KINGFISHERS contains four Old World species and six American kingfisher species, among which are: 1. The BELTED KINGFISHER *(Megaceryle alcyon;* Color plate, p. 37) from North

Fig. 1-3. 1. Common kingfisher *(Alcedo atthis)*; 2. Belted kingfisher *(Megaceryle alcyon)*; 3. Amazon kingfisher *(Chloroceryle amazona)*; 4. Giant kingfisher *(Megaceryle maxima)*.

Fig. 1-4. Lesser pied kingfisher *(Ceryle rudis)*, also in southern Asia.

Fig. 1-5. Green-and-rufous kingfisher *(Chloroceryle inda)*.

Fig. 1-6. Sacred kingfisher *(Halcyon sancta)*.

Fig. 1-7. Gray-headed kingfisher *(Halcyon leucocephala)*.

America, which has a L of 29.5-33 cm. The ♀ has a brown breast band. This species is found in central and southern North America. 2. The RINGED KINGFISHER *(Megaceryle torquata)*, which has a L of 40 cm and a weight ranging from 260-360 g. It is found in South America. 3. The GIANT KINGFISHER *(Megaceryle maxima)* has a L of 40 cm. Its breast is slate-gray with black and white stripes. The side of the neck is nut-brown, and there are reddish-brown areas on the lower body; these areas are particularly large in the female. This species is found in Africa south of the Sahara. 4. The LESSER PIED KINGFISHER *(Ceryle rudis;* Color plate, p. 38) which has an L of 28 cm. It is black and white; the ♂ has two breast bands, and the ♀ has only one. These birds are found from Asia Minor to southern China and Africa. 5. The PIED KINGFISHER *(Ceryle lugubris)* has a L of 40 cm, and is found from Kashmir to northern Japan.

The giant kingfisher is not common in its homeland. It lives along the banks of forested rivers and shaded pools, but does not hover like the lesser pied kingfisher. Instead it waits motionlessly. In Uganda it breeds from May to July, as well as in October, while in the Transvaal it breeds in September. The lesser pied kingfisher can be seen almost everywhere on the larger rivers and lakes in East Africa. When not breeding, these birds often keep together in small groups. They particularly like to perch on trees, tree trunks, or *Euphorbia* bushes along the shores. From there they plunge into the water or, even more often, hover over the water and then dive into it. They are much more companionable than the Eurasian kingfisher, and several pairs often breed together in a small colony. In Tanzania and Uganda the breeding season is from April to June; however a second brood may follow later in the year. The eastern subspecies *(Ceryle rudis leucomelanura)* is found in southeastern Asia.

The South American GREEN KINGFISHERS (genus *Chloroceryle*) are also included in the group of long-tailed kingfishers; there are four species, including: 1. AMAZON KINGFISHER *(Chloroceryle amazona;* Color plate, p. 37), which has an L of 28 cm and weighs 110-140 g. It lives along jungle rivers in South America. 2. GREEN-AND-RUFOUS KINGFISHER *(Chloroceryle inda)*, which weighs 46-62 g. The upper side of the body is a dark shiny green, while the underside is chestnut brown. These birds are found on wooded river shores from Nicaragua to southern Brazil. 3. The GREEN KINGFISHER *(Chloroceryle americana)*, which has a L of 17 cm and weighs 25-29 g. The males are green and white and have a brown breast band. This species is found from Texas to Bolivia, and in central Argentina. 4. The PYGMY KINGFISHER *(Chloroceryle aenea)*, which has an L of 12.5 cm and weighs 11-16 g. It is found in South America as far as southern Brazil. This bird is a smaller, sparrow-sized replica of the green-and-rufous kingfisher. It is particularly common in overgrown ditches of large plantations.

The TREE KINGFISHERS (subfamily Daceloninae) are not always readily separable from the true or water kingfishers. These birds are found only in the Old World. They are larger than the members of the preceding subfamily and are not restricted to water. Generally their beaks are wide and flattened; however, many species have a downward-curved beak with a slightly hook-like tip. Tree-hole nesters are confined to this group, but there are also species which make their burrows in the nests of tree termites. Since these birds are largely independent of water, they may live on seacoasts, in dry forests, city parks, and on high mountains. Their food consists predominantly of large insects, crabs, amphibia, reptiles, young birds, snakes, and when they are near the water, naturally, of fish. Their voice is far-reaching and variable and reaches its fullest development in the Australian kookaburra.

The STORK-BILLED KINGFISHER (*Pelargopsis capensis*; Color plate, p. 38), with a L of 32–35 cm, belongs to the STORK-BILLED KINGFISHERS (genus *Pelargopsis*). It has a blue back, brownish undersides, and a red beak. It is distributed from India to the Philippines. The upward-curving mandibular sheaths and the flattened culmen indicate that it is a hunter, for the fishers have laterally compressed beaks. This bird is found from mangrove forests to the well-crowned shoreline rocks of centrally flowing rivers.

The birds of the genus *Halcyon* vary in color from black to blue or green, from white to cream, and from rust to chestnut and black-brown. The beak is black, brown, or red, and sometimes bicolored. There are thirty-four species, including: 1. The WHITE-BREASTED KINGFISHER (*Halcyon smyrnensis*; Color plates, pp. 32 and 38), which has a L of 28 cm and is the best known species. Its back is pale blue, and the center of the breast is white; otherwise, the bird is brown. It is distributed from Asia Minor to Formosa, Vietnam, and the Philippines. 2. The WHITE-COLLARED KINGFISHER (*Halcyon chloris*; Color plate, p. 37) which has a L of 25 cm. It occurs in 47 subspecies from the Red Sea to Samoa and northern Australia, and has the widest distribution of all the kingfisher species. 3. The SACRED KINGFISHER (*Halcyon sancta*), which has a L of 22.5 cm and weighs 50 g. Its upper side is blue-green, and the abdomen is yellowish-white. This species is found in Sumatra and Borneo, and from Australia to New Caledonia and New Zealand. 4. The SENEGAL KINGFISHER (*Halcyon senegalensis*). 5. The STRIPED KINGFISHER (*Halcyon chelicuti*), which has a L of 17 cm. 6. The GRAY-HEADED KINGFISHER (*Halcyon leucocephala*; Color plate, p. 37) which has a L of 28 cm. 7. The MANGROVE KINGFISHER (*Halcyon senegaloides*). The last two species are found in Africa, along with other species.

WHITE-BREASTED KINGFISHERS are often encountered far from water, deep in the jungles, or on open rice fields. They feed largely on arthropods, lizards, and frogs. At the Frankfurt Zoo, these birds were

Subfamily: tree kingfishers

▷
The Eurasian hoopoe (*Upupa epops*) got its name originally from the call "wudhoop." The head carries a crest, directed to the rear, which becomes erect into a fan when the bird is excited.

▷▷
True bee-eaters (genus *Merops*) inhabit the warmer parts of the Old World; shown here is the chestnut-headed bee-eater (*Merops viridis*).

▷▷▷
The Abyssinian ground hornbill (*Bucorvus abyssinicus*) looks like a huge crow. The sides of the head and throat are bare; they are red in the male and blue in the female. The throat sacs may be greatly distended.

kept in a glass flight cage with an artificial waterfall; they dug a nest hole in a clay bank, and hatched and reared young in it for four successive years. After the fledging of each brood, the parent kingfishers cleaned out the nest hole. The parents incubated in alternation and fed the young small fish, cockroaches, maggots, and clumps of soft food. Fish which they caught in a pool were beaten against branches with such force that the head was generally flung off.

The SACRED KINGFISHER is a wanderer; thus it travels to Australia from Sumatra and the Philippines in August, and leaves there in March to travel further north in order to breed from October to January. It bores out a nest hole in termite nests and breeds there, as well as in hollow trees and river banks. It feeds on small amphibia, fish, crabs, and larger insects.

The GRAY-HEADED KINGFISHER lives in bush country and feeds almost exclusively on land animals; it may also be found near water. This common kingfisher eats grasshoppers, beetles, and other arthropods as well as small amphibia. The subspecies *Halcyon leucocephala pallidiventris*, from Africa south of Lake Victoria, has an ash-gray head; the flight and tail feathers are a brilliant violet color. This kingfisher also migrates; it appears in Uganda and Tanzania between May and September, and in southern Africa from September to March. Apparently it does not breed in South Africa.

The best known species of TREE KINGFISHERS (genus *Dacelo*) are the KOOKABURRA or LAUGHING JACKASS (*Dacelo gigas;* Color plate, p. 38) which has a L of 42–47 cm, a weight of 360 g, and a bill that is 8–10 cm long, and *Dacelo leachi* which has a L of 42 cm. Both species are found in Australia. The kookaburra is even pictured on Australian stamps. The scientific name *Dacelo* is, incidentally, an anagram of the kingfisher name *Alcedo*.

The Australian aborigines have a legend about this striking bird of their homeland: when the sun rose for the first time, the god Bayame ordered the kookaburra to utter its loud, almost human laughter in order to wake up mankind so that they should not miss the wonderful sunrise. The aborigines also believed that any child who insulted a kookaburra would grow an extra slanting tooth. The unusual laughter of this bird delights all ornithologists. This laughter is usually heard early in the morning and again shortly after sunset; therefore the bird is also known in Australia as the "bushman's clock."

Kookaburras generally live in pairs or in small groups in open woodland; they also visit city parks or gardens and will accept food from people, and even enter rooms. They feed on small amphibia, insects, and crabs. They will also rob the nests of other birds and take domestic chicks. The fact that they even kill poisonous snakes has increased their popularity in Australia. They incubate their two to four pure white eggs between September and December in hollow tree trunks, tree holes, or in excavated termite nests.

◁

Left, from top to bottom: Von der Decken's hornbill *(Tockus deckeni);* White-breasted kingfisher *(Halcyon smyrnensis);* Rufous motmot *(Baryphthengus ruficapillus);* European bee-eater *(Merops apiaster).* Right, from top to bottom: Red-billed hornbill *(Tockus erythrorhynchus); Halcyon winchelli;* Blue-crowned motmot *(Momotus momota);* Common roller *(Coracias garrulus).*

In recent years kookaburras have accepted nest boxes in the zoos of Wassenaar (Holland), Chester (England), Washington, and Basel, and they have hatched young. Both adults incubate for a period of twenty-five days. In Washington, zoologists repeatedly observed the female attracted the male to the nest for relieving her by rubbing her beak on the nest tree—with successful results. The young were fed cut-up mice, cockroaches, and worms. The first of the two youngsters left the nest thirty days after hatching, and the young were fed by their parents for an additional forty days after that. The Washington female laid another two eggs even before her first brood became independent. In Basel, a young kookaburra artificially fed with forceps uttered its first "laughing" call when forty days old.

Fig. 1-8. Caroline racket-tail *(Tanysiptera sylvia).*

The EARTHWORM-EATING KINGFISHER (*Clytoceyx rex;* Color plate, p. 38) has a L of 32 cm and lives in the high mountains of New Guinea. It is believed to dig for larvae and worms with its short shovel-like beak, but it also catches lizards. The HOOK-BILLED KINGFISHER *(Melidora macrorhina),* with a L of 26 cm, is not a fish catcher either. Its hooked beak is used to seize grasshoppers and beetles in the New Guinean jungles. Like the CAROLINE RACKET-TAIL *(Tanysiptera sylvia),* which has a L of 33 cm and is found in eastern New Guinea and northern Australia, it breeds in holes in termite nests. The Caroline racket-tail has two considerably elongated central tail feathers which are expanded at the tip. These birds are found in Australia only during the local summer; they breed from November to January.

The TODIES (family Todidae) form a small group of New World coracids. In shape and in behavior, these birds are superficially reminiscent of the kingfishers, to which they are generally believed to be related; certain characteristics, however, suggest a relationship to the bee-eaters and motmots. The L is about 9 cm; there are three strongly fused front toes. The long legs have a horny sheath in front; the tongue is long, and the intestine is very short. The long, flattened beak, which is held pointing up at a slant when the bird is perched, has cutting edges near the tip. There are short, toothed bristles at the base of the beak. The plumage is a brilliant green above, and is generally whitish below; these birds have a red throat patch which is often fluffed out when they are at rest. There is only one genus *(Todus)* with five species, all of which are found on the Caribbean Islands: 1. The CUBAN TODY *(Todus multicolor)* found in Cuba. 2. The JAMAICAN TODY *(Todus viridis)* which has a red with some green breast. 3. The NARROW-BILLED TODY *(Todus angustirostris)* and 4. The BROAD-BILLED TODY *(Todus subulatus;* Color plate, p. 48), both of which are found on Hispaniola. 5. The PUERTO RICAN TODY *(Todus mexicanus),* which has yellow flanks and is found in Puerto Rico.

Family: todies, by J. Steinbacher

Todies usually live together in pairs in territories at forest edges or on bushes where, like flycatchers, they sit and wait for arthropods.

Fig. 1-9. Motmots (family Momotidae).

They snatch their prey out of the air or occasionally pick it off leaves, bringing the insects back to their perch before devouring them. In flight the wings produce a whirring noise which, intensified, is used as a signal in territorial defense. The sexes are similarly colored. They dig nest burrows thirty to sixty centimeters long in the banks of rivers or in sandy slopes; the beak is the main digging tool. At the end of the tunnel there is a fist-sized breeding chamber which holds three or four white eggs. Both parents incubate and feed the young. These birds are often so trustful of people that one can almost touch them, and one can easily catch them with a butterfly net. Because of their high metabolism, they are difficult to keep in captivity, and any attempt to do so is rarely successful.

Family: motmots, by A. F. Skutch

The MOTMOTS (family Momotidae) resemble the kingfishers physically and in habits, although they are not found near water. The L is 16–50 cm. The beak is curved downward at the tip and has toothed cutting edges. The tarsus is short, and the middle toe is almost completely joined to the inner toe; there is only one rear toe. The plumage is soft blue or reddish-brown in color; some species have blue or emerald stripes at the side of the head. A group of black feathers at the chin and throat is characteristic of all motmots. The tail is spatulate. The central pair of feathers is elongated, and barbs near the tail fall off readily, leaving part of the shaft of these feathers bare and looking like a thin wire. This barren area gives way to an oval disk at the feather tip where the barbs are retained, this forms the spatulate tail tip. These birds exhibit a jerky tail twitching when disturbed. The ♂ and ♀ are similar in all species. There are six genera with eight species, found from Central America to Bolivia and northern Argentina.

Motmots hunt from perches, and as a result, they live on trees, although they breed in ground holes; like all original hole nesters, they have white eggs. Their food consists of caterpillars, beetles, mantises, cicadas, butterflies, spiders, centipedes, and other arthropods, as well as small lizards, frogs, and birds. The large species also take fruit.

The BLUE-THROATED MOTMOT (*Aspatha gularis*; Color plate p. 48) has a L of 28 cm; it is found from the mountains of southern Mexico to El Salvador. This bird sings at daybreak, after leaving its earth hole, and its song consists of pure full tones which rise and then fall. In the Guatemalan highlands, these birds dig their holes soon after the young are fledged, in late June or July. Each pair spends the nights in the hole during the rainy season, as well as during the dry winter months when the nights are cold. The following April, the female lays three or four white eggs in the hole, which until that time has been used only for sleeping. After an incubation period of twenty-one or twenty-two days, young hatched in the cold highlands are kept warm by their parents for a considerable period. Unlike the adults, the young do not return to the nest hole at night after fledging.

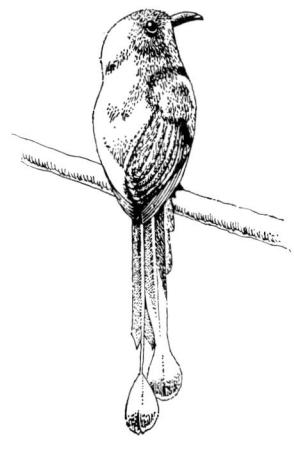

Fig. 1-10. Tail feathers before (right) and after (left) formation of the spatula in the broad-billed motmot.

The BROAD-BILLED MOTMOT *(Electron platyrhynchum)*, with its L of 33 cm, has an unusual beak form. These birds are found from Honduras to northern Bolivia and the Mato Grosso (Brazil). Their nest holes, like those of the other motmots, are one to two meters long, and occasionally they may be much longer. These nest holes may even change direction quite suddenly. The very conspicuous opening is found in vertical earth banks, on river banks, or on steep slopes beside roads or railway lines; it may also be found in fissures in caves or wells. The partners relieve one another during incubation only two times within each twenty-four hour period. As is the case with all motmots, the young broad-billed motmots are naked and blind when they hatch; they rest on their rumps and "ankles," which are protected against friction by calluses. Both parents feed the young with squashed arthropods initially; soon, however, the young are fed surprisingly large pieces of food—pieces the size of those the adults eat.

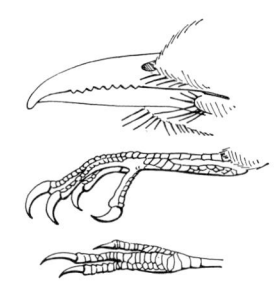

Fig. 1-11. The head and feet of the blue-crowned motmot: the beak is serrated along the cutting edge; on the foot, note the toes grown together.

The pretty TURQUOISE-BROWED MOTMOT *(Eumomota superciliosa;* Color plate, p. 48) has a L of 33 cm. It has longer central tail feathers than all the other motmots. Its only vocalizations are low wooden "kawaak" calls.

The largest species of this family are the RUFOUS MOTMOT *(Baryphthengus ruficapillus;* Color plate, p. 32), which has a L of 40 cm, and the BLUE-CROWNED MOTMOT *(Momotus momota;* Color plates, p. 32 and p. 48), which has a L of over 45 cm, and has no barbless zone in its tail feathers. Both species are found in Central America, to northern Bolivia. Besides arthropods and small amphibia, they also eat fruit, and the blue-diademed motmot even visits feeding stations to eat bananas.

▷

Kingfishers: L. White-collared kingfisher *(Halcyon chloris);* 2. Eurasian kingfisher *(Alcedo atthis);* 3. Gray-headed kingfisher *(Halcyon leucocephala);* 4. *Halcyon torotoro;* 5. Amazon kingfisher. *(Chloroceryle amazona);* 6. Belted kingfisher *(Megaceryle alcyon);* 7. Indian forest kingfisher *(Ceyx erithacus).*

At dawn, rufous motmots utter their mysterious "hoo hoo hoo" calls; blue-diademed motmot partners call each other with a soft "coot coot." Both species occasionally construct their nest burrows in the den of some animal; sometimes they build their own burrows in the ground, in which case the entrance becomes extremely difficult to find. In southern Costa Rica, the blue-crowned motmot digs its hole during the rainy months, August to October, when the soil is soft. However, the birds do not reappear in that area until the following March or April, the time of the breeding season. One adult incubates without rest from early afternoon until the following dawn, when it takes off in the twilight, and its partner takes its place for the morning. Incubation lasts for about twenty-one days. In the lowlands, motmots stop covering their young at night when they are a week old. The nest is not kept clean, yet at fledging the plumage of the young looks surprisingly fresh and clean. The young resemble the adults in their coloration, but they lack the long "racket-like" tail feathers.

BEE-EATERS (family Meropidae) are among the world's most colorful birds. The L is 17-35 cm, and the weight is about 50 g. The beak is

Family: bee-eaters, by C. König

very long, pointed, and slightly curved downwards. The legs are short, with relatively small, kingfisher-like feet. The wings are long and pointed. Bluish-green, green, or red colors predominate in the plumage; the throat is especially brilliantly colored. Some species have greatly elongated tail feathers, and one species even has a forked tail. The sexes are either indistinguishable or can be recognized only with difficulty. The young have faded plumage patterns. European bee-eater young are characterized by the lack of long tail feathers. There are seven genera with twenty-four species.

Bee-eaters are sociable and hunt in swallow-like flight. They generally perch on dry twigs or telephone wires. They often breed in large colonies. The nest burrows are placed in sand or earth banks and, if these are absent, in flat ground, where the tunnel runs downward at a slant. Such tunnels are 1.2 to 2 meters long; the area widened into the nest chamber is generally on the side of the tunnel. The females lay between two and five white eggs, or more. The remains of the pellets form the nest lining, since no nest material is carried in. The vocalizations consist of trills, whistles, or raucous calls like "oo" or "ee." Rhythmical repetition of certain calls produces a type of song.

Insect prey is generally captured on the wing. Bee-eaters also follow locust swarms and gather at the edge of grassland fires. They are called bee-eaters because of their preference for wasps, bees, bumble bees, and hornets. When a bee-eater catches one of these insects, it beats it repeatedly against a hard base until the stinger is out of action. The stinger then projects motionless from the insect's abdomen. The prey is then thrown up several times, and finally swallowed, generally head first. Non-stinging insects, even mealworms, fed to bee-eaters in captivity are treated in the same way. Most species need little water, and thus the water content of the insects suffices for their thirst. I kept bee-eaters for five years, and they did not take a drop of water during this period. We fed them mainly cooked mealworms which were sprinkled with a vitamin powder. These birds never took note of a well-filled water container, but they liked to bathe in the rain.

Bee-eaters are confined to the Old World; however, they are not present in New Zealand. Most species are found in open country, although a few are found in forests. They prefer to be near water because insects, particularly dragonflies, congregate there. These birds become migrants in the temperate zone, and winter in the tropics. The tropical species have also been known to make seasonal migrations.

The genus *Merops* has the most species, with nine, and they are found from southern Europe to Australia. Among these are: 1. The COMMON OR EUROPEAN BEE-EATER (*Merops apiaster*; Color plates, p. 32 and p. 48), which is 28 cm long and weighs 60 g. It is chestnut brown on its upper side and bluish-green below. The throat is yellow. Females have a somewhat more pale coloring. These birds are found from

◁

Kingfishers: 1. Gurial *(Pelargopsis capensis gurial)*; 2. Kookaburra *(Dacelo gigas)*; 3. Earthworm-eating kingfisher *(Clytoceyx rex)*; 4. White-breasted kingfisher *(Halcyon smyrnensis)*; 5. Lesser pied kingfisher *(Ceryle rudis)*.

Southern Europe and northwest Africa to Kashmir and southwestern Cape Province. 2. The BLUE-CHEEKED BEE-EATER *(Merops superciliosus)* has a L of 28 cm. The throat is reddish-brown, and there is a blue cheek stripe. The central tail feathers are elongated. These birds are found from the Caspian Sea to eastern China, and in parts of Africa. 3. The RAINBOW BEE-EATER *(Merops ornatus)* has a L of 25 cm. It is green with a yellow throat. It lives in Australia and migrates to Indonesia. 4. The CHESTNUT-HEADED BEE-EATER *(Merops viridis;* Color plate, p. 30) has a L of 24 cm. It is found from southeastern China to Java and the Philippines. 5. The CARMINE BEE-EATER *(Merops nubicus;* Color plate, p. 48) has a L of 33–35 cm. It is the most colorful species of all, with scarlet lower parts, a blue throat, and greatly elongated central tail feathers. It often breeds in huge colonies from Senegal to Somalia and Tanzania. 6. The SOUTHERN CARMINE BEE-EATER *(Merops nubicoides)* is the same size as the carmine bee-eater. It breeds from the southern Congo to Natal and Damaraland, migrating to the north outside the breeding season, while the carmine bee-eater wanders south.

In Europe, the common bee-eater breeds mainly in the south and southeast. There have been several successful breeding attempts in Central Europe, especially in several areas of southern Germany and at Birkensohl in Baden. In recent years, bee-eaters have nested in gravel pits near Augsburg (Germany), and in 1964 they even nested near Hamburg; migrating bee-eaters have been seen in Scandinavia.

Lilly Koenig has the following report of the breeding of the common bee-eater: "Bee-eaters about to breed land near a partner with a pronounced wing clapping; they exhibit the characteristic jerky display movement several times when perching, while at the same time uttering throaty calls. In this movement, the feathers of the back of the head are erected, and the wing tips are placed beneath the tail by means of an extreme adduction of the wrist. With each jerk, the pupil is narrowed, which allows the iris to become a bright red. From time to time the bird performs a ceremonial 'beating to death,' as if it were killing prey, against the partner's belly feathers. Both sexes display in this manner for days. Once a pair spends the night close together, they are firmly wed. Working in alternation, they dig a nest hole by vigorously hacking at the soil with their beaks, and scratching it away with their feet. My birds took eight to fourteen days, depending on the hardness of the soil, to finish the tunnel and living chamber. Then the male fed the female several times a day, and mated with her after the feedings. The adults alternate their incubation of the five (on the average) eggs which are laid at intervals of one to several days. At night they sleep in the nest. The first young hatches after twenty-two to twenty-five days, depending on the intensity of incubation after the laying of the first egg. The newly hatched young are naked, blind, and weigh only two or three grams. The first feather shafts emerge on the fifth day of life, and the young open their eyes on the sixth day.

Fig. 1-12. European bee-eater *(Merops apiaster).*

Fig. 1-13. Blue-cheeked bee-eater *(Merops superciliosus).*

"When feeding the young, the adult flies to the nest hole entrance while uttering a 'nest call.' Within the dark nest, the young orient themselves toward the call. The adult then moves into the tunnel and offers the food with gentle movements. As soon as a youngster touches something with its beak, it grabs. The hungry young await the parent in the tunnel; each withdraws when satiated, moving backwards while the next most hungry young advances. Thus each young has its turn. Bee-eaters that are thirteen to sixteen days old almost resemble hedgehogs because of their long feather keels. After about twenty days, the young take turns in peeking out of the nest hole, and as a result, they become acquainted with their immediate vicinity. While the adults weigh only about fifty grams, nestlings that are between twenty and twenty-eight days old weigh up to seventy grams; they fast during the last few days before fledging until they have reached the normal weight, which corresponds to the weight their wings can lift. The young leave the nest hole for the first time when they are about thirty days old, but they return to it for several more nights in order to sleep with the parents; later, they sleep on branches in a tight row which enables them to keep warm. The adults feed the young for about three weeks after they have left the nest. During this time, the young's ability to hunt independently develops completely."

The breeding population of bee-eaters in South Africa is probably derived from winter visitors; these birds are indistinguishable from the European birds. It was formerly believed that European bee-eaters nested for a second time in their winter grounds; this cannot be true, however, because the European birds moult during the breeding season of the South African birds, which appear in their nuptial plumage at that time.

The GREEN SWALLOW-TAILED BEE-EATER *(Dicrocercus hirundineus)* has a L of 22.5 cm. It is found in savannas from Senegal to Chad, Nigeria, and Cameroun. It has a swallow tail, and a yellow throat, bordered with blue. Birds of this species have been observed several times while sleeping in clusters in treetops.

The African genus *Melittophagus* consists of eight bee-eater species without pointed tails. Included within this genus are: 1. The LITTLE BEE-EATER *(Melittophagus pusillus)* which has a L of 17 cm. The upper part of the body is green; the throat is yellow. The tail is square and graduated. These birds are found in Africa south of the Sahara; they are quite common in open country. They generally occur in pairs; there are no large colonies. 2. The RED-THROATED BEE-EATER *(Melittophagus bulocki;* Color plate, p. 48), which has a L of 26 cm. It is found from Senegal and Cameroun to Uganda. 3. The BLACK BEE-EATER *(Melittophagus gularis),* which has a L of 20 cm. The throat is scarlet, and the rump is cobalt. This bird is found from Sierra Leone to Uganda. 4. Blue-headed bee-eater *(Melittophagus muelleri)* has a L of 21 cm. It is found in African equatorial forest areas.

The WHITE-THROATED BEE-EATER *(Aerops albicollis)* has a L of about 27 cm and is found from Africa south of the Sahara to southern Arabia. This species usually breeds in colonies and, outside the breeding season, it migrates to Tanzania and Angola. BOEHM'S BEE-EATER *(Aerops boehmi)* is rarely found in flocks. It is distributed from the southeastern Congo to Usambara (Tanzania) and Malawi.

The BLACK-HEADED BEE-EATER *(Bombylonax breweri)* has a L of 32 cm. It is black and brownish in color. This species lives in forests from Gabon to Kasai and Loango (Congo and Angola). It has a relatively thick beak. The forest-dwelling NIGHT BEE-EATERS (genus *Nyctyornis*) are active at night as well as during the day. The RED-BEARDED BEE-EATER *(Nyctornis amictus;* Color plate, p. 48) erects its head and throat feathers when calling. This species has a L of 30 cm, and is distributed over the Malayan Peninsula, Sumatra, Banka, and Borneo. The BLUE-HEADED BEE-EATER *(Nyctyornis athertoni)* has a blue throat and inhabits forests in the lower Himalayas from India to Thailand and Vietnam; a subspecies lives on the island of Hainan. *Meropogon forsteni* has a L of 29 cm. It lives in the jungles of Celebes. The upper parts of the body are mostly green, while the head and most of the lower parts are blue. This species usually keeps to the crowns of tall trees. When calling, it flicks its tail, the central feathers of which are elongated.

Family: coraciidae, by H. Schifter

The members of the family Coraciidae are brilliantly colored birds with striking vocalizations. The body is strong and the beak is partially crow-like. The rollers have a beak that is widened at the base. The tongue and caeca are long. These birds have short legs and weak feet; consequently they rarely walk or hop. The three anterior toes are joined along their basal segments. The members of this family are relatively long-winged, and are good fliers. The eggs, as is the case with almost all hole nesters, are white. Initially the young are naked; later they look somewhat like hedgehogs, because the horny sheaths of the feathers are late in rupturing. They feed on arthropods captured in flight. There are three subfamilies: 1. The Cuckoo-rollers (Leptosomatinae); 2. The Ground rollers (Brachypteraciinae); and 3. The True Rollers (Coraciinae).

Subfamily: cuckoo-rollers, by J. Dorst

The subfamily of CUCKOO-ROLLERS includes the CUCKOO-ROLLER *(Leptosomus discolor;* Color plate, p. 57), which has a L of 42 cm, and is found only on Madagascar and the Comoron Islands. Many ornithologists view this species as the most primitive representative of the whole order; it is related to both the true rollers and the cuckoos. The ability to rotate the outer toe (semi-zygodactylism) is reminiscent of the cuckoos. Other characteristics, like the soft dense plumage and the presence of powder downs, point more strongly to the rollers.

The cuckoo-roller has a large, strong beak, relatively long wings, a short tail, and very short legs. There are obvious sexual differences in plumage (in contrast to the rollers). The upper body plumage of the male, with the exception of the nape band, is dark gray with a metallic green and copper-brown gloss; the rest of the body is ash-gray. The female is somewhat larger; the upper plumage of her body is rust-brown with black bands, while the underparts have large black spots on a rust-colored background. Young cuckoo-rollers resemble the female.

These birds are found in forests with low undergrowth. They show a preference for treetops, where they are often found together in small groups. Their vocalizations consist of piercing calls and whistles. The shape of the cuckoo-roller's body enables it to catch chameleons; however, this species feeds mainly on arthropods which it finds sitting on leaves, as well as on beetles, grasshoppers, and even hairy caterpillars. The breeding period occurs during the rainy season of the year. After the display flights, which are executed high in the air, the female lays her white eggs in tree holes, and possibly even in ground holes on steep banks.

Subfamily: ground rollers, by J. Dorst

The members of the subfamily Brachypteraciinae are also confined to Madagascar. These birds are a deviant group because of their external features, and because of their biological peculiarities; however, they are closely related to the rollers. They have large heads, very strong beaks, short, rounded wings, strong feet, and relatively long tails (the long-tailed ground roller has a particularly long tail). There are three genera with a total of five species.

The SHORT-LEGGED GROUND ROLLER (*Brachypteracias leptosomus*; Color plate, p. 57) has a L of 30 cm; the head is red, the nape blue, the back green, and the undersides are white with red spots. Both this species and the SCALED GROUND ROLLER (*Brachypteracias squamigera*), which has red shoulders, green upper parts, and white lower parts covered with blackish scaly markings, provide a transitional link to the true rollers. A second genus includes the BLUE-TAILED GROUND ROLLER (*Atelornis pittoides*), which has a blue head, a green back, a white throat bordered with blue, and a pale red abdomen, and CROSSLEY'S GROUND ROLLER (*Atelornis crossleyi*; Color plate, p. 57), which is distinguished by a green back and tail, bright red underparts, and a black-and-white throat patch.

All four of the above-mentioned species inhabit dense forests; they will also take to the ground when the plant cover is not dense, generally hiding in small bushes. They are quiet birds, uttering only soft calls. They move by walking, or in low flight. Their food consists of arthropods and small vertebrates such as reptiles and frogs which they capture on the ground. These birds apparently nest in holes where they lay white eggs.

The fifth ground roller species, the LONG-TAILED GROUND ROLLER (*Uratelornis chimaera*), which has a L of 42 cm, differs from the others considerably because of its size and its long tail and legs. Its colors are also more pale. The upper part of the body is brownish-red with blackish stripes, and the underside is whitish with a black breast band. The wings are spotted azure. This roller lives in the semideserts of southwestern Madagascar. It feeds on insects and breeds in earth holes.

The TRUE ROLLERS (subfamily Coraciinae) are distributed over Europe, Africa, Asia, and Australia in two genera with eleven species. One species of the genus *Coracias* is the COMMON ROLLER (*Coracias garrulus*; Color plates, p. 32 and p. 57). This bird has a L of 32 cm and weighs 120-190 g. It is the only species in Europe, and is blue and brown in color. The INDIAN ROLLER (*Coracias benghalensis*) has a L of 35 cm. It is brownish with a pale striped throat, and is found in southern Asia. The RUFOUS-CROWNED ROLLER (*Coracias noevia*) has a L of 37 cm, and is found in mountainous country in Africa. Among the species with elongated outer tail feathers are: the LILAC-BREASTED ROLLER (*Coracias caudata*; Color plate, p. 57), which has a L of 40-45 cm. The ♂ tail feathers are 25.5 cm long, and the ♀ tail feathers measure 17.5 cm. This species is found in the lowlands of tropical Africa. The RACKET-TAILED ROLLER (*Coracias spatulata*), which has a L of 38 cm and inhabits the forests of tropical Africa. The BLUE-BELLIED ROLLER (*Coracias cyanogaster*), which has a L of 30 cm. This is a particularly beautiful bird with an ultramarine-blue abdomen and wings. It is found in West Africa.

Subfamily: true rollers, by H. Schifter

The COMMON ROLLER was once one of the most widely distributed Central European birds; however today it is found only in a few areas. In Austria, it breeds in eastern Styria and in Burgenland; in East Germany, it breeds near Berlin. This beautiful bird is common in Southern Europe and in Turkey, where it likes to sit on telephone wires. Because of their bright colors, these birds look much like tropical birds. They prefer open areas in flatlands with scattered trees. Common rollers used to land on shocks (Mandeln) of threshed grain, and as a result, they were also called "shockcrows" (Mandelkrähen) in Germany. The frequently heard "rak rak rak" calls are characteristic.

These birds first appear in Germany from mid-April to early May, at which time one may see their fluttering display flights. They nest in tree holes; in the south they also breed in holes in the ground. They use very little nest material. The clutch consists of four or five eggs. Only a single brood is raised from May through June. Incubation lasts for nineteen days, and the young are fledged in about four weeks. Otto von Frisch noted that the young opened their beaks wide when begging for food, at which time they also uttered a begging call. The parents feed their young on arthropods caught in flight or on the ground; grasshoppers also constitute a considerable part of their diet.

Common rollers migrate south from early August to mid-September, wintering in the southern half of Africa as far south as the Cape Province. They are regular visitors in Natal. Great numbers of

Fig. 1-14. Rollers (genus *Coracias*): 1. Breeding area of the common roller (*Coracias garrulus*), winter areas (arrows); 2. Indian roller (*Coracias benghalensis*); 3. Celebes roller (*Coracias temminckii*); 4. African rollers: African rufous-crowned roller (*Coracias naevia*), Senegal roller (*Coracias abyssinica*), Lilac-breasted roller (*Coracias caudata*), Racket-tailed roller (*Coracias spatulata*), Blue-bellied roller (*Coracias cyanogaster*).

Fig. 1-15. Broad-billed rollers (genus *Eurystomus*): 1. Blue-throated roller *(Eurystomus gularis)* and Broad-billed roller *(Eurystomus glaucurus)*; 2. Dollarbird *(Eurystomus orientalis)*.

them have been seen in Kenya assembling in a thornbush which they used as a communal roost.

The INDIAN ROLLER is found in Burma up to an altitude of 1600 meters. Aside from arthropods, frogs are believed to form its principal diet. This roller breeds not only in tree holes, but also in holes in rocks on cliffs; its clutch generally consists of only two eggs. In Southwest Africa, these birds breed from June until July; outside the breeding season, they, like the following species, make lengthy migrations. According to Van Someren, the lilac-breasted roller is usually found by itself in East Africa, although grass fires attract many of these birds from all directions, because they can easily hunt grasshoppers and beetles at such times. Occasionally they also take lizards. Their unlined nests are in natural tree holes or in termite hills. Sometimes these birds also take over woodpeckers' or kingfishers' nest holes. They lay two or three eggs. The RACKET-TAILED ROLLER is found in more forested country, in pairs, or in groups of three or four birds.

The three species which comprise the BROAD-BILLED ROLLERS (genus Eurystomus) all have short tails, and are somewhat smaller than the racket-tailed roller. They are all hole nesters, laying between two and five white eggs. These birds are distributed in Africa, parts of Asia, and Australia. The DOLLARBIRD (*Eurystomus orientalis*; Color plate, p. 57) has a L of 35 cm. It is so named because of a pale spot on the wing which suggests a silver coin. These birds are found from Manchuria across southern Asia to Australia. The BROAD-BILLED ROLLER *(Eurystomus glaucurus)* has a L of 30 cm. It is found in the African tropics, and in Madagascar as a migrant. The BLUE-THROATED ROLLER *(Eurystomus gularis)* has a L of 25 cm, and inhabits the jungles of West and Central Africa where it is a resident.

These birds are called rollers because of their display flights, in which they actually "loop the loop." The DOLLARBIRD is a characteristic jungle inhabitant. It chooses a perch at the tip of a dead tree, from which it makes short flights to capture butterflies and beetles (even after dusk). This bird is a migrant in Australia, and it appears there in late September or early October, leaving again in March, at the latest. The BROAD-BILLED ROLLER prefers light forests and clearings where it utters its loud calls. This very skillful flier attacks any larger passing bird. It usually breeds at the beginning of the rainy season. The BLUE-THROATED ROLLER may sit on the highest branches of trees, or it may glide about the forests making shrill calls. The nest is built high in a tree. These birds' broods may be found throughout most of the year in the Congo.

Family: hoopoes, by E. Sutter

The HOOPOES (family Upupidae) are clearly separable into two subfamilies: the short-clawed ground-dwelling HOOPOES proper (subfamily Upupinae), and the long-clawed WOOD HOOPOES (subfamily Phoeniculinae), which are reminiscent of tree creepers in their behavior. There are a total of three genera with seven species.

The EURASIAN HOOPOE (*Upupa epops;* Color plates, p. 29 and p. 57) has a L of 28 cm, and weighs 51–80 g. It is the only representative of its subfamily and genus. The plumage is a reddish-cream color, and has a black-and-white wing pattern, as well as a striking crest; the sickle-shaped bill is long and thin, the tongue is small. There are nine subspecies found from Central Europe to South Africa, Madagascar, all over Asia to eastern Siberia, China, and Malaya. The northern populations are migratory; they remain at their residence from early April to late August. These birds winter in Central Africa and Southern Asia.

The Eurasian hoopoe inhabits open, sunny parkland with pastures, moist meadows, and cultivated fields. It is found in savannas and steppes, as well as in orchards, vineyards, and stony wastelands. Like the pigeon, it walks about nodding its head and probing the soil between stones, grass clumps, or in cattle dung for its prey which consists of larvae, soil caterpillars, mole crickets, and beetles. The hard parts of large insects are removed before the insect is eaten, as the bird beats the insect on the ground until the legs, wings, and head come off the soft trunk. The hoopoe's small, rudimentary tongue is not suited for moving the pieces of food to the rear of the mouth; consequently the bird skillfully flings its prey up, and catches it with its mouth wide open, just as the almost equally tongueless hornbills do. The hoopoe's wingbeat is extraordinarily weak; it looks almost like that of a butterfly. The wave-like flight path is also characteristic.

In the Spring, the males utter a constant low "oop oop oop" sound which is difficult to localize, and which resembles the distant barking of a dog. Both the Latin generic name *Upupa,* and the English name "hoopoe" are derived from this call. The nest holes are found in rotten trees, in buildings, in holes in walls, or in stone heaps, soil banks, or cliffs. The female incubates the five to eight eggs alone, and she is fed by the male during the incubation period. Thereafter, the mother keeps the young warm for an additional eight days without interruption, and the father must continue to be the provider during this time.

Hoopoe nestlings are covered with fluffy down on the upper sides of their bodies. When they beg, they open up their beaks (like young passerines) and show a bright red palate, which is framed with thick, white ridges at the angles of the mouth. The young hiss at any approaching intruder. They then follow up this first warning by squirting thin streams of their liquid feces, which they squeeze out of their raised cloakas, at the intruder, hitting its face with excellent aim. At the same time, a cloud of offensive musk-like odor moves toward the intruder; this odor comes from the well-developed preen gland which normally secretes a yellow oil used to oil the feathers. In the female, this oil changes to an offensive-smelling brown material (during the breeding season), which is imparted to all members of the brood; the preen

▷
Hornbills: the great Indian hornbill *(Buceros bicornis)* is strikingly colored. The male's elongated beak process is grooved above; that of the female is smaller.

Todies, Motmots, and Bee-eaters: 1. Turquois-browed motmot *(Eumomota superciliosa)*; 2. Blue-crowned Motmot *(Momotus momota)*; 3. Blue-throated motmot *(Aspatha gularis)*; 4. Red-bearded bee-eater *(Nyctornis amictus)*; 5. Broad-billed tody *(Todus subulatus)*; 6. Carmine bee-eater *(Merops nubicus)*; 7. European bee-eater *(Merops apiaster)*; 8. Red-throated bee-eater *(Melittophagus bulocki)*; 9. *Melittophagus variegatus*.

Fig. 1-16. Distribution of hornbills (family Bucerotidae).

Family: hornbills, by K. Sanft

glands of nestlings also have this offensive secretion. This substance is secreted in drops when the bird is excited, and the smell then becomes unbearable to humans. Thus this bird is also sometimes called Stinkhahn (stink cock) in Germany. Little is known about the action of this evidently protective substance and its success against potential enemies, but it has been noted that cats and other predators avoid hoopoes.

The nearest relatives of the Eurasian hoopoe are the WOOD HOOPOES (subfamily Phoeniculinae), which have an L of 24–38 cm. The plumage is metallic blue or greenish-black with white marks; these birds have long, graduated tails. There are two genera with six species, found in Africa south of the Sahara.

Both nestlings and adults (possibly only the female) WOOD HOOPOES (genus *Phoeniculus*) have an offensive odor. In addition, these birds have a particular defensive behavior, similar to that of the SCIMITAR-BILLED HOOPOES (genus *Rhinopomastus*). Walter Hoesch has the following report on the SCIMITAR-BILLED HOOPOE *(Rhinopomastus cyanomelas)* from Southwest Africa: "Young disturbed in the nest erect their feathers, and slowly, like a pendulum, they move their heads from side to side; this movement is interrupted by split-second forward thrusts of the head, accompanied by hissing vocalizations. The pendular movements are more effective in the semidarkness of the nest hole, since the light-colored head stands out; the dark feathers of the juvenile plumage are still hidden by long, whitish sheaths, although they have fully unfolded on the rest of the body."

Wood and scimitar-billed hoopoes live in pairs or in small groups. They hunt for arthropods and larvae on trees. They climb tree trunks like tree creepers, and they maneuver among the branches like chickadees. Unlike the barely vocal hoopoe, they utter their loud cackling calls in a chorus of the whole group.

Amateur observers often confuse the HORNBILLS (family Bucerotidae), which belong to the Coraciiformes, with the toucans, which belong to the woodpeckers. Hornbills are tropical tree birds (ground hornbills are an exception) of the Old World. They range in size from that of a blackbird to turkey. The beak is unusually large, generally curved, and it often has ridges, a helmet, or a horn-shaped process. The hornbill gets its name from its beak. The cutting edges of the beak are toothed, particularly in adults; the beak and horn are very lightweight, despite their size, because the horn is hollow or filled with only a loose, bony network (the helmeted hornbill, with its massive horn, is an exception). The body is elongated, and the neck is very long with fourteen or fifteen vertebrae. There are brilliantly colored bare skin patches on the head and neck; the upper eyelid has long, rigid eyelashes, and the head and neck sometimes have a crest of loose feathers. The tail is long or very long, and is made up of ten feathers (the helmeted hornbill has a tail that is almost double the wing length). The wings have ten rounded primaries. The tarsus is shorter than the middle toe; the third and fourth toes are joined in tree birds, but the tarsus of ground hornbills is larger, and all the toes are deeply cleft. The plumage is usually black, and often has a metallic gloss; it may also be brown or gray with white marks. The beak and horn are often colorful. Only the left carotid is present; the crop, appendices, and powder downs are lacking. The preen gland is feathered. On the average, the female is ten percent smaller than the male. In Asia the males and females often differ in their coloration; such a differentiation is rare in the African birds, however. There are fourteen genera with fifty-five species and seventy-four subspecies.

The closest relatives of the hornbills are the hoopoes. A raven-sized bird, probably a hornbill *(Geiseloceros robustus),* was found in the coal forests of the Eocene (about forty million years ago) in the Geisel Valley near Halle/Saale (Germany); there were feather remains as well as skeletal parts. Present-day hornbills live in Africa south of the Sahara, in southwestern Arabia, southern Asia from India to Bali and Sumba, in the Philippines, the Moluccas, New Guinea, the Bismarck Archipelago, and in the Solomons. They are found in jungles, as well as in prairies. Some of their characteristics and peculiarities can be seen as adaptations to their particular habitat. Thus, most forest dwellers have a black plumage, often with a metallic sheen; even the skin of these birds has a greater amount of melanin than usual, so that it is often a glistening black, particularly the back. I never found this dark skin color in young from the tropics, or in adult birds from German zoos, however. The plumage of the steppe-dwelling birds, on the other hand, is gray, brown, or black-and-white spotted; the skin is yellow-brown, but never black.

The large hornbills feed mainly on fruit from jungle trees; smaller animals like frogs, lizards, young birds, snails and arthropods are only tidbits. The small animal life of the jungle is not nutritious enough for birds of such size, and they lack the skill, the sharp claws, and the sharp beak needed to overcome larger animals. However, the forest supplies these birds with an abundance of nutritious fruit rich in sugar (particularly figs) all year long. The desired fruit is found on the ends of twigs, and thus it is not easily accessible to the relatively heavy birds. The double-horned hornbill, for example, weighs 3.4 kilograms. Its large beak acts like a prolonged arm, enabling it to seize its food (see Toucans, Chapter 2). Nevertheless, such a meal is difficult to acquire. Potter observed how Blyth's hornbills often fell from the ends of branches in the Philippines. We can easily see why the beak of the larger hornbill forms must be longer relative to the wings. In the small Narcondam hornbill, for example, the beak is forty percent of the wing length; in the large Ceram hornbill it is fifty-three percent of the wing length; and in the great Indian hornbill of Burma it is sixty-one percent of the wing length.

The hornbill's flight is rather clumsy because of the bird's rounded wings. The large species produce a definite noise in flight, a noise which does not sound as if it were made by a bird. Oskar Heinroth describes the noise produced by one hornbill species as follows: "As I stood in the bush forest at Simpang (New Guinea), I suddenly thought I was hearing the approach of a train from the distance. ...It was not long until I saw the bird fly past and land in a tree. The noise ceased at the same moment. Only when more and more birds flew past, each again accompanied by the train noise, were my doubts stilled." This noise can be explained by examining the bird's wing structure. The coverts are too short and, as a result, part of the primary and secondary shafts are left uncovered. In the downstroke of the wing, air passes between the feather shafts and causes the upper wing coverts to vibrate.

Hornbill breeding behavior is most peculiar. All of these birds, with the exception of the ground hornbills, wall up the entry to the nest hole, leaving only a small hole through which the male feeds his family, and through which the female and young defecate. The droppings often contain seeds, which germinate below the nest tree. The natives of the area can determine the age of the young, and when they can be taken out of their nest and sold to an animal dealer, from the height of these sprouts. The dealers, therefore, usually get hornbills which have been taken directly from the nest; besides adult females, they get young which need to be hand reared. Hand-rearing may later prove dangerous for these birds because, in spite of all prohibitions, their tameness encourages zoo visitors to attempt to feed them dangerous objects. The adult females, however, remain shy, do not approach the barriers, and as a result, live to be quite old.

The nest hole is not, as Brehm wrote, "a prison in which the female must remain." Many researchers who looked in the nest holes for hornbill broods found, to their disappointment, that the female resented the intrusion, broke open the "prison door," and disappeared. The wall of the nest hole is good protection against undesirable intruders, which include the animal dealer as well as monkeys and snakes. As a result, we do not know the exact incubation and fledging periods of any of the hornbill species. The ground hornbill, which is not a hole breeder, has been found to have an incubation period of twenty-eight to thirty-one days. The red-billed hornbill apparently had an incubation period of twenty-eight to thirty days in the Frankfurt Zoo. All species, whether large or small, probably incubate just as long, but we have no knowledge on this matter.

The rainy season as a breeding releaser

In arid areas, the rainy season acts as a breeding releaser for these birds. At that time, there is moist soil for wall construction; when the young hatch, the development of small animals, notably arthropods, is at its height. Those species inhabiting tropical rain forests are not tied to a particular breeding season. This is just as well, for the larger species are evidently hard pressed for nesting habitats. Holes with a breeding chamber, like those I found used by the rhinoceros hornbills of Sumatra, are not too common. At any rate, it was noted that such a hole was reoccupied by another female of the same species after its former inhabitants had left it two months earlier.

When a pair has found a suitable nesting hole, which may be forty to fifty meters above the ground, they begin to build their nest. The female is usually the builder, while the male merely brings in the materials. The birds use droppings, food remains, and moist earth, which they mix with saliva. The female thickens this material by beating it laterally with her beak so that it forms a very hard cement upon hardening. When the female is no longer able to leave the nest, the male finishes the external work.

The small hornbills lay up to five eggs, while the large species, like the ground hornbills, lay only two. Generally only one young hatches. The uniformly white eggs are not glossy; they become more or less brown with the prolonged incubation, stained by the nest material. The size ranges from that of a pigeon's egg in the barely magpie-sized red-billed hornbill to that of a goose egg in the turkey-sized ground hornbills. After an incubation period of one month, the young of the small toko species remain in the nest for an additional one and one-half months; the young of the large species remain in the nest for two and one-half to three months. Toko mothers leave their young two to three weeks before they fledge, while it was observed that the females of the large jungle species left the nest together with the young.

The diet of the birds provides a clue to these differences. The small tokos live predominantly on arthropods, while the large jungle forms feed mainly on fruit. Male African tokos were observed flying to the nest with one single insect; they did not accumulate insects in their gullets. A male rhinoceros bird, however, a larger species, brought fifty-eight figs, which it had stored in its gullet, to the nest. In cases like the latter, the male is able to feed the family alone; this is impossible for the toko father, once the young grow beyond a certain size. When this happens, the female must leave the nest hole and participate in the feeding of the young. Incidentally, the young wall themselves in again once the mother has left them.

The female is not "inactive" while she is in the nest, for her flight and tail feathers moult at this time. Usually this process occurs so rapidly and so thoroughly that, temporarily, the female is totally unable to fly. The process may be completed in one to two weeks in the tokos. The rapidity of the moult is necessary in these species since the female spends only one and one-half months in the nest hole; afterwards she must help the male find food for the young. Thus the

Rapid molt

faster she sheds her old feathers, the more time she has for growing the new ones. This rapid moult is interrupted at once, however, if the female leaves the nest prematurely because of a disturbance. One female toko had already lost her tail feathers when she was chased out of her hole. Normally, the loss of the flight feathers would have been followed by an inability to fly, but the moult was interrupted once the bird was outside the nest hole, and she remained capable of flight. It is possible that the darkness of the nest cavity may act as a moult releaser. Non-breeding females always moult in steps, and thus they are still able to fly. Breeding ground hornbills also moult in this way, for they are not walled up in a nest hole. Females of the large hornbill species, which spend three and one-half to four months in the nest hole, have more time and are not really forced to have a precipitate moult. Nevertheless, females incapable of flight have been found in these nest holes as well; cases of the slow moult have also been reported for these species, however.

Initially the young differ from the adults in the shape of their poorly developed beaks; the color of the iris and the areas of bare skin may also be different. As yet, we know little about these color differences. Many of the rhinoceros hornbills in zoos are young birds which have been taken directly from the nest hole. Should any zoologist or interested layman collect continuous data on the colors of the bare neck and eye skin, the iris and the beak, he would contribute greatly to our knowledge of this family.

The plumage of the young generally resembles that of the adults. When the sexes differ in color, the young's plumage usually resembles that of the female. Thus there was a surprise when a yearling was examined at the Milan Zoo; it was assumed to be a male because of its whitish head and neck (the females of its species are totally black). Examination later showed that this "he" was in fact a "she." The case was particularly puzzling at first; however later study of many museum skins revealed the surprising fact that the initial plumage of the young of both sexes in the Asiatic genera where the sexes differ in color resembled that of the male adult. Female young moult within their first year of life, at which time they develop their "normal" plumage. The only African species with a sex color difference, the black-casqued hornbill, follows the general rule of the bird world: all young wear the female plumage. Observations of zoo birds can help increase our knowledge in this field as well.

The smallest hornbills are the TOKOS (genus *Tockus*), which comprise thirteen species, including: The RED-BILLED HORNBILL (*Tockus erythrorhynchus*; Color plates, p. 32 and p. 58) which has a L of 50 cm and weighs 180 g. The ♀'s bill is uniformly red. These birds inhabit the thornbush steppes of Africa. The YELLOW-BILLED HORNBILL (*Tockus flavirostris*), which has a L of 56 cm. It is found from Somalia to Kenya and South Africa.

VON DER DECKEN'S HORNBILL (*Tockus deckeni*; Color plates, p. 32 and p. 58), which has a L of 50 cm. The ♀ has a black breast. These birds are found in East Africa.

Von der Decken's hornbill has caused scientists some problems, because two forms are found in the same area, and the only difference between them is the presence of white spots on the wing coverts of one. The spotted form was originally considered as a distinct species, *Tockus jacksoni*. A study of the skins in many museums, however, showed that the spotted birds were merely young, two to four months old, of the unspotted form. Usually the young are identified by the weakly developed beak, but in this species the beak may have already reached the size and yellowish-red color of the adults before the moult into the adult plumage occurred.

Hornbills of the genus *Tockus* take both plant and animal food. They skillfully seize swarming termites in flight, and pursue grasshoppers on the ground with clumsy leaps. The onset of the rainy season stimulates breeding behavior in the RED-BILLED HORNBILL. Moist soil, which is gathered from the edges of puddles, is available at that time. The females lay up to five eggs. When the young hatch, after a month of incubation, the female has, in the meantime, shed all her wing and tail feathers, and her new feathers have reached a length of one to three centimeters. The male never fails to bring food to his brood; up to twenty feedings may occur in half an hour. The female leaves the nest about two weeks before the young do; the young wall themselves in again after she leaves. The first successful rearings of the red-billed hornbill occurred in 1926 and 1927 in the Frankfurt and Berlin Zoos.

The group of large African hornbills includes the BLACK-CASQUED HORNBILLS (genus *Ceratogymna*) and the TRUMPETER HORNBILLS (genus *Bycanistes*) which, together, have a total of five species, among which are: The BLACK-CASQUED HORNBILL (*Ceratogymna atrata*; Color plate, p. 58), which has a L of 90 cm. The WL of the ♂ is 39.5 cm, and that of the ♀ measures 35.5 cm. The beak may reach a length of 15.5 cm in the ♂, while in the ♀ it may be 13 cm. The female has a brown head and neck and narrow beak ridges. This species is found in the rain forests of West and Central Africa, as well as on the island of Fernando Po. The TRUMPETER HORNBILL (*Bycanistes buccinator*; Color plate, p. 58), which has a L of 85–90 cm. The ♂ has a WL of up to 30 cm, while in the ♀ it may reach 28 cm. The ♂'s beak may be up to 11.5 cm long, while that of the ♀ may reach 10 cm. The ♂ weighs 670–941 g, and the ♀ weighs 565–670 g. The ♀ is smaller and has a weaker beak process. This species inhabits forest edges, thornbush steppes, and mangrove forests from Angola and Kenya south to the Cape Province of South Africa.

As we mentioned previously, the young of the black-casqued hornbill resemble the adult female, unlike the young of the Asiatic hornbill

Fig. 1-17. Subspecies of red-billed hornbill (*Tockus erythrorhynchus*): 1. *Tockus erythrorhynchus erythrorhynchus*; 2. *Tockus erythrorhynchus rufirostris*; 3. *Tockus erythrorhynchus damarensis*.

Fig. 1-18. 1. Von der Decken's hornbill (*Tockus deckeni*); 2. Black-casqued hornbill (*Ceratogymna atrata*).

Fig. 1-19. Trumpeter hornbill (*Bycanistes buccinator*).

forms. Little is known about the habits of this species. The two eggs in the Berlin Museum are the only ones known to science; they have a gray shell and are quite elongated. The food of this species consists mostly of fruit, particularly that of the calamus palm and of the wine palm; however, Gerd Heinrich also saw these birds plundering a weaverbird colony in Angola. One black-casqued hornbill lived to more than fifteen years old in a zoo.

C. A. Stonor has the following report on a breeding attempt of the TRUMPETER HORNBILL: "A pair which had already been in the London Zoo for eight years examined a hole on March 29, 1936; on April 6 the female began walling up the hole entrance. The male carried the nest material, which consisted of cherry-sized balls of clay, in his throat and beak. To the great disappointment of the wardens, however, there were no eggs or young—only several feathers, for the female had moulted her large feathers in the meantime." Because the female laid no eggs, Stonor assumed that the absence of light in the nest hole caused the moult.

Included among the Asiatic jungle forms of hornbills are the following: the WHITE-CRESTED HORNBILL *(Berenicornis comatus)*, which has a L of 90 cm, and a wedge-shaped tail. This is a rare forest bird found from Malaya to Sumatra and Borneo. The RUFOUS-NECKED HORNBILL *(Aceros nipalensis)* which has a L of 1 m. This bird has no beak process, but it does have black bands across its white beak. It is found from Nepal to Laos. TICKELL'S HORNBILL *(Ptilolaemus tickelli)* which has a L of 70 cm and weighs 900 g. It is found from Assam to Vietnam. TARICTIC HORNBILL *(Penelopides panini)*, has a L of 55 cm. This species has eight subspecies found in the Philippines, all of which have cross grooves on the beak. A related species in Celebes has two subspecies with longitudinal grooves on the horn. The WREATHED HORNBILL *(Rhyticeros undulatus)*, which has a L of about 1 m. Adult ♂♂ have a WL measuring 35.5–52.8 cm. The head and body feathers of the ♀ are black, while those of the ♂ are white. The sides of the head are bare, and are blue in the ♀, yellow in the ♂. This species is found from Assam and Bengal to Java and Bali; the smaller subspecies, *Rhyticeros undulatus aequabilis*, which has less variable body and wing measurements, is found on Borneo. Other species of hornbills in the genus *Rhyticeros* are mentioned later in this chapter.

The Javanese call the wreathed hornbill "Anggang tahon" (year bird) because they believe that the cross ridges, which form the process on the beak, increase each year, so that the bird's age can be determined by counting them. Only one aspect of this belief is true: the first ridge appears before the end of the second year. The rest of the belief does not hold true, as more than one ridge may be produced in a year, or the anterior ridges may even drop off. When this happens, the supposed "calendar" shows too few years. Up to nine ridges have

Fig. 1-20. Hornbills (family Bucerotidae): 1. *Rhyticeros undulatus undulatus*; 2. *Rhyticeros undulatus aequabilis*; 3. *Rhyticeros narcondami*.

been discovered on a single bird; however, these hornbills may live to be more than twenty years old. It was formerly believed that there were two species of wreathed hornbills—one with smooth sides on its beak, the other with ridged side plates. This error was supported by the fact that birds with smooth-sided beaks were also able to breed and were, consequently, sexually mature. Only after 150 skins had been examined in museums did it become clear that the ridges and plates indicated age, and not characteristics of two different species.

H. Bartels observed these birds for several years in Sumatra; he found that a pair remained faithful to its nest hole, which had already been used by other birds for the previous twenty years, for at least nine years. Both birds inspected the hole early in January; they then occupied it three weeks later. Within two or three days, the female had walled up the entrance. The male fed her, predominantly with fruit, at intervals of one and one-half to four hours, from sunrise to sunset. When the young had hatched, the female "demanded" animal food and accepted no more fruit. The mother and her offspring left the nest hole after four and one-half months. At that time the young had the male plumage, the head and neck being white. If the young is a female, then the white feathers are lost at the age of seven to eight months, and are replaced by black ones.

BLYTH'S HORNBILL *(Rhyticeros plicatus)* has a L of up to 1 m. As an insular, helmeted hornbill, this species occurs from the Moluccas to the Solomons; the subspecies *Rhyticeros plicatus jungei* lives in New Guinea. A smaller relative, the NARCONDAM HORNBILL *(Rhyticeros narcondami)*, with a WL of up to 31 cm, inhabits only the tiny island of Narcondam in the Bay of Bengal. There are about 200 of these birds on the three by six kilometer islet. A dangerous surf has so far protected them from visitors, and thus from extinction.

The GREAT INDIAN HORNBILL *(Buceros bicornis;* Color plate, p. 47) is often kept in zoos. A larger subspecies *(Buceros bicornis homrai),* with a L of about 1.2 m, has a WL of up to 58 cm in the ♂ and up to 51 cm in the ♀. The ♂ beak may reach a length of 29.5 cm, while that of the ♀ may be up to 25 cm long. This subspecies inhabits the jungles of India and Burma. A smaller subspecies *(Buceros bicornis bicornis)* has a WL reaching 49.5 cm in the ♂ and 47 cm in the ♀. The ♂ may weigh up to 3.5 kg; the beak can reach 25 cm. The ♀ may weigh up to 2.5 kg; the beak may reach 23 cm. These birds inhabit the Malay Peninsula and Sumatra. ♂ and ♀ are easily distinguished; the ♂'s iris is red, while that of the ♀ is white. The front and back rim of the ♂'s horn is black, while that of the ♀ is reddish. Among the related species in the same genus are: the RHINOCEROS HORNBILL *(Buceros rhinoceros),* which has a L of 1.2 m and weighs 2.5 kg. This species is found from Malaya to Borneo. The PHILIPPINE HORNBILL *(Buceros hydrocorax),* which has a L of 90 cm, is a black-reddish-brown. It has a white tail and a flat, red horn. It is found in the Philippines.

▷ Rollers and Hoopoes: 1. Common Roller *(Coracias garrulus);* 2. Lilac-breasted roller *(Coracias caudata);* 3. Dollarbird *(Eurystomus orientalis);* 4. Crossley's ground roller *(Atelornis crossleyi);* 5. Short-legged ground roller *(Brachypteracias leptosomus);* 6. Eurasian hoopoe *(Upupa epops);* 7. Cuckoo-roller *(Leptosomus discolor);* 8. *Rhinopomastus minor;* 9. Green wood hoopoe *(Phoeniculus purpureus).*

Fig. 1-21. 1. Homrai or great Indian hornbill *(Buceros bicornis homrai);* 2. Lesser great Indian hornbill *(Buceros bicornis bicornis).*

Hornbills: 1. Trumpeter hornbill *(Bycanistes buccinator)*; 2. Helmeted hornbill *(Rhinoplax vigil)*; 3. Black-casqued hornbill *(Ceratogymna atrata.* Note the sex differences in size of the beak processes. 4. Red-billed hornbill *(Tockus erythrorhynchus;* 5. Abyssinian ground hornbill *(Bucorvus abyssinicus);* 6. Von der Decken's hornbill *(Tockus deckeni).*

Tickell gives us the following report of the GREAT INDIAN HORNBILL from Tenasserim: "On February 16, 1858, I heard that a large hornbill was nesting in a tree hole, and that the same spot had been used by a pair for several years. I visited the site and saw that the hole was in the trunk of an almost straight tree which had no branches for fifteen meters above the ground. The hole was closed off with a thick layer of clay, except for a small opening through which the female could pass her bill and be fed by the male. A villager climbed the tree with difficulty and began to clear away the clay. While he was busy, the male uttered loud, groaning sounds, flew to and fro, and came very close to us. When the opening was sufficiently enlarged, my man put his arm inside and was able to pull out the female. Liberated on the ground, she hopped about, unable to fly. At last, she climbed a small tree and remained sitting there. In the depth of the nest, about a meter from the hole, lay a single egg on decayed wood, pieces of bark, and feathers." Tickell tells of a hand-reared individual which became very tame, but which always remained bold, and threatened strangers with its beak. "It flew about in the garden, came down to the ground at times, hopped at a slant, and searched for food in the grass." Strangely enough, there is a Liberian stamp from 1920/21, which carries a picture of this bird; it is not known to live in Africa, but there was no ornithologist in the state printing works in Berlin, which produced the stamp.

The HELMETED HORNBILL *(Rhinoplax vigil;* Color plate, p. 58) has a L of up to 1.65 m. The WL of the ♂ reaches 54 cm, and that of the ♀ reaches 47 cm; the weight ranges from 2.5–3 kg. The tail may be up to 98 cm long in the ♂, and up to 79 cm long in the ♀. This bird is one of the strangest hornbills. Its massive horn is particularly heavy; the skull, complete with horn, weighs about 320 g, or ten percent of the bird's total weight. Thus the helmeted hornbill has the heaviest skull of any bird in the world. The horn and skull of the much larger great Indian hornbill weigh only about 200 g or six percent of the bird's total weight, while the huge bill of the marabou (Volume VII) weighs only up to one hundred g, or one and one-half percent the body weight.

Fig. 1-22. Helmeted hornbill *(Rhinoplax vigil).*

The heavy head of the helmeted hornbill presents a problem for the bird's flight. Hornbills fly with their heads extended and, consequently, this bird is too heavy in front. The two extraordinarily elongated central tail feathers help to counterbalance this. Hornbills generally moult their tail feathers in pairs; however such a moulting procedure would hinder the light ability of the helmeted hornbill considerably; as a result, rather than moulting its feathers simultaneously, the helmeted hornbill moults them in succession, so that the "balancing pole" is always present.

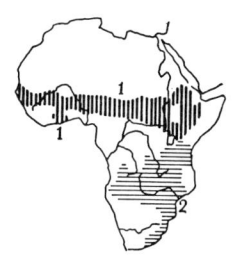

Fig. 1-23. Abyssinian ground hornbill *(Bucorvus abyssinicus);* 2. Cape ground hornbill *(Bucorvus cafer).*

The color plate on page 58 shows the striking shape and color of this bird. The bare skin areas on the heads and necks of the males are blood-red while those on the females are a dirty lilac color; in addition, the female has a spotted beak tip. This species lives in the treetops of the tropical rain forests of Malaya, Sumatra, and Borneo. It feeds mainly on fruit, as well as on an occasional animal. It breeds throughout the year; the breeding proceeds as in other hornbills, however, we are not yet familiar with the eggs of these birds. The helmeted hornbill is greatly pursued, because the Chinese use its horn for carvings, as well as for medicine and love potions, as is the case with the rhinoceros horn and with deer antlers. A large amount of "bird ivory" was exported to China during the Middle Ages; it cost twice as much as real ivory.

Gustav Schneider gives us the following report of an experience he had with helmeted hornbills: "On July 10, 1898, I followed an elephant herd deep into the jungle area of Mendaris, near Padang Bedagei Deli (Sumatra). There I heard a strange hammering. I followed this sound until I came below a waringin tree; above me I saw a number of hornbills fluttering in the air, violently flying at one another, and striking one another with their foreheads, thereby causing the mysterious hammer beats. After each strike the birds fell completely to the ground or a short distance above it depending on the intensity of the impact. Some came so close to me that I could have grasped them, and I noticed that those which fell to the ground had the greatest difficulty in getting up again. It was comical to see the long, misshapen birds rising slowly like paper dragons, and then attacking one another. I told my companions that I thought the birds were intoxicated; they said the 'Anggang tokok' had taken too much of the fruit of the waringin tree, and so were drunk. As the birds were easily caught, I took six; I started skinning them and ascertained that their stomachs were all crammed with berries, some already partly digested. There was a smell of wine-like acidity which proved that fermentation had taken place." Some time later some Malays brought him some helmeted hornbills which had been seized below a fig tree in similar condition. "When I placed them on a table, they tumbled about to and fro as if drunk. When two birds struck one another they made a great row."

The only ground dwellers among the hornbills are the GROUND HORNBILLS (genus *Bucorvus*). The L is 1.1 m; the WL reaches 62 cm in the ♂ and 55 cm in the ♀. The weight is 3.5–4 kg. The tarsus is twice as long as that of the similar-sized tree-dwelling relations. There are two closely related species (without subspecies) found in the savannas and steppes of Africa south of the Sahara. The adult ♂ ABYSSINIAN GROUND HORNBILL (*Bucorvus abyssinicus*; Color plates, p. 31 and p. 58) has a beak that may be up to 22 cm long, while the ♀ of this species has a beak that may be up to 17.5 cm long. The horn is open in front; there is a yellow spot at the base of the upper mandible. The bare skin around the eye is blue. The bare neck skin is blue with red spots in the ♂, while in the ♀ it is completely blue. The CAPE GROUND HORNBILL *(Bucorvus cafer)* has a beak which reaches a length of 17 cm in the ♂ and 15.5 cm in the ♀. This species differs from the Abyssinian ground hornbill in its anteriorly closed horn, the lack of a beak spot, a shorter beak, and the red eye skin. The neck skin is usually red in the ♂, while in the ♀ it is a bluish-red.

Ground hornbills prefer steppes and savannas with a fairly low grass growth, which makes the search for food easier. These birds follow grass fires and visit plantations. Their food consists largely of small vertebrates and larger insects. Since they hunt snakes and locusts, many tribes protect them and magnanimously overlook the fact that they also take part of their food from peanut fields.

Ground hornbills need thick trees for their nests; these trees, however, with the exception of the giant monkey bread trees, are scarce in the steppes. Consequently, it is no wonder that the distribution areas of the monkey bread tree and the ground hornbill largely coincide. These birds are quite noisy during the breeding season, which occurs during the rainy season. A deep sonorous thrumming, resembling the distant roar of lions, can be heard in the steppes at this time. Since ground hornbills are the only hornbills which do not wall in their nest holes, the female is only able to moult in installments, for she is largely self-supporting. The clutch consists of two eggs and is incubated, by the female, for a month. The young remain in the nest for three months, and are then fed by the parents for an additional nine months. The young remain with the family unit until they reach sexual maturity, after about three years. As is the case with many birds which lead their young for so long a period, the parent hornbills live together in a permanent bond. A male Abyssinian ground hornbill that came from the Rotterdam zoo in 1951 was still living at the San Diego Wild Animal Park in September of 1972 with a female that had arrived in 1961. This pair mated in the summer of 1972. She laid two eggs and incubated them for thirty-nine days.

2 Woodpeckers, Barbets, Toucans, and Related Forms

It is true that all WOODPECKERS (order Piciformes) are more or less closely adapted to arboreal life; however, of the six families included in this order, only the true woodpeckers are real tree climbers, and have stiff supporting tails. The L ranges from 8–60 cm, and the weight is between 6–300 g. The beak is strong; it is especially powerful and colorful in the toucans, while the true woodpeckers have a chisel-shaped beak. These birds have two toes directed toward the front and two (the first and fourth) directed toward the rear. The members of this order have various skeletal features in common. There are fourteen cervical vertebrae. All of the thoracic vertebrae are unattached, and there are five complete ribs. There are also four notches on the rear edge of the sternum. These birds have other common features in their musculature, digestive system, and feather pattern. The woodpeckers' food consists of insects, fruits, seeds, and, for the honey guides, of beeswax (the latter being a unique diet, at least among the birds). The white eggs are laid in holes. The young are blind when they hatch, and in most species, they are naked as well. There are a total of 383 species, distributed over the whole world, with the exception of Madagascar, Australia, New Zealand, and the South Sea Islands.

Several structural characteristics allow us to separate the six families within this order into two suborders: A. The JACAMARS (Galbuloidea) have a syrinx which is expanded into a drum. The preen gland is bare, and the appendices are well developed. There are two carotid arteries. These birds breed in self-excavated ground holes in South and Central America. There are two families: 1. The jacamars (Galbulidae) with fifteen species, and 2. The PUFFBIRDS (Bucconidae) with thirty-one species. B. The WOODPECKERS (Picoidea) do not have a drum-like syrinx; there are no appendices, and the preen gland is usually covered with feathers. Only the left carotid artery is present. The nestlings have ankle swellings. Almost all of the members of this suborder breed in

Order: woodpeckers, by G. Niethammer

tree holes, although a few families (honey guides) are brood parasites. These birds are distributed over all parts of the world, with the exception of Australia. There are four families: 1. The barbets (Capitonidae) with 76 species; 2. The honey guides (Indicatoridae) with 17 species; 3. The toucans (Ramphastidae) with 40 species; and 4. The woodpeckers (Picidae) with 209 species.

Family: jacamars, by J. Steinbacher

The JACAMARS (family Galbulidae) are slim, middle-sized, tree birds. The L is 13–28 cm, and the beak is fine and slightly curved downward. The short feet have two climbing toes. The first segments of the second and third toe are fused, and these two toes are directed forward, while the first and fourth toes are directed toward the rear. The members of the genus *Jacamaralcyon* lack the first toe. The plumage is generally loose and has a metallic green iridescence which is reminiscent of the plumage of many hummingbirds. The appendices are long. There is no gall bladder. The preen gland is bare, and the tongue is long and thin. The contour feathers have a short secondary shaft (except in the genus *Malacoptila*). The short wings have ten primaries. There are ten to twelve tail feathers.

In the primeval forests of Central and South America, the jacamar birds live singly, in pairs, or in small troops at forest edges and in clearings, in stands of trees, and along water courses. There are five genera with a total of fifteen species, of which we will mention the following: 1. The JACAMARS (genus *Galbula*) which include: The RUFOUS-TAILED JACAMAR *(Galbula ruficauda)* which has a L of 25 cm. The plumage is golden, coppery, and metallic green. The GREEN-TAILED JACAMAR *(Galbula galbula;* Color plate, p. 68), which has a L of 21.5 cm. The PARADISE JACAMAR *(Galbula dea,* Color plate, p. 68), which has a L of 30 cm; it is bluish-black with a white throat. 2. The GREAT JACAMAR *(Jacamerops aurea;* Color plate, p. 68) is the only species of its genus; it has a L of 30 cm, and is broad-billed. 3. The genus *Brachygalba,* with dull plumage colors, includes the BROWN JACAMAR *(Brachygalba lugubris),* which has a L of 15 cm. This bird is predominantly brown. 4. The THREE-TOED JACAMAR *(Jacamaralcyon tridactyla;* Color plate, p. 68) has a L of 20 cm; it is found in southeastern Brazil. 5. The WHITE-EARED JACAMAR *(Galbalcyrhynchus leucotis)* has a L of 20 cm. This species, like the other members of its genus, has many characteristics which deviate from the normal characteristics of the family. The plumage is dark brown and has a white ear spot. The white beak is very strong.

The chief prey of the jacamars are the butterflies. These birds sit on perches waiting for butterflies, which they catch in the air, like flycatchers, and which they kill by beating them against a branch. In spite of their often long and inactive waits on perches, jacamars are usually quite lively birds. They are reminiscent of hummingbirds in their plumage, as well as in their often explosively uttered sharp calls, trills, and short songs, as A. Skutch has noted. Some species look like bee-eaters when perching. The sexes are generally similarly colored, but the females of some species may have less striking head and neck colorations. These birds nest in ground holes with long entry tunnels, which they dig out with their beaks and feet, on steep river banks. One may also find these sloping tunnels some distance from the water, on soil banks, in earth masses between the roots of fallen trees, or in the ground; there is a nest chamber at the end of the tunnel.

Some jacamars even breed in termite nests. Unfortunately, we are familiar with the breeding habits of only relatively few species. The nest chamber is evidently used repeatedly, and it contains no actual nest material. In some species, both sexes work together at the time-consuming process of building the nest hole; in other species, only the female does this work. The two to four white eggs are alternately incubated by both parents during the day; during the night, however, the female incubates alone. The male feeds his partner several times during the day. The young are hatched after twenty to twenty-three days of incubation; unlike most woodpeckers, they have long, white down. The young have soft heel pads, as is the case with most birds which use no nest lining. The young are fed by both parents, and are raised in twenty-one to twenty-six days. They leave the nest hole when fully feathered, and thus, they look almost like the adults.

The family of PUFFBIRDS (Bucconidae) comprises a group of small to medium-sized birds with a compact build and a thick head. The L is 13–30 cm. The plumage is soft, and is particularly loose on the head and neck. The wings are short and round, and the tail is short. The feet are like those of the jacamars. These birds are generally dull in color and they have strong, laterally compressed beaks which are equipped with rictal bristles. The contour feathers have no aftershaft; the preen gland is exposed. There are ten primaries and twelve tail feathers. Puffbirds nest in ground holes and lay white eggs.

Family: puffbirds, by J. Steinbacher

These birds live in pairs or in loose groups on forest edges, in clearings, or in the bush of tropical jungle areas in Central and South America from Mexico to Argentina and Paraguay. The genus *Notharchus* consists of several large black and white species, one of which is the BLACK-BREASTED PUFFBIRD *(Notharchus pectoralis)* with a L of 20 cm. The COLLARED PUFFBIRD *(Bucco capensis;* Color plate, p. 68) has a L of 18 cm and weighs 35 g. This bird and the WHITE-WHISKERED PUFFBIRD *(Malacoptila panamensis),* as well as the CRESCENT-CHESTED PUFFBIRD *(Malacoptila striata;* Color plate, p. 68), which has an L of 19 cm, all have brownish-colored plumage, and there are often dark spots on the undersides of these birds as well. The BARRED PUFFBIRD *(Nystalus radiatus;* Color plate, p. 68), with a L of 20 cm, has a striking set of crossbars. The smallest puffbirds belong to the genus *Nonnula;* their beaks are longer, thinner, and have no hooks, as in the RED-THROATED NUNLET *(Nonnula rubecula;* Color plate, p. 68) which has a L of 15 cm. The NUNBIRDS (genus

Monasa), on the other hand, are the largest of the puffbirds. The BLACK NUNBIRD *(Monasa atra)* has a L of 30 cm, and weighs 90 g; it is predominantly dark-colored, like all the members of its genus. The genus *Chelidoptera* contains particularly long-winged species with narrow, downcurved beaks. Among the species in this genus is the SWALLOW-WING (*Chelidoptera tenebrosa;* Color plate, p. 68), which has a L of 16 cm and weighs 30 g.

The Germans call puffbirds "lazy birds" because of their habit of often sitting on perches for long periods of time while awaiting their prey, which consists particularly of small beetles seized in skillful flight. The swallow wings are an exception to this practice, as they hunt their prey on the wing, passing in between the branches, much in the manner of swallows. Puffbirds rarely hop about in the branches, and they almost never go down to the ground. Their calls sound like high-pitched, soft whistles, or like soft melodic trills. These birds are faithful to the same spot for years, and the name "lazy bird" may possibly have come about because of this fact.

Two nest types have been found so far: ground holes and holes in the tree nests of termites. The ground holes lead into the ground at a slant for about one-half meter; the end of the hole is expanded into a nest chamber which is lined with dry leaves. Nests which are built in termite nests are unlined. Nunbirds and the black and white puffbirds (genus *Notharchus*) hide the entrance hole with small twigs and leaves. Some species, like the black nunbird, for example, build a sort of antechamber, a tunnel which runs between leaves and twigs above the ground.

The tunnel of the swallow-wings descends steeply into the ground, and it may be up to two meters long. Both partners work on the nest hole, which may take fourteen days to complete. The two or three eggs are incubated by both parents at intervals which vary with the species; the incubation periods are largely unknown.

The newly hatched young are naked and blind; they are fed by the father, initially. After only a few days, they crawl up the tunnel to the entrance, where the mother feeds them. They leave the nest at twenty-one days of age, at which time they are in full plumage and look like the adults, as A. Skutch has noted in some species.

Family: barbets, by H. Schifter

The BARBETS (family Capitonidae) are usually colorful birds; they have compact bodies and strong beaks. They have rictal bristles which form the "beard" implied in the name barbet; this beard is responsible for the common name. These birds vary in size from that of a wren (the African pygmy barbet has a L of 9 cm) to almost that of a jackdaw (the larger species of green barbets have a L of 32 cm). The sexes are generally similarly colored; the female coloration differs only in a few species. Barbets are distributed all over tropical America, Africa, and southern Asia. There are sixteen genera with seventy-six species.

Barbets are found in large numbers only at special feeding places, such as fruit trees. Normally these birds live in pairs. A few species are social and form groups outside the breeding season. The young remain with the parents for a long time after they have fledged. They require an insect diet for at least the first two days of life; after this time they eat fruit, particularly figs, as well as insects and other arthropods.

Barbets are so well hidden in the foliage, choosing nest sites close to tree trunks, that they are rarely seen, in spite of their bright plumage. Their loud calls betray them, however. Some species have characteristic vocalizations which often consist of a single syllable repeated several times. A few of these birds, like the southern Asiatic coppersmith, and the minute African pygmy barbets, which are known as "tinker birds," have been named after their calls.

The barbets are also hole nesters, as is usual with most of the Piciformes. The holes are always built in rotted wood which the birds clear away with their strong beaks. A narrow tunnel leads to an expanded nest chamber. Outside the nesting season, this hole becomes a sleeping area; the barbets sometimes use woodpecker holes for this purpose. The two to five white eggs lie on a layer of wood chips; these eggs have longitudinal furrows on the shell, and they are incubated by both parents, although the female broods at night. The young hatch naked and blind; like woodpeckers, they have marked heel pads. They leave the nest when four to five weeks old, and their plumage colors are still dull at that time. Some species have limited breeding seasons, while others breed almost all year round in the evergreen tropical rain forests. At present we know little about the habits and reproductive behavior of many of these species.

The PRONG-BILLED BARBETS (genus *Semnornis*) have a relatively short but deep upper mandible, and an anteriorly notched or prolonged lower mandible. They inhabit the tropical highlands of Central and South America. The TOUCAN BARBET (*Semnornis ramphastinus*; Color plate, p. 97) has a L of 21 cm. The male has elongated nape feathers. This species is one of the best-known representatives of the family, yet, as an inhabitant of dense forests, it usually escapes most observers. The PRONG-BILLED BARBET (*Semnornis frantzii*), with a L of 20 cm, has an inconspicuous brown-green plumage. The female lacks the black nape band of the male.

Alexander Skutch found these birds in the moist coastal forests of Costa Rica at altitudes of up to 2000 meters. Outside the breeding season they wander in small groups, generally keeping to the scrub brush close to the ground. These birds are not shy of people. Their call is strikingly low-pitched and sounds melodious from a distance. Skutch initially thought it was a quail's call, and he described it as "cwa cwa cwa." The call can be heard for some distance, and it is usually

▷ Black-throated honey guide *(Indicator indicator)* and honey badger (*Mellivora*, Vol. XII). Above: The bird has led the badger to a bee's nest. Below: The honey guide at the wax of the combs.

The prong-billed barbets

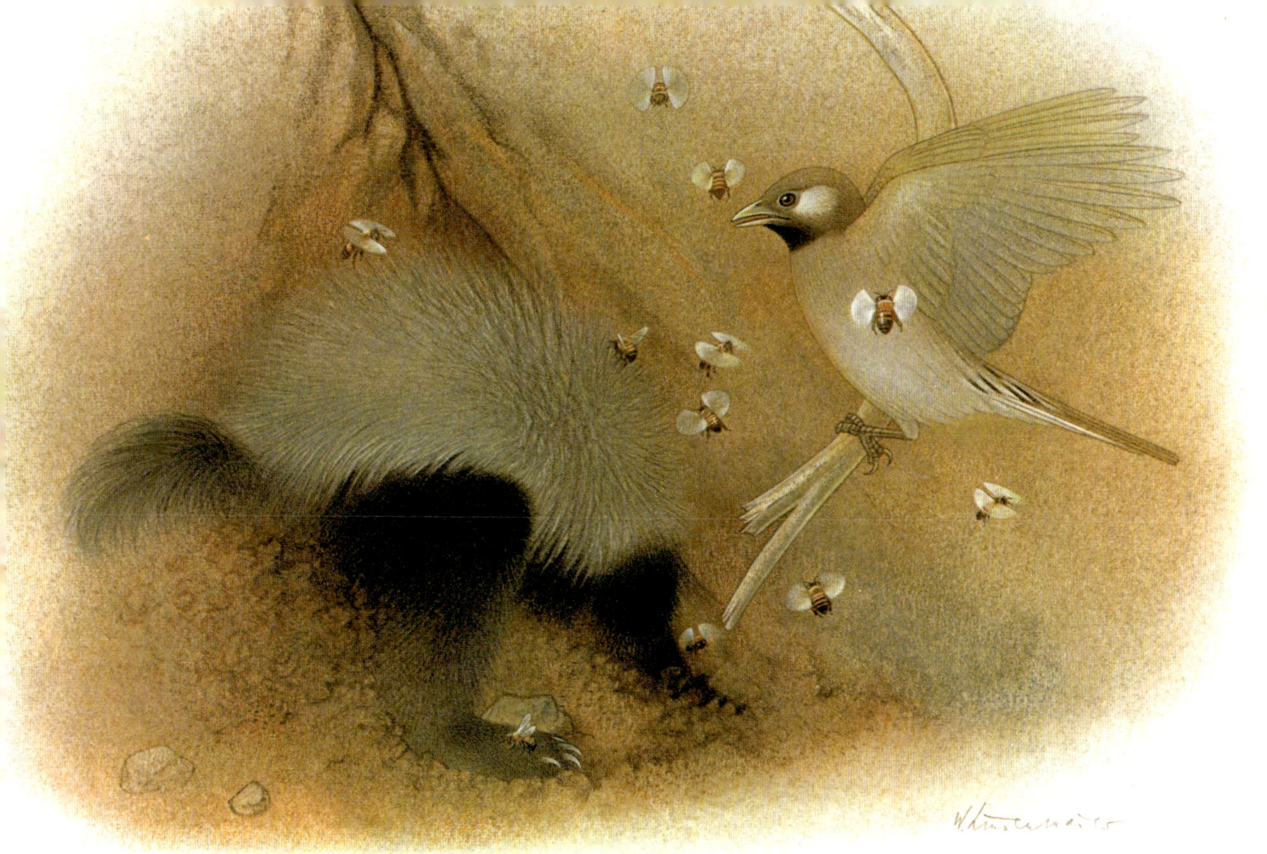

◁

Jacamars and puffbirds: 1. Green-tailed jacamar *(Galbula galbula)*; 2. White-necked puffbird *(Notharchus macrorhynchus)*; 3. Paradise jacamar *(Galbula dea)*; 4. Swallow-wing *(Chelidoptera tenebrosa)*; 5. Red-throated nunlet *(Nonnula rubecula)*; 6. Barred puffbird *(Nystalus radiatus)*; 7. Great jacamar *(Jacamerops aurea)*; 8. Crescent-chested puffbird *(Malacoptila striata)*; 9. Collared puffbird *(Bucco capensis)*; 10. Three-toed jacamar *(Jacamaralcyon tridactyla)*.

repeated several times. Prong-bills often gather in trees where they call together in chorus. Apart from fruit, they also break off and eat blossoms. They use tree holes for roosting at night. Skutch observed seven of these birds for an extended period of time; they always slept in the same hole in a rotted tree trunk. Sixteen birds emerged one morning from a hole that was even smaller than that used by the seven birds under observation.

The breeding season begins in March, at which time the pairs form. According to Skutch, both partners appear at the nest hole and work on it in the early mornings and again in the late afternoons. The birds bite off the wood with their strong beaks instead of chiseling it away like woodpeckers do. Their greatest difficulty is in starting the hole, because they must cling to the tree trunk. It took eight days for a pair to finish a nest hole; however, as soon as the hole was large enough it was used as a night roost by both partners. Barbet holes are not readily distinguishable from those of large woodpeckers when seen from a distance; close examination, however, shows that prong-bills dig much deeper into the tree so that there is always a narrow tunnel leading to the actual nest cavity.

Incubation begins when the clutch is nearly or quite complete with four or five eggs. During the day both partners share the incubation equally, but Skutch was unable to determine the role of the sexes during the night incubation, as both partners roost in the hole. The young in one nest he observed hatched thirteen days after the female had laid the last egg. The young are fed small insects during their first few days of life; after only a week, however, they eat fruit almost exclusively. The nest interior is kept scrupulously clean. The young develop slowly, according to Skutch; they are still naked and blind after nine days. It takes them a month to leave the nest, but they are able to fly at that time. The young look out of the nest hole while awaiting the parents, although they are still fed in the interior of the hole.

Skutch found that several nests were damaged as a result of bad weather, or through predators. One nest, which was only six meters above the ground, contained two almost fledged prong-bills; they had made their first short flights, and in the evening they were brought back to the nest, with some difficulty, by the parents. During the month that Skutch observed these birds, all four of them returned to the nest hole to sleep. Apparently prong-bills nest only once a year in Costa Rica.

Genus: *Capito*

The members of the genus *Capito,* which is also found in America, have a sex color differentiation. The yellow-throated, but otherwise olive-brown SCARLET-CROWNED BARBET *(Capito aurovirens)* is found in the jungles of the upper Amazon and has a L of 18 cm. The male has a red crown, while that of the female is gray. The BLACK-SPOTTED BARBET

(Capito niger), which has a L of 17 cm and weighs 50 g, is found in numerous subspecies in the northern South America; the male has a yellowish upper head and black upper parts, while the female is brown above and has light spots. This species is fairly common in many districts in Surinam; it lives mainly at forest edges. It is sometimes found singly, but it also often occurs in small groups and in greater numbers on fruit-bearing trees. A pair of black-spotted barbets nested in the Frankfurt Zoo in 1965. Ingrid Faust noted that the young left the nest hole thirty-six days after hatching. I watched a pair of them in a flight cage for some time; they seemed to enjoy calling a deep "croo croo croo" and a higher pitched "cra cra." They preferred insect food and took sweet fruit only occasionally. They held large insects under their feet, and this enabled them to work the insects over. They seemed to enjoy using their beaks on rotton wood. They were not very companionable; when the male wanted to occupy a place, the female had to yield. These birds were attentive to changes in their environment, and they examined new objects at once. They skillfully reached every corner of their cage, by climbing rather than flying, and yet they remained rather shy and wary.

The three species of the genus *Eubucco* have a sex color differentiation; however we will mention only the RED-HEADED BARBET *(Eubucco bourcierii)*. The males of this species are bright red on the head and breast, while the female is blue and yellow. This pretty barbet lives in Costa Rica, Panama, and Andean tropical areas from Ecuador to Peru. Skutch noted that its habits differ from those of the prong-billed barbet in Costa Rica. These birds live solitarily, feeding almost exclusively on insects which they take mainly from raked-up leaves and other such hiding places. Skutch found these barbets to be surprisingly quiet. They have been successfully kept in the Frankfurt Zoo (Germany) for several years. Ingrid Faust observed a successful breeding in a tree hole; the fledging period was thirty-one days.

Africa has many more barbet species than does America; there, too, these birds are adapted to the most varied environments, ranging from dense tropical jungle to almost treeless steppes. Some birds give up the usual tree-hole nesting typical of the family and nest instead in holes, which they dig themselves in banks or even in flat ground.

The members of the genus *Lybius* have tooth-like projections on the cutting edge of the beak. The BEARDED BARBET *(Lybius dubius)* is found in West Africa and has a L of 23 cm. It has a striking red and black plumage and a groove on both mandibles. The BLACK-BREASTED BARBET *(Lybius rolleti)* has no such grooves on the lower mandible. The DOUBLE-TOOTHED BARBET *(Lybius bidentatus)* from the central African jungles, has a much weaker beak. Birds of this species have red wing bands and red underparts. The L is 23 cm. The East and South African COLLARED BARBET (*Lybius torquatus;* Color plate, p. 97), with its L of 20 cm, is a particularly striking bird. It has a rust-colored head, a red throat, a black nape and breast band, and yellow underparts. The WHITE-HEADED BARBET *(Lybius leucocephalus)* has a L of 18 cm; this black and white bird is from East and West Africa.

The bearded barbet, a true tree bird, never comes to the ground in the dry bush forest areas of West Africa. This species is not as noisy as the other barbets. The breeding season begins with the onset of the rainy season; the birds choose a dead tree or branch for the nest hole site. The black-breasted barbet is said to make its holes in living trees, a habit otherwise unknown in this family. The American ornithologist J. P. Chapin found the double-toothed barbet in great numbers in many places in the Congolese jungles. These birds gathered around papaya fruit along with the mousebirds (see Volume VIII). Chapin also saw them eat red pepper fruit, its strong taste apparently not affecting them. Nestlings of this species have been found in April and also in November.

One pair of double-toothed barbets nested regularly between March 1966 and the spring of 1968 in the Frankfurt Zoo. Eleven young hatched from six broods, and ten young survived. The flight cage in which these birds were kept was as natural as possible. Old tree trunks of twenty-five to forty centimeters in diameter were set up between high rubber trees; dead leaves were scattered over the floor. The food consisted mainly of fruit like bananas, apples, oranges, cherries, and various berries, as well as insects, chopped meat, rice, salad, and ground-up turnips. There was also a basic mixture for omnivores; it contained chopped meat and rice, with cornmeal and ground-up rusks. Ingrid Faust observed the breeding behavior of these birds. She noted that the double-toothed barbets were peaceful and companionable during the breeding period. They preened and fed one another often, and even slept together, often with as many as six birds in one hole. The male showed off the white parts of his plumage during the display period. He sat beside the female, flicking his tail, calling loudly, and erecting himself as much as possible. Simultaneously he spread his white flank and back feathers. After mating, he led the female to the nest hole and tapped the floor of the entrance. The female generally followed. Newly fledged young were called to the nest hole in the same way by both parents. Both partners worked on the nest hole, and they removed wood chips from it in their beaks. Ingrid Faust's best observations were of the fifth brood. Incubation, which was done by both sexes, lasted thirteen days; the male only incubated for ten to fifteen minute periods, while the female did so for hours. Food was first carried into the nest hole fourteen days after incubation was begun. Both adults fed the young equally eagerly; the young preferred fresh insects. Other members of the species were allowed to help rear the young; a second female and the young of the preceding brood (along with the parents) fed them and also removed fecal pellets from the hole. The first young fledged twenty-four days after hatching, and twenty-six days after hatching, the second youngster fledged; it weighed seventy grams at that time, while the first young (weighed at the same time) weighed eighty grams. The young of the sixth brood weighed seventy-three and seventy-five grams on the day they fledged. The young are quite black on the breast and neck; the tail is shorter than that of the adults, and the beak is not as strong. The family continues to feed the young after fledging, and they are lured back to the nest hole to sleep.

The colorful collared barbet is the only species which has become accustomed to man; it is found in parks and gardens in Natal. These birds live in areas ranging in altitude from sea level to about 2000 meters, and according to Clancey, the only habitats they avoid are treeless country, and some mountains. A small group of collared barbets draws attention to itself by its calls. These birds breed in a small hole in a dead branch; there is sometimes an entrance tunnel some twenty centimeters long which leads into this hole. After eighteen to nineteen days of incubation, there is a thirty-three to thirty-five day nestling period. The collared barbet often rears a young honey guide along with its own brood, as these latter birds are brood parasites.

When there is fruit on the fig trees, one can be sure of finding pied barbets. These birds will often travel far in search of fruit trees. Van Someren observed a pair in Kenya making a nest hole in a dead tree; the job took ten days to complete. Another nest was three meters above the ground on the underside of a rotted branch; it had an entrance tunnel thirty centimeters long. Strangely enough, four birds worked together on this latter nest, and when the young hatched two weeks later, all four adults fed them. The young were fed by two adults only after they were fourteen days old; presumably these two adults were the parents. Van Someren assumes that the other two birds helped voluntarily, especially since these barbets are occasionally sociable.

The members of the genus *Tricholaema* are small, sparrow-sized birds with a "tooth" on each side of the upper mandible; they also have hair-like tips on the upper throat feathers. The PIED BARBET *(Tricholaema leucomelan)* from South and East Africa, weighs 29–33 g, and is a very colorful black, white, red, and yellow.

Genus: Tricholaema

Walter Hoesch and Günther Niethammer have observed pied barbets intensively during the breeding season, which lasts from October to June in southwestern Africa. They report the following: "Nest holes are made by the birds themselves, apparently only by the females. With their short, thick beaks, they are only able to tackle wood which has soft fibers. Most holes, as a result, are on the underside of rotted branches, and are only eight to ten centimeters in diameter. Incubation lasts fourteen days, and the nestling period is five weeks." These birds often roost overnight in weaverbird nests; sometimes three or more adults were found roosting together in tree holes. Hoesch took pied barbets from the nest and reared them successfully on bananas and soaked white bread. These barbets show a particular preference for green and ripe tomatoes from settlement gardens.

Pied barbets also feed on insects, especially termites. Their strong beaks allow them to open termite passages under tree holes and to remove these arthropods. These birds are quite noisy, particularly during the breeding season. Nest holes are often placed in the easily worked *Candelabra euphorbia*. The entry hole is barely five centimeters wide, while the passage may be twenty to forty-five centimeters long, leading to a nest chamber which is ten centimeters wide. As Van Someren observed in Kenya, both sexes work on the nest hole, although they may rest for several hours in between work periods. After the young have hatched, only the male brings food; he leaves the nest again at once while the female keeps the young warm.

The WHITE-EARED BARBET *(Stactolaema leucotis)* has a L of 18 cm. It is found in Kenya and Angola, and, like all of its genus, it differs from all the other barbets described so far because its beak has smooth cutting edges. This is one of the more sociable species. R. E. Moreau, an expert on the African avifauna, has seen at least five of these barbets enter a tree hole to roost together. He observed four adults bringing food for four nearly-fledged young in a nest seven to eight meters above the ground in a tree. Since this barbet species often lays only two or three eggs, Moreau assumes that in this case, two pairs used the same nest hole and shared the rearing of the young, or that some young of an earlier brood helped their parents.

Genus: *Gymnobucco*

The members of the genus *Gymnobucco* have an inconspicuous brown plumage, a tuft of rictal bristles, and a more or less bare head. All four species of this genus, which is confined to West and Central African jungle areas, are sociable. The BRISTLE-NOSED BARBET (*Gymnobucco peli*; Color plate, p. 97), which has a L of 17 cm, is found from Ghana to Gabon; according to Martin Eisentraut this species is found on Mount Cameroon only in the lower altitudes. The NAKED-FACED BARBET *(Gymnobucco calvus)*, however, occurs at altitudes of up to 1600 meters on Mount Cameroon. Several pair of these birds chop out their holes in large dead trees where they rear their three or four young. These nest colonies are visited by honey guides which lay their eggs in the barbet nests and let the barbets rear their young. Barbets congregate on fruit trees on Mount Cameroon and show little fear of man at such times.

Genus: *Pogoniulus*

The smallest barbets, which are no larger than a wren, are those of the genus *Pogoniulus*. There are nine species which are distributed widely throughout much of Africa, but which are often difficult to discover. These include: the YELLOW-THROATED TINKER-BIRD *(Pogoniulus subsulphureus),* which has a L of 9 cm, weighs 8.5–11 g, and is found from Guinea to Uganda; the YELLOW-FRONTED TINKER-BIRD *(Pogoniulus chrysoconus)* which weighs 12.5 g and is found from Angola to the Transvaal and Malawi; the SPECKLED TINKER-BIRD *(Pogoniulus scolopaceus),* which has a L of 12 cm and is found from Sierra Leone to Kenya; and the GOLDEN-RUMPED TINKER-BIRD *(Pogoniulus bilineatus;* Color plate, p. 97), which is found from Malawi to South Africa.

Martin Eisentraut has noted that the members of this genus live very secretively on Mount Cameroon, betraying themselves only by their calls. The speckled tinker-bird is said to call rather like a quail. The male golden-rumped tinker-bird has a monosyllabic call which is repeated continuously for three to four minutes. These birds feed on fruit and insects, mainly caterpillars and beetle larvae. Nest holes are made in tree trunks and branches, even outside the breeding season. The nest construction takes at least ten days; the entry is on the lower side of a dead branch. These birds incubate for about twelve days, and the young fledge in only about twenty days, according to Van Someren.

Barbets of the genus *Trachyphonus* inhabit open steppe country and, in contrast to the other barbets, are often found on the ground. LEVAILLANT'S BARBET *(Trachyphonus vaillantii)* has a L of 21 cm and is equipped with an erectile crest. It is a hawfinch-sized bird found in East and South Africa. The RED AND YELLOW BARBET *(Trachyphonus erythrocephalus;* Color plate, p. 97) is found from Ethiopia to Tanzania. The females have red heads, but the males are only red on the sides of the head. D'ARNAUDS BARBET *(Trachyphonus darnaudii)* is distributed from the Sudan to Tanzania; the sexes have a similar coloration. The YELLOW-BREASTED BARBET *(Trachyphonus margaritatus)* has a L of 19 cm. The back is decorated with white drop-like spots; this species lives in thornbush scrub from Somalia to Nigeria.

Genus: *Trachyphonus*

Levaillant's barbet feeds largely on insects, especially termites. It becomes very tame near towns and in sanctuaries, and will even come to feeding stations. O. P. M. Prozesky, of the Transvaal Museum in South Africa, observed a pair of these birds which nested in a hole in a dead quaking poplar in a park near Pretoria. They reared several broods in this hole, deepening it after each brood. They gave up the nest, however, after the sixth brood. This hole was more than one meter deep. Both adults incubated and shared the feeding of the young. During the day the parents flew to the nest between 50 to 120 times in order to feed the young. Soon the young came toward the parents when they entered the tunnel. Fledging took twenty-one days, but the parents began another brood even before the young were independent.

The red-and-yellow barbet builds its nest in the ground on steep shore banks, or on steep slopes. One nest found by A. H. Paget-Wilkes in Uganda has a tunnel thirty centimeters long and seven and one-half centimeters wide; this led to a somewhat longer nest chamber which held one egg. D'Arnaud's barbet builds a vertical tunnel, which may be one meter long, in flat ground; the nest chamber is placed laterally and at a slightly higher level so that it is protected from rainwater. Breeding begins in the middle of the rainy season in Uganda, when the ground is soft enough to be worked by the barbets. The naked-faced barbet builds tunnels in soil banks, as well as in wells and deserted houses.

Some African barbets sing duets, males and females uttering well-attuned sounds in such rapid alternation, that the whole song sounds as if it came from a single bird (antiphonal songs). These duets evidently help pair formation; they have been studied by Wolfgang Wickler. He was able to represent the contribution of one partner (in black) on a sound spectogram; the other partner's contribution was in white. This representation showed the alternation of each of the two birds, as well as the overlapping. Other species, like the naked-faced barbet and Viellot's barbet *(Lybius vieilloti)*, sing in trios; this chorus may possibly strengthen the family bond.

Genus: *Megalaima*

Barbets are found in Asia from India and Ceylon to southern China and Taiwan, the Philippines, Java, and Bali. Most of the Asian species belong to the genus *Megalaima* which has a predominantly green plumage; the head may have brown, red, yellow, and orange marks, which make many species very colorful. The sexes have a similar coloration; the young have weaker colors. Among the species of this genus are: the almost jackdaw-sized GREAT HILL BARBET (*Megalaima virens;* Color plate, p. 97), which has a L of 32 cm, and may be found from Kashmir to southern China and Vietnam; the GREEN BARBET *(Megalaima zeylanica),* which has a L of 25 cm, is mainly brown, and has a light striped head and a naked yellow eye ring; the more colorful BLUE-CHEEKED BARBET (*Megalaima asiatica;* Color plate, p. 97) found from India to Vietnam; and the GOLDEN-THROATED BARBET (*Megalaima franklinii),* which has a L of 23 cm and is found from Nepal to Laos and Vietnam.

The coppersmith

The GAUDY BARBET *(Megalaima mystacophanos)* has a large beak; the sexes differ in color. Males have red, yellow, black, and blue on the head, while females have blue and red only on the crown and not on the throat. These birds live in lowlands and moist coastal jungles from Malaya to Borneo. They feed on fruit. The COPPERSMITH (*Megalaima haemacephala)*, with its L of 17 cm, is one of the smallest barbets, being barely sparrow-sized. It is dark green with a red forehead and a yellow throat; the Javanese subspecies *Megalaima haemacephala rosea* has a red throat as well. These birds inhabit India and Ceylon as far as Java.

The great hill barbet lives at an altitude of 1000 to 2000 meters during the summer. In winter it descends to the lower hills and plains at the foot of the Himalayas. This bird is difficult to spot in the dense foliage of trees, but it gives itself away by its loud call, which Baker described as "peeo-peeo-peeo-peeo." Flocks of thirty to forty birds congregate in the winter when they are noticeably silent. The undulating flight is similar to that of woodpeckers. Three to four eggs are laid in a tree nest hole, made or merely widened by the birds. When the great hill barbet pulls arthropods out of cracks in tree bark, it clings to trees like a woodpecker, but looks much more clumsy.

The green barbet is not uncommon in the open forests, gardens, and even in the cities of Ceylon and India. It utters a monotonous call. The breeding season is from February to July; in Ceylon there are also second broods in August and September. This bird uses rotted, erect tree trunks or dead branches for its nest holes, making rather small nest chambers. The blue-cheeked barbet is also common in some areas of the plains and mountains. It was found breeding in Assam at altitudes of up to 2500 meters. Occasionally these birds will congregate in numbers on preferred fruit trees. Initially, the nestlings have a shiny green head; the bright marks do not appear until later.

The blue-cheeked barbet is one of the barbet species equipped with strongly developed rictal bristles; when these bristles are laid along the beak, they reach to its tip. Derek Goodwin, the English ornithologist, observed that when these birds were at rest, and while they accepted food, the bristles were well abducted; when the birds were excited, however, the bristles were pressed along the beak. The function of these bristles is not entirely clear; they may prevent bits of food from sticking to the head feathers. The somewhat rarer golden-throated barbet is found along with the blue-cheeked barbet; it is shyer, but betrays itself by its loud, melancholy calls. In Assam it breeds in dark gorges about 1000 to 1500 meters above a stream or river. Unlike the other barbets, it feeds almost exclusively on fruit.

In India, the coppersmith is the first bird to attract a visitor's attention. Its continually repeated "tonk" calls are uttered even during the hottest hours of the day, and even in cities. It uses any suitable tree for its nest hole, even if the tree is in a garden or on a much-used street. The entry hole is always on the lower side of a rotted branch; the two to four eggs are incubated for only twelve days.

We will mention two more rather inconspicuous representatives of the Asiatic barbets here; these birds are more sociable than their relatives, and they wander through the jungles in troops. These species are: the BROWN BARBET *(Caloramphus fuliginosus)*, which is found in several subspecies in Burma, as well as in Sumatra and Borneo; and the FIRE-TUFTED BARBET *(Psilopogon pyrolophus)*, which has a L of 28 cm and is found in Malaya and Sumatra. Its tail is longer and its beak is shorter than those of the other Asiatic species. The brown barbet, strangely enough, lacks the normally characteristic rictal bristles; it is also less vocal. This species searches tree trunks and branches for arthropods, like chickadees do; it also eats fruit. The fire-tufted barbet usually occurs in groups of five or six birds; it rarely calls, and feeds mainly on fruit.

Barbets, particularly those from India, have been exhibited in zoos for a long time. Some birds, especially those from the tropical rain forests, are delicate in their requirements, but others do well in captivity on a diet of various fruits, soft food, and insects. A blue-cheeked barbet lived at the Schönbrunn Zoo (Vienna) for eleven years. Barbets are somewhat unpleasant when kept in private homes because of their loud calls. They are often viewed as quarrelsome birds; this behavior does not apply to all species and individuals, however, although caution is advisable if other birds are to be kept with them.

Breeding in captivity has been rare; this is probably related to the difficulty of distinguishing the sexes in many species. The African Levaillant's barbet, however, bred as long ago as 1928 while under the care of an amateur in France, and later in the flight cage at the Philadelphia Zoo. Three species bred more recently in the well-arranged and well-planted flight cage at the Frankfurt Zoo.

Family: honey guides, by H. Friedmann

The HONEY GUIDES (family Indicatoridae) are the closest relatives of the barbets. The L is 10–20 cm. The tongue is short and thus can be protruded only slightly. These birds have nine primaries; the plumage is an inconspicuous brown, gray, or greenish color. Only a few species have white or yellow marks. Honey guides are brood parasites; their pure white eggs have thick shells. These birds are found in Africa, generally south of the Sahara; there are only two species in Asia, and these are found on the southern slopes of the Himalayas from Thailand and Malaya to Borneo. There are four genera with seventeen species.

Most honey guide species belong to the genus *Indicator*. The BLACK-THROATED HONEY GUIDE *(Indicator indicator;* Color plate, p. 67) has a L of 20 cm; there is a yellow ear patch. These birds inhabit most of Africa with the exception of the jungle areas, deserts, and grasslands. The SCALY-THROATED HONEY GUIDE *(Indicator variegatus)* is found from the Sudan to the Cape Province. The LESSER HONEY GUIDE *(Indicator minor;* Color plate, p. 97) inhabits various areas of Africa. The LEAST HONEY GUIDE *(Indicator exilis)* is found from Guinea and the Sudan to Gambia. The MALAYAN HONEY GUIDE *(Indicator archipelagicus)* is found from Thailand to Borneo. *Indicator xanthohotus* has a L of 15 cm; the rump is orange-yellow. These birds inhabit the Himalayas from Punjab to Burma.

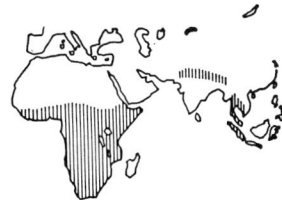

Fig. 2-1. Distribution of the honey guides (family Indicatoridae).

Members of other honey guide genera include the LYRE-TAILED HONEY GUIDE *(Melichneutes robustus),* which has a L of 17 cm and inhabits Central African forest areas; WAHLBERG'S HONEY GUIDE *(Prodotiscus regulus),* which has a L of 12.5 cm and is found in the Sudan and in East and South Africa; it is mainly a parasite of rock sparrows; and the TINKER HONEY GUIDE *(Melignomon zenkeri),* which has a L of 13 cm and inhabits the rain forests of southern Cameroun to the Congo; this bird is very rare.

The name "honey guide" comes from the peculiar habit of the black-throated honey guide, and to a lesser extent the scaly-throated honey guide, of apparently leading people to the nests of wild bees. This behavior is, of course, not purposive, nor is there any insight involved; it stems from the birds' excitement upon meeting a possible partner. Honey guides call not only to people, but also to the honey badger or ratel (see Volume XII), and possibly to other animals as well. The bird makes a continuous "churr churr' call, flies to a nearby tree, and waits to be followed by the man or badger. If this happens, the bird flies on to the next tree, and so on until it hears the bees or their humming. These sounds cause it to stop and to remain motionless on a nearby branch. The honey guide does not fly right up to the entrance of the hive, but African natives know that when the bird stops, there must be a hive nearby. Once the man or badger has opened the nest, and has taken the honey, the honey guide picks up the scattered remains. It will also feed on insects which it captures primarily in flight.

For a long time it was assumed that honey guides were interested in honey, or in the bee larvae; now, however, we know that members of the genera *Indicator* and *Melichneutes* (but not *Prodotiscus*) eat the wax of the honey combs. Such wax-eating is called "cerophagy." Two, and possibly more species of honey guides have the microorganism *Micrococcus cerolyticus* in their intestines; this microorganism decomposes wax into fatty acids which the birds can then utilize. This bacteria can decompose wax far more effectively than can the birds' digestive secretions, which are probably produced in the small intestine where wax digestion is most thorough.

According to the latest information, honey guides can, at least in part, detect bee hives by the smell of wax. They inevitably respond to the smell of a burning candle—an interesting observation, since the report of their wax eating was long overlooked. In the Sixteenth Century, a Portuguese missionary in Africa noted that these small birds flew into his church and nibbled at the candles. Today we know that honey guides have well-developed olfactory lobes in the brain.

As far as we know, all honey guides are brood parasites. We are sure of this behavior in six species, and it is unlikely to be different in the rest, although their reproductive habits are virtually unknown. All young birds generally have an egg tooth on the upper mandible, however. Black-throated honey guide young and those of two other species which parasitize hole-nesting birds, have two egg teeth; the tooth on the upper mandible is bent downward and is sharply pointed, as is the tooth on the lower mandible. This device helps young honey guides remove the host's young from the nest. The tiny, naked nestlings use the two egg teeth like scissors, destroying eggs or young birds with them. The egg teeth drop off a week after the honey guide young have hatched.

Brood parasites

The presence of such structures, developed in association with brood parasitism, suggest that this parasitism is very old. If all species prove to be parasitic, then the honey guide family would be the only avian family in which all the members have lost the original breeding pattern. If we compare the honey guides with other brood parasitic birds—like the cuckoos (Volume VIII), and a few icterids and wydahs—we find that they deviate the most from the normal breeding patterns of the bird. Their display is only feebly developed, and there is virtually no pair formation.

The nestling period in the genus *Indicator* is very long, as far as we know, lasting about a month. The young honey guides do not leave the host's nest until they are fully grown. Usually there is only one honey guide per nest, but in a few rare cases two have been observed. Occasionally these birds (particularly immatures) are found in small loose groups. Generally, however, they are found only singly. Because of their usually dull coloration, and the absence of a song, these birds are not often seen, although they are not uncommon.

Family: toucans, by A. F. Skutch

The TOUCANS (family Ramphastidae) are distinguished by their large, often colorful beaks; the ♂ Toco toucan, for example, has a bill that is 23 cm long, that of the ♀ is 21.5 cm, thereby exceeding the length of the body, not including the tail. The L is 32–60 cm. The bulk of the beak consists of a much widened upper mandible with the bird's nostrils at its base. The beak itself is not massive, but rather, consists of a network of bony septa which give it maximal strength with minimum weight. The beak is almost always brilliantly colored, often two or three different hues; the Keel-billed toucan even has five different colors. Toucans have a "feather tongue," a long, narrow, thin, flattened plate which is 15 cm long in the larger species. The beak is toothed at the edge; the indentations become deeper towards the tip, giving it a brush-like appearance. The wings are short and round. There are ten tail feathers; the tail of the aracaris is long and wedge-shaped, while it is shorter in the true toucans. The legs are powerful, and the foot has two anterior and two posterior toes. The sexes are usually similar, but ♂ ♂ have longer, thinner beaks. These birds are hole nesters, and they lay white eggs. Their diet consists of fruit and small vertebrates. There are seven genera with a total of forty species.

The huge beaks of the toucans are reminiscent of those of the Old World hornbills (Chapter 1); these two families are not related, however. The toucans' beaks make these birds the strongest and most easily recognized birds of the tropical American forests. The earlier folk name of these birds, "peppereaters" is inappropriate, because they eat various fruits and animals; only the Toco toucan occasionally eats paprika shoots.

Fig. 2-2. Distribution of the toucans (family Ramphastidae).

The GREEN TOUCANETS (genus *Aulacorhynchus*) are small, predominantly green toucans with white, gray, or blue throats, cinnamon-red lower parts, and red tips on the outer tail feathers. The EMERALD TOUCANET

(*Aulacorhynchus prasinus*; Color plate, p. 107) with a L of 35 cm, and the GROOVE-BILLED TOUCANET (*Aulacorhynchus sulcatus*) are the best known species of this genus. They live in mountain forests from southern Mexico to Bolivia and Guiana, rarely below an altitude of 900 meters. Some species spend the cold nights at altitudes of 3000 meters or more, descending to the valleys during the local winters. All of these birds have a "saw bill." They often try to dig out a hole in rotted wood, and the female does the bulk of the work. The toucanets' beaks are not suitable for such work, however, and as a result these birds usually use woodpeckers' holes for their nests. The female emerald toucanet incubates at night; incubation lasts for sixteen days. The young remain in the nest for about forty-three days.

▷ Toco toucan *(Ramphastos toco)* above. The marginal teeth on the upper mandible are clearly visible. Channel-billed toucan *(Ramphastos vitellinus)* below left. Plate-billed mountain toucan *(Andigena laminirostris)* center below. Collared aracari *(Pteroglossus torquatus)* below right.

The SHORT-BILLED TOUCANS (genus *Selenidera*) lead a secretive life in lowland forests and foothills from Honduras to Peru and Brazil. The SPOT-BILLED TOUCANET (*Selenidera maculirostris*; Color plate, p. 107) has a L of 33 cm, and weighs about 130 g. All of the six species in this genus have a sex color differentiation. The male is usually black on the head, neck, and breast, while the female is chestnut-brown or gray on the head and nape. There is a yellow tuft of feathers on each side of the head, and a yellow spot on the flanks; the under tail feathers are carmine-red. The MOUNTAIN TOUCANS (genus *Andigena*) include five species, all of which live in the Andes and the other South American mountains as far as Brazil. The BLACK-BILLED TOUCAN (*Andigena nigrirostris*; Color plate, p. 107), which has a L of 43 cm and inhabits Colombia and Ecuador, is a member of this genus. This bird is characterized by a blue head and nape, an olive-colored back, a yellow rump, blue lower parts, red under tail coverts, and chestnut-brown thigh feathers; its beak is generally more than one color. A close relative, the SAFFRON TOUCAN (*Baillonius bailloni*; Color plate, p. 107) has a L of 33 cm and is found in southeastern Brazil; it has golden underparts, but is otherwise more greenish in color.

The short-beaked toucans

▷▷ Wryneck *(Jynx torquilla)* left, at its nest hole. Green woodpecker *(Picus viridis)* right, feeding a youngster.

▷▷▷ Greater spotted woodpecker *(Dendrocopos major)* with food, feeding young.

The ARACARIS (genus *Pteroglossus*) comprise eleven species and are found from the southern Mexican lowlands to Argentina. They have a long, graduated tail, and they look lighter and slimmer than the other toucans. These birds are predominantly blackish or olive-green above, but they have a carmine-red rump. There is often a chestnut-brown or deep red collar on the nape. The breast and abdomen are usually yellow with a wide red, or black-and-red belt. The beak is smaller than that of the large toucans, but it is still huge in comparison with the bird's size. The edge of the upper mandible is often strongly toothed. The COLLARED ARACARI (*Pteroglossus torquatus*; Color plate, p. 81) are the FIERY-BILLED ARACARI (*Pteroglossus frantzii*) are included in this genus. These smaller, active toucans survive the destruction of jungles better than their larger relatives.

The aracaris

Fig. 2-3. Collared aracari.

Six collared aracaris roosted together at night in a hollow branch thirty meters above the ground in Panama. Only one adult spent the night in this inaccessible hole when brooding began, but four birds returned to their habitual roost after the young were fledged. In another instance, five adults brought food for the three young and slept together with the nestlings; they were attacked by a large white buzzard (see Volume VII) which, after killing one of the young, chased off the rest of the family. Five fiery-billed aracari adults often live together in one hole. Only one bird spent the night in the nest when it contained eggs, but both parents roosted with the young after they had fledged. Aracaris are the only toucans which roost in the nest hole outside the breeding season.

Genus *Ramphastos*

The largest toucans belong to the genus *Ramphastos*. The plumage is mostly black, but there are lighter colors in three areas; the throat and breast spots are white, orange, orange-red, or yellow, depending upon the species or subspecies, and they are often separated from the lower parts by a red band. The upper tail coverts are either red, yellow, or white, while the under tail coverts are always carmine-red. The bill is huge and ranges in color from almost pure black with a light colored root and culmen, to shades of various colors. The bare facial skin is orange, yellow, blue, or green. There are eleven species in dense jungles at lower elevations from central Mexico to Bolivia and northern Argentina. These include: the TOCO TOUCAN (*Ramphastos toco*; Color plates, p. 81 and p. 98) which, with its L of 60 cm, is the largest species; it has a white throat, an orange bill, and is found in Brazil; the CHANNEL-BILLED TOUCAN (*Ramphastos vitellinus*; Color plate, p. 81), which has a L of 45 cm, relatively short wings and a longer beak; it is found from northeastern South America to central Brazil; and the SWAINSON'S TOUCAN *(Ramphastos swainsonii)*, which has a L of 50–55 cm, and a broad yellow band on the culmen of its beak. The most colorful species of this genus are the RED-BREASTED TOUCAN (*Ramphastos dicolorus*; Color plate, p. 98), which has a L of 47.5–50 cm, and the KEEL-BILLED TOUCAN (*Ramphastos sulfuratus*; Color plate, p. 98), which has a L of 45–50 cm, and a beak with five different colors.

Toucans wander through the forests and adjacent clearings in families and flocks; such flocks rarely consist of more than twelve birds. Toucans are not intensely sociable; they never take flight in a tight group, but rather, they wander about in loose groups. The agile aracaris fly swiftly and in a straight line; the large toucans are poor fliers. After they have beaten their wings a few times, they hold them out and glide downward, as if pulled down by the weight of their large beaks. Then they begin to beat their wings again. The flight, as a result, is both undulating and brief. Toucans bathe most readily when there is a water-filled depression in the fork of a tree, or on a thick horizontal branch.

◁
Greater spotted woodpecker *(Dendrocopos major)* above, and the black-naped green woodpecker *(Picus canus)* below, taking off.

The toucans, in contrast to the "pompous" trogons, the "serious" pigeons, and the "busy" woodpeckers, are the "jokers" of the American tropical forests. They will often fly to and fro among the branches, as if in play, beating their beaks loudly against them, much as if they liked to hear the noise. They will often tease one another and use their beaks like dueling swords. One toucan may grab another by the beak and push until the other is forced off its perch. Another form of play is tossing up a berry, which another bird then snatches up.

Toucans preen one another, particularly on the head and nape, with the tips of their beaks. In captivity they are interesting and, at times, troublesome pets. They keep other birds and small animals away with their large, bright beaks, and often they become a real plague to those that live with them. They become so trusting of their guardian that they will slip inside his clothing and warm themselves against his body. Toucans are also inquisitive in the wild. Some Swainson's toucans watched a botanist from a low branch while he arranged the plants he had collected in a remote mountain forest; the birds looked as if they were interested in what the man was doing.

Toucans feed largely on fruit which they pluck while they are perched. Getting a small berry from the tip of the huge beak into the throat is quite a task; toucans perform it by jerking the head back while the beak is open. When the soft part of the berry has been digested, the hard seeds are regurgitated. These birds also eat insects, spiders, and other small invertebrates; in addition, they capture several forms of termites which fly around in the evening. Large lumps of food are held under the foot and reduced to smaller lumps with the beak. Alexander von Humboldt reported that a toucan kept on his canoe during a journey up the Orinoco River liked to catch fish on the shore. Toucans even plunder the nests of other birds for their eggs and young; they are mobbed by other birds as a result.

Naturalists have long puzzled over the significance of the toucan's beak. It was once suggested that the beak was to be used as a weapon in nest hole defense. This is not so; when toucans sense danger, they come out of the hole in a hurry, threatening the enemy only out in the open, if at all. The long beak enables these rather heavy birds to pluck berries from the tips of branches without leaving their stable perch. A thin, dark-colored beak would, however, be just as useful for this purpose. Possibly the toucan's beak plays a role in pair formation and in the social life of the group; according to E. Thomas Gilliard, it acts like a signal.

However, toucans can also use their beaks to threaten those birds whose nests they plunder. Tyrant flycatchers and even small raptors are so scared by the giant beak, which is even more effective because of its lively colors, that they fly about helplessly while the toucans devour their young or eggs. These birds will only attack the toucan

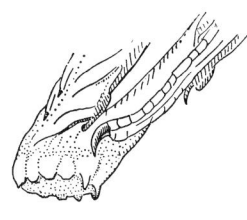

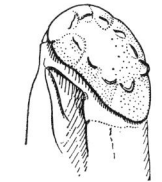

Fig. 2-4. Tarsal joint pad on the right food of a toucan nestling, seen from the side (upper drawing) and from below.

Fig. 2-5. Toucan nestling.

Family: woodpeckers, by H. Wendt

while it is in flight, because then it is unable to defend itself with its beak. When sleeping, toucans lay their beaks up over their backs, and tip up their tails—forming a roof over the back and beak. One can see how advantageous such a "feather ball" is to the black aracaris which sometimes sleep five or six together in an old woodpecker tree hole or in rotted hollow tree trunks; the last bird to squeeze in enters the hole backwards, with its tail laid over its back.

Toucans utter their melodious calls many times in the evening, continuing when most other birds have gone to rest. They nest in tree holes, usually high above the ground; the larger species prefer rotten patches in living trees, while the smaller ones often use woodpecker holes. Toucans drive woodpeckers away from newly constructed nest holes, and then enlarge the entry way if it is too narrow for them. The floor of the hole may be eight to twenty centimeters below the opening. Toucans sometimes lay their eggs in the same hole year after year. The eggs lie on wood debris, or often on a lining of seed pellets which the birds have regurgitated before egg laying. Both sexes share the incubation during the day, but they are impatient and rarely stay on the eggs for more than forty to fifty minutes at one time.

Newly hatched toucans have a bare red skin which is not covered with down. The short beak has a longer and wider lower mandible, as is the case with young woodpeckers. A thick horny swelling on the heel protects this joint from friction. These pads on the heels have sharp outgrowths which may form a ring. The parents bring some of the food in the tips of their beaks, but most of it is carried in the throat or esophagus; this is regurgitated after they have delivered the food in the beak. The parents and the older nestlings clean the nest hole and carry great beakfuls of refuse away from it.

The young toucans open their eyes only after they are three weeks old. The feathers grow so slowly that the young still show much bare skin even after they are a month old. The young of the larger species probably remain in the nest for about fifty days. The main enemies of the adult toucans seem to be the raptors; sometimes they catch toucans almost as large as themselves. Indians use the colorful toucan feathers for decoration. Both the first white settlers and the Indians kept free-flying toucans taken from the nests. Such tame toucans can even dominate chickens in a village or farmyard.

WOODPECKERS (family Picidae) are so much adapted to a climbing and pecking mode of life that a certain narrowness in their behavior has taken place. This specialization notwithstanding, these birds are found all over the world in the most diverse habitats, with the exception of Australia, New Guinea and the adjacent islands, Madagascar, and the polar areas. There are three subfamilies: 1. Wrynecks (Jynginae); 2. Piculets (Picumninae); and 3. True Woodpeckers (Picinae), with a total of forty genera and 209 species.

The WRYNECKS (subfamily Jynginae) have a relatively short beak, but the tongue can be projected as far as that of the true woodpeckers, even though it does not have barbs at the tip. These birds are somewhat larger than sparrows; the L is 16.5 cm and the weight is 30–39 g. Wrynecks nest in holes. Their behavior and structure is like that of the woodpeckers; thus these birds have climbing feet and a sticky tongue, but no supporting tail or wedge-like beak. Wrynecks are inconspicuous birds and are not really common anywhere. There is one genus with two species: the RED-BREASTED WRYNECK (*Jynx ruficollis*) in Africa and the EURASIAN WRYNECK (*Jynx torquilla;* Color plates, p. 82 and p. 108) in Europe and Asia. This latter species winters in Central Africa and southern Asia; there are six or seven subspecies.

Subfamily: wrynecks, by D. Blume

The Eurasian wryneck lives in open deciduous forests, grassy clearings, copses, gardens, and along country roads during the breeding season. It needs trees with coarse bark and holes that were once used as nest sites. It also breeds in ground holes or holes on steep banks or loess slopes in treeless areas such as in Asia. The bird gets its common name from a peculiar pendular head movement which it uses both as a defensive and as a courtship gesture. Von Lucanus, the ornithologist, describes this movement as follows: "The bird spreads its tail, erects its neck and head feathers, and then performs peculiar distortions of the head by drawing it in and then jerking it forwards or turning it from side to side." The bird hisses considerably while engaged in these unusual movements—thus increasing the serpentine effect of this behavior immeasurably. When feeding, the wryneck first exploits one source (an ant's nest for example) fully before flying on to the next one. Its sticky tongue enables it to catch ants with ease; pupae, larvae, and even adult ants stick to the tongue and are taken in to the bird's throat, which may hold up to 150 of these insects at one time.

Fig. 2-6. 1. Wryneck (*Jynx torquilla*), breeding range; 2. Winter area of Eurasian wrynecks (with the exception of the Mediterranean races); 3. Winter area of the Asiatic races; 4. Red-breasted wryneck (*Jynx ruficollis*).

As soon as they return from their winter quarters, in mid-April or May, the males occupy territories with suitable holes, and announce their presence with quacking calls. Females, attracted by these calls, reply in the same manner. The male shows the hole (which may be a natural cavity, a woodpecker hole, or a nest box) to the female by repeatedly entering it or emphatically looking out of it. The seven or eight eggs are laid on the floor of the hole. Both sexes incubate alternately when the clutch is complete; apparently the male incubates the eggs at night. The naked young hatch after eleven days; they have thick heel pads, and are also equipped with tactile swellings on the lower mandible. Before feeding the young, the adults touch these latter swellings and thus induce the young to gape. Then the parents regurgitate the food (ant pupae) into a young's throat. O. Steinfatt found that one wryneck brood requires 8,000 to 12,000 ant pupae per day. The young leave the nest when they are twenty-one days old. The

Fig. 2-7. The wryneck moves its head to and fro and twists its neck in threat or display posture. The effect of these movements is reinforced by hissing and a sort of purring.

parents continue to feed them for another fourteen days, after which they become independent. Occasionally there is a second brood, but in such cases the clutch is generally smaller. Central European wrynecks leave their breeding areas singly from July to early September.

Recently there has been a great decrease in the number of these birds in Central Europe. According to H. Menzel, many of them are shot during migration. Increasing traffic and chemical weed control are two more dangers for the wrynecks. More than half of the young only reach an age of three months; the rest of them live for an average of one to two years. The oldest age known for a banded bird was six years and four months.

Subfamily: piculets, by J. Steinbacher

The PICULETS or PYGMY WOODPECKERS (subfamily Picumninae) differ from the true woodpeckers because of their long tails with the short shafts, which cannot be used to support the body, and because of their short, laterally compressed beaks. These are small birds of the typical woodpecker shape with a L of 8–15 cm. The tongue can protrude only slightly, and the ends of the hyoid bone end at the base of the beak. There are four genera with twenty-nine species, including: the THREE-TOED PICULET *(Sasia abnormis)*, which has bare eye rings; its upper parts are olive-green. This species is found from Thailand to Borneo; the AFRICAN PICULET *(Verreauxia africana)*, which has a L of 8 cm. It is golden-olive above and dark gray below; and the HISPANIOLA PICULET *(Nesoctites micromegas)*, which has a L of 13 cm, and is found on Hispaniola and the neighboring island of Gonave. There are twenty-four species of the genus *Picumnus* which is also included in this subfamily; these birds are found in tropical America. The ARROW-HEAD PICULET *(Picumnus minutissimus)* is pictured in the color plate on page 108.

Fig. 2-8. Wryneck catching an ant with its sticky tongue. The tongue is constructed like that of woodpeckers; the keratinized tip is smooth.

The piculets, like their larger relatives, can climb in trees; however, they also sit across branches like passerines, or they can walk down tree trunks headfirst, and even hop from branch to branch like nuthatches. Their soft tails often become greatly worn, especially during nest construction. The undulating flight is short and slow. Piculets work over bark and soft wood searching for arthropods and their larvae, particularly ants, termites, and wood-boring beetles. They chip out nest holes in rotted tree trunks, or enlarge preexisting holes. The two to four eggs are incubated for twelve to fourteen days. Both parents feed the young, which fledge in twenty-one to twenty-four days. The piculets' spotty distribution in Asia, Africa, and America indicates the great age of this subfamily.

Subfamily: true woodpeckers, by D. Blume

The TRUE WOODPECKERS (subfamily Picinae) are small to medium-sized birds with strong, compact bodies; the L is 9–55 cm. Food is reached by chiseling and probing with the tongue. Besides flying, climbing is the major method of locomotion, and these birds climb on tree trunks, branches, or on cliffs. Structural adaptions for climbing include: the chisel-like beak with its hardened edges, the shock-absorbing formation of the skeleton and the muscles of the head and neck; the worm-like tongue which can extend for some distance; the climbing foot (first and fourth toe are directed to the rear, while the second and third toe are directed to the front; the fourth toe can be abducted to the side and front); the sharp claws which are curved like climbing irons; the supporting keel-shaped tail with its hard feathers; and the generally short and round wings. The small and medium-sized species have a pronounced undulating flight, while the larger birds have a shallow undulating flight or fly in a straight line. These birds are hole nesters; their holes are usually made by chiseling, or by a boring type of digging. Woodpeckers roost in the nest hole outside the breeding season. There are 34 genera with 182 species.

Fig. 2-9. Pileated woodpecker showing the characteristic woodpecker climbing posture.

The woodpeckers' diet consists of insects and their larvae, which are found in the wood of trees, in the ground, or in ant and termite nests. The abilities of climbing and chiseling have been developed to such a degree that these birds can capture their prey. The hard climbing claws hold the woodpeckers firmly on tree trunks, branches, rocks, or walls of buildings. If the surface is rough, the bird uses two toes pointing forward, and two back; if the surface is smooth, the bird rotates its outer hind toe forward and thus gains support. These birds not only climb up, but also down, with equal agility, although they do not climb down headfirst as nuthatches do. The only time the woodpecker climbs down headfirst is when it is entering into its hole. Woodpeckers climb down on trees in jerky movements, where the method of support by means of the tail is particularly evident. They climb equally readily on both the upper and lower sides of horizontal branches and twigs.

Hopping on the ground looks awkward because the tendons and muscles are particularly adapted to climbing. The characteristic jerky mode of climbing is made possible because the tail has evolved into a support structure. The twelve strong tail feathers, which are pointed at the tip, lie above one another like shingles. The keels of these feathers are very strong and elastic, and the barbs are both hard and resistant. The central pair of tail feathers are moulted last, so there are always some tail feathers to act as supports.

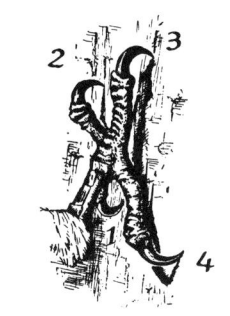

Fig. 2-10. The foot of a woodpecker, adapted for climbing. The first and fourth toes (1,4) are opposite the second and third (2,3).

Many species, like the green and gray woodpeckers and the North American flickers, which are grouped together as ground woodpeckers, cling closely to tree trunks when climbing and move easily and skillfully, although they cannot cling to one spot for long. These woodpeckers are more inclined to search and probe, and they wander from one site to another on trees in search of food. They sit on ant nests on the ground for long periods, cleaning them out thoroughly. Other species, like the European spotted and three-toed woodpeckers, hold their bodies well away from the tree trunk, and moving up and down in jerky hops when climbing. These birds can remain in one spot

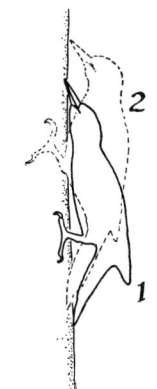

Fig. 2-11. Ground woodpeckers climb skillfully and keep the upper part of the body close to the trunk (1 and 2 are different climbing phases).

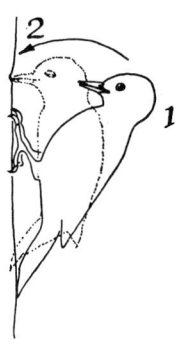

Fig. 2-12. Pecking movements of a three-toed woodpecker *(Picoides)* (1 and 2 are different phases).

Fig. 2-13. The woodpecker skull has adaptations to protect the brain against shock effects.

for longer periods, and with powerful blows with the beak they can chisel open and exploit abundant food sources. There are also species, like the ivory-billed woodpecker, which climb with legs outspread, as they need a wide base of support for the lateral blows with which they chip off the tree bark.

The beak is the woodpecker's major tool; it is used for chipping out the nest hole, to expose food sites on the ground or in wood, as well as for "drumming," a behavior equivalent to bird songs that in other species are used to attract females and to stake out territories. Because of the kind of use made of the beak, these birds have structural shock-absorbing adaptations that protect the skull and brain. What these devices are in particular, we know only from the recent studies of L. W. Spring and others. According to these studies, all woodpeckers have an ossified interorbital foramen, and certain skull muscles which are inserted in the beak and mandibular joint are so strongly developed that they take up the impact of the beak's blows by contraction and pressure. The frontal bone above the upper mandible has supporting bony septa. The horny sheath of the beak in the woodpeckers that chisel is very strong, and it is strengthened even more by ridges; the tip is wedge-shaped. The ground woodpeckers, on the other hand, often get their food by probing or light tapping, and as a result they only have thin, pointed, somewhat downward-curved beaks.

The peculiar hyoid apparatus is the final adaptation of the woodpeckers to their way of life; with it woodpeckers can extract prey from the smallest holes in trees, or from inside of ant or termite nests. Many woodpeckers can extend their tongues surprisingly far out of their beaks by means of special structures; the green woodpecker, for example, can extend its tongue as much as ten centimeters out of the beak. Very thin, flexible bony filaments project from the base of the tongue and run in two loops about the head, and come together at the top; these end on the forehead, or in the right upper mandible if they are very long. If the tongue is to be extended, certain muscles pull on the hyoid loops, pushing the base of the tongue, the hyoid horns, and the tip of the tongue forward.

The tip of the tongue is horny and has bristles. In tree woodpeckers these bristles serve as hooks on which the larvae are impaled. Some ground woodpeckers have a more spoon-like tongue which ends in a wide cluster of bristles; they can literally "spoon up" ants and termites with this tongue. An African banded woodpecker has a special fold for storing the tip of the tongue in the lower mandible. The tip of the tongue is modified to resemble a brush in the American sapsuckers, and it is used for sucking up sap.

Heinz Sielmann, the well-known naturalist and photographer, has illustrated the workings of the woodpecker's tongue in outstanding films, and he has written the following description: "Using a procedure similar to what we had tried with other species of woodpeckers, we mounted a piece of wood from an ant's nest onto a branch next to the window of the flight cage. Soon we had a rare sight: the green woodpecker was pushing its sticky tongue into the pupae chamber. The tongue snaked its way through the passages like a long worm, and because of its thick coating of mucus it caught everything it touched, ants or pupae, into the bird's gullet." Sielmann also noted that the tip of the tongue could be moved independently of the rest. "Viewing the tongue closely, we could see how the tip probed the area for food and shoveled out everything (particularly the pupae), much as one does with a little spoon, so that everything would get stuck on the long tongue." In seasons when there are few insects, however, many species of woodpeckers also take berries and fruit as a supplement or even as the mainstay of their diet. A unique habit of storing food often develops in connection with this, and the tongue plays a role in this habit as well; it is used to push fruit into hiding or out of holes and crevices.

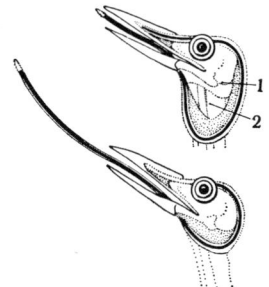

Fig. 2-14. Green woodpecker's tongue. The hyoid horns form a sling (1). Muscles which retract the tongue (2).

Usually woodpeckers chisel out their nest cavities themselves, placing them, by preference, into trees. However, a pair does not make a new nest hole every year. If there are still good nest sites available, the birds readily accept them. When trees suitable for nest holes become scarce, one may also observe the same nest hole being used for several years. Black woodpeckers merely chip out the nest interior somewhat, because wood chips serve as a base for the eggs. Both sexes work at chipping out the new cavity, although the male does the larger share of the work. Nest holes are generally made in places in the wood which are already damaged or weakened; however, a woodpecker may chip out its nest hole in a tree trunk that is completely solid. When woodpeckers use an old hole again, they dissipate their drive to chip by pecking at other tree holes in their territory.

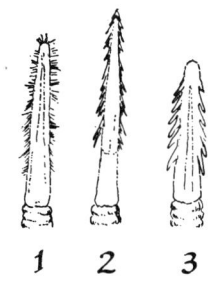

Fig. 2-15. Enlarged views of keratinized tongue tips: 1. Sapsucker; 2. Greater spotted woodpecker; 3. Green woodpecker.

There are also woodpeckers which do not nest in trees. Some South American flickers make cavities in soil cliffs. One species breeds in cacti. Some Indian banded woodpeckers breed in the nest of tree ants, and the African termite woodpecker breeds in termite nests.

The eggs of all woodpecker species are glossy white. The clutch usually consists of two eggs in the tropical species, while the nontropical species generally lay four to seven eggs. After a strikingly short incubation period, which is only eight to nine days in the greater spotted woodpecker, and only thirteen days in the black woodpecker, the blind, naked young hatch in quick succession. Their development within the nest hole lasts three weeks in the smaller, and four weeks in the larger species. There is only one brood per year in the temperate latitudes, but in warmer areas there may be two. The male does most of the incubation. The partners incubate alternately during the day,

Fig. 2-16. Nest hole of the greater spotted woodpecker.

Fig. 2-17. Black woodpecker at its hole; the entrance hole is elongated.

and they also alternate in keeping the newly hatched young warm; only the male sits on the eggs, and later on the nestlings, at night. Only in the Central American golden-naped woodpecker and a few other species do both the male and female sit together in the nest during the night.

Woodpeckers use old nest holes as roosting places after the breeding period; occasionally they construct holes just for roosting. They leave the roost hole in the morning and reenter it in the evening, according to a definite timetable specific to each species. In the Central European species, for example, the female has the longer "working day"; she rises earlier in the morning and returns later at night than does the male. The black woodpecker has the shortest "working day," while the greater spotted woodpecker has the longest. These different timetables may possibly play a useful role, since they may enable a number of different woodpecker species to spend the night in a single wooded area without disturbing one another and consequently coming into conflict. The birds arrive at their holes one after another in the evening, and they fly out again in the morning one after another.

European woodpeckers are essentially residents; they stay within their chosen areas, and only during the months after fledging do they migrate any great distance from where they were born. The northern subspecies of the greater spotted woodpecker undertakes longer migrations and the young are particularly involved in these movements. These woodpeckers do not avoid crossing the Baltic or North Sea when migrating to western and central Europe. There are also migrants among the non-European woodpecker species, including many North American sapsuckers, flickers, and acorn woodpeckers.

Woodpeckers, like other birds, communicate by calls and song-like sequences of calls. In addition, they have developed a unique system of calls which consist of drum-like rolls and taps which are made with their beaks. They not only attract partners with these sounds, but they also indicate territory ownership, nest trees, choice of nest site, and readiness for relief during incubation. Almost every species drums and taps in its own typical rhythm. To drum, a woodpecker perches on a dead branch, a hollow tree trunk, or some similar spot with good resonance, and beats the base with very rapid blows of its beak, producing a roll. The length of the individual rolls and the intervals between them enables woodpeckers to recognize members of their own species. The well-trained observer can also discern the meaning of the tapping or drumming of a particular woodpecker, including whether it is near the nest hole, whether the conspecific to whom it is showing the hole is already nearby, whether rivals are in conflict, or whether two partners are relieving one another at nest building or incubation.

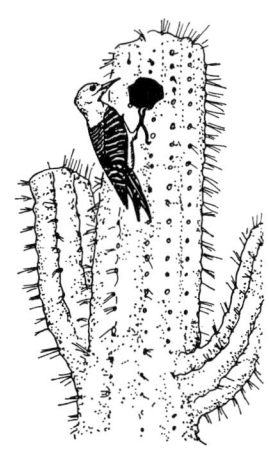

Fig. 2-18. Gila woodpeckers build their nests in cacti.

While making his films, Sielmann also made tape recordings which he played back to the woodpeckers in order to determine the specific functions of these sounds. "The four minute tape which we used contained the drumming of the greater spotted woodpecker, and at the end, the black woodpecker's phrase. When the tape was played in an area inhabited by both species, the spotted woodpecker always replied first. The black woodpecker came in only when it heard the drumming of an apparent rival of its own species. This indicates that every woodpecker can be recognized by a member of its own species because of its typical drumming rhythm."

Ornithologists have made many tape recordings of woodpecker drummings, and have made quantitative determinations by playing them back at varying speeds. According to these results, the black woodpecker's drum roll consists of thirty-eight to forty-three beats; it lasts for 2.1–2.69 seconds and is delivered three times per minute. The gray woodpecker's drumming consists of about thirty beats and lasts for just over a second; the drumming of the lesser spotted woodpecker, which also has thirty beats, can barely be distinguished from the drumming of the gray woodpecker, except that it is given twelve to fourteen times during the same time period. The greater spotted woodpecker's drum roll is shorter, with twelve to sixteen beats given in 0.6 seconds; it is repeated eight to sixteen times a minute.

Unlike the various species' drumming patterns, the tapping signals, which are used to show nest holes and to communicate with the partner at close range during nest relief, show fewer differences. The North American red-bellied woodpecker has an interesting "duet tapping" which was discovered and interpreted only a few years ago by L. Kilham. When the male and female tap by a hole together, this indicates that they have agreed to make that hole their nest site.

Woodpeckers, with a few exceptions, are unsociable and live solitarily. Males and females have separate territories outside the breeding season, or at least their activities involve different portions of their range. If in the course of the day they approach one another too closely, they soon fight vigorously. Their beaks can be dangerous weapons, and fatalities have actually resulted from such encounters. Usually, however, conflicts among woodpeckers are carried out without bloodshed—by threat calls, flights, or special sham duels with the beaks. A certain hostility always persists between the members of a pair even during the breeding season. Oskar Heinroth, the great ornithologist and co-founder of ethology, has the following report on this behavior in the black woodpecker: "When we chased the incubating woodpecker from the nest hole and its partner appeared after a while, there was always a period of aggressive pecking between them; we also noted that as the one bird flew onto the nest tree, the other would emerge and immediately rush off. One had the impression that both birds found the fact that they needed a partner for the breeding and rearing of young rather horrible." Consequently the woodpecker's

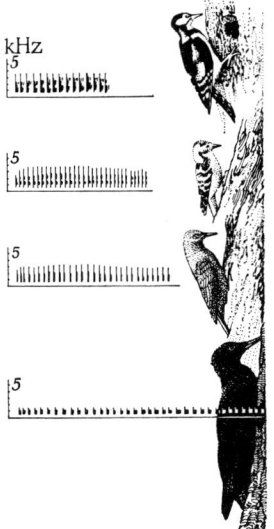

Fig. 2-19. The drumming sequences of some woodpecker species on sound spectrograms. From top to bottom: greater spotted woodpecker, lesser spotted woodpecker, gray woodpecker, and black woodpecker.

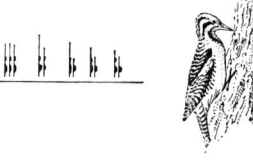

Fig. 2-20. The sapsucker's drumming consists of a short roll and distinct double or single taps.

Fig. 2-21. Ground woodpecker (Geocolaptes olivaceus).

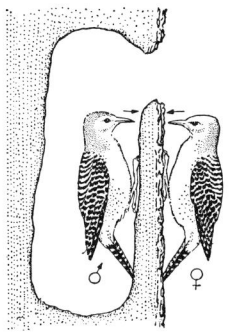

Fig. 2-22. The red-bellied woodpecker's tapping duet. Above: both partners tap at the nest hole entrance. Below: the male taps within; the female taps outside the hole.

Fig. 2-23. Greater spotted woodpecker attracts a female to the nest hole. 1. The male drums; 2. The female approaches in flight; 3. The male approaches the hole in display flight; 4. The male taps emphatically beside the hole.

courtship display looks like a threat display. Initially the future partners face one another in a threat posture, and one will not permit the other to come close. Gradually the threats decline; the birds tend to keep only slightly apart, and the threat movements are largely inhibited.

For some time there was a difference of opinion about the role of the woodpeckers in cultivated forests. Their chiseling on trees was formerly much resented. Today we know that woodpeckers are necessary for the community life of the forest, and as a result they are regarded as beneficial to forestry. G. Kneitz stated, after much thorough study, that: "The role of woodpecker holes in providing nests for hole-breeding songbirds can hardly be overestimated. The holes provide nest sites as well as refuge and roosting places for many species, particularly in winter." When tree trunks are attacked by wood-boring insects, they are regularly worked over by woodpeckers which take a portion of the pupae and larvae. Woodpeckers also completely debark diseased tree trunks, thus depriving harmful insects of a refuge. Tree stumps and fallen trees are worked over by ground woodpeckers and completely chipped to pieces in due course. Thus woodpeckers play a considerable role in the decay processes in forests. Beneficial forest ants can be protected from woodpeckers by covering their nests with protective hoods made of wood, wire nettings, or sheet metal.

There is no general agreement about the systematic arrangement of the true woodpeckers. Eight tribes are recognized at present; these are the ground woodpeckers, the flickers, the barred woodpeckers, the black woodpeckers, the banded woodpeckers, the sapsuckers, the spotted woodpeckers, and the crested woodpeckers.

The first of these groups, the GROUND WOODPECKERS, consists of only one species, the GROUND WOODPECKER (*Geocolaptes olivaceus*; Color plate, p. 108) from South Africa; the L is 25 cm. This bird is about the size of the green woodpecker. The plumage is like that of the American flickers; however, the ground woodpecker scratches more on the ground for its food than does the flicker. These birds migrate over open steppe country in groups of thirty to forty, and gather ants and termites. They build their nest holes in soil cliffs rather than in trees, and as a result they are diggers rather than carpenters. They prefer to roost in rock clefts.

FLICKERS are found in the New World in three genera with ten species. The YELLOW-SHAFTED FLICKER (*Colaptes auratus*; Color plate, p. 108) has a L of 32 cm and is found in North and Central America; the western subspecies, the RED-SHAFTED FLICKER (*Colaptes auratus cafer*; Color plate, p. 108), and its relatives have a L of 30 cm and are found from southern Alaska to Guatemala. The shafts of the flight feathers and the undersurface of the wings are coppery-red (these parts are yellow in the eastern species). The male has a brown head with a red cheek stripe, while the members of the eastern species have a black cheek stripe. The ANDEAN FLICKER *(Colaptes rupicola)* has light brown breast sides and triangular black spots which are arranged in bands; it is found on cliffs in the Andes, up to altitudes of 4700 meters. The CAMPOS FLICKER *(Colaptes campestris)* is found in southern Brazil and the Argentine pampas; it feeds on termites. FERNANDINA'S WOODPECKER *(Nesoceleus fernandinae)* has a L of 30 cm; this bird has yellow bars on a black background. The GREEN-BARRED WOODPECKER *(Chrysoptilus melanchloros)*, which has a L of 30 cm, is yellow with black bars; it is found from eastern Brazil to Chile and Argentina.

Flickers are lively birds with a colorful plumage, particularly in the golden or copper-colored feather shafts of the wing and tail. Since they are ground woodpeckers, they take the place of the European green and gray woodpeckers in America. Flickers are widely known in North America because of their striking wing and tail movements and their calls. They often hop about on the ground, probing for ants, snapping up arthropods, or picking up fallen fruit. Forty percent of their food consists of fruit and other parts of plants; sixty percent of the food consists of insects, two thirds of which are ants. Flickers migrate in large troops from the northern parts of their range southwards in the fall. Their behavior during the courting season is striking; they perform unusual dances in which they wave their heads to and fro and hold their expanded tails at a slant. These sham fights reach their climax when the tails are fully fanned out and twisted forty-five degrees to the side, so that the partner can see the bright yellow undersides of the tail feathers.

Both partners work on the nest hole, but they do not make a new one every year. As is the case with the green woodpeckers, flicker males and females try to attract one another to suitable holes by means of call sequences, marking their breeding territories by drumming. The clutch consists of six to eight eggs. The young hatch after an incubation period of fourteen to sixteen days. Both parents brood the young (as they do the eggs) during the day; only the male covers them at night, however. The young leave the nest hole when they are twenty-five to twenty-eight days old.

The tribe of BANDED WOODPECKERS, with eight genera and fifty-seven species, is widely distributed in the New World as well as in Eurasia and Africa. The birds of the genus *Piculus*, which has a L of 20-25 cm, have a greenish background color with red-brown or yellow tones on the inner edges of their flight feathers, and bands across the breast and belly. The GOLDEN-OLIVE WOODPECKER *(Piculus rubiginosus)* is a well-known representative of this genus. This bird has golden-yellow upper parts and a gray crown; it is found from Mexico to Bolivia. The African woodpeckers of the genus *Campethera* have a L of 15-24 cm and are

Fig. 2-24. Subspecies of the yellow-shafted flicker *(Colaptes auratus)* in North America: 1. the *cafer* group; 2. *auratus* group; 3. subspecies *Colaptes auratus chrysocaulosus*; 4. *mexicanoides* group; 5. *chrysoides* group; H. Center of zone of hybridization between *auratus* and *cafer* groups.

▷
Barbets and honey guides: 1. Toucan barbet *(Semnornis ramphastinus)*; 2. Great hill barbet *(Megalaima virens)*; 3. Blue-cheeked barbet *(Megalaima asiatica)*; 4. Collared barbet *(Lybius torquatus)*; 5. Red-and-yellow barbet *(Trachyphonus erythrocephalus)*; 6. Diadem barbet *(Tricholaema diadematum)*; 7. Lesser honey guide *(Indicator minor)*; 8. Bristle-nosed barbet *(Gymnobucco peli)*; 9. Golden-rumped tinker-bird *(Pogoniulus bilineatus)*.

Fig. 2-25. Dance of the yellow-shafted flicker, whereby it shows the underside of the tail. When the bird is more excited the under surfaces of the wings are also exposed.

Fig. 2-26. Termite woodpecker *(Campethera nivosa)*.

◁

Toucans: 1. Toco toucan *(Ramphastos toco)*; 2. Cuvier's toucan *(Ramphastos cuvieri)*; 3. Keel-billed toucan *(Ramphastos sulfuratus)*; 4. Red-breasted toucan *(Ramphastos dicolorus)*; 5. Ariel toucan *(Ramphastos vitellinus ariel)*.

somewhat smaller than the members of the previous genus. The TERMITE WOODPECKER *(Campethera nivosa)* has a L of 15.5 cm; this bird is so adapted to a diet of termites and ants that it sometimes makes its nest hole in tree termite nests. The GOLDEN-BACKED WOODPECKER *(Campethera maculosa)* has a special fold in its lower mandible for the tip of its tongue.

The GREEN WOODPECKERS (genus *Picus*) have a total of thirteen species (two in central Europe) found in Eurasia and North Africa. These birds are included within the tribe of banded woodpeckers. The GREEN WOODPECKER *(Picus viridis*; Color plates, p. 82 and p. 119) has a L of 35 cm and weighs 170-210 g. It has several shades of green, as the name indicates; these birds also have a black face mask and red crown. The malar stripe is red in the males and black in the females; the rump is yellow. Juveniles show strong barring, and the red cheek stripe is pale in young males. Green woodpeckers are found in Europe and West Asia. The BLACK-NAPED GREEN WOODPECKER *(Picus canus*; Color plates, p. 84 and p. 119) has a L of 30 cm and weighs 98-125 g. It is distributed in fifteen subspecies from central Europe to northern Japan, and Vietnam. Males have a red forehead, while the forehead in the females is grayish-green. These birds have no face mask and only a thin black cheek stripe. The green plumage is partially replaced by gray.

Green woodpeckers may be viewed as structural and behavioral examples of the whole tribe of banded woodpeckers. The green woodpecker settles in copses, at the edges of woods with adjacent meadows, in pastures with bushes, in open mixed woodlands, tree-lined avenues, gardens and parks. Ants of all kinds make up the major part of its diet; it captures them by means of its long sticky tongue, which ends at the tip in a spoon-like horny point. This woodpecker is often flushed from the banks along tracks in fields or in forests where it has been probing ants' nests. It regularly visits the nests of the red forest ants in the fall and winter, and it may open up passages up to seventy-five centimeters in length. The green woodpecker also eats bumble bees, flies, and beetles, and, occasionally as a supplement, wild cherries, rowan berries, and other fruits.

Green woodpeckers visit favorite feeding spots again and again, and one can find their characteristic marks on the bark of nearby trees. These marks are made when the birds land; holes are made by tapping before going down to the ground, and other marks are made as the bird whets its beak after eating. Green woodpeckers go to their sleeping places earlier in the evening than do the black woodpeckers.

Green woodpeckers utter their courtship calls in the mornings and evenings from February onward, near the tree in which they roost; the call is a ringing sequence, reminiscent of laughter, which sounds like "klew klew klew." This vocalization serves to establish contact with possible partners in the vicinity. Males and females then visit one another in their territories, look at tree holes, and keep in touch for days by calling. Green woodpeckers drum less often than do other species. The call sequences are loud and multisyllabic at the beginning of courtship; they become softer and shorter as the pair formation proceeds. The sexes recognize one another as the result of head movements to and fro, whereby the crown, face mask, and the cheek stripe are displayed.

Fig. 2-27. Green woodpecker *(Picus viridis)*.

When green woodpeckers need a new nest hole, they prefer to build it in apple and pear trees, aspen and linden trees, in alders, robinias, cherry trees, and in other soft woods. Their holes may also be found in oaks, beeches, and occasionally in conifers. Eggs are laid in early May in the mountains of central Germany. The clutch consists of five to seven eggs; these are incubated by both partners during the day, but only by the male at night. Relief of the partner during incubation, and later when the young are kept warm, takes place at intervals of one and one-half to two hours. In bad weather the interval between changeovers may last up to four hours.

Usually all the young hatch in quick succession and they are brooded by the parents for several days, to keep them warm after they are fed. The parents carry ants and ant pupae in their gullets to the young; they then regurgitate this material in lumps into the youngs' beaks. The begging call of the young is a rhythmical chattering, sounding like "akakak." The young leave the nest after a period of twenty-five to twenty-seven days; some of them then follow the father, while others follow the mother. They keep in contact with one another, and with the parent they follow, by means of "kiak" calls. The young seek out their own holes for roosting about a week after fledging. Groups of communally living green woodpeckers may be seen well into August. These birds seek food on meadows, and the young still beg successfully from the adult birds. Young which have become independent generally settle within a radius of twenty kilometers and stay within the particular area they have chosen for several years.

Fig. 2-28. Encounter of male and female green woodpeckers at onset of the breeding season at the nest hole. The male aggressively moves his head to and fro, while the female moves hers in appeasement.

Green woodpeckers reach an average age of two to two and one-half years; the maximum age is about five years. Some birds fall prey to goshawks and sparrow hawks, particularly during the breeding season. Roosting woodpeckers are occasionally taken by martens. Other green woodpeckers often die when they fly over a road and then strike a wire or a vehicle. Hard winters in many parts of Europe reduce the green woodpecker population to such an extent that it takes more than five years to recover.

The black-naped green woodpecker probably came into Europe from Siberia after the last Ice Age. It is still extending the boundary of its range westward. It reached western France in 1950, but is still absent in southern Europe. This bird lives in open mixed forests and deciduous mountain forests in western China and the Himalayas at altitudes reaching 3400 meters, as well as in woods and copses of cultivated land. According to K. Conrad's determinations, many areas of the black-naped green woodpecker's greatest abundance (such as Westphalia) coincide with those of beech forests on a stratum of lime or chalk.

The black-naped green woodpecker is mainly a ground bird in Asia; in Europe, however, it lives more in trees. It is believed that competition with green woodpeckers in Europe has forced this woodpecker, in part, to exploit other niches of its habitat. This transition was facilitated by the fact that the black-naped green woodpecker is less dependent on ants than is the green woodpecker. Because of this, they predominate over green woodpeckers in many areas, surviving hard winters much better than their close relatives and competitors. This was clearly shown in the winter of 1961/1962 when many green, but only a few black-naped green woodpeckers perished. Aside from ants, flies, and tree-dwelling arthropods and their larvae make up their diet. These woodpeckers may become regular visitors at feeding places, sometimes even in pairs. They show a particular preference for beef lard which has been filled with some soft food.

The black-naped green woodpecker is one of the "difficult" species, as Conrad puts it. Its secretive way of life, its protective color, and the rarity of its calls outside the breeding season make it difficult to observe; as a result, the true population density of these birds cannot be established. On a beech tree trunk this woodpecker is very difficult to see, and when it flies it is readily mistaken for a green woodpecker.

Black-naped green woodpecker territories become noisy from the end of April on, as the males and females drum and call continuously. The drum roll lasts a little more than a second; it consists of twenty to thirty beats, and is repeated three times a minute. Although they usually only drum near trees with holes, some will occasionally seek out more distant spots which are suitable for drumming, spots like metal roofs or towers. The call sequence, "kewkewkewkewkewkewkew," declines in pitch and, unlike the green woodpecker's "laughter," it sounds somewhat plaintive. Apart from the series of calls which are used in courtship and as call notes otherwise, variously pitched "kewk" calls and squeaky "kyaik" calls are heard all year.

Black-naped green woodpeckers become very secretive with the beginning of egg-laying. The four daily exchanges between incubating birds are announced by call sequences initially, but later they take place silently. Like their green and black relatives, these woodpeckers regurgitate food consisting almost entirely of ant pupae into the beaks of the young. According to Conrad, the young are fed once or twice per hour during the middle of the nestling period. They appear at the nest hole entrance beginning with the eighteenth day; the sensitive skin at the corners of the mouth, which releases the food-begging response when touched by the parent, has already shrunk by that time. The young are able to fly when they are twenty-two days old, but they do not leave the nest until two or three days later. Whether both parents then lead the young or only one parent does so is not yet clear. One may still occasionally see these woodpeckers in groups of two or three in early August; however, sometimes one of the young fledged in late June is already independent by mid-July. The independent young then has its own hole to roost in and looks for food within a relatively limited area surrounding this hole. According to my observations, this area averages about 200 hectares.

The BLACK WOODPECKER group includes large woodpeckers which inhabit large territories in primeval forests. The SLATY WOODPECKERS (genus *Muelleripicus*) contains four species in Southeast Asia. The BLACK WOODPECKERS in the narrower sense (genus *Dryocopus*) are native to America and Eurasia. One of the species of this genus is the INDIAN GREAT BLACK WOODPECKER *(Dryocopus javensis)*, which is found from India to Korea and the Philippines; the males have a red cheek stripe. The EURASIAN BLACK WOODPECKER *(Dryocopus martius;* Color plate, p. 108) has a L of 50 cm and weighs 250–315 g. This bird is the size of a crow. The plumage is black with a brownish sheen on the wing tips. Males have a bright red crown, while females have a triangular spot on the nape. This species is distributed in three subspecies from the Netherlands to Japan. Its North American relative, the PILEATED WOODPECKER *(Dryocopus pileatus)*, has a L of 40 cm and weighs 230 g; this bird has a white pattern on the throat, neck, and wings. Males have a large bright red crest and a red cheek stripe, while females have a small red crest and a black cheek stripe. The pileated woodpecker lives in continuous woodlands and prefers swamps and woodlands in central and eastern North America. There are four closely related South American species.

The Eurasian black woodpecker has gradually advanced into Northern and Western Europe during the last decades. Originally it was an inhabitant of large deciduous forests with many diseased and fallen trees; today, however, it is also found in coniferous and well-managed mixed forests. The territory of a pair in Central Europe is about 200 to 800 hectares. The black woodpecker seeks its food mainly on the ground, on tree stumps, on the lower segments of tree trunks, and in the nests of forest ants. It also debarks the trunks of diseased trees, removing the bark along the entire length of the trunk. Apart from ants, its diet consists of various beetle larvae and wood wasps. Its presence is indicated by worked-over tree stumps or pine trunks which have exposed ant nests in oval holes. The black woodpecker's loud flight call, "kuerrkuerrkuerr," and the penetrating call "kiyah," which is given when the bird is perched, may be heard at these locations. During March and April the black woodpeckers are particularly

Fig. 2-29. Male black-naped green woodpecker feeds a 17 or 18 day-old nestling. The touch-sensitive swelling at the corner of the mouth of the latter can still be seen.

Fig. 2-30. Black woodpecker *(Dryocopus martius)*.

Fig. 2-31. Pileated woodpecker *(Dryocopus pileatus)*.

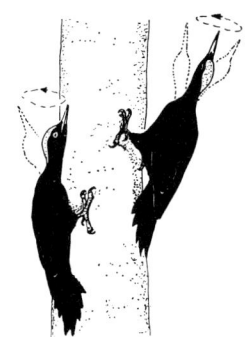

Fig. 2-32. Threat display of two male black woodpeckers with head circling and beak thrusts.

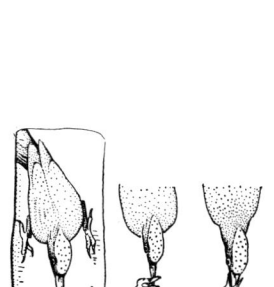

Fig. 2-33. Feeding of young black woodpeckers. Left to right: first the young sit in a clump which helps to conserve body heat. The adult touches the touch-sensitive swelling at the youngster's mouth which then opens its beak.

lively in their territories. An observer seated near one of the nest hole trees might observe the drumming, the partners flying over one another with loud "kwee" calls, the showing of nest holes with the accompanying "kiyak" calls, and the tapping at the hole entrance at that time. The repeated drumming can be heard for several kilometers; it consists of three rolls with thirty-eight to forty-three beats each, given over a period of two and one-half seconds.

When two black woodpeckers approach one another, they perform a peculiar sham fight with their beaks; when birds of the same sex are involved, this action has the effect of a mutual threat. If, however, the two birds are of opposite sexes, the sham fighting then declines, turning into a head waving which seems to indicate pacification. Either males or females may attract a partner, and offer the other nest holes. Often males and females will roost in neighboring trees during the winter; pair formation in such cases then proceeds rather inconspicuously.

An old nest hole used as a roosting place may also be repeatedly used as a nest site. Some breeding holes have been used by black woodpeckers for four to six years in succession. Generally, however, these birds make a new hole in the smooth trunk of a tree, below the first branch; the preferred trees are beech, aspen, cherry, alder, pine, and white spruce. The excavation of a new nest may take four weeks, because the nest must be large enough to accommodate the large birds. Males and females work in alternation. The entry hole soon assumes the characteristic elongated oval shape, the diameter along its longest axis being ten to twelve centimeters. The inside of the hole may be fifty-five centimeters, and in some cases up to a meter deep.

Egg-laying begins in the first half of April in Central Europe; many pairs are still laying in mid-May. The clutch usually consists of four eggs, but clutches of three to six eggs have been found, for example in Finland. The partners relieve each other very ceremoniously at one to three hour intervals during the twelve to fourteen days of incubation. The relieving bird flies in from some distance with "kuerr" calls and stays near the nest for some time. Then it flies to a tree near the one in which the nest is located, or onto the nest tree itself, and calls "ki-jak." The partner in the nest replies by tapping; this is speeded up until there are as many as forty taps in nine seconds. The relieving bird then lands some distance below the nest hole and its "ki-jak" calls become more penetrating. The tapping stops, although two or three more taps will sometimes follow, and the brooding bird takes off. The male always stays on the eggs at night; later, shortly before the young leave the nest, the male moves into his own roosting place elsewhere.

Young black woodpeckers are fully fledged after twenty-eight days in the nest. It is astonishing that they can already make all the calls of their species at that time, and they soon begin drumming. Each of the adults leads one or two of the young until the first part of August. One may hear the young keep in touch with the adult through the "kwee" calls. Although the adults are residents, the young wander about for some distance, as much as twenty, forty, or even 500 kilometers, as has been shown by banding.

The black woodpecker has few enemies. Occasionally a marten takes one in a roosting hole, and goshawks may sometimes kill one of these birds, especially during the breeding season. The maximum age known for a black woodpecker to date is four years. The roomy holes made by these woodpeckers are used by jackdaws, rollers, starlings, redstarts, nuthatches, tawny and Tengmalm's owls, stock doves, and goldeneye ducks, and many other birds.

Fig. 2-34. Rival male pileated woodpeckers threaten one another.

The PILEATED WOODPECKER of America has been more successful than the ivory-billed woodpecker in adapting to progressive cultivation of the land and in maintaining a good population density. It behaves somewhat differently from the black woodpecker in courtship and fighting. The partners or rivals ceremoniously spread their wings, displaying the black and white marks as signals of the species and the sex of the bird involved. L. Kilham has observed pileated woodpeckers in a territorial conflict near the Potomac River, and has made a drawing of the scene. Drumming is also involved in the maintenance of the territories, and it may be heard all year round as a series of rolls or a single roll. Pileated woodpeckers communicate nest hole choice and relief during incubation by taps and calls, just as black woodpeckers do. Examination of stomach contents has shown that the diet of this large American woodpecker consists mainly of large ants and beetle larvae, and a lesser but significant amount of fruits and seeds.

The TRUE BANDED WOODPECKERS, with four genera and twenty species, are confined to the New World. The RED-HEADED WOODPECKER *(Melanerpes erythrocephalus)* has a L of 21.5 cm and is found in eastern North America. The black-and-white ACORN WOODPECKER *(Melanerpes formcivorus)* has a L of 25 cm; these birds have a red crown and are found from California to Colombia. Members of the closely related genus *Centurus* include: the GOLDEN-NAPED WOODPECKER *(Centurus chrysauchen),* which has a L of 23 cm and is found in Central and South America; the RED-BELLIED WOODPECKER *(Centurus carolinus),* which has a L of 22 cm and is found in North America; the GILA WOODPECKER *(Centurus uropygialis)* from California and Central America; and the YELLOW-TUFTED WOODPECKER *(Centurus cruentatus)* from South America.

The True banded woodpeckers

Fig. 2-35. Northern and southern limits of range of the *Centurus-Melanerpes* group of woodpeckers.

Some of these woodpeckers show varying degrees of innate food storage behavior, which probably evolved from the "smithies" used by greater spotted woodpeckers. Originally, banded woodpeckers probably inserted acorns, other fruit, and an occasional large insect into a cleft in order to be able to work these objects over. Later, some species may have left some of the food as a reserve to be used when-

Fig. 2-36. Red-bellied woodpecker shows how birds of this group hide an acorn; it uses its tongue to push the acorn into its hiding place or to remove it.

Fig. 2-37. Acorn woodpecker at its storage tree, the bark of which is covered with storage holes.

ever they needed it. Finally, some species chiselled holes specifically for items to be stored (the so-called "acorn cups") and evolved a procedure to cover and hide the food to protect it from jays and chickadees. During the winter, some of these woodpeckers are very much tied to their own territories because of this food storage habit, while during other seasons, they may migrate in loose troops.

Next to the flickers, the RED-HEADED WOODPECKER is one of the most common North American woodpeckers. The northern populations are migratory, but some occasionally winter in the breeding areas when there is a good acorn harvest. About fifty percent of their food consists of acorns, berries, and other fruit; the other fifty percent of their diet consists of ants, grasshoppers, beetles, and similar insects.

European observers have been surprised to see red-headed woodpeckers remove grains of corn from the ears. These woodpeckers use natural cavities to store acorns for the winter. If necessary, they will break these acorns into suitable pieces before inserting them into a hole. According to the observations of L. Kilham, red-headed woodpeckers are very hostile toward possible competitors in their wintering areas; they try to protect their stores by covering the storage holes, which are scattered all over an individual bird's territory, with pieces of bark. This protection notwithstanding, many blue jays and chickadees succeed in finding and emptying many of the holes. The acorn woodpecker makes regularly spaced holes in the bark of an oak or conifer, and then inserts one acorn after another into these holes; however, because the woodpecker does not cover them, mice, squirrels, and jays can easily find the acorns and consume them.

Occasionally half a dozen acorn woodpeckers may be seen climbing peacefully on the same tree. The social behavior of the golden-naped woodpeckers is even more striking. A. F. Skutch provides the following description: "When the young are fledged, they return to the nest hole every evening with their parents. They find the entrance right away, often without any help from the adults. Sometimes a particular call of the parents will suffice to bring the young to the hole. The young are then fed by the parents as if they were still nestlings." Sometimes the family is also found in the nest hole during an afternoon rainstorm.

L. Kilham studied and described the tapping duet of the red-bellied woodpecker. "Males attempting to lure a female to a hole or site available for breeding use emphatic taps, call "kwarr," and drum. When a female is attracted, she flies to the tree from which the signals came, perches beside the male, and joins in his tapping. Sometimes the male will be inside the hole and the female will be outside. One phrase of the mutual tapping lasts about two seconds, and it may then be repeated several times. Besides other fruit, these woodpeckers, strangely enough, will also eat oranges from which they pick out the pulp and seeds; they generally only do this, however, with diseased or fallen oranges. The red-bellied woodpecker also stores fruit and arthropods, hiding them under pieces of wood and bark.

The gila woodpecker nests not only in trees, but also in cacti. Fifty percent of the nests counted in Arizona were in cacti, particularly in the organ pipe cactus *(Cereus giganteus)*. These holes in cacti are so durable that, like tree holes, they may be used for breeding for years.

The SAPSUCKERS, with only two species, and considered to be among the most unusual of woodpeckers, are found in North America. Some ornithologists classify these birds with the banded woodpeckers, while others consider them to be related to the spotted woodpeckers. They show certain peculiarities in structure and behavior which justify the classification of these colorful, starling-sized woodpeckers as a separate group. One species, the YELLOW-BELLIED SAPSUCKER *(Sphyrapicus varius;* Color plate, p. 119), with a L of 21 cm, has two yellow-bellied and two red-bellied subspecies which differ not only in their coloration, but also in their migratory behavior. *Sphyrapicus varius varius* is a pronounced migrant; *Sphyrapicus varius nuchalis* is less so. *Sphyrapicus varius daggetti* rarely migrates at all, and the RED-BELLIED SAPSUCKER *(Sphyrapicus varius ruber)* of California is a resident. Females predominate among the sapsuckers which winter in Mexico, Central America, and on the islands of the Caribbean; males, on the other hand, predominate in the wintering areas in the United States. Studies by T. R. Howell have shown that the subspecies with the most marked migratory behavior have the least plumage coloration. The red-bellied sapsucker is the most colorful and shows the least difference between males and females.

The second species of this group, WILLIAMSON'S SAPSUCKER *(Sphyrapicus thyroideus)*, which has a L of 20 cm, is one of the most unusual of the American woodpeckers, partially because of its coloration, but also because of the history of its discovery. The two sexes of this species show a greater difference than is known for any other woodpecker species, and as a result they were once thought to represent two different species. A female was described in 1854 under the name *Melanerpes thyroideus*. In 1857 another ornithologist named a male *Picus williamsonii*. The error was not discovered until twenty years later. Males are black with white spots on the wings and lower back; they are the only four-toed American woodpeckers which have no red on the head. The female is a grayish-brown with a brown head and white marks similar to those of the male. Williamson's sapsucker inhabits western North America, breeding in mountain forests and wintering in southern California, western Texas, and Mexico.

The most peculiar feature of these woodpeckers is the behavior to which they owe their vernacular name, the licking up of tree sap. This behavior is occasionally found in other species of woodpeckers, but no where does it play as important a role as in the sapsuckers. E. Martini has the following description: "These woodpeckers are adapted

The sapsuckers

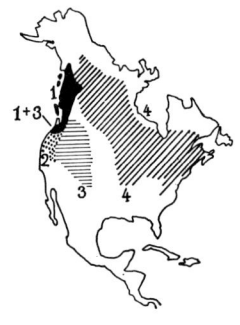

Fig. 2-38. Breeding range of the four subspecies of the yellow-bellied sapsucker *(Sphyrapicus varius)*: 1. *Sphyrapicus varius ruber*; 2. *Sphyrapicus varius daggetti*; 3. *Sphyrapicus varius nuchalis*; 4. *Sphyrapicus varius varius*.

▷
Aracaris: 1. Plate-billed mountain-toucan *(Andigena laminirostris)*; 2. Saffron toucan *(Baillonius bailloni)*; 3. Black-billed toucan *(Andigena nigrirostris)*; 4. Banded aracari *(Pteroglossus aracari)*; 5. Spot-billed toucanet *(Selenidera maculirostris)*; 6. Emerald toucanet *(Aulacorhynchus prasinus)*; 7. Blue-banded toucanet *(Aulacorhynchus caeruleocinctus)*; 8. Red-necked arecari *(Pteroglossus bitorquatus)* 9. Curl-crested aracari *(Beauharnaisius beauharnaisii)*.

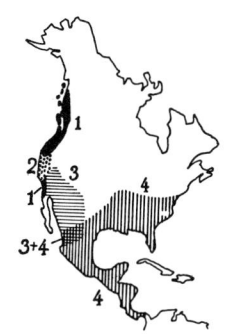

Fig. 2-39. Winter ranges of the four subspecies of the yellow-bellied sapsucker *(Sphyrapicus varius)*: 1. *Sphyrapicus varius ruber* (only part of the population migrates); 2. *Sphyrapicus varius daggetti* (only a few birds migrate); 3. *Sphyrapicus varius nuchalis* (essentially separate breeding and wintering areas); 4. *Sphyrapicus varius varius* (a definite migrant).

◁

Woodpeckers: 1. Arrowhead piculet *(Picumnus minutissimus)*; 2. Eurasian wryneck *(Jynx torquilla)*; 3. *Camphetera nubica nubica*, a spotted woodpecker; 4. Black woodpecker *(Dryocopus martius)*; 5. Ivory-billed woodpecker *(Campephilus principalis)*; 6. Red-shafted flicker *(Colaptes auratus cafer)*; 7. Yellow-shafted flicker *(Colaptes auratus)*; 8. Ground woodpecker *(Geocolaptes olivaceus)*.

to a diet of tree sap and soft bark tissue, and they only take insects and fruit on the side. Trees are ringed with holes which penetrate the bark, the sapwood layer below, and even the heartwood. Individual trees may have complete rings or three-quarter rings of holes from the foot up to the crown. The holes are often directed somewhat downwards so that the emerging sap gathers in them. If a tree is visited constantly by the birds, and if the holes are regularly renewed, the vital flow of sap into the upper parts of the tree is almost completely arrested, and the tree slowly dies. More than 275 species of trees are known to have been ringed by sapsuckers."

Even nestlings and young sapsuckers are fed primarily tree sap. The sapsucker's tongue has a relatively wide tip which helps the bird get its food; this tongue has so many horny bristles on both sides that it resembles the brush tongues of those birds which feed on nectar. The sapsucker's tongue, unlike that of most woodpeckers, cannot be extended far. Thus sapsuckers obtain insects, which are one of their main food supplements, mainly on the surface of tree trunks and branches, and just below the bark. Sapsuckers are conspicuous during the mating season by their drumming. This consists of one to two short rolls followed by very emphatic single or double taps at short intervals.

The SPOTTED WOODPECKERS are distributed over a large part of the Old and New World in thirteen genera and seventy-three species. The GREATER SPOTTED WOODPECKER *(Dendrocopos major;* Color plates, p. 83, p. 84 and p. 119) is one of the best-known and most studied species of this group. It has a L of 25 cm and weighs 74-95 g. Juveniles have a red crown until mid-September. This species is distributed in twenty subspecies from the Canary Islands and Tunisia over all of Europe as far as the Caucasus mountains and Asia Minor. The closely related SYRIAN WOODPECKER *(Dendrocopos syriacus)* is found in Asia Minor and the Balkan peninsula; it has a L of 25 cm. The black cheek stripe does not reach the nape band; the lower tail coverts are pink, and the white wing coverts are very broad. Juveniles have a red breast band. The MIDDLE SPOTTED WOODPECKER *(Dendrocopos medius)* has a L of 21 cm. This woodpecker is, in the words of the ornithologist Voous, "a characteristic bird of the European primeval forest which has now almost disappeared." The crown is red as far as the nape in males, while only the forehead is red in females; the lower belly is pink. There are stripes running lengthwise on the breast and abdomen. The largest European woodpecker of this group is the WHITE-BACKED WOODPECKER *(Dendrocopos leucotos)* which has a L of 27 cm and weighs 100-112 g. This species is distributed, with some gaps, from Greece, the eastern Alps, and Finland as far as Southeast Asia and Japan in fifteen subspecies. The LESSER SPOTTED WOODPECKER *(Dendrocopos minor)* male has a red crown, while that of the female is brownish-white. The L is 16 cm and the weight is 20-25 g. The thirteen subspecies of this species inhabit the northern and temperate zones of Eurasia and the Mediterranean area.

The American spotted woodpeckers include: the HAIRY WOODPECKER *(Dendrocopos villosus)* which, with a L of about 25 cm, looks much like the greater spotted woodpecker. This species is distributed in twenty-one subspecies over North and Central America; the DOWNY WOODPECKER *(Dendrocopos pubescens)* which, with a L of 16 cm, resembles the European lesser spotted woodpecker. There are four Atlantic and four Pacific subspecies found in open North American landscapes; and the RED-COCKADED WOODPECKER *(Dendrocopos borealis)* which is found in southeastern North America. It has a black cap and nape, and a banded back. (Syrian, lesser spotted, middle spotted woodpeckers: Color plate, p. 119.)

Fig. 2-40. Sapsucker on a tree which it has ringed repeatedly.

The European greater spotted woodpecker primarily inhabits trees and keeps almost exclusively to the trunks and branches of deciduous and coniferous trees. It is only occasionally seen on a stump or on the ground, where it hops about very clumsily. This bird lives in mixed as well as in deciduous and coniferous forests, and in parks and copses. It is resident and remains in the same general area all of its life, although it may change its territory. Usually one pair's territory covers about forty to sixty hectares, but territories only ten to twenty hectares in size may be found in deciduous forests with heavy undergrowth, or in parks.

This species feeds largely on arthropods picked from trunks, branches, and leaves, or hacked and probed out of diseased wood during the spring and summer. Wood-dwelling beetles and their larvae, which are speared in the depths of their passages with the woodpecker's harpoon tongue, make up fifty to sixty percent of the greater spotted woodpecker's diet in Germany. In addition, this species feed on ants, hemiptera (bugs), caterpillars, aphids, earwigs, and spiders. Occasionally it pulls nestling birds, which it probably views as fat larvae, out of their nest holes and devours them. Initially the young are fed aphids of many kinds; later on they are fed larger insects. Otto Steinfatt reported 665 different feedings, made up of at least 2347 individual prey animals, many of which (2200 nun moth caterpillars, for example) were harmful to trees. Fledged young are also fed plant food, including cherries and seeds like sunflower seeds which are rich in oil.

Fig. 2-41. Greater spotted woodpecker *(Dendrocopos major)*.

The greater spotted woodpecker may switch to a complete plant diet during the fall and winter. This assures the bird's survival in the hard winters of northern latitudes; it has enabled this species to maintain itself in hostile areas during the last millenium. The woodpeckers eat conifer seeds, hazelnuts, and walnuts in gardens, as well as plums and almond kernels during these seasons. The woodpecker has a special technique which it uses for eating this food. First it chisels or twists off a pine, birch, or spruce cone and brings it to the so-called "smithy." This is a crack in the bark of a tree which the bird has found, enlarged

Fig. 2-42. Conflict between two greater spotted woodpeckers over a roosting hole. The dominant hole owner threatens with abducted wings; the intruder timidly stays in a subdued posture and will shortly fly away.

somewhat, or even constructed itself. The woodpecker wedges the cone into the crack and, with hefty blows, it knocks off the scales and eats the seeds. One may find hundreds of worked-over cones beneath the "smithy." Pynnönen, the Finnish biologist, has been able to quantitatively determine the seed usage of the greater spotted woodpeckers. He found that this woodpecker eats 165 to 170 pine seeds, or 0.8 grams in an hour. Thus in an active day, which is a maximum of seventeen hours, the greater spotted woodpecker can eat fourteen grams of food. Between late February and early April these birds ring various species of trees, just as the sapsuckers do. They prefer pine, spruce, maple, birch, and willow trees. The intensity of this behavior varies in different districts; there are some areas in which ringed trees are never found.

Roosting follows a definite schedule; greater spotted woodpeckers roost in old woodpeckers' nests, in artificial holes, or, more rarely, in rock clefts or niches in walls. They go to roost at a specific time which varies according to the length of the day. The bird clings to the inside of the hole, fluffs out its plumage, puts its head under one wing, and sleeps this way until morning. Upon awakening—and the greater spotted woodpecker is an early riser—it first looks out of its hole for a while and then emerges, and preens itself before flying into its territory with a high-pitched "keex" call.

The first manifestations of courtship may be observed from mid-January onward. Males and females fly to holes and drum near them in the course of the day. The courtship proper does not begin until mid-March, at which time both partners drum persistently near trees with holes early in the morning, around noon, and again in the evening. The drum roll consists of twelve to sixteen beats; it lasts about 0.6 seconds and is repeated eight to ten times a minute. The partners gradually become more peaceful with one another; they inspect holes shown them by the other partner, or they begin to chip out new holes here and there with funnel-shaped openings. Finally the partners agree on one of these spots, or they choose an old, completed hole. If they do build a new hole, the pair works on it in alternation for ten to twenty-eight days. The entry hole has a diameter of between four and five centimeters. The diameter of the interior is ten to fifteen and the depth is twenty-three to forty-five centimeters. The courtship behavior is greatly intensified shortly before the egg laying, which occurs between the last third of April and mid-May in central Germany.

The female lays the four to eight white eggs in the early morning hours. As soon as the first egg is laid on the wood chips in the nest hole, the male roosts there overnight. Incubation begins when the last egg is laid; the parents relieve one another every thirty to fifty minutes. The young hatch in the surprisingly short time of twelve or thirteen days, all within twenty-four hours of one another. The naked young lay in a bunch over one another and thus conserve heat; however, the parents also cover them without pause for almost a week to keep them warm. The young utter a scratching "krrkrr" begging sound from the moment of hatching; this vocalization later acquires a "whirring" character.

When the young are sixteen days old, the adults feed them from outside the nest hole, although they do enter the nest hole after feeding to remove the youngs' feces. The food consists initially of small insects, including many aphids; later on they eat larger insects. The parents bring the food in their beaks, and feed the young on an average of every five to ten minutes. The young leave the nest on the twenty-third or twenty-fourth day. The family does not remain a complete unit; each parent leads some of the young. After eight days of this, the parents threaten the young, which then leave the territory. The young may still remain together for a while in sibling groups of two or three.

Banding has shown a maximal age of eight years for the greater spotted woodpecker; the average age is probably two to four years. The young are exposed to danger (especially right after fledging) from martens, sparrow hawks, goshawks, and tawny and other owls. Skeleton finds indicate that it is not rare for the adults to die of old age in their holes.

Members of the northern subspecies *Dendrocopos major major* will make extensive migrations in some years; these take them to Western and Central Europe, as well as to Southern or Southwestern Europe. Northern greater spotted woodpeckers have been found in Hungary and northern Italy in such migratory years. Most of the wandering birds are young. There have been twenty migratory years within this century and, of those, 1929 and 1962 were particularly striking. Thus in late August of 1962, 130,000 northern greater spotted woodpeckers passed through the Aaland Islands. These birds' return in the spring is inconspicuous, presumably because many participants remain in Western and Central Europe, breed there and merge with the local population. As far as we know, these migrations are all triggered by food shortages due to a poor crop of pine or spruce cones.

Fig. 2-43. Greater spotted woodpecker harvesting conifer cones and opening one which is wedged into a crevice.

Since 1890, the SYRIAN WOODPECKER has spread from Asia Minor and the Balkans in a northwesterly and easterly direction; at first the movement was slow, but since 1939 it has become more intensive. These birds are now present in Austria, Slovakia, and parts of the Ukraine. The Syrian woodpecker takes the place of the greater spotted woodpecker in large areas of Southwest Asia. When both species occur in the same area, the Syrian woodpecker avoids mountain forests and uses park and garden-like areas near settlements. It is also found in open woods along lowland rivers. It breeds with the greater spotted woodpecker at the edge of its range; a number of these hybrids have been found. The Syrian woodpecker feeds much like the greater spotted woodpecker, but according to L. Szlivka of Yugoslavia, fruit plays

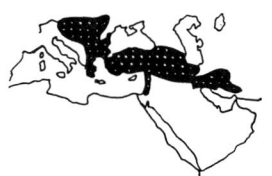

Fig. 2-44. Syrian woodpecker *(Dendrocopos syriacus)*.

Fig. 2-45. Middle spotted woodpecker *(Dendrocopos medius)*.

a more important role in its diet. The young are fed raspberries, mulberries, and strawberries to such an extent that their droppings become red, or even black.

The MIDDLE SPOTTED WOODPECKER, once so characteristic of primitive European forests, now lives in mixed deciduous woods with soft trees and diseased tree trunks along large rivers. It prefers oak and hornbeam stands, but is also found in park and garden areas with old hollow trees. This species is not common in any area, but it remains faithful to those limited areas it has occupied for decades. It is more of a prober and searcher, rather than a chiseler, and it uses its beak primarily as a pair of pincers to remove arthropods from all sorts of cracks and clefts in the rough bark, or to pick them off branches in the treetops.

This bird is a resident, and only migrates occasionally; the middle spotted woodpeckers migrate singly between July and mid-March. Drumming, so important in the courtship of other species, is replaced by a peculiar quacking series of calls. This "song" usually has six to nine syllables and is heard from the vicinity of hollow trees from mid-January onward. P. Feindt describes this vocalization as follows: "This sound pattern has quite a peculiar melody. It generally starts on a high pitch and is strongly accented; it then declines with more or less of a 'swing.' It is plaintive and even somewhat nasal in tone."

The breeding holes are chipped out of trees, especially oaks, and they are generally at the middle or higher in the tree. Incubation lasts twelve days from the laying of the last egg. The young leave the nest holes in the second third of June in Germany, and occasionally as late as early July. The family stays together for another eight days, and then it dissolves.

Fig. 2-46. White-backed woodpecker *(Dendrocopos leucotos)*.

Certain areas of the WHITE-BACKED WOODPECKER'S distribution, including the western Pyrenees, the Abruzzi, the Chinese province of Szechwan, and southeastern Europe, date back to the last glaciation; these areas were separated from the rest of the bird's present range at that time. Perhaps some of these isolated populations originated when man, through extensive deforestation, left cleared areas in regions which the white-backed woodpecker had settled. As far as we know, this woodpecker feeds on large insects and their larvae, which it finds beneath the bark of diseased trees. Thus these birds require a habitat of mixed deciduous forests on foothills which are still primeval—a habitat like the Bohemian Forest in which there are many dead or dying trees. Nest holes are usually in deciduous trees, but they may also occasionally be in coniferous trees. Incubating partners relieve each other at intervals of several hours. Pynnönen found four and one-half hour intervals in Finnish forests; this long interval may be related to the large territories of these woodpeckers. In one case, J. Franz noticed that up to sixty percent of the young's food consisted of long-horned beetles.

Fig. 2-47. Lesser spotted woodpecker *(Dendrocopos minor)*.

The lively LESSER SPOTTED WOODPECKER inhabits the crowns of mixed deciduous forest trees, or orchards and parks with some amount of old trees. These birds are active in their territories from March to May. Twelve to fourteen drumming rolls per minute resound from dead branches in the tree crowns over the area, and high-pitched "keekee-keekee" calls are heard at the same time. This woodpecker usually chips out a new nest hole every year. The nest hole is ten to eighteen centimeters deep, and ten to twelve centimeters wide; the entry hole has a diameter of about thirty-two millimeters. The nest holes are built in soft deciduous trees at heights varying from one to twenty meters. A full clutch has five to six eggs. The young hatch after twelve days of incubation, and they are tended in the nest hole for another twenty days. These woodpeckers migrate over a wider area in winter, and they are seen as "leaders" of troops of tits in large upland forests at that time. Northern members of the species may sometimes migrate to more southern areas, but they rarely reach Central Europe.

Fig. 2-48. Hairy woodpecker *(Dendrocopos villosus)*.

The European greater spotted woodpecker is replaced in North America by the HAIRY WOODPECKER. This bird is a tree woodpecker and as such, it lives on tree trunks and strong branches; seventy-five percent of its food consists of insects, of which forty-five percent come from the inside of the wood. Unlike the greater spotted woodpecker, this species also eats conifer seeds, although it does not open them at special "smithies." Aside from this, both species have a similar behavior. The hairy woodpecker, however, does exhibit a unique behavior (at least among woodpeckers) in its courtship display. Potential partners examine a nesting area with hollow trees by flying together over the tree crowns. In the threat display, they move the body to and fro and abduct the wings, effectively showing the black and white pattern of the plumage.

Fig. 2-49. Threat behavior between two male hairy woodpeckers. The upper one is pausing in its pendular threat movements; the lower one aggressively thrusts out its bill.

The DOWNY WOODPECKER prefers the dense crowns of trees and is therefore more similar to the lesser spotted woodpecker, which it much resembles in behavior. Downy woodpeckers are not always easy to observe, however. The English name downy seems to suggest that man feels a certain tenderness toward this small woodpecker.

The RED-COCKADED WOODPECKER is confined to the coniferous forests of the southeastern United States. It has also been seen visiting cornfields, where it chops insects out of the corncobs. These woodpeckers chop many small holes around their nesting holes; resin then flows from these holes. In this way the red-cockaded woodpecker is believed to produce a sticky rim around its nest, and this helps to ward off ants and squirrels which might harm the bird's young. Unlike the other spotted woodpeckers, this species is decidedly sociable, and it is often observed in troops of six to ten birds.

The THREE-TOED WOODPECKERS (genus *Picoïdes*) have heads decorated with yellow rather than red; they are regarded as tree and chiseling

Fig. 2-50. Red-cockaded woodpecker *(Dendrocopos borealis)*.

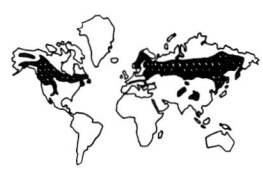

Fig. 2-51. Distribution of the three-toed woodpeckers (genus *Picoides*).

woodpeckers. These birds have no first toe; this is an adaptation to their life as chiselers. When one of these woodpeckers clings to a tree trunk to make hard blows with its head well back to begin the stroke, a toe directed to the rear and down would only be a hindrance. The BLACK-BACKED THREE-TOED WOODPECKER *(Picoïdes arcticus)* is found in North America; this species may live alongside the North American subspecies of the Eurasian NORTHERN THREE-TOED WOODPECKER *(Picoïdes tridactylus;* Color plate, p. 119), which has a L of 22 cm and weighs 63 g.

The original habitat of these woodpeckers was probably the dense, lichen-covered coniferous forest of the taiga zone, which had many dead and diseased trees. Three-toed woodpeckers inhabit a fairly continuous area in the arctic of the Old and New World. South of this area, their distribution includes isolated mountain ranges which are probably glacial relics. The black-backed woodpeckers and the northern species form twin species, as is the case with the gray and green woodpeckers. These birds prefer swampy woods burnt by forest fires and subsequently invaded by insects. Such damaged forests offer food and nest sites; they are therefore often invaded by large three-toed woodpeckers. Such invasions were particularly noted in 1924 and 1956/1957. Three-toed woodpeckers search their food trees more thoroughly than do many other woodpecker species; they remain at spots where food is abundant until they have chopped all the larvae out of the wood. Their three-toed climbing method particularly facilitates these actions.

Little is known about the habits of the several South American, African, and Asian woodpeckers which belong to the spotted woodpecker group in the wider sense. One species included in this group is the African CARDINAL WOODPECKER *(Dendropicus fuscescens)*. The back of this bird is barred; the lower parts are striped lengthwise; the forehead is brown and the crown is red. This species is thought to be common south of the Sahara, but as Mrs. G. D. Attwell writes, about its breeding habits little is known. In 1951, Mrs. Attwell was able to observe a pair of cardinal woodpeckers near her house. Both birds began building the nest hole in a dead tree on July 23. By August 17, two eggs were laid in the hole. Incubation lasted ten to twelve days, and the young left the nest hole after a twenty-seven-day nestling period. According to Mrs. Attwell's observations, the family stayed together for at least another two months. This woodpecker drums in a lively manner during the courtship season, even on the particularly resonant fruit of the monkey bread (baobab) tree.

Fig. 2-52. Cardinal woodpecker *(Dendropicos fuscescens)*.

The MAROON WOODPECKERS (genus *Blythipicus*) and the GOLDEN-BACKED WOODPECKERS (genus *Chrysocolaptes*) both belong to the Asiatic spotted woodpeckers. These birds are the size of spotted woodpeckers; they are very colorful inhabitants of the forests of Nepal, Borneo, and Sumatra. The MAROON WOODPECKER *(Blythipicus rubiginosus)* has a L of

22 cm and is characterized by its brownish-red back. This species is found from Malaya to Borneo and Sumatra. The GOLDEN-BACKED WOODPECKER *(Chrysocolaptes validus)* has a L of 30 cm and is found in the forests of Malaya, Java, and Borneo below an altitude of 1000 meters. The lower parts are reddish; the crown is red. The cheeks and back are yellow to golden-brown, and the wings and tail are brown. The BLACK-BACKED WOODPECKER *(Chrysocolaptes festivus)* is the largest woodpecker of Ceylon. It settles in coconut palm stands and chops an oval nest hole into the trunks of these trees.

Fig. 2-53. Distribution of the black-backed woodpeckers (genus *Chrysocolaptes*).

We will conclude our discussion of this order with the group of the New World *Campephilus* and the SHARP-CRESTED WOODPECKERS. The three species of the genus *Campephilus* are the world's longest woodpeckers, with a L of over 50 cm. Unfortunately all three species are threatened with extinction and can only be preserved if vigorous steps are taken. The IVORY-BILLED WOODPECKER *(Campephilus principalis;* Color plate, p. 108), which has a L of 50 cm, is almost or perhaps already extinct. The IMPERIAL WOODPECKER *(Campephilus imperialis)* is the largest woodpecker of all, with a L of 55 cm; it is found in Mexico. The MAGELLANIC WOODPECKER *(Campephilus magellanicus)* male has a red head and throat.

Fig. 2-54. Ivory-billed woodpecker *(Campephilus principalis)*. The stippled area indicates the distribution in the 19th century. The 20th century distribution is in black.

North Americans have taken some steps to protect the ivory-billed woodpecker since the beginning of this century; only the future will tell whether these measures will be successful, or whether they came too late. According to J. W. Tanner's very complete description, which he wrote in 1942, ivory-bills apparently mate with the same partner for many years and inhabit areas of six to eight square kilometers. These birds were once found in large forests in Louisiana, Florida, and South Carolina. Their food consisted predominantly of insects taken from beneath the bark of dead or dying trees. This woodpecker is deprived of food whenever such trees are cut down. In his report, Tanner recommended that diseased trees be left standing, and that other trees be cut and left to rot in order to preserve this threatened species. Since ivory-bills were sighted in Texas as recently as 1967, such measures, coupled with the strictest protection, may preserve these birds from total extinction.

Two decades after his studies on the habits of ivory-billed woodpeckers, Tanner studied the residual population of the imperial woodpecker in Mexico. His report caused consternation. Although this species was the world's largest woodpecker, and although it was still locally common in the mixed forests of the Sierra Madre Occidental about 1900, Tanner did not find any of these birds in that area after weeks of search in the 1960's. He ascertained that hunting of these birds by the natives, more than anything else, is likely to cause the complete extinction of the species. The young imperial woodpecker's flesh is regarded as a delicacy, and the feathers are used for medicinal purposes. The position of the Magellanic woodpecker of southern South America is probably similar.

Fig. 2-55. Ivory-billed woodpecker relieving its mate during incubation.

Fig. 2-56. Distribution of the sharp-crested woodpeckers (genus *Phloeoceastes*).

The SHARP-CRESTED WOODPECKERS (genus *Phloeoceastes*) are found from Mexico to northern Argentina in seven species. The CRIMSON-CRESTED WOODPECKER *(Phloeoceastes melanoleucos)* has a L of 35 cm. The male has a red crown and crest, while the female usually has a black crown and crest. These birds are somewhat smaller than the European black woodpecker. Some species have a red-brown background color, while others have a black-and-white pattern.

Woodpeckers can be kept in spacious flight cages, although trees on which they can climb and chip must be provided to satisfy their need for movement, and holes must be available for roosting. With the exception of the essentially ant-eating species, most woodpeckers can be well fed on the usual soft bird food which contains fat with added ant pupae, chopped raw meat, and some live insects or insect larvae. Species which feed mainly on plant food in the wild are particularly easy to keep. They can be well nourished on beef suet, sunflower seeds, hazelnuts, walnuts, peanuts, and conifer seeds. All woodpeckers must have water available to them.

Yellow-shafted flickers and red-bellied woodpeckers have bred in large cages. Sapsuckers and downy woodpeckers have shown the full courtship display or even laid eggs in captivity. The first successful breeding of the South American golden-crested woodpecker in captivity was achieved by H. Hillemacher, and it earned him the gold medal of the magazine "Die Gefiederte Welt" (The Feathered World). Hand-raised woodpeckers can become extraordinarily tame; those caught in the wild will lose their fear of man only slowly. In both cases, habituation to the person looking after the birds is more a matter of relying on him for food, rather than the formation of a social bond.

3 Passerine Birds

The last of the twenty-six avian orders, that of the PASSERINE BIRDS *(Passeriformes),* includes more than a fifth of all the living bird species. Unfortunately we can only allot one fourth of the bird section of this encyclopedia to the *Passeriformes,* and this is possible only because there are so many similar species in this order. All of these birds are altricial. The L ranges from 7.5-110 cm, and the weight is from 4.8-1350 g. These birds have four toes (the babbling thrush is the only member of this order with a greatly regressed first toe). The toes all originate from the same level on the tarsus, and they are generally free to the base. There is always one toe (generally the largest) directed to the rear; this toe cannot be rotated forward; the claw of this rear toe is, with few exceptions, larger than that of the middle anterior toe. Passerines have a bony palate with a design only rarely seen in other orders. There is always a distinct sternal keel, but only traces of an appendix. The gametes have a unique structure, as far as we know. The young hatch with their eyes closed; the inside of the mouth is brightly colored, often with dark spots inside the mouth as well as other juvenile developments which disappear later. The young gape, which elicits feeding by their parents. These birds have a worldwide distribution with the exception of a few very remote oceanic islands and areas near the poles. This is usually the predominant bird group in an area, both in number of species and in number of individuals. There are four suborders: 1. The BROADBILLS (Desmodactylae), in which the flexor tendons of the third toe (the middle front toe) and those of the rear toe are joined together. The front toes are fused at their bases. 2. The NOISEMAKERS (Clamatores), in which the flexor tendons of the toes are separate; the lower syrinx has one or two tensor muscle pairs inserted on the half-rings of the trachea either in the middle, through the entirety, or only at its end. 3. The LYRE BIRDS (Suboscines) also have separate flexor muscles in the toes; the lower syrinx has two or three pairs of tensor muscles inserted at both ends of the tracheal half-rings.

Order: passerine birds, by W. Meise

▷
True Woodpeckers:
1. Lesser spotted woodpecker *(Dendrocopos minor);*
2. Red-headed woodpecker *(Melanerpes erythrocephalus);* 3. Golden-backed woodpecker *(Chrysocolaptes lucidus);*
4. Crimson-winged woodpecker *(Picus puniceus);*
5. Northern three-toed woodpecker *(Picoides tridactylus);* 6. White woodpecker *(Leuconerpes candidus);* 7. Yellow-bellied sapsucker *(Sphyrapicus varius);* 8. Middle spotted woodpecker *(Dendrocopos medius);* 9. Syrian woodpecker *(Dendrocopos syriacus);* 10. Black-naped green woodpecker *(Picus canus);*
11. Greater spotted woodpecker *(Dendrocopos major);*
12. Green woodpecker *(Picus viridis).*

4. The SONGBIRDS (Oscines), in which the flexor tendons of the toes are separate; the lower syrinx has four to nine pairs of tensor muscles inserted at both ends on the tracheal half-rings. The suborder Desmodactylae differs from the other three (which are grouped together as Eleuterodactylae) because of the way the tendons of the toes are joined. However, when referring to the insertion of the song muscles, we must group the first two suborders together as Acromyodae and the last two as Diacromyodae; when we refer to the number of song muscles, the first three suborders may be grouped as "Suboscines," in a wider sense than used above, to distinguish them from the fourth suborder, Oscines.

Passerines are land birds even though some of them may get food from water not far from shore. These birds evolved from ground-dwelling forms of tree-dwelling birds, as all have the typical perching foot. Their four toes are suitable for grasping branches, stalks, wires, etc. The grasp remains firm even when the bird is asleep, because the flexor tendon and its sheath rest inside one another, and each must be freed before the toes can extend and the bird can fly, hop off, or fall down. Only a very few members of this order never fly at all.

The gaping of the young in the nest is among one of the most exciting events in nature. Young thrushes which already have their eyes open can be made to gape by fastening a cardboard circle to a stick and moving it near the young, as if it were the adult's head. Such a model, however, does not contain all the stimuli needed for directed gaping. If one adds a small projection on the front of the circle, whether it resembles a thrush's beak or whether it is merely a small circle or square, the young will gape at this projection. Thus, only a few signals are enough to release the gaping, which is necessary for survival, since without it the young are not fed, fail to gain weight, and die. If there is a long interval between feedings, young passerines gape even in the absence of the adult, i.e. the movements on the nest or the darkening of the nest entrance hole which normally release gaping. This also applies to the blind young, which erect their necks and gape in an undirected way without having been induced to do so by the parents' touching the edges of their beaks, as woodpeckers do. Once the eyes are open all the young gape in the same direction. Once the young are older they respond by gaping only to their parents. Strangers, e.g. a person, then release a fear response, i.e. pressing down in the nest silently.

Aside from the passerines, this gaping of the young when they are ready for food apparently occurs only in the cuckoos and mousebirds (see Volume VIII). However, the colorful inside of the mouth in nestlings is also found in certain other groups of birds. The function of these markings becomes obvious when we see the gaping young in nests which have only a small opening; these young have thick, strikingly

The gaping of the young

◁
Broadbills and woodcreepers: 1. Wedge-billed woodcreeper (*Glyphorhynchus spirurus*); 2. Banded broadbill (*Eurylaimus javanicus*); 3. Plain brown woodcreeper (*Dendrocincla fulginosa*); 4. Green broadbill (*Calyptomena viridis*); 5. Long-tailed broadbill (*Psarisomus dalhousiae*); 6. Cape broadbill (*Smithornis capensis*); 7. Red-billed scythebill (*Campylorhamphus trochilirostris*); 8. Narrow-billed woodcreeper (*Lepidocolaptes angustirostris*); 9. Strong-billed woodcreeper (*Xiphocolaptes promeropirhynchus*); 10. Olivaceous woodcreeper (*Sittasomus griseicapillus*).

colored edges of the beak, or light-reflecting bodies inside their mouths. The oral cavity, which is colorful in any case, is then surrounded by a yellow edge; this helps the parents place the food in the youngs' mouths, even in the dim light of the nest. The young then swallow the food and close their beaks for the time being. In many species these gape-markings are also species-typical, and parents will feed only young with the proper markings.

4 Broadbills, Noisemakers, and Lyrebirds

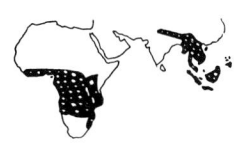

Fig. 4-1. Broadbills (family Eurylaimidae).

Suborder: Desmodactylae, by B. E. Smythies

Family: broadbills, by B. E. Smythies

Those birds which gape in the nest and which have less than four pairs of vocal muscles are considered to be primitive. They belong to the three suborders mentioned in Chapter 3: Desmodactylae, Clamatores, and the Suboscines (the primary or primitive songbirds). The 1100 species are distributed almost exclusively in the New World. The Desmodactylae are found in Africa, India and Malaya, the Clamatores in America and the primary songbirds, with only four species, in Australia. Thirty-two of the 1080 species of Clamatores, those belonging to the families of pittas, asitys, and New Zealand wrens, are an exception to this rule, as these birds are found from Africa through India to New Zealand.

The suborder Desmodactylae consists of only one family, the BROADBILLS (Eurylaimidae). These birds have large heads and are generally broadbilled. There are fifteen cervical vertebrae (other passerines have fourteen). There are small scales on the rear of the tarsus. The wings are short and round, and thus the birds only fly short distances. There are eight genera with fourteen species, two of which are found in tropical Africa, and the rest in the Orient.

This family falls naturally into two groups. The TYPICAL BROADBILLS (subfamily Eurylaiminae) have very large beaks; the GREEN BROADBILLS (subfamily Calyptomeninae), with the single genus *Calyptomena,* with a rictal brush. The size varies from that of a sparrow to that of a jay; the body is compact, and the wings are short and round. The eyes are large, and the bill is flat, wide and hooked at the tip; the green broadbill is the only bird of the family with a smaller beak, which is covered at the base with a dense hood of feathers. The sexes are usually different in appearance.

Broadbills are mainly forest birds; some prefer mountains, and some prefer plains. One can often see noisy troops of these birds. One genus commonly found in Africa, the AFRICAN BROADBILLS *(Smithornis),* with two species, was formerly regarded as a genus of flycatchers, but lately it has been classified with the broadbills. C. W. Mackworth-Praed and C. H. W. Grant consider the CAPE BROADBILL *(Smithornis capensis)* noteworthy because of its loud, explosive, grunting croaking. This sound does not sound bird-like at all; it is produced during the very short circular flights during which the bird's wings tremble. We do not know for sure whether this sound is a vocalization or whether it is produced by the wings. The only definite vocalization which the Cape broadbill makes is a fairly short, weak squeak. James Chapin, the famous ornithologist and discoverer of the Congo peacock, found similar behavior in the RED-FLANKED BROADBILL *(Smithornis rufolateralis)* in the Ituri Forest (Congo). The other African genus has only one species, the GRAY BROADBILL *(Pseudocalyptomena graueri)*: these birds are rarer and less well-known inhabitants of the eastern Congolese mountains. They behave like flycatchers. The DUSKY BROADBILL *(Corydon sumatranus)* is found in the oriental region—Burma, Thailand, eastern China, and Malaysia. The noisy calls of this blackish-brown bird, which has a hidden fiery-red spot on its back, are difficult to describe.

Fig. 4-2. Green broadbill.

The BLACK-AND-RED BROADBILL *(Cymbirhynchus macrorhynchos)* lives in the Indo-Malayan area as the sole species of its genus. This black, red, and white bird is usually silent. It lives along rivers in Borneo and is seen there much more often than the other broadbills. The genus of BROADBILLS in the narrower sense *(Eurylaimus)* consists of three species found in Malaysia and the greater Sunda Islands. These birds are mainly red, black, and white. The PHILIPPINE BROADBILL *(Eurylaimus steerii)* inhabits the southern Philippines and is notable because of the blue skin rings around its eyes and because of its blue feet. It has soft musical calls and a high-pitched whistle; on short flights it also produces a whirring noise with its wings. The long, tremulous call of the BLACK-AND-YELLOW BROADBILL *(Eurylaimus ochromalus)* consists of a sound which is repeated for about seven seconds with increasing speed; this bird is a characteristic inhabitant of the forests of Borneo and southern Asia.

The LONG-TAILED BROADBILL *(Psarisomus dalhousiae)* is a widely distributed mountain species with beautiful bright green, black, and blue colors; it has several bright yellow feathers on its head. Its loud, whistling song phrase consists of five to eight tones of equal pitch. Two species of GREEN BROADBILLS (genus *Calyptomena*) are found only in Borneo. These birds are considered members of the second subfamily. They are magnificent, predominantly green birds decorated with some black; HOSE'S BROADBILL *(Calyptomena hosii)* has a bright blue plumage also.

Most broadbills are insectivorous; only the black-and-red broadbill eats berries and even shrimp, small fish and crabs, as well as beetles, crickets, and grasshoppers. The green broadbills, however, prefer fruit. According to E. Thomas Gilliard, the display of some species is yet to be fully interpreted. These birds suddenly reveal brightly col-

▷
Horneros, spinetails, foliage-gleaners, etc:
1. Chotoy spinetail *(Schoeniophylax phryganophila)*; 2. Black-capped foliage-gleaner *(Philydor atricapillus)*; 3. Pale-breasted spine-tail *(Synallaxis albescens)*; 4. Rufous-fronted thornbird *(Phacellodomus rufifrons)*; 5. Des Mur's wiretail *(Sylviorthorhynchus desmursii)*; 6. Wren-like rushbird *(Phleocryptes melanops)*; 7. Streaked xenops *(Xenops rutilans)*; 8. Dark-bellied shake-tail *(Cinclodes patagonicus)*; 9. Sharp-tailed streamcreeper *(Lochmias nematura)*; 10. Firewood-gatherer *(Anumbius annumbi)*; 11. Striated earthcreeper *(Upucerthia serrana)*; 12. Rufous hornero *(Furnarius rufus)*.

Antbirds: 1. Scale-backed ant bird *(Hylophylax poecilonata)*; 2. Barred ant shrike *(Thamnophilus doliatus)*; 3. Variable ant shrike *(Thamnophilus caerulescens)*; 4. Bare-crowned ant bird *(Gymnocichla nudiceps)*; 5. Black-faced ant thrush *(Formicarius analis)*; 6. Spectacled antbird *(Phlegopsis nigromaculata)*; 7. Great ant shrike *(Taraba major)*; 8. White-fringed ant wren *(Formicivora grisea)*; 9. Undulated ant pitta *(Grallaria squamigera)*; 10. Short-tailed ant wren *(Myrmotherula brachyura)*; 11. Chestnut-crowned ant pitta *(Grallaria ruficapilla)*; 12. White-faced ant catcher *(Pithys albifrons)*; 13. Chestnut-backed ant bird *(Myrmeciza exsul)*; 14. Slate-crowned ant pitta *(Grallaricula nana)*.

Suborder:
Clamatores,
by W. Meise

Fig. 4-3. Noisemakers: 1. Asitys (family Philepittidae) Madagascar only; 2. New Zealand wrens (family Xenicidae).

ored spots on the plumage of the back and spread them, rather like some other birds do their colorful erectile crests. They exhibit their decorated backs from special perches as well as in flight. Quacking calls accompany this display.

All species build the same type of unusual nest, which may reach a length of two meters. It usually hangs from the tip of a branch in a shady forest glade, and it is almost always over a river or some other body of water. Its shape corresponds to that of a gigantic pear, with an elongated narrow part connecting it to the end of the branch. There is usually a large projection over the entrance. The nest is built of grass, leaves, moss, rootlets, and similar materials, and it is lined with green leaves. All kinds of strands attached with spiders' webs hang down from the nest proper. The nest exterior is often decorated with lichens and spider webs. Broadbills usually lay two to four eggs (the long-tailed broadbill lays up to five or six); the eggs are white, light red, or cream-colored, with spots of varying density. The green and African broadbills lay unspotted eggs. Nothing seems to be known about the duration of incubation, the fledging period, or other details of the nest life in any of these species.

The NOISEMAKERS (suborder Clamatores) were given their rather presumptuous name by Old World ornithologists, because the songs of most of these birds were much less pleasing to the European immigrants who traveled to the Americas than were the vocalizations of many European songbirds.

Most New World passerines are combined in this suborder. The back of the tarsus of these birds is never covered by a long shield. Noisemakers are generally distributed in America; only a few species are found in tropical Africa, Madagascar and from India to New Zealand. There are two superfamilies: A. The Furnarioidea have vocal membranes which are usually only between the trachael rings; the plumage has mainly black, brown, or brownish-red pigments. These birds are found in South American forest areas, and, to a lesser extent, in forest steppes and open country; they are also found in Central America, to a much lesser degree. There are four families: 1. Woodcreepers (Dendrocolaptidae); 2. Ovenbirds and relatives (Furnariidae): 3. Antbirds (Formicariidae); and 4. Tapaculos (Rhinocryptidae). B. The TYRANT FLYCATCHERS (Tyrannoidea) have a lower larynx which extends from the region of the trachea into that of the bronchi; the plumage has yellow and red pigments (lipochromes). These birds are found in the Americas and in warmer parts of the Old World in thirty-two species. There are eight families: 1. Pittas (Pittidae); 2. Asitys (Philepittidae); 3. New Zealand wrens (Xenicidae); 4. Tyrant flycatchers (Tyrannidae); 5. Sharpbills (Oxyruncidae); 6. Manakins (Pipridae); 7. Cotingas (Cotingidae); and 8. Plantcutters (Phytotomidae).

WOODCREEPERS (family Dendrocolaptidae) occur in all tropical areas of continental America, although they are absent on the Antilles. There are forty-eight to sixty species of these climbing birds, depending on various opinions. They resemble the tree creepers of the more northern countries in structure, color, and manner of feeding, although these two groups are not interrelated. Woodcreepers are generally larger than tree creepers; the L is about 14-33 cm. The body is slim. The tail is long and graduated, with 12 retrices; the tail feather shafts project beyond the feather vanes as sharp spines curved downward. The beak is thin; there are many beak variations, ranging from 6-7 cm long, strongly down-curved bills, to short, flattened ones. The plumage is olive, brown, brownish-red, and brownish-yellow, and it never has bright iridescent colors, but is usually striped, banded, or spotted. These birds are difficult to identify in the wild because of the uniformly colored plumage. The sexes always have a similar coloration.

Woodcreepers obtain their food, which consists of insects, spiders, and other invertebrates, from clefts in tree bark or among moss and other plants which grow on the bark. They hop up the tree trunks, often vertically, using the stiff tail as a support, while looking for food. When they have reached the top of one tree they fly to a low point on the next one. Some species, notably those of the genus *Dendrocincla*, usually follow the migrations of army ants (see Volume II). They cling to tree trunks near the ant troops, and then capture such food animals as are flushed out of their hiding places by the ants. Woodcreepers also take flying insects and many small lizards. Only rarely, if at all, do they eat fruit.

The pure, clear melodies of these industrious songsters attract attention in spite of their simplicity. Often the song is a soft trill or a long sequence of loud, ringing, very similar tones. They sing mostly at dusk while searching for bark clefts and old woodpecker holes in which they sleep, always singly. They never form flocks, but some species do live in pairs all year round. They build their nests in old woodpecker holes or in natural clefts in trees, where the nests are difficult to find. The birds carry fine rootlets, pieces of bark, lichens and strands of plant tissue to the nest site. They lay two or sometimes three white eggs which are equally rounded at both poles. Both parents of the genus *Lepidocolaptes* build the nest, incubate in alternation, and rear the young. However, in the TAWNY-WINGED WOODCREEPER *(Dendrocincla anabatina)*, only the female incubates. The SPOT-CROWNED WOODCREEPER *(Lepidocolaptes affinis)* incubates for fifteen to seventeen days, and the young stay in the nest for nineteen days after hatching. Incubation lasts twenty-one days in the tawny-winged woodcreeper, and the young are fed in the nest for twenty-four days after hatching. The female tawny-winged woodcreeper leads each of her two young to a separate tree hole for roosting, and then returns alone to a third hole.

Only six species of the OVENBIRDS AND RELATIVES (family Furnariidae)

Family: woodcreepers, by A. F. Skutch

Fig. 4-4. Woodcreepers (family Dendrocolaptidae).

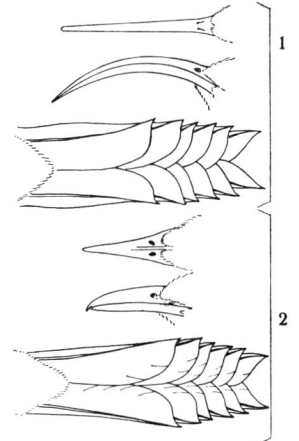

Fig. 4-5. Beak (from above and in side view) and tail of the red-billed scythebill (1) and barred woodcreeper (2).

Family: horneros and relatives,
by W. Meise

Fig. 4-6. Horneros and relatives (family Furnariidae).

Fig. 4-7. Pair of rufous horneros in alternate song on their nest.

rightfully deserve the colloquial name of "ovenbird" for only these species make "ovens." The L is 11–24 cm, and the body resembles that of the woodcreeper, although the tail feathers are not as stiff, except in the genus *Sclerurus*. These birds usually have an inconspicuous coloration; they have a superciliary stripe. The cheek and chin feathers are often light-colored, and the chin is even yellow in some species; the wing bands are often noticeably brownish-red or white. Both sexes usually have the same coloration. The food consists of insects and spiders, and occasionally small seeds. The nests are variable; these birds are hole nesters which lay uniformly white (rarely blue or greenish) eggs. The nest is almost always loose and built in holes; it may also be spherical. One species builds a sort of spherical nest suspended from reed stalks; another species builds open nests. These birds are distributed from central Mexico (seven species) to southern South America.

The fifty-eight genera and about 219 species are divided into three tribes: A. The ovenbirds are long-legged ground birds which nest in the ground or use soil for nest-building; there are forty species, found predominantly in southern South America. B. The bushcreepers are usually small birds, often with fringed or long tails; they build "ball" nests. There are ninety-five species found mainly in tropical South America. C. The leafcreepers and others are partially trunk-climbing or thrush-like birds which, like the tits and chickadees, search over leaves; they sometimes nest in excessively large nests in the open. There are eighty-four species found predominantly in tropical America.

Most TRUE OVENBIRDS resemble the European pipits and wagtails. These birds dig unusual tunnels, one to three meters long, into earth cliffs. The MINERS (genus *Geositta*) comprise ten species, among them the COMMON MINER (*Geositta cunicularia*), which has a L of 17 cm. It is earth-brown and has a rusty-brown wing area. The SHAKE-TAILS (genus *Cinclodes*) also comprise ten species, among them the DARK-BELLIED SHAKE-TAIL (*Cinclodes patagonicus*), which has a L of 21 cm; this bird has a cinnamon wing band and is gray below with white stripes. The best known representative of the OVENBIRDS (genus *Furnarius*), with six species, is OVENBIRD or RUFOUS HORNERO (*Furnarius rufus*; Color plate, p. 125), which has a L of 19 cm and weighs 75 g. It is predominantly reddish-brown.

The common miner is particularly common on the dunes of the Chilean coast. It chops worms and larvae out of the ground with its somewhat down-curved bill. It is one of the few ground birds in its inhospitable homeland where the mountains rise to 4500 meters. The shake-tails replace the true dippers in southern South America. The tunnels to their nests often wind their way into slopes covered with scattered rocks.

This ovenbird has been described by many travelers to South America. It takes advantage of man's land cultivation in the pampas and uses fence posts and masts as perches, from which it guards its territory and immediately chases off any intruder. At night, these birds prefer to withdraw into dense tree crowns even in the pampas. "As soon as it is light enough," reports Helfried Hermann, "they search the branches of the tree on which they roosted for insects. They sing all year round, and the song resembles a light bell-like laughter. It is generally given as an alternating duet between males and females. Often these birds will stand opposite one another with their necks erect, their wings slightly drooped, and their tails fanned out. A song phrase lasts three to six seconds and consists of ten to forty single syllables. The pitch is usually higher in the first phrase than in the second."

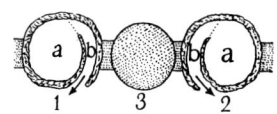

Fig. 4-8. Ground pan of ovenbird or hornero nests on a power pole. 1 and 2—nests; 3—mast or pole; a—width of the nest chamber, 13 cm; b—width of the antechamber, 5 cm; direction of take off flights (arrows).

Near Santa Fe in Argentina, choice of partners and pair formation begins as early as the local fall months. The nest is built on one of the perches at a height of two to thirty meters. It is found on the ground only very rarely. It is often built on top of another nest, and thus four-storied nests may occur. Ovenbirds can be seen nest-building all year round, except during the molt. "The mere sight of moist earth seems to stimulate the birds to build at any season." Hermann continues: "Under favorable conditions a nest is completed in ten to sixteen days. This is an astonishing performance for one pair of birds, for the nest's completion requires 1500 to 2500 lumps of clay, each three to five grams in weight. First a nest base of varying thickness is built; this depends on the nature of the material beneath, and it is often absent in the middle. Next the birds build the two short outer walls and one of the two long outer walls." According to Hermann's report, these three walls are then joined to form a roof, and the remaining almost-round opening of the other long side, which is ten to eleven centimeters in diameter, is closed up to leave an opening eleven centimeters high and 4.9 centimeters wide; this is the entrance into the antechamber. The birds then bend the outer wall inward and continue it parallel to the opposite long outer wall. The birds leave an opening in this cross wall through which they can reach the nest hole from the antechamber. Finally a nest lining of fine grass stalks is brought in (Fig. 4-8).

When the outer nest walls consist of numerous grass stalks with adherent cow manure, the nest becomes particularly rain resistant. The average weight of the nest is four and one-quarter kilograms, although nests weighing six and three-quarter kilograms have been found. One pair of ovenbirds may work on four nests at a time, either working together on all of them, or each bird working on its own. Nevertheless, at the beginning of the egg-laying season in September, only three out of twenty-six pairs in a particular area of observation had more than two nests completely ready for use. The full clutch usually consists of three or four white eggs. The incubation period is fourteen to eighteen days. The young leave the nest after twenty-one to twenty-six days, and they do not return to it for roosting after that time. This is understandable, for not only has the nest become dirty, but by December (during the local summer), the heat in the closed space would be unendurable. Many other species of birds, such as swallows and house sparrows, breed in the ovenbird nests at a later date.

"The song of the young in the nest can be heard for some distance toward the end of the nestling period," Hermann reports. "While singing, the young stretch their heads through the small exit into the antechamber. They begin to sing parts of the song phrase as early as fourteen days after hatching. After eight weeks (the end of December), the families are still in their territories, but they are only loosely held together, and often one of the members is absent for long periods." Adults are pursued by falcons and marsupial rats, while the young are hunted by a buteo, the ajaja (see Volume VII) which opens the nest chamber, and by the guira cuckoo (see Volume VIII) which goes for the eggs or young.

The BUSHCREEPERS comprise two genera, both of which are rich in species. These are: 1. The SPINE-TAILS *(Synallaxis)*, which have only ten tail feathers with closed vanes; the plumage is often a reddish-brown. There are twenty species including the RED-CAPPED SPINE-TAIL *(Synallaxis ruficapilla)*; and 2. The TWELVE-FEATHERED SPINE-TAILS *(Asthenes)*, which have twelve tail feathers; there are twenty species, including the CREAMY-BREASTED CANASTERO *(Asthenes dorbignyi)*, which has a L of 19 cm and a reddish-brown throat patch.

The red-capped spine-tail, a well-known south Brazilian species, builds small spherical hanging nests, which it enters from below. In the Peruvian Puna zone, the cold highland altitudes of over 3700 meters, the creamy-breasted canastero seeks its food on the ground among rocks or bushes. Its huge nest (fifty centimeters high, forty centimeters wide, and forty centimeters deep) is often in paya trees four meters high; the leaves of these trees project beyond the nest. Both the small nest chamber and the passage leading to it are lined with wool; the entry lies to the side and above. The spinetails take the place of the Old World leaf warblers, reed warblers, or tree creepers in South America (none of those species are found there).

The THORNBIRDS and their relatives include, among others, the following genera: 1. THORNBIRDS in the narrower sense *(Phacellodomus)*, which have seven species, including the RUFOUS-FRONTED THORNBIRD *(Phacellodomus rufifrons;* Color plate, p. 125). This bird has a L of 10 cm and is found in northern and eastern South America. 2. The TREE-RUNNERS, which have only one species, the WHITE-THROATED TREE-RUNNER *(Pygarrhichas albogularis)*, with a L of 16 cm. This species is closely related to the preceding genus, but the narrow lower mandible is bent

slightly upward. These birds are found in the Andean forests of Chile.
3. The LEAF-SCRAPERS *(Sclerurus)* comprise four species, including the BLACK-TAILED LEAF-SCRAPER *(Sclerurus caudacutus)*, which has a L of 17 cm, a slightly upward-curved lower mandible, and stiff-shafted tail feathers. This bird is found in northern South America.

Of all the birds in the ovenbird family, the thornbirds build the largest nests. The thornbird nest is one meter high, spherical, and surrounded by twigs about half the thickness of a finger. It often contains several nest chambers, all of which are entered from below. The tree-runner resembles the nuthatch in the structure of its beak and its mode of obtaining food. It chops its nest hole out of soft trees, and lays its two to three eggs on the plain wood chips. When the white-throated tree-runner searches along tree branches for food, it supports itself on its stiff tail; this bird is more usually busy on the ground where it eagerly turns over fallen leaves. It probably does not make the ground holes in which it nests.

ANTBIRDS (Formicariidae) are among the commonest birds of tropical and subtropical South America. Their characteristic features are not easily described, since there are many different forms. The L is 8.5–34 cm. There is a strong, hooked bill which, in many species, has a "tooth" as is the case with shrikes, (which are not found in South America); the smaller species have a fine, smooth beak. The wings are generally short and rounded. The tail is sometimes so short that stuffed birds seem to lack it; it is quite noticeable in the living bird, however, because it is often erected, as in the ant thrushes. The tail is very long, longer than the body in many species—like the batara which does not erect it. The lower legs of forms which live close to the ground are very muscular; the tarsus of these birds is also very long, particularly in the ant pittas, although the toes are relatively short. The plumage is full and soft, particularly on the back and sides of the body of the ant shrikes—also called woolbacks. The coloration is usually inconspicuous being darkish in the male, often black or gray, while in the female it is brown, often with paler spots on the wings and tail. The sexes usually have distinct plumage colorations (except in the ant pittas). Antbirds are found in the warmest areas of America from southern Mexico to Bolivia and northern Argentina; the greatest number of species is found in the Amazon basin.

We will only mention a few of the 226 species which are divided into 53 genera; these are: 1. The GIANT ANTSHRIKE *(Batara cinerea)* has a L of 34 cm and is the only species of its genus; 2. The PARANA SHRIKES *(Mackenziaena)* are crested birds; the SPOTTED PARANA SHRIKE *(Mackenziaena leachii)* is included in this genus; 3. The ANTSHRIKES *(Thamnophilus)* have a toothed beak. These birds sometimes exhibit tail trembling. There are 18 species including the BARRED ANTSHRIKE *(Thamnophilus doliatus;* Color plate, p. 126) which has a L of 16 cm; 4. The ANTWRENS *(Myr-*

Family: antbirds, by H. Sick

▷
Green broadbill *(Calyptomena viridis,* above left) attracts attention by its gentle call. *Cotinga amabilis* (above right). Blue-tailed pitta *(Pitta guajana,* below left) the plumage of the lower parts is surprisingly colorful.
Andean cock-of-the-rock *(Rupicola peruviana,* below right); The male is often seen in zoos, but the inconspicuous females are only rarely brought into the trade.

▷▷ and ▷▷▷
Superb lyrebird *(Menura novaehollandiae)* left, on a tree in the jungle.
Center: in display posture.
Right, from above down: seeking food, in full display, and feeding a youngster.

▷▷▷▷
Bare-throated bell bird *(Procnias nudicollis);* a gaping male.

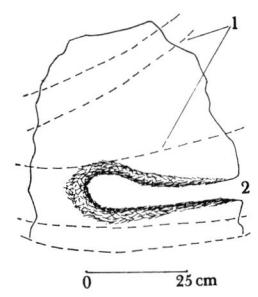

Fig. 4-9. Nest of the creamy-breasted canastero in cross section. (1) leaves which pass through the nest, (2) "hall" and nest chamber lined with soft materials.

Fig. 4-10. White-throated treerunner.

Fig. 4-11. Antbirds (family Formicariidae).

motherula) comprise 30 species, including the SHORT-TAILED ANTWREN (*Myrmotherula brachyura*; Color plate, p. 126) which has a L of 8.5 cm; 5. The FIRE-EYES (*Pyriglena*) include the FIRE-EYE (*Pyriglena leucoptera*) which has a hidden white spot on the upper back; 6. The ANT THRUSHES (*Formicarius*) include the COLMA ANT THRUSH (*Formicarius colma*); 7. The genus *Chamaeza*, also called ant thrushes in English, comprises four species, including the SHORT-TAILED ANT THRUSH (*Chamaeza brevicauda*) which has a L of 20 cm and weighs 100g; 8. The ANT CATCHERS (*Pithys*) have only one species, the WHITE-FACED ANT CATCHER (*Pithys albifrons*; Color plate, p. 126). These birds have a high pointed crest which is clearly distinct from the black head; there is a long, white, "bristly" beard; 9. The RHEGMATORHINAS (*Rhegmatorhina*) are ground birds with vividly colored bare faces which glow in the darkness of the forests as if they were phosphorescent. The size of the bare area can be altered by feather movements. There are five species, including the SPOTTED RHEGMATORHINA (*Rhegmatorhina melanosticta*); 10. the ANTBIRDS (*Hylophylax*) comprise four species, including the SPOTTED ANTBIRD (*Hylophylax naevia*); 11. The BARE EYES (*Phlegopsis*) include three species, among which is the SPECTACLED ANTBIRD (*Phlegopsis nigromaculata*; Color plate, p. 126); 12. The ANT PITTAS (*Grallaria*) comprise 27 species, including the ROYAL ANT PITTA (*Grallaria varia*; Fig. 4-13) which has a L of 21 cm. The GNAT EATERS (*Conopophaga*): see p. 139.

Antbirds live at the lower and medium levels of shady, densely leaved forests, slipping about restlessly. Only a few species have invaded the zone of the tree crowns or live in bushes exposed to the sun. The many small species may remind one of the European tits and leaf warblers which are not present in those areas, or of flycatchers, tyrant flycatchers, vireos and wrens; some antbirds resemble ovenbirds which sometimes live on the same branch with them. These birds are easily compared with the pittas from the Old World tropics because the ant pittas have a similarly compact body, an equally short tail, and long legs. Similarities between antbirds and gnat eaters are also evident; these no doubt reflect some sort of relationship. Today the gnat eaters and the tapaculos are classified in the antbird family.

Ant pittas shoot through the undergrowth like arrows when they hear imitations of their whistles in their territories. They move about in jumps. Sometimes a hunter may shoot an ant pitta in the forest twilight, thinking it to be a rabbit. The ant thrushes, which slink about on the ground, may be mistaken for small gallinaceous birds; hence the colma ant thrust is called "galinha do mato" (little forest hen) in Brazil. Sometimes these birds fly up to a branch to get a better view. The species which live low in the undergrowth prefer to use the only slightly leaf-covered vertical shoots which are so characteristic of the jungles. The birds hang on the side of the pencil-thick shoots with a grip which is perhaps particularly effective and saves energy because the outer and the middle toes are not tied at their bases.

Bright white signal spots, which are hidden deep in the plumage, particularly on the back when the bird is at rest, play a particular role in the antbird's life. During mutual threats these spots are alternately shown and hidden again, giving sort of semaphore effect which is quite an impressive means of communication in the forest twilight. Songs generally consist of short rhythmical phrases, some of which are unmelodious and quacking as in many of the ant shrikes; others are pure whistling sounds. The full, flute-like, long-lasting scales of the antthrushes are among the most beautiful bird sounds of South America; they sound as if they belonged in part to the organ or flute-like sounds of singing tinamous which inhabit the same forests. Females also sing, and sometimes a pair sings a duet. Fledged young ant shrikes occasionally make themselves noticeable by conspicuous location calls.

Fig. 4-12. Barred ant shrike.

The "tooth" on the beak is used very effectively in killing prey. Antbirds eat arthropods, crickets, bugs, and beetles, as well as spiders, centipedes, wood lice, and even snails. Larger species occasionally eat frogs, lizards, lice, helpless young birds, and (as has been ascertained in the batara) small snakes. The name "antbird" comes from the fact that many species seek out army ants, not to eat them, but in order to snatch up the numerous arthropods which are flushed by the ants and which then readily become the prey of the birds. Such ant followers include the bare eyes, the fire eyes, and the antbirds. Many of these birds are so adapted to army ants that they are rarely found far from these insects. This is also true of some other birds, like the wood creepers and the pheasant-like forest cuckoos (Vol. VIII) in the Amazon basin.

Fig. 4-13. Royal ant pitta.

The antbirds' busy life near a troop of army ants is one of the most thrilling spectacles of the South American jungle. The antbirds themselves are the surest guides to the ants, their loud calls and songs betray the insects' presence to the knowledgable observer. The birds are particularly attracted to the red army ant *(Eciton burchelli)* and the smaller blackish rain ant *(Labidus praedator)*.

During their periods of migration, hundreds of thousands of heavily "armed" ants are on the move, some in fronts several meters wide, others in narrow columns. The ground literally comes alive at the tip of the column, for it is here that most animals are flushed by the ants; the animals hop, run, and fly off every which way. Crickets, in particular, rise up in inconceivable numbers. Even larger animals like lizards drop down from the trees over which the ant army moves; these animals are then once again forced to take up the unequal fight against the insects on the ground. Caterpillars and other soft-bodied animals are immediately cut into pieces by the ants. The air is full of a ceaseless rustling and crackling. In places where certain bugs are numerous, the air becomes charged by the odor of these arthropods.

Here antbirds are in their very element. Hanging on a shoot close to the ground or from a low vine, or perched upon a stump, they await their prey. From these various perches they then snap up the organisms flushed from the neighboring leaves or twigs by the ants; then suddenly they too jump down on the ground among the swarming ants or dash off some distance. Prey is available in such numbers at the head of the army ant column that an antbird can often seize something every minute. When possible, antbirds move from the head of one ant column to another, thus continually finding food. As a result, they are nomadic, to some degree, at least outside the breeding season. A pecking order is soon established and this gives the largest and strongest species the best hunting spots along the ant train. A pecking order is also established among the individuals of any particular species of antbird. Dominant birds, particularly adult males and owners of adjacent territories, drive off immature birds which have to make do with less productive areas along the ant columns. During breeding season, antbird pairs restrict themselves to ant swarms which pass near their nests. Anting or the rubbing of live ants into the plumage, has been observed in antbirds as well as in tanagers and manakins (Vol. VII).

Antthrushes and pittas also eat seeds; they can therefore sometimes be caught in traps baited with corn which the natives set for gallinaceous birds and rails. Occasionally antbirds may be seen bathing in the shallows of quiet forest streams or in rain puddles in the jungles.

Many antbird species have nests which are deep open cups lightly fitted into the thinner branches of a shrub which may hang over water (as in ant shrikes). Other antbirds, like the fire eyes, build well-closed spherical nests on the ground; these have a lateral entrance. Still other species build what might be called woven pouch nests. Some species like the antbirds and the ant pittas breed in hollow trees. The two to three eggs are white or yellowish with fine spots; only rarely are they an unspotted white, as in the ant thrushes, or a uniform blue-green as in the ant pittas. Both parents incubate, feed the young, and lead them after fledging. Many antbirds are evidently paired all year round and remain in their territories. Some, however, become sociable after the breeding season and then wander about with ovenbirds, woodcreepers, tanager, and other small birds.

Genus: gnat eaters, by H. Sick

The GNATEATERS *(Conopophaga)* are presently included in the antbird family. There are eight species, including the RUFOUS GNATEATER *(Conopophaga lineata)* which has a L of 13 cm and weighs 21 g. The male has a long silvery-white stripe behind the eye which it can erect in excitement. The warning call is a loud "tshik," and because of this, the gnat eater is called cuspidor (spitter) in Brazil. These birds are found from eastern Brazil to Bolivia.

The second genus, which was formerly combined with the gnateaters to form a family, may possibly belong to the tyrant flycatchers.

These are the ANT PIPITS *(Corythopis)*. These are two species, including the SOUTHERN ANT PIPIT *(Corythopis delalandi)*, which has a L of 12 cm and can weigh 14 g. The breast is black striped. These birds are found from eastern Brazil to Bolivia.

The gnateaters bear some resemblance to the European robin. They are small, tame insect-eaters seen on branches in thickets. Gnat eaters are quite different from the cautious ant pipits which run about restlessly on the ground in upland forests; they are strangely reminiscent of the true pipits both in their shape and color. During the courtship season male rufous gnat eaters make a purring "brroo brroo brrroo" sound at dusk; this is a wing noise arising from special sound-producing flight feathers. The nest, an open cup made of large leaves, is on the ground. The ant pipits song is a softly whistled, but somewhat rough, three-part purr-like phrase. When the birds are excited, one can hear a crackling sound while they trip about on the ground; this sound is presumably made by the bird's beak, and it evidently acts as a warning call.

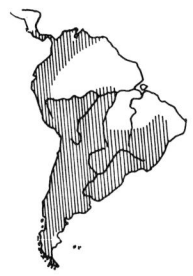

Fig. 4-14. The tapaculos (family Rhinocryptidae).

The TAPACULOS (family Rhinocryptidae) are a well demarcated group. The weight is 15.5–157 g. The nostril is protected by a raised membrane, the operculum. For example, the crested gallito is always exposed to dust storms on the ground. The tarsus is strong, and the feet are disproportionately large. The wings are short, soft and rounded; they serve as balancing aids while the bird is running on the ground or jumping about among branches—as tapaculos hardly ever fly. The tail feathers are soft and graduated, the outermost being shortest; the tail itself is rarely used as a rudder. Those species which live in dark thickets are generally dark in color, often mouse-gray. These birds are distributed from Panama southward to southern South America; their greatest density is in Chile.

Family: tapaculos, by H. Sick

We will only mention a few of the twelve genera and twenty-eight species: 1. *Pteroptochos* with three species, including the MOUSTACHED TURCA *(Pteroptochos megapodius)*; 2. *Scelorchilus* with two species including the WHITE-THROATED TAPACULO, *(Scelorchilus albicollis)* which has a L of 19 cm; 3. *Rhinocrypta* with only one species, the CRESTED GALLITO *(Rhinocrypta lanceolata)*; 4. The RUSTY-BELTED TAPACULO *(Liosceles thoracicus)* has a yellowish crop area. This is the only tapaculo of the western Amazon basin, and it lives in higher, dry areas in the jungle; 5. *Scytalopus* includes ten species among which is the MOUSE-COLORED TAPACULO *(Scytalopus speluncae)* which has a L of 10.5 cm and weighs 15.5 g. This bird is found in the upper mountain zone of southeastern Brazil, often in dense bamboo growth; 6. *Acropternis* has only one species, the OCELLATED TAPACULO *(Acropternis orthonyx)*. The legs and feet are particularly heavy. There are huge straight claws on the rear toe. This bird is found in the northern Andes.

The bamboowrens with the single species *Psilorhamphus guttatus* has a long tail and is found in southeastern Brazil.

Fig. 4-15. Crested gallito.

On his journey around the world in 1834, Charles Darwin described the tapaculos as the most interesting birds of Chile. Among other things, he wrote, "One species, called 'turco' in Chile, is not uncommon. This large-footed, rail-like creeper lives on the ground protected by thickets, which cover the arid hills here and there. It is often seen moving unbelievably fast from one thicket to another on its stilt-like legs with its tail erect. When one first sees the turco, one thinks one is looking at a badly stuffed bird skin which has escaped from some museum and come to life."

Another species has also received a Chilean name which refers to the erection of its tail; it is called "tapaculo" (cover your rear). The word "gallito," used for the single *Rhinocrypta* species means little fowl in Spanish. W. H. Hudson, the well-known English writer and naturalist, says that this bird looks "exactly like a minute cockerel." While the fowl-like tapaculos walk and run, the *Scytalopus* tapaculos move more by hopping. Though tapaculos easily escape the eye, their voices call attention to them. Their song consists of a long series of one tone syllables. Females also sing. The songs of different tapaculo species have been compared to the noise of a whetstone, snoring, grunting, sawing, flute sounds, dove calls, the barking of dogs and even to the sound of water splashing. The call is often ventriloqual. Sharp individual calls serve as warning as well as to attract other birds. The mouse-colored tapaculo silently hops about over the rock-strewn ground. Now and then it jumps into the branches to look about and sing. When the bird raises its voice, its whole body trembles and it is more easily seen at that time.

The tapaculos' food consists particularly of insects and their larvae. In some cases, plant fibers and seeds have been found in these birds' stomachs. The only tapaculo to invade the Amazon area is the rusty-belted tapaculo, which lives strictly in pairs. This species finds its food, which consists mainly of wood lice, by scratching on the ground. Occasionally it uses puddles of rain water for drinking and bathing; it roosts in dense shrubs. During the breeding season this tapaculo utters a long series of descending scales both at morning and in the evening; the female sings in a higher pitch. The nest and eggs are still unknown to us.

We do know that some tapaculos, including the tapaculos proper and the turcos, build their nests in tunnels in the ground which they excavate themselves; they may also build the nest in tall trees like some other turcos do. The *Scytalopus* tapaculos build large nests of roots and moss hidden on the ground. Two to three uniform white, non-glossy eggs are large in proportion to the birds; sometimes they may appear spotted as a result of dirt. Both parents incubate and rear the young.

Family: pittas, by N. Kuroda

The very uniform PITTA family (Pittidae) are true jewels of the bird world. The L is 15–28 cm; the body is plump, and the head is disproportionately large. The legs are long and the tail is very short and square tipped. The tarsus has a long anterior and a posterior plate. The sternum narrows towards the rear. Males and females differ in some species, but they are similar in others. Pittas are found in the Old World tropics from Africa (which has only two species) as far as Japan and the Solomon Islands.

There are twenty-three or twenty-four species, all of which belong to the genus *Pitta*. These include: 1. The BLUE-WINGED PITTA *(Pitta brachyura)*, a Japanese subspecies *Pitta brachyura nympha* with a L of 19 cm; 2. The AFRICAN PITTA *(Pitta angolensis)* is possibly only a subspecies of the blue-winged pitta; 3. The RED-BREASTED PITTA *(Pitta erythrogaster)* is small, with a bright scarlet breast and abdomen; 4. The HOODED PITTA *(Pitta sordida;* Color plate, p. 149) has fifteen subspecies of which three have brown instead of black heads; the INDIAN BLACK-HEADED PITTA *(Pitta sordida cucullata)* is one of these three subspecies; 5. STEER'S PITTA *(Pitta steerii;* Color plate, p. 149) has a black head, a green back, a white throat, and a blue breast with black in the center; the belly is bright scarlet. This species is found on the Philippines (Bohol, Samar and Mindanao); 6. The RAINBOW PITTA *(Pitta iris)* has a black face and undersides; the breast is red. This species is found in northern Australia; 7. The NOISY PITTA *(Pitta versicolor)* is found in eastern Australia; 8. The GREAT BLUE PITTA *(Pitta caerulea;* color plate, p. 149) has a L of about 29 cm, and is the largest of the pittas. The male has a blue back, while that of the female is chestnut; the undersides are yellowish-brown. These birds are found in the Malayan peninsula, Sumatra, and Borneo; 9. The BLUE-TAILED PITTA *(Pitta guajana)* Color plates, p. 133 and 149), has a longer, pointed tail. The belly is banded. This species is found from Bali to Thailand, with relatives in Burma; and 10. The SICKLE-TAILED PITTA *(Pitta phayrii)* has a L of 23 cm. The sides of the neck have long pointed feathers; these birds have striking head feathers. They are found in southeastern Burma and Thailand.

Pittas live on the ground or just above it in dense, dark forests; their habitat extends from the mangroves of the lowlands through the bamboo jungle at middle mountain altitudes up to the drooping, moss-grown forests at 2,500 meters altitude. When pittas try to fly, they tend more to hop above the forest floor rather than use their short rounded wings. In true flight (aside from migration) they cover only short distances. On the ground they have a very erect stance, although they frequently lower their heads to seek insects, worms, and snails by turning over leaves. Sometimes they even capture lizards.

The nest is a large spherical structure with a strong roof and a lateral entrance; it is built of twigs, grasses, leaves, moss, and strips of bark, all lightly worked in together and often gummed up with mud. It stands on the ground or just above it, generally in the fork of a shrub or small tree at some dark spot in the forest. The two to six rounded eggs have a rich pattern of clear as well as indistinct spots; there are

▷
Tyrant flycatchers and manakins: 1. Fork-tailed flycatcher *(Muscivora tyrannus)*; 2. Wire-tailed manakin *(Teleonema filicauda)*; 3. Eastern kingbird *(Tyrannus tyrannus)*; 4. Kiskadee flycatcher *(Pitangus sulphuratus)*; 5. Yellow-bellied elaenia *(Elaenia flavogaster)*; 6. Northern royal flycatcher *(Onychorhynchus mexicanus)*; 7. Cattle tyrant *(Machetornis rixosa)*; 8. Swallow-tailed manakin *(Chiroxiphia caudata)*; 9. Crimson-hooded manakin *(Pipra aureola)*; 10. Pied water tyrant *(Fluvicola pica)*; 11. Western flycatcher *(Empidonax difficilis)*; 12. White-bearded manakin *(Manacus manacus)*; 13. Common tody flycatcher *(Todirostrum cinereum)*; 14. Vermillion flycatcher *(Pyrocephalus rubinus)*; 15. Eared pygmy tyrant *(Myiornis auricularis)*; 16. White-headed marshtyrant *(Arundinicola leucocephala)*.

Cotingas: 1. Jamaican becard *(Platypsaris niger)*; 2. *Pachyrhamphus dorsalis*; 3. Green-and-black fruit-eater *(Euchlornis riefferi)*; 4. Black-tailed tityra *(Tityra cayana)*; 5. Black-necked red cotinga *(Phoenicircus nigricollis)*; 6. Guianan cock-of-the-rock *(Rupicola rupicola)*; 7. Amazonian umbrella bird *(Cephalopterus ornatus)*; 8. Purple-breasted cotinga *(Cotinga cotinga)*; 9. Three-wattled bellbird *(Procnias tricarunculata)*; 10. Bearded bellbird *(Procnias averano)*; 11. Bare-throated bellbird *(Procnias nudicollis)*; 12. Pompadour cotinga *(Xipholena punicea)*; 13. Rose-collared piha *(Lipaugus strephophorus)*; 14. Swallow-tailed cotinga *(Phibalura flavirostris)*.

Fig. 4-16. The pittas (family Pittidae). In eastern China, these birds are almost entirely passage migrants.

Fig. 4-17. The blue-winged pitta *(Pitta brachyura)*. This species is found in most East Chinese areas only outside the breeding season.

also wavy hair lines on a white or cream-colored background. The spots are reddish-brown to purple or blackish, depending on the species, and they are generally concentrated at the blunt pole of the egg. All the pitta calls known to date are somewhat similar, either trills or short sequences of distinct whistles.

The well-known ornithologist, Gerd Heinrich, has observed the red-breasted pitta on Celebes. He was enthusiastic about the beauty of its colors. "When one holds this wonderful bird with its bright scarlet belly in one's hand, one cannot believe how difficult it is to spot it in the jungle. Nevertheless pitta observation remains a tightly closed book. The brightly colored belly is invisible because it faces to the ground which the bird rarely leaves. The upper parts have a protective coloration which effectively conceal them in the forest twilight."

The Japanese blue-winged pitta gets its Japanese name, "Akadanna" (red lion or bodice in English) from its scarlet belly spot. This bird moves into the mountain forests of Taiwan at altitudes of 900 to 2100 meters early in April; it travels to Japan (Central Hondo, Shikoku, and Kyushu) in the end of April or mid-May. It leaves these areas at the end of September. Shortly after its arrival in Japan, one can hear its disyllabic whistles from the top branches of large trees; these syllables are reproduced as "pao-pao," "pakko pakko," "popopee popo pee," "paukai paukai," "peefai peefai," or "shiro-pen kuro-pen," depending on the particular bird, and on the individual observer's interpretation of the sound.

Blue-winged pittas breed in dense forests. Their nests are in rock clefts or forks two to eight meters above the ground in large trees. The nests are elliptical and have a lateral entrance. The most important nest materials are mosses, which are covered with dead twigs; inside, the nest floor has small roots and conifer needles. The external diameter of such a nest is 150 to 260 millimeters; the inner diameter is 160 millimeters, while that of the entrance is 60 to 130 millimeters. The depth of the nest is about 75 millimeters. When disturbed, the female covers the entrance with small leafed twigs.

The single clutch consists of four to six eggs; these are generally cream-colored or grayish-white, and they have fine pale purple-brown or gray spots. According to Kiyosu's data, the eggs measure 25 to 27.5 by 19.7 to 22.5 millimeters and weigh 5.2 grams. The young are still naked and flesh-colored just after hatching, and they carry no down. Both parents incubate and feed them.

J. T. Zimmer has often heard and occasionally seen the hooded pitta on the Philippine island of Palawan. He found it in those areas of the forests with rich undergrowth. "Most pittas of this and related species which I found," he wrote in 1918, "were on the ground or, at most, a few feet above the ground on fallen tree trunks and similar perches. One hooded pitta I saw at Brooke's Point, however, was an exception to this rule. I heard it clearly and easily reached a spot near which it had to be. Once there I had difficulty in spotting the bird, although it continued to give its explosive "wow-ha" calls, apparently only a few feet away. As the forest floor all around was more or less open, I could examine every spot, but without success; then, accidentally and surprisingly, I saw the pitta sitting on a projecting branch six meters above the ground."

Steer's pitta is found on Bohol in the Philippines; according to Hachisuka, these birds are found in forests which have small trees growing on hillocks of coral sandstone. Whitehead called this pitta the most beautiful member of the genus as early as 1899. He found these birds in the mountains of central Samar. "It feels at home in country with upland forests and fairly dense undergrowth, where the floor is covered with heaps of moss and often gigantic blocks of coral sandstone. The young were fledged in June. This pitta is said to be common on Samar from May to July; it may be a migrant."

Fig. 4-18. The Japanese blue-winged pitta. Posture when singing.

The richness of its plumage and pattern make the rainbow pitta one of the most attractive of all Australian birds. It inhabits bamboo thickets near the coast as well as mangrove forests and shrubbery, which it skillfully runs through. Its food consists of insects and snails. The nest is a large, covered structure with a side entrance—like that of other pittas.

In 1931, Cayley wrote that the noisy pitta, which is also an Australian bird, inhabits dense bush and can be attracted to within a few feet of an observer when its ringing call "walk to work," is imitated: "It lives almost entirely on the ground, eating insects and snails. Its "anvil" is a stone or a small tree trunk where it breaks the snail shells. Sometimes a stone becomes worn smooth from prolonged use. The nest is on the ground, usually among the root butresses of a fig tree in the mountain jungles. A peculiarity of pittas' nests is a sort of doormat of moist animal dung which is brought to the nest hole clinging to the bird's toes."

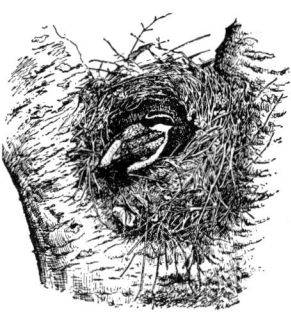

Fig. 4-19. Japanese blue-winged pitta about to slip into its spherical nest.

Davison considers the great blue pitta of Borneo to be extraordinarily shy. Unlike other pittas, it flies low and rapidly over the ground for stretches of about 180 meters when it is disturbed. The blue-tailed pitta, on the other hand, is not at all shy according to Hoogerwerf; it can be seen in the open woods and at the botanical garden at Buitenzorg (central Java). When it calls softly, the call sounds like "kurr kurr" or "purr;" louder calls sometimes sound like "pa-ok." This pitta feeds on worms, snails, ants, termites, beetles, caterpillars, and other small fry. The clutch consists of three or four eggs.

The migration of these colorful birds has not been investigated to any large extent. It is understandable that Chinese, Japanese, and southeastern Australian pittas should leave after the breeding season; however, according to Rufolf Berndt and Wilhelm Meise, there are

Fig. 4-20. Velvet asity.

Family: asitys,
by J. Dorst

Fig. 4-21. Rifleman. Above male, below, female.

Family: New Zealand wrens,
by C. A. Heming

Fig. 4-22. The bush wren.

Fig. 4-23. The New Zealand rock wren.

seasonal changes of locale in the tropical areas as well, even though the short, round wings seem unsuitable for long distance flight.

The ASITYS (family Philepittidae) are the only clamatores of Madagascar. The L is 9–15 cm. Males have a bare black eye ring. These birds are found only on Madagascar. There are two genera, each with two species: 1. The ASITYS *(Philepitta)* have a compact body with a slightly down-curved bill. The VELVET ASITY *(Philepitta castanea;* Color plate p. 150) is included in this genus; it has a L of 15 cm. The primaries have yellow edges, which are easily worn down. The females and young are predominantly olive-green. 2. FALSE SUNBIRDS *(Neodrepanis)* have a superficial resemblance to the sunbirds. The best known species is the FALSE SUNBIRD *(Neodrepanis coruscans;* Color plate, p. 150) which has a L of 9 cm. The beak is long, thin and noticeably curved downward. The female is dark green above and yellowish-gray below.

The velvet asity is an exclusive tree-dweller and never comes down to the ground; it keeps to low vegetation where it looks for fruit. These birds are usually quiet solitaries; they rarely vocalize, although they can sing pleasantly, rather like a song thrush *(Turdus philomelos)*. The nest is spindle-shaped and hangs from a twig. The clutch consists of elongated white eggs. The false sunbirds live in the moist forests of eastern Madagascar where they hunt insects on twigs and also take nectar from curved blossoms which are inaccessible to the true sunbirds.

The NEW ZEALAND WRENS (family Xenicidae) inhabit the twin islands of New Zealand. There are three genera: 1. The RIFLEMEN *(Acanthisitta);* 2. The BUSH AND ROCK WRENS *(Xenicus);* and 3. The STEPHEN ISLAND WRENS *(Traversia)*. There are a total of four species.

Like the New Zealand wattlebirds, the New Zealand wrens settled New Zealand in the Tertiary, between one and sixty-five million years ago; these birds have maintained themselves in that area since—isolated from their unknown relatives. They probably originated from some early Old World asity stock of the clamatores which need not have necessarily resembled present day pittas or asitys. All New Zealand wrens are short-tailed and soft-winged; they have strong legs and live on arthropods.

The RIFLEMAN *(Acanthisitta chloris;* Color plate, p. 150) has a L of 7.5 cm; it is the smallest New Zealand bird. The predominantly green plumage is somewhat different in the two sexes. The rifleman is well distributed in the southern beech *(Nothofagus)* forests of New Zealand. There it seeks its food singly or in pairs, moving over the tree bark like a tree creeper and uttering high pitched "see" calls. It nests in well-lined natural trees or ground holes, and lays four or five white eggs. In many places it has settled in bush areas of cut-over forest districts.

The BUSH WREN *(Xenicus longipes;* Color plate, p. 150) has a L of 10 cm.

Today, these birds are practically extinct, aside from a few remnants on smaller islands. The mountain-dwelling relative, the NEW ZEALAND ROCK WREN *(Xenicus gilviventris)* lives on sub-alpine Fjells and bush country in the mountains of the South Islands of New Zealand. It looks for food on the ground, where it also nests. Both species have a characteristic landing behavior; they dip their entire body down at the front.

Fig. 4-24. The Stephen Island wren.

The STEPHEN ISLAND WREN *(+Traversia lyalli)*, which used to inhabit Stephen Island in Cook Strait, has completely disappeared. The story of its discovery is also that of its extinction. The lighthouse keeper on the island had a cat which kept bringing small birds into the house. The keeper sent fifteen of the cat's victims to London in 1894; there it was determined that these specimens represented a new species. Shortly after this, the cat exterminated all the wrens on the island. This bird's soft, spotted, olive-colored plumage and its almost totally flightless way of life clearly distinguished it from the *Xenicus* wrens. Little is known about its way of life. According to Walter Rothschild, these birds ran like mice, kept to rock caves, and were active predominantly at night.

The TYRANT FLYCATCHERS (family Tyrannidae) form one of the largest bird families; they have more species than any other bird family in the Western Hemisphere. This family includes several rather diverse-looking forms. The name "tyrant" is derived from the kingbirds, which boldly and bravely attack and chase off raptors and other enemies from the vicinity of their eggs or young. Only rarely do these kingbirds molest their smaller neighbors. They usually have a rather simple coloration, being olive-green, gray, brown, or pale yellow; this makes specific determination in the wild difficult unless the calls, behavior, and ways of nesting are also considered. The kiskadees are more colorful and even more impressive than certain other tyrant flycatchers, for they have a red, orange, yellow, or white spot on the crown; this spot is only visible when the feathers are erected or spread out in excitement. The sexes are similar and the young resemble the adults.

Family: tyrant flycatchers, by A. F. Skutch

Tyrant flycatchers are distributed from Tierra del Fuego to beyond the Arctic Circle in Canada and Alaska, from the hottest and most moist tropical forests to the driest deserts and most inhospitable mountain heights at which insects still can live. The L is 6.5–30 cm; the third and fourth toes are joined along the most basal segment, and there are horny plates on the outer side of the tarsus. The upper mandible has a more or less distinct hook; the beak is generally short and wide, and the rictal bristles evidently help to direct flying insects into the open beak. The areas of densest distributions are in warm tropical lowlands. There are subfamilies with 115 genera and about 365 species. Some scientists include the becards with this family; we, however, include these birds with the cotingas.

▷
Pittas: 1. Hooded pitta *(Pitta sordida)*; 2. Great blue pitta or giant pitta *(Pitta caerulea)*; 3. Blue-tailed pitta *(Pitta guajana)*; 4. Elliot's pitta *(Pitta ellioti)*; 5. *Pitta superba*; 6. Blue-naped pitta *(Pitta nipalensis)*; 7. *Pitta caerula willoughbyi*; 8. Steer's pitta *(Pitta steerii)*; 9. Garnet or red-headed pitta *(Pitta granatina)*; 10. *Pitta maxima*.

Fig. 4-25. Tyrant flycatchers (family Tyrannidae).

Fig. 4-26. The long-tailed tyrant with its long tail.

◁

Scrub bird, sharpbills, asitys, New Zealand wrens, and plantcutters: 1. Velvet asity (*Philepitta castanea*); 2. Rifleman (*Acanthisitta chloris*); 3. False sunbird (*Neodrepanis coruscans*); 4. Redbreasted plantcutter (*Phytotoma rutila*); 5. Crested sharpbill (*Oxyruncus cristatus*); 6. Rufous scrub bird (*Atrichornis rufescens*); 7. Bush wren (*Xenicus longipes*).

A. The Fluvicolinae subfamily includes small to large birds; these are often ground birds with long tarsi. They are strikingly colored. There are about twenty-seven genera with seventy species, including: 1. The GREAT SHRIKE TYRANT (*Agriornis lividus*) which has a subspecies with a L of up to 30 cm; 2. The WHITE-FRONTED GROUND TYRANT (*Muscisaxicola albifrons*) which is brown with a black tail; 3. The EASTERN PHOEBE (*Sayornis phoebe*) with a L of 16 cm, and the BLACK PHOEBE (*Sayornis nigricans*) both of which are found in North America; 4. The LONG-TAILED TYRANT (*Colonia colonus*) which has a L of 23 cm, of which 16 cm is taken up by the two long, narrow central tail feathers. This bird is predominantly black. It is found at forest edges from Boliva to southern Brazil; 5. The WHITE-HEADED MARSH TYRANT (*Arundinicola leucocephala*; Color plate, p. 143) is black. The head is white and there is a white spot on the flanks. This species lives along river shores and in swampy areas; 6. The VERMILLION FLYCATCHER (*Pyrocephalus rubinus*; Color plate, p. 143) has a L of 14 cm. The male is scarlet on its crown and underparts. There is a dark brown or blackish stripe on the side of the head; the back, rump, wings and tail are also dark brown or blackish. The female is much paler. These birds are found from the southwestern United States to Argentina and the Galapagos; 7. The MANY-COLORED RUSH TYRANT (*Tachuris rubrigastra*) has a L of 11 cm. It is red, orange, yellow, bronze-green, blue, black, and white, and as a result is called "*siete colores*" (seven colors) in southern South America. It inhabits South American swamps; 8. The CATTLE TYRANT (*Machetornis rixosa*; Color plate, p. 143) has a red crown stripe. The underparts are yellow. This bird is found in South America east of the Andes.

B. The TYRANT FLYCATCHERS proper (subfamily Tyranninae) are small to medium-sized birds; usually with a concealed colored spot on the crown. There are about thirteen genera with thirty-five species, including: 1. The FORK-AND SCISSOR-TAILED TYRANT FLYCATCHERS (*Muscivora*) which have greatly elongated central tail feathers. The FORK-TAILED FLYCATCHER (*Muscivora tyrannus*; Color plate, p. 143) and the attractive SCISSOR-TAILED FLYCATCHER (*Muscivora forficata*) which has a L of 16 cm but measures 36 cm to the tip of the tail, both belong to this genus. The latter is black, white and pale gray; the sides of the body are red and yellow in color–like a sunset; 2. The KINGBIRDS (*Tyrannus*) comprise eleven species, including the EASTERN KINGBIRD (*Tyrannus tyrannus*; Color plate, p. 143) which has a L of 21 cm, and the TROPICAL KINGBIRD (*Tyrannus melancholicus*) which has a L of 23 cm; 3. The PIRATICAL FLYCATCHER (*Legatus leucophaius*) has a L of 15 cm; 4. The MYIODYNASTES FLYCATCHER (*Myiodynastes*) include five species of which the SULPHUR-BELLIED FLYCATCHER (*Myiodynastes luteiventris*) is one; 5. The BOAT-BILLED FLYCATCHER (*Megarhynchus pitangua*) has an enlarged beak; 6. The SMALL KISKADEE FLYCATCHERS (*Myiozetetes*) comprise twenty-four including: the VERMILLION CROWNED FLYCATCHER (*Myiozetetes similis*); 7. The KISKADEE FLYCATCHERS (*Pitangus*) comprise two species, including the GREAT KIS-

KADEE (*Pitangus sulphuratus*; Color plate, p. 143) which has a L of 20 cm. The crown is black with a yellow spot in the middle and white stripes on the sides; the cheeks are black. The upper side of the body is brownish-olive and the lower part is bright yellow. It is found from southern Texas to Argentina.

C. The MYIARCHUS FLYCATCHERS (subfamily *Myiarchinae*) are small to medium-sized birds, generally with a wide, often very wide, beak. There are about 25 genera with 100 species, including: 1. The genus *Myiarchus* with seventeen species, one of which is the GREAT CRESTED FLYCATCHER *(Myiarchus crinitus)* which has been found predominantly in North America. The TROPICAL PEWEE *(Contopus cinereus)* is a member of this family. 3. The *Empidonax* FLYCATCHERS which have eighteen species, several of them in North America, including the YELLOW-BELLIED FLYCATCHER *(Empidonax flaviventris)*; 4. The *Myiobius* FLYCATCHERS which comprise five species, including the SULPHUR-RUMPED FLYCATCHER *(Myiobius sulphureipygius)*; 5. The ROYAL FLYCATCHERS *(Onychorhynchus)* which include four species. The NORTHERN ROYAL FLYCATCHER *(Onychorhynchus mexicanus;* Color plate, p. 143) has a L of 16 cm. This bird has a particularly beautiful erectile crest; 6. The *Tolmomyias* FLYCATCHERS have five species including the YELLOW-OLIVE FLYCATCHER *(Tolmomyias sulphurescens)*; 7. The EYE-RINGED FLATBILLS *(Rhynchocyclus brevirostris)*;

Fig. 4-27. The river tyrannulet.

D. The TODY FLYCATCHERS (subfamily Euscarthminae), like the following two subfamilies, are small to very small birds. They include: 1. The TODY FLYCATCHERS proper *(Todirostrum)* have a bill like a minute spade. They are seventeen species, including: the COMMON TODY FLYCATCHER *(Todirostrum cinereum;* Color plate; p. 143) which has a L of 10 cm; 2. The BENT BILLS include the NORTHERN BENT BILL *(Oncostoma cinereigulare)* which has an L of 9.5 cm. The bill is somewhat curved downward. These birds are greenish-olive and yellow.

E. The TYRANNULETS AND PYGMY TYRANTS (subfamily Serpophaginae) comprise: 1. The genus *Serpophaga*, with at least four species, including the RIVER TYRANNULET *(Serpophaga cinerea)* which is white and gray and is found in the mountains from Costa Rica to Bolivia; 2. The PYGMY TYRANTS *(Perissotriccus)* with two species, one of which, the BLACK-CAPPED PYGMY TYRANT *(Perissotriccus atricapillus)*, has a L of 6.5 cm.

F. The ELAENIA FLYCATCHERS (subfamily Elaeniinae) include: 1. The ELAENIA FLYCATCHERS proper *(Elaenia)*. These birds are crested and generally have a yellow or white spot on the crown. There are twenty-three species, including the YELLOW-BELLIED ELAENIA *(Elaenia flavogaster;* Color plate, p. 143) and the LESSER ELAENIA *(Elaenia chiriquensis)*; 2. The TYRANNULETS *(Tyranniscus)* are minute birds; there are nine species, including the olive-green PALTRY TYRANNULET *(Tyranniscus vilissimus)*; 3. The PIPRA FLYCATCHERS *(Pipromorpha)* have three species, including the small olive-green OLIVE-BELLIED FLYCATCHER *(Pipromorpha oleaginea)*; one might also possibly include the two ant pipits (genus *Corythopis*) is this subfamily; these birds have already been discussed in the section on gnat-eaters included in this chapter.

Tyrant flycatchers which nest in the tropics are generally resident, and many species keep together in pairs all year round. The other flycatchers, however, are migrants. The small troops of eastern kingbirds migrate at low altitude in the daytime, when they become quite noticeable. The birds traverse the stretch between southern Canada and the United States and their winter quarters in Peru and Bolivia two times a year. Other flycatchers migrate at night.

Though predominantly insectivorous, flycatchers supplement their diet with all sorts of additional foods. Many species take berries and seeds covered with the fruit. The paltry tyrannulet has a particular preference for toe berries. The lesser elaenia sometimes comes to feeding stations to share bananas with tanagers, honey creepers, and finches; however, one of the many species of this family which visited our gardens never came to the feeding station. The largest tyrant flycatchers often catch small vertebrates like fish, frogs, lizards, and even mice; occasionally some of them, like the kiskadees and the ground tyrants, become nest robbers.

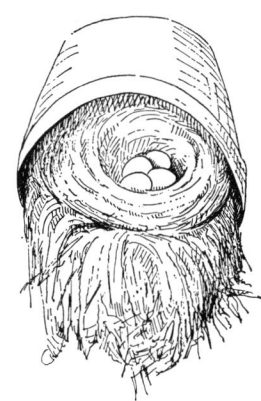

Fig. 4-28. Nest of one of the phoebe species.

The flycatcher's methods of obtaining food are quite variable. Kingbirds, peewees, and many other species rest on perches from which they seem to dive into the air to catch flying insects, returning to the perch after each flight. Whenever these birds make a catch or even just snap their beaks in vain, the movement often makes a loud noise. A large insect is beaten forcefully against a twig till it stops twitching. Occasionally a tropical kingbird may circle about in the air for a longer period, catching one insect after another like a swallow. Many small species flit unobtrusively from twig to twig through thickets or in the crowns of trees while catching insects. They often pick up insects and spiders from the foliage or the bark. The curved bill is adapted to taking insects from the undersurfaces of leaves; therefore the bird makes upward slanting flights. The Serpophaginae remain alongside swift streams, perching on projecting rocks, whipping their tails up and down, and then rising into the air to snatch up the smallest flying insects. These birds may also cling to the rocky bottom and pick up larvae from the shallow, fast-flowing water.

Fig. 4-29. Section through the spherical nest of a vermillion-crowned flycatcher.

A number of tyrant flycatchers pick up food from the ground. The black phoebe, the yellow-bellied flycatcher, and even the long-tailed scissor-tailed flycatcher, often sit on a wire or some other fairly low perch, and watch the ground beneath with care; then they fly down to catch crawling insects or worms. On the open steppes of southeastern South America and in the high Andes, ground tyrants walk or run about on the ground and pick up worms, insects, and small vertebrate, they make short flights in pursuit of flying insects during various intervals. The cattle tyrant runs alongside the heads of grazing cattle and other mammals, picking up the insects flushed from the grass, just as the anis do (Vol. VIII). When the birds' hunger is satisfied, it flies up and sits comfortably on the cow's back. The versatile kiskadee catches small fish which swim near the surface of the water; it may even stand in shallow water to catch tadpoles.

The largest tyrant flycatchers, in particular, usually make a great deal of noise; their voices may be loud and rough or soft and melancholic, depending on the species. Continuous calls are generally heard only at dawn. A few species repeat the twilight song at the end of the day. The birds rarely call or sing in full daylight unless during courtship, a territorial dispute, or the like. The twilight song is generally given for several minutes almost without interruption from some perch. The large sulphur-bellied flycatcher and the related species sing at dawn in sweet melodious tones which form a peculiar contrast to the shrill calls they utter later on during the day. A few species perform display flight. In the rising mist of dusk, the lesser elaenia flies up steeply from the thickets where it spends the day and sings a short, rough-sounding song until it is high above the crowns of the trees. Then it makes a steep dive down into the bushes and becomes silent for the night.

The northern royal flycatcher lives close to water in forests from southern Mexico to Colombia. Its crest (when not erected) is only suggested by a projection at the nape; however, it forms a marvellous head decoration after a truly magical transformation. The male's forehead is surrounded by a widespread aureole in which there are brownish-purple and velvety-black spots. This sight is a real experience for any naturalist. The females' crest is almost as wide but paler (Fig. 4-34).

Most tyrant flycatchers live in monogamy and both partners take part in nest building. In some species, however, the male does not take an active part in the building, but rather he accompanies his partner as she gathers nest material or at least he sits near the nest site and greets her when she arrives with a beakful of nest material. On the other hand, the pipra flycatchers and a few other species do not form pairs and here the female nest-builds by herself. The male pipra flycatcher sits on the same spot in the undergrowth of rain forest during the long breeding season, calling monotonously. Usually several males settle down within hearing range of one another, probably in order to better attract the females. When a female has visited the male, she leaves and lays her eggs in an attractive nest she has built by herself. Male sulphur-rumped flycatchers, eye-ringed flat bills and bent bills along with those of the related species do not help the incubating female.

The diversity of the nests in this family is almost unique. With the exception of the oven birds, no group of birds in America and possibly

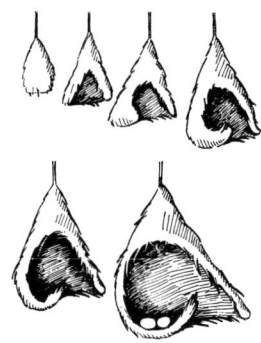

Fig. 4-30. Hanging nest of the sulphur-rumped flycatcher in six successive phases of construction.

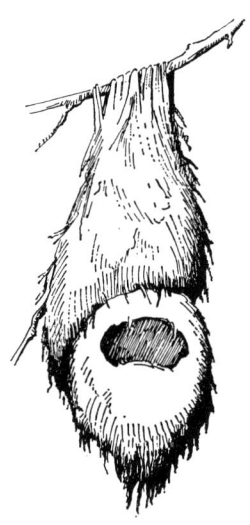

Fig. 4-31. Nest of the paltry tyrannulet attached below the hanging nest of the yellow-olive flycatcher.

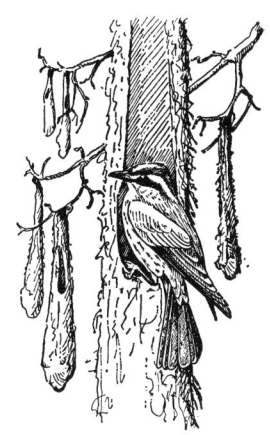

Fig. 4-32. Piratical flycatcher at its nest.

Fig. 4-33. Eastern kingbird.

the whole world breeds in so many different ways. Many tyrant flycatchers build open nests in bushes or trees and (in exceptional cases) on the ground. These open structures range from the orderly, firmly knitted and lichen-covered cup of the yellow-bellied elaenia, the nest of which is worthy of a hummingbird, to the wide, flat, shallow, disorderly cup of the tropical kingbird. Some species, like the *Myiarchus* and *Myiodynastes* flycatchers, the long-tailed tyrant, and others nest in holes in trees, posts, clefts, in buildings or in nest boxes. In these places where trees and shrubs are rare in South America, ground tyrants and the great shrike-tyrant often build their nests in holes or clefts in banks or stone walls.

All the phoebes build rather remarkable nests. They make fair-sized hemispheres of mud and plant material; then they place soft lining materials in the hollow upper part of the nest. They gum these nests onto cliffs, bridge pylons or sometimes onto the wall of a deserted house—always in a spot where the nest is protected from rain, because rain would detach the nest from a vertical wall on which it is often suspended.

There are also large nests with roofs and with a side entrance placed in the fork of two branches, or another suitable base as in the case of the kiskadee and the vermillion-crowned flycatcher. The vermillion flycatcher saves itself much trouble by simply building its nest over the old nest of some other bird. Usually the roof is dense enough to protect the bird in the nest from the whipping tropical rain bursts.

Many tyrant flycatchers build hanging nests which are suspended from thin twigs or merely hang from a single strand instead of being supported from below. These structures vary a great deal in form. They are never woven like the hanging pouches of icterids or weaverbirds which they sometimes resemble to some degree; instead, they are always built by the interlocking of the nest materials. First, the flycatcher fixes a loosely connected base of fibers onto the selected spot. Then it forces the fibers apart, and forms a nest chamber; next it lines the interior with additional material. From time to time it interwines additional strands with the twig to strengthen the support (Fig. 4-30).

The northern bent bill builds a comfortable pear-shaped structure with a small round entrance at the side. The common tody flycatcher, which often nests in gardens, builds a long, shapeless nest mass with a chamber in the center. Many tyrant flycatchers make it difficult for themselves to get into the nest. Thus the sulphur-rumped flycatcher, a forest dweller, covers the entrance to its nest with a broad apron so that it is forced to fly upward to reach the interior. The blackish nest of the yellow-olive flycatcher is shaped like a chemist's retort and is so suspended that the opening faces straight down, hence the bird must shoot upwards to enter the nest. The northern royal flycatcher's nest almost always hangs over running water in the forest. It is a loose, elongated mass of fibers sometimes one and a half meters long. The eggs are laid and the young are reared in a flat niche in the center of the nest, which is difficult to find.

The nest of the pipra flycatcher is probably the most beautiful of all. It is a long pear-shaped structure of green moss which contains a commodious chamber lined all around with thin pale fibers. The entry is at the side. This truly enchanting nest is often attached by the female to a pseudoparasitic plant which hangs down beside a moss-covered tree trunk or from a steep, rocky slope covered with ferns and orchids.

The noisy cantankerous piratical flycatcher is an exception. It takes over the nest of some other species by simply throwing out the eggs or young. As Wilhelm Meise says, according to Alexander Skutch, "When doing this, one of the partners aggressively approaches the desired nest whereby both the owners chase it; the other pirate sneaks up to the nest, throws out the eggs and henceforth uses the nest as its own." This species always chooses a roofed nest as, for example, the black retort of the yellow-olive flycatcher, the roofed structure of the vermillion-crowned flycatcher, the pouch of an icterid or even the nest chamber dug into a circular wasp's nest by a pair of purple trogons (Volume VIII). After bringing in a few dry leaves, the piratical flycatchers begin their egg-laying.

Many tropical tyrant flycatchers usually lay two eggs; some lay three or occasionally four. Two to six eggs are laid in the higher latitudes of the Northern and Southern Hemispheres, but four is probably the common number. The eggs are white, pale gray, yellowish-brown, or cream-colored; they are sometimes unspotted, but in many species they have more or less extensive brownish-red, brown, chocolate-colored or pale lilac-colored spots or blotches. The strongly spotted eggs of kingbirds and the *Myiodynastes* flycatchers are very beautiful. Only the female incubates in all the species about which we have reproductive information; females of many of the smaller species sit rather restlessly on the eggs. If the weather is good, the female flies off the nest for a short while every few minutes. Male phoebes, peewees and those of a number of other species bring food to the incubating partner, but this is exceptional in the family as a whole. Incubation varies from twelve to twenty-three days, depending on the species; it is longest in some of the smallest forms, perhaps because their nests are difficult for enemies to penetrate.

The nestlings are blind and helpless after hatching; they are only sparsely covered with down or, in a few small tropical species, are even quite naked. The interior of a nestling's mouth is yellow or orange-yellow. The young are brooded for warmth by the mother but are fed by both parents in those species which live in pairs. The female pipra flycatcher regurgitates food into the beak of the young, but most other species bring visible insects or berries to the nest in the beak. Fledging

Fig. 4-34. Northern royal flycatcher.

lasts fourteen to twenty-eight days. When we compare tyrant flycatchers which are similar in size, we find that the young of those species with open nests will leave the nest much sooner than those reared in hanging nests. Once the young are fledged their color is already much like that of the adults. The young do not return to the nest to roost; however, in some species, like the yellow-olive flycatcher and the eye-ring flatbill, the female herself continues to use the nest as a sleeping place.

A nest distraction display only occurs in these flycatchers in exceptional cases. The display consists of fluttering over the ground, apparently helplessly. Some larger species fly at raptors or toucans which come into nest view, and mob them. Even herons are chased from the territory. This behavior indicates, as Wilhelm Meise rightly observes, that these New World flycatchers are, in most cases, protectors rather than tyrants, as far as their nest neighbors are concerned.

Family: sharpbills, by H. Sick

The CRESTED SHARPBILL (*Oxyruncus cristatus*, Color plate, p. 150) is especially differentiated from the tyrant flycatchers by its straight bill with the unusual pointed tip. The sharpbill is the only member of its family (Oxyruncidae). The L is 18 cm, and the weight is 44 g. These birds have a compact body; the plumage is green. The center of the male's crown is a bright fire-red and is only conspicuous when the bird is excited. The lower parts are a light greenish-yellow or white with densely scattered dark drop-shaped spots. This species is found from Central America to southern Brazil, generally only in cool mountain areas, including the eastern slopes of the Andes (see Fig. 4-35).

Fig. 4-35. The sharpbill (family Oxyruncidae).

Sharpbills hop about among the densely leaved branches in the crowns of tall forest trees. They usually fly toward the bushy end of these branches where they find small fruit. While feeding, they cling to the branch like tits or chickadees, with the back facing downwards; the bird's weight often bends the tip of the twig down. The sharpbill's call is a long-drawn feeping, somewhat resembling the call of the three-toed sloth, which sometimes occurs in the same forests. Sharpbills are generally seen singly, but they may be encountered wandering with tanagers, woodcreepers or certain other small birds. Even today, we know little about this species aside from characteristics which place it near the tyrant flycatchers. We might be equally justified in classifying this species with the cotingas because of its behavior.

Family: manakins, by H. Sick

Even the earliest explorers of South America (Prince Maximilian of Wied-Neuwied, 1782–1867) noticed the MANAKINS (family Pipridae) because of their unique courtship dances and the "instrumental music" connected with these dances. The manakins have also been referred to as dancing birds because of their courtship. Each species has its own ritual, composed of particular modes of moving, like cowering, trembling, tripping, gliding to one side, a jump with a rotation of 180°, the raising of both wings, a display flight and others.

Manakins dance their courtship dances on thin horizontal or vertical branches without twigs or leaves. Many species even use a moving branch that begins to shake a soon as the bird sits on it, whereupon the pipra makes a characteristic movement. We will discuss the various types of dances later on in this chapter.

Manakins range from the size of a chickadee to that of the house sparrow; the L is 8.5-16 cm. The tarsal plates are like those of the tyrant flycatchers; the tarsus is almost always long. The beak resembles that of the chickadees. Males are generally very colorful, being either black with brilliant red, white, or blue on the head, or mainly red, blue, green, or white, sometimes with erectile feather "horns." The crown of one manakin species looks as if it were covered with scales of mother of pearl. The eyes are generally very brightly colored. Females, however, are dull green and, as a result, those of different species are rather similar to one another. The male's flight feathers are often greatly modified in connection with sound production; even the shortened and stiffened tail produces sounds in some species. Manakins are forest birds, and they range from Mexico to Argentina, although they are most common in Amazonia (see Fig. 4-37).

There are twenty-one genera and from fifty-two to sixty-one species, including: 1. The PIPRAS *(Pipra)* comprise about fifteen species, including the RED-HEADED MANAKIN or UIRA-PURU *(Pipra erythrocephala)*, which has a L of 9 cm and weighs 14 g. The head and thighs are a brilliant red. It is found in the Amazon region. 2. The MACHAEROPTERUS MANAKINS *(Machaeropterus)* have two species, including the FIERY-CAPPED MANAKIN *(Machaeropterus pyrocephalus)*. The crown is a golden-yellow; the secondary shafts, which are partially thickened and curved, act as resonators even in ordinary flight, which is accompanied by a loud "ringing." These birds are found in the Amazon region. 3. The HELMETED MANAKIN *(Antilophia galeata)* male has a magnificent scarlet "helmet"; this species is found in central Brazil. 4. The MANAKINS *(Neopelma)* have four species, including WIED'S TYRANT MANAKIN *(Neopelma aurifrons)*, which has a L of 13 cm. The bird is a leaf-green color with a yellow belly. The iris is a light brownish-yellow. It is found in Brazil. 5. The MANAKIN *(Manacus)* males have elongated chin feathers and pointed sabre-like outer primaries. There are six species, including the WHITE-BEARDED MANAKIN *(Manacus manacus;* Color plate, p. 143). This bird is black and white, and is common in lowland forests throughout almost all of South America. 6. The CHORUS MANAKINS *(Chiroxiphia)* comprise four species, including the SWALLOW-TAILED MANAKIN *(Chiroxiphia caudata;* Color plate, p. 143), which weighs 25 g. The male is light blue with a bright red crown; the wing and tail are black. This species is common in southeastern Brazil.

These conspicuous "dancers" live five to six meters above the ground in forests, sometimes quite close to towns. They feed on berries and small animal life which they pick off twigs. The nest, a very small, thin-walled basket, is firmly wound in the fork of a branch. The two

Fig. 4-36. Sharpbill.

Fig. 4-37. Manakins (family Pipridae).

Fig. 4-38. Dance of the red-headed manakin or uirapuru. The female is shown in white.

eggs have dark spots on a brownish background. Males leave the incubation and the care of the nestlings to the females; during the day they spend most of the time on or near the dance "arena." They also often dance among themselves uttering very peculiar creaking, whirring and clattering sounds along with the dance.

The male white-bearded manakin dances on shoots close above the ground and removes the fallen leaves, giving rise to a "dance floor" which looks as if it had been swept. In other species, like the red-capped pipra, the display consists of males chasing one another with loud calls. The *Neopelma* males jump up a few inches from a branch just above the ground, while the Wied's tyrant manakin produces a low-pitched "dop dop" sound by beating the wings together for its display. Sometimes it flings itself into the air and makes a rotation of 180° during which the bright yellow streak on the crown, which is otherwise hidden, flashes up. Males of this species always display alone, although one can sometimes hear other birds going through the same performance nearby.

Members of the genus *Pipra* have a more complex display. Thus the male uirapuru sits on the main branch with its tail trembling, calling "zlit"; it cowers, then places itself at an angle to the branch and moves to and fro along it with little, very rapid steps while shaking its tail and wings. Sometimes it spreads out the wings horizontally for a brief moment or flings them up, simultaneously uttering a gutteral "seee" or "gah." The male always turns its tail toward an approaching female; it assumes a posture of escape, whereby it shows the female a rear view of its red crown and thighs. Suddenly it takes off in its whirring display flight for about forty meters into the forest; it returns at once, calling wildly, executes an S-shaped curve in front and above the female, and then attempts to land on her back.

The fiery-capped manakin has a much different behavior. The male hangs down headfirst from a vertical branch and utters a grasshopper-like whirring call which lasts from five to twenty seconds. Every now and then it jumps rapidly around the branch. A single white-bearded manakin male jumps around on its dance floor, to and fro between two and four fixed points of a triangle. Its head is stretched forward and the impressive "beard" on its chin is erected to such an extent by special muscles that the beak disappears in it. The crackling and grunting sounds which are heard with this display are evidently produced by the modified secondaries which have thickened shafts with stiffened outer vanes. The keels are not firmly attached to the ulna, but are instead supplied by special strands of muscle.

The swallow-tailed manakin has proper communal display grounds, each used by two or three males and one female. The males perch close together on a horizontal or slightly slanting branch, all facing the same way with heads bent down; they make tripping foot movements which cause their whole bodies to tremble. They utter their chorus without interruption; it is a monotonous churring, which sounds like a distant frog chorus, and which waxes in intensity. The female sits motionless at one end of the row. Suddenly the lowest male takes wing and hovers a few inches above the group. It turns toward the female, hovers above her for a moment and utters a penetrating "dik dik dik." Then it settles beside the female at the upper end of the row of males, facing in the same direction as they are, and rejoins them in their foot tripping. The female continues to sit immobile. Then the male which is now at the lowest end of the row begins the hovering flight and so it goes, like the regular motion of a wheel. If there is no female present, her place is taken by an immature male, one that is still green and therefore like a female in appearance. The synchronization in the trembling movements in the dance of these birds is particularly astonishing. Furthermore, the males are not of equal rank; a young green male is as it were, ordered to play the female role.

Evidently the manakins of different sexes meet only on the dancing areas and their vicinity. A female arrives at a display area and watches the offerings of the male or males, often without any apparent interest, although sometimes she takes a lively part in the display "dances" herself. The arrival of a female on the "arena" stimulates the males to dance. The male approaches the female to mate only after he has danced before her for a while. Some males are very successful in attracting females. In one area where many white-bearded manakins were color-banded, it was observed that no less than twenty different banded females, as well as some unbanded ones, visited a particular male on its arena. Banding also revealed that females do not become mated to particular males but go from one to another, or rather from one arena to another. No doubt the manakin female selects her mate (or mates), and, as in the birds of paradise and some others, sexual selection prevails. This has led to an increasing emphasis of the peculiarities of males in the course of evolution, while the females retain their modest green plumage.

Manakins are part of local tradition, particularly in the Amazon basin. They are looked on as bringers of good fortune in matters of the heart. A young man who wishes to gain the favors of a girl in these areas puts the colorful skin of a manakin in his pocket; he feels then pretty sure to get the girl. Many a settler will bury a manakin under the doorstep of the new house he is building; this is to bring good fortune to the new house.

The COTINGAS (family Cotingidae) are a very multiform but nevertheless little-investigated group. Many species are highly decorative; they have striking colors, decorative plumes, crests, inflatable throat sacs, strands of skin or bare leppets on the forehead or at the angle of the beak. According to Wilhelm Meise, these forest birds, which vary from the size of a goldcrest kinglet to that of a crow and

Fig. 4-39. Swallow-tailed manakin dance directed at a female or young male. Successive phases of a particular sequence.

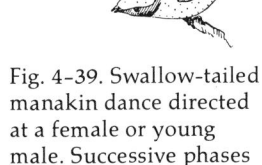

Family: cotingas, by E. Schäfer

Fig. 4-40. The cotingas (family Cotingidae).

Fig. 4-41. Green-backed becard.

which have some biological resemblances to the shrikes, show "the most colorful mixture of forms of all the clamatores families."

In spite of their multiformity, cotingas are characterized by compact bodies and a hooked bill. The L is 7.5–50 cm. The tarsi are surrounded only by band-like plates in front, but covered at the rear with very small platelets which are not all contiguous. The vocal muscles are strong and have a particular arrangement. The sexual dimorphism is usually marked. Cotingas are distributed from Central America throughout most of South America as far as Argentina.

There are about thirty genera with some ninety-five species, including: 1. The TRUE COTINGA *(Cotinga)*. The males are brilliant blue to purple in different patterns, while the females are dull brown. These birds are found in the tropical zone of the Amazon forests. There are eight species, including the BANDED COTINGA *(Cotinga maculata)* which has a L of 21 cm and is found in Brazil; 2. The FRUITEATERS *(Pipreola)* comprise eleven species, including the BARRED FRUITEATER *(Pipreola arcuata);* 3. The BECARDS *(Pachyrhamphus)* comprise eleven species, including the GREEN-BACKED BECARD *(Pachyrhamphus viridis)* which is green and black with a bright yellow breast band; 4. The TITYRAS *(Tityra)* are whitish birds with black marks. There are four species, including the MASKED TITYRA *(Tityra semifasciata)* which has a L of 20 cm; 5. The RED-RUFFED FRUITCROW *(Pyroderus scutatus)* has a L of 45 cm; 6. The CAPUCHIN BIRD *(Perissocephalus tricolor);* 7. The AMAZONIAN UMBRELLA BIRD *(Cephalopterus ornatus;* Color plate, p. 144) has a L of up to 51 cm; 8. The COCKS-OF-THE-ROCK *(Rupicola)* are plump, short-tailed, broad-footed birds. This genus is sometimes regarded as a separate sub-family. There are three species, including the ANDEAN COCK-OF-THE-ROCK *(Rupicola peruviana;* Color plate, p. 133) which has a L of 31 cm in males and 26 cm in females; 9. The BELLBIRDS *(Procnias)* have a L of 20–26 cm. There are four species: a. The WHITE BELLBIRD *(Procnias alba);* b. The BARE-THROATED BELLBIRD *(Procnias nudicollis;* Color plates, p. 136 and p. 144); c. The THREE-WATTLED BELLBIRD *(Procnias tricarunculata;* Color plate, p. 144); and d. The BEARDED (or MOSSY-THROATED) BELLBIRD *(Procnias averano;* Color plate, p. 144).

Many of the larger cotingas are distinguished not only by the gloss and brightness of their plumage and their quite unusual appendages, but also by their tuneful, far-reaching calls. Most cotingas are shy, unobtrusive avoiders of civilization, and as such they inhabit the upper and middle tree levels of continuous forest areas, as residents or wanderers. Only a few species are also found in open landscapes, in secondary forests newly grown on formerly cultivated ground; they are inconspicuous and at the same time widely distributed. Cotingas generally eat berries and tropical fruit, but the larger species and those which inhabit open country also like to take insects.

The starling-sized true cotingas are among the most beautiful of all tropical birds. The banded cotinga is predominantly ultramarine-blue with a violet throat and breast band. The quiet behavior of the members of this genus is in contrast with the wonderful colors. The predominantly emerald-green green cotingas also exhibit this quiet behavior. They live in pairs in the moist forests of the higher mountain ranges. Both the true and the green cotingas are poor fliers; they feed mainly on fruits. Their subcutaneous and perivisceral fat often takes on the blue color of the berries they prefer.

Becards and tityras, both of which generally occur in more open country, prefer a mixed diet. Most of these chickadee to thrush-sized birds are reminiscent of shrikes in their posture and behavior. Tityras nest in holes. One of the commonest species, the masked tityra, has a delicate pearl-gray coloring with a black forehead and anterior part of the crown; the bare area around the eyes is bright red.

The red-ruffed fruit-crow or pavao, meaning peacock, as it is called in South America, is reminiscent of the jackdaw; it is a good flier but shy and rare, although it lives over a great range. This species is found in closed forests of tall trees and its display call is a deep-pitched hum (because of which it is also called "trompetero"). The capuchin bird lives in the lonely jungles of the Guianas and Brazil; it owes its name, capuchin, to the naked area, which resembles a monk's tonsure, on the front of its head. Its striking calls have been likened by observers to the bleating of sheep.

The umbrella birds are distributed from Costa Rica to Bolivia; they are the largest cotingas. These birds usually inhabit tall jungle trees of the tropics. Their structure and their calls make them the most unique of the cotingas. The head carries a canopy-like metallic glistening crest along its entire length; this crest projects over the tip of the heavy beak. In addition, an apron-like feathered wattle, forty centimeters in length, hangs down from the breast. Wilhelm Meise has the following description: "The inflated throat sac, which looks somewhat like a pine cone with spread scales, is moved to and fro like a pendulum; soft sounds are heard with this movement. With the utterance of the loud, low-pitched rumbling courtship call, the head is thrown back and the wattle swings forward." The much-widened trachea enables umbrella birds to utter "terrible roaring" sounds which have earned them the name of "bull birds." As is the case with cotingas, the females are smaller and inconspicuously colored.

Fig. 4-42. Umbrella bird. *(Cephalopterus ornatus penduliger)*

The pigeon-sized Guinan cocks-of-the-rock, with their teased-out feathers on the forehead, back and wings, have a particularly striking coloration. They live in the mountainous jungle areas of northern and northeastern South America, or in the areas near the sources of the Orinoco and the Amazon Rivers. The male's plumage is a bright orange-red; the head is decorated with a helmet-like erect crest. The plumage of the back is teased out (the feather vanes are loose) as in egrets. In the courtship season, males gather on rocks amid the foam

Fig. 4-43. Head of white bellbird.

Fig. 4-44. Bare-throated bellbird.

Fig. 4-45. Head of three-wattled bellbird.

Fig. 4-46. Head of bearded bellbird.

Fig. 4-47. The plantcutters (family Phytotomidae).

Family: plantcutters, by E. Schäfer

of river rapids to display their colors in most unusual dances. Robert Schomburgk (1804-1865), the well-known South American traveler, described these dances as follows:

"A whole troop of these wonderful birds was holding their dance on the smooth, flat upper surface of a tremendous rock. About twenty admiring observers, both males and females, sat on the bushes nearby while a male moved about over the top of the rock in every direction with some rather unusual movements. It would spread its wings, toss its head in every direction, scratch the rock with its primaries, and hop upwards at varying speeds, always from the same point; again it would fan out and erect its tail and once more walk about coquettishly with proud steps. When it seemed to be tired, it uttered a different phrase from the usual call and, flying to the nearest twig, it left its place on the rock to another male. After awhile, this second bird, having first demonstrated its grace and readiness to dance, gave way to a third male."

The brown females nest in small colonies and lay both their spotted brownish eggs in rock clefts. Because of their bright plumage, cocks-of-the-rock are hunted by men of numerous Indian tribes; the natives also eat their flesh.

The bellbirds range from thrush to turtle dove-size, and are also distinguished by compact bodies, flat beaks, short tarsi and a plumage of small feathers. They are famous not only because of their truly enchanting calls, but also because of the inflatable skin appendages about the heads of the males. Males also differ from the females in their plumage coloration. Females are predominantly green. The completely white, white bellbird carries a long horn on its head; this horn is covered with small white feathers which can be erected during display. The male bare-throated bellbird is also white; it is distinguished by bare wattles and a bare, inflatable throat skin of greenish color. The three-wattled bellbird, on the other hand, has a long cone of skin on the forehead and on each side of the root of the beak; this bird's foreparts are white, and the rest of the body is chestnut-brown. The bearded bellbird has a bare throat with beard-like threads of skin set in bundles of the skin of the throat. Its crown is brown, and the flight feathers and tail are black; the rest of the plumage is a delicate pearl-gray. In display this species opens up its gape like a frog's mouth so that the threads of the "beard" (which are comparable to a wreath of tuning forks) reproduce its pure bell-like tones.

Male bellbirds prefer high perches, often on bare jungle trees, which project above the crowns of surrounding trees; they defend such places jealously against rivals. These birds feed on fruit and breed in tree holes. Their far-reaching bell-like calls are characteristic of their jungle home; those of the white bellbird sound as if an anvil were being struck with a hammer.

The PLANTCUTTERS (family Phytotomidae) are closely related to the sharpbills and cotingas. These birds live in South America and they bear a superficial resemblance to finches. There is only one genus, the PLANTCUTTERS *(Phytotoma)* with three species, including: the RED-BREASTED PLANTCUTTER *(Phytotoma rutila;* Color plate, p. 150), which has a L of 17.5 cm and is found in Argentina, Uruguay, and eastern Bolivia; and the CHILEAN PLANTCUTTER *(Phytotoma rara),* which has a L of 17 cm. The forehead is reddish-brown; the lower parts are a cinnamon color and the back is olive-green with black stripes.

The plantcutter's beak shows an adaptation for plucking fruit and cutting vegetation which is unique in the bird world; it is very short but strong and has a high culmen. Both mandibles have toothed edges. With this beak, the birds can saw off buds, fruit or juicy shoots, often staining the beak green in the process. When plantcutters visit orchards or vineyards in small troops, their visits are not exactly popular with farmers and settlers. The rarus, as they are called locally, are poor fliers. They lay three or four eggs into a cup-shaped nest in shrubbery. Their screeching calls remind one of the noise of a saw blade.

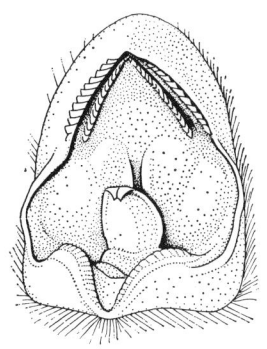

Fig. 4-48. Beak and mouth of a Chilean plantcutter.

There is no vernacular term for the PRIMARY SONGBIRDS (suborder Suboscines). The two families, the Lyrebirds and the Scrub Birds, could perhaps be called Lyrebird relatives. It is unusual that the scales on the tarsi of these birds are like those of Antbirds.

Suborder: primary songbirds, by W. Meise

The LYREBIRDS (family Menuridae) are among the most famous of the Australian birds and they have even appeared on postage stamps. The L is 75–100 cm. The male's tail consists of twelve loose feathers, two wire-like central feathers and two long but strong outer tail feathers. These two Lyre feathers are about 75 cm long and almost 4 cm wide. The underside of these feathers is a silvery-violet with half-moon-shaped golden-brown spots which, when the tail is tipped forward in display, are visible for a long distance. There is one genus, the LYREBIRDS *(Menura)* with two species: 1. The SUPERB LYREBIRD *(Menura novaehollandiae;* Color plate, p. 134), which has a L of up to 100 cm and is found in eastern Australia. It has also been successfully introduced in Tasmania; 2. PRINCE ALBERT'S LYREBIRD *(Menura alberti),* which has a L of 75 cm. The lyre feathers are black. This species is found along a 240 km-long stretch of coastline in New South Wales and Queensland, and inland to a distance of about 100 km.

Family: lyrebirds, by A. H. Chisholm

Fig. 4-49. 1. Superb lyrebird *(Menura novaehollandiae)* and 2. Prince Albert's lyrebird *(Menura alberti).*

These giants in the order of passerine birds might, because of their shape, long legs and strong tarsi, be taken as pheasants on superficial examination; even experts formerly classified them with the gallinaceous birds. However, their loud melodious voices, their marked ability to imitate other bird calls and their breeding habits, which they share with their less conspicuous compatriots, the scrub birds, led to their classification as a subgroup of the passerines.

The first lyrebird was seen in 1798 by an ex-convict near the then penal settlement of Sydney, the capital of New South Wales. He called

Fig. 4-50. Displaying lyrebird.

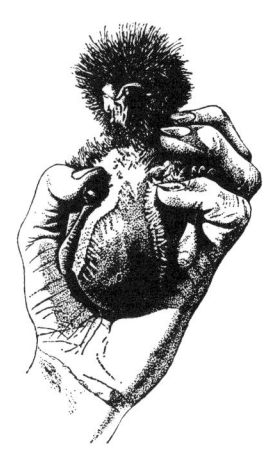

Fig. 4-51. Three-day-old lyrebird chick.

the bird the native pheasant of New South Wales or bird of paradise; the latter name probably came about because of the tail which the male tips forward in such a surprising manner that the rest of the bird, as it were, disappears beneath it.

During the courtship period in autumn, but also during the winter, the male lyrebirds are mainly occupied in singing and showing themselves off, a display which has been enthusiastically described by every traveler to Australia. The displaying cock may stand on the stump of a tree, or on a branch, but most often on a mound of soft earth about one meter in diameter which it builds in some forest clearing. Here the majestic bird sings with a full voice, mingling its own calls in a deceptive manner with those of other birds; it also gives very good imitations of locomotive whistles, the sound of car horns, barking of dogs and other sounds. In Sherbrook Forest near Melbourne one can find a number of lyrebirds which have lost their usual shyness of man as a result of constant exposure to people. Consequently, these birds are readily photographed and their calls recorded. Even though the lyrebird is not readily kept in captivity, superb lyrebirds have bred twice under such circumstances.

The lyrebirds nest is located on the ground, in a rock crevice or in a forked branch; it consists of a large pile of twigs and branches, and soft building materials, and it has a side entrance. The nest-building begins in late fall (May in Australia) or in early winter. Only the plain-colored female does the building. She incubates the single, dark-colored egg, which is the size of a hen's egg, and raises the young, which is still naked several days after hatching. Incubation lasts six weeks, and the young stay in the nest for another six weeks. With her strong feet the mother scratches the moist subsoil to find all sorts of insects which she feeds her young; she also looks for food in decaying wood and other types of ground covering. She carries feces of the young to the water or buries them in the soft ground. This precaution is biologically meaningful. Besides this, the young is capable of uttering a penetrating cry when danger threatens. The female adult's call is more subdued, but she is still just as good a mocker as the male.

Fig. 4-52. Three-week-old lyrebird chick.

Prince Albert's lyrebird, which is confined to the tropical rain forest area, builds no special display hillocks, but usually performs on a tree stump. It is as good a singer and sound imitator as the other species, although it has been hard hit by changes man has made in its environment. The superb lyrebird can still be found near large cities such as Sydney, Canberra, and Melbourne; Prince Albert's lyrebird has become rare and its habits are largely unknown.

Family: scrub birds, by A. H. Chisholm

The small SCRUB BIRDS' (family Atrichornithidae) relationship to the large lyrebirds is evidenced by their structure. The L ranges from 17-23 cm. These birds are ground dwellers and they rarely fly. There is no furcula. The sternum is unusually large. The vocal muscles are very strong and the voice is loud. Scrub birds can also imitate other birds. The position of the intestinal loops is similar to that of the lyrebirds. The tail is long. Scrub birds have various tones of brown in their coloration. There is one genus, *Atrichornis*, with two species: 1. The WESTERN SCRUB BIRD *(Atrichornis clamosus)* has a L of 23 cm. It is dark reddish-brown above with narrow blackish bands. The throat and breast are white with a dark gray spot below which the breast is yellowish-brown; the female does not have this dark breast spot. These birds are found in scrubland near the coast of southwestern Australia; 2. The RUFOUS SCRUB BIRD *(Atrichornis rufescens;* Color plate, p. 150) has a L of 17 cm. It is reddish on the belly. This species is found in the coastal rain forests from New South Wales to Queensland. The rufous scrub bird is separated from the other species by about 4,800 km; this separation probably occured in a post-glacial dry period. Both species are ground dwellers which can barely fly. The smaller species lives in the dense coastal jungles of eastern Australia; it probably never comes out into the sunlight. Both species run along the ground (often with raised tail) in a lively manner and scratch up worms, snails and snail eggs; however, they also eat insects and small seeds.

John Gilbert, a collector for John Gould (1804-1881), the famous illustrator of Australian bird life, discovered the western scrub bird in 1842. Since then, only a few males have been seen and because none were observed after 1889 the species was believed to be extinct. In late 1961, however, a male was seen near Albany. As a result the government of West Australia, which had originally planned a town park for that very area, magnanimously dropped the plan—an example which is unique in the history of Australia. A nest with one egg was found in June 1967 and again in June 1968. It was in dense grass almost on the ground and was constructed of dry leaves and bark. It had a lateral entrance, and a central chamber which was lined with moist wood debris.

The major distinctive feature of this endangered species is its unique, very melodious song, which is reminiscent of that of the nightingale. This song often has ventriloqual qualities and it contains imitations of the calls of other birds. The singer generally remains concealed in the dense bush, and as a result, its song has a mysterious and, as it were, disembodied effect.

The rufous scrub bird has been spared the fate of its larger relatives, for it ranges over a more extensive area covered with tropical rain forest. It lives secretively and when disturbed it will even hide beneath the bark of trees near the ground. As is the case with its relative, the male rufous scrub bird's voice is extraordinarily loud and it too has ventriloqual qualities and includes imitations. As far as we know, the rarely encountered female only produces soft ticking and squealing calls. It is to be hoped that this species will be preserved, particularly in national parks and other sanctuaries.

Fig. 4-53. 1. Rufous scrub bird *(Atrichornis rufescens);* 2. Western scrub bird *(Atrichornis clamosus).*

Fig. 4-54. Western scrub bird.

5 Songbirds

Suborder: songbirds, by W. Meise

The SONGBIRDS (suborder Oscines) comprise approximately 4,000 species, almost half the number of all avian species; this group of birds is the one most different from the primitive birds. The length varies from 7.5-110 cm, and the weight ranges from 5-1350, exceptionally to 1750 g. The anterior tarsal plates cover only the anterior ridge of the tarsus. There is a longitudinal plate behind the tarsus on each side; the lower part of this plate is often subdivided along its whole length (as in larks and a few other groups). The ring-like plates on the front of the tarsus are occasionally fused into a semi-lunar horny groove. These birds have more than three pairs of vocal muscles.

All songbirds do not have a true song, in spite of the number of vocal muscles; many of them do not sing, or do so only poorly. The natural classification of the songbirds has not yet been satisfactorily clarified. In the past the best singers, the thrushes and thrush-like birds, were viewed as the most highly developed songbirds; seed eaters, like the finches, were placed in the middle, and the medium-sized hunters of large insects, like the corvids, were regarded as the more primitive representatives of the suborder. Later, many ornithologists began to regard the finches as the most advanced songbirds, and these "songsters" were placed almost at the beginning of the series as the so-called "primitive insect eaters"; the shrikes and corvids, which had originally been placed in that position, were moved to the middle. In this book we will follow the most recent viewpoint, outlined in the *Check List of Birds of the World* by Peters, Mayr, Greenway and Paynter. They place the larks and swallows at the beginning of the suborder; these birds are well differentiated in their structure, but they certainly do not belong together and have not remained most similar to the original songbirds. This list places the corvids at the end, at the supposed apex of the avian world, because these birds have a relatively large brain and its performance is correspondingly outstanding.

Altogether, we differentiate forty-five songbird families; other ornithologists differentiate between forty-two to fifty-three families, not including the fossil families of which, unfortunately only two are known so far. Only in a few cases can these families be arranged in groups with a certainty of common origin. Two different arrangements are shown in the systematic review. A comparison of these two arrangements reveals that there is no general agreement in the arrangement of the orders of the Class Aves (Volume VIII), or in the sequence of families among the songbirds. Thus, much thorough study is yet to be done before we can further clarify the true relationships among these birds.

▷
The barn swallow (*Hirundo rustica*, above), feeding its young. It particularly prefers to nest in barns. The tail is deeply forked, the outer feathers are longer than the others and pointed. Well-known is the skylark (*Alauda arvensis,* below), whose song can be heard from March until summer from above the fields.

6 The Larks

Family: larks, by M. Abs

The Lark family (Alaudidae) is a well demarcated family of songbirds. Because of their apparently primitive characteristics, they are presently placed at the beginning of the group of songbirds. They range in size from that of a finch to that of a thrush; the L is 11.5-23 cm. The tarsus is covered with individual horny plates even on its rounded rear side; the borders of these plates become indistinct only in very old birds (in nearly all other songbirds, these plates have become fused into single pieces like greaves which form a sharp angle toward the rear). The syrinx lacks an ossified pessulus. The claw of the rear toe is long and straight; occasionally it is lightly curved. The sexes have a similar coloration, although there are a few exceptions; the general pattern is brown above, while the under parts are a lighter color with spots on the breast. Males are larger than females. The shape of the beak varies a great deal and is, in every case, adapted to the particular diet. There are ten primaries; the outermost is smaller in some species and sometimes it is even reduced to minuteness. There are fifteen genera with over seventy species. Larks have a worldwide distribution with the exception of South America and the oceanic islands; man has even introduced these birds to certain islands. The main area of distribution is Africa, and that is where larks are believed to have originated.

Larks owe their name to their rich and long-lasting song. Their inconspicuous earth-colored plumage characterizes them as ground birds of steppes and deserts; this designation persists in spite of the fact that larks can fly, although their flight performances are neither facile nor elegant. Many larks are long distance migrants; others are residents. These birds can run skillfully on their large feet and they can develop a ground speed of seven kilometers an hour. Larks are not choosy about food and drink. Many satisfy their thirst with dew and can dispense with water in other forms. They never bathe in water, only in sand or dust like chickens do. They sleep in small ground

◁ The barn swallow (*Hirundo rustica*, upper) in flight. The cliff swallows (*Petrochelidon pyrrhonota*, below; also illustrated p. 180) once built their hemispherical nests in huge, dense colonies on overhanging cliffs; these nests have protruding entrance passages. Today cliff swallow nests are primarily found on the sides of buildings.

depressions which they scratch out themselves and which they use repeatedly. After the breeding season, many larks undergo a complete moult. The young have a spotted juvenile plumage which is replaced by the adult plumage within the first year of life. We will discuss eleven representatives of the most important lark genera in this chapter.

The FLAPPET LARK *(Mirafra rufocinnamomea)* is one of the short and fine-billed larks of the genus *Mirafra* which are perhaps the most primitive larks. It has a wing length of 70 to 90 mm and is medium sized, compact, and generally has pronounced rust-colored edges on the flight feathers. It occurs in bush steppes and savannas from the Kalahari to the eastern Sudan and Gambia.

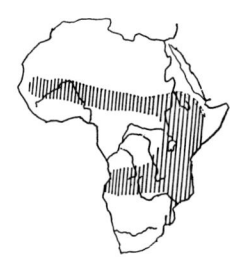

Fig. 6-1. Flappet lark *(Mirafra rufocinnamomea).*

When traveling through the southwestern African steppe, one may hear a rhythmically repeated clattering sound, "rrr-rrr-rrrrrr." At first one suspects a cicada as its source, but then one notices a flappet lark high up in the sky. The male utters its clattering sound at the end of a rising flight; he planes downward a little after each phrase and then rises again to clatter for a second, third or more times. Finally he flies on for a stretch and then begins his clattering again. The sound is not a vocal one but rather is produced by the flight feathers. Flying and clattering, the bird moves over its territory, which it marks out and defends against rivals in this way. This species has developed the "clattering song" particularly well. The CLAPPER LARK *(Mirafra apiata)* also has an upwards-clattering flight, but it finishes its song flight with loud, long-lasting, flute-like whistles. Other larks of this genus distributed over Africa, southern Asia and Australia, have a flute-like song according to Wolfgang Wickler; thus one can find all the transitions between singing and clattering in this genus.

The FINCH LARKS (genus *Eremopterix*) are probably also close to the most primitive larks. The WHITE-FRONTED LARK *(Eremopterix nigriceps)* which has a wing length of 72–82 mm, is also an African bird; the sexes of this species are strikingly different. The male has black lower parts and a black head which contrasts against the white forehead and cheeks; the female has the usual brown lark plumage. This bird is finch-sized and has a heavy conical beak with which it removes husks from millet grains in a finch-like manner.

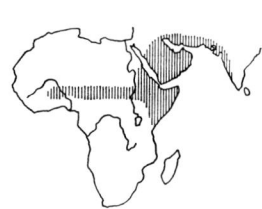

Fig. 6-2. White-fronted lark *(Eremopterix nigriceps).*

This species needs only little water and is even said to drink sea water on the coast of Somalia. Males rise in tight spirals on their song flights, constantly repeating a dry-voiced "deedl eedl tu"; they soon return to the ground again in a spiral path. In display, the male circles about the female, lowers his head, partly raises his wings and fans out his tail. This lark lives in savannas and steppes. The clutch consists of two eggs. After the breeding season, these larks associate with other birds and wander far in search of suitable food.

The DESERT LARK *(Ammomanes deserti),* which has a L of 16 cm, and four

Fig. 6-3. Desert lark (*Ammomanes deserti*).

other representatives of its genus live in the midst of the desert; nevertheless, this bird is a strict resident. Its plumage is sand-colored above, and it adapts to the local color of its habitat, which, for the genus as a whole, ranges from Morocco to Somalia and northwestern India.

Many desert lark subspecies have been distinguished according to the grayer or more reddish tint of the plumage. Thus, a light colored subspecies lives in Arabia in areas of pale sand; right beside it is a dark subspecies, which lives on black basalt. A pale bird will never move onto the dark ground even if one attempts to chase it over. This adaptation is the result of selection by predators of the larks; falcons, in particular, can spot a pale bird on the dark background from afar. In those areas where raptors are scarcer, however, for example in Tibesti, dark desert larks look for food in dried-out valleys filled with pale sand. It is strange that desert larks do not behave as their concealing color would lead one to expect; they do not crouch and freeze at the approach of danger; instead, they always try to escape by running or flying. Their flight song has five parts; the bird flies upward with each phrase, and at the end of it, glides down a little. Nests are placed in the shelter of large stones and are surrounded at the side of the entry by a small wall of pebbles, which many observers believe is meant to provide protection from the wind. Along the edge of the desert the clutch size is four, but in the harsher conditions of the true deserts three constitute a full clutch. Desert larks eat seeds and beetles which have hard carapaces; they drink dew if no open water is available.

The BIFASCIATED LARK (*Alaemon alaudipes*; Color plate, p. 179) has a L of 23 cm and a wing length of 11 to 13 cm. This bird looks like a small hoopoe when its two large, white wing bars flash out in flight. It has a long curved beak, but its hind toe is short and thick for a lark. It is found from northwestern Africa to northwestern India.

Fig. 6-4. Bifasciated lark (*Alaemon alaudipes*).

These larks run about swiftly on their tall legs over the sandy ground. Their plumage color is adapted to this ground, and yet, they too, do not crouch when danger approaches. They are strict residents; each individual defends a territory of about a hectare in size. Their favorite foods are the larvae of certain beetles and the birds dig holes up to five centimeters deep with their beaks in the loose sand which forms drifts in the lees of small trees in search of these larvae. The larks pursue this job so intently that they almost disappear in a cloud of sand. How the bird detects a beetle larva before it starts to dig is a mystery, for it is almost always successful. The male rises only a few meters into the air for its song flight; then he loops the loop and returns to his starting point. During this time, he utters his beautiful plaintive whistles. The female often builds her nest above the ground in a shrub and fastens it with mud; such a nest resembles that of a shrike. Only two eggs are laid, but there is often a second brood during the season.

The THICK-BILLED LARK (*Rhamphocorys clotbey*; Color plate, p. 179) has a L of 16 cm and a wing length in males of 12.5 cm. It is only as large as the skylark, but because of its thick head and heavy beak it looks larger than it actually is. The beak is deep, has a gap in the middle, and reminds one of a parrot's beak. The plumage is sandy-gray above, and white with large black drop-shape spots below. There is a white feather ring around the eye and a small black spot on the white cheeks. This species is found in the pebble deserts "hammada" from southeastern Morocco to Syria and northern Arabia.

Fig. 6-5. Thick-billed lark (*Rhamphocorys clotbey*).

The contact call of the lark is a soft "djoo" but the song sounds rough and abrupt. In display the male holds a pebble in its beak and offers it to the female; these pebbles are gathered about the nest entrance. The eggs of this species differ from those of other larks because they have a reddish background color. These larks wander about irregularly, after the breeding season.

The CALANDRA LARK (*Melanocorypha calandra*; Color plate, p. 179) has a L of 19 cm and a wing length of 11.3 to 11.5 cm. It gives the effect of a large skylark but one notices the large black spots on each side of the breast at once. The rear edge of the wing is white and this is readily noticeable in flight; the male performs circles high up in the air during its song flight. This bird is often caged because of its melodious song; many calandra larks include bits of imitations of other birds in their song. This species is found from the Mediterranean eastwards as far as Turkestan and northern Afghanistan, on dry cultivated lands or sage steppes.

Fig. 6-6. Calandra lark (*Melanocorypha calandra*).

Mauersberger has described the ground display of the calandra lark in Bulgaria: "The male stands fully erect with loosely hanging but tightly closed wings and an erected tail. It sings softly in this posture and nods its head." The usual clutch consists of five eggs. After the breeding season, calandra larks come together into huge flocks which are often associated with corn buntings. This bird is a rare vagrant in Britain and Germany; the last observation in Germany was on Helgoland in 1933.

In spite of its finch size, the SHORT-TOED LARK (*Calandrella brachydactyla*; Color plate, p. 179), which has a L of 14 cm and a wing length of 8–10 cm, gives an impression of smallness; it resembles a small calandra lark and has similar breast spots. This species is found from Southern Europe northwards as far as Brittany, as well as in Africa and Central Asia. The LESSER SHORT-TOED LARK (*Calandrella rufescens*), found in the Canary Islands, and from southern Spain to eastern Africa and Afghanistan, with twin species ranging as far as Manchuria, is closely related to the short-toed lark.

Fig. 6-7. Short-toed lark (*Calandrella brachydactyla*).

When short-toed larks migrate towards their breeding areas in troops of six to twelve birds on a sunny April morning, their white underparts show up brightly against the blue sky. The male defends its small territory in waste lands or in dry, infertile, cultivated, ploughed land vigorously, as these birds live in colonies. It performs several song flights daily in spirals over its "own land." As it rises, it sings a phrase of its clattering song, which is followed by a short, steep downglide; then rises again, but remains mute until it flies still higher to sing another phrase. Over twenty song phrases are sung in flight of only a few minutes. The song reminds one of that of the black redstart. Often songs of other birds are imitated. Short-toed larks get along peacefully with larks of other species, and they may live so close together that individual territories intersect; in such cases, one may hear the various species of larks imitating one another's songs.

Ground display of the short-toed lark

Guichard describes the ground display of the short-toed lark as follows: "On returning from its song flight, the male perches beside the female, which acts as if uninterested; the male trips about in a proud-looking posture with his tail erect, his neck held up vertically and the feathers of the crown erected." The female chooses the nest site and builds the nest, which is lined with plant wool; the entrance to the nest is lined with small pebbles. The young leave the nest only when they are fully fledged. The short-toed lark has disappeared from central France in the course of the last century and at the same time vagrants to Germany have been seen more rarely. Single birds are occasionally seen in Helgoland.

The CRESTED LARK (*Galerida cristata*; Color plate, p. 179) has a L of 18 cm and a wing length of 9.4–11.4 cm; it only immigrated into northern Central Europe within the last decades of the 19th Century. This bird is like the skylark in size, although plumper. It has a spotted breast, like many species of larks, and a crest on the head; this crest is noticeable even when the bird's feathers are not erect, because they project beyond the back of the head. The outer tail feathers are rust-colored. The crested lark is distributed in many subspecies over Africa, Europe, and Asia.

Fig. 6-8. Crested lark *(Galerida cristata)*.

The crested lark became familiar to many Germans after the last World War in the ruins and wasted areas. One can still see a pair of these birds today near the Bochum, Germany railway station, even though the ruins have long since disappeared from that area. Nevertheless, this lark has not become any more common since the war. It follows civilization, and, as such, it settles at the edges of towns and villages, garbage dumps and factory sites; however, nowadays it finds fewer means of subsistance in these areas. All roads and squares are paved, weeds are controlled, and since the triumph of the car, there is nary a horse dropping to contribute undigested oat grains to the larks' food. The crested lark rises high into the air for its song flight; it utters a phrase of its song with fluttering wing beats, and then moves on in wavy flight and begins a new phrase. The bird usually stays up in the air for three minutes and circles over its territory. The longest song flight observed lasted twenty-five minutes and was performed by a male which had remained unpaired. Rivals are driven out of the territories (which average four hectares in area) in wild chases. Crested larks utter a flute-like and emphatic "tritritree" as a contact and alarm call; they can imitate other whistles to perfection, as Erwin Tretzel was able to show particularly well. Crested larks near Erlangen have learned to imitate a shepherd's whistle so accurately that the sheep dogs obey their signals just as if the shepherd has given them. In addition, a comparison of recordings showed that the whistles of the birds were purer than those of the shepherd; the larks had a sense of rhythm and pitch which the shepherd lacked. Tretzel, therefore, surmises that the larks had grasped the ideal motif and whistled it accurately, while the shepherd had it in mind but could seldom perform it in reality.

W. Sudhaus reports the following on the display and mating of these birds. The female crouches low on the ground with her tail lightly fanned out and then begins to whirr her wings ever more violently. The male hops about in a particularly lively manner and briskly flaps his wings while singing loudly. He jumps onto the female and presses his tail beneath hers. Then he hops about the female again, while the latter remains, wings whirring, in the mating posture. Again the male mounts the female and copulation occurs. Then the male flies away and begins to sing softly.

The female usually builds the nest in the protection of a shrub or a clod of earth, in railyards and even between tracks which are in use. She brings many grass stalks, threads, shreds of paper and bits of textile material to the nest, heaping some of this outside the edge of the nest. The three to five eggs are laid at twenty-four-hour intervals in the early morning hours. There are usually two broods per season; each clutch of eggs is incubated for eleven to thirteen days by the female alone. Both parents, however, feed the young, which leave the nest nine days after hatching. Initially they only hop about a little before they begin to walk; after a week they are fully fledged. The young are fed caterpillars, grasshoppers and earthworms and usually only one youngster is fed at one feeding.

The WOODLARK (*Lullula arborea*; Color plate, p. 179) has a L of 14 cm and a wing length of 8.8 to 10.1 cm. It is a characteristic bird of the north German pine heaths. It has a pale stripe above the eye and a black-and-white pattern on the wing edge. It lays three to five eggs per clutch and incubates for thirteen to fifteen days. Its breeding range is from Europe as far as eastern Iran and the Urals.

Woodlarks which breed in Germany leave their winter quarters in France and other Mediterranean countries as early as February or March. At first the male defends a territory many hectares in size, but by the time of the breeding season this area is reduced to a few hectares. The male rises at a slant to deliver its song flight, and performs

Fig. 6-9. Woodlark (*Lullula arborea*).

irregular circles and curves while delivering its fluty song, "üdlüd-lüdl." It also often uses the tips of pines as its song perch. The female builds a neat nest of moss and rootlets; both parents bring food, consisting of caterpillars, beetles and flies, to the young. Once at the nest, they put down their food "packets" and feed each youngster in succession. Nine to eleven days after hatching, the young leave the nest. A second brood generally follows. The families stay together after breeding and wander south in small troops.

The song of the SKYLARK (*Alauda arvensis;* Color plates, p. 169 and p. 179), which has a L of 18 cm and a wing length of 9–12.4 cm, is a symbol of spring in much of Europe. This bird differs from the crested lark in the white outer border of its tail and because it has only a mere suggestion of a crest. It lays three to five eggs per clutch and incubates for eleven days. There are two or even three broods per season. This species is widely distributed in Europe and Asia, and has been introduced in one locality in Canada (Victoria, B.C.) and New Zealand (see Fig. 6-10).

Fig. 6-10. Skylark (*Alauda arvensis*).

Soon after the males arrive in February, they start establishing the boundaries of territories in which they tolerate no conspecifics; the smallest territories are about one acre in size. Females arrive ten days later. These birds occupy grain and clover fields, pastures and long meadows in Central Europe. Most of them return year after year to the same spot. About 10% of the skylarks have no territories, but they may take over those of other birds which die in the course of the breeding season.

J. D. Delius has the following description of the skylark's flight song: "After a silent takeoff, the bird rises steeply, into the wind if possible, with short wing beats and outspread tail, singing continually. On the average it flies to a height of fifty meters above the ground. Then there is a pause during which the male remains more or less level without changing its manner of flight or the song; often the bird then circles over its territory during this time. This pause lasts for about two and a half minutes. Then the bird slides down slowly with almost motionless, fully extended wings, uttering long drawn flute-like tones. It either glides right down to a landing or becomes silent when still ten to fifteen meters high, closes its wings and plunges down with body bent forward, opening the wings to act as a brake just before landing."

The duration and frequency of the song flight reaches several peaks, namely just before pair formation and during the female's two periods of incubation. Delius also described the display and breeding. Males display before rivals by standing upright with fluffed-out breast feathers, raised crest and loosely held wings. However, when they threaten, they stretch out horizontally, hold the wings distinctly off the body (but not extended), and with their beaks wide open, they utter the hostile call, "gee gee." If a female lands in a male's territory, the male flies to her and begins to display. It stretches itself upright, points the beak up and makes hops, one and two centimeters long, either up and down or toward the female. If the female remains, a pair bond is formed; the partners can soon recognize one another at a distance of thirty meters. The "tail up" display of the male is also part of its display repertoire; it turns its slightly raised tail toward the female and at the same time trembles its wings. In the "tank display," shown mainly against strange females, the male enlarges itself by fluffing out its plumage. It draws in its neck, holds the tail slightly fanned out and at an upward slant; it droops the wing tips and trembles them. It walks about the female with short steps and sings loudly with open beak.

During the pre-nuptial display, according to Delius, the female walks towards the male in a crouch, with horizontally held body and lightly fluffed out plumage; the tail is cocked up, the neck drawn in, the drooping wing tips tremble. The male approaches by hopping and stands alongside the female touching her back with one trembling wing. Then it jumps onto the female, keeps its balance by whirring its wings, and flicks its tail under the female's; she, in turn, facilitates mating by moving her tail sideways. Then the male jumps off and quickly moves away. The laying of the first egg follows three or four days after mating.

Delius writes that "the female chooses the nest site. She visits many ground depressions and indicates a preference only after some time. She deepens this depression by stamping in it with her feet. During nest building, some males will accompany their females on every step, while others show no interest in the female at all at this time. The female incubates alone and she interrupts this more often during the late morning hours to go off to seek food. The young hatch after eleven days. The empty egg shells and the fecal sacs are removed from the nest and the young are fed regularly and brooded right after hatching. The male too brings in food for the young right away. The young are initially blind and they gape in response to sound signals. On the fourth day their eyes open and then they gape towards the feeding parents. After six days the young hop out of the nest for the first time; they leave the nest on the eighth day, and on the fifteenth day they become fully fledged. After the young leave the nest, the male carries the main burden of feeding them, particularly when the female again lays eggs and incubates. Tending of the young ceases after about four weeks at which time the male begins to attack them."

According to Delius, young skylarks join into small bands and wander about until they migrate in the fall to winter quarters. Many migrate southwestward and winter mainly in southwestern France. The young return to their birth place in the following year. As the adults too are loyal to their breeding sites, the same pairs often form in successive years in the same territory. In winter skylarks feed on parts of green plants; during the breeding season they feed on insects;

▷
Larks: 1. Horned lark (*Eremophila alpestris*); 2. Red-winged bush lark (*Mirafra hypermetra*); 3. Short-toed lark (*Calandrella brachydactyla*); 4. White-winged lark (*Melanocrypha leucoptera*); 5. Thick-billed lark (*Rhamphocorys clotbey*); 6. Woodlark (*Lullula arborea*); 7. Crested lark (*Galerida cristata*); 8. Calandra lark (*Melanocorypha calandra*); 9. Bifasciated lark (*Alaemon alaudipes*); 10. Skylark (*Alauda arvensis*); 11. Thecla lark (*Galerida theclae*).

and in the fall they feed on weed and grass seeds. Crows, weasels, hedgehogs and foxes cause great losses on skylark eggs and nestlings. Adults fall prey particularly to the Eurasian sparrowhawk, the hobby hawk and merlins.

The HORNED LARK (*Eremophila alpestris*; Color plate; p. 179) has a L of 17 cm and a wing length of 9.4–12.1 cm. It is a characteristic bird of the Holarctic tundras. The male has black ear tufts, the "horns"; these are merely indicated in the female. The head is yellow and black, and a black crop band runs across the upper chest. This species is found in the Old and New Worlds from near the North Pole, south as far as Colombia; it replaces the larks of Europe in North America as there are no other larks on that continent.

Fig. 6-11. Horned lark (*Eremophila alpestris*).

In North America the horned lark is one of the first song birds to begin breeding, as early as March, before the land has been ploughed. Thus, many broods are lost during ploughing. The male rises silently for its song flight, then flies in irregular circles while singing metallic-like sounds which are distantly reminiscent of the squeaking of an old gate. The song flight is ended by an abrupt downward swoop. Neighboring males chase one another in territorial boundary disputes, flying very low over the ground. Display takes place in the usual manner of larks in general, on the ground. The male walks about in front of the female in an imposing posture with drooped wings, fanned out tail and erect ear tufts. When the median daily temperature on two successive days has exceeded 5°C, the female starts nest building. She digs out a depression using her beak and feet, and constructs a nest from stalks, laying out the entry with small pebbles and clumps of mud. Nest building takes two to four days; three to four eggs are then laid and incubated (by the female only) for eleven days. At the approach of danger near the nest, the female decoys the intruder away by stretching out one wing, letting it hang, and walking away from the nest in this posture (nest distraction display).

◁
Swallows: 1. Barn swallow or swallow (*Hirundo rustica*); 2. House martin (*Delichon urbica*); 3. Bank swallow or sand martin (*Riparia riparia*); 4. Purple martin (*Progne subis*); 5. Black rough-winged swallow (*Psalidoprocne holomelaena*); 6. Crag martin (*Ptyonoprogne rupestris*); 7. Rough-winged swallow (*Stelgidopteryx ruficollis*); 8. Cliff swallow (*Petrochelidon pyrrhonota*); 9. Tree swallow (*Iridoprogne bicolor*); 10. Red-rumped swallow (*Hirundo daurica*); 11. White-headed rough-winged swallow (*Psalidoprogne albiceps*).

Like the European larks, young horned larks have an orange gape with black spots on the tongue and lining of the lower mandible. The female only brings enough food to the nest at one visit as will enable her to satisfy one youngster's hunger. The male, however, drags in so much food that two youngsters can be fed with it. The young are reared on insects, especially grasshoppers (in the lower latitudes) when the season is advanced. The young leave the nest ten days after hatching and hop about in the vicinity. They are fully fledged after fifteen days.

The horned lark probably became extinct in Europe during the Ice Ages, and only reinvaded that area during the 19th Century. Apart from the tundra, it has also settled in mountains above the tree line. It wanders about in small flocks and in Europe, it winters on the North Sea coasts, rarely being seen inland. American horned larks, however, congregate into huge migrating flocks.

7 The Swallows

Family: swallows
by D. St. Peters

The form which gave rise to the swallows probably separated from the primitive insectivorous birds in the Eocene (i.e., in the Lower Tertiary about fifty million years ago). In spite of many special developments connected with their mode of hunting on the wing, the swallows still show many primitive features, such as the formation of the bronchial rings, so that they are placed fairly low down in the evolutionary system of the song birds.

The SWALLOWS (family Hirundinidae) are delicately built birds which fly swiftly with great endurance and great maneuverability. Their body form, which is "drop-shaped," is an adaptation to the requirements of flight; swallows range from 10–23 cm in length. The wings are long and pointed; there are nine primaries, and the tail has twelve feathers. The tail is usually indented or even deeply forked; it may also occasionally be straight-ended. The body plumage is short and pressed close to the skin. The predominant color is earth-brown or black and white often with an attractive green or purple gloss; rust-brown or rust-red markings are not infrequent. The sexes do not usually differ externally. The flight muscles are strong, and the legs so short that they permit only a clumsy waddling gait. Swallows are insect hunters, catching their prey almost exclusively in flight. The beak is small but its opening at the base is wide and runs as far back as the eye. Nest construction is variable. Nests are rarely built of grasses and twigs alone; most often they are built partially of moist earth in tree or ground holes, on walls, often on or in buildings, or in passages dug into the soil by the birds. Their distribution is worldwide with the exception of Australia, New Zealand, and certain small islands. There are two subfamilies: 1. The AFRICAN RIVER SWALLOWS (Pseudochelidoninae); and 2. The TRUE SWALLOWS (Hirundininae), made up of nineteen genera and about seventy-five species.

A comparison of swallows and swifts (Volume VIII) enables one to understand quite well the phenomenon known to biologists as con-

▷
The gray wagtail (Motacilla cinerea) lives exclusively near moving waters, singly, or in pairs; it is a resident.

Left from top to bottom: White-throated bulbul (*Criniger flaveolus*); Yellow wagtail (*Motacilla flava*); Blue-backed fairy bluebird (*Irena puella*); Red-whiskered bulbul (*Pycnonotus jocosus*).

Right from top to bottom: Gray wagtail (*Motacilla cinerea*); White wagtail (*Motacilla alba*); Water pipit (*Anthus spinoletta*); Gold-fronted leafbird (*Chloropsis aurifrons*).

vergent evolution. Such convergence is manifest when, in otherwise not closely related organisms, similar body forms develop as a result of a similar environment and mode of life. Adaptations in the same direction account for the fact that swifts and swallows are often confused and were once regarded as close relatives. In fact, they belong to different orders of birds; the swifts, as described in Volume VIII, are not even passerine (or perching) birds. In spite of all external similarities, swifts and swallows are readily distinguished. Swifts fly much faster, in straighter lines, and at a greater height than do swallows. Their largest primary feather is three times as long as the secondaries, while in the swallows it is only about twice as long. The tail of the swift has only ten feathers but the swallow's has twelve. While swallows often interrupt their flights to perch horizontally on roofs, branches, or wires, swifts fly with more endurance; when swifts perch, they suspend themselves vertically, with the head up, on vertical walls or similar surfaces such as cliffs. In addition, there are, of course, structural and behavioral differences.

Subfamily: Pseudochelidoninae

The subfamily of Pseudochelidoninae consists of two species, one of which is the AFRICAN RIVER MARTIN *(Pseudochelidon eurystomina)*, with a L of 14 cm. This bird differs in many characteristics from the true swallows and, because of this, it has sometimes been considered a member of a separate family or placed with the wood swallows. It is glossy black and has a red beak and feet. It breeds from December to April on the shores and islands of the Congo River. It digs slanting passages into the flat ground for its nests. After breeding it migrates to the West African coast. Another species, the WHITE-EYED RIVER MARTIN *(Pseudochelidon siantarae)*, which is distinguished by its white rump and thread-like elongated tail feathers, was only recently discovered in Thailand and was first described in 1968.

Subfamily: true swallows

All other swallows are grouped together in the subfamily of the TRUE SWALLOWS (Hirundininae) and the genera characteristics given above apply to these birds as well. According to their nest types, true swallows can be divided into four types which do not, however, necessarily correspond to their true relationships: 1. African river swallows; 2. Ground-hole nesters or bank swallows; 3. Tree swallows; and 4. Mud-nest swallows.

The MASCARENE MARTINS (genus *Phedina*) occupy a special position, for they do not build mud nests nor do they dig nest passages; rather, they build single twig nests. One species, the MASCARENE MARTIN *(Phedina borbonica)* inhabits the island of Madagascar; another species, *Phedina brazzae*, lives in the Congo. Many ornithologists believe, and probably rightly so, that these swallows (the words swallow and martin imply no significant distinction) represent the primitive type of the family and that they are still close to the original swallows.

Some species of swallows (those classified above as bank swallows) nest in a manner which one would hardly expect in these short-billed, weak-footed, and rather delicate birds. They build their nests at the end of passages which they dig into the soil and which are often one meter long. Among these is the BANK SWALLOW (*Riparia riparia*, called the sand martin in Europe), the smallest of the central European swallows; it does not confine itself to the banks, but also breeds far from water in such places as sand or gravel pits. In Lusatia, (area in southeastern East Germany and southwestern Poland), I have even come to know it as a peaceful inhabitant of deserted World War II trenches. The clutch consists of five to six white eggs. There are two broods per season. The bank swallows' commonest call is a hoarse, whetstone-like sound. (Color plate, p. 180.)

Fig. 7-1. Bank swallow or sand martin (*Riparia riparia*).

The ROUGH-WINGED SWALLOWS, represented by only one species of that name in America *(Stelgidopteryx ruficollis)*, and the AFRICAN ROUGH-WINGED SWALLOWS *(Psalidoprocne)*, which occur in nine species, are also ground-hole nesters. In both genera the males show a more or less hook-like thickening of the outer vane of the first primary, whose function is still unknown. According to G. Steinbacher, it may play a role in mating, enabling the male to anchor itself to the female's plumage. Possibly it provides a means of clinging on a vertical wall.

The AFRICAN GRAY-RUMPED SWALLOW *(Pseudhirundo griseipyga)*, with a L of 15 cm, is also included with the ground-hole nesters. It breeds not only in passages which it has dug out itself, but also primarily in holes made by rodents in the dry parts of its range. There are also a number of other swallow genera, each of which contains only one or two species; the generic distinctiveness of these does not appear to be altogether certain.

Swallows of the TREE SWALLOW group are also hole nesters. In America some species have attached themselves closely to man. Among them is the TREE SWALLOW *(Iridoprocne bicolor)*, which has a L of 15 cm. This species readily accepts nest boxes, as does the PURPLE MARTIN *(Progne subis;* Color plate, p. 180) which has a L of 20 cm. Indians used to hang up hollowed-out pumpkins as nest sites for this bird. This large species is noteworthy because of the glossy blue-black males, which, in contrast to those of most swallows, are differently colored than are the females.

Tree swallows

All the remaining swallows are mud nesters. One species in this group nests in tree holes; this is the TREE MARTIN *(Petrochelidon nigricans)*, which has a L of 13 cm and is found in Australia, Tasmania, and the Lesser Sunda Islands. However, tree martins make the nest hole entrance smaller by using mud; thus, it reveals itself as a mud hole nester. The AMERICAN CLIFF SWALLOW *(Petrochelidon pyrrhonota;* Color plates, p. 170 and p. 180); originally nested on cliffs where its colonies sometimes numbered many thousands of pairs. Nowadays, this very attractively colored species has become a follower of civilization and

Mud nest swallows

Barn swallows

Fig. 7-2. Red-rumped swallow *(Hirundo daurica)*.

Fig. 7-3. Swallow or barn swallow *(Hirundo rustica)* breeding area black, wintering areas vertical lines.

often nests on buildings with projecting roofs like the European swallows, or under bridges.

The swallow genus with the most species is *Hirundo*. Its nineteen species have been divided by several authors into two or more different genera on the basis of plumage patterns and above all on the basis of differences in the construction of their mud nests. Thus those birds which resemble the cliff swallow in plumage, and which build retort-shaped nests, have been combined in the genus *Cecropis*. This would include the STRIPED SWALLOW *(Hirundo abyssinica)* which has a L of 16.5 cm, is widely distributed in Africa and often nests colonially on buildings. The RED-RUMPED SWALLOW *(Hirundo daurica;* Color plate, p. 180), which is divided into 12 subspecies and distributed from central Spain and northern Africa as far as eastern and southeastern Asia would also be included in this group. Although it occasionally nests on a house, it breeds incomparably more often away from settlements. The closer relatives of the barn swallow (swallow in Europe) build mud nests which are open above (i.e. cup as opposed to retort-shaped, liked those mentioned previously). We will mention here the WHITE-THROATED SWALLOW *(Hirundo albugularis)*, which has a L of 16.5 cm, from South Africa, the ANGOLA SWALLOW *(Hirundo angolensis)*, which is very similar to the barn swallow but gray-bellied, the SOUTH SEA SWALLOW *(Hirundo tahitica,* called the Welcome Swallow in Australia) from southeastern Asia and Australia, and the BARN SWALLOW *(Hirundo rustica;* Color plate, p. 169, p. 170, and p. 180), which has a L of 19 cm and has nine subspecies in the northern parts of the Old and New Worlds.

Together with the stork, the cuckoo, and the nightingale, the swallow is one of the most popular European birds. It is appreciated as an omen of good luck and a harbinger of spring. It has lived for millenia in the closest association with man. Originally, it probably attached its nests to cliffs, cave walls and clefts in the ground, as it still does in a number of areas, but nowadays most of its nests are in buildings, including even the tents of nomads of the Asiatic steppes and living rooms of farm houses. The swallow is a marked follower of civilization and it has colonized many areas only as a follower of man. It prefers rural habitats with stables and dung heaps where its prey, flying insects, readily develop. It has therefore long disappeared from the densely populated districts of central European cities. It is only found in those areas when there are green open spaces and water. Even in villages equipped with modern technology, it seems to be decreasing.

However, in areas where the swallow finds suitable conditions, it shows great tameness. Arnold Baron Vietinghoff-Riesch in his well-known book on the swallow lists a number of unusual nest sites. Even in inns, the smoke, music and noise of dancers does not distract them from breeding.

In Central Europe swallows arrive from about the end of March onward but the date depends largely on the temperature. Even within one village, weeks may elapse between the arrival of the first and the last pair. Climatic conditions limit their altitudinal distribution. While they are found up to about 1,000 meters in Central European mountains and in the Riensengebirge of Germany they breed just up to 1,400 meters, they nest at higher elevations in the south. In the Alps, they breed to about 1,800 meters. I found a barn swallow's nest at the southern edge of its North American breeding range in the Mexican state of Puebla, on an arch in the city of Cholula at 2,200 meters. However, when migrating, swallows cross mountain ranges even higher than this.

The swallow's mud nest is shaped like an open quarter-sphere and is usually placed in the inside of buildings, although nests placed on the outside may be found on some occasions, particularly when projecting roofs or balconies provide enough shade. It usually takes the birds eight days to build a new nest, but in unfavorable weather this period may be as long as four weeks, according to von Vietinghoff-Riesch. The time given includes that taken to line the nest, for which mainly feathers are used. A nest may be used every year for a decade or longer. Naturally the birds have to repair it from time to time. In Central Europe the clutch consists of five to six eggs; the later in the season it is started, the fewer eggs. Generally, only the female incubates in Central Europe, although the male sometimes takes part. In the American subspecies, the barn swallow, it seems to be normal for both partners to incubate. The young hatch after eleven to eighteen days of incubation and they are fledged after fifteen to twenty-three days. During this time their weight increases from 1.6 to 23 grams; after fledging they rapidly lose weight again and soon weigh the same as the adults, namely 18 to 20 grams. In the last days before fledging the length of the last primary increases by five mm a day. Generally swallows breed twice per season, rarely even three times.

Swallows migrate south again from the end of August on, but particularly in the first half of September. The old German peasant rule "on virgin Mary's birthday (September 8) the swallows leave," is therefore fairly accurate. Before the birds depart and while they rest in the course of migration, large flocks often form, frequently together with house martins, in the reed beds of ponds and lakes. Late migrants can still be seen in Germany in October or even during the first days of November. According to my observations, the last swallows in the southernmost American breeding area in the Mexican plateau disappeared during the first ten days of October.

The swallow's voice, no doubt, contributed to its popularity; in this regard it is the most talented of its tribe. Its contact call, which is also used to attract others, can best be described as a high-pitched, fairly loud, often repeated "weet." The song consists of a leisurely purring

▷
Pipits: 1. Cape long-claw (*Macronyx capensis colleti*); 2. Tree pipit (*Anthus trivialis*); 3. New Zealand pipit (*Anthus novaseelandiae*); 4. Red-throated pipit (*Anthus cervinus*); 5. Tawny pipit (*Anthus campestris*); 6. Water pipit (*Anthus spinoletta*); 7. Meadow pipit (*Anthus pratensis*); 8. Gray wagtail (*Motacilla cinerea*); 9. Tree wagtail (*Dendronanthus indicus*); 10. Yellow wagtail (*Motacilla flava*); 11. White wagtail (*Motacilla alba*).

twittering which strikes many people as "homely." In many countries the swallow's song has suggested the words of certain songs.

The CRAG MARTINS (genus *Ptyonoprogne*) belong exclusively to the Old World but are joined into one genus with the swallows by a number of authors. The CRAG MARTIN *(Ptyonoprogne rupestris)* inhabits mountainous areas in southern Europe, northwards as far as the Bavarian Alps. Its nest, which resembles that of the swallow, is found suspended from overhanging rocks. This species is found on middle elevations and does not migrate from suitable breeding areas. Related species are found in Asia and Africa.

The HOUSE MARTIN *(Delichon urbica)* has a L of 14 cm and is found in six subspecies in the Old World southwards as far as Asia Minor, northwestern India and southeastern China. It belongs to another genus. In central Europe the house martin inhabits settlements alongside the swallow. It is readily recognized by its entirely white lower parts and its white rump. Its nests, which are closed except for the round entry hole, are usually placed on buildings, but it still breeds locally on rocks in Germany as well. It is more frequently found in towns than is the swallow; the newer, more open settlements particularly seem to suit it. Both sexes incubate the four to five eggs and there are two broods per season.

Fig. 7-4. House martin *(Delichon urbica).*

◁
1. Yellow-streaked greenbul *(Phyllastrephus flavostriatus);* 2. White-vented bulbul *(Pycnonotus barbatus);* 3. *Spizixos semitorques cinereicapillus;* 4. Red-whiskered bulbul *(Pycnonotus jocosus);* 5. Common iora *(Aegithina tiphia);* 6. Blue-backed fairy bluebird *(Irena puella);* 7. Golden-fronted leafbird *(Chloropsis aurifrons);* 8. Long-billed marsh wren *(Cistothorus palutris);* 9. Winter wren *(Troglodytes troglodytes);* 10. Carolina wren *(Thryothorus ludovicianus);* 11. Cactus wren *(Campylorhynchus brunneicapillus);* 12. Eurasian dipper *(Cinclus cinclus);* 13. Canyon wren *(Salpinctes mexicanus).*

8 Wagtails, Cuckoo-Shrikes, Bulbuls, Fairy Bluebirds and Leafbirds

The WAGTAIL family, which is difficult to place, will be discussed here, for the sake of space, along with three other songbird families, which though possibly related to one another, are not related to the wagtails. These other families are the CUCKOO-SHRIKES, the BULBULS, and the FAIRY BLUEBIRDS and LEAFBIRDS.

WAGTAILS (family Motacillidae) are lively, generally attractive inhabitants of grass and wetlands. They also live on river shores or in rocky areas, but four species live in wooded country. In southern areas, they inhabit mountains up to 4,000 meters. They avoid areas far from water more than do larks, which often occur in deserts. As insect-eaters, they are migrants in northern latitudes.

Family: wagtails, by W. Meise

These small to medium-sized birds (the L is 12.5 to 23 cm) have an extended body and soft plumage. They have ten primaries; the outermost primary is very short. The young undergo a partial molt. The tail ranges from medium length to very long and has narrow feathers; the outermost feathers always have a white or pale edge and are rarely completely white or yellow. The body is generally held horizontally; these birds move by walking rather than hopping. The tarsus and toes are long; the claw of the rear toe is often prolonged. The beak in contrast to that of larks is narrow. The food consists mainly of insects, spiders, small mollusca and crustacea, also of soft seeds and other plant materials. It is generally taken from the ground, and more rarely while walking on water plants, from trees, or sometimes after a chase on the wing or while fluttering. Wagtails breed in cup-shaped nests on the ground, a little above it or in shallow cavities (the tree wagtail is an exception). The song is generally a sequence of mono- or disyllabic calls which are often sharp and hissing, but melodious in a few species. The clutch size in middle latitudes is three to eight eggs; in the tropics it may be up to four. The eggs are more or less spotted and are incubated by the female only, or by both sexes; the incubation period is twelve to sixteen days. The young leave the nest before they

Fig. 8-1. Meadow pipit (*Anthus pratensis*).

Fig. 8-2. Water pipit (*Anthus spinoletta*).

Fig. 8-3. Red-throated pipit (*Anthus cervinus*).

are fledged at eleven to fifteen days. Many species breed several times a year. The wagtail's distribution is world-wide; they are absent only from Oceania and the vicinity of the poles. Fossils dating back to the Upper Oligocene (about 30 million years ago) have been found. There are two groups of genera: wagtails and pipits, with a total of six genera and about fifty species.

PIPITS have the same distribution as the family as a whole. They are often ground-colored above with indistinct stripes below; they generally show dark spots on the throat. Their tail is relatively short and their flight course is not markedly wavy. At night they rest on the ground. There are four genera with thirty-nine species, of which we will mention the following: 1. The PIPITS in the narrower sense (*Anthus*). The sexes have a similar plumage; the breeding and off-season plumage are similar in the tropics, and in Europe they are only rarely markedly different. The MEADOW PIPIT (*Anthus pratensis;* Color plate, p. 189), which has a L of 15 cm and weighs 19 g, and the WATER PIPIT (*Anthus spinoletta;* Color plates, p. 184 and p. 189), which has a L of 17 cm and weighs 24 g, are often to be seen at the same time on wet mountain meadows or on sea coasts. The meadow pipit also occurs on moors, heaths, pastures and waste land, and in winter along ditches and on fields. The RED-THROATED PIPIT (*Anthus cervinus;* Color plate, p. 189), of the tundra, resembles the meadow pipit and is a passage migrant in Central Europe. The TREE PIPIT (*Anthus trivialis;* Color plate p. 189) has a L of 16 cm, and is distinguished by a markedly curved claw on the hind toe. It lives on forest edges and clearings in open parklands and often sits on branches. The TAWNY PIPIT (*Anthus campestris;* Color plate, p. 189) has a L of 17.5 cm and is more sandy in color; it occurs locally in bare soft ground on wasteland areas and dunes. The NEW ZEALAND PIPIT (*Anthus novaeseelandiae;* Color plate, p. 189) has a L of 18 cm and weighs 33 g. It is distributed in at least twenty-nine subspecies from eastern Asia and Siberia to New Zealand, Australia and Africa. It has a very long hind claw and is occasionally seen in Germany as a vagrant from Siberia. 2. The LONG-CLAWS (*Macronyx*) have a L of 17 to 21 cm and a particularly long rear claw. The males generally have a dark border around the throat; below they are partially or nearly totally yellow to red. These birds are found in seven to eight species in Africa. Among them is the CAPE LONG-CLAW (*Macronyx capensis;* Color plate, p. 189) with an orange-yellow throat.

Most pipits, at least all Central European species, have display flights during which they sing in the breeding season. They begin their song up in the air and return to the ground after a few turns. The tree pipit returns to its elevated perch after its song. In flight pipits look more delicate than larks. The song, which is also given by perched birds, is generally more monotonous and contains more hissing sounds than that of the lark. The tree pipit's song is particularly melodious and variable, and as a result, this bird has been called the "woodland canary." When rising and particularly when descending, each pipit flight proceeds in steps, not at a continuous slant. Pipits that migrate far have more pointed wings than do those birds which do not. Long-spurred pipits resemble quails in flight.

Although both pipits and wagtails wag their tails (the former only slightly), they are easily distinguished, because pipits have a more or less spotted plumage, a shorter tail, a monosyllabic flight call and a less wavy flight than do wagtails. In the wagtail group, the winter plumage differs from the breeding plumage except in the tree wagtail and in most tropical species. The back is uniform in color; the flight path shows a marked undulating (up and down) pattern. These birds are generally sociable; they are found only in the Old World and are not found in Australia. There are two genera with eleven species: 1. The TREE WAGTAILS *(Dendronanthus)* with the single species the TREE WAGTAIL *(Dendronanthus indicus;* Color plate, p. 189) which has a L of 14 cm, and is an intermediate between the pipits and the wagtails. It is a forest dweller of eastern Asia, and does not move its tail up and down like the true wagtails, but rather from side to side. It nests on horizontal branches four to ten meters from the ground. 2. The TRUE WAGTAILS *(Motacilla)* have a noticeable tail. There are ten species, among them the YELLOW WAGTAIL *(Motacilla flava;* Color plates, p. 184 and p. 189) with a L of 17 cm and a weight of 17 g. There are eighteen subspecies, one of which *Motacilla flava tschutschensis* also breeds in Alaska and migrates to Asia for the winter. The YELLOW-HEADED WAGTAIL *(Motacilla citreola)* is a vagrant in Central Europe from eastern Russia and Siberia. The GRAY WAGTAIL *(Motacilla cinereca;* Color plates, pp. 183, 184 and 189) has only penetrated the northern German plain since the middle of the last century; it prefers clear streams and dams. The best-known species is the WHITE WAGTAIL *(Motacilla alba;* Color plates, p. 184 and p. 189) which has a L of 19 cm and has eleven subspecies, of which the PIED WAGTAIL *(Motacilla alba yarrellii),* found in Britain and occasionally on the southern North Sea coast, is striking because of its black back (see Fig. 8-13).

The yellow wagtail inhabits not only pastures and meadows, but in some places, also fields. It spends the winter in tropical Africa where it is found in congregations, often with several subspecies together on the shores of rivers and lakes. It returns to its breeding areas in March. Soon after this, each male "owns" a territory of grassland or part of a field. Initially the territory holders will still feed together and follow ploughing farmers, although now and then they fly back to their territories. In territorial disputes they face one another, the head thrown back, chest puffed out and fluffed up, wings lowered and tail laid on the ground. Ten seconds may elapse before a fight begins. Then they look like a ball of feathers, fluttering up in the air clinging to one

Fig. 8-4. Tree pipit *(Anthus trivialis).*

Fig. 8-5. The tree pipit wags its tail like the wagtails.

▷
Shrikes: 1. Red-backed shrike *(Lanius collurio collurio);* 2. Northern shrike *(Lanius excubitor);* 3. Gonolek *(Laniarius barbarus);* 4. Magpie shrike *(Urolestes melanoleucus);* 5. Rufous-backed shrike *(Lanius schach);* 6. Lesser gray shrike *(Lanius minor);* 7. Isabelline red-backed shrike *(Lanius collurio isabellinus);* 8. Brown shrike *(Lanius cristatus);* 9. Woodchat shrike *(Lanius senator);* 10. Masked shrike *(Lanius nubicus);* 11. Black-backed puffback *(Dryoscopus cubla);* 12. Redwing shrike *(Tchagra tchagra).*

Fig. 8-6. Display flight of the tree pipit.

Fig. 8-7. Tawny pipit (*Anthus campestris*).

1. Phainopepla (*Phainopepla nitens*); 2. White-browed wood swallow (*Artamus superciliosus*); 3. Blue vanga (*Leptopterus madagascarinus*); 4. Palm chat (*Dulus dominicus*); 5. White-helmeted shrike (*Prionops plumata*); 6. Bohemian waxwing (*Bombycilla garrulus*); 7. Helmet-bird (*Euryceros prevosti*); 8. Hook-billed vanga (*Vanga curvirostris*); 9. White-crowned shrike (*Eurocephalus anguitimens*); 10. Pollen's vanga (*Xenopirostris polleni*); 11. Retz's red-billed shrike (*Sigmodus retzii*).

another, or suddenly they release each other and chase one another through the air.

When the females arrive, which is about three weeks after the males, the most intense fights break out. The future partner is, however, immediately treated differently than rival males. While the female stands still, the territory owner trips about her with erected breast feathers and lowered wings. When a pair has been formed, the increased arousal is shown by the fact that the male now also fans out its tail and performs hovering flights up to three meters above the female. After such a flight it lands beside its partner before copulation. Other courtship flights are made from some raised perch. With this, as in displays generally, one hears a series of robin-like "tsip" calls. The wagtail does not fly downwards to land during its song flight; instead it flies along an undulating course and then rises up once more, not singing while ascending. It then descends singing, but does not land. It does not fly up and down only once as do pipits.

The female selects the nest site within the territory; the male accompanies her when she brings in the nest material and this may take them both over distances of up to one kilometer. The nest depression is made on a slope, in the cover of grass or heather clumps, in stands of vegetables or in other places. The nest consists of grass with a lining of hair. Only rarely is the nest in shrubs or bushes up to 1.2 meters above the ground. Nest-building takes from four days to three weeks. The female generally lays five to six (in Spain four to five) eggs, which are only incubated from the laying of the last egg for twelve to thirteen days by both partners; the male, however, only participates during the daytime.

Stuart Smith describes the nest relief of English yellow wagtails as follows, "The partner which was to feed flies to the territory and lands in it. It looks about as if to see that all is in order, then flies and lands usually about fifty meters from the nest. Then it calls and walks about aimlessly, also pecking. Soon the bird which had been incubating appears, having first walked some distance away from the nest. The relieving bird now makes its way toward the nest, although this seems painfully slow (to the observer) as the bird often feeds en route. The calls only cease when the nest has been almost reached." Ludwig Schuster noticed, on the other hand, that the female usually descended directly onto the nest from a hovering flight. The young are covered with down upon hatching and their eyes open at about four days. When disturbed, they are able to fly after ten days: normally, however, they fly at eleven to thirteen days. They are fully fledged after the sixteenth to seventeenth day of life. They are fed by the adults for several weeks after that. In rare cases, a second brood may follow.

The popular, but less well-studied white wagtail, returns as early as March. It is also seen in villages where its lively "tseesees" call is heard from the roofs. It catches insects in flight, as well as by other methods, and it likes to follow the plough, tripping along and wagging its tail. Its nests stand in any kind of cleft, thirty centimeters to six meters above the ground. They often do not build a new nest for the second and third broods. Outside the breeding season it roosts in flocks in reed beds, and in city trees in London beneath the shades of street lights.

The CUCKOO-SHRIKES (Campephagidae) are a largely tropical, Old World family. Their German name, Stachelbürzler, spine rumps, is due to the widened, then suddenly pointed, shafts of their dense rump feathers. In size they range from that of a sparrow to that of a dove; the L is 14-36 cm. They are predominantly pale and dark gray, more rarely black and white, and occasionally black and red. The wings are fairly long and pointed, the tail moderately long, rounded and generally provided with the spiny rump mentioned above. The rictal bristles are well developed and in many species, they cover the nostrils. The sexes differ in the degree of color intensity; females are pale, bleached-looking editions of the males. Nestling young often have a white plumage which is exchanged in the juvenile moult for one like that of the female. There are two groups of genera: cuckoo-shrikes and minivets.

Family: cuckoo-shrikes, by B. E. Smythies

The CUCKOO-SHRIKE group (Campephagini) has eight genera, three of which have only one species: 1. The GROUND CUCKOO-SHRIKE *(Pteropodocys maxima)* of Australia; 2. The ORANGE CUCKOO-SHRIKE *(Campochaera sloetii)* of New Guinea; and 3. The BLACK-BREASTED TRILLER *(Clamydochaera jefferyi)* of Borneo. The remaining five genera are: 4. The CUCKOO-SHRIKES PROPER, *(Campephaga)* which include the WATTLED CUCKOO-SHRIKE *(Campephaga lobata)* and the BLACK CUCKOO-SHRIKE *(Campephaga flava)*, both from Africa: 5. The genus *Coracina* with the BLACK-FACED CUCKOO-SHRIKE *(Coracina novaehollandiae)*; 6. The FLY-CATCHER SHRIKES *(Hemipus)* with the PYGMY TRILLER or PIED SHRIKE *(Hemipus picatus)*; 7. The TRILLERS *(Lalage)* with the VARIED TRILLER *(Lalage leucomela)* from northern Australia and Melanesia; and 8. The genus *Tephrodornis* with the HOOK-BILLED GRAYBIRD *(Tephrodornis gularis)*.

Fig. 8-8. Tree wagtail.

In contrast to the rather dark and pale colored cuckoo-shrikes, the MINIVETS (Pericrocotini) are considerably more colorful. There is only one genus, MINIVETS *(Pericrocotus)*, which is absent from Africa and Australia, but widely distributed in the far east of the Old World. From there, the ASHY MINIVET *(Pericrocotus roseus divaricatus)* migrates in the winter to the southeast; it is the only long distance migrant of the family, although some of the cuckoo-shrikes, particularly in Australia, wander about at certain seasons.

All cuckoo-shrikes are tree birds; only the deviant ground cuckoo-shrike spends most of its time on the ground. The varied triller, when searching for food, also comes down often to the ground for a few moments. Some species live only in dense forests; others are characteristic of forest edges, secondary growth, gardens and coastal vegetation. Their calls are loud, either flute-like, or to our ears, rather unpleasantly screeching.

The minivets form a clearly distinguished group of birds about the size of our wagtails, with fairly long graduated tails. In most species, the male is strikingly red or black while the female is yellow, orange and black, or gray. Minivets live mainly in the crowns of trees; they are generally seen in flocks of a dozen or more birds. When hunting for insects in the foliage through trees, each bird follows its predecessor and group members maintain contact by musical calls. When these flocks whirr about in the tree tops or fly over a clearing, they give the effect of scarlet-red and yellow streaks. They are one of the characteristic sights of the teak forests of Burma and are a source of delight for naturalists. Babblers (Chapter 11), Old World warblers (Sylviidae, Chapter 11) and flycatchers (Chapter 12) and members of still other families are often found with these birds.

The display of some of the larger cuckoo-shrikes is quite unusual. The male alternately raises a wing without opening it and repeats this about six times while calling loudly. Both sexes build the nest; in many species only the females incubate. The nest is usually high up in a tree fork and difficult to see from below. It is a small flat cup with soft lining, and is constructed of twigs, roots and grass stalks or mosses and lichens. The two to five eggs are often pale green with a brown, gray, or purple pattern, although they generally become black on exposure to light.

The African members of the genus *Campephaga* are quite attractive and striking. The black cuckoo-shrike is distributed from the Senegal to the Cape, and lives mainly in savanna forests. The male has a glistening blue-black color and sometimes has scarlet shoulder spots. The rare and little-known wattled-cuckoo-shrike (*Campephaga lobata*) inhabits only a narrow range in western Africa. The male has a yellow lobe of skin beneath his scarlet eye, near the gape.

The black-faced cuckoo-shrike (*Coracina novaehollandiae*) is widely distributed in southeastern Asia, Australia, and Melanesia. This thrush-sized, gray bird has a black band through the eye region. Its clamorous troops are often seen in tree crowns. The pygmy triller is only the size of a flycatcher; it is well known in India and Burma and since it catches flying insects, behaves much like a flycatcher.

The black-breasted triller is a mountain bird of Borneo which inhabits a very small area there from the Kinabulu to the Dulit. It was first discovered by J. Whitehead in 1885 during the exploration of Kinabulu and has since been seen and collected by several observers, although it is still little known. It has a striking plumage pattern of black and pale brown and feeds only on fruit. Deignan has observed the ground

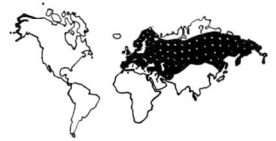

Fig. 8-9. Yellow wagtail (*Motacilla flava*).

Fig. 8-10. Yellow wagtail. Above: threat display of two males. Below: display flight of the male above female on the ground.

cuckoo-shrike in the Australian interior. According to his report, it inhabits treeless plains, perches on fence posts, flies after insects like a swallow, and, in contrast to all other cuckoo-shrikes, walks about readily on the ground.

Throughout the tropics and subtropics of Africa and Asia, the bulbuls have attached themselves to man, in the words of E. Thomas Gilliard; they inhabit settlements, gardens and parks, arenas and cemeteries in considerable numbers. These beautiful songsters can become extraordinarily tame; some species have therefore been introduced in Australia, the Mascarene Islands and in other areas where they did not originally occur.

The BULBULS (family Pycnonotidae) are sparrow to thrush-sized birds with a L of 14–28 cm. The neck and wings are short; the tail is straight or slightly indented, and the legs and feet are relatively weak and small. The body feathers are soft, long and loose, particularly on the lower back; there are always a few hair-like feathers, which are often long and striking in the nape, (hence the German name, Haarvögel, in English, "hairbirds"). Some species have a distinct crest. The color is generally olive-green, yellow or brown; some birds have striking red, yellow, white or reddish-brown lower tail coverts. The sexes are similar in color and so are the young, though they are sometimes paler or darker. The earliest fossils are from the Middle Pliocene (about one million years ago). All bulbuls are residents and move about locally in search of fruit. Ten bulbul genera are confined to Africa and Madagascar with only two in southern Asia. The BULBULS PROPER (genus *Pycnonotus*) and those of the genus *(Criniger)* with the major species the BROWN WHITE-THROATED BULBUL *(Criniger ochraceus)*, living in Africa as well as Asia. The genus *Hypsipetes* occurs from Madagascar to eastern Asia. *Pycnonotus melanoleucos* from Malaysia, Sumatra, and Borneo, is unusual insofar as its plumage is black. The feathers do have some white spots; these, however, are not present in the young. The BLACK-HEADED BULBUL *(Pycnonotus atriceps)* is mainly olive-yellow with a black head, although in some parts of its range the olive-yellow is replaced by gray, as on the island of Maratua between Celebes and Borneo.

The voice of most bulbuls is loud and lively rather than musical. The YELLOW-CROWNED BULBUL *(Pycnonotus zeylanicus)* of Malaysia and the Greater Sunda Islands is an exception in this regard, for it sings long melodious songs which are better than those of the nightingale or the European blackbird. Its phrases, however, always remain the same and are merely performed at maximum loudness. If one travels down the rivers of Borneo in a canoe, the voice of this bird is one of the most striking animal sounds one can hear. The gay sounds of the RED-VENTED BULBUL *(Pycnonotus cafer)* are also very attractive. The species ranges over India, Burma and Ceylon; it is a well known garden bird and gives no real song but merely repeats two calls. The STRIPE-THROATED BULBUL

Family: bulbuls, by B. E. Smythies

Fig. 8-11. Yellow-headed wagtail *(Motacilla citreola)*.

▷ Contrary to earlier opinion, the northern shrike *(Lanius excubitor)* actually does impale its prey.

Fig. 8-12. Gray wagtail *(Motacilla cinerea)*.

Fig. 8-13. White wagtail *(Motacilla alba)*.

◁

Here the northern shrike *(Lanius excubitor)* has wedged its prey into the fork of a branch.

(Pycnonotus finlaysoni) of Burma has a very characteristic loud and melodious call as has the BLACK BULBUL *(Hypsipetes madagascariensis)* which in spite of its scientific name, is also found in southern Asia.

Some bulbuls are among the most common inhabitants of the gardens of villages and towns; they sit on the tips of bushes and draw attention to themselves by their loud calls. Thus, the RED-WHISKERED BULBUL *(Pycnonotus jocosus;* Color plates, p. 184 and p. 190) and the YELLOW-VENTED BULBUL *(Pycnonotus goiavier)* are generally known in southern Asia. Other species, however, live on forest edges or even in dense forests and are difficult to observe. The rarest of all bulbuls is the NIEUWENHUIS BULBUL *(Pycnonotus nieuwenhuisi),* a dark olive-brown bird with wide white tailtips and pale blue fleshy eyelids. Only two individuals from two of its subspecies in Sumatra and Borneo are known, and in spite of much search, no further members of these subspecies have been found there.

Bulbuls generally inhabit lowlands; only a few prefer mountainous country. The YELLOW-STREAKED GREENBUL *(Phyllastrephus flavostriatus;* Color plate, p. 190) of Central Africa, utters its piping chatter in the mountain forests of its home; unlike the other bulbuls, it moves along branches like a tree creeper. The OLIVE-BREASTED MOUNTAIN GREENBUL *(Pycnonotus tephrolaemus)* is one of the most common birds in the forests of the Ruwenzori, a mountain range in eastern Africa. It is olive-green with a gray head and upper breast. BLYTH'S BULBUL *(Pycnonotus flavescens)* in Borneo inhabits the mossy forests of mountains and has been reported up to 3500 m. The black bulbul also lives at medium elevations in Burma and descends to lower levels in the winter. Its flocks in the tree tops often include up to one hundred birds. The birds rarely descend to the undergrowth and rarely stay long on one tree, but they fly in loose troops from one group of trees to another.

Most bulbuls feed mainly on fruit and berries. *Pycnonotus gracilirostris* of Africa often wanders over considerable distances in large flocks in search of wild figs. In some parts of their range, bulbuls are therefore considered harmful in orchards. In its morning visits to the fruit trees, the yellow-vented bulbul of Borneo likes to join a shining starling (Chapter 13) in search of fruit fifty meters high or higher. It is, however, very adaptable in its search for food and, in contrast to other bulbuls, regularly seeks food on the ground where it takes large worms, dung beetles and crickets. Before swallowing such prey, it beats it vigorously against the ground or a branch. It also often visits low shrubs and there consumes caterpillars and grasshoppers; on sandy shores it even catches small flies. The yellow-crowned bulbul also has a varied diet; on small rivers or in the shoreline thickets of larger ones it takes freshwater snails.

Bulbuls build their open, cup-shaped nests from twigs and root fibers; the nests usually give a somewhat coarse and disorderly impression and are lined with finer materials. They stand or hang in the forks of branches in the lower levels of trees or in bushes. The two to three (rarely four to five) eggs are pink or white with various patterns. Van Someren has observed the reproduction of the WHITE-VENTED BULBUL (*Pycnonotus barbatus;* Color plate, p. 190), the most common African bulbul in Kenya. This bulbul probably mates for life. The female incubates for twelve to thirteen days; the male brings her food and occasionally relieves her. Both parents feed insects to the young and the male helps in covering them for warmth. By the end of the first week the young are well feathered, and they can leave the nest when sixteen days old.

The genus *Spizixos* includes a very unusual bulbul, the FINCH-BILLED BULBUL *(Spizixos canifrons),* which lives in the mountains of southern Asia. It has a peculiar, short, thick finch-like pale yellow beak which looks almost white in the field, and a thin crest which can be erected toward the front. It is found often in swarms of a hundred in the bush jungle which surrounds the mountain villages and on the fallow, strongly sloping fields which are cleared and burnt over at certain distances from one another for the planting of mountain rice and other fruit.

Those bulbul species which sing well are popular cage birds; they are kept not only in India and other parts of their native range, but are also regularly imported to Europe.

Systematists are not in agreement about the LEAFBIRDS and FAIRY BLUEBIRDS (family Irenidae). 1. The LEAFBIRDS *(Chloropseinae)* with two genera and eleven species, and 2. The FAIRY BLUEBIRDS (Ireninae) with one genus and two species. However, recently, ornithologists have placed the fairy bluebirds as a separate subfamily among the orioles (Chapter 18). We will not consider this unsettled question further, but follow the well-known French ornithologist, Jean Delacour, who places the leafbirds and the fairy bluebirds into the family Irenidae. The IORAS (genus *Aegithina*) are generally included with the LEAFBIRDS (genus Chloropis) in the subfamily of leafbirds.

Family: leafbirds, by S. Ali

The eight species of the LEAFBIRD genus *(Chloropsis)* are closely related to the bulbuls and have sometimes been placed in the same family. Their length is from 17-20 cm; the wings are like those of the babbling thrushes, but less rounded and longer. The beak is slender, and slightly curved down; the tarsus is short and fairly strong. The plumage is mainly bright grass-green, long and loose (particularly on the lower back). The sexes differ; males in many species have black or bright blue on the chin, throat and beard stripe and an area of malachite-green above the band of the wing. One species of these lively tree birds, the ORANGE-BELLIED LEAFBIRD *(Chlolrorpsis hardwickei),* is distinguished by a bright orange-brown lower body; another, the well-known GOLD-FRONTED LEAFBIRD *(Chloropsis aurifrons;* Color plate, pp. 184 and 190), is marked by a bright yellow forehead. The leafbirds are confined to the oriental region; Malaysia is particularly rich in leafbird species.

Fig. 8-14. Display postures of three female and two male white wagtails.

These birds are at home in the evergreen crowns of the jungle trees and hardly ever come down to the ground. Their green plumage provides effective camouflage. They stay in pairs or small troops and keep particularly to flowering trees, e.g. the species of *Salmalia* and *Erythrina*, where they take nectar which is a major part in their diet. Their beaks and tongues are very well adapted to nectar feeding. Thus, the leafbirds are included among the most important "flower birds" on which the fertilization and reproduction of many species of plants depends. In their behavior towards other birds, they are anything but friendly. They are cantankerous and chase away other birds, even from distant trees. Being imitators, they do not confine themselves to their own rough, partly musical calls, but imitate exactly the calls of many other birds as well. The phrases of imitations follow one another without interruption and can begin to confuse the listener.

The nest is a cup loosely made up of runners, moss and rootlets poorly held together with spiderwebs. The lining consists of soft grass and bask fibers. As the nest is carefully hidden high up on a tree, at the end of a projecting branch, it is difficult to find; however, the fearful behavior of the owners usually betrays its location. Usually two, sometimes three eggs are laid. Leafbirds are popular cage birds; particularly the gold-fronted species is kept by many bird fanciers. Because of the bellicosity, these amusing birds cannot, unfortunately, be kept in community cages.

The IORAS (genus *Aegithina*) are somewhat reminiscent of tits and chickadees in their appearance and behavior. The L varies from 15-18 cm; the plumage is largely yellow or olive-green on the wings and the tail is black with a little white, as in the leafbirds. The plumage is dense and loose, particularly on the lower back. The beak is straight and fairly strong; the tarsi are longer than in the fruit-suckers. These birds are tree dwellers and are mainly insectivorous. There are about four species in southern southeastern Asia, one of which, the COMMON IORA (*Aegithina tiphia*; Color plate, p. 190) has many subspecies. The sexes are superficially similar; males have a brightly colored plumage during the breeding season.

Ioras inhabit lightly wooded dry and moist habitats in which leaves fall from time to time. They not only keep to forests, but also to gardens and copses not far from settlements. They hunt in the foliage in pairs for caterpillars, moths and spiders, hopping from twig to twig and often hanging upside down or in other acrobatic postures among the leaves. Members of a pair maintain contact with soft flute-like tones and short musical phrases. In display the male chases his partner, then perches with wings lowered, fluffs up the loose lower back feathers, erects the tail and utters melodious, long-drawn hissing phrases. From time to time it jumps up one or two meters above its perch, fluffs out the glittering white back feathers and glides slowly and gracefully back to its perch in a spiral, looking like a floating ball of down as it does so. The nest is a deep, well-constructed cup covered on the outside with spider webs, sits in a fork three to four meters above the ground and contains two to four eggs.

The FAIRY BLUEBIRDS (genus *Irena*) are about the size of a starling. In the BLUE-BACKED FAIRY BLUEBIRD (*Irena puella*; Color plates, p. 184 and 190), the males are light blue and velvet-black; the females, however are greenish-blue. In the PHILIPPINE FAIRY BLUEBIRD (*Irena cyanogaster*) the sexes are more or less similar; both are blue and black.

Family: fairy bluebirds, by B. E. Smythies

These beautiful, lively fruit-eaters live in pairs or small troops and keep to the evergreen lowland forest. There, one occasionally hears their very loud contact calls and their short sharp whistles. In Sarawak (northern Borneo), the fairy bluebirds have also been known to eat insects. The nest is a fairly flat, loose cup of roots and other materials; it is lined with rootlets. The female lays two pale blue eggs with brown, gray or purple markings.

9 Shrikes and Waxwings

Shrikes and waxwings, by G. Niethammer

The groups of birds discussed in this chapter consist of at least four families, all of quite different sizes. The interrelationships between these families will require further clarification. However, for practical reasons, all four are considered here together. They are: 1. Shrikes (Laniidae); 2. Vangas (Vangidae); 3. Waxwings (Bombycillidae); and 4. Palmchats (Dulidae).

The shrikes are small to thrush-sized birds with a beak curved downward with a hook at the tip and a sharp horny "tooth" at the edge of the upper mandible on each side of the beak. The shrikes are able to crack insects with hard carapaces with the help of this "tooth". All true shrikes (subfamily Laniidae) impale their prey onto thorns or wedge it into forks of branches, thus catching it or holding the prey so that pieces can be torn off and eaten. This is an innate behavior typical of these birds. These are four sub-families, although the Pityriasinae can only be included with the shrikes with some reservations. The vangas, which only occur on Madagascar, are probably the closest relatives of the shrikes in external appearance and mode of life.

The waxwings, however, are short-legged tree birds with straight beaks which are widened at the base and hence they look more like flycatchers. The palmchats of the Antilles hold a much disputed systematic position for they differ from the waxwings in their harder, i.e. not so silky, plumage and particularly in their habit of building large communal nests.

Family: shrikes

The SHRIKE family, Laniidae, is divided into four subfamilies: 1. Helmeted Shrikes (Prionopinae); 2. Bush shrikes (Malaconotinae); 3. True shrikes (Laniinae); 4. Wood shrikes (Pityriasinae).

Subfamily: helmeted shrikes, by C. W. Benson

The HELMETED SHRIKES (subfamily Prionopinae) are confined to Africa south of the Sahara. The L varies from 20–25 cm. The feathers on the forehead are stiff and lie forward over the nostrils. There are two genera: 1. *Prionops,* including the WHITE-HELMETED SHRIKE (*Prionops plumata*; Color plate, p. 196) which has a L of 20 cm and is black and white.

This species is found in open woodland from Transvaal to Eritrea and Senegal; other species are distributed in part over large areas of Africa, while still others are found only in restricted habitats. 2. *Eurocephalus*, which includes the WHITE-CROWNED SHRIKE (*Eurocephalus anguitimens*; Color plate, p. 196) which has a L of 25 cm and is predominantly brown with striking white areas on the crown, throat and breast. This bird has no eye lappets or crest; it is found in the dry bush of southwestern Africa north to Rhodesia. There is one sub-species in eastern Africa which ranges as far as Ethiopia.

All helmeted shrikes are probably sociable and always form small troops which search the ground beneath trees for insects. They build an open-shell nest. The German name "Brillenwürger" (spectacled shrike), is derived from the fact that the eye is usually surrounded by light colored skin lobelets in the genus *Prionops*. The white helmeted shrike always occurs in groups of six to twelve birds, even while breeding. Members of the group also roost together. Even the nests are located in groups of five or six in neighboring trees. Usually four eggs are laid, sometimes two clutches laid by different females are found in the same nest. Other helmeted shrikes are less sociable. The white-crowned shrike, one of the largest of all shrikes, generally perches at the tip of one of the outermost branches of a tree, from which it makes its straight flights with short, rapid wing beats.

The BUSH SHRIKES (subfamily Malaconotinae) have many more species and are, with the exception of one species, also confined to Africa south of the Sahara. The L ranges from 15-27 cm; these birds resemble the true shrikes in the hooked tip of the beak, but they are clearly different in their mode of life. There are six genera with thirty-six species, of which we will mention only the following: 1. The CHAT SHRIKE *(Lanioturdus torquatus)* has a L of 15 cm. The beak has only a slight hook; the wings are pointed and the legs are very long. The plumage is black, gray and white; the tail is almost all white. This species occurs in the acacia forests of northern southwest Africa and southern Angola. 2. The BRUBRU *(Nilaus afer)* has a L of 15 cm. The plumage is black and white with brown spots; the bird is found in forested areas from Senegal and Eritrea to southern Africa. 3. The BLACK-BACKED PUFF-BACK *(Dryoscopus cubla;* Color plate, p. 195) has a L of 18 cm. It is also black and white, and marked like other species of this genus by notable feather formations on the rump (white in the male, gray in the female) which look like powder puffs when the feathers are erected. This bird is found in forests from western Africa to the Cape area. 4. The REDWING SHRIKE *(Tchagra tchagra,* Color plate, p. 195) has a L of 21 cm, and like the others of its genus, it is decorated with head stripes and rust color on the wings. It is found mainly in the undergrowth of forest thickets in the South African coastal country. 5. The CRIMSON-BREASTED SHRIKE *(Laniarius atrococcineus)*, has a L of 22

Fig. 9-1. Limited areas of species distribution of certain helmet shrikes (genus *Prionops*): 1. Yellow-crested helmet shrike *(Prionops alberti)*; 2. Gray-crested helmet shrike *(Prionops poliolopha)*; 3. Chestnut-fronted shrike *(Prionops scopifrons)*; 4. Angola red-billed shrike *(Prionops gabela).*

Subfamily: bush shrikes, by C. W. Benson

Fig. 9-2. Distribution of red-bellied species of the genus *Laniarius*. Crimson-breasted shrike *(Laniarius atrococcineus)* occurs separately in southwestern Africa.

Fig. 9-3. Distribution of bush shrikes of the genus *Malaconotus*.

cm. It is particularly colorful with its black upper parts and deep red lower parts. It is found on thorn bushes of the Kalahari Desert and adjacent areas. 6. The GRAY-HEADED BUSH SHRIKE *(Malaconotus blanchoti)* has a L of 27 cm. It has a very deep beak, and is found from Senegal and Eritrea to the Cape Province and northwards again to Angola.

In contrast to the true shrikes, bush shrikes are more inclined to be "stealthy" birds, which live largely in dense undergrowth and seek their food there. One never finds them on a perch in open country. The smaller species presumably live only on insects, while the larger ones, which have unusually strong beaks, also take small vertebrates. The redwing shrike is said to live mainly on berries. The nests of all these birds are cup-shaped. The nests of the redwing shrike and its relatives are very lightly built.

The two or three eggs are white or bluish with brownish spots. In some bush shrikes the sexes differ in color. All the species of this sub-family probably have distinct calls which often sound like whistles and are sometimes sung in duets. Bush shrikes are strict residents and are rather poor fliers.

Subfamily: true shrikes, by W. Münster

The characteristic feature of the TRUE SHRIKES (subfamily Laniinae) is the strong beak which resembles that of a raptor. The upper mandible has a hooked tip and behind this is a strong "tooth" on each side which fits into an equally sharp-edged indentation on the lower mandible; this is not a true tooth formation as in the primitive birds (see Volume VII), but rather a modification of the horny sheath of the beak in which a projection of the bone participates. Insects are the principal food of these birds but the larger species, particularly, will also eat various vertebrates. There are three genera: the SHRIKES PROPER *(Lanius)*; the MAGPIE SHRIKES *(Urolestes)*, and the YELLOW-BILLED SHRIKES *(Corvinella)* with a total of twenty-five species.

The decorative-looking shrikes were formerly judged in a prejudicial manner because of their supposed lust to kill. Even their German name, Würger, which means strangler has contributed to give these birds a bad reputation among uninformed people, yet shrikes are often attractively colored and sing pleasantly. Their habit of impaling insects or small vertebrates on thorns, sharp twigs or more recently even on barbed wire, is due to an innate drive which has long been incorrectly interpreted. Thus, the German name, Neuntöter, for the red-backed shrike (English),"killer of nine" originated from the belief that the bird had to kill at least nine animals before it could start to feed. Pax has reported that 141 of these shrikes were killed in a city park of 6.5 hectares between 1898 and 1900 because it was believed that the shrikes pursued small song birds and expelled them from their breeding areas.

Fig. 9-4. Looped migration route and wintering area of the red-backed shrike *(Lanius collurio)*.

As a result, shrikes have undergone a strong diminution in Central Europe. Only the RED-BACKED SHRIKE *(Lanius collurio;* Color plate, p. 195) which has a L of 18 cm, is still a widely distributed breeding bird. Its population is, however, subject to fluctuations which Peitzmeier attributes to climatic influences. The deterioration of the climate since 1950 has led to a slight decline in the number of red-backed shrikes, but to a marked decline in the case of the WOODCHAT SHRIKE (*Lanius senator*; Color plate, p. 195) which has a L of 18 cm. The clearing of hedges in many places has surely also contributed to the decrease of shrikes.

The red-backed shrike settles in sunny districts with more or less extensive bushy areas. The male arrives in the breeding area a few days before the female and at once betrays his presence by striking demonstrations of his claim to a territory. He makes straight-line flights from the tops of trees or bushes and often utters his contact call. In flight display, the male throws himself from side to side, showing his striking colors (Color plate, p. 195). He achieves the same effect when he performs all sorts of bows and contortions in front of his chosen partner. Courtship feeding is another display, whereby the female cowers trembling her wings, and utters a soft "wree wree" call, which resembles the food begging calls of the young.

The song of the red-backed shrike is rarely heard, mostly only during the first few days after its arrival from wintering areas. If one does hear this bird's song it is an unusual experience. Aside from its own twittering, this shrike combines imitations of the songs of its neighbors into a first-rate composition, which it delivers with great endurance. This imitation is characteristic of many shrike species.

Nest-building is preceded by calls to the nest and by the nibbling of thorns. Initially, the male flies again and again to the chosen bush, uttering repeated "tschoog" calls. If the female follows, both birds settle and creep about with loosely held wings. The birds use their beaks to shorten twigs and thorns which interfere with the nest site. As they often return for years to the same breeding place, the partners of the preceding year often come together again. Oskar Heinroth calls this phenomenon "Ortsehen", translated, "marriages dependent on locality." The completed nest may be found as early as a week after the female's arrival. The male generally brings nest material to the site while the female performs the actual nest construction. The nest is generally placed low in bushes or small trees. The outside consists of twigs and strong grass stalks; the middle layer consists of grass stalks, webs and moss; while the lining always consists of rootlets.

The eggs can vary greatly, but those in one nest are always similar. The background color is yellowish, greenish or reddish-brown; the spots, which vary in shape and color are joined into a sort of wreath about the blunt end. The clutch generally consists of five or six eggs which are incubated for fourteen to sixteen days by the female only. After hatching, the young remain in the nest for about fourteen days, and they are then fed and guarded outside the nest for another two or three weeks. Red-backed shrikes are very watchful parents and they vigorously chase away any intruders. At such times, one can hear their defensive call, a sequel of "tshek tshek." Strong excitement is shown by a lively tail twisting. While families of this species soon break up, young northern shrikes (*Lanius excubitor*; Color plates, pp. 195, 201 and 202) called great gray shrikes in Europe, are sometimes still with the parents even in the late fall. Fledged red-backed shrikes show the curved lines on the plumage of the upper and lower parts which are characteristic of shrikes. In many areas, the cuckoo *(Cuculus canorus)* (See Volume VIII) shows a particular preference for the red-backed shrike as a host. Around Leipzig, Germany, Schlegel found 803 eggs and young of this brood parasite, of which 613 were in red-backed shrikes' nests.

Although insects are the shrikes' main food, these birds also take other invertebrates to a lesser extent and occasionally even vertebrates, particularly mice, amphibians, lizards, blind worms (a legless lizard) and smaller birds. Only rarely does the red-backed shrike eat berries. An individual keeps watch from the top of a tree, bush, or a wire, and catches its prey in the air or on the ground. Only once was I able to observe a female red-backed shrike catching a songbird, a woodwarbler *(Phylloscopus sibilatrix)* which the shrike seized on the ground, killed with a few blows of its beak and carried off in its feet in clumsy flight. The "specialists" among shrikes which hunt birds are an exception. According to human standards their utility far exceeds their harmfulness. In addition, shrikes are valuable to us because their lively ways and the beauty of their plumage provides a pleasing enrichment to the countryside.

Red-backed shrikes work over captured prey in an unusual manner. Smaller insects are squashed in the beak and larger pieces of chitin are then picked off. The gut content or even the whole gut of caterpillars is flung out. Stinging insects are rendered harmless of their sting by being rubbed on a branch. Larger prey is impaled on thorns or other sharp objects so that they can be more readily worked over. There may also be a storage function in this, keeping food for use during periods of poor weather. In fact, in England, for example, where rainy or foggy weather is frequent, impaled prey is found more frequently than in climatically more favorable countries. The woodchat shrike, also impales prey, while the LESSER GRAY SHRIKE (*Lanius minor*; Color plate, p. 195), which has a L of 22 cm, rarely does so. One also seldom finds prey impaled by the northern shrike; instead this species wedges its prey into the forks of branches.

If breeding runs its normal course and there are no markedly late replacement broods, the red-backed shrike leaves Central Europe between the end of August and early September. The adults begin the

Fig. 9-5. Red-backed shrike. From the top down: courtship posture, alarm posture when there is a disturbance near the nest, tail turning in excitement, another courtship posture.

migration. Red-backed shrikes migrate singly and, if possible, all night long. The circular migration, in the fall over the eastern Mediterranean, in spring returning over the Red Sea, Syria and Asia Minor, is interesting. Verheyen explains it as being due to the different direction of prevalent winds in the fall and spring, for these shrikes fly mainly with the wind (i.e. a tail wind). While the flight to southern Africa takes them about three months, the return migration is made in about two months, ending between the end of April to mid-May.

Towards the east, the red-backed shrike is joined by several other forms: some are only subspecies, while others are distinct species. The ISABELLINE SHRIKE (*Lanius collurio isabellinus*; Color plate, p. 195), prefers the dry steppes of the Tarim basin. The BROWN SHRIKE (*Lanius cristatus*; Color plate, p. 195), is found breeding between Tomsk (Siberia) and northern Japan. The lesser gray woodchat shrikes are among the rarer, locally breeding birds, in Germany. The northern shrike is more widely distributed but also occurs only very sparingly. While the three smaller German species of shrikes are migrants, the thrush-sized northern shrike remains there all year in its breeding area. This species does not undergo a complete winter moult (as its relatives do); it renews its plumage in November as an adaptation to its residency. In winter it is not uncommon to see the attractive northern shrike leading a flock of green finches or killing a mouse.

Only the red-backed shrike has a sex color differentiation; the sexes in the other Central European species are very similar. There are no essential differences among these species of shrikes in breeding behavior, but both partners alternate in incubation in the lesser gray shrike and the male northern shrike also participates in incubation. In contrast to the smaller species, the northern shrike breeds very early; in western Germany, it breeds in late March, while in central Germany it breeds in the second half of April. The display of the lesser gray shrike is particularly striking. The male follows his partner for up to thirty minutes, flying half a meter lower than she does. The tendency of the light-colored lesser gray and northern shrikes to place white strips of materials on the outside of their nests may be a form of camouflage.

The colorful MASKED SHRIKE *(Lanius nubicus)*, found from southern Yugoslavia to southeastern Iran and Egypt, distinguishes itself by its melodious song. The RUFOUS-BACKED SHRIKE *(Lanius schach)* breeds locally on the southern coast of the Sea of Aral, on the eastern coast of central China and as far as southeastern New Guinea. The MAGPIE SHRIKE *(Urolestes melanoleucus)* has a L of 45 cm and belongs to a different genus. It resembles a magpie with its black and white color, its long tail, its body form and flight. It inhabits southern Africa as far as Angola and Lake Victoria. In the bush and tree steppe of the Kruger National Park, one can observe these attractive birds hunting in

Subfamily: wood shrikes, by B. E. Smythies

Fig. 9-6. Distribution of the bald-headed wood shrike (subfamily Pityriasinae).

Family: vangas, by J. Dorst

groups. The closest relative of this species is the LONG-TAILED SHRIKE (*Corvinella corvina*) which has a L of 31 cm and is found from Senegal to the Nile.

The BALD-HEADED WOOD SHRIKE (*Pityriasis gymnocephala*) which has a L of 25 cm and only occurs in Borneo, represents a separate subfamily, Pityriasinae. The beak is massive, and bent like a hook at the tip; the tail is short, and the head is partly bare, warty, and bright yellow in color. The feathers on the covered part of the head and on the neck are transformed into stiff bristles which are more reddish in color. The body color is dark gray. Females differ from males in having red spots on the flanks.

The bald-headed wood shrike is an unusual bird without close relatives and systematists are still uncertain about where to place it; possibly it is most closely related to the Australian butcher birds (Cracticidae, see Chapter 19). It lives in the upper part of forest crowns and takes beetles, orthoptera and similar small animals; it has not been known to take any plant food. Unfortunately, its nest, which perhaps might throw some light on its relationships, has never been found. This bird's plaintive, nasal, high-pitched calls, interspersed with sequences of almost crow-like sounds are most peculiar. When these birds join together in a flock, they begin a choral song consisting of trills with subsequent "weeping" sounds.

The VANGAS (family Vangidae) have a L of 12–30 cm and occur only on Madagascar and in one species on Grand Comoro. They vary so much in color and particularly in the shape of the beak that they were formerly allocated to different families. However, their striking resemblance to other groups of birds is only the result of superficial adaptation in the same direction (convergence). Some remind one of shrikes, others almost of swallows, bulbuls or even the nuthatches. Nevertheless, their skull shape and particularly the formation of the bony palate reveal that they are a single group of birds.

When one looks at the individual forms on page 196, one can hardly believe that one is looking at members of a distinct family. Compare the HOOK-BILLED VANGA (*Vanga curvirostris;* Color plate, p. 196) which has a L of 25 cm and a marked hook at the tip of a strong beak, with the HELMET BIRD (*Euryceros prevosti;* Color plate, p. 196) which has a L of 27 cm, whose big, narrow beak is dome-shaped, the *Xeno pirostris* with the POLLEN'S VANGA (*Xeno pirostris polleni;* Color plate, p. 196), which has a L of 23 cm and strong, high and laterally narrow beaks and finally with the delicate BLUE VANGAS (genus *Leptopterus*) which have a L of 15–19 cm and include the WHITE-HEADED VANGA (*Leptopterus viridis*) with a L of 19 cm, the CHABERT VANGA (*Leptopterus chabert*) with a L of 15 cm, and the BLUE VANGA (*Leptopterus madagascarinus;* Color plate, p. 196) which has a L of 17 cm.

The SICKLEBILL or FALCULEA *(Falculea palliata),* which has a L of up to

32 cm including the curve of the bill, is found in the semi-deserts of western and southwestern Madagascar; this bird looks particularly unusual as its very long, thin beak is curved like a sickle. The RED-TAILED VANGA *(Calicalicus madagascariensis)* has a L of 13 cm. This is a delicately shaped bird with a short, hook-like bill. The male has a gray back, a red rump, a black throat, and a white belly suffused with red. The female is gray above and has beige-colored underparts. The RED VANGA *(Schetba rufa)* has a L of 19 cm and is surprisingly brilliantly colored. The male has a black spot on his head and a black breast; his back is a pale red color, and the belly is white. The female has whitish-gray underparts which look washed out. The straight, conical beak of BERNIER'S VANGA *(Oriolia bernieri)* resembles that of a starling, only it has a hook at the tip. This species has a L of 21 cm; the male is black and the female is reddish with black stripes.

Today we include two additional species in the family of vangas; the systematic position of these two species has been obscure until recently. They are: 1. The GRAY KINKIMAVO *(Tylas eduardi)* which has a L of 22 cm. This bird was formerly classified with the bulbuls. It is surprisingly similar in color to the members of the genus *Xenipirostris*. 2. The MADAGASCAR NUTHATCH *(Hypositta corallirostris)* has a L of 13 cm. The beak is bright red; the male is azure-blue, while the female is blue above and blue-gray beneath. This bird lives like a nuthatch insofar as it climbs on tree trunks, although not head downwards. Its feet are well adapted to hold onto the rough bark.

Tylas and Madagascar nuthatch

The multiformity of the vangas is due to their special development in a geographic area which, because of its insular nature, was settled by only a few groups of birds. Next to the honeycreepers of the Hawaiian Islands, the vangas provide one of the best examples of radiating evolution. Although all vangas are predominantly tree dwellers which eat insects and other small fry including small vertebrates, they were able to occupy quite different habitats and ecological niches because they lacked competition. Even though we still do not know much about the biology of most vangas, much of their adaptation to the environment has been ascertained.

The blue vanga species live chiefly in the crowns of trees where they hunt insects on the twigs. The largest members of the family, particularly the hook-billed vanga and the helmet bird, fill in their diet with small vertebrates and tree frogs; they behave much like true shrikes or like small raptors. On the other hand, Pollen's vanga and Bernier's vanga sit immobile on a twig and plunge down on passing insects like flycatchers. The red-tailed vanga collects insects and their larvae from twigs in the manner of a warbler *(Sylvia)* which it biologically replaces along with other birds of its genus on Madagascar. The sicklebill and its relatives can be regarded as "representatives" of the woodpeckers, which are completely absent on Madagascar; they search tree trunks with their long sickle-shaped beaks for insects hidden in clefts.

The sociable vangas live in troops of several dozen birds; these include not only members of the same species but also companions chosen from species of other families. They breed in open shell nests and have clutches of three to four eggs which carry brown spots on a white or green background.

Family: waxwings, by H. Bub

The waxwing family (Bombycillidae) contains three separate subfamilies. These are: 1. Waxwings proper (Bombycillinae) with one genus and three species; 2. Phainopeplas (Ptilogonatinae) with three genera and four species; and 3. Hypocolius (Hypocoliinae) with one genus and one species.

The WAXWINGS PROPER (subfamily Bombycillinae) are distributed in three species over the northern parts of the Old and New Worlds. They are birds of starling-size (the L is 16-22 cm) with a decorative crest and soft brownish plumage, the striking beauty of which is a delight to any naturalist. There are bright, more or less numerous, red, horny, wax-like platelets on the secondaries of the wing. The CEDAR WAXWING (*Bombycilla cedrorum*) of America is most plainly colored; it lacks white wing coverts. The JAPANESE WAXWING (*Bombycilla japonica*), which breeds in northern Manchuria and parts of eastern Siberia, has a red tail band and carries no horny platelets on the wing.

Fig. 9-8. 1. Bohemian waxwing (*Bombycilla garrulus*); 2. Japanese waxwing (*Bombycilla japonica*); 3. Cedar waxwing (*Bombycilla cedrorum*).

The BOHEMIAN WAXWING (*Bombycilla garrulus*; Color plate, p. 196) is distributed circumpolarly almost all the way around the Northern Hemisphere in both the Old and the New Worlds. This bird has caused excitement over and over again during past centuries in Europe. Its irregular mass invasions during the six winter months in certain areas gave rise to all sorts of superstitions and gave it the name "pest bird," a name which it still carries in the Netherlands. Thus, Aitinger wrote in 1631: "Many people believe that when these birds are seen with us, it is always a particular omen and has a special meaning; people even believe that these birds bring the three principal punishments of man—war, pestilence and famine or hunger with them, though they have not been seen in many parts of the country in fourteen or more years." Later such opinions were rarely taken seriously.

During the breeding season, waxwings prefer tall, dense coniferous forests with an undergrowth of berry-bearing bushes. They also inhabit the edges of moors and river shores where there are isolated birches or poplars. The Bohemian waxwing is a characteristic bird of the Taiga (the Siberian forest area between the tundras in the north and the steppes to the south; this corresponds to the forest-zone in North America.) It took a long time for the breeding behavior of this well-known bird to become known. Several famous explorers searched for its breeding places in vain until in 1856 the Englishman, John Wolley, with the help of his Lapp collector, saw the first nest with eggs. These waxwings often breed fairly close together. Both partners share in nest building but only the female incubates. The clutch consists of three to five eggs which hatch after twelve to fifteen days. The young are fledged in fourteen to sixteen days. During the summer, insects, caught after the manner of flycatchers, form the major part of the diet of these birds, while during the rest of the year they feed mainly on berries.

The display of these birds is a real experience. Meadows, the Englishman, has given us a description of this behavior and the German ornithologist, Heinz-Sigurd Raethel has passed it on. Meadows was the first to succeed in rearing these birds in captivity in 1964. In display, the male fluffs out the feathers of the lower back (which causes an exaggerated humped appearance) and the belly plumage. He erects his tail and holds it pointing downward, at a perpendicular to his body axis. In this posture he first turns his head away from the female. If she is in the same mood, she adopts a similar but less pronounced posture. The male now passes her a "present," some small object which need not be edible. He offers it to his partner at the tip of the beak and she accepts it. The "symbolic food" is passed back and forth several times between the two, but the object, whether a berry or an ant pupa, is never swallowed. The American ornithologist, Sidney Porter, has made similar observations in cedar waxwings; however, these birds eat the berry at the end of the display sequence.

Display

The migrations of the Bohemian waxwing are a particular phenomenon. The Finnish zoologist, Siivoven, defined three main forms of movements in 1941: first, the annual winter migration which extends, for example, as far as Hungary, but does not consist of many birds and is therefore not very noticeable; second, migrations which occur at intervals of several years and which are set off by a lack of berries; and finally the large irruptions which occur at roughly ten year intervals. During these large migrations, the birds travel in masses, often very rapidly, to the south and they may even reach Algeria. In the winter of 1932/33 Warga caught and banded 1,371 waxwings within 12 days in Budapest. A bird banded in Poland was found the next winter 5,700 kilometers away in eastern Siberia. These irruptions are always preceded by an increase in the breeding population and an enlargement of the breeding range.

Migrations

The PHAINOPEPLAS (subfamily Ptilogonatinae) are confined to southern parts of the North American continent and Central America with three genera and four species. These are: 1. The LONG-TAILED SILKY FLYCATCHER *(Ptilogonys caudatus)* which is slender and gray with a pointed upright crest, and bright yellow under-tail coverts, inhabiting unplanted areas from Costa Rica to western Panama. Males have a L of 24 cm and females have a L of 20 cm. 2. The GRAY SILKY FLYCATCHER *(Ptilogonys cinereus)* has a L of 21 cm and differs from the preceding species mainly in the absence of projecting central tail feathers. It

Subfamily: phainopepla, by A. F. Skutch

Fig. 9-8. Range of the phainopepla group (subfamily Pitlogonatinae).

occurs in the mountains of Guatemala and Mexico. 3. The BLACK-AND-YELLOW SILKY FLYCATCHER *(Phainoptila melanoxantha)* has a L of about 21 cm. The male looks thrush-like and is largely black with bright yellow on the lower back and sides of the body. The female is predominantly olive-green, and has yellow flanks. This species occurs in the mountains of Costa Rica and western Panama. 4. The PHAINOPEPLA *(Phainopepla nitens;* Color plate, p. 196) has a L of 18 cm. The male has a tall crest, and is glossy bluish-black with a striking white area on the open wing. The female is also crested but is olive-gray in color. This species is found in semi-deserts from central Mexico to the southwestern United States.

These lively birds seek out exposed perches on branches at some height from which they catch insects in playful-looking swoops. They also eat many small fruits. The phainopepla has a melodious song; the gray and the long-tailed silky flycatchers, though they are noisy and chatty, sing only in whispers. The nest, an open cup in a tree or in bushes, is built mainly by the male, but in the long-tailed silky flycatcher both sexes share the work almost equally. The eggs, only two in the long-tailed silky flycatcher, two to four in the phainopepla, are gray to greenish-white, and marked in various ways with brown, purple and black tones. The male phainopepla takes over a large part of the incubation during the day, but at night only the female is on the eggs; she is fed by her partner, however. The young remain in the nest for eighteen to twenty-five days, depending on the species.

Subfamily: hypocoliinae, by W. Meise

A southwestern Asiatic waxwing, the GRAY HYPOCOLIUS *(Hypocolius ampelinus)* with a L of 27 cm, which occurs from the Red Sea to the Caspian and Sind Seas (although only in small breeding areas), is regarded as the only representative of the subfamily Hypocoliinae. This bird is gray with a black and white pattern. The beak is wide at the base, and the legs are flesh-colored. The female has much less black. Nests have only been found in Traque to date.

Factors in favor of the gray hypocolius' relationship with the waxwings are its sociable life outside the breeding season, its feeding on fruit of the nightshade and mulberry, figs and dates (less on insects), and the straight line flight which takes it directly into the foliage where its cup-shaped nest is located. "Both adults," writes S. Marchant, "build and bring nest materials from some distance directly to the nest sites, usually flying high up together and calling." These birds desert their nests, even after only a slight disturbance. Marchant assumes that the nests are destroyed by their builders and then, according to his observations, built eight days later, fifty to one hundred meters away. A strange influx of almost a hundred paired birds in the Soviet Union east of the Caspian in May, 1960, is reminiscent of the mass irruptions of waxwings described above; this influx did not lead to a definite colonization, however.

Fig. 9-9. Gray hypocolius.

The PALMCHATS (family Dulidae) are distantly related to the waxwings. There is only one species, the PALMCHAT (*Dulus dominicus;* Color plate, p. 196) which has a L of 17.5 cm. It is found on Hispaniola and Gonave (Greater Antilles). It is distributed over all of Haiti and the Dominican Republic except for the higher mountainous areas. The upper side of the body plumage is olive-green; the head is darker at the side. The rump is greenish, and the wings and tail have greenish edges; the lower parts are pale yellowish-white with dense dark stripes. The sexes have an external resemblance.

In French, this bird is called "oiseau palmiste;" in Spanish, "sigua de palma." It occurs in pairs or in groups of two to five pairs which together inhabit a large, unskillfully built nest of thin dry twigs. Such a structure has a diameter of a meter or more. All nest materials are brought to the site in the beak not, as has been reported on occasions, with the feet. Each pair has its own compartment in this communal nest; the eggs lie on a thin layer of grass and shreds of bark. They are white with a slight gloss and thickly covered with dark gray to blackish-gray spots, particularly about the wider pole. In the lowlands, nests are generally in Royal Palms; in the mountains coniferous trees are occasionally used, but only single or double nests are built there.

Outside the breeding season, the group uses the communal nest as a roosting and resting place. When the supporting palm fronds tear or fall off, the structure is lost and must be replaced by a new one. Palmchats, as indicated by their breeding and sleeping habits, are very sociable. Males and females often rest side by side. They sometimes utter their calls in chorus, but they do not have a true song. They live on flowers which they swallow whole or in pieces, as well as palm berries and other kinds of berries. The rough dense plumage distinguishes the palmchat from the waxwings, as does its somewhat different form, and, above all, its pronounced social life.

Family: palmchats, by A. Wetmore

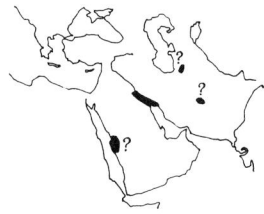

Fig. 9-10. Gray hypocolius *(Hypocolius ampelinus.* The only nests found so far have been in Iraq. This bird is found from Mosul to the northern coast of the gulf. The breeding area probably extends to southern Afghanistan and western Arabia. The arrival in Bagdad in mid-April indicates migration, but no notable occurrences in southern Arabia, Pakistan or India have been determined.

10 Wrens and Related Forms and Accentors

Superfamily: primitive insectivores

The families dealt with in this and the next two chapters are combined as primitive insectivores or flycatcher-like birds in the superfamily Muscicapoidea. We will begin our discussion of these birds with the DIPPERS (Cinclidae), the WRENS (Troglodytidae), the MOCKINGBIRDS and THRASHERS (Mimidae), and the ACCENTORS (Prunellidae), the first three of which at least are closely related and possibly of common origin.

Family: dippers, by O. Jost

Of all the passerine birds, the dippers (family Cinclidae) are the only true inhabitants of water. The body size is that of a starling with a L of 14-19 cm. The body itself is compact, the wings are short and strong, and the tail is short. The male is a little larger than the female, although otherwise the sexes are very similar. Dippers show close ties to specific habitats, streams and rivers with swift flowing water, stony beds with rocks and shore line shrubbery. During the breeding season, pairs inhabit segments of water courses several hundred meters long and defend these against others of the same species. There is one genus, *Cinclus*, with five species distributed in the mountainous areas of tropical America and North America and in the middle range and high mountains of Eurasia; the species differ mainly in their color patterns.

The EURASIAN DIPPER (*Cinclus cinclus*; Color plate, p. 190) inhabits Europe, northern Africa and Asia. The Central European subspecies shows a rusty-brown band between the white breast and the darker belly; this band is absent in the North European subspecies. The subspecies of southeastern Europe and Italy has a lighter, reddish-brown color on the side of the belly, and the upper parts are paler. The subspecies from Central Asia has a completely white belly. The BROWN DIPPER *(Cinclus pallasii)* from Central and Eastern Asia, is uniformly brown. It may occur together with the Eurasian dipper on the rivers of the Hindukush. The gray AMERICAN DIPPER *(Cinclus mexicanus)* inhabits the mountainous country of Western, Northern and Central America. Of the South American species, we will mention primarily the

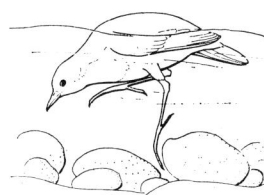

Fig. 10-1. Dipper walking on a stream bed.

WHITE-CAPPED DIPPER *(Cinclus leucocephalus)* from the mountains of Venezuela to Bolivia.

Some ornithologists assume that this very old family originated from thrush-like ancestors in Central Asia, which settled America when there was still a land bridge between northeastern Asia and Alaska. Others believe there was a development from wren-like ancestors in the American Cordillera and that the path of distribution led in the opposite direction, to the west, over the land bridge to Asia and Europe.

The naturalist always finds it exciting to observe the lively Eurasian dipper in its habitat. In swift whirring flight it comes along just above the water and lands on a rock amidst the foaming stream. Then it bobs. Swift movements of the legs at the ankle joint lower and raise the body. The eye seems to light up when it blinks. The upper eyelid is moved swiftly over the eye revealing its white outer surface. The dipper's song may be heard even in winter when it is cold and there is snow. It is an alternately soft and loud twittering of trills and sequences of whistles and screeches. It sounds much like the splashing and rushing of a mountain stream, hence one of the German names for this bird is in translation, "water chatterer."

The Eurasian dipper feeds mainly on water insects and their larvae, also on sand hoppers which it knows how to catch in flight, by walking, swimming or even by diving. The very thick plumage is waterproofed with a secretion from the preen gland. There are no webs on the feet; in swimming and diving, dippers use their strong legs and toes to push off and to walk on the bed of the stream. The dipper can also hunt under water, moving quickly by rowing skillfully with its short wings and legs drawn up. Since they dive most readily in fast-flowing water, it is assumed that the water pushes the body, which is held at a slight slant to the current, downward. It is believed that the dipper can walk about the stony stream bed in this posture and seek its food, diving for up to thirty seconds (See Fig. 10-1). Indigestible hard parts of the diet are eliminated as pellets. Numerous studies of its food have shown that the dipper does no harm to fishery interests, as was long assumed; only occasionally does it take small fish of no economic importance.

These otherwise unsociable birds begin to build their nests early in spring. The nests are usually built right beside fast-flowing deep water, in holes in walls, below bridges, on tree trunks, below overhangs of the shore or even on rocks, with moist moss and other plant materials. Characteristic nest sites are behind waterfalls, which the adults can only reach by flying through the down-rushing water. The young hatch from the four to six white eggs after sixteen days of incubation. At the approach of danger, they jump into the water and they prefer to swim and dive rather than fly. Sometimes there is a second brood. In Central Europe, this species is a resident; the young wander

▷
Mockingbirds, thrashers, and accentors: 1. Brown thrasher *(Toxostoma rufum)*; 2. Mockingbird *(Mimus polyglottos)*; 3. Galapagos mockingbird *(Nesomimus parvulus)*; 4. Catbird *(Dumetella carolinensis)*; 5. Black-capped mocking thrush *(Donacobius atricapillus)*; 6. Blue mockingbird *(Melanotis caerulescens)*; 7. Woodthrush *(Hylocichla mustelina)*; 8. Orange-billed nightingale thrush *(Catharus aurantiirostris)*; 9. Eastern bluebird *(Sialia sialis)*; 10. Orange-headed ground thrush *(Zoothera citrina)*; 11. Hedge sparrow *(Prunella modularis)*; 12. Hermit thrush *(Hylocichla guttata)*; 13. Alpine accentor *(Prunella collaris)*.

▷▷
Bubbling thrushes:
1. Sooty-headed babbler *(Malacopteron affine)*; 2. Bearded tit *(Panurus biamicus)*; 3. Red-capped babbler *(Timalia pileata)*; 4. Silver-eared mesia *(Leiothrix argentauris)*; 5. Striped jungle babbler *(Pellorneum ruficeps)*; 6. Hume's tree babbler *(Stachyris rufifrons)*; 7. Nepal babbler *(Alcippe nipalensis)*; 8. White-crested laughing thrush *(Garrulax leucolophus)*.

(Continue on p. 223)

up or down-stream and into other rivers. Northern European dippers migrate from Scandinavia over the sea to Britain and Germany.

Floods, weasels, rats, cats, jays and many other enemies regularly destroy many broods. Increasing pollution, unnatural shores on flowing waters, drainage and destruction of the shore-line vegetation leads to a diminution of the animal life of streams and causes dippers to move away. However, the lack of safe nest sites on many suitable waters is the main reason of the decline of this species. Nest boxes have been found to be very effective in the preservation of dippers; they should be placed below bridges or be provided with a sloping roof and fitted on walls.

Wrens (family Troglodytidae) have a L of 9.5-22.2 cm and are chestnut to gray-brown or blackish-brown, often with whitish, pale yellow, dark brown or black bands, stripes or spots, particularly on the wings and tail. The lower parts are pale. The sexes look similar. The wings are short and rounded, an adaptation to the thickets; the tail may be short. The beak is fairly long, slim and slightly curved; the legs and feet are strong with long claws which facilitate clutching twigs, tree trunks, marsh plants or rocks.

Family: wrens, by E. A. Armstrong

When we look at the WINTER WREN (*Troglodytes troglodytes;* Color plate, p. 190), the common wren of Europe, with a L of 10 cm, we may wonder at the fact that this species is the only representative of its family in the Old World. The rest of the twelve to fourteen genera, with sixty-two species, inhabit the New World and species are particularly numerous in the American tropics. Apparently this wren came across the Bering Strait during the Glacial Period into Asia and then spread to Europe and northwestern Africa. It has also settled islands in the northern Pacific and Atlantic Oceans as well as in the Mediterranean; on the mainland it is particularly at home in forest and mountainous districts.

◁
9. Pekin robin *(Leiothrix lutea);* 10. Southern chowchilla *(Orthonyx temmincki);* 11. Gray-necked rockfowl *(Picathartes oreas);* 12. White-necked rockfowl *(Picathartes gymnocephalus);* 13. Chestnut quail thrush *(Cinclosoma castanotum);* 14. Rail babbler *(Eupetes macrocercus);* 15. Rusty-cheeked scimitar babbler *(Pomatorhinus erythrogenys);* 16. Gray-crowned babbler *(Pomatostomus temporalis);* 17. Ceylon jungle babbler *(Turdoides sommervillei rufescens).*

These restless birds busily seek their food, which consists mainly of invertebrates, particularly insects and spiders, and occasionally of molluscs. The smaller species can reach into cracks and clefts and seize prey there which is inaccessible to other birds. The Icelandic subspecies of the winter wren, *Troglodytes troglodytes islandicus,* enters houses and picks up bits of dry meat. CACTUS WRENS (*Campylorhynchus brunneicapillus,* Color plate, p. 190) with a L of 22 cm, have been observed performing acrobatics over stones and mud crusts. The LONG-BILLED MARSH WRENS (*Cistothorus palustris;* Color plate, p. 190) sometimes sucks the eggs of other birds nesting in its territory, while the HOUSE WREN (*Troglodytes aedon),* in competition for nest sites, may destroy the eggs of its own or of other species. Some wrens, like the CAROLINA WREN (*Thryothorus ludovicianus;* Color plate, p. 190) with its L of 13 cm, and the winter wren, eat a certain amount of supplementary plant food.

The unusual urge to sing and the tendency to roost collectively is characteristic of many species of wrens. Though most members of this family live and breed in low plant growth, some species have adapted themselves to desert and rocky country, among them the ROCK WREN (*Salpinctes obsoletus*) which has a L of 14 cm, and the CANYON WREN (*Salpinctes mexicanus*; Color plate, p. 190) which has a L of 13.5 cm. The cactus wren occurs in scrub and cactus growth. Other species inhabit marshes of northern areas; many settle in moist forest. The BAND-BACKED WREN (*Campylorhynchus zonatus*) differs from the cactus wren and most other members of its genus in its preference for moist areas. It is distributed from southeastern Mexico to northwestern Ecuador and occurs from sea level to almost 3,300 meters. Southern subspecies of the house wren vary almost as much in their altitudinal distribution. In Europe, the winter wren may nest in areas hardly one meter above sea level; in Tibet and Kashmir, it breeds at 4,000–4,200 meters. Though the winter wren in Europe is generally an inhabitant of gardens, parks and woodland, it has adapted itself to the storm beaten cliffs of St. Kilda off the northwestern coast of Scotland at an altitude of about 400 meters and to the heather-grown moors of the Shetland Islands.

Wrens can only be seen with difficulty in their habitat; most species indicate their presence by frequent singing and calling. In many, the warning call is an alarm clock-like purring. Components of the highly developed song range from fast, sometimes coarse but often very musical tone sequences to slow phrases of pure whistles. Some species are among the most accomplished avian songsters. Henry Walter Bates (1825–1892), an explorer of South America and contemporary of Darwin, wrote the following about the MUSICIAN WREN (*Cyphorhinus aradus*) which has a L of 12 cm: "When its unique tones first strike a listener's ear, he can only suspect a human voice as their cause. Some musical fellow must be searching for fruit in the thickets and be singing a few tones to cheer himself up. Then the sounds become softer and more plaintive; they sound as if produced by a small flute and although it makes no sense, one is still convinced for a moment that someone must be playing an instrument....The musician wren is the only songster which makes an impression on the natives of Amazonia; sometimes they rest their paddles and interrupt the journey in their small canoes on the shady tributaries, as if deeply struck by the puzzling sounds."

The black-capped wren

According to the reports of Alexander Skutch, the calls and the song of the BLACK-CAPPED WREN (*Thryothorus nigricapillus*) are easily heard even in noisy surroundings. "That its voice can successfully compete with a rushing mountain stream is proved by the fact that I have often heard them singing on the shore from over sixty meters away, and their voice overcame the splashing sounds of the falling water. Apart from these songs which are to be heard throughout the year, it also has softer and purer bell-like calls and similar pure tones that are made in flight."

Mated pairs keep in touch while seeking food in the dense foliage with contact calls. The song, too, has a contact function, for males and females sing in alternation, and sometimes the phrase of one partner follows so exactly that of the other that a listener may only accidentally discover that actually two birds are involved. This duet singing is characteristic of those species in which the males and females live close together and maintain the pair bond from one brood to the next and from year to year. High development of the song in wrens is connected either with their strict and long-lasting occupation of a territory or in some non-tropical species with polygamy. For example, some populations of certain subspecies of the winter wren, the house wren and the long-billed marsh wren, sing as long as they are capable of reproduction. In suitable areas, the winter wren occupies territories throughout the year and consequently also sings throughout the year.

A number of species, among them the winter wren, the house wren and the long-billed marsh wren, change the song when they wish to intimidate a rival, court a partner or wish to invite it to the nest. As the male winter wren approaches the nest, he becomes even more lively; he utters song fragments and moves jerkily from twig to twig. Then he stands on a fern frond by the nest with outspread tail and trembling, lowered wings and sings softly and invitingly. Meanwhile, the female comes closer, with small hops, apparently "unconcerned." Her suitor flies to a branch, trembles the half-open wings, chatters softly, bows far forward and observes the reappearance of the female. Then he flies back to the perch by the nest, sings soft, short phrases and spreads out his trembling wings like a mantle. Suddenly, he shoots down to the nest and slips into it. In this way, he induces the female to look inside the nest. If the female being courted accepts the nest, she begins to line it with feathers; then the male has time to invite another female to accept one of its many other nests. Long-billed marsh wrens, when showing their nests, behave in a similar way.

Polygamy The long-maintained pairs of many tropical and sub-tropical species, and even of some subspecies in colder areas (as, for example, those of the Carolina wren) are in contrast to the complete or partial polygamy of the northern species, described above. The extent of polygamy varies according to conditions, particularly the abundance of food from place to place. As these wrens have an urge to build several nests and as the male usually completes the raw nest before the female arrives, the latter have a choice among future partners. This is, therefore, a sexual selection which works against unsuccessful males, for the females often reject poorly built or poorly situated nests. Naturally, in such polygamy, the female has the main responsibility in caring for the young. Although the male in such multiple matings generally concerns himself but little with the young, the winter wren is a notable exception. Here the male helps to feed the young of one of his partners on a signal from her; he even looks after the young if the female dies.

Many species of wrens use the nest not only for breeding, but for sleeping as well. Males or females may roost in the nest before egg-laying. The fully fledged young are often led into the breeding nest or some other nest of the male to sleep, and the female sometimes spends the night with them. In some tropical wrens, both the family bonds and pair bonds are very strong; parents and young then roost together for weeks. Such close and long-lasting sociability often leads to a youngster of an early brood helping its parents to feed the young of a later brood. Some wrens build sleeping nests which may be more loosely built and placed higher than breeding nests. Bent has written the following report of his observations in the United States: "The nest of the cactus wren is more than a mere gathering place for eggs and young; it serves throughout the year as a real home, for protection against cold and rain and as a refuge from enemies at night. When necessary, it is repaired or rebuilt and every young bird as soon as it is capable of reproduction, builds its own home and thus a sleeping place for the coming winter."

In the polygamous species, the male usually builds either the entire or nearly all of the nest; however, in the monogamous wrens, both partners as a rule work together. The lining is inserted by the female. Exceptionally, female winter wrens may also build nests. Some species, the MOUNTAIN WREN *(Troglodytes solstitialis)* for example, erect an open cup nest. The house wren brings a layer of plant material into a cavity while the rock wren walls up the entrance to the rock cleft in which it breeds with small stones.

In cold weather, winter wrens in Europe sleep communally in a nest or some other cavity; up to sixty-four birds have been counted in such a sleeping place. Communal sleeping developed in a warm climate and initially served as protection against enemies, but it has probably enabled the band-backed wren to survive in high altitudes and helped the winter wren to adapt to the northern winter.

The eggs of all wrens are white or almost white, and in many species they have some brown or purple spots. Tropical species lay two to five eggs; in higher latitudes the clutch size increases, according to a rule known to ornithologists for some time. Thus, the winter wren in Britain lays five to six eggs, but in Iceland and on the Faroes it lays seven to eight eggs. Central American wrens incubate for fifteen to twenty days and feed the young in the nest for sixteen to eighteen days. In harsher northern areas, wrens can only rear one brood per season, but in more favorable areas, two broods are the rule. In warmer zones there may be three broods. This is facilitated in the Carolina wren in the south of its range as the female leaves the care of the young to the male and meanwhile lays and incubates in one of the other nests built by the male. Mammals are the principal enemies of wren broods, and birds to a lesser extent; some species are much persecuted by snakes.

Breeding nests and sleeping nests

Family:
mockingbirds and
thrashers,
by E. O. Höhn

Fig. 10-2. Breeding area of the mockingbirds and thrashers (family Mimidae).

Fig. 10-3. California thrasher.

Most wrens do not migrate, although species of northern continents may seek out warmer regions. Thus, some European populations of the winter wren wander to southwestern Europe for the winter. During hard winters, this species suffers great losses in Europe, but after one or two mild winters the former population density is usually restored. In general, wrens are interesting and attractive birds, volatile in their behavior and adapted to their particular environment in many ways. The winter wren in the Old World has long attracted human attention. The tale of a contest between the wren and the eagle as to which of them could fly higher, attests to this. The wren won, having hidden in the eagle's plumage, then flying higher still when the eagle thought he was high enough and had won.

MOCKINGBIRDS AND THRASHERS (family Mimidae) though reminiscent of the slim, long-tailed thrushes, probably share a common ancestry with the wrens, the larger American species of which they resemble. The L varies from 20–30 cm; the beak is narrow and strong, straight or curved and, with the exception of one species, there are bristles at its base. The tarsi show horny plates in front (sometimes only indistinctly); the middle toe is joined to the outer toe for about the length of the first segment. The color is generally an inconspicuous gray to reddish-brown; lower parts are pale to white, and in some species they are spotted, while in others they are uniformly dark. The tail is long, rounded and graduated, and is often moved in a lively manner. There is very little sex differentiation, or there may be none at all. Juveniles are spotted above. These birds are found only in the Americas from southern Canada to southern Chile and Argentina; those which breed in Canada or in the northern and middle states of the United States are migrants.

There are thirteen genera with thirty-one species, of which we will mention the following: 1. MOCKINGBIRDS *(Mimus)* including the MOCKINGBIRD *(Mimus polyglottos;* Color plate, p. 221) has a L of 24 cm; 2. GALAPAGOS MOCKINGBIRD *(Nesomimus parvulus;* Color plate, p. 221) with a L of 21 cm, has nine subspecies on the Galapagos Islands; these are probably on their way to evolving into species: 3. THRASHERS *(Toxostoma)* including the CALIFORNIA THRASHER *(Toxostoma redivivum);* 4. SAGE THRASHER *(Oreoscoptes montanus);* 5. CATBIRD *(Dumetella carolinensis;* Color plate, p. 221) has a L of 22 cm; 6. The BLACK-CAPPED MOCKINGTHRUSH *(Donacobius atricapillus;* Color plate, p. 221) has a L of 22 cm and is brown with brownish-yellow lower parts and a black head; the tail is particularly long. This bird is distinguished by a yellow spot of thickened skin between the feather tracts on the side of the neck. The black-capped mocking thrush occurs from Panama to northern Argentina; it sings loudly with clear whistles and also utters a whirring chatter reminiscent of the wrens.

These birds feed on invertebrates, mainly arthropods, which they take almost exclusively from the ground, as well as on fruits and berries. The two to five eggs are thrush-like with a beige, green or blue background color; they may be unspotted or covered, from lightly to densely, with darker spots. The nests, which are deep cups, are in bushes, in dense tree growth, or, in the case of the thrashers, also in cacti; as far as known, they (unlike the thrush nests) are never reinforced with clay. Both parents feed the young. These birds live either singly, or, during the breeding season, in pairs; they never gather into flocks.

Many mockingbirds and thrashers sing outside the breeding season as well; even females do. They often sing at night, particularly by moonlight. MOCKINGBIRDS (genus *Mimus*) not only sing from a high perch, but also when they fly up from it briefly. The sage thrasher even sings most of its songs on the wing. The song is long and very variable in the mockingbird and in the catbird; song imitation is characteristic of these two genera. Mockingbirds and catbirds not only imitate the songs and calls of other birds, but the croaking of frogs, the chirping of crickets, the screeching of wheels and many other sounds.

The greatest number of species and individuals are found in the semi-deserts of the southwestern United States and northern Mexico with their scant vegetation of thorn bushes, cacti and sage. Here, there are eleven species of thrashers alone. The California thrasher is particularly notable among them, as the partners of this species remain paired for the whole year and they are able to breed from November to July although they generally do so in February and March.

The MOCKINGBIRD *(Mimus polyglottos),* the best known species, is one of the most varied songsters. It inhabits the width of North America in two subspecies, from northern California to New York State in the north and southwards as far as Mexico. The mockingbird has also occasionally bred in Canada. Mockingbirds strongly defend not only their breeding territories, but also their winter territories as well; they are not afraid to attack domestic dogs, snakes and people. In the winter, females often have their own territories and they then sing to mark and defend these areas. On the borders of territories, there are often threat displays in which two birds face one another with raised head and tail and then hop up and down as if dancing.

Both partners build the nest in only about two days. It is placed in bushes or in dense trees from one to two and one-half meters above the ground. The female incubates her three to six (usually four to five) eggs for nine to twelve days. The adults may raise up to three broods a year.

Most of the long, melodious song consists of phrases which are repeated four to five times; it generally contains imitations of other bird songs. Chapman recorded imitations of no less than thirty-two other bird species in a ten minute song of a single mockingbird. On the other hand, Miller in 1938, found that only 10% of a mockingbird song consisted of imitations. When some European nightingales were kept in a flight cage in Florida, the native mockingbirds soon imitated the nightingales' song, and they did this so often and for so long, that the latter gave up singing altogether (possibly because of their moult, but the "mockers" continued their nightingale imitations for some time). In contrast to its melodious song, the calls of the mockingbird are rough, consisting of "tshur, tshook, tshik" and a long, drawn "whee."

The CATBIRD *(Dumetella carolinensis)* breeds from southern Canada to the southern United States; it winters on the Atlantic coast of North America from New York State southwards to Texas. It derives its name from its long-drawn warning call, which resembles a cat's screech and which is also often woven into its otherwise pleasant song. It too is a great imitator, and it defends its territory, for example against squirrels and orioles (Icterids, see Chapter 15). Males display with lowered wings and raised tail showing up the rust-brown lower tail coverts. Both partners build the nest in five to six days; it is placed in bushes. The four glossy eggs are bluish-green, only rarely spotted, and are incubated by the female for twelve to thirteen days. A pair usually raises two broods per season.

In one case, a catbird pair was observed to stay together not only for the two broods of the same year, but also for the first brood of the subsequent year. Another ornithologist was able to show that a particular male was mated to six different females over a five year period and that in twelve pairs under observation there was a change of partners between the first and second brood. Banding results have shown that catbirds reach an average age of two and a half years.

Family: accentors, by H. Schrifter

The ACCENTORS (family Prunellidae) with their delicate but hard beaks which are somewhat strengthened as in finches, look somewhat like finches. The L varies from 12–18 cm; the weight is 18–40 g. The plumage is inconspicuous, and predominantly brown. Males and females are similar. The nests are deep cups carefully constructed, often strongly formed with moss and are usually built at a slight height above the ground (up to a maximum of two meters), or even on the ground. There are three to six bluish-green eggs per clutch and generally there are two broods a year. Both partners share in the care of the young. There is only one genus *(Prunella)* with twelve species distributed over Europe, northern Africa and the non-tropical areas of Asia.

As accentors keep largely to the ground or in cover, they are not often seen in the wild. The HEDGE SPARROW or DUNNOCK *(Prunella modularis;* Color plate, p. 221) has a L of 15 cm and weighs 19 g. It prefers woods with plenty of undergrowth, hedges and shrubbery at forest edges. In many areas it has become a characteristic, but little-known

Fig. 10-4. Distribution of the accentors (family Prunellidae).

garden bird, because of its secretive ways. In Central Europe, it wanders as a migrant to southern France and Spain, although in more southern areas it is a resident. Its song, which it likes to deliver from the tips of conifers of medium height, is a short bright twittering. Breeding takes place from the end of April to the end of July. The incubation period is twelve to fourteen days, and the young are fully fledged at thirteen to fourteen days of age.

The ALPINE ACCENTOR (*Prunella collaris*; Color plate, p. 221) has a L of 18 cm and weights 40 g. It is a mountain bird and inhabits high mountain meadows above the tree line, but also rocky slopes. It builds its nest of moss, grass stalks and roots in holes and clefts. Clutches of four to five eggs may be found from late May to July. Its song resembles that of the skylark. This bird is a resident or short distance migrant which withdraws to the valleys in winter. The SIBERIAN ACCENTOR *(Prunella montanella)* has a L of 15 cm and is found from the Urals to the Bering Sea; it winters particularly in China. The RUFOUS-BREASTED ACCENTOR *(Prunella strophiata)* on the other hand, breeds in the Himalayas at altitudes of 2,500 to 4,500 m, only wanders down into the valleys in the winter.

▷
Sylviidae warblers:
1. Aquatic warbler *(Acrocephalus paludicola)*; 2. Great reed warbler *(Acrocephalus arundinaceus)*; 3. Sedge warbler *(Acrocephalus schoenobaenus)*; 4. Fantailed warbler *(Cisticola juncidis)*; 5. Cetti's warbler *(Cettia cetti)*; 6. Palla's grasshopper warbler *(Locustella certhiola)*; 7. Moustached warbler *(Acrocephalus melanopogon)*; 8. Savis' warbler *(Locustella luscinioides)*; 9. Grasshopper warbler *(Locustella naevia)*.

11 Babbling Thrushes or Babblers and Old World Warblers

Family:
flycatcher-like birds,
by J. Steinbacker

◁
Sylvidae warblers:
1. Blackcap *(Sylvia atricapilla)*; 2. Barred warbler *(Sylvia nisoria)*; 3. Garden warbler *(Sylvia borin)*; 4. White-throat *(Sylvia communis)*; 5. Lesser white-throat *(Sylvia curruca)*; 6. Reed warbler *(Acrocephalus scirpaceus)*; 7. Icterine warbler *(Hippolais icterina)*; 8. Marsh warbler *(Acrocephalus palustris)*.

Subfamily:
babbling thrushes,
by B. E. Smythies

The many groups of birds to be described in this and the following chapter are presently united in the large family of OLD WORLD FLYCATCHER-LIKE BIRDS, Muscicapidae, which includes not only the Flycatchers, but the Thrushes, the Old World Warblers, the Babbling Thrushes, and a number of other forms like, for example, the Kinglets, which up to a decade ago were considered related to the tits and chickadees. In this, the first chapter devoted to this giant family, we will discuss the closely related babbling thrushes and warblers, as well as some subfamilies which stand close to them. These are: 1. BABBLING THRUSHES OR BABBLERS (Timaliinae); 2. PARROTBILLS (Panurinae); 3. AUSTRALIAN QUAIL THRUSHES (Cinclosomatinae); 4. OLD WORLD WARBLERS (Sylviinae); 5. AUSTRALIAN WRENS (Malurinae) 6. KINGLETS (Regulinae); 7. HYLIAS (Hyliinae); 8. GNAT CATCHERS (Polioptilinae). These subfamilies are considered to be the most primitive songbirds, those least provided with special features, and with a few exceptions, they inhabit the Old World.

Of all passerine birds, the babbling thrushes (subfamily Timaliinae) show the greatest differences in appearance and shape. Many resemble thrushes, wrens, chickadees, and Old World warblers or flycatchers; perhaps they share a common origin with these birds. The great multiformity and difficulty of defining subgroups among the babbling thrushes in particular was once humorously expressed by the well known ornithologist, Ernst Hartert: "What can't be classified is regarded as a babbling thrush."

Babbling thrushes are small to fully thrush-sized birds, with a L of 9–42 cm. The wings have ten primaries, the three outermost ones being shorter than those following. The tail has twelve rectrices; the plumage is loose. The front of the tarsus has scales running across it. The legs and feet are strong. These birds exhibit a great jumping ability, but they have little flight capacity. The beak is generally strong and curved. Babbling thrushes are residents or partial migrants, and they live mainly on the ground or in thick undergrowth. In contrast to most other songbirds, they scratch their heads by bringing the foot up over the front of the wing. There are about 245 species.

The range of babbling thrushes includes most of Africa (only in oasis in the Sahara), as well as Madagascar, the Arabian coast, southern Asia, (where by far the greatest number of species live), China, the Philippines and Australia. In the New World, there is only the wren tit on the western coast of North America. There is also a fossil babbling thrush from the Middle Pleistocene, about one million years ago.

Fig. 11-1. Distribution of the babblers (subfamily Timaliinae).

In 1964, Jean Delacour gave a general characterization of the babbling thrushes: "They move restlessly among twigs and on the ground; they hop about and dig among fallen leaves. Usually they live in the undergrowth, sometimes on the ground among dense plant growth, fallen branches, climbers and evergreen trees, where they can be observed searching for berries and insects. While doing so, they move busily, flutter the wings a great deal, wag their tails and utter noisy calls. As a rule, they are loud and varied vocally, hence the name babbler, for they are virtually never quiet. Some sing very well and their full melodies ring out far. Outside the breeding season, they move about in small troops. Often they join with other birds into the mixed flocks characteristic of tropical forests, all seeking food together."

In contrast to the flycatcher and thrush-like birds to be described in the next chapter, young babbling thrushes never have a spotted first plumage. Many species have what Delacour calls a "nobly beautiful" pattern, sometimes with scarlet, yellow or green hues, but only rarely with metallic colors. However, there are also more or less uniformly brown-colored species. In Burma and Borneo, babbling thrushes occur everywhere in the undergrowth of the tropical forests; most of the peculiar twittering calls one hears can be attributed to one of them with certainty. Nevertheless, they are difficult to observe as they are quite shy and live secretively.

We can distinguish at least six groups of genera in this great subfamily: 1. JUNGLE BABBLERS (Pellorneini); 2. SCIMITAR AND WREN BABBLERS (Pomatorhinini); 3. TIT BABBLERS (Timaliini); 4. WREN TITS (Chamaeini); 5. SONG BABBLERS (Turdoidini); 6. ROCKFOWL (Picathartini).

The jungle babblers (genera group Pellorneini) live in Africa and southern Asia; they are particularly numerous in Burma and Malaysia. They keep to the forest undergrowth near the ground. Both sexes are similar, generally being simply brown with paler lower parts, and sometimes chestnut-brown with a gray or black pattern. These birds are predominantly insectivorous. They build spherical nests and lay spotted eggs. There are five genera: 1. Birds of the genus *Pellorneum* are uniformly brown above and white below; they are sometimes spotted. The legs are long, and the tail is short. One of the best known

Jungle babblers

species is the STRIPED JUNGLE BABBLER (*Pellorneum ruficeps*; Color plate, p. 222) with a L of 16.5 cm. 2. Birds in the genus *Trichastoma* are similar to the first genus described, but have a thicker beak. Included in this genus are ABBOTT'S BABBLER *(Trichastoma abbotti)*, BLYTH'S JUNGLE BABBLER *(Trichastoma rostratum)* and the BLACK-BROWED JUNGLE BABBLER *(Trichastoma perspicillatum)*, which is very rare, and is known only from a specimen collected in Sumatra and one in Borneo. 3. The THRUSH BABBLER *(Ptyriticus turdinus)* has a L of 20 cm and is the sole species of its African genus. 4. *Leonardina woodi* is a rare thrush-like bird of the Phillipines. 5. Birds of the genus *Malacopteron* have longer wings and tails. They live higher above the ground in the lower story of tree trunks and the lower branches of great trees; the GREATER RED-HEADED BABBLER *(Malacopteron magnum)* belongs to this genus.

One of the best known sounds of the teak forests is the bright call of the striped jungle babbler, which sounds like "pretty dear". More often, one bird after another utters quite a different whistle, like a group of schoolboys loudly whistling in turn. Davison has the following description of this species: "It lives exclusively on insects and their larvae, but particularly on ant eggs. On sunny days, many species of ants bring out all their eggs, particularly when there has been much rain before. When our 'pretty dear' has found such an ant drying place, it quickly picks up all the eggs before the ants know what is going on."

H. C. Smith describes Abbott's babbler in the dense evergreen undergrowth: "The bird calls continually, particularly at dawn. It has a loud tri-syllabic call, whose pitch drops with the second syllable. Sometimes it also has a four-syllable call, whereby the first sound is low pitched, and the other three are higher." Blyth's jungle babbler is found on the shores of the upper reaches of rivers in Borneo. A traveler going downstream hears its tri-syllabic call. The musical fluting of the greater red-headed babbler, which also lives in Borneo, sounds very pleasant; it contains five or six tones of different pitch, each clearly separated from one another.

The Pomatorhinini include two groups of birds which look quite unlike and that are, however, joined by intermediates. Thus, the scimitar babblers have long, curved beaks and fairly long tails; the wren babblers, on the other hand, have short tails and a nearly straight, fairly short beak. Two species, however, one of which is the LONG-BILLED WREN BABBLER *(Rimator malacoptilus)*, are exactly halfway between the two groups. Included among the genera of the Pomatorhinini group are the following: 1. The genus *Pomatorhinus* from southern Asia; the beak is long, straight, and usually bright yellow or scarlet. 2. The genus *Pomatostomus* from Australia and New Guinea; they are only slightly different from the first genus, having a thicker beak and different in mode of life and breeding habits. 3. The SLENDER-BILLED SCIMITAR BABBLER *(Xiphirhynchus superciliaris)*, found in the Himalayas and the mountains of northern Burma, has the longest, thinnest and most curved beak of this group. 4. LONG-BILLED WREN BABBLER *(Rimator mala coptilus),* from the Himalayas, northern Burma, Tongking and Sumatra, has the long, curved beak of the scimitar babblers, but the very short tail and striped feathers of the wren babblers. 5. The genus *Ptilochichla* has three species in Borneo, Palawan and the Philippines. 6. STRIPED WREN BABBLER *(Kenopia striata)* has white stripes above, and inhabits lowland forests. 7. The genus *Napothera.* 8. The genus *Pnoepyga.* 9. The genus *Spelaeornis.* 10. The WEDGE-BILLED WREN *(Sphenocichla humei)* from the Himalayas and northern Burma has a clearly wedge-shaped beak.

The widely distributed YELLOW-BILLED SCIMITAR BABBLER *(Pomathorhinus montanus)* is characteristic of its group. Its usual call consists of three to six calls which sound like "hoot" and are delivered with varying rapidity. Another call sounds like "cow cow kejit" and the response to it is a long drawn "tshioo tshiai". The warning call is a rough "scree tshit, tshit, tshit". While gathering food, the members of a flock utter soft contact calls. All species of scimitar babblers are sociable birds of very similar behavior; they hide themselves extremely well on the ground or in dense growth and only rarely rise to a higher tree level. As they are very shy and know how to hide themselves so well, one rarely sees them, although their loud characteristic calls do betray their presence.

Gerd Heinrich, who observed these birds on Mount Victoria in Burma, has written the following: "The restlessness, swiftness and maneuverability of these birds in slipping through the forest thickets is quite astonishing. Not for a second do they stay put; they slip along horizontal branches like mice, hop here and there through the ground cover like shadows and fly on to another thicket before the observer's eye can quite make out what it was dealing with."

H. Whistler reports the following on the behavior of the RUSTY-CHEEKED SCIMITAR BABBLER *(Pomatorhinus erythrogenys;* Color plate, p. 222): "This bird inhabits the dense undergrowth of forests, as well as areas of grass and bush on treeless mountain slopes. It seeks most of its food under the ground cover, walking slowly over the layer of fallen leaves; occasionally it may develop greater speed and hop away with long jumps. Usually one finds these sociable birds in small troops which betray their presence by their sounds. The best-known vocalization is often a concert of soft 'cor quee oh' whistles to which a far-ranging light 'quiop' may be added. This species also has a harsh scolding call which is reminiscent of the jay thrushes or the tree pies. Only when the birds are quite close does one hear their soft feeding call, 'tep tep.' This scimitar babbler immediately replies to imitations of its calls, and it can be attracted in this way. These birds rarely leave cover, but when they fly up, their flight is swift and strong, although the short wings in combination with the heavy beak and tail give the bird an unusual appearance. It has even been claimed that this species performs a regular dance."

The RUFOUS BABBLER *(Pomatostomus isidorei)* of New Guinea does not build its nest on the ground like the true scimitar babblers, but, rather, suspends it. Thomas Gilliard saw such a nest which had a roofed side entrance. "It hung from a single strand of a climbing plant. According to the reports of the usually well informed natives, a whole crowd of these builders gathers each night to sleep in the large nest pouch. Sometimes, it is said, the group needs three months to complete the nest. I find it somewhat surprising that only two or three brown, striped eggs are laid in such a nest."

Only a few naturalists have observed the slender-billed scimitar babbler *(Xiphirhynchus superciliaris)* in its habitat. Like other scimitar babblers, this bird has two calls of attraction which are apparently used indifferently: a soft "hoot", sometimes composed of seven to eight syllables; and a rough, whetstone-like call which is like that of the RUFOUS-NECKED SCIMITAR BABBLER *(Pomatorhinus ruficollis)*. Incidentally, these two species sometimes join together and there is little difference in their way of life. I have encountered the slender-billed scimitar babbler on a bamboo in hilltop forests where it was apparently picking up ants and other small arthropods, and where from time to time, it flew over to another stalk. I could never discover the purpose of its highly developed beak.

We will describe only three species of wren babblers. The GREATER SCALY-BREASTED WREN BABBLER *(Pnoepyga albiventer)* has a lively chocolate brown color; the lower parts are white in one color phrase and in the other, they have a rusty and black color pattern. The tail is so short as not to be visible on the living bird. This babbler utters its warning call, "tsik," at intervals of four to five seconds for several minutes at a time. Gerd Heinrich has written the following about the behavior of this bird on Mount Victoria in Burma: "The breeding area is in the peak zone between about 2,500 to 3,000 meters. The moist, shady floor of the closed leaf forest, particularly thickly overgrown ravines and the shores of streams with low vegetation in the dark of the forest, constitute this bird's habitat. It shows a preference for the vicinity of some large tree in the forest, whose bent and half rotten branches are overgrown with herbs and moss." The bird moves on the ground almost without a sound and only rarely does it climb up a little in the vegetation. When it notices a human's presence, it usually utters its warning call. If one then waits quietly for some time, one may see the dark-brown bird hopping across a gap in the undergrowth, lifting its stumpy wings regularly and jerkily with each "tsik." This gives a particularly droll effect, with the bird's spherical tailless form.

The much smaller LESSER SCALY-BREASTED WREN BABBLER *(Pnoepyga pusilla)* has a L of 9 cm and is strikingly similar. Its characteristic call is a loud, shrill whistle and this is followed after about a second by a softer tone. Usually this call is heard from the moist dark ravines of teak forests, and it is so penetrating that it can be heard above the rushing sound of the waters. The bird calls for several minutes, at intervals of ten to twenty seconds, as it moves about. The warning "tsik" sounds sharper and softer than it does in the greater scaly-breasted wren babbler. The lesser scaly-breasted wren babbler is easily overlooked because it likes to stay in dense fern growths and among rich, mossy cushions on the shores of streams; there it runs about like a mouse and the observer can see nothing of it aside from a slight movement of the plants now and then.

The small, dark brown LONG-TAILED WREN BABBLER *(Spelaeornis chocolatinus)* is easily distinguished from the other wren babblers by its long tail. According to Gerd Heinrich, this bird is an inhabitant of the densest undergrowth on Mt. Victoria in Burma, particularly at the edges and in clearings of the evergreen forests. It prefers the almost impenetrable bamboo thickets which are permeated with thorny brambles in many places. It is not as exclusively a ground dweller as is the lesser scaly-breasted wren babbler, for example. Its song consists of a strong, ringing phrase of generally three, often only two, syllables which are repeated many times without interruption. The nest, a roofed-over oval structure of dead leaves, is lined with a water-proof layer which probably consists of the ribs of leaves and some soft fibrous materials. The birds works both up to a pulp which it applies to the nest interior, producing a proper cup.

The BARRED-WING WREN BABBLER *(Spelaeornis troglodytoides)* lives in the mountains of western China and also in Yunnan. So far it has been seen by only a few ornithologists. In 1953, I reported on a pair which hopped about in the bushes on a dry slope. These were certainly not ground dwellers like the other wren babblers.

TIT BABBLERS (tribe or genera group *Timaliini*) are small birds with fairly short, pointed beaks and tails of medium length. The plumage is soft and fluffy, often with attractive patterns and prolonged, sometimes disintegrated feathers on the back. These birds are found mostly in the undergrowth and lower levels of forests; thus, in general, they are not such pronounced ground dwellers as the first two tribes. There are six genera with thirty-five species, found in the Orient and Madagascar.

The MADAGASCAR JERYS (genus *Neomixis*) are very small yellowish-olive colored birds with thin pointed beaks like those of the tree babblers. There are four species, of which we will mention only the GREEN JERY *(Neomixis viridis)*. These birds live only on Madagascar and they resemble the small species of tree babblers (genus *Stachyris*). There are twenty-four species of tree babblers, and these vary a great deal in size and color; they live in the Orient and are particularly

Tit and tree babblers

numerous in Burma, Thailand and Malaysia. Although one often hears these generally common birds, they are seen only rarely. The GOLDEN-HEADED BABBLER *(Stachyris chrysaea)* is a characteristic species. All one usually sees of this minute bird is a glimpse of its head and breast, which give the effect of a golden veil. The call consists of seven or eight tones of the same pitch, "pee---pee-pee-pee-pee-pee-pee," with a marked pause after the first. Outside the breeding season these birds form flocks of up to forty or fifty birds and then show a preference for dense stands of bamboo. This babbler rarely descends to the ground if at all. It hops or flits from one twig to another. and like many other babblers of the undergrowth, it occasionally enters a higher forest level in search of food.

The RUFOUS-BELLIED BABBLER *(Dumetia hyperythra)* and the BLACK-HEADED BABBLER *(Rhopocichla atriceps)* are predominantly brown; each is the sole representative of its genus. The TRUE TIT BABBLERS, genus *Macronous*, with four species, have very long back feathers, particularly the FLUFFY-BACKED TIT BABBLER, *(Macronous ptilosus)* is the best known species of the genus; it occurs in several subspecies throughout almost the whole range of the tribe. As a pronounced inhabitant of undergrowth, it is one of the most numerous and most characteristic birds of the teak forests in Burma.

The RED-CAPPED BABBLER *(Timalia pileata;* Color plate, p. 222) lives in high cane grass; several subspecies inhabit a large part of southern Asia. These are small birds which occur in flocks in dense cover just like the other tree babblers. Each bird works downward in the thicket from stalk to stalk picking up arthropods and other food from branches and leaves; then it moves to the top of the bush, suns itself for a few seconds and sings a short song before it plunges down into cover once more. A very characteristic call consists of a trill and a whistle of descending pitch, which differs from all other bird calls. This bird is closely related to the rufous-crowned babbler and thus forms a link to the next tribe.

Two eastern Asiatic genera *(Moupinia* and *Chrysomma)* together with the NORTH AMERICAN WREN TIT (genus *Chamaea)* form another tribe, the WREN TIT AND ALLIES (Chamaeini). The WREN TIT *(Chamaea fasciata)* has a L of 16 cm and is the only New World species; it is distributed from Oregon to Baja California in North America. This predominantly brown bird is reminiscent of both wrens and titmice. Chaparral, low hardwood thickets, are its favorite nesting habitats. When hopping through low bushes, it holds its long, faintly banded tail upright. The plumage is dense and fluffy. Its calls are harsh and the song is loud. The wren tit feeds on insects and berries.

According to the ornithologist E. Thomas Gilliard's report: "Wren tits probably pair for life. The two sexes participate about equally in nest construction, incubation, and feeding of the young. The nest is a firm cup of twigs, strips of bark, feathers and animal hair. It is built in a fork in dense scrub. The female lays three to five greenish-blue, unspotted eggs."

The RUFOUS-CROWNED BABBLER *(Moupinia poecilote)* of western China, is very similar. The ORIENTAL GOLDEN-EYED BABBLER *(Chrysomma sinense)* differs from most other babblers in that it avoids forests and wanders about in open country. It inhabits tall grass areas, low scrub and groups of bushes. It is also quite common in stands of cane grass.

The babbler tribe with the greatest number of species is that of the SONG BABBLERS *(Turdoidini)* with 17 genera and no less than 140 species. Song babblers are found throughout Africa, and above all in southern Asia. Jean Delacour has written the following description: "This tribe includes all the more vividly colored and the best songsters among the babblers. In most species the voice is loud, full and melodious. All colors are represented and the plumage pattern is often striking. The beak, legs and feet are strong; the wing is round, loosely held, and is often adorned with beautiful colors. The tail is broad, moderate to very long and usually hangs between the wings. Many of these birds are adapted to an arboreal life, although they often feed on the ground. They have a peculiar way of jumping around, of moving brusquely and of making jerky, short flights from limb to ground and back. They go about in flocks often associated with other birds and they always keep to the cover of brush, creepers and trees, usually in woods and forests; only a few come to savannas and gardens, where they keep to thickets and hedges. They feed on berries and insects. Their nests are cup-shaped (with rare exceptions), and are placed on low trees, in bushes and creepers; only a few birds build on the ground or in holes, making covered nests of moss and grass."

The most primitive genus of these highly developed babblers is that of the SONG BABBLERS in the narrower sense, *Turdoides,* with twenty-six species. These fairly large birds (the L is about 25 cm) are usually brown and may be striped; they inhabit scrub in dry areas. The SEVEN SISTERS *(Turdoides somervillei)* is commonly seen in India on the edges of towns and villages. In Burma, it is replaced by the WHITE-THROATED BABBLER *(Turdoides gularis),* a characteristic bird of the bungalow districts in the dusty city of Mandalay. Small troops of these birds look busy indeed, hopping about hunting insects, turning leaves over while holding their tails in various directions and, now and then, making short flights. Their characteristic habitats are the thornbush hedges along the edges of fields in the semi-desert plains and hills of the dry zone, where cultivated fields alternate with wasteland. There they move from hedge to hedge or from one thornbush to another, spreading their small, round wings and fanning out their tails.

One of the better known species of this genus is the SPINY BABBLER *(Turdoides nipalensis)* of Nepal, which has a L of 22 cm. It is generally brown striped, but has spiny tipped feather shafts and a relatively long

▷
Southern yellow robin *(Eopsaltria australis)* at the nest with its young.

tail. At one time there were only four specimens of this bird and it was regarded as extinct. Ripley then found another individual. Later, it was discovered that this species can be found in some numbers in the scrub right around Katmandu, the capital of Nepal.

The CHINESE BABAX *(Babax lanceolatus)* is a mountain species which links the song babblers to the LAUGHING THRUSHES (genus *Garrulax*). There are about fifty species of laughing thrushes, all of which are fairly large, and some of which are crested. Almost all of these birds utter loud musical calls and some sing very well. They are among the most colorful, interesting and striking birds of the mountain forests and the forests at the foot of mountains in southeastern Asia. The WHITE-CRESTED LAUGHING THRUSH *(Garrulax leucolophus;* Color plate, p. 222) is one of the most characteristic birds of the Burmese teak forests. In 1940, Gerd Heinrich wrote the following description of this species: "Dense shadowy forests, particularly those rich in stands of bamboo, seem to be the favorite habitat of this stately bird. It is very striking in its colors, its behavior and above all, in its voice. These feathered goblins wander far through the forest thickets in small troops or more often in large flocks." Small reddish-brown "ghosts" with bright white crests slip about the stands of bamboo. They perform acrobatics in loops of creepers and hop about, rustling the fallen leaves on the ground. If the observer emerges from the tree trunk behind which he has been concealed, everything becomes quiet after a brief scurry of the escaping birds. In the background, three or four of the birds with fully erected white crests may perhaps be seen sitting side by side, almost motionless on a broken bamboo stem.

After a short, soft grumbling, "a sudden, ringing, almost threatening laughing or neighing is heard," as Heinrich puts it. "Right away all is quiet again; then after the same introduction, the noise of a new salvo of laughter rings out again." So it goes on until a movement of the observer puts the laughing goblins to hurried flight, whereby their neighing dies down in the far off dark of the forest.

Two other species of this genus, the BLACK-GORGETED LAUGHING THRUSH *(Garrulax pectoralis)* and the NECKLACED LAUGHING THRUSH *(Garrulax moniliger)* look almost alike, except that their sizes are not the same; they are both olive-brown with a black chin spot.

According to C. V. Ticehurst, these two species are one of the most notable examples of parallel evolution in birds. "Where one species occurs, the other is often found as well, even in the same mixed flock. Nests, behavior, eggs, display and environment are almost identical and both share the same habitats. When one species shows a geographical variation, the other in the same area shows the same thing."

A number of small genera include some beautiful and fascinating species which make the hills and mountains of southeastern Asia more interesting for ornithologists. The LAUGHING THRUSHES of the genus *Liochichla,* which include *Liochichla phoenicea* resemble the laughing

◁
Those who have the patience to sit for long periods in an observation blind can see the great reed warbler *(Acrocephalus arundinaceus)* feeding its young. The female builds the nest by herself.

thrushes of the genus *Garrulax*, but are smaller, lighter in build, and brightly colored in red, yellow or green. The two species of the genus *Leiothrix* are closely related to the birds of the *Garrulax* genus. The PEKIN ROBIN (*Leiothrix lutea;* Color plate, p. 222) is found from the Himalayas to southern China and has been introduced on the Hawaiian Islands; it is known in Germany as the Chinese nightingale. Bird fanciers also keep the somewhat larger SILVER-EARED MESIA (*Leiothrix argentauris;* Color plate, p. 222), which occurs from the Himalayas to Sumatra and is one of the most beautiful and common birds in the higher mountains of Burma. It has a black head with silvery ear coverts, a golden-yellow body plumage with green lines in certain parts, red upper and lower tail coverts and a red spot on the wing.

H. Whistler reports the following on the habits of the pekin robin: "It is a bird of the moderate elevations in mountain forests; it is found in jungles of every kind but, by preference, in spruce or pine forests with undergrowth. As a lively, gay and uncommonly sociable bird, it searches through the undergrowth in small troops for insects, although it occasionally also shows itself high up in the trees. Usually it calls 'tee tee tee tee tee.' During the breeding season, one hears charming songs from the males, songs with a considerable range of tones and variety of melody. While singing, the male sits with fluffed-out plumage on the tip of a branch and flutters its wings. The nests are cups which vary in depth and firmness of construction and are usually not well hidden. The lining consists of fine fibers or hair-like moss roots. The nest site is variable, but is always less than three meters above the ground. Some nests hang like oriole nests in a horizontal fork, others are attached to a vertical fork, like those bulbuls favor; still other nests are built among several vertical shoots like the nests of reed warblers. The clutch usually consists of three eggs." This beautiful bird is one of the few babblers to have entered the animal trade and the flight cages of European and American fanciers in some number. It does not, however, breed very readily in confinement.

The small and very beautiful FIRETAIL *(Myzornis pyrrhoura)* of the higher forests in the Himalayas, has been called a "living emerald." Like the CUTIA *(Cutia nipalensis)* it is the only representative of its genus. The five species of SHRIKE BABBLERS (genus *Pteruthius*) belong here; like the cutias, they are tree dwelling mountain birds of similar habits and very colorful plumage. The sexes look different. The RED-WINGED SHRIKE BABBLER *(Pteruthius flaviscapius)* is strikingly tame and is often seen peering about as if short sighted. It moves rather slowly and gives the impression of being stolid. When looking for food, it works its way up a tree in a leisurely manner and takes insects and berries principally from the main branches, often hopping along while athwart the branch. When it has reached the top of the tree, it may stay for half an hour or longer, calling continually and then resuming its search for food on another tree.

The WHITE-HEADED BABBLER *(Gampsorhynchus rafulus)* resembles a small laughing thrush and is the only species of its genus. It has a markedly hooked beak and a striking color as the white head contrasts with the light brown of the rest of the plumage. It inhabits bamboo thickets and the undergrowth of evergreen forests, but only rarely comes down to the ground. Two additional genera of song babblers are the NUN BABBLERS (genus *Alcippe*) and the SIBIAS (genus *Heterophasia*) (for more species, see the Systematic Review). The NUN BABBLER *(Alcippe poioicephala)* is common at the foot of mountains in teak forests where its restless flocks slip about through the bamboos and the undergrowth. The LONG-TAILED SIBIA *(Heterophasia picaoides)* has a very attractive plumage marked mostly with black, white and gray. It has a melodious song, part of it giving a plaintive effect. As a true tree bird, it spends the greatest part of its life in the tree crowns, where it feeds on nectar and pollen, but also on berries and insects.

Fig. 11-2. Distribution. of rockfowl (genus *Picathartes*).

The last babbler tribe, the ROCKFOWL or BALD CROWS *(Picathartini)* consists of only one genus *(Picathartes)* with two species. As the only bald headed babblers, they look quite different from their relatives; some ornithologists consider them a separate family or subfamily. These birds are blackish-gray, white below and have exceptionally long thrush-like legs. The two species can be distinguished particularly by the color of the bald head and the neck feathers. The WHITE-NECKED ROCKFOWL *(Picathartes gymnocephalus;* Color plate, p. 222) of Sierra Leone, Ghana and Togo, has bright orange-yellow on its bare head; the GRAY-NECKED ROCKFOWL *(Picathartes oreas;* Color plate, p. 222) of the Cameroons, is red on the back of the head and has a blue forehead.

The gray-necked rockfowl leads a very secretive life on the ground, where it pursues insects and amphibia with long jumps. For this purpose, it often seeks out the vicinity of small, deep forest rivers, and also trees. Its habit of hastily flinging leaves aside and quickly picking up fallen fruit is reminiscent of the laughing thrushes. Some English ornithologists found a breeding colony of these birds among granite rocks in the depths of a hill forest. The cup-shaped mud nests were attached to the rock in clefts and below overhangs, two to five meters above the ground. Some were even attached to vertical rock faces. The nests were open on top and lined with grass and thin fibers so that the eggs did not touch the mud walls. Clutches of two brownish eggs with fine brown marks were found. Nestlings were fed by regurgitation. Today, both species can be seen in zoos where they may even breed, as in Frankfurt, Germany.

The Frankfurt Zoo has successfully kept white-necked rockfowl since 1962 and first bred this species in captivity. According to Richard Faust, the oldest of the Frankfurt birds has lived there since 1964 and is therefore at least seven years old. These birds, which came from Liberia, are fed a coarse soft food, fruit and insects at Frankfurt. They have been observed catching and eating mice several times. P. Mackrodt measured jumps of 1.2 meters, which the birds performed without using the wings. He was impressed by their striking curiosity and surprising knowledge of locality. Of sixteen different bird species, the white-necked rockfowl were the only ones which learned to slip through the curtain of threads of the flight cage when they saw the food cart approaching behind it. They are not shy of visitors and may even undo people's shoelaces or examine their handbags. Their method of head scratching is also notable, for they scratch with the foot raised from behind over the lowered wing, as observed by Mackrodt, while their relatives, the babblers, do so by bringing the foot up directly. This observation may suggest the need for a check of their systematic position. White-necked rockfowl look for food on the ground, turning over fallen leaves with the beak or seizing them and flinging them aside. They kill living prey or reduce the size of large pieces of food by flinging them against the ground. The only vocalization Mackrodt heard was a soft croak used as a contact call among these sociable birds. Among themselves, they kept individual distances of at least thirty centimeters.

Ingrid Faust observed the breeding behavior in two instances. The pair, after various trials, selected the nest site in accordance with two features: ready access in flight and cover above. As nest materials, they used moist earth mixed with dry stalks and small leaves. The earth was formed into small lumps with the beak; these were then rolled on the ground before being taken to the nest in the beak. Stalks and leaves were carried in bundles. As the nest contents could not be checked, the clutch size and incubation period are unknown, but the incubation period was at least nineteen days and one young was reared in each case. Both parents incubated alternately, although the male did so longer than the female. When relieving the incubating bird, the partner brought soil or stalks in the beak. They killed insects by beating, shaking and rolling them on the ground before offering them to the young. During the first sixteen days the parents removed fecal pellets from the nest. The fledging period differed in the two cases; one young left the nest after twenty-five days, the other after nineteen days. After the young had left the nest, the parents fed them for an additional eighteen days; twenty-five days after that, the parents already had a new egg in the nest. Thus, breeding as a whole took fifty-six days. The young of the second brood is still alive and is now four years old.

The Frankfurt Zoo was the first German zoo to import the gray-necked rockfowl in the summer of 1968 and it has been keeping them successfully since. Although according to observations in the wild,

Subfamily: parrotbills

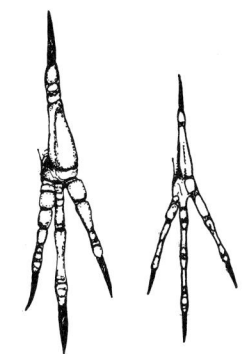

Fig. 11-3. Ball of the foot (or footpad) of "branch singers." It is widened in the great reed warbler (left) and only slightly developed in the grasshopper warbler (right).

Fig. 11-4. Distribution of the yellow-chested apalis (*Apalis flavida*).

these birds are said to sleep only in rock clefts, in the zoo they roost on branches in bushes, in spite of available clefts.

The PARROTBILLS (Panurinae) are a separate subfamily of the flycatcher-like birds. They have a short, deep, laterally compressed, parrot-like beak. In most species the tail is longer than the wings. These birds inhabit reed beds, tall grass, bamboo thickets, hedges and broad-leafed forests with undergrowth in southeastern Asia from India through China, and north as far as Manchuria; one species, the bearded tit, also occurs in Europe. There are two genera; the PARROTBILLS in the restricted sense *(Paradoxornis)* and the BEARDED TITS *(Panurus)*.

The best known representatives of the genus *Paradoxornis* are: the VINOUS-THROATED PARROTBILL *(Paradoxornis webbiana)* which has a L of 12 cm; *Paradoxornis paradoxa* which has a L of 15 cm, and is from southern China; and the GRAY-HEADED PARROTBILL *(Paradoxornis gularis)*, with a L of 17 cm which occurs from Sikkim to Assam. The vinous-throated parrotbill, which occurs from Ussuriland to Taiwan and western China, resembles the long-tailed tit and, like the bearded tit, was formerly classified with the tits or chickadees, in spite of the difference in moult and other characteristics.

The BEARDED TIT *(Panurus biarmicus;* Color plate, p. 222) has a L of 16.5 cm and weighs 14 g; it is the only representative of its genus. Those families with older bird books may be surprised to find that this bird belongs not with the tits but the parrotbills. In contrast to the parrotbills in the narrow sense, it has a weak, pointed beak; the tenth primary is degenerate. The bearded tit lives in extensive reed beds on the shores of lakes and in densely grown reed marshes. It is distributed from Manchuria over central and western Asia to southern Russia and southeastern Europe, but it also breeds in southeastern England, Holland, France, central and eastern Spain, northern and southern Italy, Austria and northern Germany, even in the former eastern Prussia. Since 1959 new breeding places have been observed increasingly in Germany, eastwards at least as far as Sylt, Hildesheim and Munich.

The bearded tit is a partial migrant, and wanders great distances outside the breeding season. It is tame and sociable, and moves skillfully among the reed stalks, sometimes grasping a different stalk with each foot. It also comes down to the ground, where it walks about and scratches for insects in the summer; in the winter, reed seeds are its main food. The flight is undulating and slow, and usually only covers short distances. The nest, a deep cup, is placed low down, and concealed among the water plants; it is made from reed and grass stalks. Both males and females build the nest and incubate the four to seven eggs, which are whitish and have dark markings, for twelve to thirteen days. The young leave the nest after thirteen days and are cared for as a group for an additional period after that. There are often three, and rarely four, broods per year. The bearded tit's voice is a soft "tsit,

tsit" or a nasal "ping ping" and a purring "churr"; the song is a soft twitter.

The well-known Austrian ethologist, Otto Koenig, has observed this species for many years in the wild at Lake Neusiedel (Austria) as well as in large flight cages. He probably observed over a thousand of these birds in the field and he reared about sixty young. According to his reports, bearded tits form a permanent pair bond, after an "engagement" as immatures. Several groups of siblings form a flock until the "engagements" take place; afterwards the pairs separate, although they form flocks once more during the juvenile moult. In the spring, the flocks of adults form loose breeding colonies. There are no territorial boundaries although at the beginning of the breeding season the birds keep others of their species away from their nests.

Fig. 11-5. Long-billed crombec *(Sylvietta rufescens).*

Bearded tits do not inhabit particularly large territories, but rather several small ones and they move from one to the other on the wing. As Otto Koenig points out, these birds are very well adapted to life in reed-covered, lagoon-like marshy areas. These adaptations include the straight line flight, the ability to hop in the reeds like reed warblers, the ability to walk on the ground, the ability to perch with each foot on a different reed stalk so that the body is suspended between them, and lastly, the so-called flutter-swimming, should a bird happen to fall into the water. Bearded tits eat soft food (i.e. mainly arthropods) during the breeding season. The fledged young prefer a vegetarian diet, e.g. unripe seeds and parts of reed blossoms blown off by the wind. After the moult, the adults too eat plant materials. As sociable birds, bearded tits have many related calls and movements of expression. Koenig has even been able to demonstrate learning abilities which can be very different in one individual as compared to another.

Fig. 11-6. Gray-backed camaroptera *(Camaroptera brevicaudata).*

The RAIL-BABBLERS (subfamily Orthonychinae) are generally considered close to the babblers systematically. They are ground dwellers; the L ranges from 16–29 cm. The legs are relatively long but the toes are short; the sexes often differ in color. Their range is Australia and New Guinea, with one species in Malaya, Sumatra and Borneo. There are nine very different genera with twenty-three species in all.

Subfamily: rail-babblers, by H. Th. Condon

The QUAIL THRUSHES (genus *Cinclosoma*) resemble many true thrushes *(Turdus)* in size, shape, plumage and the spotting of the young. They are brown or reddish-brown in color with striking white eye and cheek stripes in the males; the males also have much black on the throat and breast. There are also white spots on the black wing coverts of these birds, except in the New Guinea species. Quail thrushes are found on the Australian mainland, and also in Tasmania and New Guinea. There are seven species, including: the SPOTTED QUAIL THRUSH *(Cinclosoma punctatum);* both sexes have a gray breast band and coarse black spots on the flanks; the CINNAMON QUAIL THRUSH *(Cinclosoma cinnamomeum)* which is a desert species, and has bright reddish-brown on the

Fig. 11-7. An apalis.

Fig. 11-8. The crombec.

Subfamily:
Old World warblers,
by B. Leisler

Fig. 11-9. Eremomela.

back; and the AJAX SCRUB ROBIN *(Cinclosoma ajax)* which has bright-orange-brown spots, and is found in the lowland forest of New Guinea.

Other genera and species of this subfamily are also found in New Guinea: these are: 1. The genus *Ptilorrhoa,* which has a L of 20-23 cm. The tail is long; the plumage of the upper parts is predominantly blue and chestnut brown. There are three species, including the LOWLAND EUPETES *(Ptilorrhoa caerulescens).* 2. The MELAMPITTAS (genus *Melampitta*) have a short tail. The plumage is all black. There are two species, including the LESSER MELAMPITTA *(Melampitta lugubris)* which has a L of 17 cm. 3. The IFRITA *(Ifrita kowaldi)* has a L of 17 cm and is the only species of its genus. The tail is of medium length. The plumage is dark olive, and the crown is blue with a black center.

The RAIL BABBLER *(Eupetes macrocercus;* Color plate, p. 222) has a L of 29 cm. It is a widely distributed inhabitant of lowlands in Malaysia. Its plumage is brown, and the head, neck, and breast are orange-brown.

The QUAIL THRUSHES are the best known birds of this subfamily. They are reminiscent of quail in flight. These birds are easily recognized by their alarm call which consists of a single high-pitched whistle. They generally lay two or three brown spotted eggs into a coarse, cup-shaped nest placed on the ground. The eggs have unusually thin shells. All seven species inhabit particular habitats from rain forests to deserts. Of the other species mentioned above, the ifrita stands out because of its habit of hopping up tree trunks.

The WARBLERS, subfamily Sylviinae, are among the most characteristic songbirds of the Old World. They range from the size of a leaf warbler to that of a thrush; the L is 7.8-29.2 cm. They have narrow pointed beaks, often with bristles at the gape. The wings are generally of medium length, and predominantly rounded (only the species of northern Eurasia had to develop the pointed wing characteristic of migrants); there are ten primaries and, generally twelve tail feathers, although some species only have ten and one genus only has eight. The plumage is usually brown, olive-green and gray; only a few African representatives are more colorful. The sexes are generally similar. The legs are short to medium long. The foot has expanded soles on the toes ("branch singers", Fig. 11-3); the balls of these toes are less well developed in grass and ground dwelling species and in the leaf warblers. These birds are insect eaters which search out their prey; only a few species hunt passing insects from a perch. Warblers are good nest builders; they build in the open and lay two to twelve eggs per clutch. Incubation is by the female or by both partners. The young, which are born naked or covered in down, are generally reared by both parents. They are excellent songsters; many species have a display flight. Warblers live in open habitats in the Old World except in the extreme north and certain oceanic islands. One species, the ARCTIC

WARBLER *(Phylloscopus borealis)* has immigrated into Alaska over the Bering Strait; the BUSH WARBLER *(Cettia diphone)* was introduced into Hawaii from eastern Asia. There are altogether about 65 genera and 312 species.

As pronounced insect-eaters, the Old World warblers are among the most ancient songbirds. A fossil sylvid from the Upper Oligocene of France (about 28 million years ago) has been found. The group evolved into an unusually large number of species which now belong to the most abundant and striking songbirds. Because of this, representatives of this subfamily often serve as hosts to cuckoos and other brood parasites in a wide variety of areas. The European cuckoo (Volume VIII) particularly, tends to lay its eggs into the nests of birds of the genus *Sylvia,* and those of reed warblers. Asiatic and African cuckoos, and certain African weaverbirds which parasitize leaf warblers and grass warblers, act similarly.

Fig. 11-10. Camaroptera.

Most Old World warblers are marked quite inconspicuously; distinguishing marks are often found only on the head, throat, and breast. More rarely, the wing will also carry a mark. Knowlege of these birds' behavior and songs often gives the best material for identification. Many of the species native to Germany belong, because of their artistic songs, to the best feathered songsters. Those birds of the genus *Sylvia* are typical territorial birds, and their territories can readily be ascertained. It was therefore no accident that the British ornithologist Howard based his territory theory particularly on the observations of these birds. Old World warblers can be divided into six groups: A. Wren warblers and relatives; B. Grass warblers; C. Reed warblers; D. Typical warblers; E. Leaf warblers; F. Tailor birds.

The group of WREN WARBLERS AND RELATIVES consists of five genera with about sixty-five species, all of which are found in Africa: 1. The APALIS (genus *Apalis*) have twenty-two species, including the YELLOW BAR-THROATED APALIS *(Apalis thoracia)* which has a wing length of 5.1–5.4 cm, and the YELLOW-CHESTED APALIS *(Apalis flavida)* which has a L of 16 cm and a wing length of 4.9–5.3 cm. 2. The CROMBECS (genus *Sylvietta*) include the LONG-BILLED CROMBEC *(Sylvietta rufescens)* which has a L of 8.8 cm and is found from Senegal to the Cape. 3. The EREMOMELAS (genus *Eremomela*) include the BROWN-CROWNED EREMOMELA *(Eremomela badiceps)* with a L of 11 cm, which is found from Ghana and the southern Sudan to Angola, the Congo, and Rhodesia. 4. The CAMAROPTERAS (genus *Camaroptera*) include the East African GREEN-BACKED CAMAROPTERA *(Camaroptera brachyura),* and the GRAY-BACKED CAMAROPTERA *(Camaroptera brevicaudata)* with a L of 12.5 cm, both of which are now often regarded as mere subspecies of the single species. 5. The BUSH CREEPERS (genus *Macrosphenus*) include the YELLOW LONGBILL *(Macrosphenus flavicans),* which has a L of 12.5 cm and is found from Sierra Leone and northern Angola to Uganda (see Fig. 11-4 through 11-6).

Fig. 11-11. Bush creeper.

Fig. 11-12. Fan-tailed warbler *(Cisticola juncidis).*

Several species of the genus *Apalis* have a characteristic black band across the chest. They are small, rather long-tailed birds. The yellow-chested apalis busily searches leaves and small branches for arthropods, like a leaf warbler, and, like birds of this group, also builds a nest which is covered with a roof. The slightly curved bill of the long-billed crombec is suitable for probing in clefts in tree bark, hence the German name of the genus—nuthatch warbler. These birds inhabit dry forest and bush areas. The eremomelas, apart from their marked pattern and color, resemble the leaf warblers, with which they are probably in competition in Africa. The gray-backed camaroptera makes itself conspicuous by erecting its tail and by its peculiar nest structure; it is widely distributed in tropical Africa. It builds its nest between two leaves which it has sewn together in a dense bush or in the fork of a branch near the ground; it often sews on another leaf as a roof. The long-necked and long-beaked yellow longbill presents an unusual appearance with its fluffy back feathers. It often seeks food in mixed flocks of birds.

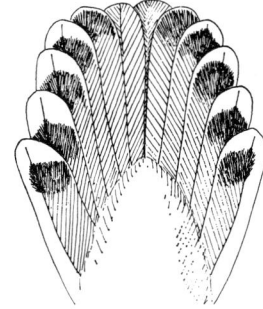

Fig. 11-13. Tail spots of the fan-tailed warbler.

The GRASS WARBLER group consists of small, round-winged forms with unobtrusive habits which inhabit Africa, southern Asia and Australia. These are: 1. CAPE GRASS BIRD *(Sphenoeacus afer)* which has a L of 21 cm. The tail is wedge-shaped. This is the largest grass warbler, and it is distributed in southern and eastern Africa. 2. FAN-TAILED WARBLERS (genus *Cisticola*) comprise forty-five species, forty-three of which occur in Africa. The following species occur outside Africa. The STREAKED FAN-TAILED WARBLER *(Cisticola exilis)* with a L of 11 cm, and the FAN-TAILED WARBLER *(Cisticola juncidis;* Color plate, p. 231) which has a L of 10 cm. 3. PRINIAS (genus *Prinia*) are from Africa and India. They include the TAWNY-FLANKED PRINIA *(Prinia subflava)* which has a L of 14 cm and is found in Africa from the Sudan to southern Africa and from India to Szetchuan, Taiwan and Java. 4. EMU-TAILS (genus *Dromaeocercus*) includes the GRAY EMU-TAIL *(Dromaeocercus seebohmi)* of Madagascar; this bird has eight tail feathers with some of the feather barbs not connected to one another.

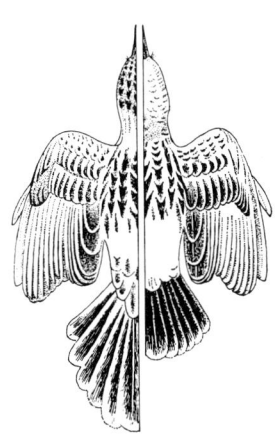

Fig. 11-14. Differences in the tail length (dimorphism) in the streaked fan-tailed warbler. Left: Nuptial plumage. Right: Non-breeding plumage.

The predominantly tiny fan-tailed warblers form one of the genera with the greatest number of species; they are characterized by generally short but well-rounded tails, by a double moult in many species, as well as by a very marked sex difference (males are much larger), and lastly by their markings (Fig. 11-13). Many males have black beak chambers in the breeding season. Apart from the special environmental requirements of particular species, grasses of a particular type are a prerequisite for the presence of all these warblers. During the mating season, only the males are conspicuous through their songs, which they sing from perches or in flight.

The most striking display flight is probably that of the WING-SNAPPING CISTICOLA or AYRE'S CLOUD WARBLER *(Cisticola ayresii)* which has a L of 8.5

cm and is distributed from southern Africa to Kenya and Gabon. It rises rapidly so high that it sometimes becomes invisible to an observer. Then, while singing and making snapping noises with its wings, it cruises about, and finally descends vertically producing a single tone in great excitement. Just above the grass tips it arrests its descent, rises up again briefly and performs a short horizontal flight. Only then does it drop down into the grass. Similar peculiarities are shown in the display of closely related species, all of which are inhabitants of treeless grassland. The display of the DESERT CISTICOLA *(Cisticola aridula)* is not so extraordinary; this bird builds a ball-shaped nest with a lateral entrance (Fig. 11-16). A still less striking display without wing-snapping is shown by the fan-tailed warbler (in the restricted sense) which includes southern Europe in its range. It flies at a height fifteen to thirty meters, or only just above its breeding area, calling "tsip tsip tsip" while flying to and fro and rising and dropping in flight. It builds a pouch-shaped or even bottle-shaped nest (Fig. 11-17) and in this connection it occasionally perforates a few leaves of grass with its beak and pulls threads of spider web through the holes. Like other fan-tailed warblers, it lines the nest interior with plant wool. This bird is widely distributed and has penetrated eastward from Indonesia to northern Australia.

Fig. 11-15. Desert cisticola *(Cisticola aridula)*.

On the rest of the Australian continent, it is replaced by the streaked fan-tailed warbler, which towards the northeast also inhabits New Guinea and the Bismarck archipelago. This species has also spread to southern China and India, into the range of the fan-tailed warblers; it occupies dry areas and builds a ball-shaped nest. The still more artful nests of the tailor bird type (Fig. 11-18) are built by the RED-FACED CISTICOLA *(Cisticola erythrops)* which has a L of 13 cm and is found in southeastern Africa, and the SINGING CISTICOLA *(Cisticola cantans)* of eastern and southeastern Africa. Like many of the fan-tailed warblers, the small, slim, long-tailed and delicate-looking WREN WARBLERS (genus *Prinia*) also build skillfully woven or sewn roofed-over nests. Both the fan-tailed warblers, which can run very well, and the wren warblers feed on the ground.

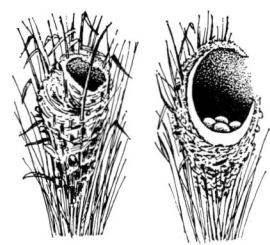

Fig. 11-16. Ball or sphere type nest of the desert cisticola.

The REED WARBLER group, which is not sharply demarcated from the preceding group, consists of several forms, many of which use reed beds. The genus *Acrocephalus* (for distribution see Fig. 11-19 through 11-23) contains eighteen species, including the one after which the whole group is named. The most well known species are: The SEDGE WARBLER *(Acrocephalus schoenobaenus;* Color plate, p. 231) has a L of 13 cm. The AQUATIC WARBLER *(Acrocephalus paludicola:* Color plate, p. 231) has a L of 13 cm. The MOUSTACHED WARBLER *(Acrocephalus melanopogon;* Color plate, p. 231) has a L of 12.5 cm. The AFRICAN REED WARBLER *(Acrocephalus baeticatus)* has a wing length of 5.3–6.3 cm, and is found from the southern Sudan and Lake Chad to the Cape. It is cinnamon-brown

Fig. 11-17. The fan-tailed warbler builds a flask-shaped nest.

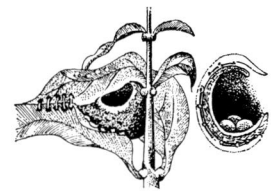

Fig. 11-18. Tailor bird-type nest of the red-faced cisticola.

Fig. 11-19. Sedge warbler (*Acrocephalus schoenobaenus*).

Fig. 11-20. Aquatic warbler (*Acrocephalus paludicola*).

Fig. 11-21. Reed warbler (*Acrocephalus scirpaceus*).

and builds a nest of the reed warbler type. The INDIAN REED WARBLER (PADDY FIELD WARBLER) *(Acrocephalus agricola)* of Russia and Asia, resembles the reed warbler, but is dull rust-colored and has a very pronounced eye stripe. The REED WARBLER *(Acrocephalus scirpaceus;* Color plate, p. 232) has a L of 13 cm. The MARSH WARBLER *(Acrocephalus palustris;* Color plate, p. 232) has a L of 13 cm, BLYTH'S REED WARBLER *(Acrocephalus dumetorum)* has a L of 13 cm and ranges from southeastern Finland and Estonia to Lake Baikal. The GREAT REED WARBLER *(Acrocephalus arundinaceus;* Color plate p. 231 and p. 242) has a L of 20 cm. The EGYPTIAN REED WARBLER *(Acrocephalus stentoreus).* The IRAQ GREAT REED WARBLER *(Acrocephalus griseldis)* has a wing length of 7.7–8.4 cm. It is a smaller bird and is found in Mesopotamia. The THICK-BILLED WARBLER *(Acrocephalus aedon)* is a somewhat divergent species; its close relationship with the other species is doubtful.

We will mention only the following of the other species and genera included in this group: 1. The RUSH WARBLERS (genus *Bradypterus*) with the LITTLE RUSH WARBLER *(Bradypterus baboecala)* which has a L of 15 cm and is round-winged. It is a dirty brown color; the eye stripe and lower parts are cream-colored. These birds perform short display flights with whirring wings; they live in lowlands and moist mountain valleys in Africa. 2. The SWAMP WARBLERS (genus *Calamocichla*) include the GREATER SWAMP WARBLER *(Calamocichla gracilirostris),* which has a wing length of 6.9–8 cm (see Fig. 11-24). 3. The CETTIS' WARBLERS (genus *Cettia*) are related to the leaf warblers. CETTIS' WARBLERS *(Cettia cetti;* Color plate, p. 231) has a L of 14 cm. This species is from the Mediterranean area, but it has wandered as far as Belgium. 4. Warblers of the genus *Locustella* (Fig. 11-26) with seven species: SAVIS' WARBLER *(Locustella luscinioides;* Color plate, p. 231) has a L of 14 cm. The GRASSHOPPER WARBLER *(Locustella naevia;* Color plate, p. 231) has a L of 13 cm. PALLAS' GRASSHOPPER WARBLER *(Locustella certhiola;* Color plate, p. 231) has a L of 13 cm and a plumage like that of the sedge warbler. The STREAKED or TEMMINCK'S GRASSHOPPER WARBLER *(Locustella lanceolata)* with a L of 11.5 cm, has a collar of stripes with a light-colored throat below. The TAIGA GRASSHOPPER WARBLER *(Locustella fasciolata)* has a wing length of 7.4–8.3 cm. The RIVER WARBLER *(Locustella fluviatilus)* has a L of 14.5 cm. MIDDENDORF'S WARBLER *(Locustella ochotensis)* has a wing length of 6.5–7.5 cm, and a markedly graduated tail; it is found in eastern Asia, and is possibly a subspecies of Pallas' grasshopper warbler.

That birds of the genus *Hippolais* are related to the reed warblers is indicated by their almost continuous, suppressed-sounding songs. There are six species: The ICTERINE WARBLER *(Hippolais icterina;* Color plate, p. 232) which has a L of 13.5 cm and is found from continental Europe eastward to western Siberia. The MELODIOUS WARBLER *(Hippolais polyglotta)* which has a L of 13 cm and occurs in the western Mediterranean area, France, and the Ticino (Switzerland). The OLIVE-TREE WAR-

BLER *(Hippolais olivetorum)* has a L of 15.5 cm and is from the eastern Mediterranean. The BABBLING WARBLER *(Hippolais caligata)* is from Russia, east as far as the Yenissei and south to Iran. UPCHER'S WARBLER *(Hippolais languida)* is found in the Near East and east from there as far as Tianchan. The OLIVACEOUS WARBLER *(Hippolais pallida)* has a L of 13 cm (see Fig. 11-27 and 11-28).

The warblers of the genus *Calamocichla* are round-winged and have large, long feet with well-developed claws. They inhabit reed-covered shores of lakes and rivers, as does the greater swamp warbler, and also moist thickets, sedges and tall elephant grass. They often weave their nests over water among reeds and papyrus.

Cettis' warbler is found in Europe only on the three southern peninsulas, in France and locally in Belgium, where its pentration northward is often arrested by high mortality in long, hard winters. It is a most secretive bird of the densest network of climbing plants, brambles and tamarisks on the shores of rivers, forest streams and irrigation ditches. Like so many inhabitants of thickets, it twitches its wings and erects its tail. Often only its song, which is sudden in onset, loud and ringing, betrays its presence. The nest is placed in dense vegetation and is not anchored at the points of support.

In the reed warblers proper, we are able to separate striped and unstriped forms. Those with striped heads have finer beaks, eggs of a different color and do not build such complex nests as do the unstriped species. The sedge warbler, which arrives in central Europe about mid-April from its winter quarters in tropical and southern Africa, is the most common of the three European striped reed warblers. It inhabits the marginal zone of swampy areas where there are scattered reeds and much shorter vegetation. Males often begin their short song flight from the willow bushes found there. Both partners build the simple nest with only little anchoring to the vegetation about it (Fig. 11-29). The female incubates for thirteen days and the young are fed by both parents. The closely related aquatic warbler is becoming progressively scarcer in Europe due to displacement by the preceding species and drainage of its habitats, large meadows with tall sedges. It also performs a short song flight and differs from the sedge warbler in appearance and by its simpler song. This song consists of a mixture of churring and whistles which sounds somewhat like "arr arrr dee dee dee dee". The aquatic warbler is also more secretive than the sedge warbler and, when searching for food, it remains hidden among the sedges and often walks on the ground like the warblers of the genus *Locustella*. It likes to place its nests over marshy ground or in a clump of sedge in shallow water without anchoring it to a great extent (Fig. 11-30). The brownish eggs, densely spotted with yellowish gray, are remarkably similar to those of the sedge warblers.

The peculiar moustached warbler *(Acrocephalus melanopogon)* differs from

Fig. 11-22. Marsh warbler *(Acrocephalus palustris)*.

Fig. 11-23. Great reed warbler *(Acrocephalus arundinaceus arundinaceus)*; 2. Eastern great warbler *(Acrocephalus arundinaceus orientalis)*; 3. Iraq great reed warbler *(Acrocephalus griseldis)*; 4. Egyptian great reed warbler *(Acrocephalus stentoreus)*.

Fig. 11-24. Greater swamp warbler *(Calamocichla gracilirostris)*.

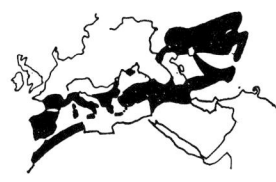

Fig. 11-25. Cetti's warbler (*Cettia cetti*).

Fig. 11-26. Grasshopper warbler (*Locustella naevia*).

Fig. 11-27. 1. Melodious warbler (*Hippolais polyglotta*); 2. Icterine warbler (*Hippolais icterina*).

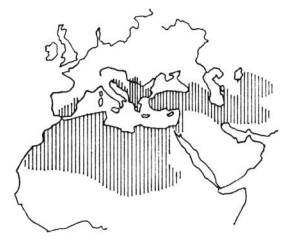

Fig. 11-28. Olivaceous warbler (*Hippolais pallida*).

the reed warbler especially in its habit of holding its tail boldly erect. It is distributed over the European Mediterranean area, also eastwards as far as Austria, Hungary and Rumania and again in the Kirgise Steppe, Turkestan, Transcaspia and southwards as far as southern Iran. As a partial migrant, it appears as early as the first days of March at Lake Neusiedel (Austria and Hungary). It enlivens the reed beds, even on cold pre-spring days with its beautiful reed warbler type song in which "lu lu lu" phrases of rising pitch are often interwoven. It is found particularly where there are stands of reed mace and *Cladium mariscus*, the scientific name of the latter plant being responsible for the bird's German name, Tamarisc Warbler. The female also builds the frequently extensive nest of thread-like algae and plant wool in these plants. The moustached warbler feeds mainly on aquatic insects and their larvae, and spiders.

The uniformly colored GREAT REED WARBLER (*Acrocephalus arundinaceus*) also obtains most of its food from the surface of the water; its diet includes large dragonfly larvae and the larvae of the diving beetle and other water beetles. In Hungary a pair of these birds was even observed feeding small fish to its young. The female also takes the dripping wet nest materials from the water and with them she skillfully weaves the nest between upright reed stalks. The Dutch ornithologist, Kluyver, has filmed and described this process (Color plate, p. 242). The loud rattling, very characteristic song, which is striking because of its considerable jumps in pitch, is widely known. The closely related Egyptian great reed warbler inhabits papyrus swamps from Egypt and Israel eastwards, while the smaller, long-billed, gray Iraq great reed warbler is only found in Iraq. The thick-billed warbler which is often placed in a separate genus (*Pragamaticola* as *Pragamaticola aedon*) occurs in southern Siberia, northern Mongolia, Manchuria, the Amur and Ussuri lands and northeastern China, in contrast to the other reed warblers.

The reed warbler, marsh warbler, Blyth's reed warbler and the Indian reed warbler are very similar to one another, but they can be distinguished, among other things, by details of the wing pattern and on the basis of their songs. The native European reed warbler (*Acrocephalus scirpaceus*) is a skillful hunter of mosquitos. Its small hanging nest is often suspended from only two reed stalks, and is composed mainly of dry, very fine materials; the edge of the nest is often covered with spider webs and its cavity is often lined with the ribs of reed leaves which the bird breaks off itself (Fig. 11-32). The naked young hatch from the densely spotted eggs after twelve days of incubation. They leave the nest when they are ten to twelve days old, at which time they are still unfledged and move around the vicinity of the nest by climbing about the reeds.

The marsh warbler (*Acrocephalus palustris*; Color plate, p. 232) of most of Europe and Blyth's reed warbler of eastern Europe and Asia, which occurs towards the west as far as southern Finland, are twin species. The marsh warbler inhabits marshy areas and nettle thickets along ditches and shore lines, as well as willow stands with dense undergrowth. Only more recently has it become a bird of cultivated areas and now it often breeds in wheat fields and as a result is called "cereal warbler" in German. Because of its fast song which often sounds as if given by two voices, and its great imitative ability (in song), the marsh warbler is one of our best songsters. Its characteristic loose nest is often hung among nettles or among the stalks of meadowsweet. The nest is built by both partners in a matter of five to six days, and the interior is only lightly lined with rootlets and hairs. The attractive bluish-white or greenish-white eggs have olive-brown spots and fine blackish dots as well as ash-gray blotches. They are incubated mainly by the female; both parents rear the young, however.

Fig. 11-29. Nest of the sedge warbler.

While the naked, newly-hatched young of all grass and reed warblers always have two dots on the tongue, young grasshopper warblers have three such spots, one at the tip of the tongue and one on each side of the back of the tongue. Grasshopper warblers owe their name to their more or less grasshopper-like whirring or stridulating song; only the taiga grasshopper warbler has a loud, tuneful song. The group is also characterized by a very markedly graduated tail, very long under-tail coverts and the absence of rictal bristles at the base of the very narrow beak. The grasshopper warbler's song is usually delivered in twilight; it is reminiscent of the green grasshopper's song and is responsible for the bird's English name. The grasshopper warbler lives in various habitats, but particularly moist meadows, swampy areas (which in America would be called a form of muskeg), and areas of bush and shrubs; there it walks about on the ground, as skillfully as a mouse in the grass, for it is a characteristic ground bird. Only when the males sing do they move to more elevated situations. Then one can see them with wide open beak; they move the head from side to side as they sing, thereby achieving ventriloquial song effects. Because it is a ground bird, the male also has a tripping display on the ground. The young leave the artless, cup-shaped nest early, when eleven days old, and can walk well from the first day. The grasshopper warbler is replaced in central and eastern Asia by Pallas' warbler which occurs in open swampy areas, and the streaked grasshopper warbler *(Locustella lanceolata)*, which lives in swampy woodland.

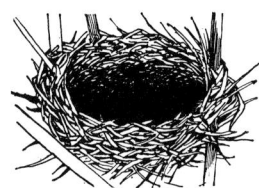

Fig. 11-30. Nest of the aquatic warbler.

The purring song of Savis' warbler *(Locustella lusciniodes)* is heard from large reed swamps. This bird often begins its whirring song phrases with a tick sound. Like the moustached warbler, it too occurs in particularly great density in stands of the common reed, which because of the dense network of leaves cannot be inhabited by other reed warblers. Here, or in the stalks of large sedges, or on broken reed

Fig. 11-31. The nest of the moustached warbler.

Fig. 11-32. The nest of the reed warbler shows marked angles at the points of attachment.

maces or reed stalks, the pair builds its often large, spherical cup nest, the only part of which looks at all "orderly" being the nest depression, which is often lined with very wide reed leaves. The eggs, like those of other grasshopper warblers, have fine spots, and are dull grayish brown or strongly rusty-red on a whitish background. As Oscar Heinroth reports, the young cannot walk properly for a while after fledging, and during this time they show a peculiar "polka hop", an intermediate phase between hopping and walking.

Recently, the river warbler *(Locustella fluviatilis)* has been spreading westwards from eastern Europe. It inhabits wooded swamps and wet woodlands along rivers, and is detected by its characteristic whetstone-like song. It wanders the furthest south on its migration (among the warblers in this group) to Zambia and the Transvaal. In densely forested areas of central and eastern Asia, it is replaced by the taiga grasshopper warbler; the males of the latter species utter a bell-like song which sounds like "tooti rooti...rooti tooti."

The HIPPOLAIS WARBLERS (for which there is no English group name) differ from the reed warblers in their non-graduated tail, the wide flat beak and the egg markings. The fact that the icterine warbler *(Hippolais icterina;* Color plate, p. 232) has been able to settle central European parklands is due to its ability to fix its nest independently of low shrubbery high up in the forks of tree branches. It weaves the nest to its surroundings with webs, stalks and bast. This bird's restless behavior while feeding is characteristic, and in the search for food it always covers considerable distances on the wing. The many fluting tunes and imitations of the sounds of other birds woven into its own abrupt and, as it were, compressed song, have earned it its German name which in translation means "yellow mimic." It sings its characteristic song with wide open beak, with accompanying head movements which look almost cramped, and with great endurance. As in all the warblers of this group, the eggs, usually five, are strikingly colored; there are sparse black dots, spots and hair lines on a deep rose-colored background. Both partners incubate the clutch for thirteen days, the female doing the greater share. The icterine warbler leaves its breeding area early in August to winter in tropical and southern Africa.

Its nearest relative is the short-winged melodious warbler *(Hippolais polyglotta)* of southwestern Europe, which prefers moister habitats. Its song, a flowing, enduring, twittering warble, is more melodious than that of the icterine warbler. The olive-tree warbler, the largest Hippolais warbler, is an inhabitant of olive tree stands, oak forests and almond tree plantations. It inhabits only a small area in the eastern Mediterranean region. Its song resembles that of the great reed warbler; it winters in southern Africa. The olivaceous warbler *(Hippolias pallida)* is extending its breeding area northwards from the Balkans. It builds its deep cup-shaped nest in bushes or trees, often also in

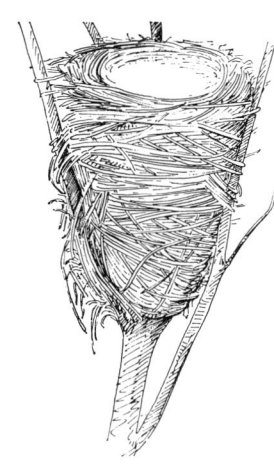

Fig. 11-33. Nests of the melodious warbler are generally in the forks of branches of shrub.

raspberry bushes, and covers it on the outside with lichens, pieces of bark, spider webs and cocoons as do the other hippolais species. It, like *Hippolais caligata* of Russia and Siberia, lacks the yellow and green tones in the plumage of others of this genus.

The warblers of the genus *Sylvia* (see distribution maps) are insect, berry, and fruit eaters. The two sexes often differ. Many species have a song flight. The dots on the tongue of the young are only occasionally present; many species have a red or rose-colored gape. Males build several nests of which the female chooses one, which is then completed by both of them. There are nineteen species distributed mainly in Europe, northern Africa, and western Asia. Included in this genus are: the BARRED WARBLER (*Sylvia nisoria;* Color plate, p. 232) which has a L of 17 cm and occurs eastwards from Europe to northwestern Mongolia; the ORPHEAN WARBLER (*Sylvia hortensis*) which has a L of 15 cm and is found around the Mediterranean and Asia Minor eastwards to Afghanistan; the GARDEN WARBLER (*Sylvia borin;* Color plate, p. 232) which has a L of 14 cm; the BLACKCAP (*Sylvia atricapilla;* Color plate, p. 232) with a L of 15 cm; the LESSER WHITE-THROAT (*Sylvia curruca;* Color plate, p. 232) which has a L of 13.5 cm. There are seven subspecies, particularly in the Transcaspian-Turkestan area where they are readily separable. This bird is found from Europe to Transbaikalia; the WHITE-THROAT (*Sylvia communis;* Color plate, p. 232) with a L of 14.5-15 cm; the SPECTACLED WARBLER (*Sylvia conspicillata*) with a L of 12.5-13 cm; MARMORA'S WARBLER (*Sylvia sarda*) with a L of 12.5 cm; the DARTFORD WARBLER (*Sylvia undata*) with a L of 13 cm; the SUBALPINE WARBLER (*Sylvia cantillans*) with a L of 12.5 cm only inhabits the Mediterranean area; RUEPPEL'S WARBLER (*Sylvia rueppeli*) of the eastern Mediterranean has a L of 14 cm; the SARDINIAN WARBLER (*Sylvia melanocephala*) has a L of 14 cm; the DESERT WARBLER (*Sylvia nana*) has a L of 11.5 cm and is a pale sand-color. It occurs from Astrakhan to eastern Persia and the Sahara; TRISTRAM'S WARBLER (*Sylvia deserticola*) has a L of 12 cm. The head and back are blue gray; the back is also somewhat brownish. The cheeks are blackish, and there is a white malar stripe, the underparts are a terracotta color. This species inhabits bush areas in northern Africa (distribution maps see Fig. 11-34 through 11-37).

Together with white-throats and red-backed shrikes, the largest of the sylvia warblers, the barred warbler inhabits thorny, bushy areas, particularly in eastern Europe. It betrays its presence by its striking song flight and by its habit of erecting its tail when excited. It delivers its fast, rich, pure-toned song in short phrases often including harsh "arr" sounds and phrases of imitations of other bird sounds. Among European Sylviidae it is the best imitator. Both partners build the nest, a fairly large flat cup in hedges or in rather large isolated bushes like the white thorn. In certain circumstances it uses berries to a considerable degree to rear its young. It leaves its breeding area early and migrates in a southeasterly direction to winter in eastern Africa.

Fig. 11-34. Garden warbler (*Sylvia borin*).

▷
1. Willow warbler (*Phylloscopus trochilus*); 2. Chiffchaff (*Phylloscopus collybita*); 3. Wood warbler (*Phylloscopus sibilatrix*); 4. Greenish warbler (*Phylloscopus trochiloides*); 5. Goldcrest (*Regulus regulus*); 6. Bonelli's warbler (*Phylloscopus bonelli*); 7. Firecrest (*Regulus ignicapillus*); 8. Tailor bird (*Orthotomus sutorius*); 9. Ruby-crowned kinglet (*Regulus calendula*); 10. Variegated wren (*Malurus lamberti*). 11. Yellow-browed warbler (*Phylloscopus inornatus*); 12. Yellow-tailed thornbill (*Acanthiza chrysorrhoa*); 13. Blue-gray gnatcatcher (*Polioptila caerulea*); 14. Black-throated warbler (*Gerygone palpebrosa*); 15. Southern emu-wren (*Stipiturus malachurus*).

Fig. 11-35. Spectacled warbler *(Sylvia conspicillata).*

◁
1. Willie wagtail *(Rhipidura leucophrys);* 2. Asiatic paradise flycatcher, with the fairly frequent white form *(Terpsiphone paradisi);* 3. Black-headed pitohui *(Pitohui dichrous);* 4. Japanese paradise flycatcher *(Terpsiphone atrocaudata);* 5. Narcissus flycatcher *(Ficedula narcissina);* 6. Golden whistler *(Pachycephala pectoralis);* 7. Collared flycatcher *(Ficedula albicollis);* 8. Pied flycatcher *(Ficedula hypoleuca);* 9. Blue flycatcher *(Cyanoptila cyanomela);* 10. Black-naped flycatcher *(Hypothymis azurea);* 11. Least flycatcher *(Ficedula parva);* 12. Spotted flycatcher *(Muscicapa striata);* 13. Rufous-bellied niltava *(Niltava sundara);* 14. Blue-throated flycatcher *(Cyornis rubeculoides);* 15. Puff-back flycatcher *(Batis capensis).*

The orphean warbler has a slower, melodious, almost thrush-like song. Adults, like those of the barred warbler, have a yellow iris. The songs as well as the habitats show geographical differences. This bird is found in open oak woods, cork oak woods, olive hedges, pine forests, parks and plantations of citrus fruits, as well as in the steppe of low dwarf bushes of the Mediterranean. It winters in Africa and India.

The blackcap is probably the most familiar western and central European warbler of this genus. It belongs to the mixed partial migrants; those which breed nearer to the pole migrate over the breeding range of the species as far as tropical Africa, while those which breed nearer to the equator (in the Mediterranean area) are residents and partial migrants. This species, with its beautiful song, which consists of a soft warbling prelude and a loud fluting main song, belongs to the most popular and earliest arriving bird harbingers of spring. It keeps higher up in the trees than do other Sylviids: it also keeps high in the tall undergrowth of mixed deciduous woods and gardens and, correspondingly, places its nest high up as well. Although the male upon arrival at the breeding place builds several nests in the territory, the later arriving female has the task of completing the chosen nest. Both partners incubate the clutch, which generally consists of five eggs, for thirteen to fourteen days. They also share the feeding and rearing of the young even when they leave the nest, after they are ten to thirteen days old. The blackcap's strong beak enables it to eat soft juicy fruit from which it knows how to peck off pieces. Like the garden and barred warblers and the common and lesser white-throats, blackcaps congregate in numbers in the fall on elder bushes to eat the ripe berries. This enables them to lay down fat for the coming migration.

The insignificantly colored garden warbler uses similar habitats, but prefers deciduous or mixed woods which must have dense undergrowth. With its uniform "organ-like" song delivered in long phrases, it belongs to the most attractive songsters. Sauer has described its simple display. When a female appears, the male with nest material in his beak shows her the nest site he has chosen; he sings while doing so, spreads his tail, moves the outspread wings and presses himself deeply into the nest site with his head trembling. A long migration takes this warbler to southern Africa where it moults the larger feathers. Most subspecies perform a migration of about 9,000 kilometers to winter in Africa.

On the other hand, European white-throats, which do not have so far to go, complete their main moult in the breeding area before beginning the fall migration. The common white-throat is a well distributed bird of low bushy areas and hedges; it also occurs in the taller weeds and in bramble tangles along ditches, roadsides and railway embankments where, more than other Sylviidae, it seeks food near the ground. From the tip of a bush or a thick stalked weed, the male delivers its rough, hurried-sounding, twittering and rises to a song flight. The nest is placed close to the ground in dense vegetation and is generally lined with spider webs and plant wool.

In contrast to the white-throat, the lesser white-throat leaves its European breeding area in autumn in a southeasterly direction. Its song, often with a soft twittering prelude, is clattering, a feature which has coined it one of its German names, "little miller." This species is found in the most varied bushy areas with higher shrubs or low conifers, particularly on the edges of woods, in gardens or parks, in the mountains even above the upper zone of the forests, in the zone of *Pinus montana*. The male begins the construction of the loose, cup-shaped nest, which is completed by both partners. Both partners incubate the four to six eggs in alternation for eleven to twelve days and brood the young for another eleven to twelve days in the nest.

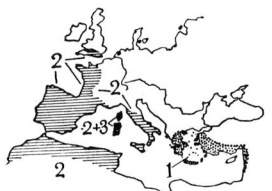

Fig. 11-36. 1. Rueppel's warbler *(Sylvia rueppelli);* 2. Dartford warbler *(Sylvia undata);* 3. Marmora's warbler *(Sylvia sarda).*

The Sylviidae of the Mediterranean lands are unbelievably fast at slipping about; they often cock up their tails boldly. The smaller the species is, the more hastily it delivers its twittering song and the more rapid are the sequences of its alarm calls. The Sardinian warbler has the widest distribution of these species. In suitable areas this warbler is a common bird of the evergreen bush lands, the so-called Macchia. There it delivers its rattling song often from pistachio bushes or rises in its short dancing song flight. It is a resident in the Mediterranean area and is therefore, like the Dartford warbler, subject to mass deaths in severe winters. The Dartford warbler has also settled southern England and there it inhabits stands of heather and broom.

Fig. 11-37. Sardinian warbler *(Sylvia melanocephala).*

Rueppel's warbler with its black head, white cheek stripe and red eye ring is one of the most beautiful of the Mediterranean warblers of this genus. More than in the other species, the males sit out in the open on the tips of shrubs and bushes in the rocky areas of the Agaean to sing their loud song which has clattering phrases. The widely distributed subalpine warbler is less choosy in its selection of habitats. It is found breeding at altitudes of up to 1,100 or 2,000 meters in mountains. The small spectacled warbler in Spain is found even higher, up to 3,000 meters, although otherwise it keeps to lower levels, often the Salicornia steppes of coastal lagoons.

The LEAF WARBLERS (genera *Phylloscopus* and *Seicercus*) occur in all parts of the Old World except the Australian continent, New Zealand and Madagascar. They are insect-eaters which build oven-like nests, lay eggs which are white or brown with spots, and have downy young which have no tubercles on the tongue. With almost fifty species, *Phylloscopus* is one of the largest genera of song birds.

Of the Asiatic species, we will mention the following: The KINGLET WARBLER *(Phylloscopus proregulus)* with a L of 9 cm and the YELLOW-BROWED WARBLER *(Phylloscopus inornatus;* Color plate, p. 259) which has a L of 10

Fig. 11-38. White-throat at the nest with young.

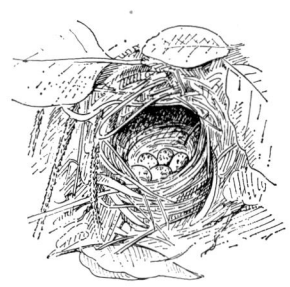

Fig. 11-39. The "oven-like" nest of the wood warbler.

Fig. 11-40. Greenish warbler *(Phylloscopus trochiloides).*

Fig. 11-41. Arctic warbler *(Phylloscopus borealis).*

Fig. 11-42. Chiff-chaff *(Phylloscopus collybita).*

cm: both species have distinct wing bands, dark stripes at the side of the head, a paler stripe through the center of the head and a yellow rump. They occur in Europe as vagrants. The MOUNTAIN LEAF WARBLER *(Phylloscopus trivirgatus)* is found in Malaya, the Philippines and the Moluccas as far as New Guinea and the Solomon Islands. The GREENISH WARBLER *(Phylloscopus trochiloides;* Color plate p. 259) with a L of 11 cm, and the ARCTIC WARBLER *(Phylloscopus borealis)* with a L of 12.5 cm, both have no head pattern apart from the pale eye stripe, but they have two wing bars of which only one is distinct: the BRIGHT-GREEN LEAF-WARBLER *(Phylloscopus nitidus).* Both the BROWN WILLOW WARBLER *(Phylloscopus fuscatus)* with a L of 11 cm and RADDE'S BUSH WARBLER *(Phylloscopus schwarzi)* with a L of 13 cm, inhabit low shrubbery areas; both are dark with a very prominent pale superciliary stripe and a dark eye stripe. The latter species also occurs in trees at the edges of the Taiga.

Of these eastern leaf warblers, the greenish and arctic warblers which belong to the subgenus *Acanthopneuste* are spreading westwards and southwards. They are readily distinguished by their songs. The Arctic warbler delivers a pleasant sounding whirring, reminiscent of the cirl bunting; the greenish warbler sings a phrase comparable to the common wren's song.

The central European and British leaf warblers (see Figs. 11-40 through 11-43 for distribution) are small, very swift, inconspicuously colored birds which live in the network of leaves and branches of tree crowns and the woodland undergrowth. There are four species: the CHIFF-CHAFF *(Phylloscopus collybita;* Color plate, p. 259) with a L of 11 cm; the WILLOW WARBLER *(Phylloscopus trochilus;* Color plate, p. 259) with a L of 11 cm; the WOOD WARBLER *(Phylloscopus sibilatrix;* Color plate, p. 259) with a L of 13 cm, and BONELLI'S WARBLER *(Phylloscopus bonelli:* Color plate, p. 259) with a L of 11.5 cm.

Finally, we will mention the mainly Indo-Malayan genus of FLYCATCHER WARBLERS *(Seicercus)* with nineteen species, of which a few occur in Africa such as the YELLOW-THROATED WOODLAND WARBLER *(Seicercus ruficapillus)* which is found from southeastern Kenya through Tanzania and eastern Rhodesia as far as the east of the Cape Province of South Africa. The YELLOW-EYED FLYCATCHER WARBLER *(Seicercus burkii)* has a L of 11.5 cm and is found in the Himalayas and eastwards to western and southern China, Burma and Indochina. The flycatcher warblers are small birds which occur particularly in evergreen forests. Generally they are more strongly colored and in more lively patterns than are the true leaf warblers (genus *Phylloscopus*) and they have wider beaks with which they often catch arthropods on the wing.

Here it is only possible to consider the life of the central European and British species of leaf warblers in some detail. As in all of this genus, the female chiff-chaff builds its oven-shaped nest without much help from the male. The nest is very well hidden, a little above the ground in low shrubbery or heather. The willow warbler, twin species of the chiff-chaff, builds its nest right on the ground; the nest itself has a very small opening so that one can hardly see the eggs. Both species line the nest amply with feathers. However, as Aschenbrenner was able to show, the wood warbler and Bonelli's warbler do not line their nests. All four species are easily distinguished by their songs, but less so by their habitat requirements, so that several species often occur side by side. The chiff-chaff prefers areas with taller trees and less undergrowth in mountains; it occurs at much higher levels than does the willow warbler, which is typically found in willow and poplar thickets of shores and further north in the subarctic birch woods and in the bush tundra. The wood warbler is a characteristic bird of beech woods; the warmth-loving Bonelli's warbler occupies dry warm oak, beech and pine woods as well as subalpine mixed beech and spruce woods on slopes with southern exposures in central Europe. Further south, however, it is also found in cork oak stands and sycamore woods.

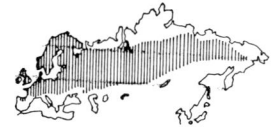

Fig. 11-43. Willow warbler *(Phylloscopus trochilus)*.

The wood warbler performs a slow song flight from twig to twig between low placed twigs. In display, it shows a bat-like flutter flight as it follows the female which has entered its territory. If the female stays in the territory, pair formation follows. The breeding behavior of these four leaf warblers shows considerable similarity. The very passive role of most males is striking; they keep in touch with the female during the breeding period only by contact calls. It is therefore the female who chooses the nest site. The eggs are incubated for about thirteen days and the young stay in the nest for eleven to thirteen days after hatching. The part taken by the male in feeding the young varies individually a great deal. The female has to be particularly busy when the young are about to leave the nest, at which time they are the most endangered. The young have a great desire for contact at that time. When they wish to rest, they line up together on a perch in a characteristic "sitting" and later "sleeping in contact." This urge disappears after they have become independent; it lasts longest in the chiff-chaff, and as a result the young of this species keep together in sibling groups for the longest period. The distances of the winter quarters from Europe differ in the four species. This distance is related to differences in wing shape, length of stay in the breeding area, the breeding cycle, the development of feathers in the young, in the moult, and weight changes before and during the period of migratory restlessness.

Fig. 11-44. Bonelli's warbler *(Phylloscopus bonelli)*.

Gwinner reports on these matters as follows: "The willow warbler, which winters in tropical and southern Africa, has the longest migration route. That of the wood warbler, which winters in the northern savanna belt of Africa, is shorter. Directly north of the wood warbler's winter range is that of Bonelli's warblers from western Europe. Finally, the chiff-chaff winters as close to its breeding range as the Mediterranean and northern Africa."

BABBLING THRUSHES OR BABBLERS AND OLD WORLD WARBLERS 265

Fig. 11-45. A tailor bird at its nest.

Subfamily: wren warblers, by H. Th. Condon

Fig. 11-46. Distribution of the wren warblers (subfamily Malurinae).

We have already noted a tendency to sew nest material among various representatives of the subfamily Sylviinae. The TAILOR BIRDS (genera *Orthotomus* and *Phyllergates*) have developed this ability to perfection. There are seven species including the TAILOR BIRD (*Orthotomus sutorius*; Color plate, p. 259) has a L of 14-17 cm in males and a L of 13 cm in females. Half of the male's length is due to the strongly graduated tail (9 cm), the central feathers of which are prolonged during the breeding season. This species is distributed in India, Ceylon, southern China, Malaya and Java. The other species, too, inhabit the Malayan area.

The tailor bird inhabits gardens, orchards, hedges, reed beds and forests with trees of medium height, preferably near settlements. All tailor birds are real artists. They bend together one or more living leaves to make a sort of cone, puncture the leaves near the edge with the beak and sew them up with spider webs or threads they have picked up. Into the pouches so formed, they build the nest of plant wool, sheep's wool and other animal hairs. Tailor birds generally lay three eggs.

The WREN WARBLERS, also known as AUSTRALIAN WARBLERS or AUSTRALIAN WRENS (subfamily Malurinae) include both highly colored as well as insignificant-looking birds. The L ranges from 9-26 cm, but is generally 9-14 cm. The legs are thin and rather long; the tail is often very elongated. The nest is enclosed all around and has a side entrance (one genus builds open nests). The distribution is mainly in Australia, but also in the Papuan area eastwards as far as Polynesia and New Zealand; only a few species are distributed westwards as far as the Malayan region. There are two groups of genera: A. THORNBILLS or THORNBILL WARBLERS (*Acanthiza*) and related genera; and B. SUPERB WARBLERS and relatives.

Included among the insignificantly colored thornbills and relatives are the following: 1. The ACANTHIZAS or THORNBILLS proper (genus *Acanthiza*) generally have spots of light color on the forehead. There are eleven species, including the LITTLE THORNBILL (*Acanthiza nana*) with a L of 11 cm, which is found in Australia. 2. The FLY-EATERS (genus *Gerygone*) have no light spots on the forehead. There are about fifteen species, including the green and yellow-colored MANGROVE FLY-EATER (*Gerygone sulphurea*) found from southwestern Thailand to the Philippines and the island of Alor (north of Timor). 3. The SCRUB WRENS (genus *Sericornis*) are generally longer billed. There are eleven species, among them the ARFALE BUFF-FACED SCRUB WREN (*Sericornis rufescens*) which is olive-green above and brownish-white below. It has a pale orange color about the eye, and is found in New Guinea. 4. The EMU WRENS (genus *Stipiturus*) have only six long and fringed (emu-like) tail feathers. The throat is grayish-blue, and the sides of the abdomen are rusty yellow. There are two species, including the SOUTHERN EMU WREN

(*Stipiturus malachurus;* Color plate, p. 259) of Australia, which has a L of 18 cm.

In the group known as the superb warblers (and by other names) there are some very colorful species. We will mention the following: 1. The warblers of the genus *Malurus* have feathers with a velvety and enamel-like effect. The males are magnificently colored, being blue, violet, red, black, and white; the females are more insignificant. The male's out of season plumage is partially like that of the female. The tail is long and not fringed. There are eleven species in Australia, and one in New Guinea. These include the VARIEGATED WREN (*Malurus lamberti;* Color plate, p. 259) of Australia, which has a L of 13 cm. 2. The GRASS WRENS (genus *Amytornis*) have a long tail; their coloration is generally brownish with whitish stripes. There are seven species including the WESTERN GRASS WREN (*Amytornis textilis*) which has a L of 18 cm and is found in central and western Australia. 3. The AUSTRALIAN CHATS (genus *Ephthianura*) are reminiscent of the wheatears and chats. They should probably be considered as belonging to a separate subfamily. They are patterned in white, red, yellow, and orange, and are more short-tailed than the others in this group. There are five species, including the CRIMSON CHAT (*Ephthianura tricolor*) which has a L of 10 cm and is found in Australia.

Most wren warblers resemble the leaf warblers of Europe in their behavior, insofar as they too work over the foliage for insects and spiders. Many of these birds can be found in the driest areas of the Australian interior, in thorn scrub, in spinifex grass, and on the ground where species of the genus *Amytornis* hop about with constantly erected tail. Many species are good songsters; the mangrove fly-eater, for example, sounds a sequence of seven tones of descending pitch. Nests are covered on top and have side entrances except in the genus *Ephthianura,* the members of which build cup-shaped nests. The *Acanthiza* and *Gerygone* species suspend their nests from a branch, often near the nests of pugnacious wasps. The crimson chat wanders about in central Australia. Like many of the birds of that area, it occasionally breeds socially, yet each pair maintains its own territory within the breeding area of the group.

In spite of its display of colors, the variegated wren gives the effect of an inconspicuous and generally silent bird. It builds its well-hidden nest close to the ground. In September to December, the female lays three or four white to reddish eggs with red spots and dots. Only she incubates. In the species so far studied, only the female builds the nest. In other *Malurus* species, the young, presumably those of the first brood, help the parents to feed the young of the later broods. A male SUPERB BLUE WREN (*Malurus cyaneus*) has lived in the Frankfurt Zoo since 1965; it is therefore at least six years old, a remarkable success of keeping in captivity. It may be noted that in spite of several partial moults, this captive male never developed a full nuptial plumage.

Subfamily: kinglets, by E. Thaler-Kottek

Fig. 11-47. 1. Goldcrest (*Regulus regulus*); 2. Firecrest (*Regulus ignicapillus*).

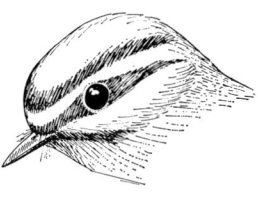

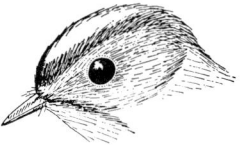

Fig. 11-48. Facial pattern of the firecrest (top). The erection of the feathers of the crown which indicates an aggressive mood (middle) and the simpler facial pattern (at rest) of the goldcrest (below).

The KINGLETS (subfamily Regulinae) are the smallest European or North American song birds. They used to be regarded as related to the tits and chickadees. The L is 8.5-9.5 cm. The beak is fine and straight with a bristle feather covering the nostril; the plumage is dense, brownish-grayish-green with a bright yellow or orange colored area on the crown which is surrounded by black. The tail tip is indented. These birds are inhabitants of coniferous forests; they are very lively, and not at all shy, but because of their minuteness (and agility) are difficult to observe. There are two genera, one in central Asia and western China, another with five species in North America, Eurasia and Taiwan. The European species are: 1. GOLDCREST (*Regulus regulus*; Color plate, p. 259) has a L of about 8.5 cm. It has a yellow crown mark, which in the male is orange-colored in the center and is yellow only in the female. The colored feathers are erected in excitement and they then take up the whole width of the crown; this serves as a signal. 2. FIRECREST (*Regulus ignicapillus*) is more highly colored than the former; it is almost orange-yellow at the sides of the neck with a wider crown-patch, which in the male is more deeply orange in the center than in the goldcrest. The pale superciliary stripe reaches forward to the root of the beak and below it runs a dark cheek stripe. This species is only a vagrant in the British Isles while the goldcrest, as in most of Europe, is a resident breeding bird.

Goldcrests live in dense coniferous forests during the breeding season. They can be recognized by their contact call—a fine, high pitched "see see see" and by the song of territory claiming males; this is a twitter, which falls and rises slightly in pitch and ends with a stronger trill of several syllables. Both partners build the small, skillfully-made nest which hangs on the down-slanting side branches of thick branches, generally those of spruce. The edge of the nest is somewhat drawn-in relative to the rest of the cup. Nest materials include mosses and liches, matted together with the webs of various arthropods; the nest cup is lined with feathers and hair. Such a nest is a good heat insulator and the small bird is able to incubate its seven white eggs which are clouded with gray (and which it cannot all cover with its body). The incubation periods lasts twelve to sixteen days. Both parents rear the young, which stay in the nest for fifteen to seventeen days and are fed and led by the parents for an additional two weeks after fledging. The young lack the colored spot on the crown until a month after fledging, when there is a juvenile moult of the small feathers.

Goldcrests have two broods a year—the first in April to May, the second immediately thereafter. They often begin the annual moult during the year's second brood and thereafter gather into flocks. They are wanderers; goldcrests which breed in central Europe, winter in southern Europe and are replaced in their breeding range by others from more northern breeding areas. During the migration periods they wander together with troops of tits, even in parks and reed beds and ceaselessly seek, sometimes by hovering, arthropods which form the bulk of their food. Even in winter they can find such prey and in a snow-covered coniferous forest they know how to climb along the underside of branches without coming into contact with the snow. Their need for food is extraordinary. To keep their body temperature at 40°C, they must eat several times their own body weight per day during the migration periods. In accordance with this intake, they weigh about five grams in the daytime but one gram more at the end of the day. In frosty nights they form overnight associations of fifteen to thirty birds. These formations are preceded by complex pacification ceremonies for these aggressive birds will not otherwise approach one another closely.

Fig. 11-49. Ruby-crowned kinglet *(Regulus calendula)*, vertical lines; golden-crowned kinglet *(Regulus satrapa)* horizontal lines.

The firecrest too, prefers coniferous forests, but also light mixed forests, and it occasionally nests in ivy or in elderberry bushes. The eggs have reddish spots and correspond in number, incubation period and fledging period of the young to those of the goldcrest. This species also has a similar contact call, but the song which is used in territory claiming is different; it is a fast "see see" which swells in volume about the middle of the phrase and then suddenly breaks off in a strong final tone. Fledged young firecrests already have the colored crown spot. In Germany, the goldcrest breeds up to altitudes of 2,000 meters in mountains; the firecrest only breeds up to altitudes of about 1,000 meters.

The two other members of the genus, *Regulus goodfellowi* of the mountains of Taiwan and the GOLDEN-CROWNED KINGLET *(Regulus satrapa)* of North America, are much like the goldcrest. The RUBY-CROWNED KINGLET *(Regulus calendula;* Color plate, p. 259) which is also North American, deviates from the others of the genus in some respects; its wine-red crown is not surrounded by black and is found only in the male, the female having no crown spot at all. The song is also quite different, being richer and more melodious; it also includes typical flute-like tones.

Fig. 11-50. Green hylia *(Hylia prasina)*.

The position of the insufficiently known genus *Leptopoecile* with *Leptopoecile sophiae* in central Asia and western China, and *Leptopoecile elegans* of western China, relative to the kinglets, is still uncertain. Even the systematic position of the kinglets themselves is by no means finally ascertained, as their behavior shows certain peculiarities. They are reminiscent both of the Sylviidae warblers and of tits and chickadees.

The GREEN HYLIA *(Hylia prasina)* with a L of 13.5 cm and a weight of 14 g, is the sole representative of its subfamily (Hyliinae) and belongs to the birds related to the sylviid warblers. This inconspicuous bird, which is olive green above and gray below, inhabits creepers and densely leaved bushes, as well as tree crowns in the African rain forest from Guinea to Angola. In a hidden but busy way it searches for

Subfamily: hyliinae, by W. Meise

▷
Eurasian robin *(Erithacus rubecula)* feeding its young.

Fig. 11-51. Green hylia.

Fig. 11-52. Blue-gray gnatcatcher *(Polioptila caerulea)*.

arthropods like a leaf warbler. Its full double whistles, "woe woe" betray it from afar. There is a pause of about half a second between the two tones and the pitch falls only at the end, and then rarely. The song is a chattering "tshet-sheri." This bird avoids human habitations. Its nest is about fifteen centimeters high and nine centimeters wide; it is placed in forks of branches, sometimes in shrubs. Apart from a side entrance some two centimeters in diameter, it is completely closed in. Two white eggs are laid per clutch.

The GNATCATCHERS (subfamily Polioptilinae) have a L of 9-12 cm. They are restless, very thin-billed hunters of arthropods. They generally pursue very small prey. The tail is very mobile, narrow and long, in many species giving the effect of a little stick; it is flicked up, down or sideways. The nests are cup-shaped and are generally placed in horizontal twigs. There are three genera with thirteen species in the Americas.

The BLUE-GRAY GNATCATCHER *(Polioptila caerulea;* Color plate, p. 259) has a L of 11.5 cm. It is common in tree crowns in moderately moist forests, but also inhabits open, dry mixed forests, oakwoods, chapparal, pinon pine woods and elder bushland. Its striking nest, clothed with lichens, is easily found before the leaves emerge. In any site all pairs are never at the same phase of the breeding cycle, hence territories in this species are not very rigid. Often pairs which are just rearing young are forced back into the nucleus of their original territory, while other members of the species settle the rest of it. Wintering may take these birds as far as Honduras, Cuba and Mexico, but many of them winter in southern parts of the breeding range. The BLACK-TAILED GNATCATCHER *(Polioptila melanura),* on the other hand, is a resident. It has a L of 11 cm and lives in the semi-deserts of the western United States and northern Mexico.

As representatives of the two other genera, we will mention the LONG-BILLED GNATWREN *(Ramphocaenus melanurus)* which has a L of 12 cm and a strikingly long beak, and the small COLLARED GNATWREN *(Microbates collaris)* which has a L of 9 cm. Both of these species inhabit warmer areas of the Americas; the long-billed gnatwren is found from southern Mexico to the state of Sao Paulo in Brazil, and the collared gnatwren lives in southern Venezuela and adjacent areas. Whether these birds really do belong to the gnatcatchers is questionable.

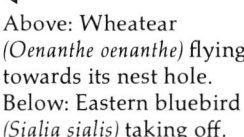

◁
Above: Wheatear *(Oenanthe oenanthe)* flying towards its nest hole. Below: Eastern bluebird *(Sialia sialis)* taking off.

12 Flycatchers and Thrushes

The presentation of the giant family of FLYCATCHER-LIKE BIRDS, Muscicapidae, begun in the preceding chapter, is concluded in this chapter with a description of the remaining members of the family. Most of the birds to be mentioned in this chapter are readily divisible into two groups which correspond to adaptations to two different modes of feeding. In the flycatchers (in the extended sense), we find a wide, fairly flat beak surrounded by bristles which make it easy for the bird to guide flying arthropods and other very small flying animals into its gape. Flycatchers also eat berries, fruit and other types of food. Thrushes, in contrast to this, use a narrower and generally not very long beak to pick up their insect food; they also feed on berries, fruit and other foods.

Nevertheless, there are so many exceptions to the basic structure in both groups that ornithologists have once more given up the attempt to draw a sharp line between flycatchers and thrushes. The flycatchers, in the extended sense of the word, include the FLYCATCHERS PROPER (Muscicapinae), the FANTAILS (Rhipidurinae), the PARADISE FLYCATCHERS (Monarchinae), and the THICKHEADS (Pachycephalinae), while the THRUSHES are grouped into one subfamily, Turdinae. The WREN THRUSH is rather distinct from these two groups and, according to the latest studies, it should be placed in its own family. Both the flycatchers (in the extended sense) and the thrushes share one common feature—the presence of light-colored spots in the plumage of the young.

The flycatchers proper are the most markedly typical of the flycatchers in the wider sense. They feed predominantly in flight. The distribution of the subfamily Muscicapinae is shown on the map in Fig. 12-1. The L ranges from 9–22 cm; the wings are generally more pointed than in the other groups. There are ten primaries, the outermost being very short; the tarsi are shorter than in the other groups. The snapping beak is pointed and more or less hooked at the tip; it is flattened at the base and surrounded by bristles (vibrissae). These

Flycatchers and thrushes, by W. Meise

Subfamily: true flycatchers, by R. Berndt

birds perform catching flights from a perch into the air, into the network of branches or down onto the ground. Their food consists of medium-sized insects, caterpillars, spiders, centipedes, wood lice, the smallest snails and other similar fare. Some flycatchers nest in the open, some in holes. The juvenile plumages are almost always spotted. These birds are distributed from northern Eurasia over the world to southern Africa, Australia and Hawaii. There are almost 60 genera with an even 200 species. These are divisible into four groups: 1. TYPICAL FLYCATCHERS, 2. PALE FLYCATCHERS, 3. PUFF-BACK FLYCATCHERS AND RELATIVES, and 4. FLAT-BILLED FLYCATCHERS.

Most members of the subfamily, including all the European species, belong to the most highly specialized groups of the typical flycatchers, of which we will first consider the hole-breeding flycatchers (genus *Ficedula*). The PIED FLYCATCHER (*Ficedula hypoleuca;* Color plate, p. 260) has a L of 13 cm and weighs 13 g; it is one of the most adequately studied west and central European birds. Its breeding range extends from northwestern Africa and western Europe as far as central Siberia. The males in the central zone of this belt are only a little darker than the females, even in full maturity, but in northern Europe, on the British Isles and in the Mediterranean, including the Alps and the Atlas Mountains, the males are almost black and white, at least in maturity.

Although this flycatcher occurs only locally in western Europe and western Germany, it is a widely distributed forest and park bird in the rest of central Europe. As a hole-breeder, it depends on suitable natural cavities or nest boxes. Up to fifteen pairs may settle one hectare if nest sites are available. It finds its best habitats in sunny, open mixed oak forests, but it also likes to live in birch and alder stands and, in much lesser density, in stands of red beech, pine and spruce forests. It is a migrant, as are all flycatchers in the north of the Old World, and it arrives in central Europe from its central African winter quarters from mid-April onwards, in central Siberia from early May onwards, and in Lapland not before mid-May. In the fall northern migrants pass through Germany as late as October, while those birds which have bred there have already left in July. Pied flycatchers generally live in the crowns of tall trees and one hardly notices more about them than the male's song, which begins with the phrase "tee whoo tee woo." Immediately after their return in spring, the males seek a nest hole, create a territory around it, and show it to the females, which arrive a few days later. If the female accepts the hole, she builds the nest within it. The nest itself is loosely constructed of dry grass stalks, leaves and glistening pieces of pine bark, a particularly favorite material. Many males also display before additional holes which may be up to several hundred meters from the first, and they may succeed in attracting other females to these holes. Thus, ten to fifteen per cent of males have several territories and are either bigamous or polygamous. The five to seven pale blue eggs measure 18 by 13 mm and weigh 1.6 g. The female alone incubates for thirteen to fourteen days. The young, which hatch with only a scanty covering of down, are tended by both parents for fifteen to sixteen days in the nest. A few days later, the family has usually already wandered away from the breeding area and scarcely a week after fledging, the young are independent. A successful first brood is only extremely rarely followed by a second one in central Europe, but in England and Russia this happens somewhat more frequently. The clutch size increases from the west to the east of the range by one egg. In addition, more eggs are laid in lowlands than in mountains, and more in favorable as opposed to less favorable habitats. Aside from this fact, we must realize that one year old females lay fewer eggs than do older ones, and clutches laid late in the season are smaller than those laid earlier.

Fig. 12-1. Distribution of the flycatchers proper (subfamily Muscicapinae) and of the fantails (subfamily Rhipidurinae).

When they reach sexual maturity, at just under a year old, some of the young of the year before settle at and around the site of their birth; however more than half of the yearlings breed at a distance of one to one hundred kilometers from where they were hatched. In all subsequent years, the males, which have become very loyal to the site of their own first breeding, return to this spot or to places within an area of at most 1,000 meters from it. Females, on the other hand, often settle in new breeding places in successive years, or return to the site of their birth, so their breeding sites may be up to 150 kilometers apart in successive years. Even within one breeding season, a female which has lost her eggs or young may settle down to a replacement breeding only a week later at a place up to fifty kilometers away, a mobility known in no other bird species.

Fig. 12-2. 1. Pied flycatcher *(Ficedula hypoleuca)*; 2. Collared flycatcher *(Ficedula albicollis)*; 3. Half-collared flycatcher *(Ficedula semitorquata)*.

As in the case of most nest-hole breeders, only a few nests, eggs or clutches fall victim to enemies or disasters; thus fledged young survive from eighty-five per cent of the eggs laid. Between the time the young fledge and the time they reach the age of one year, the mortality is very high—seventy per cent; in later years it lies below and finally above the average annual mortality of forty-five per cent. The median age of the population is between two and three years; the maximal age of certain individuals is at least eight years. Losses from nests, as is understandable in a hole-nester, are mainly due to pine martins, weasels and stoats, squirrels, dormice, the yellow-necked mouse, the great spotted woodpecker and the wryneck. Newly fledged young are particularly preyed upon by the common (Eurasian) jay. The chief enemies of the adults are the Eurasian sparrowhawk and the tawny owl.

The COLLARED FLYCATCHER *(Ficedula albicollis;* Color plate 260) and the HALF-COLLARED FLYCATCHER *(Ficedula semitorquata)* are so closely related to the pied flycatcher that hybrids are found in the areas where their ranges overlap. The three species are therefore called triplet species.

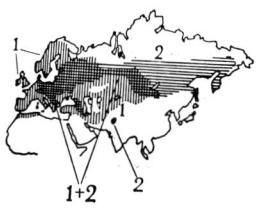

Fig. 12-3. 1. Spotted flycatcher *(Muscicapa striata)*; 2. Least flycatcher *(Ficedula parva)*.

The collared flycatcher lives locally in Italy, in eastern France, western and southern Germany, on Swedish islands of Oland and Gotland, as well as from Czechoslovakia and Yugoslavia as far as the middle Volga. The half-collared flycatcher is distributed from the Balkans and the Caucasus as far as central Asia.

The last central European nest-hole breeding flycatcher is the LEAST FLYCATCHER (*Ficedula parva;* Color plate, p. 260) which has a L of 12 cm, and weighs 10 g. Its striking song, "tink tink tink aida aida aida weed weed weed" is heard only in the late spring, at which time the bird hunts unobtrusively in the crowns of trees of deciduous, mixed and coniferous forests for insects and spiders, after it has arrived from its Indian winter quarters, generally in the second half of May. Its breeding area extends from Kamachatka and the northwestern Himalayas over Siberia and northern Iran to Europe, where it occurs locally as far as the Black Forest, the Teutoburger Forest, the Luneburg Heath (all in Germany) and in Denmark. It builds its mossy nests in semi-cavernous depressions on tree trunks, branches or behind the bark, in cracks in rocks and similar places, but sometimes also in proper holes or in the open in the forks of branches, or a whirl of branches. The four to seven eggs of the single annual brood are yellowish to greenish-white with a pink hue. Heinrich Frieling has the following description of this delicate little bird in his book, 'The Singing Bush': "The least flycatcher has the orange-colored throat of the (European) robin, the brown of the chaffinch, the rhythm of the wood warbler in song and the softness of tone of the willow warbler. It is the living sprite of the cathedral-like beech forest."

The following are hole-nesting flycatchers of the Asiatic-Indo-malayan region: the KOREAN FLYCATCHER *(Ficedula zanthopygia),* in which the male takes a share in incubation; the NARCISSUS FLYCATCHER *(Ficedula narcissina;* Color plate, p. 260), a colorful species of Japan and Hopeh (China); the BLACK-AND-ORANGE FLYCATCHER *(Ficedula mugimaki);* the ORANGE-GORGETTED FLYCATCHER *(Ficedula strophiata);* the RUSTY-BREASTED BLUE FLYCATCHER *(Ficedula hodgsonii);* the WHITE-BROWED FLYCATCHER *(Ficedula superciliaris);* and the SLATY BLUE FLYCATCHER *(Ficedula tricolor).* All five species are found in the forests of the Himalayas. The BLUE FLYCATCHER *(Cyanoptila cyanomelana)* has a L of 14 cm and forms a separate genus; it occurs as a nester in niches in forests, on cliffs and precipitous shores in eastern Asia.

The NILTAVAS (genus *Niltava)* also live in Asia; included among them is the largest of all flycatchers, the LARGE NILTAVA *(Niltava grandis)* of the Himalayas with a L of 21 cm, and the smaller RUFOUS-BELLIED NILTAVA *(Niltava sundara;* Color plate, p. 260) from western China. In southeastern Asia there are eighteen species of the genus *Cyornis,* including the beautiful BLUE-THROATED FLYCATCHER *(Cyornis rubeculoides)* which occurs from the Himalayas as far as Ceylon, and the colorful *Cyornis rufigastra* of western Indonesia. The inconspicuous OLIVE-BACKED JUNGLE FLYCATCHER *(Rhinomyias olivacea)* has a L of 14 cm and is found on Indonesian and Malayan plantations and along forest edges with others of this genus. The AFRICAN TIT FLYCATCHERS (genus *Parisoma*) look more like tits or chickadees than like flycatchers when they seek food among branches and foliage; in addition, most of them do not have a spotted juvenile plumage. Included in this genus are the GRAY TIT FLYCATCHER *(Parisoma plumbeum)*, which has a L of 14.5 cm and inhabits the savannas right around the rain forest area, and the TIT BABBLER *(Parisoma subcaeruleum)* which has a L of 14.5 cm and lives in the southern steppes. The brownish-white MOUNTAIN NEWTONIA *(Newtonia brunneicauda)* has a L of 12 cm and lives in a similar way.

The genus of FLYCATCHERS proper *(Muscicapa)* with its fifteen species, consists entirely of birds which hunt insects on the wing. One of these is the SPOTTED FLYCATCHER *(Muscicapa striata)* which has a L of 14 cm and weighs 17 g. It is a well known bird in Europe, and occurs in seven subspecies from northwestern Africa over Eurasia as far as Transbaikalia and the western Himalayas. It is a confiding bird of gardens, parks, cemeteries and all, not overly dense, deciduous forests, and, to a lesser degree, of coniferous woods. It builds its nest in semi-cavities on buildings, rocks and trees, as well as occasionally in the forks of branches, in which case it likes to use an old nest of some other bird as a foundation. The spotted flycatcher returns to central Europe in early May from its wintering area which extends from northwestern India over Arabia as far as southern Africa. The male sings only little; his song consists of monotonous sounds like "pst." Four to six cloudy brownish-grayish-white and rust-red eggs are laid and generally incubated only by the female. The young hatch after twelve to thirteen days of incubation and are fully fledged after being tended by both parents for fourteen days. There may be only one or two broods per season.

Fig. 12-4. Display behavior of the pied flycatcher.

The African species, *Muscicapa gambagae,* from the savanna south of the Sahara, is possibly the resident-remaining ancestral form of the spotted flycatcher. Other similarly inconspicuously colored species include the SIBERIAN FLYCATCHER *(Muscicapa sibirica),* the smaller BROWN FLYCATCHER *(Muscicapa latirostris)* from eastern Asia, and the DUSKY FLYCATCHER *(Muscicapa adusta)* from southern Africa. The LEMON-BREASTED FLYCATCHER *(Microeca flavigaster),* which pursues its insect hunting on isolated trees of the savannas of northern Australia and New Guinea, is distinguished by its particularly short legs, and has a L of 11.5 cm.

The five species of the related genus *Eumyias,* which includes the colorful INDIGO FLYCATCHER *(Eumyias indigo)* with a L of 14 cm, are particularly characteristic of the Malayan insular area.

The second main group of the true flycatchers is formed by the PALE FLYCATCHERS of tropical Africa with eight genera and fifteen species.

Fig. 12-5. Display of the collared flycatcher.

These birds are typical insect hunters, descending from a perch to the ground to seize their prey in semi-open country. Included in this group are the MARIQUA FLYCATCHER *(Bradornis mariquensis)* which has a L of 18 cm and is mainly in the southwest African acacia savannas, and the PALE FLYCATCHER *(Bradornis pallidus)*, which has a L of 17.5 cm and is found mainly in the East African acacia savannas; the South African BLACK FLYCATCHER *(Melaenornis pammelaina)*, which has a L of 20 cm, is also found in such habitats.

The group of relatively long-tailed PUFF-BACK FLYCATCHERS and relatives includes eleven genera and thirty-four species in Africa, as well as the PUFF-BACKS (genus *Batis*). The PUFF-BACK FLYCATCHER *(Batis capensis;* Color plate, p. 260) has a L of 11 cm and is common in southern Africa; it generally lives in pairs, often in great density, and in association with other species in the moister forests. These birds hop from branch to branch in search of food. When excited, they produce a purring sound by flapping the wings rapidly. This is probably an intensification of the wing flicking which is characteristic of most flycatchers. Puff-back flycatchers have lived for over two years in the Frankfurt Zoo. Two other puff-backs inhabit drier areas of southwestern and southeastern Africa, respectively; they are twin species: the PRIRIT PUFF-BACK FLYCATCHER *(Batis pririt)* has a L of 12 cm and the CHIN-SPOT PUFF-BACK FLYCATCHER *(Batis molitor)* also has a L of 12 cm. Most male puff-backs have the black breast bands; these are colored in the females. The ample plumage of the back can be erected to look like a ball of wool.

Fig. 12-6. Puff-back flycatcher.

The three species of WATTLE-EYED FLYCATCHERS (genus *Platysterira*) are closely related to the puff-backs; they have a pale plumage and a wattle of bare red skin above the eye. One species in this genus is the BLACK-THROATED WATTLE-EYED FLYCATCHER *(Platysterira peltata)*, which has a L of 12 cm and is found in river-side meadows, dry stream beds and mangrove swamps in eastern Africa. Representatives of these species are rarely seen in zoos. The *Dyaphorophya* flycatchers, or wattle-eyed flycatchers, are closely related to the *Platysteira* wattle-eyed flycatchers. They have toothed yellow, blue, or red areas of spectacle-like skin about the eyes. The CHESTNUT WATTLE-EYED FLYCATCHER *(Dyaphorophya castanea)* has a L of 9 cm and is characteristic of central African forests. The LITTLE YELLOW FLYCATCHER *(Erythrocercus holochlorus)* and LIVINGSTONE'S FLYCATCHER *(Erythrocercus livingstonei)* are both almost as small with a L of 9.5 cm; both occur in a similar habitat near the East African coast and they move about in the foliage in pursuit of insects like leaf warblers.

Fig. 12-7. Head of the chestnut wattle-eyed flycatcher.

The fourth and last main group of flycatchers is that of the FLAT-BILLED FLYCATCHERS with seventeen genera and forty-nine species, all of which live in the Australian-Polynesian area. The black, white and yellow PAPUAN FLYCATCHER *(Machaerirhynchus flaviventer)* from New Guinea and northern Australia has a L of 12 cm; this bird has a partic-

Fig. 12-8. The papuan flycatcher.

ularly wide bill provided with long bristles at the base. The genus *Myiagra* includes the MICRONESIAN BROAD-BILL *(Myiagra oceanica)* which has a L of 11 cm and occurs in four very differently colored subspecies.

Both the MIROS (genus *Miro*) which live in two species in New Zealand, and the PETROICAS (genus *Petroica*) which inhabit the Australian-Polynesian area in eight species, are ground dwellers and correspondingly longer-legged, longer and softer-winged than the preceding groups. The SCARLET ROBIN *(Petroica multicolor)* has a L of 11.5 cm and is distributed from Australia to Samoa. Both sexes of the TOMTIT *(Petroica macrocephala)* from New Zealand, both have a black and yellow nuptial plumage in the Aukland Islands but on the southern island only the males show such a coloration. These birds have a L of 13 cm. The FRILLED FLYCATCHER *(Arses telescophthalmus)* has a L of 16 cm and is from northeastern Australia and New Guinea; it has blue or yellow rings of bare skin around the eye, and the plumage is black and white. One flycatcher species has even reached the Hawaiian Islands; this is the ELEPAIU *(Chasiempis sandvicensis)* which has a L of 14 cm, and gets its name from its call. This bird is found in mountain forests and has a fairly long tail which it often erects.

Although flycatchers inhabit most of the earth, not one species of this successful subfamily has gained a foothold anywhere in the New World, perhaps because its habitats there were already occupied by tyrant flycatchers (Chapter Four).

The FANTAILS (subfamily Rhipidurinae) differ markedly from the flycatchers proper in the structure of the palate. They are also quite different from the paradise flycatcher group which we will discuss later. Fantails have a L ranging from 12 to 22 cm. They often carry their bodies horizontally with the wings slightly raised; the tail is very long, graduated at the sides and is often fanned out. The juvenile plumage shows no obvious light-colored spots. The nest consists of strips of bark, grass and spiderwebs and looks like a wineglass suspended upside down. These birds are found in the southeastern part of the Old World from the Himalayas as far as Australia, New Zealand and Polynesia with the greatest abundance of species being in New Guinea. There are two genera: 1. The PIED FANTAILS *(Rhipidura)* with twenty-three species including: The WILLIE WAGTAIL *(Rhipidura leucorphrys)* (see Color plate, p. 260) has a L of 22 cm and is black with a stripe above the eye; the breast and belly are white. This bird is found from Australia and New Guinea as far as the Solomons. The RUFOUS-FRONTED FANTAIL *(Rhipidura rufifrons)* has a L of 15 cm and is found from eastern Australia to Sumba, Celebes, the Marianas and the Solomon Islands. 2. The YELLOW-BELLIED FANTAIL FLYCATCHER *(Chelidorynx hypoxantha)* has a L of 12 cm and is the only representative of its genus. It is found from North Vietnam to the Himalayas and southwestern Szechwan.

The willie wagtail is one of the most common inhabitants of Australian settlements. It must have trees only in the vicinity of its habitat,

▷
Thrush-like birds:
1. White-rumped shama *(Copsychus malabrirus)*;
2. White-capped redstart *(Chaimarronis leucocephalus)*;
3. Orange-flashed bush robin *(Tarsiger cyanurus)*;
4. Eurasian robin *(Erithacus rubecula)*;
5. Redstart *(Phoenicurus phoenicurus)*;
6. Black redstart *(Phoenicurus ochruros)*;
7. Black redstart, subspecies *(Phoenicurus ochruros aterrimus)*;
8. Robin chat *(Cossypha caffra)*;
9. Bluethroat *(Luscinia svecica)*;
10. Ruby throat *(Luscinia calliope)*;
11. Nightingale *(Luscinia megarhynchos)*;
12. Thrush nightingale *(Luscinia luscinia)*.

Subfamily: fantails, by W. Meise

Fig. 12-9. Rufous fantail.

◁
1. Purple cochoa *(Cochoa purpurea)*; 2. Stonechat *(Saxicola torquata)*; 3. Blue whistling thrush *(Myiophoneus caeruleus)*; 4. Wheatear *(Oenanthe oenanthe)*; 5. Pied chat *(Oenanthe pileata)*; 6. Leschenault's forktail *(Enicurus leschenaulti)*.

Subfamily: paradise flycatchers, by W. Meise

Fig. 12-10. Distribution of the paradise flycatchers (subfamily Monarchinae).

for it is really a bird of open woodlands. It hunts flies and other arthropods on the wing from a perch, which may even be the back of a cow, but it does not, like the European flycatchers, usually return to its original perch from its hunting flights. It also picks up food from the ground and from trees. During the breeding season the willie wagtail is closely tied to its territory, from which it chases intruders, both large and small. Its simple sequence of tones is heard night and day during the breeding season. The nest is suspended from a horizontal branch and contains three to four gray to pale brown eggs, the darker spots of which often form a wreath about the blunt pole. The incubation and fledging periods both last thirteen days. While the willie wagtail is a resident, the rufous-fronted fantail wanders from southeastern Australia to New Guinea outside the breeding season.

The PARADISE FLYCATCHERS AND THEIR RELATIVES (subfamily Monarchinae) are certainly the most colorful and impressive group of the flycatcher-like birds. They are smaller, medium-sized birds with a L of 14-53 cm (including the tail). The beak is generally wide but not particularly flat and often has a keeled ridge on the upper mandible. The legs are short; the tail can be short to very long. The wings are relatively long. Unlike most flycatcher-like birds, the members of this subfamily do not hunt flying prey but rather search for insects in the foliage of trees and shrubs. The nest is a deep but minute cup fastened in a horizontal or vertical tree fork. The highly decorative males also incubate. The young are not spotted in the first plumage. The distribution is from Senegal across Africa, southern Asia and Indonesia as far as Tasmania and Polynesia. There are eleven genera with fifty-two species, including: 1. The MONARCH FLYCATCHERS *(Monarcha)* often have a velvet-like plumage on the forehead and chin or the whole head. The colors are often very strongly demarcated from one another, being reddish-brown and blue-gray, or black and white. The distribution is from Celebes to the Carolinas, the Bismarck Archipelago and Tasmania. Members of this genus include the ISLAND GRAY-HEADED MONARCH FLYCATCHER *(Monarcha cinerascens)*, the BLACK-FACED FLYCATCHER *(Monarcha melanopsis)*, which has a L of 15 cm, and the YAP MONARCH FLYCATCHER *(Monarcha godeffroyi)*.

2. The MARQUESAS FLYCATCHERS *(Pomarea)* comprises five species, including the MARQUESAS FLYCATCHER *(Pomarea mendozae)* which is found in seven subspecies only on the Marquesas; some of the subspecies show different degrees of sex differences.

3. The BLACK-NAPED FLYCATCHERS (genus *Hypothymis*) are silky-blue colored birds with black and white plumage areas. They include the BLACK-NAPED FLYCATCHER *(Hypothymis azurea;* Color plate, p. 260) found from India as far as Taiwan, the Philippines and Java. It has a narrow black stripe at the nape of the throat and above the root of the beak on the forehead. A Philippine species of this genus has magnificent crest feathers.

4. The PARADISE FLYCATCHERS (genus *Terpsiphone*) are larger and have larger beaks. One species from the Philippines is a fiery cinnamon-brown and has a tail of normal length. In many other species, the adult males have very long tails. Members of this genus include: The ASIATIC PARADISE FLYCATCHER (*Terpsiphone paradisi*; Color plate, p. 260) which, with a L of 53 cm (42 cm of which is taken up by the tail), is the largest of all flycatcher-like birds. It is almost pure white with a black head. The JAPANESE PARADISE FLYCATCHER (*Terpsiphone atrocaudata*; Color plate, p. 260) has white only on its lower parts; the rest of the plumage is black with a dark purple gloss. The AFRICAN PARADISE FLYCATCHER *(Terpsiphone viridis)* has a long tail, most of which is black and white. The back may be brown, gray, black or white, all within the same distribution area; however, in other areas it is uniformly colored, being dark north of the Congo forests and paler in southern Africa.

The island gray-headed monarch flycatcher has split into many subspecies on the minute, mangrove-surrounded islands which lie between Celebes and New Britain. This gray bird has cinnamon-brown lower parts. The east Australian black-faced flycatcher migrates as far as New Guinea and Timor. It has a white eye-ring, a black mask and brownish-red lower parts; otherwise it is gray. The Yap monarch flycatcher was discovered on Yap, one of the Caroline Islands, by collectors traveling for the businessman, Jean Cesar Godeffroy (1813-1885), who was interested in discovery. The bird is white except for its head, throat, wings and tail.

Fig. 12-11. Asiatic paradise flycatcher.

The African paradise flycatcher has penetrated the whole of the Congo jungle starting from the surrounding ring of steppes, and it has even reached the area around the mouth of the Congo River. In so doing, it entered the range of short-crested paradise flycatchers which, themselves, had split long ago into a red-bellied and gray-backed, and gray-bellied, red-backed species. Occasional hybrids of the three present-day species occur in these areas where the populations, which had formerly become separated, meet once more (just as in central Europe where the carrion and hooded crow meet).

The Japanese paradise flycatcher prefers dense forests. Here, beneath the roof of foliage, it can find leafless branches to perch on, branches around and through which it can fly freely and, much further down in the undergrowth, it can find nest sites. Only reluctantly does this bird leave the forest cover to fly in the open. Its flight is swift and skillful. H. Jahn, giving his impressions of this bird, writes: "The male offers a wonderful sight when, in pursuit of an insect, it performs a rapid turn in the air and swings its long tail about elegantly, thus emphasizing the turn."

Fig. 12-12. Black-faced flycatcher.

The two-part song "teeay tleeooioo—woit woit woit" is tittered by both males and females, and imitations of it can be used to attract the birds. Jahn notes that "to construct a nest, the birds prefer a trunk in the undergrowth varying from finger to thumb thickness which has a fork about two to six meters above the ground."

The nest, with an external diameter of about eight cm is strikingly small, which may explain why the three young leave it when only ten days old at which time they have hardly attained half the size of their mother. Their flight feathers are fully developed at this stage, but the rest of the plumage is still incomplete and consists largely of down. The young are fed by the parents for three weeks after fledging and by then they are as large as the adults. The striking-looking male takes part in incubation and feeding of the young, but he does little feeding once his moult has started and he has already lost the long tail feathers. One year old males which still lack the long tail only breed exceptionally; they can be distinguished from females only by the wider blue ring around the eyes.

The Frankfurt Zoo has kept paradise flycatchers successfully for two years and the following data from there contributes to our knowledge of the longevity of these birds. Indian paradise flycatchers have lived there for over seven years and thus are at least eight years old. African paradise flycatchers have been kept since 1965. The males of both species renew their plumage excellently and continue to form nuptial feathers. Both species are kept at an artifically prolonged day length of twelve hours, in well planted flight cages at 24 to 26°C in air which is kept moist. They are given soft food mixtures supplemented with live insects such as flies and moths.

Subfamily: thickheads, by W. Meise

The THICKHEADS (subfamily Pachycephalinae) have a L of 12-28 cm, look almost like shrikes, and were for a long time classed with them. They are compact, generally large-headed birds, with a beak which bends down at the tip. The upper mandible has a notch behind the tip. The juvenile plumage is unspotted, as in the paradise flycatchers, and in most of the fantails. Thickheads are found mainly in Australia and from New Guinea east as far as Polynesia; westwards, only one species is found as far as Bengal. There are eleven genera with forty-six to fifty-four species. These include: 1. The EASTERN SHRIKE-TIT *(Falcunculus frontatus)* which, with a L of 17 cm, is the only species of its genus. It has a small keeled beak, and no rictal bristles. The head and crest are black with two white lateral stripes. The upper body is predominantly yellowish-green, while below it is almost entirely bright yellow. 2. WHISTLERS (genus *Pachycephala*) are often predominantly yellow. There are twenty-seven species, including the GOLDEN WHISTLER *(Pachycephala pectoralis;* Color plate, p. 260), which has a L of 18 cm and is green above and yellow below. It has a white throat with a black gorget below. The female is more significantly colored. The distribution range is from Lord Howe Island, the Santa Cruz Islands, New Caledonia, the Fiji Islands, and as far as Australia and eastern Java. There are at least seventy-three subspecies. 3. PITOHUIS (genus *Pitohui*)

Fig. 12-13. Distribution of the thickheads (subfamily Pachycephalinae).

comprises seven species, including the BLACK-HEADED PITOHUI (*Pitohui dichrous;* Color plate, p. 260) which has a L of 25 cm and is reddish-brown, with a black head, wings and tail. It is found in the mountain forests of New Guinea.

The shrike-tits feed mainly on berries. Whistlers live in bush-grown areas, in the lower levels of forests or in the undergrowth where they search the foliage for arthropods and, to a lesser extent, for berries and seeds. They live alone or in pairs. Their loud song sounds pleasant and varied; their flutelike call sounds like "we we we weet." The song of these birds can also be imitated so as to attract them. A female whistler sometimes sings a duet with the male. The strikingly bright yellow males not only defend their territories energetically, but also help build the nest and take part in incubation. The cup-shaped nest is placed in a fork. Two or three yellowish eggs with brown or gray spots are laid and incubated for fifteen days.

Fig. 12-14. Eastern shrike-tit.

For a long time ornithologists were uncertain about whether or not the THRUSHES (subfamily Turdinae) form a group which is clearly distinct from the flycatchers and Old World warblers. Nowadays most ornithologists do not separate the thrushes from the latter group, particularly as there are many transitional forms among them. At best, the large number of these birds might well justify such a separation.

Subfamily: thrushes, by J. Dorst

Thrushes are medium-sized birds with a L of 11.5–33 cm. The beak is strong and slightly decurved, sometimes with an only slightly developed hook; it either has no rictal bristles at its base or those present are few in number, and small. The tarsi bear horny plates rather than scale-like plates; these are often stronger and larger than in related birds, and are thus in keeping with their mode of life, which is carried out more on the ground than in trees. The wings, which have ten primaries, are somewhat rounded except in species which are true migrants. The tail has twelve feathers. Color and plumage patterns are highly variable as is the absence or presence of plumage sex differences. The juvenile plumage always shows light color spots on a dark background. In contrast to the Old World warblers, thrushes only have one annual moult. There are about sixty genera with an even 300 species, most of them of the chat, the wheatear or the thrush type. The wren thrushes (Chapter 15) are a deviant group, and, according to the latest views, they should not be placed here but near the wood warblers (Chapter 15), which is where we will discuss them.

Fig. 12-15. Distribution of the thrushes (subfamily Turdinae).

The thrushes as characterized above are similar to several families of song birds. The dippers (Chapter 10) are close to them although they are more specialized in structure and habits. The mockingbirds (Chapter 10) also resemble the thrushes in appearance and mode of life, but they do not have a spotted juvenile plumage. The accentors (Chapter 10) also show many of the thrush characteristics, but they differ in their thin, straight beaks, the structure of the plumage, and their color, which is reminiscent of that of finches. The groups which

Fig. 12-16. Rufous chat or rufous warbler *(Erythropygia galactotes)*.

Fig. 12-17. Robin *(Erithacus rubecula)*.

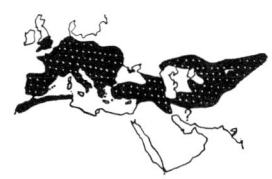

Fig. 12-18. Nightingale *(Luscinia megarhynchos)*.

have just been named are so close to the thrushes that many ornithologists include them among the relatives of the latter in the great family of flycatcher-like birds from which they have separated as forms which have evolved in their own particular directions. The thrushes are very widely distributed; one might call them inhabitants of the entire world. They have even penetrated to many isolated islands; they are absent only from Antarctica and from the most remote insular areas.

It may be assumed that thrushes originated in the Old World, where they have achieved the greatest variety; perhaps more specifically in the more northern areas. Africa then became a focus of further marked diversification, at least among the chats. The New World was probably colonized by the immigrants from Eurasia in at least three independent waves. Although most thrushes will use trees, they are still predominantly ground-dwellers. This plasticity enables them to settle the most diverse habitats, ranging from deserts and tundra to tropical rain forests. Their preferred habitats are areas where trees, bush, clearings and meadows are mixed. In general, they use a mixed diet, consisting of fruit and berries, but particularly of arthropods, larvae, mollusca and worms. Most of them are excellent songsters. They use cup-shaped nests placed in a tree or bush, but sometimes on the ground or in a cleft in a rock. The three to six eggs are either uniformly colored or show dark spots on a paler background. In warm countries thrushes are resident, but many northern forms travel very far as more or less pronounced migrants.

A large proportion of the thrushes are separable into two well characterized groups: 1. CHATS (AND WHEATEARS): 2. THRUSHES in the narrower sense. The chats are undoubtedly more primitive, and among them we find the greatest number of species which form transitions to other groups, particularly the Old World warblers and the flycatchers. They are small birds, generally with relatively weak beaks and are well represented in the tropical zones. The SHORTWINGS (genus *Brachypteryx*) and the group which includes the genus *Erythropygia* are probably the most primitive; they form a transition between the Old World warblers and the thrushes. Old World ROBINS *(Erithacus)* and related genera, like the MAGPIE-ROBINS *(Copsychus),* as well as the REDSTART GROUP *(Phoenicurus* and related genera) have evolved from relatives of the two groups. The FORKTAILS *(Enicurus)* and COCHOAS *(Cochoa)* and the SOLITAIRES *(Myadestes)* deviate rather more. At the end of this large group are the CHATS proper *(Saxicola)* and related forms, as well as the ROCK THRUSHES *(Monticola).*

The shortwings (genus *Brachypteryx*) have a L of 12 cm. They are very short-tailed and have markedly rounded wings. They range from India to the Philippines and the Lesser Sundra Islands. The LESSER SHORTWING *(Brachypteryx lecuophrys)* is olive-brown above and whitish below; it is a shy inhabitant of mountain forests rich in undergrowth.

Aside from one Australian-Paupuan genus *(Drymodes)*, the fork-tail group is purely African, although one species ranges as far as western Asia. There are six genera, including: 1. The FORK-TAILS proper *(Erythropygia)* are medium-sized birds. In build and behavior they are intermediate between sylviid warblers and the robinchats of Africa; however they differ from these birds in having less contrasting colors and white tips on the outer tail feathers. There are nine species including the RUFOUS WARBLER *(Erythropygia galactotes)* which has a L of 15 cm and a bright brownish-red tail which has a terminal black and white band. 2. The BLACKBUSH ROBIN *(Cercotrichas podobe)* has a L of 21 cm. It is blackish in color, and is found in deserts and semi-deserts from Senegal to Somalia and Arabia. 3. The RUFOUS ROCK JUMPER *(Chaetops frenatus)* has brown stripes on the back, a white stripe at the side of the throat and a brownish-red or brownish-yellow rump and lower parts. It is found in southern Africa. 4. The SCRUB ROBINS (genus *Drymodes*) have a longer tail. There are two species, which occur in arid areas in Australia. One of these, the NORTHERN SCRUB ROBIN *(Drymodes superciliaris),* is also found in New Guinea.

Fig. 12-19. Thrush nightingale *(Luscinia luscinia).*

The rufous warbler *(Erythropygia galactotes)* inhabits bush steppes, as well as areas converted by human cultivation, like gardens and orchards. It is a migrant and spends the winter at the edge of the Sahara.

The black bush robin, the only representative of its genus, inhabits deserts and semi-deserts from Senegal to Somalia and Arabia. It has a black plumage with white tail spots. It is found in thorn bushes and hunts arthropods on the ground but nests in trees or in holes in walls.

The Old World robins are marked as a group by weak beaks and long tails. Besides the Eurasian robin, this group includes the bluethroat and nightingale, and their relatives. There are five genera including: 1. The ROBINS *(Erithacus)* with the EURASIAN ROBIN *(Erithacus rubecula;* Color plates p. 269 and p. 279) which has a L of 13 cm and a reddish-orange breast. It is widely distributed from the British and Canary Islands as far as western Siberia, with related species in northern Asia and Africa. 2. The NIGHTINGALES *(Luscinia)* are very closely related to the robins, with many species in the north of the Old World, including the NIGHTINGALE *(Luscinia megarhynchos;* Color plate, p. 279) which has a L of 16 cm and is found from Great Britain to northern Africa, southwestern Siberia and central Asia. In northern and eastern Europe, as well as northwestern Siberia, it is replaced by the THRUSH NIGHTINGALE *(Luscinia luscinia;* Color plate, p. 279) which has a shorter outer primary feather, is more olive-brown in color and uses moister habitats. The RUBY THROAT *(Luscinia calliope;* Color plate, p. 279) male has a bright red throat and a white stripe above the eye and below the cheeks. This species is found in northeastern Asia and in southeastern Asia in the winter. The BLUETHROAT *(Luscinia svecica;* Color plate, p. 279) has a L of 14 cm. The male has a black-bordered blue throat with a white or reddish spot in the center; the female has a more whitish black-bordered

Fig. 12-20. Ruby throat *(Luscinia calliope).*

Fig. 12-21. Bluethroat *(Luscinia svecica).*

Fig. 12-22. Threat display of the Eurasian robin.

throat. The tail in both sexes is largely reddish-brown. This species occurs from northern and western Europe to eastern Asia. 3. ROBIN CHATS or FOREST ROBINS *(Cossypha)* have areas of bright brownish-red and black and often a white pattern on the head. Members of this genus include the SNOWY-HEADED ROBIN CHAT *(Cossypha niveicapilla)* which has a L of 18 cm. The top of the head is white and the sides are black. This bird ranges from Senegal to Kenya. 4. The MAGPIE ROBINS *(Alethe)* include the FIRE-CREST ALETHA *(Alethe diademata)* which has a L of 16 cm and is chestnut brown above with an orange striped crown, and whitish below. It occurs from Sierra Leone to Uganda.

The Eurasian robin prefers to live in deciduous or coniferous forests with undergrowth, up to an altitude of 2,000 meters. In Great Britain, it also inhabits gardens and is found near houses all the year round; however, in continental Europe, this applies only in the winter. It feeds on arthropods, larvae, and worms which it takes particularly from the forest litter.

The robin has long been the subject of investigation by a number of ornithologists; the best known of these zoologists is David Lack. He found that robins are—in human terms—very tolerant; they defend a territory of six to eight thousand square meters with emphasis. The song is full and includes short but unusually varied phrases so that it may be likened to pearls on a rope. Its main function is to delimit territories, and this is also aided by the bird's striking colors. In threat display, the red of the throat and breast is emphasized by erection of the feathers; lateral body movements further emphasize this coloring. Such threat movements are generally sufficient to cause intruders to flee from a territory-owning male. The brightly colored breast thus serves as a signal of imminent attack. This can be proved when a robin is shown a stuffed robin or even a mere bundle of red feathers, or a small piece of red cloth.

An abbreviated form of the same behavior is used in sexual display; this enables the two sexes to recognize one another. The female builds the nest in a depression in the ground, on banks, below tree stumps or roots, even in mouse holes, sometimes in a hole in a wall, but only rarely in a tree. Six eggs are laid on the average; they have reddish-brown spots on a whitish background. They are incubated by the female for twelve to fifteen days. Both parents feed the young, which leave the nest two weeks after hatching and are then fed by the parents for another two to three weeks.

Depending on the intensity of the local winter, the robin is either a resident or a migrant. In warmer areas of its range, it stays the year round (as for example, in western Europe where other robins from colder areas also winter). In the north, it is a migrant and wanders to the Mediterranean and Near East. The marked maintenance of territories persists in the winter quarters.

Of all central European song birds, the nightingale has become most famous on account of its song. It utters one of the most melodious songs of all thrush-like birds, rich and full, consisting of clear flute-like tones, but also often mixed with harsh rough sounds. The song shows a great variety of phrases and is delivered during the day as well as at night, when it is particularly impressive.

The nightingale, by preference, stays close to or on the ground, preferring low, dense bushes. In such places, it searches among the fallen leaves for worms, arthropods, larvae and spiders. The male has a territory which is sometimes very small, hence its frequent song. The female builds the nest in the center of the territory from dry leaves. The five olive-brown eggs are incubated by the female only for thirteen days. The male, however, shares in the rearing of the young for two weeks, after which time the young disperse before they can fly properly. As migrants, nightingales cover great distances because, since they depend mainly on insects for food, they cannot, like the robin, replace this by berries in winter. They migrate at night as early as September to the areas between the west and east coasts of Africa.

The bluethroat is a pronounced ground bird. It is found singly among dense vegetation and prefers moist areas, swamps and flooded areas, even when they are brackish. It feeds on arthropods and their larvae, mollusks, and worms, and occasionally on berries. The male has a loud song, in which the same tones are often repeated; it also imitates the songs and calls of other birds and even insect sounds. Unlike robins, which display secretively, the bluethroat performs conspicuous leaps from which it drops down again with widely spread wings and tail. Males also impress the females by displaying their blue breasts and simultaneously fanning out the tail and drawing the head back. The female builds her nest in holes in the ground or among roots, and incubates the five to six grayish-green, reddish spotted eggs for thirteen days. Both parents rear the young. Although individual bluethroats winter in western Europe, in places like Belgium or Switzerland, the majority migrate to northern Africa; they also cross the Sahara or wander to the Near East and southeastern Asia.

The snowy-headed robin chat lives like all robin chats, in dense forest undergrowth among the shrubs or on the ground in pairs. Here it seeks worms and insects including, among others, ants and termites. It is a good singer with a tuneful voice and utters a sequence of chattering sounds. The nest is close to the ground in a hole and is remarkably well camouflaged. The two to three bright olive-green eggs have fine brown spots.

The fire crest aletha is also at home in the dense undergrowth of primeval African forests; it is a bird which is inconspicuous, both in its mode of life and in its appearance. The song is a plaintive whistling. Its food, which it seeks on the ground, consists largely of ants. The

The nightingale

▷
Thrushes: 1. Cuban thrush *(Turdus plumbeus)*; 2. Bare-eyed thrush *(Turdus nudigenis)*; 3. Wren thrush *(Zeledonia coronata)*; 4. Orange-headed ground thrush *(Zoothera citrina)*; 5. Townsend's solitaire *(Myadestes townsendi)*; 6. Blue rock thrush *(Monticola solitarius)*; 7. Rock thrush *(Monticola saxatilis)*.

fire crest aletha is one of the most important species of the mixed bird societies of tropical forests, along with bulbuls and other thrush-like birds. All these birds gather in troops to hunt ants in the rain forests.

The MAGPIE ROBINS belong to a genus *(Copsychus)* which stands alone. It has eight species of moderate-sized Asiatic birds, the best known being the MAGPIE ROBIN *(Copsychus saularis)* with a L of 22 cm, and the WHITE-RUMPED SHAMA *(Copsychus malabaricus;* Color plate, p. 279) which has a L of 23-27 cm.

The magpie robin is distributed from India and southern China to Indonesia and the Philippines. This black bird with a bluish gloss has a white spot on the wing and a pure white abdomen; the outermost tail feathers are also white. The female has gray and blue replacing the male's black. This is one of the most common birds of India. It is found especially in gardens and on cultivated land. Its song, one of the most beautiful in the family of flycatcher-like birds, has made it world famous and provided a reason for its protection by man. The nest is placed in a hole in a wall or tree and the eggs are greenish-blue with reddish spots.

The shama is also distributed from India to Indonesia. It has a long graduated tail. The plumage is black with a white rump and outer tail feathers and a reddish-brown belly. This bird looks for food on the ground in dense jungles, particularly near water courses. The nest holds four to six greenish, dark brown spotted eggs. Like the magpie robin, the shama is also an excellent songster, which imitates other birds. It is a popular cage bird because of its readiness to sing and its liveliness.

Members of the redstart group do not always have a brownish-red tail, but these birds always depress their tails with a characteristic tremor. They are inhabitants of the northern Old World, including northern Africa, southern Asia and North and Central America. There are eight genera, including: 1. REDSTARTS *(Phoenicurus)* have brownish-red tails; there are several species, two of them in Europe. These are the REDSTART *(Phoenicurus phoenicurus;* Color plate, p. 279) which has a L of 13 cm and which ranges from western Europe to central Siberia and Afghanistan, and the BLACK REDSTART *(Phoenicurus ochrurous;* Color plate, p. 279) which is darker and occurs from western Europe and the Mediterranean as far as the Near East, northern China and the Himalayas, but not in Scandinavia. MOUSSIER'S REDSTART *(Phoenicurus moussieri)* lives in northern Africa; its eggs are all blue in some clutches, in others they are all white. 2. WHITE-CAPPED REDSTART *(Chaimarrornis leucocephalus;* Color plate, p. 279) has a L of 18 cm. It is the sole species of its genus. It has a blue-black breast and back, a white crown and a reddish-brown belly, rump and tail. It occurs in the Himalayas up to an altitude of 4,500 meters. 3. *Rhyacornis fluiginosus* is the only species of its genus. It is a glistening blue and is found from the Himalayas to China. 4.

◁
Thrushes: 1. Fieldfare *(Turdus pilaris);* 2. Mistle thrush *(Turdus viscivorus);* 3. Song thrush *(Turdus philomelos);* 4. Ring ouzel *(Turdus torquatus);* 5. Redwing *(Turdus iliacus);* 6. American robin *(Turdus migratorius);* 7. Blackbird *(Turdus merula).*

BLUEBIRDS *(Sialia)* are mainly blue and resemble the redstarts in build; there are three species, including the EASTERN BLUEBIRD *(Sialia sialis;* Color plates, p. 221 and p. 270) which has a L of 16 cm and is indigo-blue above, and reddish-brown below, except for the white belly. The females are darker and more grayish above. This species is found from Canada south to Central America.

Fig. 12-23. Redstart *(Phoenicurus phoenicurus).*

The redstart is one of the most common and most conspicuous European birds. It inhabits mainly treed areas, but these must be open; thus it never lives in dense forests. It catches flying arthropods along walls, rocks, or in the foliage of trees, as well as larvae, spiders, and worms, nor does it despise berries and other fruit. Its song consists of short phrases in which sequences of long tones alternate with shorter ones in a sort of twittering; it sometimes imitates other birds. The male performs a series of display flights in its territory, in which tail movements are conspicuous. The female builds her nest in holes, tree holes and holes in walls, including house walls. The five to seven bright blue eggs are incubated for twelve to fourteen days. Both parents feed the young for two weeks and then tend them after fledging for an additional two to three weeks. Unlike the other thrush-related birds, the family unit in the redstarts persists. As definite migrants, redstarts leave Europe in winter except for a few in the Mediterranean area. They migrate to Africa between its Mediterranean coast and the tropics, but rarely pass south of latitude 9°N. They migrate over western Europe and not the southeast.

Fig. 12-24. Black redstart *(Phoenicurus ochrurous).*

The black redstart was originally a mountain bird and it is therefore found up to an altitude of 3,000 meters in the Alps, where it lives mainly in rocky areas and on talus slopes, but also around walls and buildings. However, nowadays, it is also found in other areas right down to the sea, where it prefers cliffs. It nests in caves in rocks and holes in walls and builds a substantial heap of grasses, weed stalks and moss. It lays an average of five eggs which are white with a pink flush and sometimes spotted at the blunt pole. The female incubates for thirteen days. The reproduction is like that of the redstart.

The black redstart's song is poorer than the redstarts'; it consists of a rapid sequence of short, rough sounds which can be likened to that of paper being crumpled up, or the rubbing together of bits of broken glass. In winter this bird migrates to the Mediterranean, to the Near East and India; a few birds, however, stay in central Europe. From western Europe the black redstarts migrate to the west, although one central European population migrates southeast over Greece and Turkey to Egypt and the Near East. The establishment of the black redstart as a breeding bird in the south of England dates mainly from the period during and since the last war.

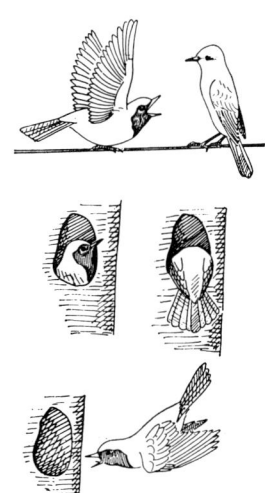

Fig. 12-25. The redstart displays in front of the female (top) and shows her the nest hole (middle and below).

Bluebirds are familiar American birds. Eastern bluebirds are frequently found in orchards, clearings, open forests and parks. These

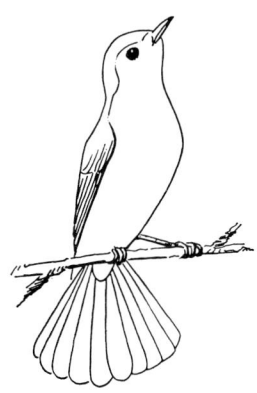

Fig. 12-26. Black redstart (in display).

Fig. 12-27. Leschenault's forktail.

Forktails and cochoas, by B. E. Smythies

birds prefer to eat grasshoppers and beetles and will also take fruit. They seize arthropods in flight, but are not as skillful at this as flycatchers; often they seek their food in dense vegetation or on the ground near selected areas. The vocal talent of the eastern bluebird is not remarkable; its song consists of only three to eight tones forming short, hardly tuneful, phrases.

However, the marked display flights of this bird are striking. The male spreads its tail and holds the wings half open. Males and females build the nest in a hole in a tree or some other cavity, the female doing most of the work. The five eggs, on the average, are glossy pale blue and are incubated for twelve days; the young remain in the nest for about two weeks. Since this bird is mainly an insect eater, it is a migrant. Nevertheless, a proportion of the population winters relatively far north, in New England for example, where it feeds mainly on berries. The wintering area extends from Arizona, Illinois and New York State to the southern United States.

The FORKTAILS *(Enicurus)* form a strongly deviant genus with seven species, of which we will mention three: LESCHENAULT'S FORKTAIL *(Enicurus leschenaulti;* Color plate, p. 280) has a L of 28 cm. It is found in southeastern Asia, and is long-legged, and black and white. The CHESTNUT-NAPED FORKTAIL *(Enicurus ruficapillus)* has a chestnut-brown crown and nape; in the female this color extends down to the back and the wing coverts. Six forktail species have long forked tails which are continually twitched up and down in the manner of the wagtails. The entire body and tail may also be swung from side to side. The mountain-dwelling LITTLE FORKTAIL *(Enicurus scouleri),* found from southeastern Russia to North Vietnam, has a relatively short tail; it can fan out and close its tail like the plumbeous redstart.

Forktails live on small, permanent water courses, usually on streams in dense forest. They take insects from the water and the stream beds. Their nest is an orderly but massive cup-shaped structure of moss mixed with roots and mud; as a result, it is almost always wet and heavy. It is placed on rocks or similar situations, generally by a water course, or even behind a waterfall, and occasionally at some distance from water.

The cochoas (genus *Cochoa*) of southeastern Asia are marked by widened beaks, and strongly developed rictal bristles. There are three species: The PURPLE COCHOA *(Cochoa purpurea;* Color plate, p. 280) the GREEN COCHOA *(Cochoa viridis),* both originally from the Himalayas and more rarely found in the mountains of Burma and Thailand, and the MALAYAN COCHOA *(Cochoa azurea)* of Java and Sumatra.

All three species of cochoas are rare and little is known of their behavior. They have brilliant colors. Thus, the purple cochoa is a dark purple-blue with black sides of the head, a black stripe above the eye, a pale blue crown and a tail which is black at the tip; females are reddish-brown rather than purple. The green cochoa is green, but blue on the crown, nape and wings with a black tip on the tail. Females differ in a somewhat more yellowish-brown wing color. The Malayan cochoa is mainly black and bright blue. These birds are exclusive tree-dwellers and fruit-eaters. They have long wings and short legs and toes. They live in evergreen mountain forests, build cup-shaped nests and are generally not found much below altitudes of 1,000 meters.

The solitaires (genus *Myadestes*) are also wide beaked, like the cochoas, but they are far more widely distributed than the latter. They were formerly often placed with the flycatchers which they resemble both in color and mode of hunting. However, certain peculiarities of structure, appearance and the song indicate their relationship to the thrushes. There are seven species, all found in America, including TOWNSEND'S SOLITAIRE (*Myadestes townsendi;* Color plate, p. 289) which has a L of 20 cm. This bird has an elongated shape. It is brownish-gray in color, has white outer tail feathers, and occurs from Alaska to Mexico.

Solitaires, chats and rock thrushes, by J. Dorst

The mountain haunts of Townsend's solitaire may extend above an altitude of 3,000 meters in Colorado. Its preferred habitats are light coniferous forests, particularly on steep and rocky slopes. There it lives singly, outside the breeding season (hence its English name), or in pairs. Its loud melodious song consists of clear tones delivered in a rapid sequence. This bird feeds on arthropods, caterpillars, ground insects, and all sorts of wild fruit. It often hunts insects like a flycatcher, as well as looking for them on the ground or among vegetation. It builds its nest on the ground, in cavities between roots, banks or among rocks. Here the female lays her four dirty-white, white or bluish eggs with small brown spots. In the north of its range, this solitaire is a partial migrant; otherwise it undertakes only local movements. It can endure the winter in areas which insect-eaters must leave at this season because it feeds on berries to such a degree.

The flycatchers of the genus *Stizorhina* have short, wide beaks. In contrast to young of the solitaires, the young of this genus are unspotted. These birds inhabit the African rain forests from Sierra Leone to the Congo and Angola. There are two species, one of which was formerly classed with flycatchers proper. This is the RUFOUS FLYCATCHER (*Stizorhina fraseri*). Strangely enough, the more ground-dwelling and slender-billed ANT THRUSHES (genus *Neocossyphus*) have the same colors. There are two species, one of which, the RED-TAILED ANT THRUSH (*Neocossyphus rufus*), lives from the Cameroons to Kenya and Tanzania. It is brownish-red above, pale on its outer tail feathers, and has a L of 18 cm.

Many systematists classified the *Stizorhina* flycatchers with the solitaires. These birds live in trees in the African rain forests while the RED-TAILED ANT THRUSH is more a ground dweller and prefers low shrubs where it hunts mainly termites and ants.

Fig. 12-28. Whinchat (*Saxicola rubetra*).

Fig. 12-29. Stonechat (*Saxicola torquata*).

Fig. 12-30. Wheatear (*Oenanthe oenanthe*).

Fig. 12-31. Black-eared wheatear (*Oenanthe hispanica*).

The CHATS proper form a large collection of species, many of which have calls resembling the sound of lip smacking. Although they are perching birds, the majority of them keep to the ground or in low vegetation. With the exception of the wheatear, these birds are confined to the Old World; in America they are replaced by other birds, similar in appearance, but belonging to other families such as the horneros.

The chats proper include seven genera; among them: 1. BLACKSTARTS (*Cercomela*) have longer tails than do the wheatears. They are rather short-winged. There are nine species including the BLACKSTART (*Cercomela melanura*) found from the Niger to Arabia, and the RED-TAILED CHAT (*Cercomela familiaris*) which has a brownish red tail with a black border, and is found from Nigeria to Kenya and southern Africa. 2. STONECHATS and WHINCHATS (*Saxicola*) have relatively long wings and a shorter tail. Apart from blackish brown feather edges, they have a paler color. They are very widely distributed in the north of the Old World. There are ten species, including the WHINCHAT (*Saxicola rubetra*) which has a L of 12 cm, and the STONECHAT (*Saxicola torquata*). These birds are widely distributed in Eurasia, Africa, Madagascar and even on the island of Reunion in the Indian Ocean. 3. ANTEATER CHATS (*Myrmecocichla*) have a very dark, often even black, plumage, although in flight their snow white wing coverts appear. This genus includes the ANTEATER CHAT (*Myrmecocichla aethiops*) which is rust colored and occurs from the Senegal to eastern Africa. 4. WHEATEARS (*Oenanthe*) are white, black and pale brown with very variable plumage patterns; they are found in the northern Old World, and one species is also found in North America. There are eighteen species; these include: The WHEATEAR (*Oenanthe oenanthe*; Color plates, p. 270 and p. 280) which has a L of 14 cm and is found from arctic Canada and Greenland over Europe to Siberia to Alaska. The DESERT WHEATEAR (*Oenanthe deserti*) has a black ear region, a gray back, and a white belly; it occurs in semi-deserts from the Sahara to Mongolia. The BLACK-EARED WHEATEAR (*Oenanthe hispanica*) is white with an orange tinge; it has a black eye stripe and wings, and is sometimes also black throated (the form "*strapazina*"). This species is found in the Mediterranean area as far as Iran. The PIED CHAT (*Oenanthe picata*) is found from Iran to the Kirgisian steppe and Pakistan. The MOURNING WHEATEAR (*Oenanthe leugens*) is found from Morocco to the Near East. The WHITE-CROWNED BLACK WHEATEAR (*Oenanthe leucopyga*) has a black plumage. The belly, rump and outer tail feathers are white; the crown may be white too. This species occurs from Morocco to southwestern Siberia and Ethiopia. The MOUNTAIN CHAT (*Oenanthe monticola*) is from southern Africa. The RED-RUMPED WHEATEAR (*Oenanthe moesta*) is found from Morocco to Iran. 5. The INDIAN ROBIN (*Saxicoloides fulicata*) has a L of 16 cm and is the only species of its genus. It has a different pattern of horny plates on its tarsus than do other thrush-like birds; this pattern is more like that of the accentors. The beak is clearly decurved. The plumage is black with a brown belly and a white wing spot; the female is browner. This bird is found in India.

The blackstart inhabits talus slopes and rocky mountain gorges on the edges of the desert. It hunts for arthropods on the ground or among the dry vegetation; its song is very modest. The red-tailed chat is also an inhabitant of rocky areas, mainly of mountains, where it lives a solitary life or occurs in pairs, and rarely in small flocks. This bird is very lively; its tail is always in motion, and it betrays its presence by almost uninterrupted hoarse calls. It seeks its food on the ground. The nest is placed below a stone in a cleft on the ground, on slopes. The four eggs are blue with red and brown spots.

The whinchat and stonechat inhabit green lands but do not enter areas of tall vegetation. The whinchat keeps mainly to meadows in swampy areas and weedgrown banks. It is found on mountain meadows up to an altitude of 2,400 meters. When the male has occupied a territory, it performs display flights with an expanded tail and puffs out its breast and head feathers in an imposing manner. It often sings, shrilly but pleasantly; its song consists of short phrases in which rough sounds alternate with whistling tones. The female builds the nest by herself; it is on the ground hidden in the vegetation, often with a sort of entry from the side. The five to six turquoise blue eggs are incubated for two weeks by the female; the young stay in the nest for two weeks after hatching. The whinchat leaves Europe in early September and migrates to tropical areas, mainly the African savannas.

Fig. 12–32. Whinchat.

The behavior of the stonechat is very similar, but this bird holds itself more upright and is even livelier. It preys on insects from the ground or from a perch. Its preferred habitats are dry, partially plant-free areas which facilitate the hunting of arthropods. Nevertheless, the stonechat does not altogether avoid swamps and other moist areas; it is also found in areas densely grown with bushes. Unlike the whinchats, this species is not found above 1,000 meters. When both species occasionally occur together, their relationship to the environment remains different. Both clear and rough tones alternate in the stonechat's short, often regularly repeated song phrases. During his display the male spreads his wings and tail, showing off the contrastingly colored areas. The nest is on the ground in a depression or a bank well hidden among the plant growth. The four to six pale bluish-green eggs with reddish-brown spots are incubated for fourteen days on the average. As a partial migrant, the stonechat winters on the British Isles, in western France and in the Mediterranean area. In these areas, it occupies territories and also takes vegetable food and even grains. Many birds, however, migrate into the Mediterranean area and even into tropical Africa as far as Ethiopia, or to India and southeastern Asia.

Fig. 12–33. Stonechat perched on a stuffed cuckoo placed near its nest plucks feathers from it.

The anteater chat lives in open African areas, often in semi-deserts where it feeds on beetles, ants and termites. It has a tuneful song consisting of flute-like tones. In display the male performs a regular dance before the female. The nest has a horizontal passage which the birds dig into the banks of loose soil. Apparently the birds use these cavities outside the breeding season, for protection.

The wheatears live in very open countrysides. Many of them have entered deserts where they use quite varied habitats. They show the phenomenon of two different color phases in the same general area; this is noticeable in six species. The head is either white or black as in the mountain chat and in the pied chat so that two color phases of these two species may occur in the same area. The phenomenon of color phases is also exhibited by the white-crowned black wheatear, but not quite so obviously, for the young have a black head which only becomes white in adults. The lower parts of the mourning wheatear and of the pied chat and the mountain chat may be light colored or black. In the black-eared wheatear the throat may be white or black. This remarkable diversity of color makes it possible to study suddenly produced color phases in birds, relative to particular genes. One can also determine different variations occurring within different populations of the same wheatear species.

The wheatear has a gray back and a rust-red band below the eye; the female, with less contrast, is more rust-colored. Wheatears perform conspicuous bowing movements which display their white rump. These ground dwellers are in ceaseless motion; they like to perch in bushes. The wheatear lives in pastures and other open areas with rocks, areas of rubble and dunes. Sometimes it hunts on the wing, but usually it feeds on ground-dwelling insects such as beetles, diptera, butterflies and caterpillars. Its calls are brief and harsh; the song consists of alternations of tones, as well as whistles mixed with harsh sounds. In display it performs leaps and bows, dances with open wings and also performs true display flights. The nest is hidden under a large stone, among rock debris, or in a rabbit burrow. The five to six eggs are pale blue, often with brown or blackish dots; they are incubated for two weeks by the female. Both parents feed the young for fourteen to sixteen days.

The wheatear winters in tropical Africa, preferably in open savannas; the various wheatear populations reach this area by routes which converge in a fan-shaped manner. Members of the large subspecies *Oenanthe oenanthe leucorrhoa,* from northeastern Canada and Greenland, cross the Atlantic twice a year and pass over western Europe to tropical Africa and back. On the other hand, Alaskan and east Siberian wheatears pass over all of Asia to reach eastern Africa. They make no use of the closer tropical areas, such as southeastern Asia, which would be suitable as winter quarters. These peculiar mi-

Fig. 12-34. Wheatear.

gration routes are presumably to be interpreted as routes by which the species has distributed itself since the Pleistocene (the Ice Age).

The black-eared wheatear lives in dry rocky areas of the Macchia (dwarf bush areas); in winter it passes over the Sahara into the African tropics. The desert wheatear is an inhabitant of semi-deserts; it keeps to plains avoiding dunes and rocks. It breeds on the ground among Macchia bushes, or beneath stones, and only wanders a short distance to the south in winter. The white-crowned black wheatear inhabits the same areas. Its habitats are arid and it nests by choice in rodent holes. The red-rumped wheatear on the other hand, prefers rocky areas and the steep shores of dry water courses where it breeds in holes among rocks. It avoids the desert as well as areas with trees.

The white-crowned black wheatear, is a characteristic desert bird. It hunts for arthropods, particularly lizards on the rocky or sandy banks of dry water courses as well as in true deserts; it also takes small lizards. This diet furnishes the bird with its water requirements for there is no free water in its habitat. In spite of its marked adaptation to life in the desert, shortage of food affects the white-crowned black wheatear's reproduction. It often lays only two to three eggs while those species which dwell in less desert-like habitats have clutches of four to five eggs. Wheatears have adapted extremely well to open areas of the most diverse habitats, including the barest deserts.

The Indian robin, which belongs to the genus *Saxicoloides,* is very tame. It searches the ground and shrubs in arid areas for insects and other small animals. It breeds in ground holes and even holes in walls.

The rock thrushes form a link between the chats and the redstarts; they are all included in the single genus *Monticola.* These birds resemble both the redstarts and the chats in behavior, however they also show several of the characteristics of the true thrushes. They are found in the northern Old World but some species also occur in southeastern Asia and Africa. Their bodies are compact; the tail is relatively short and often moves rapidly. There are nine species, including: The ROCK THRUSH (*Monticola saxatilis;* L 19 cm; Color plate, p. 289) male has a blue head, white rump, reddish tail and rusty-brown underside; female is reddish-brown and distinctly spotted; the BLUE ROCK THRUSH (*Monticola solitarius;* Color plate, p. 289) males have blue gray wings and a black tail; females are brownish with a bluish gloss, and are paler beneath with dark marks over the whole body. Both species range from Spain as far as China.

The rock thrush likes to sit on rock perches; it even flies skillfully, very close to the ground. It likes sunny, dry, rocky slopes up to 2,700 meters altitude. Its song consists of a sequence of short, repeated phrases which end abruptly and sometimes have a fluty quality; it is reminiscent of the songs of the blackbird and the mistle thrush. In display, the male rises ten meters up in the air, or higher, and then

Fig. 12-35. Rock thrush (*Monticola saxatilis).*

Fig. 12-36. Blue rock thrush *Monticola solitarius).*

descends with outspread wings and tail as with a parachute. The nest is placed in a cleft among rocks or in rock debris. It is built by the female alone. The four or five pale blue eggs sometimes have brown spots. In winter the rock thrush migrates to tropical Africa where it is found in the open savannas.

The blue rock thrush is reminiscent of the redstarts and of the blackbird because of its lively manner. It generally keeps to rocky dry areas and perches on a lookout post from which it hunts for food in the vicinity of plants. It lives at lower levels than does the rock thrush, as well as in more shaded locations. The melodious fluty song reminds one of an inexperienced blackbird. The nest is built in rock clefts and generally holds five pale blue eggs which often have reddish-brown spots. Some spend the winter in their breeding areas along the Mediterranean; others migrate to central Africa and southeastern Asia.

In comparison with the large group of chat-like birds, the thrushes are larger and have larger beaks. They inhabit all parts of the world. Of the two subgroups, many ornithologists regard the whistling thrushes as the most primitive and the true thrushes as the most highly developed members of the whole subfamily.

Whistling thrushes, by B. E. Smythies

The whistling thrush group, with only one genus, *Myiophoneus,* which comprises seven species, inhabits southeastern Asia and the south of the northern Old World (e.g. Turkestan, Tibet, China and Taiwan). These birds are large, strong-legged, dark thrushes; the sexes are identical in color. They are almost water birds as they live along small water courses, usually in mountainous areas. The young are unspotted or only slightly spotted. The BLUE WHISTLING THRUSH (*Myiophoneus caeruleus;* Color plate, p. 280), which has a L of 30 cm, and the SUNDA WHISTLING THRUSH *(Myiophoneus glaucinus)* are included in this group.

A characteristic call which has already been attributed to whistling thrushes resembles the shrill tone made on a slate by a slate pencil. However, these birds also utter pleasant flute-like tones and extended songs. In India the whistling thrush has the nickname "whistling schoolboy." Its diet is variable; in Burma it feeds on berries and other plant material as well as water beetles and snails, while the Sunda whistling thrush in Borneo feeds on frogs, crickets, beetles, snails, woodlice, and berries. The nest is usually close to water; it is a deep massive cup with a lining of fine grasses and other materials.

Whistler described the life of the whistling thrush as follows: "This common and characteristic bird of the Himalayas can be looked on as a water bird in a certain sense for it is a bird of rivers and other water courses. It is found everywhere in the mountains, where one sees it shooting among steep cliffs, between the trunks of the dark pine forests, or seeking food on open slopes. However, a brief period of observation always shows that its home range is in some ravine with a stream or a raging river. Its song and its call give further evidence of its ties to the water. The call is a loud, melodious whistling and the song is long-lasting like that of most thrushes; however in both the call and the song, one hears something harsh and unharmonious, an almost uncanny squeaking element, which clearly serves to be audible above the sound of rushing waters. If one hears the bird deep down in a ravine where the rushing of the water dampens all sounds, one realizes how useful these squeaking tones are.

We will begin our description of the true thrushes with the GROUND THRUSHES (genus *Zoothera*). These birds have very rounded wings, the backs of which, as well as several entire feathers of the secondaries, are white or rust colored; this is particularly noticeable on the lower wing surface. The axillaries have two colors; the color pattern in which individual feathers are deeply demarcated varies so much from species to species that these birds were formerly divided into five genera. They inhabit warm areas of the Old World, as well as Siberia, and two species live in western North America as far south as Mexico. There are twenty-nine species, including: WHITE'S THRUSH (*Zoothera dauma*) which has a L of 28 cm. This bird has marked dark half-moon spots on the light background of its upper and lower parts; the tail is brown with a whitish tip. It ranges from Siberia and India as far as Indonesia, New Guinea and Australia. The SIBERIAN GROUND THRUSH (*Zoothera sibirica*) male is uniformly blackish-blue; the female is olive-brown with a white belly. It breeds in eastern Siberia and Japan, wintering in India and southeastern Asia. The ORANGE-HEADED GROUND THRUSH (*Zoothera citrina*; Color plates pp. 221 and 289) has a L of 22 cm. It is more brightly colored; the male has an orange head and lower parts, a blue-gray back, and white wing spots, while the female is olive-brown with pale feather edges. This species is found in India and China as far as Indonesia. The VARIED THRUSH (*Zoothera naevia*) has a L of 23 cm. It is dark gray above, and reddish-brown below with a black breast band. It breeds from Alaska to California.

White's thrush (also called the golden mountain thrush) shows great adaptability towards its environment for it lives in Siberian conifer forests, as well as in tropical jungles. It prefers rocks covered with moss, or dense plant growth, in otherwise quite different habitats. This secretive bird nests one to four meters above the ground and lays three or four yellowish-green eggs which are almost completely covered with reddish spots. In the north of its range it is a migrant which wanders into tropical areas. Occasionally it strays into Europe (to Germany, Austria, Italy and Belgium) as a vagrant.

The Siberian thrush is an inhabitant of eastern Asiatic forest areas while the orange-headed ground thrush lives in the ferns and undergrowth of the dense jungles in the warmer zones. It hunts over moss-covered rocks for insects and spiders and utters a song which sounds "unfinished" for it consists of soft fluty tones and harsh calls. Its tree nest holds three or four spotted greenish or cream-colored eggs.

Thrushes proper, by J. Dorst

▷
The nest hole of a nuthatch *(Sitta europaea)* holds a considerable number of young. The feeding adult clings skillfully to the wall of the cavity.

Two ground thrush species have settled western North America. The best known species of the two is the varied thrush. These colorful birds lead a secretive life in dense, moist coniferous forests particularly at lower levels. The melodious song consists of a sequence which gives a melancholy effect. They feed on ground-dwelling insects and on fruit. In winter, acorns may form up to seventy-six percent of its diet. This ability to vary its diet enables the varied thrushes to survive the winter with only insignificant migratory moves. The nest is placed in a tree up to five meters above the ground and it holds three blue dark spotted eggs, on the average.

Several other genera are placed close to the ground thrushes. These are: 1. HAWAIIAN THRUSHES *(Phaeornis)* occur only on Hawaii; the Oahu subspecies of the HAWAIIAN THRUSH (+ *Phaeornis obscurus oahuensis*) has become extinct. 2. NEW WORLD NIGHTINGALE-LIKE THRUSHES *(Cathurus)* have a long outermost primary and are dark in color. Some have a black crown. The song is of medium quality. They occur from Mexico to Peru. One member of this genus is the ORANGE-BILLED NIGHTINGALE-LIKE THRUSH *(Cathurus aurantiirostris;* Color plate p. 221) a long legged, short-winged bird which is brown above with a darker head, and pale brown below; the belly is whitish, and the beak is a lively orange color. It occurs from Mexico to Colombia and Venezuela. 3. WOOD THRUSHES *(Hylocichla)* with the exception of the hermit thrush, have a short outer primary. They occur in five species in North America. The HERMIT THRUSH *(Hylocichla guttata;* Color plate, p. 221) has a L of 18 cm, and is brown on the back, brownish-red and white with black spots below. It breeds from Alaska to California and Virginia. SWAINSON'S THRUSH *(Hylocichla ustulata)* is more olive, and has no reddish-brown color. The GRAY-CHEEKED THRUSH *(Hylocichla minima)* is olive-brown with gray cheeks; it has spread into eastern Siberia from North America.

Wood thrushes are at home in the most varied types of North American forests where they are among the best songsters; the hermit thrush is particularly so and, as a result, it has been called the "American nightingale." All these birds have a mixed diet, consisting of mainly arthropods in summer, and berries in winter. They are migrants, many of them moving to South America for the winter, a practice which is not all that common among North American birds.

The true thrushes (genus *Turdus*) are rather large, bulky, and strong legged, in comparison with the forms just discussed. Their habits also differ from those of the ground thrushes, nightingale thrushes and their relatives. In display and when feeding they keep mainly to the ground. These are spotted or otherwise contrastingly colored species, as well as uniformly brown or black thrushes, and they were formerly divided on this basis into "thrushes" and "blackbirds". However, such a division is artificial and not in accord with true relationships. Thrushes of the genus *Turdus* are found everywhere except Madagas-

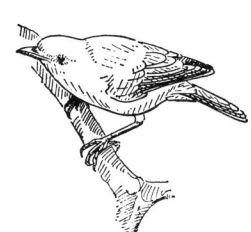

Fig. 12-37. Hawaiian thrush.

◁
Tits are beneficial in combatting insect enemies of useful plants. A marsh tit *(Parus palustris)* is holding a beakful of larvae of such insects.

car, the Mascarene Islands and Australia. The sixty-two species amount to one fifth of the entire subfamily. We will mention the following here: 1. GROUND-SCRAPER THRUSH *(Turdus litsipsirupa)* is the only species with white on the lower wing surface (therefore reminiscent of the ground thrushes). It is found from Ethiopia to South Africa. 2. MISTLE THRUSH *(Turdus viscivorus;* Color plate, p. 290) has a L of 26 cm. It is found all over Europe, except for part of Norway, as far as Siberia and northern India; in winter it migrates to southern Europe, to a lesser extent to northern Africa. Like the next four species, it has distinctly spotted lower parts. 3. REDWING *(Turdus iliacus;* Color plate, p. 290) has a L of 21 cm and a whitish stripe above the eye; the flanks are reddish-brown. It is found further north, from Scandinavia to northern Russia and Siberia; it winters in western Europe and the Mediterranean area. 4. SONG THRUSH *(Turdus philomelos;* Color plate, p. 290) has a L of 23 cm. It has a yellowish underwing and is found from Europe to western Siberia, although it is absent in part of the Mediterranean area. 5. FIELDFARE *(Turdus pilaris;* Color plate, p. 290) has a L of 25 cm. The head and rump are pale gray; and are distinctly separated from the brown back and the black tail. This bird is found from central Europe as far as Lake Baikal and, since 1937, it has also colonized Greenland. 6. RED-CAPPED THRUSH *(Turdus ruficollis)* is from eastern Siberia; it has one subspecies, the DARK-CAPPED THRUSH *(Turdus ruficollis atrogularis)* in eastern Russia and western Siberia, and another red-capped subspecies *(Turdus ruficollis ruficollis)* in eastern Siberia. This latter subspecies has a reddish crop region. 7. The BLACKBIRD *(Turdus merula:* Color plate, p. 290) has a L of 24 cm. 8. The RING OUZEL *(Turdus torquatus;* Color plate, p. 290) is largely black with paler feather edges and a wide white band across the breast; the female is mainly brownish-black with a less distinct breast band. This bird is found in Scandinavia and in central and southern European mountains. It has a L of 27 cm. 9. The AMERICAN ROBIN *(Turdus migratorius;* Color plate, p. 290) was called robin by the first settlers because of its brownish-red lower parts. Otherwise it is uniformly dark. It occurs from the North American treeline south as far as Guatemala, and migrates mainly to the Gulf States where the climate is pleasant and mild.

Several of the South American thrush species are uniformly colored with black streaks on a white throat; however, they are unspotted further back. These species include: 10. The WHITE-NECKED THRUSH *(Turdus albicollis)* is found from Venezuela to Argentina. 11. The RUFOUS-BELLIED THRUSH *(Turdus rufiventris)* has a bright brownish-red belly and is found from Brazil to northern Argentina and Bolivia. It resembles the olive thrush. 12. The OLIVE THRUSH *(Turdus olivaceus)* is from Africa. Divergent insular forms of this species live in the gulf of Guinea, west of Africa and on the Antilles.

The preferred habitats of the mistle thrush are meadows and pas-

Fig. 12-38. Mistle thrush *(Turdus viscivorus).*

Fig. 12-39. Redwing *(Turdus iliacus)* [note the British Isles should not be blacked in; this thrush does not breed there].

Fig. 12-40. Song thrush *(Turdus philomelos).*

Fig. 12-41. Fieldfare *(Turdus pilaris).* This bird has been in Greenland since 1937.

FLYCATCHERS AND THRUSHES 305

Fig. 12-42. Red-capped thrush *(Turdus ruficollis).*

Fig. 12-43. Blackbird *(Turdus merula).* There are several species in southern Asia.

Fig. 12-44. Ring ouzel *(Turdus torquatus).*

tures mixed with woods, as well as hedges and fields. In spite of its liveliness, it is quite shy. Like other thrushes, it feeds on arthropods, larvae, and snails from the ground during the summer; in winter it feeds partially or entirely on fruit, preferring the berries of the mistletoe. As a result, it can winter in areas with a hard climate. Since it sheds the seed kernels of the mistletoe in its excrement, it plays an important role in the distribution of this plant. When the seeds of this tree parasite fall on a branch, they take root there. The contact call of the mistle thrush is loud and rolling; its song is reminiscent of that of the blackbird but is less variable and only consists of short, rather chopped-off phrases with simple motifs.

This bird generally breeds in forests up to altitudes of 2,000 meters, but it may make do with a copse in a park. The territory is restricted to the area immediately around the nest and is of no significance as a feeding area; the birds feed equally outside it. The nest is fixed in a tree fork up to twenty-five meters above the ground. It includes a layer of mud, and consequently resembles the blackbird's nest; it holds four greenish-blue to bluish-gray eggs with brown spots. Incubation lasts twelve to fourteen days; the young stay in the nest two weeks after hatching and remain with the family for some time after that. Large flocks form in the winter.

The song thrush, which owes its name to its beautiful song, is also well known because of it. The song is tuneful, melodious and very variable. The bird repeats the one to four syllable phrases two to four times, makes a pause, and then sings on; the whole song is delivered rapidly. The song thrush haunts dense vegetation where trees stand on moist ground; it also penetrates mountainous areas, particularly in the south of its range. Like most true thrushes, it feeds on the ground and, apart from arthropods, it shows a particular preference for snails which it cracks open on stones—the so-called "thrush smithies". Fruits complete its diet, particularly in the fall and winter. Like the redwing and others of its relatives, it is particularly attracted by grapes.

The female song thrush builds the nest alone, in a tree and inside she adds a layer of mud and pieces of rotting wood, forming a sort of papier mâchée. The four to five (or sometimes fewer) eggs, are greenish-blue with a few brown spots. They are incubated for twelve to fourteen days. Both parents feed the young, which leave the nest after thirteen days. After breeding, song thrushes move to the moistest areas of undergrowth in their habitats. This thrush is a migrant in the north of its range. It winters in western Europe and the Mediterranean area, occasionally in Iran and India.

Fieldfares were formely caught in countless numbers with baited snares made of horse hair; they were then brought to market to be sold as delicacies. The fieldfare is easily recognized by its pale gray head and rump, which contrasts with the chestnut-brown of the back and the blackish tail. It does not seek to hide in its preferred haunts, which are open country with hedges and bushes. It belongs to those thrushes which like the society of others, hence it breeds in small colonies of three to ten pairs or sometimes more. Although each pair defends a territory around its nest, the birds also feed in neutral areas. In summer they feed on arthropods, spiders, worms and small snails; in winter they feed largely on fruit, berries and grains. The harsh and hard call is uttered often; the song has a remote resemblance to that of the blackbird and consists of a sequence of calls and whistles.

Fig. 12-45. Mistle thrush in aggressive poses.

The nest is placed in a tree seven to eight meters above the ground, and sometimes higher; it is lodged in a fork and has a layer of mud. On the average there are five eggs; these are greenish to greenish-blue with red or brownish-red spots and are incubated for thirteen to fourteen days. The young stay in the nest for two weeks after hatching and after fledging they remain together with the adults in small flocks. A plant diet in winter enables the fieldfare to distribute itself all over Europe and to seek out mainly the western and the Mediterranean areas during the cold season.

The blackbird is one of the song birds which has adapted itself completely to the changes wrought by man in its habitats. Originally a forest dweller and still found in forests, it also lives in great numbers in parks, orchards and even in the interior of great cities, provided there are a few trees there. One can really speak of two blackbird populations, forest blackbirds and town blackbirds, which have adapted to the vicinity of man. The two differ not only in their behavior towards man, but also in certain biological features, as, for example, the course of breeding, and migration—for the climate of cities is favorable to them. Although the blackbird seems to be a bird that keeps to itself because of its territorial behavior, outside the breeding season it shows a social tendency. It was once a tree dweller but is now found more in the undergrowth and on the ground where it seeks worms, arthropods, subterranean insect larvae and other small animals; it also eats many berries and fruit, particularly those of the ivy. One can hardly imagine summer evenings in green parts of European cities without the varied, slowly delivered song of these birds. This song contains melodious flute-like tones which may be modified, almost infinitely, with harsh twittering phrases in between. The composition of the song shows marked individual differences, and an expert can easily identify individual birds by their song.

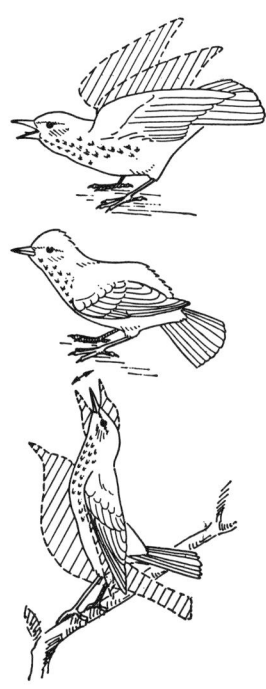

Fig. 12-46. Display postures of the song thrush.

▷
True tits: 1. Great tit (Parus major); 2. Coal tit (Parus ater); 3. Blue tit (Parus caeruleus); 4. Siberian tit (Parus cinctus); 5. Crested tit (Parus cristatus); 6. Marsh tit (Parus palustris); 7. Southern black tit (Parus niger); 8. Willow tit (Parus montanus); 9. Azure tit (Parus cyanus).

In the relatively rare display, the male erects his feathers and spreads his tail. The nest may be in a tree, generally in a fork, or in a shrub; it is often in a hole in a wall, on wire netting, or on beams close to human habitation. The blackbird lays three to four (occasionally five) eggs which are bluish-greenish-gray and densely covered with red spots. They are incubated by the female but occasionally also by the

male as well for an average of twelve to thirteen days. Both parents feed the young, which leave the nest thirteen days after hatching. The fertility of the blackbird (it rears three to four broods a year) must have contributed to its wide distribution. A marked loss of eggs or of young brought about by magpies or other nest robbers is thus soon made up. Only in the most northern populations do the males also migrate to western Europe during the winter; in central Europe only some of the females and the young migrate.

The ring ouzel is found along the upper tree line in mountains and in light mountain woods, as well as on moors at lower levels. Its calls are harsher than those of the blackbirds. Its song reminds one of that of the song thrush, but it is slower and less varied. Its nests are in conifers and they hold four to five bluish-green, brown spotted eggs. In winter, this bird migrates to southern Europe and northern Africa.

The American robin is a very common species among the American thrushes. It replaces the European blackbird in North America for it is equally tame and lives in a similar way. Its song is shallower and has shorter phrases than does that of the wood thrushes, but it still forms quite a melodious sequence. Like other members of its genus, it lives on insects, snails and other small animals but it also takes fruit. The female builds the nest and she alone incubates the four uniformly blue eggs. Although their habitats are very different, the rufous-bellied and the white-necked thrush, and other South American species have the same habits as their relatives. However, at least one species lives mainly on fruit and only occasionally supplements its diet with arthropods. The African species are much the same although they inhabit the continent from the rain forests up to high mountains.

The American robin

1. Verdin *(Auriparus flaviceps)*; 2. Long-tailed tit *(Aegithalos caudatus)*; 3. Red-headed tit *(Aegithalos concinnus)*; 4. Sultan tit *(Melanochlora sultanea)*; 5. Tufted titmouse *(Parus bicolor)*; 6. Gray tit *(Parus afer)*; 7. Varied tit *(Parus varius)*; 8. Short-toed tree creeper *(Certhia brachydactyla)*; 9. Tree creeper *(Certhia familiaris)*; 10. Penduline tit *(Remiz pendulinus)*; 11. Mouse-colored penduline tit *(Anthoscopus musculus)*.

13 Tits (Chickadees), Nuthatches, and Tree Creepers

This chapter brings together several groups of small song birds whose general origin is generally assumed nowadays to be from the flycatcher-like birds. Nevertheless their relationships to one another and their systematic position are still often disputed. TITS, NUTHATCHES, TREE CREEPERS and their relatives show much in common in their way of life, but they are more or less divergent in many features of body conformation and structure. Some forms are particularly specialized and therefore difficult to compare with others.

All birds of this group have short, rounded wings and, therefore, a limited capacity for flight. They are small, with a L of 8.5 to 20 cm, lively and mobile; some have penetrating voices but they only rarely show a true song. The plumage is soft and loose with long aftershafts; there are ten primaries and twelve tail feathers. These birds live in or on trees (one species lives only on cliffs) and are generally sociable outside the breeding season. They feed mainly on arthropods (on the resting phases of these in winter), spiders, snails and also seeds, fruit, grains and nuts. Their climbing skill is used in the search for food. They are largely residents or mere wanderers.

By arranging the group in eight families, we are able to compare the first four (LONG-TAILED TITS, PENDULINE TITS, TRUE TITS, and NUTHATCHES) with the last four (AUSTRALIAN NUTHATCHES, TREE CREEPERS, PHILIPPINE CREEPERS and SPOTTED CREEPERS). In this case, the last four families do not appear as different from one another as do the first four, and as a result the former are often classed as subfamilies of the tree creepers. The last four families are separated mainly because of the very prolonged hind toe, the strongly curved claws and the finely pointed, generally decurved beak (the beak is curved upwards in the Australian nuthatch).

The LONG-TAILED TITS (family Aegithalidae) differ from the true tits in the following behavioral features: they do not keep breeding territories. Both parents share in nest building. They breed in the open.

Tits, nuthatches, and tree creepers, by J. Steinbacher

Family: long-tailed tits by R. Berndt

Fig. 13-1. 1. Long-tailed tit *(Aegithalos caudatus)*; 2. Red-headed tit *(Aegithalos concinnus)*; 3. Java long-tailed tit *(Psaltria exilis)*.

There is no joint behavior in nest defense. Several adults often take part in rearing the young of one pair. Neither young nor adults maintain individual distances; there is a special call to induce hunching up together. They are very sociable towards members of their species, but not towards other birds. They never take food under their toes. There are only three genera with eight species and about forty-five subspecies. This marked formation of subspecies is due to the resident nature of these birds.

The only European species of this family is the LONG-TAILED TIT *(Aegithalos caudatus;* Color plate, p. 308) which has a L of 14.5 cm (9.5 cm of which is due to the tail), and weighs 9 g. It is one of the most delicate European birds, and is called "little panhandle" in German because of its long tail. It avoids conifer woods and mountains, but otherwise occurs moderately commonly in woods, parks and large areas of shrubs. It feeds on small and even the smallest arthropods, particularly on mosquitoes and aphids. Its nest, which is a highly skillfully woven longish oval pouch, closed apart from an entry hole at the side, is placed among dense twigs. The clutch consists of seven to fourteen eggs. Incubation lasts twelve to thirteen days, the fledging period fifteen to seventeen days. The absence of breeding territories is connected with the great sociability towards members of the same species and the frequent participation of these other birds in rearing the young and in keeping warm at night. The birds sleep bunched up in groups.

There are twenty subspecies in the large range of the long-tailed tit; this includes Europe and northern Asia. Western, southern and eastern subspecies have a dark eye-stripe, the others all have white heads. As Germany is an area of transition for these birds, striped, stippled and white-headed forms occur there, the last particularly as winter visitors. There are four other species in this genus, and we will mention only the RED-HEADED TIT *(Aegithalos concinnus;* Color plate, p. 308) which has a L of 10 cm. It occurs in six subspecies from the Himalayas to Burma and Taiwan. The JAVA LONG-TAILED TIT *(Psaltria exilis)* has a L of 8.5 cm and is the only species of its genus. It has the shortest beak, apart from swifts, of all birds; its tail is only 6.5 mm long. The third genus is that of the BUSH TITS of southwestern North America. The twin species, the COMMON BUSH TIT *(Psaltriparus minimus)* which has a L of barely 9 cm, and the BLACK-EARED BUSH TIT *(Psaltriparus melanotis)* which is the same size, live in semi-deserts, where they suspend their cucumber-shaped nests in bushes.

Family: penduline tits, by R. Kinzelbach

The penduline tits (family Remizidae) were formerly included with the tits. They too have common ancestors, but the penduline tits have diverged to such a degree, that they are now separated as a distinct family. These birds have a L ranging from 8–11.5 cm; they have short wings and tails, and delicate heads. The culmen of the beak, unlike that of the true tits, is not curved but straight. Their spherical nests, which are, in part, skillfully made pouches, are the basis for their German name, which means pouch tit. There are four genera with about eleven species in a large part of Eurasia, Africa and southern North America.

The PENDULINE TIT (*Remiz pendulinus*; Color plate, p. 308) has a L of 11 cm and weighs about 10 g. The two sexes are similar, although the female is somewhat paler. The crown and sides of the head are brown outside the breeding season, instead of pale gray and black. The fully fledged young look similar. Penduline tits look for food in shore line vegetation, particularly in reed beds where they show themselves to be skillful climbers. They feed largely on arthropods and spiders; in winter they also eat grass seeds (particularly those of reeds and reed maces). In addition they feed on Compositae seeds which they often take like goldfinches while sitting on the flowerheads. Larger food particles are held between the toes in the manner of tits, so that the bird can reduce them with its beak. When they wander about alone or in small troops, in winter and on migration, penduline tits often utter their high-pitched contact call, "tsee". Their song is a continuous twittering often rendered as "seeseesee".

The male begins nest construction alone usually in April; the nest is reminiscent of that of many weaver birds in shape and manner of construction. It is placed in willows, poplars, tamarisks or birches, generally on a fork hanging down vertically, often over water. Some groups regularly build in reeds. The nest consists of long fibers, grasses, willow roots, bast, nettles and a short-fibered, wooly component of seed hairs from willows, poplars, reed maces and spider webs. In many districts, penduline tits even use sheep wool instead of plant fibers.

The male weaves these long fibers into a ring of twenty-five to thirty centimeters diameter using both limbs of the fork as supports; more rarely the nest may be suspended from a single, unforked twig. The ring forms the base and side walls of the final nest. The floor is widened until a structure resembling a hanging basket is produced. Next a rear wall is constructed so that the nest becomes a hanging pear. The front wall is raised and finally the builder reduces the size of the entry hole and provides it with a somewhat prolonged entry tube. While still building, the male looks for a partner who soon takes part in the completion of the nest. She is especially involved in taking over the lining of the nest skeleton with plant wool, and the lining of the nest cup. Often even before the entry tube is finished she lays five to eight, rarely ten, white eggs and begins to incubate. The male, meanwhile, sometimes will start even earlier to build another nest and to look, generally successfully, for a second or even greater number of females.

The young hatch after twelve to fourteen days of incubation; they are then tended in the nest generally by the female alone for another

Fig. 13-2. Penduline tit (*Remiz pendulinus*).

Fig. 13-3. Nest construction in the penduline tit.

A hanging "ring" in an early stage of nest construction.

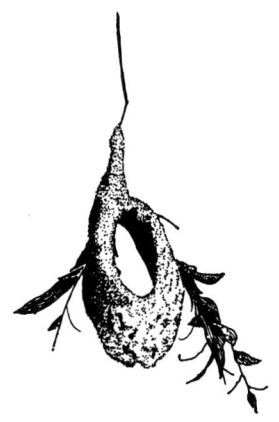
Fig. 13-4. A "hanging basket" in a later stage of nest building.

fifteen to twenty days. The faeces of the young are not completely removed so that the nests in which there has been a brood are easily recognized. In many eastern European districts, children used to gather empty nests of the penduline tit for use as slippers. This tit is not to be found throughout the area shown on the distribution map; it requires a habitat which is only locally available—marshes, low trees near sweet or brackish water, combined with reed beds, as well as meadows along rivers. Local changes, such as floods, may cause temporary breeding in places where in other years this species occurs as only a rare visitor. In addition, penduline tits make irregular advances into areas at the edge of their normal distribution—advances comparable to the invasions of the slender-billed nutcracker (Chapter 19) and the sand grouse (Volume VIII). Thus, the penduline tit has been seen in the Rhineland in some numbers only in 1820, from 1880 to 1900, about 1935 and from 1960 to 1965; it has only rarely bred in that area. Its migratory behavior is also variable; it may, according to locality, be a resident, a wanderer or a real migrant. A whole population may migrate for some distance in one year, while in another year they may not travel as far.

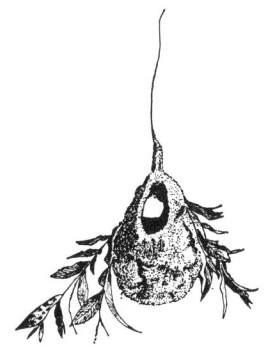

Fig. 13-5. Almost completed nest, front view.

The systematics of the penduline tits in Eurasia are involved. It is not yet clear whether the eastern representatives present one or more distinct species. At any rate, two groups of these tits found in the area between the Sea of Aral and Lake Baikal behave like two species; the western one breeds in reed beds, and the eastern one in the shore vegetation; no intermediates are known. Curiously enough, at Lake Neusiedel in Austria, in the area of the western group, there are also bush and reed nesters, although there they intermingle and, quite clearly, all belong to a single form.

There are other species of penduline tits in Africa. The AFRICAN PENDULINE TIT *(Anthoscopus caroli)* of southeastern Africa builds a nest which is perhaps even more astonishing than that of the Eurasian species. The adult can shut the nest by pinching the entry tube with its beak upon leaving. Probably this keeps off nest robbers. The CAPE PENDULINE TIT *(Anthoscopus minutus)* uses its nest very fully. The young of the first brood still spend the night in it while the next brood is already growing up. Up to six fully fledged young have been found in a nest which also held the smaller young of the next brood. In winter, the young of the second brood also use the nest to sleep in. The FIRE-CAPPED TIT *(Cephalopyrus flammiceps)* of the Himalayas, in contrast to its relatives, nests only in tree holes where it builds quite a simple nest.

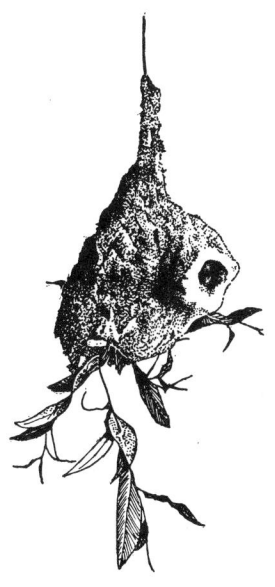

Fig. 13-6. Typical tree nest suspended on a single willow twig.

The only representative of the penduline tit family in the New World is the VERDIN *(Auriparus flaviceps;* Color plate, p. 308). Its nest is a hemisphere of thorn twigs with a side entrance, placed in shrubs of semi-deserts. The eggs, in contrast to those of the Old World relatives, are greenish and spotted. As in the penduline tit, the male builds several nests for several females. The young are fed for the first five days of their life with regurgitated food.

The TRUE TITS (family Paridae) are a group rich in species; they include the chickadees. The L ranges from 9-20 cm; all the members of this family are small, lively tree and bush birds with a short, hard conical beak and short strong legs suitable for climbing. The coloration may be very contrasting, involving yellow, blue, green, white or black, or it may be dominated by gray or brown. Almost all these birds are hole nesters; they lay large clutches of white, red-spotted eggs. Outside the breeding season they live socially. The food consists of arthropods in all life stages, but mainly of caterpillars and butterfly eggs and, in part, of oily seeds. These birds are distributed over the entire northern hemisphere and Africa. Many species can endure the severest winters and therefore are able to live as residents in the far north; others, as partial migrants, perform mass movements. There are three genera with about forty-five species and about 230 subspecies. The marked subspecific differentiation is related to the resident habits.

The genus, *Parus*, with over forty species, includes all European tits, among them the six central European species. The most common is the well known GREAT TIT (*Parus major*; Color plate, p. 307) which has a L of 14 cm and weighs 18-20 g. It can be found breeding in every treed area, but it prefers deciduous and particularly mixed oak woods. It is far more common in these areas than in coniferous woods. As a hole nester, its density depends on the number of natural or artificial cavities. It uses woodpecker holes, holes left where branches break off, and nest boxes. Great tits prefer holes with a diameter of about thirty-two millimeters at the entry. When suitable holes are scarce, they use mail boxes, pump pipes, etc. Very rarely they build a nest in the open in the fork of a branch.

As early as autumn the males divide the area into territories by threatening one another in a head-up pose, and by chasing off rivals. After this autumnal period of hostility, they join peacefully into the mixed winter flocks of climbing birds. From mid-winter onwards, males utter their rhythmical spring call, particularly near feeding tables; this call sounds like "spitz dee sha". In March, the final territories are laid down. As great tits sleep in holes, the teritories which usually are 0.3 to 3 hectares in size, are centered around the male's sleeping holes; the pair then uses the female's sleeping hole as a nest site. The female builds early in April and because there are still night frosts, the nest is particularly warm; rootlets, grass stalks and much moss are used in its construction, and it is lined with hairs, wool and feathers. After the daily egg laying, the female always covers the incomplete clutch with nest material to protect it from cold and the sight of enemies.

The great tit mother incubates for about fourteen days; the young

Family: true tits, by R. Berndt

Fig. 13-7. The African penduline tit at a closed nest; left; sections through the opened (1) and the closed (2) nest.

Fig. 13-8. Great tit (*Parus major*). Subspecies groups *major* vertical lines, *minor* group of subspecies horizontal lines, *cinereus* groups dots.

Fig. 13-9. Postures of the great tit.

Two males threaten one another in the head-up pose in the breeding season dispute.

Fig. 13-10. Great tit in the head forward threat pose used towards a food competitor.

are tended by both parents and take eighteen to twenty-one days to reach the fledging stage. After an additional one to two weeks, they are independent. Great tits only occasionally breed twice per season in deciduous woods, but in pine woods, they breed two to three times a year. The clutch size depends on food availability. In deciduous woods, first clutches vary from eight to fourteen eggs. They are smaller in pine woods, although second clutches in these habitats are larger than first ones. One year old females lay fewer eggs than do older females. The number of second broods and the clutch size diminishes with increasing population density.

Thus, a female great tit may lay up to three dozen eggs a year, and a pair may rear up to two dozen descendents. This high reproductive rate is accompanied by a high rate of mortality. Eighty-five percent of the fledged young die during the first ten months, and only fifteen percent survive to breed next spring. About half of the adults die in the course of a year. The median life expectancy of a great tit is only one and a half years, and the average age of the population is only two years. Nevertheless, individuals may reach an age of ten or twelve years.

The main enemies of the great tit at all seasons are the Eurasian sparrow hawk in the daytime, and the tawny owl at night. Many adults, nests, and nestlings are also destroyed by pine martens, stoats, weasels, dormice, squirrels, the yellow-necked mouse, the greater spotted woodpecker and the wryneck. Jays, in particular, prey on the newly fledged young. In severe and long winters very marked losses occur due to lack of food when, in addition to a snow cover, frost and ice cover all branches. If, after a long winter night, a tit finds no food for half a day, it dies of starvation. Feeding stations can reduce such losses only locally. However, natural circumstances restore the population fairly rapidly so that usually the former population is restored or exceeded after a year and a half.

Without these periodic losses, the high reproductive rate would produce an overpopulation in a few years which might lead to shortage of food. Overpopulation releases an urge to wander in many young tits, as early as mid-summer or the early fall. They travel hundreds to thousands of kilometers, from the northern into the southwestern half of Europe. Such mass movements have occurred every three to four years in this century. Some of the wanderers die on route, others settle permanently in a new area up to 500 kilometers away from their origin. Only a few return, so that the overpopulation generally becomes an under-population. Since all the great tits of any particular area never migrate, the bird is called a partial migrant, and since their irregular wanderings depend on population density, great tits are "density dependent irrupters". Their migratory behavior is probably the major device which stabilizes the population density.

Great tits feed on insects in all life stages, as well as on other arthropods; hence their great utility for man (this applies to all tits, tree creepers and nuthatches). As supplements, they take berries, fruit, half-ripe peas and, through the winter half of the year, oily seeds. At feeding tables these birds are given poppy seeds, oats, sunflower seeds, nuts, oatmeal, unsalted meat and fat. When seeking food, tits test all possibly edible objects by hammering them with the beak. Some adults sleep all year round in holes, but most sleep among dense twigs during the six months of summer. Nevertheless, they take to holes about the time the leaves fall. Although many pairs keep together for the whole year, in winter (during the season when birds are almost sexless) the male takes over the best available hole for himself, without considering the female; only as the display season approaches does he gallantly leave the best accommodation to his partner.

Most species of tits greatly resemble one another in their social life, food and reproduction, population fluctuations, migratory behavior and other features; this is also true of the six central European tits. However, each species also shows certain characteristic differences. Thus, the BLUE TIT (*Parus caeruleus*; Color plate, p. 307) with a L of 11.5 cm and a weight of 11 g, is somewhat smaller than the great tit, and though almost as common as the latter in deciduous woods, does not nearly approach its total abundance. Unlike the great tit, with its black reminiscent of conifer forests, the blue tit with colors more in harmony with deciduous woods (blue, yellow and white) is only abundant in oak-dominated deciduous woods. It is scarcer in mixed woods, parks and gardens, rare in pine woods and almost absent from spruce woods. Its diet is more strongly biased in the direction of animal food, particularly insect eggs, so that, from the human point of view, it is outstandingly beneficial. Birch and alder seeds predominate in its plant diet. The blue tit's call is finer than that of the great tit, and its song is marked by a high-pitched trilling final phrase. An entry hole of twenty-six millimeters suffices for its nest holes. Aside from moss and hairs, the nest contains many stalks, bast fibers and feathers, preferably blue ones. Reproduction is high, with eleven to fourteen or more eggs, although only a few pairs, mostly those living in pine woods, have two broods a year. On the other hand, this tit, more than others, is affected by severe winters so that the population density has a particularly marked fluctuation from year to year. Blue tits, too, show mass wanderings usually every two to three years.

In addition to these two tits from deciduous woods, we will discuss two species from coniferous forests of which the dark COAL TIT (*Parus ater*; Color plate, p. 307) with its L of 11 cm and its weight of 10 g, is the smaller and more abundant. It prefers spruce woods to pine woods. This bird is most readily noticed by its song, which sounds like "stee-fle". As a markedly early nester, it builds a particularly thick-walled

Fig. 13-11. Male in nest showing display; he shows the female a nest hole by looking into it, tapping the edge of the entry hole and turning his head about.

Fig. 13-12. A breeding female attempts to scare off a nest robber with a threat display.

moss nest in tree holes, nest boxes, holes in the ground or in holes among rocks. The first clutch generally consists of ten eggs; there may be two or three broods per year. This high reproduction rate soon leads to overpopulation in the absence of unfavorable circumstances, and great movements from northern to southern and western Europe take place.

The inconspicuously gray-brown colored CRESTED TIT (*Parus cristatus;* Color plate, p. 307) has a L of 12 cm and weighs 11 g. It lives in pine woods. Its call, "tseegurr gurr gurr" is unmistakable. It builds a very warm nest which is narrowed at the top so that it is almost pouch-like. The nest building is begun as early as the beginning of April, and the nest itself is made of moss, wool, spider webs and the like. Nests are placed, in part, in holes which the birds excavate themselves in trees, in rotten tree stumps or posts, as well as in nest boxes, and occasionally in the heaps of twigs of the nests of much larger birds. Crested tits lay only five to six eggs on the average. As second broods are rare, the rate of increase is low, and overpopulation and consequent wanderings occur only exceptionally.

The remaining two central European tit species form a pair, the two members of which are not so easily distinguished. They are the MARSH TIT (*Parus palustris;* Color plates pp. 302 and 307) which has a L of 12.5 cm, weighs 11 g. and which is markedly common everywhere in deciduous and mixed woods, and the WILLOW TIT (*Parus montanus;* Color plate, p. 307) which has a L of 13 cm, weighs 11 g. and is generally considered to be conspecific with the North American BLACK-CAPPED CHICKADEE *(Parus atricapilla).* The willow tit is found in woods along large rivers, in willows, but mainly in oak, birch, pine or spruce woods. While the marsh tit has a glossy black cap, a grayer area on the side of the head towards the back, a smaller chin spot and uniformly colored wings, the willow tit has a dull black crown, whiter sides of the head towards the back, a smaller chin spot which forks toward the bottom and secondaries with pale feather edges. The best aid to identification, however, is the voice: Marsh tits have a loud clattering, "dyigs dyigs dyigs" song and a short toned "tsi da da da dad" call. The willow tit, on the other hand, whistles its song, a three to five syllable "tsee" and calls a prolonged "tsee tsee daa daa." The marsh tit hardly ever excavates its nest hole itself and always breeds only once a year; the willow tit almost always digs out its own nest hole and may have two broods a year at least in pine woods. As both species lay clutches of only six to nine eggs, overpopulation is rare. As a result, only occasional mass movements affecting the willow tit are known. Both these tits, as well as the coal and crested tits, and the Siberian tit, hide stores of food, particularly in the north (seeds and dead arthropods are hidden beneath bark, under moss or lichens on branches and tree trunks).

Additional relatives of these two species include the SOMBRE TIT *(Parus lugubris)* found from the Balkans as far as Iran; the NORTH AMERICAN BLACK-CAPPED CHICKADEE *(Parus atricapillus)*; the CAROLINA CHICKADEE *(Parus carolinensis)*; the GRAY-SIDED CHICKADEE *(Parus sclateri)* from Mexico; and the MOUNTAIN CHICKADEE *(Parus gambeli)* from the Rocky Mountains, the last four being from North America. In addition, we will mention: *Parus superciliosus* from the central Asiatic mountain tundra, which breeds in holes in the ground or in spherical nests built in shrubbery; the SIBERIAN TIT or GRAY-HEADED CHICKADEE *(Parus cinctus;* Color plate, p. 307) which has a L of 12 cm, is found from Norway to Alaska, and is extremely tame; the BROWN-HEADED CHICKADEE *(Parus hudsonicus)* which has a L of 11 cm and is from Canada; and the CHESTNUT-BACKED CHICKADEE *(Parus rufescens)* which occupies the Pacific coastal strip of North America and has a L of 11 cm.

Relatives of the crested tit are: the BROWN-CRESTED TIT *(Parus dichrous)* which has a L of 11.5 cm and is found in the mountains of Central Asia; the BRIDLED TITMOUSE *(Parus wollweberi)* which has a L of 11.5 cm and is an inhabitant of the dwarf oak and elberberry stands of northern Mexico; the PLAIN TITMOUSE *(Parus inornatus)* which has a L of 12.5 cm and which lives north of the preceding species; the TUFTED TITMOUSE *(Parus bicolor;* Color plate, p. 308) which has a L of 14 cm and is found from eastern North America as far as the Atlantic coast. This latter species is well known as a guest at feeding tables and a breeder in nest boxes.

The following species are related to the coal tit: the RUFOUS-BELLIED CRESTED TIT *(Parus rubidiventris)* which has a L of 12 cm and is found in the central Asiatic mountains; the CRESTED BLACK TIT *(Parus melanolophus)* which has a L of 11.5 cm; *Parus venustulus,* which has a L of 10 cm; and the VARIED TIT *(Parus varius;* Color plate, p. 308) which has a L of 12.5 cm, is colored like a nuthatch and occurs in Japan and Taiwan.

The great tit is distributed in over thirty subspecies from northwestern Africa over almost all of Eurasia as far as the Lesser Sunda Islands. Close relatives of this species are: the BUKHARAH GREAT TIT *(Parus bokharensis)*; the GREEN-BACKED TIT *(Parus monticolus)*; and the BLACK-SPOTTED YELLOW TIT *(Parus xanthogenis)*, all from Asia. Also in this group are three central African species, the AFRICAN GRAY TIT *(Parus griseiventris)* [now regarded as a subspecies of the gray tit]; the DUSKY TIT *(Parus funereus)*; and the BLACK TIT *(Parus leucomelas)*; as well as the SOUTHERN BLACK TIT *(Parus niger:* Color plate, p. 307) which has a L of 16 cm; the GRAY TIT *(Parus afer;* Color plate, p. 308) which has a L of 14.5 cm and is from southern Africa. The latter breeds from September to December and does not lay more than three to four eggs.

The blue tit group includes, aside from the blue tit which ranges from the Canary Islands over northwestern Africa and all of Europe as far as southwestern Asia in about fourteen subspecies, the very closely related, long-tailed, more blue and white-colored AZURE TIT

Fig. 13-13. 1. Southern black tit *(Parus niger)*; 2. Gray tit *(Parus afer)*.

▷
1. Velvet-fronted nuthatch *(Sitta frontalis)*;
2. Pygmy nuthatch *(Sitta pygmaea)*; 3. Turkish nuthatch *(Sitta krueperi)*;
4. Orange-crowned sitella *(Neositta chrysoptera pileata)*; 5. Chestnut-bellied nuthatch *(Sitta europea castanea)*; 6. Nuthatch *(Sitta europea)*;
7. White-breasted nuthatch *(Sitta carolinensis)*;
8. Corsican nuthatch *(Sitta whiteheadi)*; 9. Striped-headed creeper *(Rhabdornis mystacalis)*; 10. Madagascar nuthatch *(Hypositta corallirostris)*;
11. Red-breasted nuthatch *(Sitta canadensis)*; 12. Wall creeper *(Tichodroma muraria)*; 13. Rock nuthatch *(Sitta neumayer)*;
14. Gray spotted creeper *(Salpornis spilonotus)*;
15. Red-browed tree creeper *(Climacteris erythrops)*.

◁

Sunbirds and honey eaters: 1. *Anthreptes collaris;* 2. Lovely sunbird *(Aethopyga shelleyi);* 3. Noisy friarbird *(Philemon corniculatus);* 4. *Nectarinia venusta alliventris;* 5. Tui *(Prosthemadera novaezealandiae);* 6. VanHasselt's sunbird *(Nectarinia sperata);* 7. Royal sunbird *(Cinnyris regius);* 8. Cape sugarbird *(Promerops cafer);* 9. *Myzomela cardinalis.*

Family: nuthatches

Subfamily: true nuthatches, by R. Berndt

◁

10. Crescent honey eater *(Phylidonyris pyrrhoptera);* 11. White-naped honey eater *(Melithreptes lunatus);* 12. *Notiomystis cincta;* 13. *Xanthomyza phrygia;* 14. *Phylidonyris undulata;* 15. Lesser yellow-eared spider hunter *(Arachnothera chrysogenys);* 16. Cinnamon-breasted wattle bird *(Melidectes torquatus);* 17. Ooaa *(Moho braccatus).*

(*Parus cyanus;* Color plate, p. 307) which has a L of 13 cm, and adjoins the blue tit's range from eastern Europe as far as Ussuria, in seven subspecies.

The tit family also includes two additional genera, each of only one species. The SULTAN TIT (*Melanochlora sultanea;* Color plate, p. 308) has a L of 20 cm and is the largest of the tits; it is found in four subspecies at lower levels from Nepal and Burma to Sumatra. The YELLOW-BROWED TIT *(Sylviparus modestus)* has a L of 9 cm and is the smallest tit. It is yellowish-green and like the leaf warbler in both plumage and behavior. It lives in the crowns of mixed deciduous and coniferous forests in the mountains of central and southeastern Asia. There are three subspecies.

In many parts of the world, tits, together with long-tailed tits, tree creepers and nuthatches, form a group particularly rich in species and numbers of individuals; this group often, particularly in winter, forms the very basis of the local bird life. Tits are generally known and are popular as indicators of spring and as destroyers of noxious insects; they have also been extensively studied.

The NUTHATCHES PROPER (subfamily Sittinae) are presently united with the wall creepers (subfamily Tichodrominae), which were formerly included with the tree creepers, in the family of nuthatches (Sittidae).

The nuthatches proper are generally blue-gray above and white below with some chestnut brown; they have a white eye stripe. They are the only birds which customarily climb head downwards instead of only upwards while seeking food. The compactly built birds do this by reaching back down with one foot while holding themselves with the other foot, and then continuing this practice with alternate feet. When climbing, movements are generally at a slant so that a spiral or zigzag pathway is followed. Using their long, strong, and pointed beak (which is often turned slightly upwards at the tip), they fetch all life stages of arthropods from cracks and clefts in tree trunks and rocks, or chisel them out of the surface layer of the bark—from the covering layer of moss and lichens. Sometimes these birds also pick up their prey from leaves, and occasionally from the ground, or they may even catch it in flight. During the winter half of the year they also take a variety of oil-rich tree seeds and other seeds, such as walnuts, conifer seeds, the fruit on the linden tree, sunflower seeds and oats. In order to hammer the seeds open they wedge them into suitable clefts; the same practice is adopted for breaking open small snails. When there is an abundance of food, they store it in clefts and use it in times of need.

The EUROPEAN NUTHATCH (*Sitta europaea;* Color plates, p. 301 and p. 319) has a L of 14 cm and weighs 23 g; it is roughly sparrow-sized. When it does not use up all its food stores, one may see sunflower or hazelnut bushes growing out of cracks in tree bark, walls or rocks. This bird is a regular guest at feeding tables in wooded districts; it generally appears in pairs. It is quite noticeable from the very beginning of winter because of its loud whistles and trills.

As the nuthatch's food on tree trunks is easily reached even in winter, the bird is an early breeder. Its conspicuous song therefore ceases early in the year, and, since second broods are almost unknown, the song does not recur till the nest year. Old pairs live in permanent monogamy and unless disturbed, occupy the same territory as long as they live. The normal foundation of territories by males begins in mid-July, at the age of two and a half months, and after intensive fights, particularly in September, against both old established pairs and young rivals, it is generally concluded by February or March.

Fig. 13-14. When climbing downwards, the nuthatch uses its feet in alternation by reaching down with one foot while clinging with the other foot above its body.

Early in the breeding season, nuthatches, being strict hole nesters, narrow the entry to their nest holes with moist clay-like soil carried in the beak; this is a protection against nest robbers and rival would-be occupants of the nest hole, particularly starlings. The opening is narrowed until it barely admits the bird's body. Occasionally game droppings or plant fibers are used for this purpose. This habit of "gumming up" the nest entrance accounts for the German name for nuthatches, meaning "gummer." The female further rounds off the nest cavity by placing small pieces of wood in the corners; she also plugs any cracks and adds more material to that around the nest entrance. The nest itself, however, is merely a loose structure of bark flakes and dry leaves.

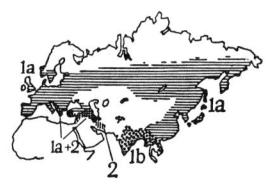

Fig. 13-15. 1. Nuthatch (*Sitta europaea*); 1a. Main group of subspecies; 1b. Chestnut-colored nuthatch group of races; 2. Rock nuthatch (*Sitta neumayer*).

Usually six to eight white red-spotted eggs are laid and incubated only by the female. The young hatch after fifteen to eighteen days and are fed in the nest for twenty-three to twenty-four days by both parents, which then tend them out of the nest for another ten days. Then they are independent and become sexually mature in the following spring. The main enemies of the adults are the Eurasian sparrowhawk, the tawny owl and, where it occurs, the pygmy owl; greater spotted woodpeckers and pine martens occasionally plunder the nests as well.

In Germany the nuthatch prefers deciduous woods, mainly oak woods, but in other areas it also uses mixed and conifer woods. It is a pronounced resident; only occasionally in certain years does it undertake migrations which are probably released by high population density. It is found from Morocco and Asia Minor over almost all of Eurasia—in some areas up to latitude 68°N. It has formed about twenty-five geographic subspecies, including the seven subspecies of the entirely chestnut colored southern Asiatic CHESTNUT-BELLIED NUTHATCH (*Sitta europaea castanea*: Color plate, p. 319), which is often regarded as a distinct species.

Fig. 13-16. Eastern rock nuthatch (*Sitta tephronota*).

There are roughly fifteen species in the single nuthatch genus (*Sitta*),

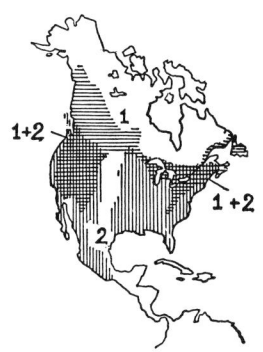

Fig. 13-17. 1. Red-breasted nuthatch *(Sitta canadensis)*; 2. White-breasted nuthatch *(Sitta carolinensis)*.

Subfamily: wall creepers, by H. Psenner

of which the ROCK NUTHATCH *(Sitta neumayer;* Color plate, p. 319) which has a L of 15 cm, and the EASTERN ROCK NUTHATCH *(Sitta tephronota)* which has a L of 15 cm, show an extraordinary nest building ability. These nuthatches, which live between the Himalayas and the Adriatic and are unable to climb head down, build spherical nests out of soil moistened with rubbed arthropods or berries; the nests have entry tubes several centimeters in length. Most other nuthatches do not have the habit of using mud or similar building materials; they either excavate their breeding holes in rotted wood or they enlarge holes already present. The WHITE-BREASTED NUTHATCH *(Sitta carolinensis;* Color plate, p. 319) has a L of 14.5 cm, is found in North American deciduous woods, and rubs arthropods around its entry hole, while the RED-BREASTED NUTHATCH *(Sitta canadensis;* Color plate, p. 319) has a L of 11.5 cm, is from North America, and smears resin about its nest entry holes. The former species is closely related to the central Asiatic WHITE-CHEEKED NUTHATCH *(Sitta leucopsis),* while the latter species belongs to a widespread group of nuthatches from coniferous woods which, outside of North America, occur in isolated populations in Eurasia. These are: the KOREAN NUTHATCH *(Sitta villosa);* the KANSU NUTHATCH *(Sitta bangsi);* the YUNNAN NUTHATCH *(Sitta yunnanensis);* the TURKISH NUTHATCH *(Sitta kreuperi;* Color plate, p. 319); and the CORSICAN NUTHATCH *(Sitta whiteheadi;* Color plate, p. 319) which has a L of 11 cm and weighs 13.5 g.

In North America, there are two additional small species, the PYGMY NUTHATCH *(Sitta pygmaea;* Color plate, p. 319) from the western conifer woods, and the BROWN-HEADED NUTHATCH *(Sitta puscilla)* from the southeastern pine woods. These two forms may possibly be two subspecies of the same species and, thus, closely related to the Asiatic WHITE-TAILED NUTHATCH *(Sitta himalayensis).* The GIANT NUTHATCH *(Sitta magna)* which has a L of 18.5 cm, is also found in this area. Included among the southeastern Asiatic species are: the BEAUTIFUL NUTHATCH *(Sitta formosa)* which has a L of 16.5 cm, is particularly colorful, and is found in Sikkim and Assam; the VELVET-FRONTED NUTHATCH *(Sitta frontalis;* Color plate, p. 319) which has a L of 12.5 cm, and the AZURE NUTHATCH *(Sitta azurea)* which has a L of 12.5 cm, both are the only members of the subfamily which are found as far as Indonesia. A velvet-fronted nuthatch has been living in the Frankfurt Zoo for five and a half years.

The WALL CREEPERS (subfamily Tichodrominae) consist of only one genus and species, the WALL CREEPER *(Tichodroma muraria;* Color plate, p. 319) which has a L of 16 cm. It occurs in two subspecies from the Pyrenees, Alps, northern Apennines and Carinthian mountains over the Balkans to Syria, the Himalayas and China.

Many descriptions lead one to believe that the wall creeper is a real mountain bird. This is not correct. If this bird is found on the rocks of European mountains other than in the lower-lying Alpine areas, this is only because it finds the sort of habitats it prefers at higher elevations. The wall creeper is at home in ravines and rock cliffs, even when they lie along low lying rivers. It does not like bare rock walls—there must be some scattered vegetation. Even its name indicates that it is a rock dweller and seeks its food by climbing. The claw of its hind toe is particularly long, and its feet can grasp any slight projection on a rock wall. While climbing, it opens its wings in a jerky manner showing the observer a flash of the red band of the wing. This color becomes intensified in males during the breeding season.

On a cliff called the Martinswand, near Innsbruck, Austria, I was able to see five wall creepers in display at once. They clearly showed the ability to descend from the top to the base of the cliff in batlike flight. Wall creepers often appear on churches of mountain villages if these are built of natural stones. In the past, such birds were regarded in Innsbruck as escaped exotics and they were much admired. Only since the Alpine Zoo in that city, which was the first and only zoo to exhibit this species, have many mountaineers come to know a bird they did not notice before.

A wall creeper nest which I observed in Pitztal was at arm's length inside a rock cleft through which a trickle of water ran. The base of the nest was so wet that I could wring out the nest material like a wet sponge. Nevertheless, hairs which formed the lining of the relatively large nest were quite dry. The fully fledged young withdrew into the furthest crack of the cleft and were still being fed by the adults when I took out the nest to examine it. The adult birds were quite tame, and continued feeding the young, even though I was only a few meters away; they brought their offspring arthropods and spiders which they found in fine cracks in the rock. Wall creepers only live socially while breeding and until the young are independent; otherwise they are solitary birds, a stronger one chasing away a weaker one from its territory.

Two bird groups sometimes considered subfamilies among the relatives of the tits and nuthatches are the TREE RUNNERS (Neosittinae) and the Australian TREE CREEPERS (Climacterinae); here we will regard them as one family, the TREE RUNNER-LIKE BIRDS (Climacteridae). G. F. Mees says the following about these forms: "I am convinced that the Australian nuthatches are the closest relatives of the tree runners, and that neither have anything to do with the tree creepers or the nuthatches."

Family: tree runner-like birds, by H. Th. Condon

The tree runners or SITELLAS (subfamily Neosittinae) have a L of 10–12 cm; they resemble the true nuthatches in size, shape and movements, as well as in their ability to hop downwards, head down, on tree trunks. However, tree runners are more colorful and do not nest in tree holes or rock cavities, but rather, build deep cup-shaped nests. There are two genera with a total of three species. Members of the genus *Neositta*, which has two species found in Australia and New Guinea, have a beak which points upwards at the tip; the rump is

Subfamily: tree runners

white. The ORANGE-CROWNED SITELLA *(Neositta chrysoptera)* is the best known species of this genus; it has an orange or white wing spot and occurs in Australia but not Tasmania. It has well recognizable subspecies which interbreed at the borders of their ranges.

The orange or white wing spot of the orange-crowned sitella is conspicuous in flight. The birds move about in small flocks of up to twenty or more. Their call is an excited-sounding soft twittering which is uttered in flight, while feeding, and at the nest. The nest consists mainly of spider webs and cocoons; pieces of bark camouflage its exterior so that it can hardly be told from the branch which supports it. There are usually three eggs which are almost the same color as the lichen lining of the nest. Not only the pair concerned but other members of the species, as well, take part in nest building and in the feeding of the young.

Subfamily: Australian tree creepers

The Australian tree creepers (subfamily Climacterinae) are small resident birds with a L of 16 cm. They resemble the true tree creepers in their habit of climbing trees with rough bark on a spiral course while in search of insects; however they have softer tails. As a result, they depend on their strong feet which have long curved claws for climbing. They too, inhabit Australia and New Guinea in one genus *(Climacteris)* with seven species. The BROWN TREE CREEPER *(Climacteris picumnus)* is grayish-brown with a bright yellowish-brown wing band and a tail with black cross bars; below it has conspicuous white stripes with black edges. The male has a blackish throat spot while that of the female is reddish. This species is found in eastern and southern Australia.

The call of these birds is a harsh twittering and a loud "pink pink". The soft nest holds two or three eggs which have intense reddish-brown spots on a pale background. Nests are found in tree holes six to fifteen meters above the ground; occasionally they may be found in a hole in a post. The nests are constructed from pieces of bark bound together and lined by hairs and feathers.

There is still considerable uncertainty about the correct systematic position of the PHILIPPINE CREEPERS (Rhabdornithidae) and the SPOTTED CREEPERS (Salpornithidae). In 1963, the ornithologist E. Mayr expressed the view that these forms should be considered separate families since they show no unequivocal relationship to other groups of birds.

Family: Philippine creepers, by J. Steinbacher

The PHILIPPINE CREEPERS were formerly generally treated as a mere subfamily of the true tree creepers. There is one genus with two species, the STRIPED-HEADED CREEPER *(Rhabdornis mystacalis;* Color plate, p. 319) with a larger beak, found on the Philippine Islands of Luzon, Negros, Panay, Mindanao, Leyte and Samar and the PLAIN-HEADED CREEPER *(Rhabdornis inornatus)* which has a shorter, thicker beak and is found on Samar and Mindanao.

These birds are obviously larger and stronger than the true tree creepers. They are grayish-brown above and white below; on the sides they have characteristic black and white stripes. Their beak is much more decurved than in the tree creepers. When the ornithologist Whitehead first saw one of these birds in the wild he thought he was looking at a modified sunbird. The Philippine creepers are, in fact, able to lick up nectar by means of their tongue, which has a brushlike tip, although usually they search for insects in the bark of trees. They inhabit dense forests in the Philippines and are not common in any area. Their nests are placed in tree holes.

The family Salpornithidae consists only of one species, the GRAY SPOTTED CREEPER (*Salpornis spilonotus*; Color plate, p. 319) which has a L of about 15 cm and is found in parts of India and in Africa south of the Sahara. It is blackish above with white spots, and reddish-yellow below.

Family: spotted tree creeper, by C. W. Benson

This bird resembles the tree creepers in its slender and decurved beak and its habit of climbing trees to search for arthropods. However, its tail is not strong and pointed so that it cannot use it as a support when climbing. In addition, this creeper does not place its nest behind loose bark, but rather builds an open cup-shaped nest in a tree fork. It lays two or three bluish eggs with lilac or brown spots. It occurs mainly in open woodland where it is generally found in groups. Its flight from tree to tree is undulating, but swift. The most frequent call is a rapid sequence of shrill but not very conspicuous whistles.

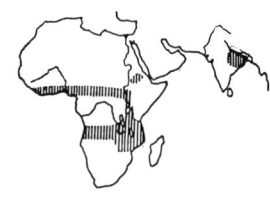

Fig. 13-18. The gray spotted creeper (*Salpornis spilonotus*).

The European representatives of the TRUE TREE CREEPER family (Certhiidae) are the only passerine birds in Europe which climb tree trunks upwards like the woodpeckers, to which they are not, however, related; this climbing habit evolved independently in the two groups in association with a similar mode of life. There is only one genus (*Certhia*) with five species: 1. SHORT-TOED TREE CREEPER (*Certhia brachydactyla*; Color plate, p. 308) has a L of 13 cm and weighs 9 g; it occurs from western (not the British Isles) and central Europe eastwards as far as Asia Minor. 2. TREE CREEPER (*Certhia familiaris*; Color plate, p. 308) differs from the above species in being more white below, and having a wider white stripe above the eye; it is also browner above. The call is a tremulous "sirrl" and its song is longer than that of the other species; it is a soft twittering of whistles and little trills of fluctuating pitch. This bird occurs from Europe as far as eastern Asia and in North America from Alaska to Newfoundland, south as far as Nicaragua. 3. HIMALAYAN TREE CREEPER (*Certhia himalayana*) occurs in four subspecies from Baluchistan and Turkestan to southern Burma and central China. It prefers conifers, elderberry bushes, and rhododendrum. 4. BROWN-THROATED TREE CREEPER (*Certhia discolor*) is long-tailed; it occurs in five subspecies from Nepal to Burma and is found more among oaks, alders and other deciduous trees. 5. STOLICZKA'S TREE CREEPER (*Certhia nipalensis*) is equally long-tailed but has a short straight beak. It occurs in the Himalayan

Family: true tree creepers, by R. Berndt

Fig. 13-19. 1. Short-toed tree creeper (*Certhia brachydactyla*); 2. Tree creeper (*Certhia familiaris*); 2a. *Familiaris* group of races; 2b. *Americana* group of races.

forests between altitudes of 2,500 and 3,500 meters and in winter it wanders down to elevations of 1,500 meters.

The two species found in Germany are among the most common and widest distributed birds of that area; as a result one can see at least one species, not both, every day in woods and parks. Nevertheless, the novice birdwatcher is rarely familiar with these birds. They are both small birds with a mouse-gray coloring and a type of bark-like pattern on the upper parts of the body, a pattern which serves as an excellent camouflage, making these birds almost invisible on a tree trunk. As their calls are also insignificant, tree creepers attract attention only in the spring through their song—a song which Hermann Löns has called a "merry silver song."

The short-toed tree creeper generally inhabits lowland deciduous woods, and often parks with older trees as well. With its fine, sharp and curved claws, it clings to the bark while hopping up a tree, often taking a spiral course around the trunk. At the nest tree trunk, it begins once more at the bottom. Like a fly on a ceiling, it can also creep along the underside of thick horizontal branches with its back facing downwards. It can even work its way downwards for short stretches with the tail leading, but only rarely in a head down position. It uses its tail for support just as woodpeckers do. For this reason, in the moult of the stiff shafted tail feathers, the central pair are shed last so that there is always some tail support. With its fine long, curved, tweezer-like beak, it pulls insect eggs and spiders out of cracks in the bark of trees.

As tree trunks support a rich fauna of small invertebrates, a supply which is largely independent of the weather and is already available in early spring, tree creepers are among the earliest breeding birds. Pairs generally remain together through the winter and keep in contact by means of "titt" calls. From January onwards the male delivers its rhythmical song which can be rendered as "teet teet teeteroiteet". Nest building may begin before the end of March. Nest sites are in wind tears of tree bark or in places where the bark has ruptured, as well as in stacks of wood, behind walls of wood paling, and even in occupied nests of much larger birds, for tree creepers breed in clefts. In such spaces, which usually have no floor and often two side entrances, the bird first throws in coarse, dry twiglets, often larger than its own body; as soon as these twiglets have caught, the tree creeper builds the nest cup which consists of moss, fibers, wool, spider webs and such-like materials on this base.

There are usually five or six white eggs with brownish-red spots. The young hatch after two weeks of incubation and leave the nest after an additional two weeks. Only the female incubates, but both parents feed the young. A second brood is usual. When the family has dispersed, tree creepers often join tit flocks, singly or in twos. The young may then wander some distance; they are sexually mature after one year. Adults seem to keep more strictly to the same locality and to breed year after year in the same small area.

According to Thielcke, certain short-toed tree creepers and certain tree creepers learn the song of the other species and deliver it mixed in with their own song, or alternate between the two songs. The two species differ only slightly in their requirements and adaptations. The short-toed tree creeper with its very long beak and much curved claw prefers older stands of oaks, elms, alders and other trees with coarse bark. The tree creeper is apparently slightly weaker and inferior in competition with the other species. It has a shorter beak and a flat rear claw. It has to make do with younger trees with smooth bark like spruce, pines and beech trees.

The different sensitivities of the two species, which becomes apparent in their distribution (the short-toed species is absent from Britain and northern Europe), is also shown in their sleeping habits. Short-toed creepers sleep, if possible, in holes often with several birds (up to fifteen) together. Tree creepers sleep only in pairs, even at less than minus 12°C; otherwise they sleep alone in self-dug depressions in the bark. The main enemies of tree creepers are the Eurasian sparrow hawk in the daytime and the tawny owl at night as well as the pygmy owl, where it occurs.

14 The Sunbirds, Flower Peckers, Honey Eaters, and White-eyes

The honey eaters and relatives, by J. Steinbacher

This chapter deals with four families of song birds, whose preferred food is the nectar of flowers and blossoms of every kind. Because of their probable origin, they are placed close to the tits and their relatives. The four families are remarkably uniform among themselves, but are separated from one another by a number of striking characteristics involving shape, structure and plumage. The FLOWER PECKERS (mistletoe birds), the SUNBIRDS and the HONEY EATERS resemble one another considerably in life style; as their names indicate, they are dependent on frequent visits to blossoms to obtain nectar of fruit juice. The modification of the tip of the tongue, which ends in a bundle of fibers or which can become a sucking tube, enables them to feed this way. Their feeding in many cases also results in the fertilization of the flowers they visit. In the white-eyes family, a few species with only slightly fringed tongues are transitions to the more specialized families. Evidently the white-eyes have developed rather away from the other three families; the white-eyes have only nine primaries (the tenth is quite degenerate) while the members of the other families (apart from the genus *Dicaeum*) have ten.

Aside from the small and very small forms which are among the very tiniest of birds, there are, among the sunbirds, larger species up to the size of a jay, but much more slender. All sunbirds are tree dwellers and are sociable. They are good fliers, are always in motion, and have striking melodious or sharp calls and in some cases, chattering songs.

Family: flower peckers (mistletoe birds), by B. E. Smythies

The flower peckers or mistletoe birds (family *Dicaeidae*) are distributed over all of southeastern Asia in seven genera and fifty-five species; they are found as far east as Australia, New Guinea and the Solomons. Two species, the BRONZE-BACKED FLOWER PECKER *(Dicaeum ignipectus)* and the YELLOW-BELLIED FLOWER PECKER *(Dicaeum melanoxanthum)* are found as far as southwestern China and the Himalayas. All are residents, apart from the AUSTRALIAN MISTLETOE BIRD *(Dicaeum hirun-*

dinaceum), which has long, swallow-like wings and, as a strong flier, leads a nomadic life.

Most of the species belong to the genera *Dicaeum* and *Prionochilus.* They are small birds with wing lengths from 4 to 7.6 cms, and short square-ended tails. They have short beaks, with toothed edges in the third towards the tip. The beaks vary from the short blunt tit-like type to the thin pointed, decurved beak present in short-beaked sunbirds. The tongue is short, the edges of the deeply cleft half towards the tip are rolled inwards forming two semi-tubular endings (similar to those of sunbirds), probably an adaptation to feeding on nectar. In some species, the plumage is simple and the same in both sexes; in other species, the males are brightly colored and the females are generally duller. The outermost (tenth) primary is well developed in the flower peckers, but only suggested in the *Dicaeum* in the flower peckers, with the exception of the yellow-bellied flower pecker.

Fig. 14-1. Distribution of the flower peckers or mistletoe bird, family (Dicaeidae).

The New Guinean genera of the BERRY PECKERS *(Melanocharis)* and SPOTTED BERRY PECKERS *(Rhamphocharis)* are primitive with a simple tongue structure. The berry peckers of the genera *Oreocharis* and *Paramythia,* each with only one species, also found in New Guinea, deviate more. The DIAMOND BIRDS or PARDALOTES *(Pardalotus)* are found in New Guinea as well with seven species, including the SPOTTED PARDALOTE *(Pardalotus punctatus)* of Australia and Tasmania. Evidently, these birds are only distant relatives of the flower peckers and only share with them the features of an almost atrophied first primary. However, berry peckers have simpler tongues and no "teeth" on the beak.

Flower peckers of the genus *Dicaeum* generally inhabit evergreen forests of lowlands as well as mountains; some species, like the SCARLET-BACKED FLOWER PECKER *(Dicaeum cruentatum)* and the ORANGE-BREASTED FLOWER PECKER *(Dicaeum trigonostigma)* also visit nearby gardens and areas of second growth. They are usually seen singly. These minute, compact but very lively birds slip from one blossoming tree to another and, when crossing over trails, demonstrate their zig zag flights and sudden changes of direction. In flight they utter sharp "tshit tshit tshit" calls which sound much like two pebbles being struck together. The Australian paradolates behave in a different manner; they feed much like tits on the bark and leaves of trees from the crown down to the ground.

The rather different, very sociable CRESTED BERRY PECKER *(Paramythia montium)* of New Guinea is as large as a thrush and is blue with much white in its erectile black and white crest. It is found in mountains up to the tree line. The related TIT BERRY PECKER *(Oreocharis arfaki)* has a bright yellow breast and looks somewhat like a tit and is often encountered in flocks. In the SPOTTED BERRY PECKER *(Rhamphocharis crassirostris)* the female, which is yellowish white below, is much larger than the male.

There is a close relationship between certain flower peckers and particular mistletoe berries. Many species feed almost exclusively on these berries, apart from arthropods, and they are the most effective agents for the dispersal of mistletoe seeds on their home grounds. Thus, a mutual dependence between the birds and the plants has developed. F. Salomonsen has written the following description: "The manner of eating the berries varies with the beak structure of the bird. Thick-billed species like the STRIPED FLOWER PECKER *(Dicaeum agile)* use the beak to detach the pulp of the fruit from the seed. Then they swallow the fruit and get rid of the seed by rubbing it off the beak on a twig. The thin-beaked forms, like TICKELL'S FLOWER PECKER *(Dicaeum erythrorhynchos)* from India, swallow the whole berry and shed the seed from the cloaca in a matter of minutes. In both cases, the seeds retain the ability to germinate. The speed with which mistletoe berries pass through the intestine is partly due to their laxative action, but also to a peculiar structure of the stomach. The gizzard (ventriculus) has become an evagination of the main digestive tract with a sphincter opening. Readily digestible berries therefore pass directly from the gullet into the intestine without entering the gizzard. On the other hand, arthropods and spiders enter the gizzard where they are ground up and thoroughly digested. The digestive tracts of the BLACK BERRY PECKER *(Melanocharis nigra)* and of the crested berry pecker *(Paramythia montium)* do not show these features and are more like those of other fruit-eating songbirds."

According to Barbara and Tom Harrison, the YELLOW-RUMPED FLOWER PECKER *(Prionochilus xanthopygius)* of Borneo feeds mainly on blossoms, pollen and nectar; only in a quarter of birds they examined did they also find minute beetles and spiders. The PLAIN FLOWER PECKER *(Dicaeum concolor)* apparently eats only arthropods, while the SCARLET-THROATED BLACK-SIDED FLOWER PECKER *(Dicaeum celebicum)* takes large fruit and has a varied diet which also includes aphids, the smallest flies, insects and seeds. The Australian pardolates eat arthropods almost exclusively; the four genera which occur only in New Guinea are fruit-eaters.

The flower peckers build small, pear-shaped nests into which they can slip through a lateral entrance near the upper end and which resemble those of sunbirds. These nests are beautifully woven of various plant materials, are suspended from twigs and are lined with additional plant material. The pardolates, on the other hand, nest in tree holes or in holes in the ground which they dig themselves; their nests are cup-shaped or spherical. The only New Guinean bird of this type whose nest man is familiar with is the crested berry pecker. The nest is cup-shaped and stands in dense bushes. In most flower peckers, the eggs are white; only a few members of the genus *Dicaeum* and the crested berry pecker have spotted eggs.

It is generally accepted that the sunbirds (family Nectariniidae) take much the same place in the Old World tropics as hummingbirds do in those of the New World. Certainly these birds play a role in the fertilization of flowers and have a certain number of structural features in common with the hummingbirds—features concerned with adaptation to life with flowers and others like the reduction in size and the changes in feathers to make them iridescent. However, these similarities are superficial. The basic structure of both groups is very different and they are not related.

Sunbirds are a relatively uniform group. They are small or very small, the size of a starling at most; the L is 9-25 cm. The beak and tongue are reminiscent of those of hummingbirds, the beak being long and thin, tube-shaped and narrow right from the base. It is almost always somewhat decurved—so much so in the GOLDEN-WINGED SUNBIRD (*Nectarinia reichenowi*) that it looks like a sickle. It is always pointed and black, and is only a little paler at the base of the lower mandible in some species. The tongue is long and its tip can be projected forward into two tubes and is useful for sucking up nectar or for catching arthropods. The limb structure is quite different from that of hummingbirds. The tarsi are fairly long and generally thin. The toes have curved claws of medium length. The foot is built on the usual passerine pattern. The wings, too, conform to the passerine pattern—the elbow portion is well developed and the upper wing surface is rounded; there are ten primaries of which the fourth and fifth are the longest. The tail always consists of twelve well-developed feathers; it is never forked and is always either square-ended or graduated. The central tail feathers are sometimes very much prolonged. There are 8 genera with 108 species.

In adult males the dense plumage often has a metallic iridescence or bright colors—features which are reminiscent of the hummingbirds. Sunbirds even exceed the latter in the magnificence of their colors as they have not only iridescent colors, but a non-glossy bright red and yellow, or other colors as well. In general, plumage differences between the two sexes are much more frequent than in hummingbirds as the females do not have the iridescence or the striking colors of the adult males. Less colorful sunbirds often have yellow or red tufts of feathers on their flanks. These are concealed by the wings when the birds are at rest; such tufts only rarely occur in females.

As the beak and the tongue structure suggest, the food of sunbirds is similar to that of hummingbirds. It consists of small insects and nectar, which the projectile tongue sips up from the base of flowers; occasionally fruit pulp or juice are also taken. Certain large species, like the SPIDER HUNTERS (genus *Arachnothera*) are regarded as being entirely spider and insect eaters. To get nectar, the sunbirds cling to flowers; they take arthropods on the wing. Although incapable of the stationary hovering of the hummingbirds, they can hang in the air for

Family: sunbirds, by J. Berlioz

▷
Above: Yellow-throated miners (*Manorina flavigula*); Below: The woodpecker finch (*Cactospiza pallida*).

▷▷
Left, from above down; Goldfinch (*Carduelis carduelis*); Yellow cardinal (*Gubernatrix cristata*); Right, from above down: Cardinal (*Cardinalis cardinalis*); Painted bunting (*Passerina ciris*).

a few seconds by means of rapid wing beats. They often "stand" in front of flowers or spider webs as hummingbirds do. They are particularly interested in tubular flowers and in South Africa, for example, the numerous aloes are one of their favorite habitats.

Sunbirds are swift and lively in all their movements and show unfailing energy. Their flight is swift and straight-lined, but they show little endurance in flight, nor are they able to fly backwards like hummingbirds. They can readily take off with their feet from twigs and small branches without any help from their wings. Although they seem to like to quarrel, they are by far less cantankerous than hummingbirds. They are generally seen singly or in pairs, but they often join into small troops which visit flowering shrubs and large trees together. When doing this, they follow one another in flight with short, sharp calls, but show no real hostility toward one another.

In their reproduction, the sunbirds differ considerably from the hummingbirds. Pairs form for the whole breeding season; both sexes take part in nest building, and the feeding of the young, and are thus monogamous. However, only the females incubate. The nest is always closed and has a side entrance. Usually it is pouch-shaped and suspended from a branch. It is carefully constructed out of a variety of materials which are generally held together by spider webs. The clutch usually consists of two eggs which are very pale with fine brown or black spots. Strangely enough, these small spherical nests are often visited in tropical Africa and Asia by small species of cuckoos such as the Klaas and the emerald cuckoo (Volume VIII); while the young cuckoo is being fed by the sunbirds, the rapidly developing parasite often bursts the nest.

Sunbirds are found throughout the tropical world on the mainland as well as on islands, wherever there are flowers and insects in abundance; there are sunbirds from the West African coast as far as the Australian Pacific coast. The greatest number of species are found in Africa. Like the New World hummingbirds, sunbirds have adapted to the most varied habitats. Some are at home along the edge of the desert or in the semi-deserts of tropical and southern Africa; others in the damp forests of the lowlands; still others prefer higher elevations in mountains, particularly the zone where frequent mists ensure a rich vegetation. As is the case of hummingbirds, certain human activities, such as the spread of gardens favor their distribution. However, in contrast to the hummingbirds, they have not invaded the temperate zone as the distribution map shows. As in the hummingbirds, unfavorable circumstances release local movements, which in the case of mountain forms, lead them to lower elevations, and in other species, over greater distances. The largest genus is *Nectarinia* which can be divided into several sub-genera, including the LONG-TAILED SUNBIRD (subgenus *Nectarinia*) to which the African species belong. These birds

◁
From above down: Bullfinch *(Pyrrhula pyrrhula)*, female; Brambling *(Fringilla montifringilla)*, male; Bullfinch, male.

◁◁
Above from left to right: Seven-colored tanager *(Tanager fastuosa)*; Blue-winged mountain tanager *(Anisognathus flavinucha)*; Center from left to right: Brazilian tanager *(Ramphocelus carbo bresilius)*; Red-legged honey creeper *(Cyanerpes cyaneus)*; Blue-and-yellow tanager *(Thraupis bonariensis)*; Below left to right: Purple honey creeper *(Cyanerpes caeruleus)*; Glistening-green tanager *(Chlorospingus phoenicotis)*.

have both central tail feathers prolonged, as in the MALACHITE SUNBIRD *(Nectarinia famosa)* a beautiful, fairly large bird common in eastern and southern Africa. The male has a greenish-bronze plumage and bright yellow flank tufts. The subgenus *Cinnyris* includes many African species as well as others from islands in the Indian Ocean. The males of this group have either straight-ended or graduated tails and iridescent feathers on the upper parts and the breast as in *Nectarinia coccinigaster.* One of the largest and most iridescent species is *Nectarinia superba* of the lowlands about the Gulf of Guinea. In the subgenus *Chalcomitra,* the males have a matte, often black and brown back. *Nectarinia amethystina* is a beautiful species found in several subspecies in open areas of eastern and southern Africa. One of the most inconspicuous species is *Nectarinia olivacea* which has a completely olive-green plumage on which the males have yellow side tufts. Members of the subgenera *Hermotimia* and *Cyrtostomas* do not have iridescence in the plumage of the back. The first subgenus includes the smallest of all sunbirds, the SMALL SUNBIRD *(Nectarinia minima)* of southern India, which in its minuteness and striking, partly iridescent plumage, can rival the small hummingbirds.

In the sunbirds of the genus *Aethopyga,* the male's plumage includes blood-red or bright yellow feathers which are made more prominent by the other areas where there are scale-like metallic green or bluish purple feathers. The two central tail feathers are more or less prolonged. These very small birds are particularly attractive. They all inhabit southeastern Asia where most of them live in wooded zones of the mountains. There are fifteen species, but we will mention only the widely distributed YELLOW-BACKED SUNBIRD *(Aethopyga siparaja).*

The sunbirds of the genus *Anthreptes* generally have shorter and less curved beaks than do those of the genus *Nectarinia.* Their size varies from species to species; some are minute. The plumage is generally inconspicuous and loose; only a few species are made more attractive by bright metallic-looking upper parts. In their habits, they are generally less impetuous and cantankerous than the typical sunbirds; they are also shyer. These birds are generally found singly or in pairs in fairly dense vegetation. Of the dozen species about three quarters are restricted to Africa. One of the most common sunbirds of Malaysia, the Philippines and the Sunda Islands as far east as Alor, is the PLAIN-THROATED SUNBIRD *(Anthreptes malacensis).* The RUBY-CHEEKED SUNBIRD *(Anthreptes singalensis),* also found in the Indo-Malayan region, is separated as a distinct subgenus *Chalcoparia.* This little bird, which has a fairly short tongue which ends in a brush-like tip, is sometimes regarded as a link with the honey eaters.

The ten species of spider hunters (genus *Arachnothera*) can also be regarded as transitional forms to certain New Guinea honey eaters. All are fairly large and massive in structure and show no bright colors or decorative features apart from the side tufts so characteristic of sunbirds, found in some species. The tail is fairly short and cut off straight; the sexes look almost identical. The beak is strong, decurved, and very long; it is highly suitable for catching spiders. All spider hunters inhabit the Indo-Malayan mainland and its islands; thus the LESSER YELLOW-EARED SPIDER HUNTER (*Arachnothera chrysogenys*; Color plate, p. 320) is found from the Greater Sunda Islands and Malacca north as far as southern Burma.

For some time it has been possible to import sunbirds, particularly the African varieties, to Europe and to keep them in captivity. If they are provided with the same living conditions, manner of feeding and food as hummingbirds, they are relatively easy to keep. Fanciers often prefer them to hummingbirds as they are more lively and attractive, although hummingbirds are more unique and more varied.

Family: white-eyes, by G. F. Mees

The WHITE-EYES (Zosteropidae) are distributed over Africa, southern and eastern Asia, Australia and western Oceania, and form a very uniform songbird family. They owe their English and scientific name to the ring of small white feathers around the eye; this is usually very pronounced and is absent only in a very few species. It is not known whether this eye ring has any biological significance; it seems reasonable to suggest that it aids mutual recognition.

All white-eyes are small; the L is 10–14 cm and the wing length is 5–7.6 cm. The plumage is fairly inconspicuous; generally the upper parts are yellowish green, and the lower parts are yellow, gray or whitish. A few species, usually inhabitants of small islands, are brown or gray. The beak, as in other birds which visit flowers or eat insects, is fine; the tongue is adapted to the sucking of nectar. The first (outermost) primary is atrophied to a minute remnant. The sexes are similar but males are somewhat larger. There are twelve genera with about ninety species.

The white-eyes proper (genus *Zosterops*) with over sixty species, form one of the largest of all avian genera. A series of species are extraordinarily similar to one another. Thus, *Zosterops maderaspatana* is only distinguishable from the NEW GUINEA MOUNTAIN WHITE-EYE (*Zosterops novaeguineae*) with difficulty; *Zosterops griseovirescens*, which only occurs on the west African islet, Annobon, resembles *Zosterops natalis* of Christmas Island, south of Java. A subspecies of the southern Indian ORIENTAL WHITE-EYE (*Zosterops palpebrosa nilgiriensis*) can hardly be differentiated from *Zosterops citrinella citrinella* of Timor.

Another genus of this family is that of *Lophozosterops*. Its six species are mostly mountain birds of islands in the Indo-Australian area (Java, Sumbawa, Flores, Celebes, Ceram and Mindanao). These birds are larger than the white-eyes and have a gray or brown crown which is often striped or has small white droplet-like spots. *Lophozosterops dohertyi* of Sumbawa has a crest.

The MOUNTAIN BLACKEYE *(Chlorocharis emiliae)*, the sole representative of its genus, lives on high mountains in Borneo. Its eye ring consists not of white, but of black feathers. It is never found below 1,300 meters, and on Kinabalu, the highest peak of Borneo, it even lives on the bare rock mountain top at over 4,000 meters.

Here we will consider only the white-eyes proper *(Zosterops)* in more detail. The various species resemble one another not only in shape but also in habit. They feed on arthropods, fruit and berries as well as a considerable amount of nectar. Outside the breeding season they wander about in small flocks of three to twenty members through the woods, much as tits do in Europe. Their sociability is also evident when they often sit tightly pressed together in sleeping places where they also regularly preen one another. Most species have a pleasant song reminiscent of that of the canary or the hedge sparrow.

The nests and eggs of over half of the species have not yet been described. As far as is known, the nests are small baskets suspended in tree forks. The two to five eggs are greenish-blue to milky-blue or white, and they are only spotted in two species of the Lesser Sunda Islands, *Zosterops wallacei* and *Lophozosterops dohertyi*. Both sexes share in incubation and the rearing of the young. As far as is known, the incubation period generally lasts eleven to thirteen days. The PALE WHITE-EYE *(Zosterops pallida capensis)*, according to the careful researches of Skead and Ranger, incubates only for ten days and three to four hours. Presumably this applies also to the ORIENTAL WHITE-EYE *(Zosterops palpebrosa)* and the GRAY-BREASTED SILVEREYE *(Zosterops lateralis)*. This gives these birds the shortest incubation period of all birds. The fledging period in some species is ten to twelve days.

Although these birds have been known to science for more than two hundred years, they were only "elevated" to a genus of their own in 1826. In subsequent years they were allocated to the most varied songbird families until 1888 when the British zoologist, A. Newton, separated them as a single family.

Nevertheless, their position in the system has remained uncertain; because of their flower sucking and their tongue adapted to nectar sucking, they are generally placed near the other sunbirds.

The white-eyes are very successful colonizers of small and remote islands and groups and islands where separate subspecies, species or even genera have often evolved. Thus, in southwestern Oceania, almost every island has its own subspecies or species. On the other hand, large land masses are poor in white-eyes species. The whole of Australia has only two species, while there are three on minute Norfolk Island east of Australia. The greatest wealth of species is found in the Indo-Australian island area. In Africa, the YELLOW WHITE-EYE *(Zosterops senegalensis)* inhabits an area which extends from the Senegal over central and eastern Africa, to Zululand. It is replaced in southern

Fig. 14-2. Gray-breasted silvereye *(Zosterops lateralis)*.

▷
Old World buntings:
1. Yellowhammer *(Emberiza citrinella)*; 2. Black-headed bunting *(Emberiza melanocephala)*; 3. Cirl bunting *(Emberiza cirlus)*; 4. Siberian meadow bunting *(Emberiza cioides)*; 5. Rock bunting *(Emberiza cia)*; 6. Reed bunting *(Emberiza schoeniclus)*; 7. Corn bunting *(Emberiza calandra)*; 8. Ortolan bunting *(Emberiza hortulana)*.

Africa by the PALE WHITE-EYE *(Zosterops pallida)* and northeastern Africa and Arabia by the ABYSSINIAN WHITE-EYE *(Zosterops abyssinica).*

The best known species in captivity is probably the ORIENTAL WHITE-EYE *(Zosterops palpebrosa)* which occurs in eleven subspecies from Afghanistan and India to Burma, Malaya, Sumatra and Java to the Lesser Sunda Islands. Another Asiatic species, the CHESTNUT-FLANKED WHITE-EYE *(Zosterops erythropleura)* is marked by dark red-brown sides. It breeds along the lower Amur—further north than any other members of the family. It winters in mountain forests in Burma, Thailand and southern China. On migration this species covers a distance of 3,500 kilometers, twice a year. The JAPANESE WHITE-EYE *(Zosterops japonica)* lives not only in Japan, but in a series of subspecies on Taiwan, the northern Philippines and a large part of China. Unfortunately, it has only rarely been bred in captivity.

The MOUNTAIN WHITE-EYE *(Zosterops montana;* Color plate, p. 372) resembles these Asiatic species, but differs from them in having white, not brown eyes. It has never been found below 1,000 meters in its home—the high mountains of the Sunda Islands, Moluccas and Philippines (but not on Borneo). Usually it is found from 2,000 meters up to the bare mountain tops.

With the GRAY-BREASTED SILVEREYE *(Zosterops lateralis)* found in ten subspecies from western Australia to the New Hebrides and Fiji Islands, we will begin our discussion of the series of Australian-Polynesian species. This bird is characterized by a gray middle of the back; in other species the middle of the back is always yellowish-green, as it is in two subspecies of the species just mentioned. The subspecies *gouldi,* which has an entirely green back, is abundant in southwestern Australia, where it avoids the dry areas; thus it is generally found within 100 to 150 kilometers from the coast. This small, pretty bird has been persecuted without mercy in Australia for decades; in 1900, 20,000 birds were shot about the city of Perth alone. They are regarded as harmful in Australia because they take soft fruit and often visit orchards in swarms. On the other hand, they have great economic significance for man as destroyers of insects.

There is hardly a bird book which does not mention how the TASMANIAN GRAY-BREASTED SILVEREYE *(Zosterops lateralis lateralis)* which settled New Zealand in the thirties of the last century. This subspecies found originally only in Tasmania, visited southeastern Australia annually as a migrant. About 1835, small troops were first seen on the western coast of the south island of New Zealand; apparently they had drifted across the Tasman Sea. Twenty years later, in 1856, there was an invasion of the northern island. Since then this form has spread all over New Zealand and the adjacent islands, and it is now one of the most common birds there.

Tiny Norfolk Island, between New Zealand and New Caledonia, was

◁
1. Orange-billed sparrow *(Arremon aurantirostris);* 2. Saffron finch *(Sicalis flaveola);* 3. Song sparrow *(Melospiza melodia);* 4. Red-crested cardinal *(Paroaria coronata);* 5. Ruddy-breasted seedeater *(Sporophila minuta);* 6. Yellow-faced grassquit *(Tiaris olivacea);* 7. White-throated sparrow *(Zonotrichia albicollis);* 8. Rufous-sided towhee *(Pipilo erythrophthalmus);* 9. Yellow cardinal *(Gubernatrix cristata);* 10. Tree sparrow *(Spizella aborea);* 11. Ringed warbling finch *(Poospiza torquata);* 12. Oregon junco *(Junco oreganus);* 13. Blue-black grassquit *(Volatinia jacarina);* 14. Seaside sparrow *(Amnospiza maritima);* 15. Snow bunting *(Plectrophenax nivalis);* 16. Lapland longspur *(Calcarius lapponicus).*

settled by this species in 1904, probably from New Zealand. There it met two native species, *Zosterops tenuirostris* and *Zosterops albogularis* both of which were probably earlier immigrants also derived from this species, and which had meanwhile developed a few differences. *Zosterops albogularis*, in contrast with all other white-eyes, is not sociable but lives singly in the forest; the total population nowadays probably does not exceed thirty to forty birds.

Three forms of white-eyes have become extinct in very recent times. *Zosterops semiflava* of the Seychelle Island, Marianne, was last seen in 1890. *Zosterops everetti everetti* on Cebu in the Philippines has not been seen since 1906. *Zosterops strenua* of Lord Howe Island east of Australia was still frequent there until 1920 but disappeared soon thereafter. In all three cases man was the cause of extinction. On Marianne and on Cebu, deforestation was the direct cause. In the case of Lord Howe Island, it was the accidental introduction of rats when the steamer "Makambo" stranded there in 1918. The rats increased in a few years and exterminated the white-eyes and a whole series of other native birds.

The great family of the honey eaters (Meliphagidae) with the one exception of the CAPE SUGARBIRD, is so characteristic of the Australian fauna that the famous ornithologist, John Gould said in 1865: "they are without question the most peculiar and most striking figures in Australian ornithology". The size is very variable; the length, including the long beak ranges from 10 to 45 cms. They have a projectile brush tongue which is cleft near the tip, each half of which has a number of stiff horny projections—adaptations to taking up nectar. The nostrils usually lie in a deep groove in the base of the beak, which generally has a leather-like lid bare of feathers.

Family: honey eaters, by D. L. Serventy

Plumage colors are usually largely greenish or olive-brown with yellow, rust color, black and white; only in the genus *Myzomela* is there a more or less extensive area of scarlet. There is often a marked head pattern, particularly on the ear coverts, which contrast in color to the rest. Blue is found only occasionally on bare areas of the head. Certain parts of the head are bare of feathers in several forms and such skin areas are colored; in a few species there are flaps of skin or other projections on the bare areas; these are a striking red, yellow or blue color. Other species have feather tufts on the head. There are two subfamilies: 1. Meliphaginae with 38 species and about 170 species found with few exceptions in Australia and Oceania. 2. The SUGARBIRDS (Promeropinae) of southern Africa with one genus and one species which, in my own opinion and that of some others, should not be placed in this group.

As the scientific and vernacular names indicate, these birds prefer nectar as food and their beaks and tongues are adapted for obtaining it. Occasionally nectar ferments when strong rains have increased its water content; in such cases honey eaters have been observed in a state of intoxication and unable to fly. This can also happen to bees under certain circumstances. John Gould, over a hundred years ago, surmised that the shape of the beak and tongue of honey eaters and the blossoms of eucalyptus were adapted to one another. Quite independently, the Australian botanist, O. H. Sargent expressed the opinion in 1962 that honey eaters were the fertilizers of eucalyptus flowers and that adaptation to the needs of these birds played a role in the phylogeny of these characteristic Australian plants. In 1967, G. F. Mees found further proof of the correctness of Sargent's views. He regarded the peculiar form of a western Australian plant, the kangaroo paw *(Anigozanthos manglesii)* as "inconceivable without a bird pollen carrier". The bird responsible for this is the WESTERN SPINEBILL *(Acanthorhynchus superciliosus)*. However, not all honey eaters are linked with certain species of eucalyptus; they also visit the flowers of many other plants, among them acacias, the shrub *Eremophila, proteacea* and *epacridizea* for food. Artifical mixtures of honey and sugared water attract these birds and feeders containing these mixtures are used as "bait" in banding and observation stations in Tasmania and western Australia for honey eaters of the genus *Melithreptes.*

[margin: Honey eaters as flower pollinators]

In spite of these adaptations, we do not know of any species which feeds exclusively on nectar. The PAINTED HONEY EATER *(Conopophila picta)* seems to dispense with nectar altogether by living mainly on mistletoe berries. Many species take both native and introduced fruit; they also take insects, perhaps accidentally while seeking for nectar, or on hunts directed at insects. They can catch insects on the ground as well as in flight; they also attack swarms of flying termites and search for insects beneath bark or among foliage. Some rapacious species even take the eggs and young of small passerine birds. Thus, the SINGING HONEY EATER *(Meliphaga virescens)* tears a hole in the nests of zebra finches to get at its prey. The larger LEATHERHEADS (genus *Philemon*) also rob nests as well as devouring small lizards and frogs.

[margin: Adaptations to nectar feeding]

As an adaptation to nectar feeding, the lower opening of the oesophagus and that of the duodenum lie side by side. Thus, nectar and other easily digested food constituents can pass directly from the gullet into the intestine. However, arthropods and the like are dealyed in the intestinal tract for a longer time, according to Salomonsen.

Males and females sing busily although differently in the individual species. Thus, the melodious song of the BROWN HONEY EATER *(Lichmera indistincta)* of Bali is reminiscent of the nightingale's, while the harsh calls of the WATTLEBIRDS (genus *Anthochaera*) are unpleasant to human ears. The songs are part of territory marking flights in many species; the bird rises vertically in the air, then descends performing various evolutions. Duet songs also occur and several birds often cooperate in delivering what sounds like a single song. When the YELLOW-TUFTED HONEY EATER *(Phylidonyris melanops)* and the singing honey eater perform such a group song in a wood densely inhabited by these birds, these choirs form unforgetable experiences.

Only one species, the BELL MINER *(Manorina melanophys)* lives strictly speaking in colonies, but all have a marked social tendency. Members of some species often nest close together. After the breeding season they form small flocks which may occasionally fuse into large swarms. Nevertheless, according to K. Immelmann, they lack many forms of behavior of social birds. They do not preen one another, but the young will beg from any member of the species.

The nest is almost always cup-shaped; it is either fixed in the fork of a branch or hangs from a twig. Only honey eaters of the genus *Ramsayornis* build roofed-over nests and even these have evolved from cup-shapred nests in which the edges were raised up, with the exception of a lateral opening. The BLUE-FACED HONEY EATER *(Entomyzon cyannotis)* often uses the nests of other birds, particularly that of the gray-crowned babbler which it merely lines anew with strips of bark or into which it builds its own nest. Sometimes it does build its own nest. Usually only the female honey eater builds the nest.

The eggs are generally flesh-colored with reddish-brown markings; in Australia the clutch size is usually two. The LITTLE WATTLEBIRD *(Anthochaera chrysoptera)* lays only one egg in western Australia, but two in the east of the continent. In the western spinebill, early laying females lay only one egg, but those which lay later, lay two. The EASTERN SPINEBILL *(Acanthorhynchus tenuirostris)* lays either two or three eggs. There are often two broods and the breeding season may be long. Eggs of the YELLOW-WINGED HONEY EATER *(Phylidonyris novaehollandiae)* have been found every month of the year around Sydney, although this species breeds mainly in spring and summer. Usually only the female incubates, but the male takes part in feeding the young and in caring for them. The young hatch after thirteen to sixteen days. Honey eaters are important foster parents for those of the Australian cuckoos which are parasitic birds and which build open nests. The very common brush cuckoo (Volume VIII), uses the nests of quite a series of large and small honey eater species. Honey eaters breed when they are still in their first year.

These lively birds are quite cantankerous towards one another as well as towards other birds. In contrast to most other birds, the antagonists become involved in real fights as Immelmann emphasizes. This is probably due to the fact that in this family threat and submissive postures are only feebly developed. A rank order was found among several species living in the same area. Thus, the large WHITE-GAPED HONEY EATER *(Meliphaga unicolor)* chases smaller species of its genus away from feeding and drinking places. The YELLOW-PLUMED HONEY EATER *(Meliphaga ornata)* is particularly full of fight; it attacks other birds even when they are as large as itself. All honey eaters rarely accept strange members of their species; instead they chase them away.

Although they generally inhabit loose or dense stands of trees and keep to them, there are a few honey eater species which occur in open sandy heaths and dry desert shrubbery. Honey eaters are found in all the climatic zones of Australia from the very damp rain forests to the temperate zone, to the deserts of the interior. The MANGROVE HONEY EATER *(Meliphaga fasciogularis)* can be regarded as the most pronounced "sea bird" among the sunbirds. On Moreton Bay and other areas of its habitat in Queensland, it constantly lives in mangrove flats which are surrounded by the open sea for miles at high tide.

Most honey eaters in tropical and temperate areas of Australia are more or less resident; at any rate they only undertake local travels. On the other hand, in the irregular climatic conditions of the dry deserts and semi-deserts, they, like other birds, can only maintain themselves by leading a largely nomadic life. The only marked migrations are found in two southeastern species, The YELLOW-FACED HONEY EATER *(Meliphaga chrysops)* and the WHITE-NAPED HONEY EATER *(Melithreptus lunatus)* Color plate, p. 320); both migrate north in the local spring and return in the fall.

Most honey eaters live in Australia (sixty-seven species) and New Guinea (sixty-three species). Three species are confined to New Zealand and various additional species occur on islands of the western Pacific. The Bali honey eater has crossed Wallace's line (see Volume VII) between Bali and Lombok and settled on Bali. The APALOPTERON *(Apalopteron familiaris)* lives only on the Bonin Islands near Japan. Among the five Hawaiian species which belong to the genera *Moho* and *Chaetoptila,* four are very probably extinct.

Subfamily: sugarbirds, by J. Steinbacher

The SUGARBIRDS (subfamily Promeropinae) have only two species, the CAPE SUGARBIRD *(Promerops cafer;* Color plate, p. 320) which is distinguished by a strikingly long tail. It occurs in southern Africa far from the range of all other honey eaters. The L, including the tail, is 45 cm in males and 28 cm in females; the male's tail measures 32 cm. The second species is the NATAL SUGARBIRD *(Promerops gurneyi).*

This South African bird has the same tongue structure and habits as do the other honey eaters. It visits the flowers of the tree *Protea,* takes nectar and while doing so fertilizes the flowers. It lays two yellowish-brown, dark spotted eggs in its open cup nest. A number of ornithologists regard it as the result of special evolution in the same direction as the honey eaters from quite different ancestors. Thus, its similarities to them would be the result of convergence. However, most still include the sugarbirds with the honey eaters either as a deviant genus or as a separate subfamily.

15 Buntings and Their Relatives

Next to the roughly 1,400 to 1,500 species of Flycatcher-like birds (Chapters 12 and 13), the approximately 950 species in the groups to be dealt with in this and the following chapter, makes them the second largest unit among the songbirds. Buntings, vireos, wood warblers, New World orioles, finches and their relatives are small to medium-sized birds of divergent modes of life. The L varies between 9 and 55 cm, but is generally between 11 and 25 cms; the majority of these birds are native to America. There are seven families: 1. The BUNTINGS (Emberizidae); 2. The WOOD WARBLERS (Parulidae); 3. The WREN THRUSHES (Zeledoniidae); 4. The HONEY CREEPERS (Drepanididae); 5. The VIREOS (Vireonidae); 6. The NEW WORLD ORIOLES (Icteridae); and 7. The FINCHES (Fringillidae).

Buntings and their relatives, by W. Meise

Opinions about the origin and systematic position of these birds have changed repeatedly; even today the matter is not settled. Probably the original members of this group of families were mainly seed eaters which reared their young on insects, as most of them still do, while other forms specialized in a diet of arthropods, fruit, the soft parts of plants or nectar. In contrast with earlier systems, the cardinals and tanagers are placed with the buntings, on the basis of the structure of their palates and jaw muscles. The earlier family, Coerebidae, most members of which suck nectar was, again in accordance with palatal and jaw muscle structure, divided up among the tanagers and wood warblers. In distinction from older systems, the insect and fruit eating wood warblers and honey creepers are placed here with an otherwise seed eating group of families. The vireos, which are also insectivorous, suggest, by their hanging nests and the occasionally very short tenth primary, a connection with shrikes and orioles.

Fig. 15-1. Distribution of the buntings (subfamily Emberizinae), some species introduced in New Zealand.

As we already mentioned, according to their palatal structure, buntings, cardinals and tanagers belong together and are grouped here to form the family BUNTINGS *(Emberizidae)* with the subfamilies BUNTINGS (Emberizinae), CARDINALS (Cardinalinae) and TANAGERS (Thraupinae).

Family: buntings

Subfamily: buntings, by E. O. Höhn

Fig. 15-2. Corn bunting *(Emberiza calandra).*

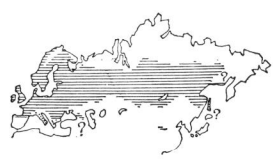

Fig. 15-3. Yellowhammer *(Emberiza citrinella)* and pine bunting *(Emberiza leucocephalos).*

Fig. 15-4. Ortolan bunting *(Emberiza hortulana).*

To the tanagers are added the PLUSH-CAPPED FINCHES and HONEY CREEPERS as groups of genera even though they are often treated as subfamilies or even families.

It is probable that the buntings (subfamily Emberizinae) originated in North America because only about one sixth of the species (the so-called Old World buntings) occur in the Old World. They are small to scarcely medium-sized birds; the L is 9.5-21 cms. Their legs are of medium length, the feet are usually large, the tail is short to medium length, and the beak is short, conical and pointed. The plumage of the majority is brown, gray or olive-colored, and rarely largely reddish-brown or quite black; sometimes it is patterned in black, white, yellow, greenish or reddish-brown and/or striped. Short crests are not uncommon, but only one species has a long crest. These birds lay two to six eggs with a pale background color with, in most cases, darker spots; they are incubated by the females only. Initially, at least, the young are fed by regurgitation from the crop of the parents. There are buntings all over the world except the Indo-Australian region, Madagascar and Australia; there is only one species in southern Asia. Two species have been introduced in New Zealand.

There are three groups of genera: 1. The BUNTINGS, in the restricted sense (Emberizini) most of which have a downward kink at a blunt angle on the edge of the lower mandible of the beak. They are found in the New and Old Worlds. 3. DARWIN'S FINCHES (Geospizini) resemble the grassquits but have various types of beaks and a very densely feathered lower back; they are restricted to the Galapagos Islands and Cocos Island.

The buntings in the restricted sense are further divisible into two groups: OLD WORLD BUNTINGS and NEW WORLD "SPARROWS". In Old World buntings, the edge of the lower mandible is slightly overlapped by the edge of the upper mandible when the beak is closed. There is often a horny protuberance on the palate which helps in getting seeds out of the birds' coats and in cracking them. In the New World sparrows the beak is generally slimmer and many species scratch the ground with both feet while looking for food. Although these are mainly New World birds, two species are circumpolar and thus also occur in northern Eurasia, while three species are found on southern Atlantic islands which lie nearer to Africa than South America.

The Old World buntings have a L of 13.5-19 cms. They comprise three genera with about forty species.

1. The TYPICAL BUNTINGS (genus *Emberiza*) include: the CORN BUNTING *(Emberiza calandra;* Color plate p. 341); the PINE BUNTING *(Emberiza leucocephalos)* has a white head with black stripes and replaces the yellow bunting in eastern Siberia—in western Siberia both species occur, generally each on its own in adjacent areas but hybrids are frequent in areas where both species occur; the ORTOLAN BUNTING *(Emberiza hor-*

tulana; Color plate, p. 341); the BLACK-HEADED BUNTING *(Emberiza melanocephala;* Color plate, p. 341) with a black cap, chestnut brown back and yellow lower parts, is found from southeastern Italy eastwards as far as the lower Volga; the RED-HEADED BUNTING *(Emberiza bruniceps)* occurs from the Volga to Mongolia, and hybridizes with the black-headed species in the area where their ranges border one another; the REED BUNTING *(Emberiza schoeniclus;* Color plate, p. 341) with subspecies which differ a great deal in the beaks, among them the BULLFINCH-BILLED REED BUNTING *(Emberiza schoeniclus pyrrhuloides)* of the Caspian Sea area.

2. The STRIPED BUNTINGS (genus *Fringillaria*, but often included in the genus *Emberiza*) occur in Africa and western Asia. Members of this genus include: the CINNAMON-BREASTED ROCK BUNTING *(Fringillaria tahapisi)* which is cinnamon colored with a white crown and has three white stripes on each side of an otherwise black head. It occurs in southern Arabia and Africa; the PALE ROCK BUNTING *(Fringillaria impetuani)* of southern Africa; the CAPE BUNTING *(Fringillaria capensis),* another southern African species which is gray below; and the HOUSE BUNTING *(Fringillaria striolata)* which is reddish-brown with a black throat and wide black stripes below. It is found from Morocco to northern India.

The CRESTED BUNTINGS (genus *Melophus*) have a single species *(Melophus lathami)* which has the same vernacular name as the genus. It has a long pointed crest. The male is a glossy blue-black with a largely reddish-brown wing and tail; the female is brown. It occurs in northern India and Indochina and is the only bunting of the subtropics and tropics of Asia.

The corn bunting is the largest species of the Old World buntings. Its song ends in a noisy jangling which can be rendered as "tsick tsick tsick shnirrrps". A male may be mated to as many as seven females. The best known species of this group is the yellowhammer. It inhabits open landscapes with bushes and is generally a resident, although in the north of its breeding area it is a migrant. Its modest song is easily recognized; a traditional rendering is "little bit of bread and no cheese". The commonest call is "tsreek". In central Europe, males take up territories in February and pair formation soon follows. The nest is a cup, generally thirty to fifty centimeters high in bushes and occasionally placed on the ground. The white eggs, usually three to six per clutch, have gray or brown spots and fine wavy "hair lines" as well. The female incubates for twelve to fourteen days; generally there are two to three clutches per year. Both parents feed the young, and consume the egg shells as well as the feces of the young in the first few days. At first the young are fed from the crop of the parents; later the parents bring food in the beak. The young are given arthropods, particularly caterpillars. In contrast to most finch-like birds which first hold the food in the crop, young buntings' food slips into their stomachs right away. Once they have left the nest, young buntings pick up

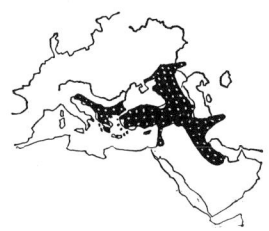

Fig. 15-5. Black-headed bunting *(Emberiza melanocephala).*

Fig. 15-6. Reed bunting *(Emberiza schoeniclus).*

▷

Darwin's finches: 1. Vegetarian tree finch *(Platyspiza crassirostris)*; 2. Warbler finch *(Certhidea olivacea)*; 3. Cactus ground finch *(Geospiza scandens)*; 4. Woodpecker finch *(Cactospiza pallida)*; 5. Small insectivorous tree finch *(Camarhynchus parvulus)*; 6. Cocos finch *(Pinaroloxias inornata)*; 7. Large ground finch *(Geospiza magnirostris).*

very small stones which help grind up their subsequent, mainly vegetarian, food in the stomach.

The ortolan bunting is an inhabitant of quite varied habitats and has been highly regarded since Roman times as a bird easy to fatten; this has, of course, been harmful for it. The black-headed and red-headed buntings are noteworthy because of their molts. Initially, the males undergo a partial molt in June and July and thereafter they resemble the females; then in November to December a complete molt follows whereby the nuptial plumage is restored, but its bright colors are hidden at first by pale feather edges. These are worn off in the course of the winter. It is not quite certain whether the male red-headed bunting also has an eclipse plumage like that of the female, but it is known to have its full molt in mid-winter (unlike the other species of buntings, in late summer). The reed bunting is a common inhabitant of reed beds. It feeds on animal food, more than other buntings. Its nest is placed low above the ground or occasionally over water, attached to reeds. Its usual call is a long-drawn "tsee." Its short little song can be rendered as "tsia tit tai tseesees." The house bunting is a characteristic bird of settlements in northwestern Africa, as the house sparrow is in so many parts of the world; however, in the eastern part of its range the house bunting also lives in stony deserts.

The New World sparrows and relatives (L 9.5 to 21 cm) form a much larger and more diversified group of genera; some of the more diverse genera are not readily separable from other finch-like birds like the Darwin's finches or cardinals. Altogether there are 54 genera with 162 species. In the northern circumpolar belt two genera, that of the SNOW BUNTINGS *(Plectrophenax)* and the LONGSPURS *(Calcarius)* link the present group with the Old World buntings. Other genera in this group are: 1. The CROWNED "SPARROWS" *(Zonotrichia)* with five species, among them the WHITE-THROATED SPARROW *(Zonotrichia albicollis;* Color plate, p. 342) and the RUFOUS-COLLARED SPARROW *(Zonotrichia capensis).* 2. The JUNCOS (genus *Junco*) which have gray heads and upper parts but often have a different color on the back and white outer tail feathers. There are seven species, among them the SLATE-COLORED JUNCO *(Junco hyemalis)* and the OREGON JUNCO *(Junco oreganus:* Color plate, p. 342). 3. The SAVANNAH SPARROWS (genus *Passerculus*) with two species, the SAVANNAH SPARROW *(Passerculus sandwichensis)* and the IPSWICH SPARROW *(Passerculus princeps).* 4. The SONG SPARROWS (genus *Melospiza*) with three species, among them the SONG SPARROW *(Melospiza melodia;* Color plate, p. 342) which is grayish brown above with dark brown spots and a reddish-brown rump, and is white below with dark brown stripes and a larger spot in the middle of the breast. It breeds from southern Alaska and Newfoundland as far as Kansas and Georgia. 5. The CHIPPING SPARROWS (genus *Spizella*) with seven species, among them the CHIPPING SPARROW *(Spizella passerina)* with a reddish-brown crown and a white and black

◁
1. Ultramarine Grosbeak *(Cyanacompsa cyana);*
2. Pyrrhuloxia *(Pyrrhuloxia sinuata);* 3. Cardinal *(Cardinalis cardinalis);*
4. Yellow grosbeak *(Pheucticus chrysopeplus);* 5. Blue grosbeak *(Guiraca caerulea);* 6. Black-cheeked ant tanager *(Saltator atriceps);*
7. Painted bunting *(Passerina ciris);* 8. Indigo bunting *(Passerina cyanea).*

line on the sides of the head. 6. The SHARP-TAILED SPARROWS (genus *Amnospiza*) with three species, including the SEASIDE SPARROW (*Amnospiza maritima*; Color plate, p. 342) which is olive gray, whitish below and has small stripes just about all over, plus a yellow spot in front of the eye and painted tail feathers. 7. The BRUSH FINCHES (genus *Atlapetes*) with twenty-seven species in Central and South America, form the genus richest in species. They include the CHESTNUT-CAPPED BRUSH FINCH *(Atlapetes brunneinucha)* which is olive-green above, mainly white below and has a black forehead, cheeks and breastband. It is found from Mexico to Peru, particularly in the Andes. 8. The CRESTED FINCHES (genus *Coryphospingus*). 9. The MELANODERA FINCHES (genus *Melanodera*) with two species in South America. One of these is the BLACK-THROATED FINCH *(Melanodera melanodera)* which is greenish, bright yellow below and has a white stripe above the eye and in the malar region. 10. The TRISTAN BUNTINGS (genus *Nesospiza*) with two species on the island of Tristan da Cunha, about 3,200 kms from South America; their closest relatives are those of the preceding genus, hence they have presumably emigrated from South America. They include the CUNHA BUNTING *(Nesospiza acunhae)* which has a L of 18 cm and weighs 30 g; it is olive-gray, paler and yellower below and has become extinct on the main island of the group. Other genera and species of this group are listed in the systematic review.

The SNOW BUNTING (*Plectrophenax nivalis*; Color plate, p. 342) which weighs 40 g, is, in contrast to others of this group, also found in the Old World; it breeds on Iceland, the Faroes, in northern Scandinavia and the U.S.S.R. as well as many far northern islands. It breeds further north than any other land bird—to about 670 kilometers from the North Pole, as well as in all of arctic North America and northern Greenland. It nests in very small numbers on the highest peaks of Scotland. In winter, flocks of these birds occur on the German North Sea coast. Their white wing areas are particularly conspicuous in flight. In the male's breeding plumage, only the back, the primaries and the central tail feathers are black, the rest of the plumage is snow white. MCKAY'S BUNTING *(Plectrophenax hyperboreus)* which only breeds on two Bering Sea islands and is sometimes regarded as a mere subspecies of the snow bunting, has even more white on the back.

Fig. 15-7. Snow bunting *(Plectrophenax nivalis)*.

The most common call of the snow bunting is a jangling "tshurrt" and a long drawn whistle, "deeoo." The song is lark-like and consists of a pleasant "tshuree, tshuree, thsuree, ta teeoo"; it is delivered sitting or in a circling display flight. In the breeding season, snow buntings are tied to rocks for they breed in rock clefts and are absent from tundras which lack rocks or cliffs. They show a predilection for settlements and may build their nests on buildings. I have even found their nests in old rusty tin cans lying on the ground. In the Arctic, males appear in spring several weeks before the females. Usually a pair rears

Fig. 15-8. Lapland longspur *(Calcarius lapponicus)*.

Fig. 15-9. Male Lapland longspur in nuptial plumage.

Fig. 15-10. Head of the white-throated sparrow.

only one brood of four to six young per year, but D. Nethersole Thompson, who observed these birds for several years in Scotland, found second broods in nine out of nineteen cases. Although snow buntings are monogamous, he observed one male which, while its first female was incubating, paired itself with another female. He also saw a few cases where the female was paired to more than one male.

When in mid-summer, only a few snow fields are to be found at lower levels in the Arctic, snow buntings congregate in these areas to search for food. Presumably areas of snow attract them because they can more readily spot insects (from adjacent areas) on this background and because these are soon paralyzed by the cold. In Greenland, snow buntings and rock ptarmigan are the only birds which breed on the so-called nunataks, rocky peaks which project over the inland ice like islands and which are often far from the ice edge.

The LAPLAND LONGSPUR (*Calcarius lapponicus;* Color plate, p. 342) has a L of 15 cm. Its breeding range and general distribution is similar to that of the snow bunting, but it does not nest in Iceland, Scotland or Spitzbergen. In winter it too occurs on stubble fields near the German North Sea coast and most regularly on Helgoland. As it is not tied to rocks like the snow bunting, its distribution on the tundra is much more even and it is much more numerous. It sings almost always on the wing, in part probably because on flat open tundra there are hardly any elevations which could serve as song perches. When these birds nest in the course of migration in April, still over a thousand kilometers from their breeding areas, their song is often to be heard. In this season, huge flocks of Lapland longspurs and snow buntings may be seen on stubble fields, especially in southern Canada. Related North American species are listed in the systematic review.

The North American white-throated sparrow is worthy of mention because of its song which is simple, and much like a human whistle; the pattern is suggested by the words "hard times Canada" with the last three syllables prolonged. The closely related rufous-collared sparrow is a bird of open country; it shows a predilection for settlements where it may become very tame. Its tropical subspecies breeds in every month of the year but mainly about the time of the two equinoxes. The two annual dry seasons usually induce a molt in females and so ends egg laying. However, males have two annual breeding seasons, both of four months each followed by a full molt which takes about two months. If exposed to artificially prolonged daylengths, the young, which are normally sexually mature at five to eight months old, can breed when only three to four and a half months old. In males of the tropical subspecies, there is no season when development of the testes cannot be induced by prolonged daylength. In contrast to this, North American sparrows have a refractory period in late summer and fall, when artificial illumination cannot stimulate the testes. Experiments in which the testes and ovaries of birds were stimulated by artificial daylength prolongation were first carried out in 1929 by W. Rowan on captive slate-colored juncos.

Both the slate-colored junco and the Oregon junco are breeding birds of the western North American coniferous forests. They range from southern Alaska to southern California. In winter some migrate too far north; thus I have encountered Oregon juncos in the Arctic on Banks Island and another well north of the tree line on Hudson Bay.

Fig. 15-11. Oregon junco.

Among the North American "sparrows" (which are really buntings), the savannah sparrow is a modestly chirping songster of grasslands. It is grayish-brown above, white below and striped both on its upper and lower parts; it also has a yellow spot above the eye. It breeds from Alaska to Labrador and southwards as far as the central states of the U.S. Its close relative, the Ipswich sparrow, is interesting because of its limited breeding area. It nests only on Sable Island, a sand island about thirty kilometers long, but only one and a half kilometers wide, which lies in the northern Atlantic 168 kilometers from the mainland of Nova Scotia. As the island is slowly being reduced in size by the action of currents, the eventual extinction of the Ipswich sparrow is to be expected.

The song sparrow is the one North American sparrow whose life we know a great deal about. It breeds in hedges and semi-open country with bushes where the birds live in small territories and often form loose colonies. Its song can be suggested by the words, "pres pres pres presbyterian"; it is not inherited but acquired by each individual by imitation of other members of the species. The female too has a short song. Margaret Morse Nice studied these birds very thoroughly by means of color banding. According to her findings in Ohio, fifty per cent of males and twenty per cent of females were residents, which in most cases, used the same territories year after year. Those which were migrants also generally settled within 100–1400 meters of where they had been hatched. Song sparrows are almost always monogamous but occasionally a female deserts its male in the course of the mating season.

Fig. 15-12. Song sparrow.

In Ohio, egg laying takes place in April and begins six days after the temperature has been above 20°C for at least two days, although most females only lay after an additional four successive warm days.

The eggs of one year old females are narrower and more elongated than those of females two or more years old. While young females lay four clutches, each of four eggs, per year, older females lay two five-egg clutches and two of four. Only the female incubates and during the daytime it leaves the eggs every twenty to thirty minutes on the average for eight minutes. Incubation lasts twelve to thirteen days. About five per cent of the eggs are lost through the actions of brown-headed cowbirds which remove some eggs and stab others with their claws. Another thirty-three per cent of the eggs or young are lost to other animals, while human interference accounts for a loss of four to seven per cent; parental maladjustments cause a further loss of nine per cent and finally, extraordinary droughts account for two per cent loss. The young are fed by both parents from their first to third day of life every thirteen to thirty-two minutes. During the first five to six days, the parents brood the young during the day more than half the time for as in all nidiculous birds, the body temperature of the young is not initially independent of that of the environment. When they are eight to twelve days of age, the young leave the nest but are still fed by the parents for they are not fully fledged until thirty days old.

The chipping sparrow makes itself conspicuous by its song—a monotonous trill of unchanging pitch. Among other habitats, it lives in numbers in rural and urban settlements. The seaside sparrow as the name suggests, is an inhabitant of the salt marshes of the North American eastern and southern coasts. Of the many Central and South American species (see the systematic review) we will mention the tame chestnut-capped brush finch. It nests in bushes and lays pale brownish eggs. The RED-CRESTED FINCH *(Coryphospingus cucullatus),* found from Guyanna to northern Argentina is more colorful. The female is brown-backed with a whitish throat and lower parts which are otherwise rust-colored, while the male has a fiery red crown with a black stripe below. The sides of the head, back and tail are brown, the throat is pale red and the underparts from the breast downwards and the rump are reddish-brown. This bird lays white eggs and lives in open areas with some tree growth. The black-throated finch which is also South American, is found from southern Argentina to Tierra del Fuego and the Falkland Islands. Among its close relatives are the Tristan buntings, already mentioned, of the Atlantic Island Tristan da Cunha, which are so round-winged they can hardly fly and the GOUGH ISLAND BUNTING *(Rowettia goughensis)* of Gough Island which is still nearer to Africa.

The grassquits are small, generally very thick-billed birds which form the group of genera called Tiaridini. The L is 9.5–18 cm. The plumage is gray, grayish-green, greenish, yellow or black with additionally black and white or dark brown to cinnamon-brown areas. There are fifteen genera with forty-nine species, including: 1. The YELLOW FINCHES *(Sicalis)* are largely yellow, somewhat like the Old World siskin or serin finch but with larger inner primaries and a quite different palatal structure; there are eleven species including the SAFFRON FINCH *(Sicalis flaveola;* Color plate, p. 342) found from Guyana to northern Argentina and introduced in Panama and Jamaica. 2. The SEED EATERS *(Sporophila),* comprise twenty-eight species. The VARIABLE SEED EATER *(Sporophila americana)* may be taken as an example; it is small, has a black plumage with a white mark on the wings, a white rump in some areas of its range and is paler below with a white band at the side of the neck. It is found from Mexico to Amazonia. 3. The ST. LUCIA BLACK FINCH *(Melanospiza richardsoni)* has a L of 13.5 cm. It has pale reddish feet and is found only on Santa Lucia, one of the Lesser Antilles. This bird and the yellow-faced grassquit may look somewhat like the ancestors of the Darwin finches once did. 4. The BLUE-BLACK GRASSQUIT *(Volatinia jacarina;* Color plate, p. 342) has a L of 10 cm. Males are glossy blue-black above. This bird is found in Central and South America. 5. The TIARIS GRASSQUITS *(Tiaris)* have four species, among them the YELLOW-FACED GRASSQUIT *(Tiaris olivacea;* Color plate, p. 342) which has a L of 11 cm. It is olive green, more grayish below, has a white stripe above the eye and a yellow throat; it occurs in the Greater Antilles, Central America and northern Colombia. The CUBAN GRASSQUIT *(Tiaris canora)* which has a L of 9.5 cm and is from Cuba, has a yellow collar which rises at the sides up to behind the eyes.

Many of the grass finches and seed eaters are among the most long-lasting and popular cage birds, both in their home countries and in Europe. The saffron finch lives in South American thornbush forests. Its cup-shaped nest is found in tree holes or holes in walls as well as in nest boxes. The seed eaters, on the other hand, build their loose cup-shaped nests low in vegetation and at least one species suspends its nest from reeds. Seed eaters live in grasslands, swamps, as well as in bushy areas and on the edges of woods. They have a pleasant-sounding song. To ensure a sufficiency of grass seeds for the young, egg laying has to coincide with the onset of the rainy season. For this reason, the variable seed eater, in Costa Rica alone, breeds from May to September or even from April to January as there are great local differences in the times of the rainy season. The female incubates alone for twelve to thirteen days; she is sometimes fed in the nest by the male. The young generally fly after thirteen days in the nest.

The grassquits, by W. Meise

The blue-black grassquit is of interest because its display includes little jumps. The yellow-faced grassquit searches for minute grass seeds in flocks. In contrast to the other species of the tribe Tiaridini but like the St. Lucia black finch, it builds a spherical nest with a side entrance in the grass or low down in a bush. The two to four young are dark skinned and almost naked when they hatch; the parents feed them by regurgitating softened food. The Cuban grassquit has long been popular as an easily reared cage bird. Its spherical nest is placed higher in shrubs and trees; it also inhabits bushes on the edges of fields and even pine woods.

Fig. 15-13. Yellow-faced grassquit.

Darwin's finches of the Galapagos (tribe Geospizini) are nowadays included in the bunting subfamily. They have played a particular role in the history of biology and they first gave a clue to the process of species formation. When the great naturalist, Charles Darwin, visited

Darwin's finches, by E. Curio

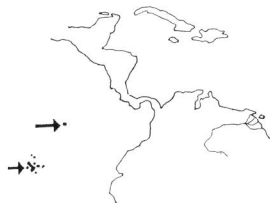

Fig. 15-14. Position of Cocos Island and the Galapagos Islands (arrows).

the Galapagos, west of Ecuador, in the fall of 1835, on his famous journey around the world, he discovered there, among other wildlife, the birds now named after him. When he saw very similar but distinct forms living together, the novel idea of the mutability of species and of the derivation of similar species from a common ancestor, came to him. All Darwin's finches have probably originated from a single bunting-like ancestor from the Central or South American mainland, which was related to the present day grassquits or black finches. Since, apart from mockingbirds, only few other songbirds have reached the remote Galapagos, these finches were able to settle in the most diverse habitats and to adapt to the most diverse modes of feeding Some became seed eaters with more or less conical beaks, others became leaf and soft food eaters with far more delicate beaks, still others became insect eaters, and the woodpecker finch has even taken over the role of a woodpecker.

Long separation on the various islands of the group has resulted in the formation of different species, but these, in spite of many differences, have not differentiated enough to abolish interbreeding, for hybrids between different species and even different genera are frequent. Also even the most pronounced insect eaters in the group still feed their young by regurgitation, as seed eaters do, and not from the beak as is the case in all other families of insect eaters.

In beak structure and the structure of the skull, head muscles and stomach muscles, the tongue and the size of the stomach and heart, the various Darwin's finches differ a great deal. The L ranges from 9.5 to 13.5 cm, the weight is 9 to 38 gm. The tail is short, and the lower back is densely feathered. These birds have a relatively low heart weight; in the heaviest species the heart weight amounts to 5.4 per cent of the body weight, less than it is in almost all other birds. There are six genera with fourteen species:

1. The GROUND FINCHES (*Geospiza*) have a thick beak and are the most finch-like. Young males and all females are grayish-brown with blackish spots; males become deep black from the forehead, breast and belly, until finally only the lower tail coverts retain pale brownish edges. Possibly some males may remain female-like in plumage all their lives. There are six species, including: the LARGE GROUND FINCH (*Geospiza magnirostris*; Color plate, p. 351) which weighs 27 to 39 gm and has a heavy conical beak; the SMALL GROUND FINCH (*Geospiza fuliginosa*); the SHARP-BEAKED GROUND FINCH (*Geospiza difficilis*); and the CACTUS GROUND FINCH (*Geospiza scandens*; Color plate, p. 351) in which not only males but females and young are adapted to a lava terrain by a dark plumage. 2. The VEGETARIAN TREE FINCH (*Platyspiza crassirostris*; Color plate, p. 351) has a L of 13.5 cm and weighs up to 41 gm. It is the sole species of its genus. The beak is a bullfinch-like "fruit-squeezer"; this bird has the longest intestinal tract of all Darwin's finches. The black

Fig. 15-15. Three phases in the development, with age, of the male plumage of ground finches.

plumage extends as far as the breast in mature males. 3. The INSECTIVOROUS TREE FINCHES (Camarhynchus) resembles the preceding species. They are olive-greenish above, pale brownish below and have drop-shaped spots on the upper back and breast; the crown and forepart of the head are black in older males. There are three species, including the SMALL INSECTIVOROUS TREE FINCH (*Camarhynchus parvulus*; Color plate, p. 351) which has a L of 10 cm and the LARGE INSECTIVOROUS TREE FINCH *(Camarhynchus psittacula)*. 4. The WOODPECKER FINCHES *(Cactospiza)* are olive-brownish above and yellowish-gray below; the sexes are similar. There are two species, the WOODPECKER FINCH (*Cactospiza pallida*; Color plates, p. 333 and p. 351) and the MANGROVE FINCH *(Cactospiza heliobates)*. 5. The WARBLER FINCH (*Certhidea oliveacea*; Color plate, p. 351) has a L of 9.5 cm. It has a longish awl-shaped beak and the shortest intestine of all Darwin's finches. 6. The COCOS FINCH (*Pinaroloxias inornata*; Color plate, p. 351) has a L of 11.5 cm and is colored like the ground finches. It is the only Darwin finch not found on the Galapagos; instead it lives on Cocos Island which is about 180 kilometers northeast of the Galapagos and belongs to Costa Rica.

Fig. 15-16. The large insectivorous tree finch.

Darwin's finches are clumsy flyers with the exception of the warbler finch which is lively and can maneuver readily. However, even the plump species which fly with whirring wings find their way through the network of branches with which the Galapagos are endowed. Members of the same genus have similar songs, but there are considerable individual song differences. Often the only more or less reliable specific distinction is the volume.

Most of these birds breed during the hot season, January to May, when the heavy annual rains begin. Ground finches then move from the evergreen uplands to the dry zone nearer the coast which begins to become green at that time. Some of the small insectivorous tree finches and vegetarian tree finches do this too, but most of them breed on the uplands. The cactus ground finch, on the other hand, never leaves the arid zone. Males take up territories in advance of the females and defend them against other small birds. Within the territories, spherical nests made out of grasses with an entrance at the side, are built. They are placed in tree forks, or more often, among cactus shoots. Males which still lack females often steal nest material from the nests of their neighbors to use in their own nest building.

The female lines the nest with grass, feathers and orchilla lichens, and the male, at least in the ground finches, helps. The eggs, one to five but generally three, are incubated by the female alone for eleven to fourteen days. During this time the male feeds the female on the nest or near it, as Orr observed in captured Darwin's finches.

Both parents feed the young on chewed-up arthropods from the crop by regurgitation; however, some ground finches also feed them milky seeds. Only the male leads the young when they are fledged. In any case, the small ground finch raises at least two broods a year as it often

▷
Tanagers: 1. Crimson-collared tanager *(Phlogothraupis sanguinolenta)*; 2. Masked tanager *(Tangara nigrocincta)*; 3. Blue-gray tanager *(Thraupis episcopus)*; 4. Chestnut-breasted chlorophonia *(Chlorophonia pyrrhophrys)*; 5. Silver-beaked tanager *(Ramphocelus carbo)*; 6. Red-crowned ant-tanager *(Habia rubica)*; 7. Blue-winged mountain tanager *(Anisognathus flavinucha)*; 8. Flame-faced tanager *(Tangara parzudakii)*; 9. Magpie tanager *(Cissopis leveriana)*; 10. Paradise tanager *(Tangara chilensis)*; 11. Flame-crested tanager *(Tachyphonus cristatus)*; 12. Summer tanager *(Piranga rubra)*; 13. Scarlet-rumped tanager *(Ramphocelus passerinii)*; 14. Diademed tanager *(Stephanophorus diadematus)*.

◁
1. Red-legged honey creeper *(Cyanerpes cyaneus)*; 2. Plush-capped finch *(Catamblyrhynchus diadema)*; 3. Slaty flower piercer *(Diglossa baritula)*; 4. Swallow tanager *(Tersina viridis)*; 5. Green honey creeper *(Chlorophanes spiza)*; 6. Blue dacnis *(Dacnis cayana)*; 7. Rose-breasted thrush-tanager *(Rhodinocichla rosea)*.

Heads of Darwin's finches. Species which take a mixed diet with a preference for plant food (genus *Geospiza*).

Fig. 15-17. Large ground finch.

Fig. 15-18. Medium ground finch *(Geospiza fortis)*.

Fig. 15-19. Small ground finch.

continues to breed in the cool season, the season of the light rains in the uplands, from May to December. A female kept by Orr laid ten clutches at monthly intervals, a total of thirty-one eggs. Other species, like the woodpecker finch, which feeds only insects to its young, may skip breeding for a year when drought reduces the insect supply—much as the tawny owl (Vol. VIII) does in poor mouse years in central Europe. Many eggs, particularly those of ground finches, are infertile.

Darwin's finches molt during the cool season. There are various reports on this. Snow found that the ground finches of Indefatigable Island begin the molt of their large feathers before the breeding season; the molt was resumed after the breeding season, at exactly the point on the wing where it had ended before the interruption. However, on Tower Island, the German zoologist, Peter Kramer, and I found that these finches were able to molt in one uninterrupted process. Young ground finches first molt the contour feathers when they are two months old and some weeks later they molt the flight feathers; they may still begin a second molt of the large feathers in the same year. As the breeding season approaches, the beak color, which is yellowish pink in immatures and brown-colored in adults, becomes black.

After the breeding season, young and old of species with similar food requirements form loose mixed flocks. Up to 400 small ground finches, medium ground finches with a few cactus finches and vegetarian tree finches, may form such a flock. The other ground finches, particularly the insectivorous ones, generally keep to themselves.

The Galapagos Archipelago has twenty-four named islands and many additional islets, and it is understandable how thirteen species of Darwin's finches could develop there. This number of species is also due to the adaptations at various altitudes and the many habitats of the islands. Thus on some islands, there is a coastal dry zone, a more elevated woodland zone and an upland pampa. There are about 100 island populations of Darwin's finches; some are similar to others, others are insular subspecies, each confined to one island. Still others of these birds, as indicated above, have evolved into full species and even genera.

A consideration of the shapes of the beaks helps us to understand the evolution of the various Darwin's finches. The seven species of ground finches alone form a varied assembly. They like a mixed diet with a preference for plant food. The size of the beak of these closely related birds corresponds very closely to the size and hardness of the seeds they crack, according to the research of Bowman. This is true for the various species and even for individual birds. The two species of cactus finches feed mainly on various parts of prickly pear cacti (opuntia), but they also detach pieces of the bark to get at insects. The sharp-beaked ground finch and the small ground finch eat a good deal of animal food, besides seeds, berries and nectar; at low tide they pick up small marine organisms on the shore and they take ticks off iguanas (Vol. VI). Sometimes there is a marked difference in diet between insular races; the reasons for this are still unknown. Thus, the sharp-beaked ground finch of Tower Island, *(Geospiza difficilis acutirostris)* is almost entirely a plant-feeder, while its larger representative on Wenman Island *(Geospiza difficilis septentrionalis)*, which at twenty-four grams is almost twice as heavy, prefers a varied animal diet, but does not despise certain plants. Bowman and Billeb observed how these finches pecked at the skin over the elbow of breeding boobies to obtain blood. A comparable consumption of blood is known for only a very few other birds, such as the African oxpeckers. The Wenman Island sharp-beaked finch is also eager to get at the flesh of dead animals—crabs which have been cracked open, and bits of fish dropped about booby nests to which its attention is called by the food begging of the young. This finch also cracks open the eggs of sea birds to drink their contents.

Fig. 15-20. Large cactus ground finch.

Fig. 15-21. Cactus finch.

Fig. 15-22. Sharp-beaked ground finch.

When seeking food on the ground, ground finches turn over leaves and scratch outwards with both legs simultaneously. They also lever up small stones with one foot while supporting themselves with the other. They hold pieces of food on the ground with one or both feet, while working them over with the beak just as the woodpecker finch does on branches. On Hood Island where there are large areas of gravel, the cactus finch, and occasionally also the small ground finch, expose seeds hidden between the pebbles by supporting themselves with the ridge of the beak against a piece of rock so that they can use both feet alternately to push pebbles aside. In this manner the cactus finch, which weighs only 25 gm, can tip over stones weighing up to 358 gm.

Plant-eater tree finches (genus *Platyspiza*).

Fig. 15-23. Vegetarian tree finch.

Three tree finches eat a mixed diet but prefer arthropods. The larger a species is, the more it pursues insects beneath torn-off strips of bark, during which time the birds often hang upside down on the branch like chickadees. They also take buds, fruit and seeds. The related vegetarian tree finch feeds on buds, leaves, blossoms and fruit. It only rarely takes arthropods and works them over without the help of its feet. The small warbler finch is almost entirely insectivorous. Like a chickadee, it examines leaves and small twigs and also catches flying insects; it holds larger arthropods with its feet like the tree finches do. Judging by its beak, the Cocos finch has a similar diet.

Finches which take a mixed diet with a preference for arthropods (Tree finches of the genus *Camarhynchus*).

The woodpecker finch and the mangrove finch, which is restricted to the mangrove belt of Albemonte and Norborough Islands, have attracted the particular attention of investigators. They too are insect hunters, but also take fruit and mangrove leaves. Both capture arthropods in the manner of woodpeckers. They hack off pieces of rotten wood, detach bark from tree trunks, break off thin twigs and then

Fig. 15-24. Large insectivorous tree finch.

Fig. 15-25. Medium insectivorous tree finch.

Fig. 15-26. Small insectivorous tree finch.

Insectivorous finches (genera *Cactospiza*, *Pinaroloxias* and *Certhidea*).

Fig. 15-27. Woodpecker finch.

Fig. 15-28. Cocos finch.

Fig. 15-29. Warbler finch.

probe the insects out of their hiding places. However, they accomplish the latter not with their beaks but by using a cactus thorn or a small stick. This unique use of a tool among birds was discovered in the woodpecker finch by the American ornithologist, Gifford, in 1905. Later others like Lack, Bowman, Eibl-Eibesfeldt and Sielmann described the use of small sticks; finally Kramer and I described this habit in the mangrove finch.

The woodpecker finch breaks off its tool, spine, small piece of stick, or a fork of a twig, and then probes with it in tree holes or among cactus shoots which are in contact. Beforehand, it checks visually and possibly also by ear whether the hole is occupied; it may also feel along the passage with its thorn and finally it levers out its prey. Woodpecker finches kept in a cage by Bowman and Millikan never tried to spike a mealworm which has kept hidden in wood, with a thorn. A male kept by Eibl-Eibesfeldt hid unwanted mealworms in clefts and later fetched them out again; it thus created the probing situation itself. Woodpecker finches usually only go to fetch a thorn when they have spotted prey. There are indications that this behavior is not simply inherited but also has to be learnt, but how this happens has not been studied. Such an investigation would be well worth-while for so few other animals use tools; other examples are certain bower birds (Chapter 19), the Egyptian vulture among the raptors (Vol. VII) the sea otter (Vol. XII) and a number of species of monkeys (Vols. X and XI).

Two very different answers have been given to the question of the possible evolution of the different beak shapes, but they are not necessarily mutually exclusive. Thus, David Lack in his book on the Darwin's finches, believes that related species on a particular island compete with one another for food. In times of food scarcity, it must be advantageous for one species to use, even to a small extent, other foods than the competing species and it can do this with a somewhat different beak. Lack's view is supported by the distribution of species with corresponding diets and their beak shapes. In contrast, Bowman suggests that beaks were adapted to different food plants. He points out that the plant communities often differ in half of their species, even on islands close to one another and this would produce different adaptations in beak shape. According to Bowman, the multiplicity of plant species and their densities in particular islands often mirrors closely the number of finch species. However, the matter is so complex that possibly neither of these suggestions is able to give a final explanation of the evolution of different forms.

The main enemies of Darwin's finches in the daytime are the Galapagos hawk, a buteo (Vol. VII), the short-eared owl (Vol. VIII) and a snake; at night the barn owl (Vol. VIII) is an additional enemy. Thus, the Galapagos are not a "paradise" without enemies for these birds, as was believed up to twenty years ago. In adaptation to predator pressure, Darwin's finches have developed a number of protective adaptations. Thus, they recognize both the buteo and the owl and avoid them while uttering hostile calls. Experiments with model predators which Kramer and I carried out on ten species on four different islands showed that the intensity of fear these induced corresponds to the degree the finches are pursued by predators. On Indefatigable Island, where all the predators occur, fear responses are most intense while, for example, on Wenman, which has no predators, they are weak but still recognizable.

Reactions to snakes are similarly related to their distribution. On snakeless Tower and Abingdon Islands, the ground finches show less fear of a large living snake *(Dromicus dorsalis)* than do those on the Indefatigable Island where the snake occurs. Not one of the finch species has an inborn fear of cats or men since neither originally occurred on the islands. On the mainland, however, where there are many predatory mammals, many birds have an inborn fear of such enemies. Thus, the fear of predators in the Darwin's finches mirrors the predator threat in their particular environment quite accurately.

So far not one species of this group of birds so significant for the understanding of evolution has become extinct as the result of human interference; only one subspecies on one of two islands, the MEDIUM SHARP-BEAKED GROUND FINCH *(Geospiza difficilis debilirostris)* has been lost on Indefatigable, but survives on James Island. Probably this species, which is very much a ground bird, became the victim of the many feral domestic cats on the larger island.

Cardinals (subfamily Cardinalinae) are not readily placed either among the buntings or the tanagers, although they have often been classified with the latter. The L is 12.5–23 cm. The beak is clumsy and somewhat distended at the base. The plumage is generally very colorful, often with red or yellow pigments. They breed in North and South America, and generally visit the West Indies only as migrants. There are seventeen genera with forty-five species.

A. The CRESTED GENERA: 1. The CARDINALS PROPER *(Cardinalis)* comprise two species, among them the CARDINAL *(Cardinalis cardinalis* [scientific name used in America is *Richmondena cardinalis*], Color plates, pp. 334 and 352) which has a L of 21 cm and weighs 35–50 g. Males are red with a black ring around the base of the red beak. The female is mainly brown with a red beak. The range has recently been expanding northwards, particularly in the Mississippi Valley. This bird has been introduced in Hawaii and is locally common there. 2. The PYRRHULOXIAS comprise two species, among them the PYRRHULOXIA *(Pyrrhuloxias sinuata;* Color plate, p. 352) is mainly brown but has red on the crest, breast and wings. It has a curved beak which is bright yellow in the breeding season but corn colored or brown in the fall. It is found in Mexico and the southwestern United States. 3. The PAROARIA CARDINALS

Fig. 15-30. Distribution of races of the sharp-beaked ground finch *(Geospiza difficilis)* on the Galapagos Islands. 1. Culpepper sharp-beaked ground finch *(Geospiza difficilis nigrescens);* 2. Large sharp-beaked ground finch *(Geospiza difficilis septentrionalis);* 3. Sharp-beaked ground finch, nominate race *(Geospiza difficilis difficilis);* 4. Small sharp-beaked ground finch *(Geospiza difficilis acutirostris);* 5. Medium sharp-beaked ground finch of James Island *(Geospiza difficilis debilirostris);* 6. Medium sharp-beaked ground finch of Indefatigable Island *(Geospiza difficilis debilirostris),* presumably extinct.

Subfamily: cardinalinae, by H. Schifter

Fig. 15-31. Sharp-beaked ground finch pecking between the secondaries of the masked booby.

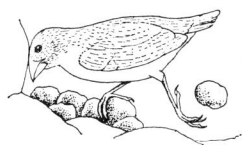

Fig. 15-32. Large cactus ground finch of Hood Island using its beak for support while scrambling among pebbles for seeds.

Fig. 15-33. Medium ground finches in silent hostile display at a snake model in the snake area of Indefatigable Island. In this display, they fan the tail and wings, stretch the neck and now and then fly up.

(Paroaria) are red only on the head; otherwise they are gray above and white below. There are five species, including the RED-CRESTED CARDINAL *(Paroaria coronata;* Color plate, p. 342) from southern Bolivia, and Brazil to Argentina and the RED-COWLED CARDINAL *(Paroaria dominicana)* which is also red-headed but with a very short crest—it replaces the preceding species in northern Brazil. 4. The YELLOW CARDINAL *(Gubernatrix cristata;* Color plates, pp. 334 and 342) has a L of 21 cm and is predominantly green; it breeds in southernmost Brazil, Uruguay and a large part of Argentina.

B. The CRESTLESS SPECIES: 1. The GROSBEAKS *(Pheucticus)* comprise four species of which we will mention two. The ROSE-BREASTED GROSBEAK *(Pheucticus ludovicianus)* has a L of 19 cm. The male's breast is red, the head and upper parts are black, and the belly is white. The female is brown above and white below with brown stripes. Birds of this species have lived up to twenty-four years in captivity. The BLACK-HEADED GROSBEAK *(Pheucticus melanocephalus)* males have a brown breast and yellow abdomen; the female is much like that of the rose-breasted grosbeak. This bird is found in the midwest of the United States; hybrids with the preceding species have been known to occur and the two are possibly not distinct species. 2. The BLUE GROSBEAK *(Guiraca caerulea;* Color plate, p. 352) is the only species of its genus. The male has a beautiful blue color with some brown marks on the wing, while the female is an inconspicuous brown bird. This species breeds in the southern United States, showing some spread to the north recently; it winters in Central America and Cuba.

C. PAINTED BUNTINGS (genus *Passerina*) have a L of 12.5-14 cm. They are much smaller than the preceding forms and constitute a separate group; they are popular cage birds. They are found in the United States and Mexico and winter south as far as Panama. There are six species; we will mention five: The INDIGO BUNTING *(Passerina cyanea;* Color plate, p. 352) males are bright blue in their breeding plumage; in winter they are dark brown like the female with only the wings and tail bluish. The LAZULI BUNTING *(Passerina amoena)* is blue like the former species but with white wing bands, a brown breast and a white belly. Its range is the western United States and it hybridizes in the zone of overlap with the eastern indigo bunting. The PAINTED BUNTING *(Passerina ciris;* Color plates, pp. 334 and 352) male is perhaps one of the most colorful of all birds with its red, blue, green and yellow. It occurs in the southeastern United States and winters south as far as Panama. The VARIED BUNTING *(Passerina versicolor)* is rather darker in color and occurs from the southwestern United States to Guatemala. The ORANGE-BREASTED BUNTING *(Passerina leclancherii)* is blue and green above the yellow and orange below; it occurs in Mexico.

The cardinal has a preference for thickets along rivers, but also likes hedges and bushes in the immediate vicinity of houses; however, it avoids open spaces. Nowadays, it can be found in towns and villages in North America where it shows little shyness and where it is well liked because of its colorful appearance and pleasant song. Its flutelike, descending song can be heard all year round in the south, and in the north mainly from March to August. This bird is regarded as one of the best North American songsters. Both sexes sing; the female's phrases are said to sound softer than those of the male's.

These resident birds wander but little, although banded cardinals have been recovered up to three hundred kilometers from where they were banded. In the north they withdraw into hedges and thickets in the winter, and their bright plumage forms a particularly attractive picture in a snowscape. At this season, they also come to feeding tables where they are well able to assert themselves against other birds. Occasionally they form loose flights of sixty to seventy birds.

Nests are rarely placed in trees but normally in bushes or hedges at a height of one to two meters. The nests are cup-shaped, often rather carelessly constructed, from flexible twigs, including willow twigs, grass and roots into which the birds like to weave leaves and shreds of paper. The female builds alone and takes three to nine days to complete the task. Clutches consist of two to five pale green or bluish eggs with brown spots of variable density. The female incubates for twelve to thirteen days until the mouse-gray, down-covered young hatch. They develop rapidly and only remain in the nest for nine to ten days; on disturbance they can even leave it when only seven days old and they then make their first attempts to fly. Both parents feed the young, at first by regurgitation, but soon from beak to beak. While the young are reared exclusively on insects, adults subsist on reeds and fruit to a greater extent than they do on animal food. At feeding tables, they eagerly take sunflower seeds.

Juveniles are brown with a red flush on the forehead, crest and tail, but they assume the nuptial plumage in their first spring and then soon start to sing. There are often three or four broods a year so that, for example, in Michigan, cardinals breed from mid-April to mid-September. In the favorable climate of Hawaii, they breed all the year round. The greatest enemy of cardinals is the brood parasitic brown-headed cowbird; blue jays and squirrels also occasionally plunder nests. However, a pair of cardinals has been seen successfully driving away a pair of the much larger jays from their nest. In the wild, cardinals live to an average age of three to six years but one banded bird was observed for thirteen and a half years and it still reared a brood in its last year. The cardinal was formerly a very popular cage bird in Europe, but its export from the United States is now forbidden as a protective measure.

The Pyrrhuloxia's song resembles that of the cardinal, but the calls are distinctly different. In addition, this equally resident bird lays

Fig. 15-34. Cardinals (Subfamily Cardinalinae).

Fig. 15-35. Cardinal (Cardinalis cardinalis).

Fig. 15-36. Red-cowled cardinal *(Paroaria dominicana)*, horizontal stripes. Red-crested cardinal *(Paroaria coronata)*, vertical stripes. Yellow cardinal *(Gubernatrix cristata)*, dots.

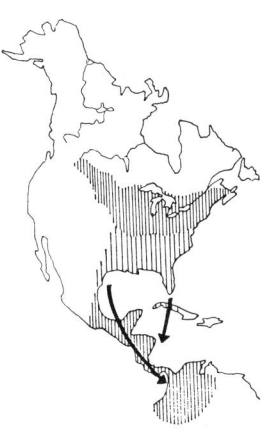

Fig. 15-37. Rose-breasted grosbeak *(Pheucticus ludovicianus)*. Northern breeding area, areas of passage, wintering area in South America.

clutches of only two to three eggs and usually only rears one brood of young a year. The red-crested cardinal and the red-cowled cardinal are frequently imported into Europe and kept as cage birds, but they are not compatible with one another in the same cage.

The rose-breasted grosbeak is one of the best known song birds of North America. It is a migrant which appears in its breeding area in May. Its choice is woodlands, preferably near rivers or marshes, but nowadays it is also found in parks and gardens. The song is very variable; in addition males court the females in peculiar postures and with strange movements. Unlike the cardinal, male grosbeaks of this species take part in nest building as well as in incubation and the rearing of the young. Nestlings are fed mainly on caterpillars and insect larvae; this bird is altogether one of the most efficient destroyers of harmful insects. There is only one brood per year. Migration begins soon after the molt and brings the birds to Central America from mid-October onwards. There they keep mostly to the uplands above 1,000 meters in farming areas or in open forests, but rarely use the dense jungles. They seldom sing in their winter quarters and they migrate northwards in March.

The blue grosbeak does not reach its breeding areas until May. It keeps to bushes and often sings on electric wires, but avoids the interior of woods. Its cup-shaped nests are often decorated by bits of paper or even snake skins. The clutch is usually four bluish-white, unmarked eggs. Only the female incubates, but both parents feed the young. The fledged young are often fed mainly by the male, particularly when the female has in the meantime started to build a second nest. This grosbeak has two annual broods in a large part of its range. The young gather in large flocks and visit rice fields especially.

Turning to the painted bunting group, we will discuss only the indigo bunting here. Its behavior is much like that of the painted bunting and the other species of this genus. Its preferred habitats are in open dry woodlands, and it builds its nest in bushes only a few feet above the ground. The strongly built, bowl-shaped nest consists of roots, dry grass and leaves lined with plant wool, feathers and hairs. Two broods are reared in the same nest between May and August. There are three to four eggs per clutch; these are pale bluish with only faint brown spots. The female incubates for twelve to thirteen days and also feeds the young alone. The busily singing male only participates in defense of the nest and the territory, within which it tolerates no others of its own species. The young are naked for the first few days after hatching, but they develop rapidly and can leave the nest when eight or nine days old, although they can only fly poorly at that time. The juvenile plumage is like that of the female.

Indigo buntings are frequent victims of cowbirds. Their fall migration begins in the last days of August and lasts until November. In their winter quarters they often mingle with seed eaters and American goldfinches. They are found on the coast and also in the mountains up to two thousand meters. According to Alexander Skutch, they particularly favor coffee plantations in Costa Rica. They leave their winter quarters up to mid-April and soon after this their song is to be heard in the breeding areas. They prefer projecting branches or telephone wires as song perches.

The tanagers (subfamily Thraupinae) have often been placed in a family of their own. The L is 9-25 cm. The beak is usually short and fairly thick, but the swallow tanagers and honey creepers are an exception to this. Some few species are inconspicuous in plumage, but far more of them are colorful. They occur in the Americas, including the West Indies. There are four tribes: A. The TRUE TANAGERS (Thraupini); B. The SWALLOW TANAGERS (Tersinini); C. The PLUSH-CAPPED FINCHES (Catamblyrhynchini), and; D. The HONEY CREEPERS (Dacnidini). Altogether there are 73 genera with 236 species.

The true tanagers, with their 62 genera and 207 species, form the bulk of the subfamily, and there are many forms which occur in great numbers and which contribute more to the colorfulness of the bird life in the American tropics than any other group, the hummingbirds not excluded. They cause admiration more because of their colors than for their voices as only few of them are good songsters. As warmth-loving birds they are most numerous in the moist lowland forests. Relatively few species have adapted to the lower temperatures of high tropical mountains, and even fewer migrate into the temperate zone to breed. Almost all of them live in stands of trees and bushes where they take a mixed diet of fruit and insects. Exceptions to this rule are the CHAT-TANAGER *(Calyptophilus frugivorus)*, a bird with a pointed beak, olive-colored above and white below, from the West Indian islands of Hispaniola and Gonave, where it lives largely on the ground in dense mountain woods or thickets in semi-deserts and the ROSE-BREASTED THRUSH-TANAGER (*Rhodinocichla rosea;* Color plate, p. 362) of Mexico to Venezuela, which is gray-black above and a beautiful pink below. It neither looks nor behaves like a typical tanager and searches over fallen leaves below densely growing bushes.

One of the largest and most widely distributed genera is that of the EUPHONIAS *(Euphonia),* which rarely exceed 11 cm in length. Their beaks and tails are short. The males are shiny black with blue or purple on the upper parts and sometimes on the throat; they are bright yellow on the forehead, on part of the crown, and on the lower parts. They inhabit both the American mainland and the West Indies. Some species have bright blue, chestnut brown or orange colored areas of plumage. The females are olive-colored above and a muddied-yellow beneath. The CHLOROPHONIAS (genus *Chlorophonia*) are closely related to them and have beautiful bright green, yellow and blue plumages.

Subfamily: tanagers, by A. F. Skutch

▷

Wood warblers: 1. Black-throated blue warbler *(Dendroica caerulescens)*; 2. Magnolia warbler *(Dendroica magnolia)*; 3. Golden-winged warbler *(Vermivora chrysoptera)*; 4. Black and white warbler *(Mniotilta varia)*; 5. Ovenbird *(Seiurus aurocapillus)*; 6. Red warbler *(Ergaticus ruber)*; 7. Hooded warbler *(Wilsonia citrina)* 8. Rose-breasted chat *(Granatellus pelzelni)*; 9. Collared redstart *(Myioborus torquatus)*; 10. Bananaquit *(Coereba flaveola)*; 11. Prothonotary warbler *(Protonotaria citrea)*; 12. Painted redstart *(Myioborus pictus).*

Fig. 15-38. Distribution of the tanagers (subfamily Thraupinae), introduced on Hawaii and Bermuda.

◁
1. Mistletoe bird *(Dicaeum hirundinaceum)*; 2. Scarlet-backed flower pecker *(Dicaeum cruentatum)*; 3. Tit berry pecker *(Oreocharis arfaki)*; 4. Spotted pardalote *(Pardalotus punctatus)*; 5. Mountain white-eye *(Zosterops montana)*; 6. Red-eyed vireo *(Vireo olivaceus)*; 7. Yellow-throated vireo *(Vireo flavifrons)*; 8. Gray-headed greenlet *(Hylophylus decurtatus)*; 9. Green shrike vireo *(Vireolanius pulchellus)*; 10. Rufous-browed peppershrike *(Cyclarhis gujanensis)*; 11. Crested honey creeper *(Palmeria dolei)*; 12. Apapane *(Himatione sanguinea)*; 13. Mamo *(Drepanis pacifica)*; 14. Akepa *(Loxops coccinea)*; 15. Akiola *(Hemignathus obscurus)*; 16. Iiwi *(Vestiaria coccinea)*.

While the chlorophonias inhabit the cool upland forests, the euphonias generally live at lower levels in warmer areas. They eat mainly mistletoe berries and travel great distances to gather on trees covered with these parasites. The seeds remain covered by a glutinous mass as they pass through the stomach of the birds so that when they are passed, they stick to branches and germinate. However, euphonias are not the only birds which aid in the dispersal of mistletoe. While seeking food, euphonias utter soft whistles and chatterings. Most species are not good songsters.

Unlike the other tanagers, euphonias and chlorophonias build closed nests. These are small spheres composed of dry plant materials, or fresh moss, and they have a round entrance at the side. They are placed in the moss growth of a tree, in clefts in the bark, in holes in fenceposts, clefts in rocks or sometimes under an orchid growing in a hanging basket slung from a garden tree. Both sexes build these comfortable dwellings. Euphonias lay two to five minute white, spotted eggs; the number is unusually high for a neotropical passerine. The female incubates and she is often accompanied by the male when she returns from an absence from the nest. The male then flies so close behind his partner that it looks like a race for the nest entrance. However, the female always wins and enters the nest while the male swerves aside and flies away.

The almost naked young hatch after fourteen to eighteen days of incubation. Their gapes are red, as in other tanagers. Usually both parents come to the nest together. They carry no visible food in their beaks but regurgitate it for the young. The male enters the nest to feed the young first, then the female; she then stays to brood the young, as long as they are bare. The young remain in the nest for a surprisingly long period—twenty to twenty-four days. Both euphonia and chlorophonia males often breed before they have acquired the full adult plumage.

The tanagers of the genus *Tangara* have a L of 11–15 cm. They show a great variety of plumage colors, usually arranged in complicated patterns. Even in the most colorful species, the sexes are quite or almost the same in color and these colors are retained right through the year. One of the most famous species is the PARADISE TANAGER *(Tangara chilensis;* Color plate, p. 361) which, as can happen in zoology, has a misleading scientific name for it does not occur in Chile. It is widely distributed in the densely forested parts of the South American tropics. It deserves its vernacular name, for in the northwestern subspecies, the plumage has red, orange, yellow, golden-green, turquoise-blue, purple and black elements. The GREEN-HEADED TANAGER *(Tangara seledon)* looks equally beautiful; it occurs from southern Brazil to northern Argentina. The cinderella of this genus is the PLAIN-COLORED TANAGER *(Tangara inornata)* of southern Central America and

Colombia in which the gray plumage is enhanced only by a blue suffusion of the wing coverts.

Tanagers of this species are very numerous at the forested base and on the lower slopes of tropical mountains. However, they also occur in the lowlands and some wander up into the cooler heights of the mountains. All tanagers like to eat berries and all sorts of fruit, and they are easily attracted to feeding tables with bananas. Four species came to our feeding table in southern Costa Rica. They supplement their diet with insects and spiders. A characteristic mode of feeding involves hopping along a thin, more or less horizontal branch and then leaning far over to one side or the other to spot small prey in the moss or lichens on the lower side of the branch. These birds live in permanent monogamy and are thus found throughout the year in pairs or small family troops.

As far as is known, no tanager of this genus is an outstanding singer; some species do not even utter the most modest tune. Their cup-shaped nests are built by both partners (more rarely by the female alone) in trees or bushes. The female lays two strongly spotted eggs which she alone incubates, being fed by the male while doing so. After thirteen or fourteen days, the young (which are initially red-skinned) hatch; they carry only scanty down and are fed by both parents on food carried in the beak. If the young are not disturbed, they usually remain in the nest for fourteen to sixteen days by which time they can make short flights. In some species like the SILVER-THROATED TANAGER *(Tangara chrysophrys)*, the plain tanager and the MASKED TANAGER *(Tangara nigrocincta;* Color plate, p. 361), one or two "outsider" birds in adult plumage may help the parents feed the young. Sometimes young masked tanagers still in the juvenile plumage bring food to their siblings of a later brood.

The THRAUPIS TANAGERS *(Thraupis)* are a little larger than the euphonias and the tangara tanagers, with a L of 15-18 cm. One of the best known birds of gardens and shady plantations in the American tropics is the BLUE-GRAY TANAGER *(Thraupis episcopus)* found from Mexico to Bolivia and Brazil. Both sexes have the same blue-gray plumage with bright blue wings and tail. Birds once paired remain together for the whole year and, in towns, they often visit feeding tables to eat fruit. Their breeding behavior is like that of the tangara tanagers. Sometimes they steal the nest of some other bird, such as that of the smaller masked tanager. They may then incubate the eggs of the former owner and later feed its young along with their own.

In the area of Rio de Janeiro, the very similar SAYACA TANAGER *(Thraupis sayaca)* replaces the blue-gray species of the more northern towns. The PALM TANAGER *(Thraupis palmarum)* which is olive-green with blackish wings and tail, is widely distributed. It often builds its cup-shaped nest among the lowest parts of palm fronds where it is difficult to reach; however it may also breed in holes in rotting tree trunks or clefts in the walls of buildings. The only tanager to reach Chile (only its most northern part) is the BLUE-AND-YELLOW TANAGER (*Thraupis bonariensis;* Color plate, p. 336) an attractive bird with a blue head and neck, green back, a yellow rump and belly, and a black breast band which separates the blue throat from the orange-colored breast.

The widely distributed RAMPHOCELUS TANAGERS (genus *Ramphocelus*) are fairly large birds, with a L of about 18 cm. They inhabit thickets, open woods and plantations. In contrast to the preceding two tanager genera, pairs are maintained only during the breeding period. They move about in loose flocks wherein birds in female plumage exceed those in the adult male plumage. The SCARLET-RUMPED TANAGER (*Ramphocelus passerinii;* Color plate p. 361) of southern Mexico and Central America is velvety black with a fiery red lower back and upper tail coverts; the female is olive-colored and yellowish and, in a subspecies from Costa Rica, she often has an orange spot on the breast and rump. The YELLOW-RUMPED TANAGER *(Ramphocelus icteronotus)* is found from Panama to Peru. It is black with a lemon-yellow lower back and upper tail coverts; the female is yellow on the rump and on the lower parts, and has a dark olive-colored back and sides of the head. The SILVER-BEAKED TANAGER (*Ramphocelus carbo;* Color plate, p. 361) is widely distributed in northern South America; the appearance of the male's plumage varies from almost pure black to the brightest red, according to the direction of the light.

These tanagers place their large open nests in thickets and bushes, usually not very high. In the scarlet-rumped species, only the female builds, but she is often accompanied by the male. The two eggs are pale blue with black and pale lilac spots and hair lines. They are incubated by the female only, for twelve days. The male has not been observed to feed the incubating female, but he does help to feed the young at most nests. However, it also happens that in some broods in this species as well as in the silver-backed tanager, the male does not visit the young at all. It is probable that because of the scarcity of adult males, some females must rear their young unaided. The young leave the nest at eleven to thirteen days of age, much earlier than in the other genera described so far.

The PIRANGA TANAGERS (genus *Piranga*) generally live at medium to higher elevations. Males usually have a certain amount of red in the plumage and females are olive-green and yellow. Four migrant species of this genus breed in the United States and, before the blue-gray tanager was introduced in Florida, they were the only tanagers in North America (north of Mexico). The most pronounced migrant of all the tanagers in the SCARLET TANAGER *(Piranga olivacea).* This colorful black-winged bird breeds in southern Canada, in the northern and middle states of the United States, and winters far to the south of Peru and Bolivia. Males show a marked seasonal plumage change and wear a yellowish plumage during the southward migration.

On the other hand, the SUMMER TANAGER (*Piranga rubra*; Color plate, p. 361), like the tropical members of its subfamily, keeps its bright red plumage all year round. It breeds in the middle and southern United States, and winters from Mexico to Bolivia and Brazil. It snaps up many insects on the wing, like a flycatcher, and often opens wasp nests to devour the white larvae and pupae. These tanagers of the North American temperate zone, like many other birds of northern countries, rear larger families than do most of their tropical relatives; they lay three to five eggs. They are also better and more enduring singers than many of their tropical relatives.

In the dense lowland forests of the tropics, the colorful tanagers are more numerous in the sunny crowns of the trees than in the twilight of the undergrowth. Nevertheless, a few interesting species also live in the lower forest levels, species like the shy GRAY-HEADED TANAGER (*Eucometis penicillata*) which has a L of about 17 cm. Both sexes are yellowish-olive-green above and lemon-yellow below. They usually live in pairs and follow the swarms of wandering ants thereby meeting antbirds (Chapter 4), woodcreepers, and other small and medium-sized birds. Like most birds which accompany army ants, they catch insects and spiders which have been flushed from their hiding places beneath the ground cover by the hunting ants; the tanagers do this from perches one or two meters high. A male sometimes gives his female the prey it has seized. In the dry season when army ants rarely appear, I watched a pair of these tanagers intensively; they caught insects disturbed by a group of scratching hens at the forest edge. Although gray-headed tanagers usually build their simple, open nests in shrubbery or in thorn palms right in the forest, they sometimes breed in coffee plantations near the forest, where there are fewer enemies. They lay two or three bluish-gray, spotted eggs. They are among the best songsters of the whole subfamily.

ANT-TANAGERS (genus *Habia*) are found in the lower forest levels. In RED-CROWNED ANT-TANAGER (*Habia rubica*; Color plate, p. 361) males are dull red and have an erectile, scarlet crest, which is only displayed on special occasions. The female is olive-colored. This species occurs from Central America to Paraguay and northern Argentina. They accompany the mixed flocks of small birds which noisily move through the undergrowth in search of arthropods and fruit in the bushes and the lower branches. Their rough calls make them the most striking members of these colorful mixed troops. In the breeding season, males repeat their loud, clear song for minutes on end about dawn, but they rarely sing later in the day.

If, in a tropical mountain region, one climbs above 1,500 to 2,000 meters, the lovely tangara tanagers become progressively rarer, but one encounters BUSH TANAGERS (genus *Chlorospingus*) instead. They are about the same size (with a L of 11–15 cm) but are less colorful than

Fig. 15-39. Grass-green tanager.

the tangara tanagers. The COMMON BUSH TANAGER *(Chlorospingus ophthalmicus)* is found from southern Mexico to Bolivia. Both sexes are olive-green with a dark head and a conspicuous white spot behind the eye. In contrast to many other tanagers, bush tanagers do not make up for their dull plumage by a melodious voice; their song is little more than an oft-repeated twittering. The female, with only slight help from her mate, builds the large nest in an astonishing variety of places. It may be hidden beneath bromelias, under ferns or other tree-growing epiphytic plants or sunk in the mossy pad of the tree trunk, on a grass-grown slope or even on the ground beside a bush. Although the nest is open at the top, it is often covered by the plants among which it is placed. The female incubates the two eggs for fourteen days. The male takes part in the feeding of the young, which when they leave the nest resemble the adults in their inconspicuous plumage.

In addition to the dull-colored bush tanagers, there are also a few wonderfully colored species in the tropical Andes. There, above 2,000 meters, one finds the GRASS-GREEN TANAGER *(Chlorornis riefferi;* see Fig. 15-39) which has a L of 21 cm. This bird has a lively green colored plumage with reddish chestnut sides of the head and lower tail coverts. Another beautiful inhabitant of the Andes is the SCARLET-BELLIED MOUNTAIN TANAGER *(Anisognathus igniventris),* which is mainly black but has a chestnut red triangle at the side of the neck. Its lower back and a spot on each wing are light blue.

Our knowledge of the behavior of the tanagers that live on the flower-covered slopes of the Andes is only sketchy and this is also true for the MAGPIE TANAGER *(Cissopis leveriana;* Color plate, p. 361) which, with its tail, is 25 cm long. Both sexes are black and white in a pattern surprisingly like that of a magpie, but they have bright yellow eyes. These birds live in pairs or flocks in the rain forest areas of South America. Probably many years will elapse before ornithologists find their nests and probe into their habits.

The SWALLOW TANAGER *(Tersina viridis;* Color plate, p. 362) has a L of 16 cm and is the sole representative of the tribe Tersinini. It has a black forehead and throat, but the rest of its plumage, depending on the angle of the light, looks deep aquamarine-blue or light turquoise-blue. Females are banded on a pale green background. The wings are long and pointed; the beak is wide, flat and hooked at the tip. The gullet is very distensible. This bird feeds on insects, which it often catches on the wing, and also on fruit and sprouts. Its song is a tinny clatter and a monotonous chirping. It ranges from Panama and Trinidad to northern Argentina.

I have watched male swallow tanagers in Venezuela defending their territories, of which they are very conscious, with determination. In display they sit opposite one another in a stiff pose and then bow to one another with peculiar hops. Favorite nest sites are holes in steep

Fig. 15-40. Distribution of the swallow tanagers (tribe Tersinini).

The swallow tanagers, by E. Schäfer

banks, below bridges, or the eaves of roofs or in clefts in the walls of old buildings. The cup-shaped nest is built mainly by the female which incubates the three glossy white eggs for fourteen to seventeen days. The fledging period is twenty-four days.

The PLUSH-CAPPED FINCH (*Catamblyrhynchus diadema*; Color plate, p. 362) has a L of about 15 cm and is found in the subtropical zone from Colombia to Bolivia. It is regarded by some ornithologists as the sole representative of a family, others place it among the tanagers, while still others place it with the finches. Here we regard it as representing a tribe Catamblyrhynchini of the tanagers. The common name refers to the short erect golden-yellow feathers of the forehead and crown which give a plush-like effect. Otherwise this thick-billed bird is dark gray above and reddish-brown below. Almost nothing is known of its way of life, only that it is seen singly or in pairs in dense moist shrubbery or low bamboo growth where it is difficult to observe.

The classification of the DACNIS (tribe Dacnidini) has also caused uncertainty. For a long time, these birds were aligned with the conebills in a family of sugarbirds; this has now been abandoned. Studies of their anatomy by W. J. Beecher indicate that the conebills are derived from wood warblers and that the dacnis are close relatives of the tanagers. Both groups have come to resemble one another in certain characteristics, particularly in those of the beak and tongue; this is connected to their shared habit of sucking nectar from flowers.

Dacnis are small, rarely longer than 13 cm. The plumage ranges from dull black to bright blue, green, rarely with a little reddish-brown. The shape of the beak is variable. The tongue is split into two or three strips near the tip which has a fringe or is brushlike. There are nine genera with about thirty species; only two occur in the West Indies, most of the rest are found from Mexico to northern Argentina. We will mention the following: 1. FLOWER PIERCERS (genus *Diglossa*) are generally a fairly inconspicuous black to dark blue; they are rarely found below 1,200 meters from Mexico to northern Argentina. 2. The ORANGEQUIT (*Euneornis campestris*), the only member of its genus, is found in Jamaica. 3. The HONEY CREEPERS (genus *Cyanerpes*) include the RED-LEGGED HONEY CREEPER (*Cyanerpes cyaneus*; Color plates, pp. 335 and 362) which is found from southern Mexico to Brazil and Bolivia as well as on Cuba and Tobago. 4. The GREEN HONEY CREEPER (*Chlorophanes spiza*; Color plate, p. 362) is the sole species of its genus. 5. The DACNIS in the restricted sense (genus *Dacnis*) have five species.

Most or probably all dacnis drink nectar, but they obtain it from flowers by different means. They supplement their diet with fruit and arthropods; some are fruit eaters to a great degree. They sing either poorly or not at all. One of the most widespread species is the veritably diamond-like red-legged honey creeper. At the end of the breeding season, the male molts from a turquoise-blue, dark blue and black

The plush-capped finch,
by A. F. Skutch

Dacnis,
by A. F. Skutch

Fig. 15-41. Distribution of the plush-capped finch (tribe Catamblyrhynchini).

Fig. 15-42. Distribution of the dacnis (tribe Dacnidini).

nuptial plumage into a greenish one, like the female's. This makes it one of the few South American male passerine birds which shows a striking seasonal plumage change. Some males even start becoming greener while they are still feeding their fledglings. These birds are sociable and sometimes move in great flocks through the flowering shade trees found in coffee plantations. They like fruit and if offered bananas, come to feeding tables right away. Our feeding table in southern Costa Rica also attracts the dark blue, yellow-footed SHINING HONEY CREEPER *(Cyanerpes lucicus),* the green honey creeper and the black-throated BLUE DACNIS *(Dacnis cayana;* Color plate, p. 362).

Nothing is known about the nest of the shining honey creeper, but we do know that others of this group build open cup-shaped nests on trees. The green honey creeper uses fairly large, dry leaves in its nest, but the other species use finer material. The females build the nest without help from their mates, but these follow them as they gather nest material. The three white, spotted eggs are incubated by the female only. In one case an incubating female dacnis was observed being fed at long intervals by the male; however, this has not been seen at the nests of the other species.

Both parents feed the nestlings, bringing the food in their slim beaks. Usually, however, the female bears the greatest share of this burden too. Two males in full breeding plumage have been observed helping a female feed the young at a nest in a guava tree. As they arrived at the nest, the red seeds of a cusia tree in their beaks contrasted vividly with their blue plumage so that they looked uncommonly colorful. Dacnis seem to like such covered seeds, (which are distasteful to humans) above all other food.

All dacnis have pointed beaks which vary in length from the short, straight beak of those of the genus *Dacnis,* to the long, slender, downward curved beak of red-legged honey creepers. To reach the nectar they dip their beaks down to the base of the flower, in the manner characteristic of hummingbirds, but while perched instead of while hovering.

A number of species of flower piercers, however, proceed differently. These generally inconspicuous birds vary in plumage from black to dark blue. The short, somewhat upcurved bill, ends with a marked hook on the upper mandible, and behind this hook there are several flat "teeth" on the cutting edge. The lower mandible comes to a sharp point. When a flower piercer wants to get at the nectar in a tubular flower, it holds the flower by grasping the upper part of it with the upper mandible, while the lower mandible stabs a minute hole, through which, apparently, the feather-shaped tongue draws out the nectar. In this way, flower piercers can deal with surprisingly large flowers. Like hummingbirds, they supplement their diet with small arthropods, which they catch in the air. Bristles at the gape aid them in this latter activity.

Fig. 15-43. Slaty flower piercer *(Diglossa baritula).*

Flower piercers are a good example of the dependence of the breeding season on food suitable for the young. In the high mountains of Central America, they breed early in the dry season, which corresponds to the northern hemisphere winter. The nights are cold and frosty, then there are more bright flowers than at any other season. The hummingbirds breed during the same months, however most of the other small birds which do not sip nectar postpone their breeding until April or May when the nights are warmer and fruit and insects are more abundant, although flowers are fewer. The female SLATY FLOWER PIERCER *(Diglossa baritula)* builds an open cup nest in low bushes; this nest is larger than that of most dacnis. She alone incubates the two light blue eggs which have fine brown spots. The young hatch after fourteen days and are fed by both parents. The food is regurgitated, as in hummingbirds conebills, and not directly passed from beak to beak as in most other dacnis. The young remain in the nest for about sixteen days. Slaty flower piercers are more eager songsters than most other species, but their voices sound thin and weak.

The wood warblers (family Parulidae) are related to the tanagers. They vary in length from 11 to 19 cm and have slender or flat beaks without the "tooth" of vireos. The tongue is, at most, slit and thus unlike the brush tongue of the dacnis. In contrast to many tanagers, the plumage is never glossy. There are two subfamilies: 1. The WOOD WARBLERS PROPER (Parulinae) have 25 genera and about 115 species distributed over almost all of the New World from Alaska and northern Canada to northern Chile and Argentina; they are most numerous in North America. 2. The BANANAQUITS (Coerebinae) have two genera and ten species and are found on the Bahamas and from southern Mexico to Argentina.

The wood warblers proper are characterized by very fine thin beaks and well developed muscles for closing the beak. The rear of the palate has a striking ridge. They generally have yellow, red, black, gray or green areas of plumage. Males are usually much brighter in color and have sharper patterns than do the females, but they resemble the latter in fall and winter when they are more inconspicuous. We will mention only the following genera: 1. The PROTHONOTARY WARBLERS *(Protonotaria)* have only one species. 2. The BLACK-AND-WHITE WARBLERS *(Mniotilta)* have only one species. 3. The VERMIVORA WARBLERS *(Vermivora)* have eleven species. 4. The DENDROICA WARBLERS *(Dendroica)* have twenty-six species. 5. The WATERTHRUSHES *(Seiurus)* have three species. 6. The WILSONIA WARBLERS *(Wilsonia)* have three species. 7. The AMERICAN REDSTARTS *(Setophaga)* have two species. 8. The ERGATICUS WARBLERS *(Ergaticus)* have two species. 9. The BASILEUTERUS WARBLERS *(Basileuterus)* have twenty-three species. 10. The MYIOBORUS REDSTARTS *(Myioborus)* have eight species. 11. The GRANDELLUS CHATS *(Granatellus)* have three species. 12. The YELLOW-BREASTED CHATS *(Icteria)* have one species.

Most North American wood warblers are long distance migrants.

▷
Icterids: 1. Spotted-breasted oriole *(Icterus pectoralis)*; 2. Green oropendola *(Psarocolius viridis)*; 3. Montezuma oropendola *(Gymnostinops montezuma)*; 4. Russet-backed oropendola *(Psarocolius angustifrons)*; 5. Wagler's oropendola *(Psarocolius wagleri)*; 6. Yellow-rumped cacique *(Cacicus cela)*; 7. Boat-tailed grackle *(Cassidix mexicanus)*; 8. Common grackle *(Quiscalus quiscula)*; 9. Yellow-headed blackbird *(Agelaius icterocephalus)*; 10. Giant cowbird *(Scaphidura oryzivora)*; 11. Baltimore oriole *(Icterus galbula)*; 12. Western meadowlark *(Sturnella neglecta)*; 13. Long-tailed meadowlark *(Sturnella loyca)*; 14. Bobolink *(Dolichonyx oryzivorus)*; 15. Troupial *(Icterus icterus)*; 16. Shining cowbird *(Molothrus bonariensis)*.

Family: wood warblers, by Lester L. Short, Jr.

Wood warblers proper

1. *Serinus tristriatus;* 2. Red crossbill *(Loxia curvirostra);* 3. Pine grosbeak *(Pinicola enucleator);* 4. Long-billed rosefinch *(Uragus sibiricus);* 5. Yellow-fronted canary *(Serinus mozambicus);* 6. Serin *(Serinus serinus);* 7. Yorkshire canary; 8. German-hooded canary; 9. Harz canary; 10. Wild canary; 11. Agate lipochrome pastell canary; 12. Lizard gold canary; 13. Gold-fronted finch *(Serinus pusillus).*

Their regular passage flocks are one of the most striking features of North American bird life. During the height of the spring migration in May, it is possible to see up to thirty species in a day in eastern North America. The North American species winter mainly in Central America and northern South America, but a few resistant species winter in the United States, while others move as far as Argentina. They migrate during the day and insert pauses for feeding and resting of several days duration.

The YELLOW-BREASTED CHAT *(Icteria virens)* with a L of 19 cm is the largest of the family; it is also one of the most peculiar wood warblers. It deviates entirely from the other species both in its behavior and its plump, conical beak. The songs of wood warblers are often only weak and buzzing. One exception is the LOUISIANA WATERTHRUSH *(Seiurus motacilla),* an insignificant-looking ground dweller which often utters its loud, pleasant tune to a background of rushing water. The yellow-breasted chat, on the other hand, has a loud, flute-like and gurgling song and it often imitates other birds. It sometimes sings in the middle of the night. In contrast, the feeble "heee-bsss" song of the BLUE-WINGED WARBLER *(Vermivora pinus)* which sounds like an insect, is readily overheard by inexperienced human observers and even by members of the same species. Songs are uttered during spring migration as well as in the breeding territories.

Wood warbler nests may be on the ground, among grass, in bushes or on trees, often more than fifteen meters high. They are usually firm, densely woven structures, generally with an inner lining layer of rootlets or moss. The clutch size varies a great deal within the family. The eggs are usually white with pale reddish-brown or black markings, sometimes arranged as in a wreath. Wood warblers feed mainly on insects which they seek busily everywhere, much like the Old World leaf warblers. While searching for food, they move on the ground among grass, on bushes or in the foliage—in some cases also on the tree bark. According to Robert McArthur, five species breed very close together in an evergreen forest. Nevertheless, the nesting sites and manner of feeding of these five are different. Thus, they divide the available food among themselves and there is almost no competition. Although small arthropods are the main food of all wood warblers, some species occasionally take berries. The MYRTLE WARBLER *(Dendroica coronata)* can only survive the winter in eastern North America by feeding on berries and sometimes seeds.

The beautiful PROTHONOTARY WARBLER *(Protonotaria citrea;* Color plate, p. 371) has a L of 14 cm. It is a swamp dweller in eastern North America. It has a long, thin beak, a bright golden yellow and gray plumage and a fairly short tail. The female is similar but less brightly colored. This warbler builds its nest mainly of moss in tree holes such as old woodpecker holes, or in nest boxes; this is remarkable as hole breeding is rare among the wood warblers. The BLACK-AND-WHITE WAR-

Fig. 15-44. Area of distribution of the wood warblers (family Parulidae).

BLER *(Mniotilta varia;* Color plate, p. 371) has a L of 13 cm. It pecks its food out of clefts in the bark of trees which it searches carefully. It is one of the earliest wood warblers to return to North America in the spring and can be recognized by its soft, often repeated "wee ee" song. It usually works over one tree trunk after another from near the ground upwards. Its rear toe and claw are unusually long; this is probably connected with its creeper-like habits. This species is common in northern North America. It builds its nest of strips of bark, moss, grass, and other materials, behind a piece of bark on the lower part of a tree trunk or more frequently on the ground beneath a tree where it covers it partly with leaves.

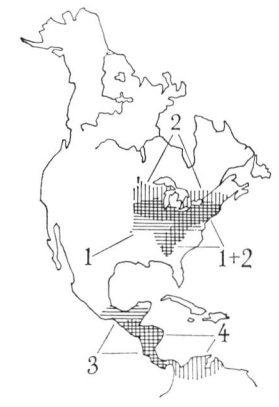

Fig. 15-45. 1 and 2, breeding areas of the blue-winged warbler *(Vermivora pinus)* and the golden-winged warbler *(Vermivora chrysoptera)* and 3, winter range of the blue-winged warbler and 4, of the golden-winged warbler.

The very thin beaked GOLDEN-WINGED WARBLER *(Vermivora chrysoptera;* Color plate, p. 371) has a L of 12 cm and inhabits forest edges and swamps. It and the blue-winged warbler are very closely related and interbreed very frequently along a zone which extends through the whole of eastern North America. Both species and their hybrids, some of which incidentally have been described as distinct species, may occur in the same stands of bushes or in wooded areas side by side. Although the buzzing songs of the two species sound different, some of these birds sing the songs of both species. Evidently the more southern blue-winged warbler has taken the place of the golden-winged warbler in some areas during the last century. These two species are very similar in behavior; they breed in low bushes. I have observed an adult male hybrid helping to feed the young in the nest of another male hybrid and a female blue-winged warbler.

One of the dendroica warblers, the YELLOW WARBLER *(Dendroica petechia)* looks entirely yellow at first glance. It breeds mainly in willows and other bushes. The black, yellow and gray MAGNOLIA WARBLER *(Dendroica magnolia;* Color plate, p. 371) has a L of about 13 cm. Even though this bird is seen in a variety of habitats while on migration, during the breeding season it keeps entirely to conifers of medium height in the northern United States and southern Canada. Near the home of my parents in New York State, it used to appear in plantations of Douglas firs and hemlock when the trees were three to five meters high. In the gloom of the dense trees, the brightly colored males would sing their "weetee weetee tee tee" songs not far from their bulky nests, which stood low in the trees. After rearing the young, the adults molt into an inconspicuous winter plumage.

Fig. 15-46. Black-and-white warbler.

The very common OVENBIRD *(Seiurus aurocapillus;* Color plate, p. 371) nests in mixed and deciduous woods lacking dense bushy undergrowth. In such places it runs about the fallen leaves and tosses them aside in its search for hidden arthropods. It responds to the presence of an intruder, particularly when the latter is near the nest, by flying up, uttering "tick" calls or even by erecting the orange-colored feathers of its crown patch. Its song is one of the best known North American

Fig. 15-47. Ovenbird *(Seiurus auricapillus)*. breeding area, lines; wintering area, dots.

Fig. 15-48. Hooded warbler *(Wilsonia citrina)* breeding area, vertical lines; wintering areas, dotted. Rose-breasted chat *(Granatellus pelzelni)*, horizontal lines.

Family: bananaquits-conebills, by W. Meise

bird songs, and consists of a sequence of "teach-er, teach-er teach-er" which gets progressively louder. The popular name "teacherbird" derives from the song which is delivered from the ground or from a low, up to a fairly high-placed branch. The nest, which is covered above, is responsible for the name "ovenbird."

The HOODED WARBLER *(Wilsonia citrina;* Color plate, p. 371) has a L of 14 cm and is a bright yellow and black inhabitant of the undergrowth of mature deciduous woods. There are conspicuous bristles at the angles of its fairly wide beak and this is probably connected with the fact that, like a flycatcher, it often catches flying arthropods. Its loud ringing song, "wee te wee tee o" proclaims its presence in thickets, often near water. It builds its nest among bushes and climbers; the nest itself is firmly built of leaves and grasses and lined with rootlets or fine grass. The female lays three or four eggs which resemble those of other wood warblers.

The PAINTED REDSTART *(Myioborus pictus;* Color plate, p. 371; scientific name used in North America is *Setophaga picta*) has a L of 13 cm and is a strikingly colored black, white and red bird. It flits about in the oak forests of northern Central America as it chases arthropods on the wing or searches leaves in rapid movements. Alexander Skutch renders its song as a fluted "witsher witsher witsher". The RED WARBLER *(Ergaticus ruber;* Color plate, p. 371) which has a L of 13 cm and is a brilliantly red bird with silvery white cheeks, is found in the mountain country of western and central Mexico. This little known species lives in forests of tall pines or oaks; it probably nests on the ground, in which case the nest may be presumed to be of the roof-over type.

The tropical lowlands of Amazonia are the home of the highly colored ROSE-BREASTED CHAT *(Granatellus pelzelni;* Color plate, p. 371) which has a L of 14 cm. This bird has relatives only in northern Central America. It probably seeks its food in the dense undergrowth in the same manner as its northern relatives. The COLLARED REDSTART *(Myioborus torquatus;* Color plate, p. 371) is restricted to the high, cool and moist mountain forests of Costa Rica and western Panama. Because of its fearlessness, it has received there the nickname "amigo de los hombres", friend of man. It builds a roofed-over nest on the ground in which two or three eggs are laid. According to Skutch, its loud song exceeds that of other Central American wood warblers in fullness and softness.

The bananaquit-conebills (subfamily Coerebinae) are mainly nectar feeding relatives of the wood warblers. The tongue is cleft near the tip. These birds generally show no sex difference in the plumage. There are two genera: 1. The CONEBILLS *(Conirostrum)* have a short, pointed beak and a blue-gray coloring with black, brown and/or white areas. There are nine species including the BICOLORED CONEBILL *(Conirostrum bicolor)* which has a L of 12 cm. 2. The BANANAQUITS *(Coereba)* have only one species, the BANANAQUIT (*Coereba flaveola*; Color plate, p. 371) which has a L of 10 cm. It has a beak that is curved downwards, blackish upper parts and throat, a yellow rump and lower parts, and a pale stripe above the eye with pale areas on the wing and tail.

Birds of both genera stab flowers to get at the nectar, but they also feed on berries and insects. Bananaquits are usually very common but they do not live in aggregations greater than pairs or family parties and are not conspicuous as they are peaceable and rather silent. Males utter a modest song and take part particularly energetically in the construction of the low-placed spherical nest; they even build separate sleeping nests. The female incubates and both parents regurgitate food for the young. On several West Indian islands, usually non-adjacent, the bananaquit has acquired an entirely black plumage.

Fig. 15-49. Distribution of the bananaquit (tribe Coerebinae).

The wren thrushes (family Zeledoniidae, genus *Zeledonia*) are represented by only one species, the WREN THRUSH (*Zeledonia coronata*) which has a L of 11 cm, and is sufficiently distinctive when compared to other birds to deserve family rank. The only contrasting color in this bird (which is olive-brown or olive-green above and ash gray below) is a smoky-brown or orange-brown spot on the crown which is bounded by a black stripe on each side. The flanks and lower tail coverts are olive-green. The wren thrush has a small, fairly flat, blackish beak and long, dark legs, a short tail with only ten feathers, and rounded wings in which the tenth (the outermost) primary is very small. The sexes look alike. This species was formerly placed with the thrushes (Chapter 12) as a primitive form of that group, but it seems to belong among the relatives of the wood warblers.

Family: wren thrushes, by A. F. Skutch

The wren thrush is found in the moist mountain forests from Costa Rica to western Panama at altitudes from about 1,500 to 3,000 meters, becoming more frequent at the higher levels. It keeps to the densest and most impenetrable undergrowth of the mountain slopes, ravines, or hides in the bamboo thickets near the higher peaks. There it hops or creeps about on the ground and the moss-grown rotting tree trunks, and hunts arthropods or other small prey. It often raises both wings while searching for food but rarely does it actually take flight. The wren thrush is extraordinarily difficult to observe; its nest was recently discovered, but not described.

Fig. 15-50. Bananaquit.

In museums of ethnology, one may occasionally see exhibits of chieftains' war cloaks, headgear, neckwear, or other exhibits from the Hawaiian Islands which consist of the bright yellow, red or green feathers of honey creepers. A full third of the distinguishable forms of these small birds are now extinct but this is probably less due to their having been hunted for feathers than to the reduction of the forest area of the islands to a quarter of the former extent. Honey creepers are not merely of interest to ethnologists, economists and conservationists, they also represent an attractive family of birds for biologists,

Family: Hawaiian honey creepers, by A. J. Berger and W. Meise

Fig. 15-51. Distribution of the wren thrush (family Zeledoniidae).

Fig. 15-52. Distribution of the Hawaiian honey creepers (family Drepanididae) only the Hawaiian Islands.

Subfamily: parrotbills

for on the remote Hawaiian islands, they have taken the places occupied elsewhere by sunbirds, tree creepers, woodpeckers, finches and tanagers.

Honey creepers (family Drepanididae) have only nine primaries, and, with the exception of two species, they are inhabitants of trees and bushes. The L is 11-21 cm; the tail ranges from fairly short to medium length. The legs are relatively short but strong, and this also applies to the toes and claws. The tongue is tubular in most birds, with a fringed tip. There is quite a variable degree of dilatation of the gullet to form a crop. Honey creepers are found only on the Hawaiian Islands in two subfamilies: 1. The PARROTBILLS (Psittirostrinae) and; 2. The BLACK-AND-RED HONEY CREEPER (Drepanidinae).

The parrotbills are green or grayish-green, particularly the females and young; males are yellow, orange, red, or in some species, even dull in color. The plumage is full and bushy. There are six genera with fifteen species: 1. *Viridonia* or *Loxops* males are green with yellow beneath. The beak is almost straight. There are three species, including the AMAKIHI (*Viridonia* or *Loxops virens*) which has a L of 12.5 cm and weighs 13 g. This species occurs in four subspecies on six of the generally large eastern islands of the group. The LARGE AMAKIHI (+ *Viridonia* or *Loxops sagittirostris*) became extinct because its habitat, on the slopes of Mauna Kea on Hawaii, was transformed into sugar cane plantations. 2. The HAWAIIAN CREEPER (*Paroreomyza* or *Loxops maculata*) has a less tubular tongue. It is basically an insect searcher on trunks and branches at middle levels of trees. There are six subspecies, some of which are almost entirely red. 3. The AKEPA (*Loxops coccinea;* Color plate, p. 372) is short-billed; the tip of the mandibles are slightly crossed. There are four subspecies. Males are reddish-orange (in one form often yellowish, in another only yellow to olive-green). 4. The HALFBILL *(Hemignathus)* has a long thin, strongly down-curved upper mandible; the lower mandible is typically shorter. The plumage color is olive-green to yellow. There are three species, one of them the AKIALOA (✝ *Hemignathus obscurus;* Color plate, p. 372) has a L of 18 cm. The lower mandible is almost as long as the upper and equally curved. At least one of the four subspecies is not yet extinct. The AKIAPOLAAU *(Hemignathus wilsoni)* has a straight, thick lower mandible only half as long as the upper. This bird hammers audibly on the tree bark with its beak open and fetches insects out of holes with its upper mandible; it never takes nectar. 5. The MAUI PARROTBILL (✝ *Pseudonestor xanthophrys*) has a L of 14 cm. The beak is much thicker and shorter than that of the preceding species. The tongue is less tubular. This bird is a very rare, endangered species. 6. PARROTBILLS *(Psittirostra)* have even thicker beaks. There are six species, including: the OU *(Psittirostra psittacea)* which has an upper mandible reaching far beyond the lower, and a somewhat tubular tongue; the LAYSAN FINCH *(Psittirostra cantans)* which has a L of 18 cm.

Its tongue, as in the following species, is more fleshy. This bird has strong legs, weak wings and a yellow breast. It occurs only on western atoll-like Laysan Island and on rocky Nihoa. Of the other four species of parrotbills, three were always rare, and two are now extinct. All four species occurred only on the main island of Hawaii. The two extinct species are: the GREATER KOA FINCH (+ *Psittirostra palmeri*) which had a L of 21 cm; and the GROSBEAK FINCH OR KONA FINCH (+ *Psittirostra kona*) which had a thick, grosbeak-like bill.

Fig. 15-53. Beak of the amakihi enlarged.

Although so many forms of honey creepers have become extinct, the mode of life of those which have become known is similar, in spite of the variety of shapes. All these birds breed between December and July; pair formation begins in October when the testes of the male enlarge. Their simple cup nests are built on branches except on Laysan, where they are placed in grass. The first young leave the nests at the end of January but young still appear in July and August, although second broods have not yet been proved to occur. The two to three white or bluish eggs which have fine reddish-brown spots, are incubated by the females only. Both parents feed the young, which scatter in the late summer and fall; many of the young settle new areas of forest. While adults go through a complete molt after the breeding season, the young have a slow partial juvenile molt.

Fig. 15-54. Beak of the Hawaiian creeper enlarged.

Most honey creepers inhabit tropical rain forests in which the ohia (*Metrosideros collina*), a tree related to the myrtle, predominates, and where the undergrowth of tree ferns may be four to six meters high. On Waialeale, a mountain on Kauai slightly over 1,500 meters high, an annual rainfall of over 15 meters has been recorded. These rain forests form ideal haunts for nectar and insect-eating birds like the apapane, akepa, iiwi and akiola. The two mamos,to be dealt with later, also used to occur in this type of forest. In the northwest of this area, there is an upland surrounded by densely wooded ridges and valleys. In this area, the Alakai Swamp, which is still relatively untouched, there are no less than nine species of honey creepers, as well as other rare varieties like two of the Hawaiian thrushes (Chapter 12) and members of the only surviving species of the ooaa (*Moho braccatus*, Peterson's field guide to western birds).

Fig. 15-55. Akepa.

Nowadays there is only one largely dry type of forest on Hawaii which serves as a habitat for many honey creepers. This is the "Mamane Naio Forest", the main trees of which are the mamane (*Sophora chrysophylla*) and the naio (*Myoporum sandwicense*). It lies between 1,800 and 3,000 meters on the southwestern slope of Mauna Kea, and thus extends to the present day tree line at its upper limit. Earlier, the forest extended higher but feral horses, cattle and goats have destroyed it there. Recently the akiapolaau was found in that area as an extremely rare bird.

The unusual amakihi is found not only in the dry area just de-

Fig. 15-56. Akiola.

Fig. 15-57. Nukupau *(Hemignathus lucidus).*

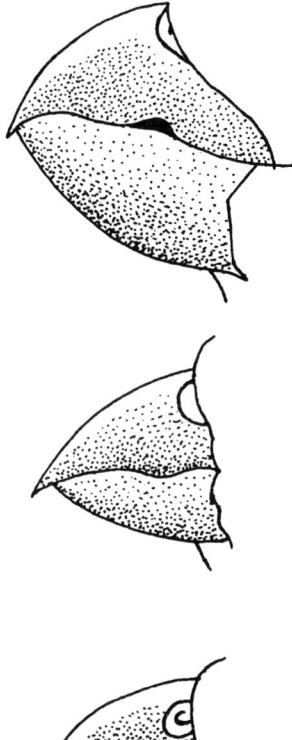

Fig. 15-58. Beak shapes of parrotbills, enlarged. Top—Koa finches; Middle—Laysan finch; Bottom—Ou.

scribed, but also in wet rain forests, mainly between 1,200 to 2,000 meters. It is found particularly in acacia parkland and it is still the most widely spread and thus the most successful species. In addition to arthropods, it also takes nectar, for which it flies not only into the ohia (the preferred nectar tree of all honey creepers, with its flat, usually red flowers), but also uses lobelias, the mamane, bananas, canna, pritchardia and the introduced fuchsias. It stabs long tubular flowers like those of the breadfruit tree *(Bryophyllum),* as its beak is short. To get at other plant juices, it bores into the cherry-sized fruit of the caroll tree *(Solanum pseudocapsicum)* but never swallows the whole berry. It searches for soft-skinned arthropods, generally of two to ten centimeters in length (especially caterpillars and plant lice, as well as spiders) in all layers of the bushy growth, but generally at middle height where it searches the leaves and twigs for prey. Sometimes it hops along the branches in the manner of a tree creeper.

As soft-skinned arthropods are available all year round, the amakihi is more resident than other honey creepers and thus several subspecies could readily evolve. Nevertheless, a male which had been moved experimentally found its way back to its territory from a distance of almost fifteen kilometers. This bird chases neighbors out of its territory very energetically, even when it has been away for weeks. Amakihis stake out territories of 0.4 to 1 hectar as early as the fall although they do not breed until rather late in the spring and early summer. Only males sing, with particular endurance in January and February; when excited they often insert the loud territorial call, a slow "tshe" trill into the song. The nests are sometimes only a meter above the ground.

The akepa with its slightly crossed bill, deals with the scales of buds in order to get at the juices in their interiors much like crossbills deal with the cones of conifers; however it feeds mainly on caterpillars. The akiola has a particular prediliction for the long curved flowers of the lobelia, but it also often looks for arthropods beneath bark or in rotting wood. Its pecking at bark often betrays its presence. The Maui parrotbill can effect a strong grasp with its beak to open the hard chambers of long-horned beetle larvae on the Koa acacia by crushing or rotation. It is found at about 1,700 meters on the island of Maui; it probably does not take nectar.

The ou, though now very rare, still occurs in the moist tree fern forests at altitudes of 800 to 1,200 meters on Hawaii where it gathers fruit in the tree tops. A good flier, it can visit trees in fruit, a stage which varies from tree to tree. Perhaps it frequently moves from one island to another; at any rate it has not formed any subspecies. Apart from fruit, it likes the female flowers of a climbing tree *(Freycinetia arborea)* which hold nectar. The lifestyle of the Laysan finch differs considerably from that of other honey creepers. It does not hop, but walks instead; it only flies for short distances and cowers beneath grass tufts. Its food is very varied, consisting of carrion, insects, roots, sprouts, soft parts of plants and seeds, but often also of bird eggs into which it bites a hole with its beak. Its cup-like nest contains no moss (there is none on Laysan) but grass and twiglets; it is placed in the grass or in small bushes. Like the ou, the Laysan finch sings pleasantly. In 1923 its population had been reduced to several dozen pairs. This remnant survived the destruction of the vegetation by introduced rabbits and guinea pigs, and the destructive storm of 1923, no doubt because this finch can use tern eggs as an alternative food supply. The Nihoan Laysan finch *(Psittirostra cantans ultima)* lives only on a volcanic islet of only 62.4 hectars in area. It has a special way of breaking tern eggs. The islet rises to 270 meters and has only very little flat ground; there are volcanic rocks of all sizes everywhere. The birds often peck at the undersurface of tern eggs; this sets them rolling and soon they fall thirty centimeters or more on to a rock and break. When this happens, a dozen or more of the birds fly to the spot, and since they all cannot eat the egg at the same time, there is often a fight. The now extinct greater Koa finch used to cling to the fruit of the mamane trees and get at the interiors through the dry husks; this caused a rustle audible from afar.

The black-and-red honey creepers (subfamily Drepanidinae) are mainly red and black; the young also have black spots or a black wash. Parts of the plumage are stiffer than in the other honey creepers; sometimes feathers on the head and other parts are longish and stiff. The primaries have square-cut tips in some species. There are five genera with six species. They include: 1. The APAPANE *(Himatione sanguinea;* Color plate, p. 372) which has a L of 13 cm. Males weigh 16 g while females weigh 14.5 g. The beak is much like that of the parrotbills; the plumage is red with black wings and tail, the belly is whitish. The Laysan subspecies (+ *Himatione sanguinea freethii)* had a shorter bill and a lighter shade of red; it became extinct about 1923. 2. The ULA-AI-HAWANE (+ *Ciridops anna)* now extinct, had a short, fairly thick beak and a gray nape; it used to occur on Hawaii. 3. The IIWI *(Vestiaria coccinea;* Color plate, p. 372) has a strong, beak which is down-curved, the sexes are similar in color, being red above and on the belly, and white on the inner wing. It is fairly common on all the main islands. Males weigh 20 g while females weigh 17 g. 4. The MAMOS *(Drepanis)* lack lancet-like feathers. There are two species. The MAMO (+ *Drepanis pacifica;* Color plate, p. 372) had a long down-curved beak, black plumage; it was yellow on the lower back and in other areas. This bird used to live on Hawaii; it is now extinct. The BLACK MAMO (+ *Drepanis funerea)* also extinct, used to inhabit Molokai; it had an even longer beak and an almost entirely black plumage.

Fig. 15-59. Apapane.

▷
Goldfinches and linnets: 1. Red siskin *(Carduelis cucullata);* 2. American goldfinch *(Carduelis tristis);* 3. Common redpoll *(Acanthis flammea);* 4. Linnet *(Acanthis cannabina);* 5. Siskin *(Carduelis spinus);* 6. Twite *(Acanthis flavirostris);* 7. House finch *(Carpodacus mexicanus);* 8. Greenfinch *(Carduelis chloris);* 9. Scarlet grosbeak *(Carpodacus erythrinus);* 10. Goldfinch *(Carduelis carduelis).*

Subfamily: black-and-red honey creepers

Fig. 15-60. Beak of the ula-ai-hawane (enlarged).

Fig. 15-61. Beak of the iiwi (enlarged).

◁
1. Evening grosbeak (Hesperiphona vespertina); 2. Hawfinch (Coccothraustes coccothraustes); 3.. Chaffinch (Fringilla coelebs); 4. Japanese grosbeak (Eophona personata); 5. Japanese bullfinch (Pyrrhula pyrrhula griseiventris); 6. Chinese bullfinch (Pyrrhula erythaca); 7. Brambling (Fringilla montifringilla); 8. Gray-crowned rosy finch (Leucostycte arctoa); 9. Azores bullfinch (Pyrrhula pyrrhula murina).

Apart from the ula-ai-hawane, the black-and-red honey creepers live mainly on nectar. The apapane is presently the one member of this family which is most conspicuous on Hawaii. It is not the most common of these creepers, but its flocks are very lively when they are busy on flowering trees. The blunt tips of their wings give rise to a noise in flight which probably supports the cohesion of the flock. These flocks often cover considerable distances between one flowering tree and another, flying high above the tree tops. When such a flock approaches a human observer, he will note a musk-like effluvium which belongs to all honey creepers and which persists in museum skins for decades. Sometimes one may see forty to fifty apapanes per hectar, particularly in the ohia forests, where they often join into flocks of thousands when the young have flown. In addition to ohia blossoms, they also work over fuchsia and mamane trees. Arthropods play only a minor role in their diet; they are not caught in a tree creeper-like manner, but picked off flowers and branches. After storms, not infrequent on the islands, one occasionally finds dead apapanes washed up on the shores. An inclination to travel probably also explains why there are no subspecies of this bird. Apapanes sing the whole year through, but most intensely in December. The song is long and although monotonous, sounds pleasant. The male also sings near the nest and thus drives away rivals, yet it permits several members of its own species to search for food on the same tree.

The extinct Laysan apapane used to sip nectar from particular flowers on the ground, but it also took moths, which it held with one foot, as well as caterpillars. Like the Laysan parrotbill, it was a peculiar inhabitant of this island, a paradise for albatrosses and other sea birds. In 1923 the tanager expedition found that only three of these birds were left and all of them perished in a sand storm which lasted three days. The ula-ai-hawane, which is also extinct, used to feed on the seeds of the pritchardia palm and also on arthropods.

The iiwi behaves like the apapane and also produces a sound by its flight. Its long beak permits it to get nectar even from deep bell-shaped flowers, which it does not need to stab, although it occasionally does so. Because of its long beak, it visits sophora flowers more frequently than do the apapane and the amakihi. Possibly its down-curved beak was an adaptation for use on bell-shaped flowers; these have greatly declined in numbers on Hawaii. Sometimes the iiwi's stomach is found to contain only nectar. This bird prefers cool, moist forests, particularly between two and three thousand meters on Mauna Loa and Mauna Kea, where it also stays through the winter. It is rather hostile against others of its species or even family which visit the same tree. Throughout the year, both sexes utter a sharp, penetrating croak which is only rarely repeated to form a song.

P. H. Baldwin describes an unusual observation of iiwis. "Several times I have noticed a loud, continuous noise from flocks of these birds which were flying into a sophora tree just breaking into bloom. About two or three hundred of the birds gathered in a copse of only twenty hectars. It seemed that display and pair formation were in progress. The gathering was maintained for the whole breeding season until March, but I found no nests in the copse or near it." This was probably an exceptional case; possibly the population density was so high that breeding was inhibited. In other areas, breeding pairs of these birds keep apart.

Fig. 15-62. Beak of the black mamo, (above) and the mamo (below), enlarged.

The mamos had almost the largest beaks of the honey creepers, beaks with which they were able to dip into the deepest flowers, like those of the lobelia which are five centimeters deep. They evidently obtained water from the moss on trees with rapid tongue movements while searching for insects there.

Along with Darwin's finches and a few other groups, the honey creepers are an outstanding example of a branching evolutionary tree developing from a bird which reaches a group of islands not yet occupied by comparable species. Certainly the Hawaiian Islands have been forested since about five million years ago and the original honey creeper must have been a forest bird because its present day descendants still mainly inhabit upland forests and only rarely visit the warm dry lowland forests, in spite of the occurrence of ohia trees there. The original honey creeper probably had a tubular tongue with a fringed tip and only nine primaries. Such birds are only found among the bananaquits and the dacnis; possibly the icterids might also have given rise to such a type. As bananaquits are American and their tongues are most like those of the honey creepers, they are, in the opinion of certain ornithologists, their closest relatives.

Fig. 15-63. Mamo.

Immigration was followed by the formation of subspecies and species, although the eastern islands, which were presumably settled first, are uniform in the composition of their forests and probably offered similar foods. The differences which evolved on different islands, or different sides of mountain ranges, became intensified when the various forms met again. Foods and the shape of the beak and tongue (which are related to food preferences) became different, as did colors of plumage, which plays a role in preventing hybridism. Thus the greater koa finch had a bright orange head while the lesser koa finch, which lived alongside it, was yellow-headed. When the evolutionary tree forked, the ancestors of both subfamilies were still short-billed; long-beaked species arose in both branches but only in the parrotbills branch did they become pure insect eaters, in part, or thick-billed species able to crack the hardest seed. Thus, in the honey creepers, evolution proceeded from thin-billed to thick-billed birds, the reverse direction from that shown by the Darwin's finches; it also differed from events in the formation of the latter in that no hybrids of honey creeper species are known.

Fig. 15-64. Distribution of the Vireos (subfamily Vireoninae), genus *Vireo* (vertical lines) genus *Hylophylus* (horizontal lines).

Family: vireos, by H. Schifter

Fig. 15-65. Peppershrikes (genus *Cyclarhis*), horizontal lines. Shrike-vireos (genus *Vireolanius*), vertical lines.

Fig. 15-66. Red-eyed vireo.

Because of the hook at the tip of the beak of most vireos (family Vireonidae), these birds were formerly called leafshrikes and placed close to the shrikes; however, more recently, they have been regarded as related to the wood warblers. The L varies from 10 to 17 cm. The plumage is inconspicuous in color, being generally greenish or gray, and there are no plumage sex differences. The food consists predominantly of arthropods. The nests are open cups fastened in the forks of branches. Vireos lay two to five eggs, and these are either white all over or show spotting in varying intensity. The incubation period is twelve to sixteen days, the fledging period ten to fifteen days. Fully fledged young resemble the adults. Vireos occur only in the Americas.

We distinguish three subfamilies with four genera: A. The VIREOS (Vireoninae) include: 1. The VIREOS in the restricted sense *(Vireo)* with twenty-five species, and 2. The GREENLETS *(Hylophylus)* with thirteen species. B. The SHRIKE VIREOS (Vireolaniinae) with one genus, SHRIKE VIREOS *(Vireolanius)*, and three species; C. The PEPPERSHRIKES (Cyclarinae) with one genus, PEPPERSHRIKES *(Cyclarhis)*, and two species.

The vireos in the restricted sense are the best known and are at home in Central and North America. In these birds it is predominantly the female which builds the nest, but the male sometimes helps; he takes part even more in the incubation and feeding of the young, which are born naked. The RED-EYED VIREO (*Vireo olivaceus*; Color plate, p. 372) is the most common North American vireo of the eastern woodlands. The more grayish WARBLING VIREO *(Vireo gilvus)* is also widely distributed but in mighty trees to which it keeps; it only gives itself away by its constantly delivered song. All vireos which nest in Canada and the United States winter in Mexico and Central America, but the red-eyed vireo wanders as far as the upper Amazon and has been found as a vagrant in Germany. As Alexander F. Skutch has observed, Central American vireos return to Costa Rica as early as February and the males begin to sing right away. In March nest building is already in full swing. Vireos do not generally arrive in the United States until April and they stay until September. After the molt in summer, they often resume their song until their southward departure.

The greenlets are more delicate (the L is 10–12 cm) and are predominently green. Only three species range northwards as far as Central America. One of these is the GRAY-HEADED GREENLET (*Hylophylus decurtatus*; Color plate, p. 372) which wanders restlessly through the foliage like a chickadee. It prefers the upper levels of rain forests. While searching for food it sometimes hangs head down from leaves or climbers. Its song is pleasant but weak in volume. The clutch evidently always consists of only two eggs.

The shrike vireos are marked by laterally compressed beaks. The CHESTNUT-SIDED SHRIKE VIREO *(Vireolanius melitophrys)* is a little-known bird of oak forests from Mexico to Guatemala, where it is found above 3,000 meters. It moves at a leisurely pace through the dense foliage, now and then seizing an arthropod which it shreds with the beak while it holds it clamped to a twig with a foot.

The olive-green RUFOUS-BROWED PEPPERSHRIKE (*Cyclarhis gujanensis*; Color plate, p. 372) is found from Mexico to northwestern Argentina in a number of subspecies. Pairs of these birds are characteristic of open areas and gardens in many places in Brazil where, according to E. G. Snethlage, they breed all year round. The nest is a little pouch woven from very fine fibers and suspended from the fork of a branch. With their strong beaks, these peppershrikes seize not only the most varied arthropods, but also their cocoons. This is probably the only vireonid which is occasionally kept in zoos or by cage bird fanciers.

The icterids (Icteridae) are a characteristic New World family of birds of unusual diversity. They are of tropical origin, but belong to the everyday features of American bird life in general, although most species still occur in South America. They probably only invaded North America in the Pleiocene (in the late tertiary about one to eleven million years ago) when the Central American land bridge became complete; they then developed a number of new forms in North America. The oft made statement that icterids replace the Old World starlings in North America is applicable to only a few genera.

Family: icterids, by E. Schäfer

Icterids or American orioles and blackbirds, range in size from that of a finch to that of a crow; the L is 17–55 cm. The beak is conical, pointed, laterally compressed and often shows a straight line along the ridge of the upper mandible, and along the lower edge of the lower. The ridge of the upper mandible is often prolonged onto the forehead, sometimes with a frontal shield; only rarely is it somewhat curved downward or short as in a finch. The legs are strong and the wings are of medium length. Many species have some bright yellow feathers, hence the family name (*icterus* means jaundice). The most common main color is black, but large areas of red, orange or rust color often produce beautiful effects. Males are larger than the females, but often similar in color; in such cases, the females sing as melodiously and variably as do the males and often in alternation with them. Generally, these birds are tree dwellers and are very energetic and sociable. The few species which live on the ground do not hop but walk instead. Their diet is mixed; nestlings are fed mainly on arthropods but adults are mainly plant feeders.

There are approximately twenty genera with about ninety-three species; we will mention the following: 1. The OROPENDOLAS (genera *Psarocolius, Gymnostinops, Cacicus* and *Amblycercus*) have a L of 38–55 cm. These are almost the largest song birds and they have a horny frontal plate. 2. The GIANT COWBIRD (*Scaphidura oryzivora*; Color plate, p. 381) has a purple-black iridescent plumage. 3. COWBIRDS (genus *Molothrus*) are starling-sized birds; the males are generally black, and the females are brown. The beak is finch-like. 4. GRACKLES (genus *Quiscalus*) vary

Fig. 15-67. Distribution of the icterids (family Icteridae).

in length from 25 to 43 cm; males have a metallic gloss, and females are dark brown. The iris is bright yellow. 5. The AMERICAN ORIOLES (genus *Icterus*). 6. AMERICAN BLACKBIRDS (genus *Agelaius*) range in length from 20 to 22 cm; the males are colorful, and the females brown and striped. 7. MEADOWLARKS (genus *Sturnella*) are long-beaked, and partridge-like in color with a light yellow breast. 8. The BOBOLINK (*Dolichonyx oryzivorus*; Color plate, p. 381) has a L of 18 cm and is short-beaked. The male is mainly black with a light colored nape and yellowish stripes on the back; however outside the breeding season it is colored like a sparrow, as are all female plumages.

In North America, many of these birds have adapted to cultivated landscapes, and because of the harsher climate, they are generally migrants there. Polygyny is frequent in icterids; a number of other species live in loose reproductive associations with several females. Only a few genera have monogamy. The brood parasitic cowbird species usually do not have a pair bond. Courtship displays include deep bows and erection of plumage. Nests are very variable; some are flat saucers placed on the ground; others are disorderly domed structures, and still others are well-woven pouches. In contrast to the Old World weaver-finches, wherein the males are nest builders, only the females work on hanging icterid nests, with the exception of a few North American species in which the male also helps. With the exception of the brood parasites, to be mentioned later, the females of tropical species only lay two eggs; however, those which breed in temperate North America lay four to six eggs. The eggs show dark spots and hairlines on a background which is generally greenish or bluish. In those species which breed colonially, only the females are concerned with rearing the young, while the larger males serve only as guards.

The most colorful group of icterids is the Central and South American oropendolas. Their graduated tails, which can be fanned out, are decorated with yellow signal feathers. Many of these birds exude a strong, crow-like odor. Their breeding colonies of up to fifty pouch-shaped nests are often arranged on isolated trees like the figures on a merry-go-round. The displaying males are fascinating to watch as they bow deeply, raise their rapidly trembling wings rhythmically, and beat them together over the back. A number of them perform cartwheels hanging from nests already started, in order to urge their numerous females to more energetic nest building. The gutteral sounds and abruptly ended fluted fanfares of the oropendolas are almost symbolical of the magic of the New World tropics. The males on guard relieve one another, give warnings in croaky voices, or follow the females into the undergrowth where, far from the breeding colony, matings take place. A strict rank order is maintained among the males and only this guarantees their peaceful coexistence in the colony.

Construction of the nests, which are woven from plant fibers and may be up to a meter and a half long, takes fifteen days on the average. The incubation period is ten to fourteen days, the fledging period lasts nineteen to thirty days, and the entire breeding season of a colony lasts four months. After leaving the colony, the family groups wander about the country like noisy gypsies. Their favorite haunts are coffee and cocoa plantations with tall shade trees, as well as recently cleared land and moist forest edges. Adults feed mainly on tree fruit, but they also sip nectar.

The widely distributed CRESTED OROPENDOLA *(Psarocolius decumanus)* is glossy black with an ivory-colored bill, a bright blue iris and lemon yellow outer tail feathers. In the higher mountain forests, it is replaced by the inconspicuous greenish-yellow RUSSET-BACKED OROPENDOLA *(Psarocolius angustifrons;* Color plate, p. 381). The smaller YELLOW-RUMPED CACIQUE *(Cacicus cela;* Color plate, p. 381) is rather different in its nest building and display behavior. It lives in drier areas and places its breeding colonies by preference on the shores of rivers or in the protective vicinity of ant or wasp nests. Where the range of the three species overlap, there are mixed breeding colonies, which probably arise because the sight of the pouch nests of one species attracts prospecting males of another. Hybrids have not so far been observed.

The giant cowbird is often a brood parasite of oropendolas. Female cowbirds enter the breeding colonies of their hosts with great obstinacy for they are received with marked hostility. The males of the two species, however, take no part in these altercations. Giant cowbirds free cattle of ticks or roll over stones and cattle droppings in their search for insects. Males walk about, as if very pleased with themselves, in imposing display postures in which the down-pointing beak is largely hidden by a collar of erected feathers and the birds bob up and down and bow. Females are more numerous than males in these species as well; the sex ratio is 4:1. So far, giant cowbirds are known to parasitize eight species of birds as hosts for their eggs. The female lays five to ten eggs, one after the other, into nests of different hosts.

Brood parasitism in cowbirds is evidently a relatively recent development. Evidence for this comes from various stages of its development in the five species shown by Friedmann's studies of their life histories. All are highly sociable birds which have evolved in South America. At the start of the series leading to parasitism is the BAY-WINGED COWBIRD *(Molothrus badius)* in which the tendency to parasitism is merely suggested. These birds live in monogamy and in exceptional cases, still build their own nests, but they do not breed until plenty of the nests of other species are available. They acquire such nests by fighting for them; they also defend them and then lay their eggs in them. However, they incubate their eggs in them and rear their young without any help from host birds.

Fig. 15-68. Bay-winged cowbird *(Molothrus badius).*

Fig. 15-69. Bay-winged cowbird.

Fig. 15-70. Brown-headed cowbird *(Molothrus ater)* interrupted lines indicate wintering areas.

Fig. 15-71. Brown-headed cowbird.

The next stage is that of the SHINING COWBIRD (*Molothrus bonariensis*; Color plate, p. 381), a widely distributed follower of cultivation in South American cattle breeding areas. In this species, too, some evidence of the nest building instinct can be shown, although it is subthreshold and leads to no definite result. Most females simply lay many eggs, which are highly variable in color and shape, on the ground, where they perish. Others, however, lay several eggs into the nest of a host bird. In one case, as many as thirteen different females laid a total of thirty-seven eggs into a single "strange" nest. Female shining cowbirds often destroy not only host eggs, but those of their own species too, with blows of the beak. Even though they have been shown to parasitize over a hundred species of other birds, their brood parasitism is ill-organized. Once the young parasite hatches, however, it has a very healthy appetite and doubles its weight on the very first day. Usually it becomes the sole occupant of the nest because it displaces its nest mates. It leaves the nest after ten to fifteen days, but is fed for a further two weeks by its host parents.

The highest stage of cowbird brood parasitism is shown in temperate North America by the widely distributed BROWN-HEADED COWBIRD *(Molothrus ater)*. Display behavior and song are most highly developed in this species. As the ratio of males to females is 3:2, or an even greater excess of males, the most variable relationship between the sexes ranging from strict monogamy to mere mating contacts without any pair bond are found. In contrast to the European cuckoo this cowbird has not specialized itself to use particular host species. According to Friedman, no less than 185 species have acted as hosts, but those which build open nests like tyrant flycatchers, vireos, wood warblers and New World sparrows are preferred. The female cowbird usually lays only one egg in any one host nest. She lays at an interval of twenty four hours and lays four to six eggs altogether, but this number may be increased if suitable host nests are available. Cowbird eggs are always larger than those of the host and hence are more easily incubated. As the incubation period is only eleven to fourteen days, the cowbird often hatches before the young of the host, which are therefore often squeezed to death or starved. In some cases, the host birds seem to realize something is wrong, for they may build a new nest floor over the cowbird egg and a repetition of this process may lead to a sort of "high rise" nest. The prairies were the favorite haunt of these cowbirds, which used to work over buffalo for their external parasites as they do now on cattle. Because these birds occur in great flocks, which in the fall roost in reed beds and are partial or true migrants, they have been compared with the Old World starling.

Grackles are rather similar to cowbirds and are also widely distributed from South to North America. Breeding colonies of these noisy birds are often right in settlements, in gardens, parks or tree-grown open squares. While the males are engaged in noisy display, the females build disorderly nests from twigs and plant fibers to which they sometimes add cow dung or mud. Nest construction takes five to ten days, incubation takes thirteen to fourteen days, and the fledging period takes eighteen to twenty-three days. The COMMON GRACKLE (*Quiscalus quiscula*; Color plate, p. 381) of eastern and central North America invades corn and grain fields, in its fall mass migration, much to the chagrin of the owners. The SOUTH AMERICAN CARIB GRACKLE (*Quiscalus lugubris*) is a resident which follows cultivation and makes free use of banana plantations. Its breeding colonies of thirty to a hundred nests are started before the beginning of the rainy season and they are often in the immediate vicinity of human habitations, on squares or in yards surrounded by walls. The "period of engagement" of these birds may be long, for egg laying does not start until the rainy seasons with their abundance of arthropods, set in.

The colorful American orioles are widely distributed over the American mainland in about fifty species and subspecies. Most species live in monogamy and the males take part in feeding the young. The most common North American species east of the Rockies is the BALTIMORE ORIOLE (*Icterus galbula*; Color plate, p. 381) which has a L of 20 cm. Males are deep black on the back, neck, and throat, but fiery orange below and on the lower back. They also have a white wing bar. Near human habitations, Baltimore orioles often use textile fibers in their nest construction. Females lay four to six eggs; incubation and fledging periods are each fourteen days. On migration, these orioles often invade orchards and they are, therefore, regarded as pests. Their wintering area extends to northern South America. In the western United States, this species is replaced by the more simply colored BULLOCK'S ORIOLE (*Icterus bullockii*).

However, most American orioles are native to Central and South America. The gold and black TROUPIAL (*Icterus icterus*; Color plate, p. 381) has a L of 30–35 cm. It is a favorite cage bird because of its ability to imitate other birds and even the human voice. Many tropical orioles, such as the ORANGE-CROWNED ORIOLE (*Icterus auricapillus*), the YELLOW-BACKED ORIOLE (*Icterus chrysater*) and the YELLOW ORIOLE (*Icterus nigrogularis*) hang fine-meshed pouched nests near water from the central rib of a banana leaf into which they have pecked holes. Consequently, corrugated iron roofs and shiny tarred roads, which suggest water to these birds, often act as foci for nest building and nests are often placed on houses or above roads. All tropical orioles lay only two eggs.

American blackbirds are the most sociable of all the icterids. The RED-WINGED BLACKBIRD (*Agelaius phoeniceus*) is widely distributed in eastern North America. Males are black with bright red wing coverts, while females are uniformly striped brown. Males first occupy territories early in the spring, females follow later. They breed in colonies,

Fig. 15-72. Eastern meadowlark *(Sturnella magna)*.

probably generally in pairs but also in polygyny. Their cup-shaped nests are often attached to reeds over water. A full clutch consists of four or five eggs; both the incubation and fledging periods last eleven to twelve days each. Males neither build nor incubate, but they take part in feeding the young. After the breeding season they migrate to the south in huge swarms and often invade grain fields. The SOUTH AMERICAN YELLOW-HOODED BLACKBIRD *(Agelaius icterocephalus)* and the YELLOW-WINGED BLACKBIRD *(Agelaius thilius)* lead very similar lives. According to Chilean tradition, the country owes its name to the characteristic call, "tshee-lai" of the latter bird.

While some of the icterids described so far are regarded as harmful to human interests, meadowlarks are appreciated as insect destroyers. They too occur both in North and South America. They favor prairies and moist meadows; only the WESTERN MEADOWLARK *(Sturnella neglecta;* Color plate, p. 381) of western North America, west to the Pacific, favors arid country. Meadowlarks are decidedly ground birds. They live mainly in monogamy, but a male may have several females in its territory. Both sexes give jubilant whistles responsible for the name meadowlark. Females build roofed-over nests with a side entrance in tall grass. In contrast to other North American icterids, meadowlarks are only partial migrants, and move south only in case of frost or deep snowfalls.

The bobolink, on the other hand, is a pronounced migrant which also lives predominantly in monogamy. However, when the males appear in the nesting areas in advance of females early in spring, they perform collective display flights in which they utter their simple but highly variable song. Nests are well hidden and are often placed close together in high grass. The full clutch consists of five to seven eggs. Bobolinks climb about in the grass stalks as skillfully as do Old World reed warblers. On their migration, they cross the Caribbean in untold millions, invade the rice fields in Colombia and Venezuela, and then move on to Brazil and Argentina. As early as the nineteenth century, laws for their destruction were passed in the United States. Once they had laid down their premigratory fat in the fall, these birds used to be caught in the southern United States and offered on local markets as "reed birds."

16 The Finch Family

Finches (family Fringillidae) differ from buntings in most cases by the absence of an angle in the lower mandible of the beak. Males feed the incubating females. They are found naturally in all parts of the world except Madagascar and Australia-Oceania, where a few species have, however, been introduced. There are few finches in southern Asia. There are two subfamilies: 1. Chaffinches (Fringillinae), and 2. Relatives of the Old World Goldfinch (Carduelinae).

The first differ from birds of the second tribe only slightly in structure and behavior, but a major difference is that the chaffinches feed their young entirely on invertebrates, mainly caterpillars which they bring to the nestlings in the beak, instead of by regurgitation of plant food. Their L is about 15 cm and they range in weight from 17 to 30 g. They are found only in the northern Old World. There is one genus with three species: 1. The CHAFFINCH *(Fringilla coelebs)*; 2. The CANARY ISLAND CHAFFINCH *(Fringilla teydea)*, bright lead-blue above and pale ashy grey below, it occurs on the Canary Islands of Teneriffe and Gran Canaria. 3. The BRAMBLING *(Fringilla montifringilla)* a pronounced migrant.

The chaffinch occurs throughout the western part of the northern Old World and is constantly spreading further eastwards. It is one of the best known European birds and breeds wherever there are large trees in deciduous and coniferous woods, parks, gardens and hedges with trees. The nest is often placed on a branch of a tree or shrub and is a firm cup, often camouflaged externally by lichens. In almost all of Europe the usual clutch is three to six eggs; second broods in the same year are frequent. In winter chaffinches feed mainly on cultivated fields, on grain and weed seeds and in some years also on beech nuts in the woods. When the ground is snow-covered, they often gather in farm yards where they make full use of food put out for birds. Many chaffinches from the northern and eastern part of their range migrate for the winter to southern and western Europe. In some areas the mi-

The finch family, by I. Newton

The chaffinch tribe

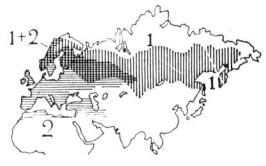

Fig. 16-1. 1. Brambling *(Fringilla montifringilla)*; 2. Chaffinch *(Fringilla coelebs)*.

gratory habits of the two sexes differ. The scientific name, *coelebs* (without marriage), given to the bird by the Swedish systematist Linnaeus, derives from the fact that almost all chaffinches wintering in Sweden are males. Banding has shown that the females of the German and Scandinavian breeding populations migrate further than the males; thus in the winter there are more males in Scandinavia, Britain and some parts of central Europe. However, more females than males spend the winter in Ireland.

Chaffinch songs, by H. Wendt

Young male chaffinches tend to imitate the songs of adults in their neighborhood. Because of this, breeders and fanciers of finches, formerly more numerous than now, would distinguish different songs, each of which ends in a particular sound pattern. As Brehm stated in the last century, "this matter has become a science of its own which has its own devotees and which will always remain mysterious to one not instructed in its secrets." At one time finch fanciers in Thuringia, the Harz Mountains and upper Austria were famous because they could identify more than twenty song patterns. One finch variety with a particularly loud song was called "blower out of the light". Chaffinches thus have song dialects. A finch which sang excellently and distinctly in one of these dialects fetched a good price.

Fig. 16-2. Display behavior of the chaffinch.

The Canary Island chaffinch is restricted to the mountain pine forests of its home islands and has already become rare. It usually breeds in June and lays only two eggs. It feeds on seeds, particularly those of the pines, and on invertebrates, particularly butterflies. It was probably the first type of chaffinch to settle the islands and was later displaced by immigrant common chaffinches into its present mountain home. Descendants of the more recently arrived common chaffinches now inhabit the uniform deciduous lowland forests of the islands.

The brambling takes the place of the chaffinch in the almost polar, subarctic birch woods and in the northern part of the Taiga. Both species have similar breeding habits, but bramblings usually lay six or seven eggs. It winters in the same area as the chaffinch, with their flocks often mixing. It has a stronger preference for beech nuts and tends to occur in great numbers in different areas in different years, depending on the supply of beech nuts.

In central Europe, bramblings on migration are often seen together with chaffinches, linnets, buntings, and greenfinches. According to snow conditions, they wander irregularly from one feeding place to another. In the hard winter of 1946/47 an estimated eleven million bramblings were feeding on the abundant beech mast at Porrentruy in Switzerland. Every night they gathered in a small valley to roost. Like all bird wanderers from the north, they are at first very tame towards people, but, like chaffinches, rather cantankerous towards members of their species.

The second tribe of finches, those related to the Old World goldfinch *(Carduelinae)* also have, with only one exception, nine primaries. In L they range from 10 to 23 cm and in weight from 8 to 100 g. Their beaks are conical, suitable for the cracking of seeds, for they also have strong jaw muscles. The gullet is readily stretched and seeds are stored in the crop. The gizzard, where seeds are ground up with the help of small stones, is strong. In contrast to the chaffinches, they feed their young by regurgitation, mainly of seeds. Plumage colors are variable; red, green, yellow and brown prevail. Females are usually the same size as males, but duller in color. The distribution is the same as that of the whole family. There are twenty-eight genera with about 122 species. These include: A. The Scarlet Grosbeak tribe, which includes SERIN FINCHES (genus *Serinus*); GOLDFINCHES and SISKINS *(Carduelis)*; LINNETS *(Acanthis)*; ROSY FINCHES *(Leucosticte)*; TRUMPETER BULLFINCHES *(Rhodopechys)*; SCARLET GROSBEAKS *(Carpodacus)*; BONIN FINCHES *(Chaunoproctus)* with only one species, the BONIN FINCH *(Chaunoproctus ferreirostris)*; PINE GROSBEAKS *(Pinicola)*; and CROSSBILLS *(Loxia)*. B. BULLFINCHES (genus *Pyrrhula*). C. HAWFINCHES (the genera *Coccothraustes, Hesperiphona, Mycerobas,* and *Eophona.*

Tribe of the goldfinches and relatives

Fig. 16-3. Distribution of goldfinches and related finches (tribe *Carduelinae*), introduced among other localities in Australia and New Zealand.

Many of these have a pleasant melodious song which they utter almost all year round. A few are widely distributed and are also common near human habitations. Because of their pleasant ways and attractive songs, they are favorite cage birds, particularly in Europe and Africa. Almost all species live in areas where there are trees and bushes; only a few live in deserts, in rocky country or in the tundra. They are almost always seen in flocks, even in the breeding season. They fetch their food either from the ground or, more frequently, directly from plants. A number of species are adept at clinging to even the thinnest twigs or weed stalks, even hanging upside down from these.

Some species breed in loose colonies, while others do so in pairs. Group breeding is commonest in the linnets, serin finches, goldfinches and siskins. Thus, in the case of the COMMON LINNET *(Acanthis cannabina),* several dozen pairs may breed together, several nests being only a few meters apart from one another. In England, over twenty pairs of HAWFINCHES *(Coccothraustes coccothraustes)* were once found breeding in a single loose colony. GREENFINCHES *(Carduelis chloris)* and TWITES *(Acanthis flavirostris)* often breed in groups of up to six pairs, but in the other European species, these groups rarely exceed three pairs.

The nests of the birds of this tribe are generally placed in trees or bushes at varying heights, depending on the species. Nests of moorland dwelling twites are often very near the ground, while those of woodland dwelling SISKINS *(Carduelis spinus)* are in the crowns of tall trees. Those species which live in deserts or rocky areas generally place their nests on the ground between rocks. In all species, the nest is an open cup and is built only by the female, although the male often keeps

Fig. 16-4. Goldfinch in defensive posture.

her company. Nest materials vary with the species, but are usually grass, roots, and moss. The lining always consists of finer material than the outer layer of the nest wall. The nests of some GOLDFINCHES *(Carduelis carduelis)* and siskins are beautiful little constructions of moss, lichens and the down from thistle seeds woven together with the help of spider webs.

In most species, the usual clutch consists of four to six eggs with a whitish, pale blue or green background color, with reddish, brown or blackish spots. Incubation lasts eleven to sixteen days and is followed by a fledging period of the same time, unless the young are disturbed; then they leave the nest earlier. Only females incubate, depending upon the males to feed them on the nest. During the first few days after hatching, the male parents are still able to bring all the food required, but both parents feed the young when they are older and already have some feathers. Almost all species breed in spring and summer and rear more than one brood a year.

The young are fed mainly on seeds, but caterpillars and some small invertebrates are added. As the young grow up, the proportion of animal food is increased. However, crossbills and linnets feed their young only on seeds, which may also happen occasionally in other species. The young receive a considerable amount of food every twenty to sixty minutes, which the parent bird regurgitates and which the young store in their capacious gullets. The skin of the neck of the young is so transparent that one can readily see and identify the food in their gullets without harming them. This stored food does not seem to have been predigested by the parents; the individual particles are merely joined to one another by a little mucus. Together with food, the nestlings are also given gravel and water by the parents.

In at least four genera, the bullfinches, trumpeter bullfinches, rosy finches and scarlet grosbeaks, during the breeding season adults have a special pharynegeal pouch beneath the floor of the mouth in which they store food for the young. It is not known whether the others have such a pouch or merely use the distensibility of their gullets for this purpose. Possession of a storage organ which can hold a considerable amount of food enables the birds to gather food at some distance from the nest. This would not be possible if, like the chaffinches, they could only bring a few items at a time in the beak.

In most species, breeding is followed by a full molt in the adults. At the same time, the contour feathers and small feathers of the wings and tail are replaced in the young, but they keep the large wing and tail feathers for a further year. In California, the HOUSE FINCH *(Carpodacus mexicanus)* takes 105 days to complete its molt, but in England, according to my observations, the greenfinch and bullfinch take eighty-five days, the goldfinch eighty, the linnet and twite seventy and the COMMON REDPOLL *(Acanthis flammea)* fifty days. In England, species which are migrants tend to start molting earlier and to complete the molt in a shorter period than those which are residents. Thus, the migrants are ready for their move about the end of September or early October. In some species, including the goldfinch, adults have a partial molt in spring and thus acquire the nuptial plumage.

The wanderings of finches of this tribe are often grand spectacles. Bullfinches banded in Russia have been recovered over 1,000 kilometers away. In the northern hemisphere most species migrate in at least part of their range, although no place is entirely deserted by all members of the population. In the northeast, where the winters are more severe, the proportion of migrants is greater. Some species of the boreal forests of Eurasia and North America migrate only every few years at irregular intervals brought on by food shortage in the north. During such wanderings, they often form huge flocks and reach districts outside their normal range. Crossbills are particularly well known for these irruptions, but in the northern Old World they have also been noted in common redpolls, the pine grosbeak and the siskin and bullfinch and in the New World in PINE SISKINS *(Carduelis pinus)* and EVENING GROSBEAKS *(Hesperiphona vespertina)*. All these species live mainly on the seeds of conifers except the pine grosbeak, which lives on those of the mountain ash; the amounts of these seeds fluctuate from year to year.

The finches are a highly interesting group for the student of evolution, for they show considerable differences in beak structure and feeding behavior. The greenfinch, linnet and redpoll have relatively short, wide beaks and they feed largely on plants which have exposed seeds, grasses, or seeds which are in capsules. They also pick up many fallen seeds from the ground, but each species prefers seeds of a certain size. On the other hand, the goldfinch and siskin have relatively long, narrow beaks and get much of their food by searching the cones of conifers and the flower heads of compositae. The goldfinch, which has the larger beak, eats more seeds of composite flowers, the siskin more tree seeds. The bullfinch and hawfinch get much of their food by breaking open fruits and seeds, discarding the outer covering in order to get at the seeds. There are also differences in feeding postures; goldfinches, siskins, redpolls and crossbills are best at clinging and hanging; accordingly they feed mainly on trees and weeds. Others which are less efficient at clinging obtain almost all their food on the ground or on firm horizontal branches.

Some of the goldfinch tribe use their toes to clamp food to any available base while they work them over with the beak. Siskins, redpolls or goldfinches searching for food in alders demonstrate this behavior. People formerly used this behavior as the basis for a form of "amusement". Suitable food was tied to the end of a long thread. The birds grasped the end of the thread with the beak and pulled on it, then held

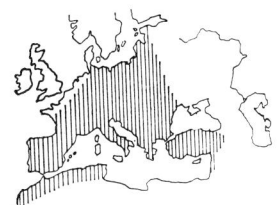

Fig. 16-5. Serin *(Serinus serinus)*, 1967 first breeding in England; also in northern France and northern Germany. Has recently advanced as far as southern Denmark.

Fig. 16-6. Citril finch *(Serinus citrinella)*.

the section pulled in under a foot and pulled in another section with the beak until they pulled up the food. Thorpe writes about this, "goldfinches are so good at this trick that for centuries they were kept in special cages so constructed that the birds could only survive in them by pulling up and holding onto the threads. One thread was tied to a little wagon loaded with food, which stood on a sloping tract, the other ended in a thimble containing water. In the sixteenth century, this sort of contraption was so widespread that the goldfinch was called 'dipper' in several European languages…"

The serin finches form one of the genera with the greatest number of species. They are small, usually greenish, yellow and brownish, often striped on the back with a short, strong beak and a slightly forked tail. They are mainly African, for of the thirty species, twenty-seven are African, many of them widely distributed and with many subspecies. We will describe the following: 1. The SERIN *(Serinus serinus;* L 11 cm, wt. about 12 g). 2. The CANARY *(Serinus canaria)*, often considered conspecific with the serin; males yellow breasted with an ash gray back, females brown and generally duller; native to the western Canary Islands, Madeira and the Azores, introduced in Bermuda. 3. The CITRIL FINCH *(Serinus cintrinella)* of the mountains of central and southern Europe, rarely found below 1,700 m. 4. The WHITE-RUMPED SEEDEATER *(Serinus leucopygius)*, common from the Senegal to Eritrea. 5. The YELLOW-FRONTED CANARY *(Serinus mozambicus)*. Subspecies *caniceps* occurs from Senegambia to the northern Cameroons; other subspecies are found south of the Sahara. 6. The YELLOW CANARY *(Serinus flaviventris;* L 13 cm), bright yellow, occurs mainly in South Africa.

Serins in general are found in open areas with bushes or cultivated land with groups of trees, particularly fruit trees. There are only two species in Europe, the serin and the citril finch. The serin usually occupies the fringe of cultivated lands and is often seen in parks and gardens of great cities. Its northward spread was already mentioned in Vol. VII, Chapter One. Males like to sing from tree tops or telephone wires. They often feed on the ground, usually not far from trees, and they live largely on weed seeds. The nest is placed in a tree or bush, usually two or three meters above the ground. Eggs are laid from March onward in North Africa, from April on in southern Europe, and in May in central Europe. There is more than one brood per year.

The wild canary, which is very closely related to the serin, is the ancestor of all domestic canaries. D. Bannerman writes about it; "The well-known song of the canary usually betrays its presence, though the songster remains hidden…even if one spends only a few hours on Gran Canaria or Tenerife and makes an outing to Santa Brigida or Tacoronte on Tenerife, one can be sure of hearing canaries singing in the eucalyptus trees…In order to hear and to see them, one should visit the orchards early in the spring: The birds are then in good plumage and males sing on every side. They are most frequent on cultivated land and small trees, bushes and fruit trees are indispensible at the approach of the breeding season...It seems to be certain that in the coastal areas all the birds have eggs towards the end of March, while those in the high mountains do not lay before June or even July."

The Spaniards conquered the Canary Islands in 1478, and they soon brought canaries to Europe in great numbers. A lively trade in this ever-more-popular cage bird developed rapidly and the well-known bright yellow domestic canary and several other races and color phases were bred. Red canaries owe their origin and their color to interbreeding with the red siskin. Nowadays, canaries are found as eager songsters in homes all over the world, often in small rooms.

As a result of selective breeding, all sorts of changes in color and shape have been produced. Thus, there are bright yellow, golden yellow, straw-yellow, beige, orange-yellow, reddish-yellow and red canaries, as well as spotted, patchily colored forms and others with lizard-like stripes, as well as the cinnamon colored breed formerly popular in England. There are smooth headed, hooded and crested forms, others with a collar of prolonged feathers and a mantle-like plumage as well as birds classified according to their type of song. Of these, the "Harz Mountain Roller" was the most famous in Germany. Increasing tourism in the Harz area has led to a decrease in the breeding of this strain of canaries since the tourist trade proved more profitable.

Yet even now, in the age of radio and T.V., the strong, loud, rolling, whistling and twittering song of the canary contributes much to the character of many a home, as E. Thomas Gilliard and Georg Steinbacher put it. In the words of these two zoologists, the canary was the first passerine bird to be domesticated and its breeding became the hobby of an untold number of bird fanciers.

In contrast to the serin, the citril finch has only a limited distribution. It is usually encountered on sunny, open spots in the coniferous forest belt. It finds its food above the tree line on meadows or in rocky areas. It feeds on conifer seeds which it is able to extract from the cones and on weed seeds. Its nest is placed high in a conifer; eggs are found from April to August, which suggests that many pairs breed more than once per season. In winter, citril finches descend to lower levels. The numbers caught for banding purposes at the Col de Bretolet in Switzerland indicate a considerable migration in the fall.

Two of the commonest and most widely distributed species of this genus in Africa are the WHITE-RUMPED SEEDEATER *(Serinus leucopygius)* and the YELLOW-FRONTED CANARY *(Serinus mozambicus).* As D. Bannerman writes, the latter is probably the most common cage bird in Africa. There is a lively trade in these birds in port cities. Originally this serin was a parkland bird, but it is now found in the bush and also in fields,

The domestic canary, by H. Wendt

Fig. 16-7. Greenfinch *(Carduelis chloris).*

Fig. 16-8. Siskin *(Carduelis spinus).*

Fig. 16-9. Goldfinch *(Carduelis carduelis).*

gardens and villages. The YELLOW CANARY *(Serinus flaviventris)* comes from south and southwestern Africa; it is rarely seen as a cage bird. It is mainly an inhabitant of the bush and bush-covered mountain slopes, but is also found in small towns.

The genus *Carduelis* includes some of the smallest finches. Many have green and yellow tones and usually show only slight sex differences in color. There are twenty-four species, seven of which occur in the northern Old World as far as southern Asia. We will describe the following: 1. The GREENFINCH *(Carduelis chloris;* L 15 cm, wt 24 to 30 g), the largest species, a denizen of the northern Old World. 2. The SISKIN *(Carduelis spinus;* L 11 cm, weight 11 to 15 g). 3. The PINE SISKIN *(Carduelis* or *Spinus pinus).* Both sexes resemble the female siskin, being brownish-green and striped with yellow wing and tail coverts. It is found in the northern and western United States. 4. The RED SISKIN *(Carduelis* or *Spinus cucullatus;* L 11 cm), females reddish-brown, salmon brown below, tail dark brown. It is found in northern South America. 5. The AMERICAN GOLDFINCH *(Carduelis tristis),* females are dull olive colored with blackish wings; in winter both sexes are colored like the female. It is found in North America from coast to coast and migrates south in the winter. 6. The LESSER GOLDFINCH *(Carduelis psaltria;* L 10 cm) has a dark back and a bright yellow breast. It breeds from the southwestern United States as far as northern South America. 7. LAWRENCE'S GOLDFINCH *(Carduelis lawrencei;* wt 8 to 11 g), has a yellow band on the wing and black cheeks. It is only found in southern California except during winters when some visit western Mexico. 8. The GOLDFINCH *(Carduelis carduelis).* The female is only slightly less colorful and smaller than the male. It occupies the whole western half of the Old World. There are two groups, BLACK-HEADED GOLDFINCHES *(Carduelis carduelis carduelis)* mainly in Europe and GRAY-HEADED GOLDFINCHES *(Carduelis carduelis caniceps)* in Turkestan.

The greenfinch is a frequent breeder in parks, gardens and farmland; it feeds mainly on weed seeds and scattered grain. When there is snow on the ground, it often lives in farmyards and comes to feeding tables in gardens. Its disorderly-looking nest is built in a large bush. In most parts of Europe, it rears up to three broods a year.

Siskins are found in summer mainly in conifer woods where they feed on seeds, small shoots and arthropods. In winter, however, they are far more widely distributed and feed mostly on alder seeds. Their wanderings are somewhat "gypsy-like" and depend largely on available food supplies. Siskin nests are placed on conifers and in May or June contain four or five spotted eggs. According to the reports of G. Svärdson, siskins in Sweden lay eggs earlier than usual in years when there is a good supply of conifer seeds, even while the snow is still on the ground and in this way manage to raise one more brood than in average years.

The call of the North American pine siskin resembles that of the Old World siskin, as do its habitats and its behavior. In winter, pine siskins often visit garden feeding places, and, as A. A. Allen reports, they are unusually tame, even bold, and many perch on people. He also tells how a group of pine siskins changed the sleeping habits of a Massachusetts bird lover one winter. The birds became used to the breakfast of grains which he put out for them and they gathered for it daily at dawn. When their host was not yet awake, they flew into his bedroom, and pulled at his hair or nipped him in the ear. At feeding places pine siskins take not only the grains, but also cling onto suspended food such as is sold for chickadees. This habit has recently also begun to show itself in siskins in England.

The brightly colored American goldfinch is one of the most striking and familiar North American birds. Like the Old World goldfinch, it breeds in open stands of trees, in parks along roadsides and in orchards, and has a preference for thistles. Apart from this it feeds on other weed seeds and in winter on conifers. In the northern part of its breeding range it arrives in April or May, about the time dandelions and some other plants begin to seed, but it does not breed until mid-July. Occupied nests may be found as late as October, but most pairs probably rear only one brood per year. The breeding season is related to the development of thistle seeds on which the young are fed. The nest is placed in thistles or other tall weeds, but also high up in bushes or trees. The eggs are pale blue. The lesser goldfinch occurs in open country with scattered trees, in scrubland or along wooded watercourses and in gardens. It builds its nest in a small bush or tree. In the Hastings reservation (California), Lindsdale noted that breeding lasted from early March to July or exceptionally into November. In summer, the birds were mostly in pastures where Napa thistles *(Centaurea melitensis)* were growing. They also showed a predilection for seeds of the common groundsel. In winter they feed mainly on seeds on adenostoma bushes.

Lawrence's goldfinch lives in hotter, more arid country than its relatives. It breeds by preference on oak-grown slopes where it seeks food on nearby open ground. The eggs, in contrast to those of other goldfinches, are unspotted white. According to Lindsdale, this goldfinch has a particular preference for the seeds of borages; in winter, it, like the lesser goldfinch, eats mainly adensotoma seeds.

There are also many goldfinches in South America, of which the red siskin is one of the most striking. It inhabits the dry tropical areas of northern Venezuela and the island of Trinidad, being found mainly in dry scrub country and pasture land.

The goldfinch is one of the most characteristic European birds which has also long been a popular cage bird. It thrives in captivity and interbreeds with canaries and other related species. Its food preference

Fig. 16-10. Threat behavior of the greenfinch.

Fig. 16-11. Common redpoll *(Acanthis flammea)* and hoary redpoll *(Acanthis hornemanny).*

Fig. 16-12. Twite *(Acanthis flavirostris).*

Fig. 16-13. Linnet *(Acanthis cannabina).*

is for the seeds of thistles, dandelions, groundsels, burdocks and other Compositae, but in winter it also takes alder seeds and those of birches and various conifers. It is the only European finch with a beak long and narrow enough to be able to reach the seeds of the teasel, which lie at the base of a long tubular structure. Darwin had already noted that male goldfinches must have slightly longer beaks to be able to reach the seeds of the teasel, than do the females, which rarely feed on this plant. The small but strongly built nest is placed high in a tree, in a settlement, park, garden or roadside. The clutch consists of four or five eggs and up to three broods a year are raised. Goldfinches thrive on warmth, so they rear more young in warm, dry English summers than they do when the weather is cold and wet. The goldfinch has been introduced in Australia, New Zealand and on Bermuda, as well as in several places in North America, but most of the American introductions have been unsuccessful.

LINNETS (genus *Acanthis*) are generally brown or gray with a flesh-colored or pink flush on some parts of the male's plumage. This color is intensified in the breeding season by the rubbing off of brownish feather tips revealing the red below. There are six species: 1. The COMMON REDPOLL (*Acanthis flammea*; L 13 cm, wt 14 g) is red on the crown, breast and rump and is mainly distributed in the Taiga zone. 2. The HOARY REDPOLL (*Acanthis hornemanni;* L 15 cm), larger and paler, sometimes mainly white, is mainly a tundra bird. 3. The TWITE *(Acanthis flavirostris)* is red only on the rump. 4. The LINNET (*Acanthis cannabina;* L 13 cm, wt 16 to 20 g) is red on the crown and breast. 5. The YEMEN LINNET *(Acanthis yemensis),* similar to the linnet and possibly only a subspecies of it, is found in the mountains of Yemen and other Arabian countries at 2,000 to 3,000 m. 6. The WARSANGLIA LINNET (*Acanthis johannis;* L 12 cm) has black wings and tail with much white at the base of the secondaries and the sides of the tail. It occurs in elderberry stands at 2,000 to 4,000 m in Somalia.

The ranges of the common and the hoary redpoll touch in several areas, e.g. in northern Norway, and the two freely form mixed pairs in such areas. This gives rise to populations containing individuals with wide differences in appearance. South of the tree line, the wanderings, breeding and population density of common redpolls depend on their ability to get enough bird seeds, which are their main diet. The common redpoll was introduced in New Zealand in the last century and has greatly increased there. It is not well liked there now as it feeds on the buds of orchard trees. The twite inhabits desolate cold areas at high altitudes; in Tibet it is found to 4,500 m. The linnet, on the other hand, breeds in bushy areas in large parts of the northern Old World. It feeds mainly on weeds on farmlands. Its German name, which means blood linnet, comes from the bright scarlet of its breast. Like so many other finches, it was a popular cage bird before the more colorful exotic finches became available. It is one of the most easily satisfied domestic birds, singing very busily in its cage and liable to become closely attached to its owner. Brehm wrote about it in the last century, "It is rarely missing in the room of the true bird lover". Now, however, it is not often found in captivity.

ROSY FINCHES (genus *Leucosticte*) breed in high mountains. They are sparrow-sized with a fairly dark brown or blackish plumage with attractive red or silvery spots. There are five species, three in North America and two in northern Asia. The GRAY-CROWNED ROSY FINCH (*Leucosticte arctoa;* L 15 cm) lives in both hemispheres. It is a migrant in part of its range.

Rosy finches usually breed in the transition zone between the trees and the open country above the tree line, at places where there are scattered small bushes. In winter they descend to lower levels. The life of most species has not been extensively studied, although A. A. Allen has given a good picture of them in North America. "As true alpinists, rosy finches in their western domain keep largely to the areas of snow fields and glaciers. If one climbs far enough above the tree line in spring, one may see a flock of them at the edge of the retreating snow walking about picking up seeds as well as arthropods. Their chirping is reminiscent of the house sparrow's. They build their nests in clefts in rocks or beneath large rocks, often at 3,800 to 4,000 meters. The nests are firm cups of grass or moss. The four or five eggs are as white as the snow on the peaks. The young hatch after about two weeks of incubation. In winter these birds wander in flocks far out on the prairies. There they may feed at doorsteps and spend the night beneath the eaves of houses and farm buildings."

The TRUMPETER BULLFINCHES (genus *Rhodopechys*) are brownish or gray birds with red, black and white spots. They inhabit the deserts and other desolate areas of North Africa and southwestern and central Asia in four species. The TRUMPETER BULLFINCH (*Rhodopechys githaginea* L 12.5 cm) occurs from the Canaries to northwestern India. The male is grayish-brown with a marked pink flush. His beak is coral red and is brightest in the spring. The female's has only a very pale pink cast.

It lives on hot dry mountain slopes and ravines and stony edges of deserts. Its color blends so well with that of the desert ground that often its presence is announced only by its calls. D. Bannerman has compared this call with "the sound of a child's trumpet sounded without cessation". He writes about its behavior on the Canary Islands: "For the greater part of the year, the trumpeter bullfinch wanders over the plains in small flocks of ten to fifteen birds, but when the breeding season approaches, the flocks disperse and one usually meets the birds in pairs. They feed largely on seeds, apparently those of desert plants, but they also eat maggots and adult arthropods. Though a bird of desolate spaces, it cannot hold out for long without water; thus small

▷
Top: Gape of the melba finch *(Pytilia melba),* the fire finch *(Lagonisticta senegala)* and the *Erythrura cyanovirens.* Nests with eggs. Upper middle row: Lapwing *(Vanellus vanellus);* Tree sparrow *(Passer montanus);* Greenfinch *(Carduelis chloris).* Lower middle row: Song thrush *(Turdus philomelos);* Golden oriole *(Oriolus oriolus);* Willow warbler *(Phylloscopus trochilus).* Lowest row: One cuckoo egg *(Cuculus canorus)* in a nest with eggs of the robin *(Erithacus rubecula);* Blackcap *(Sylvia atricapilla);* Hedge sparrow *(Prunella modularis).*

▷▷ & ▷▷▷
Top left: Weaver nest (subfamily *(Ploceinae)* in the African tree steppe. Lower left: Cinnamon weaver *(Textor badius)* at its nest. Right: Nests of the golden weaver *(Textor subaureus).*

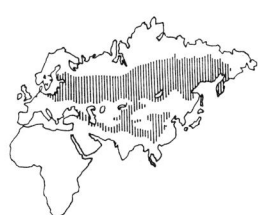

Fig. 16-14. Scarlet grosbeak (*Carpodacus erythrinus*).

family troops visit the deep foundations of the eastern island and quench their thirst there. The nest is usually built on the ground under a rock or a desert plant and the eggs, four or five in number, can be found on the Canaries from mid-February to mid-May."

The SCARLET GROSBEAK-PURPLE FINCH group (genus *Carpodacus*) consists of brown or gray birds, mostly broad striped, much like sparrows. Males have a red or reddish-purple wash and in some species the male has feathers which have a silvery shimmer on the head and throat; there are about twenty species, seventeen of them in northern Eurasia and three in North America. They include: 1. The SCARLET GROSBEAK (*Carpodacus erythrinus*; L 15 cm, wt about 20 g and up). Its breeding range is wider than that of other Old World finches, from Sweden to Japan between latitudes 25° and 68°; in winter it is found from Iran to China and Burma. 2. The PURPLE FINCH (*Carpodacus purpureus*; L 13 cm) has a purple-red rump and breast, and is found only in the conifer belt of Canada and the northern United States and the mountains of the western U.S. In winter it occurs further south. 3. CASSIN'S FINCH (*Carpodacus cassinii*; L 14 cm) is similar but lives at higher elevations in the mountains of the western U.S. and comes down to lower levels in winter. 4. The HOUSE FINCH (*Carpodacus mexicanus*; L 12 cm) has the brightest red on the breast and rump of all the birds of this genus. 5. *Carpodacus puniceus* is dark brown with black spots on the cheeks, a stripe above the eyes, with the rump and much of the lower parts and some other areas red to reddish. It is found in mountains of central Asia up to 6,000 meters.

These birds generally inhabit woodlands and bushy areas of high-lying mountain districts. The scarlet grosbeak is the only member of the genus in Europe. Its wide distribution may be due to the fact that it is not tied to a specific habitat and was able, more than other northern finches of its type, to settle farmed land and other habitats created by man. Since 1930, it has been spreading westwards in Europe, having settled by 1960 in parts of Poland, eastern Germany, southern Finland and southern Sweden. Such a spread had occurred once before at the beginning of the century, but most of the colonies had disappeared later. Scarlet grosbeaks usually breed in swampy bushland with willows, alders or birches, but also on cultivated land with hedges and gardens. The nest is placed low in a bush and the eggs, usually five, are easy to recognize as they are dark blue with coarse dark brown spots. The scarlet grosbeak is one of the few finches which undergoes a full molt during its stay in its winter quarters.

Though the purple finch breeds only in woods, it visits densely populated areas in winter and then feeds in gardens and city parks. It often comes to feeding tables, particularly when sunflower seeds are offered, and may then be seen with Cassin's and house finches, which breed in different habitats.

◁
Sparrows: 1. Snow finch (*Montifringilla nivalis*), male in breeding plumage; 2. *Montifringilla theresae*; 3. Yellow-throated sparrow (*Petronia xanthocollis*); 4. Desert sparrow (*Passer simplex*); 5. Chestnut sparrow (*Passer eminibey*); 6. House sparrow (*Passer domesticus*); 7. Spanish sparrow (*Passer domesticus hispaniolensis*); 8. Dead Sea sparrow (*Passer moabiticus*); 9. Cape sparrow (*Passer melanurus*); 10. Golden sparrow (*Passer lutcolus*).

In the western U.S., the house finch is a highly common bird, often breeding in city gardens and even on buildings; hence its vernacular name. Hundreds of these birds may roost in the climbers on the walls of a single house. It is particularly numerous in California, where it shows a liking for fruit. Its main food, however, consists of weed seeds, although in the breeding season it may also take some arthropods. The BONIN FINCH (+ *Chaunoproctus ferreirostris*) of one of the Bonin Islands near Japan, had a very thick bill. It was brown with long dark stripes above and on the lower tail coverts. It became extinct as early as 1854.

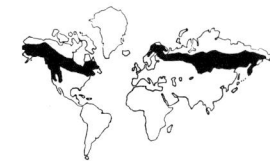

Fig. 16-15. Pine grosbeak *(Pinicola enucleator)*.

The PINE GROSBEAK (*Pinicola enucleator*; L 21 cm, wt 60 g and up) is also the only species in its genus. It is one of the largest finches, found mainly in open conifer forests. There it lives on various seeds and berries and also on shoots and buds of coniferous trees and in summer also on arthropods. In invasion years, it occurs in numbers outside its normal range. In its normal habitats it is strikingly calm and tame. Some photographers have even been able to photograph it at close range on the nest without having to use a blind.

CROSSBILLS (genus *Loxia*), as the name indicates, have the tips of the bills crossed. There are three species: 1. The PARROT CROSSBILL *(Loxia pytyopsittacus)* has a very large beak used to work over pine cones. It is found only in northwestern parts of the Old World. 2. The RED CROSSBILL *(Loxia curvirostra)* has a medium-sized beak adapted to the softer cones of spruce. 3. The WHITE-WINGED CROSSBILL *(Loxia leucoptera)* has a still smaller beak adapted in particular to larch cones. The red crossbill and white-winged crossbill occur in coniferous woods of the Old and New Worlds.

Fig. 16-16. Parrot crossbill *(Loxia pytyopsittacsus)*.

Because of the astonishing adaptations to their mode of feeding, crossbills have always attracted the attention of zoologists. The crossed points of the beak enable them to pull the seeds out of the cones of conifers; they hardly feed on anything else, although occasionally birds of any of the three species take seeds from a tree other than the one mentioned above as characteristic of its species. They usually breed early, from January to April, before the seeds fall out of the cones. The days then are still short and cold and the ground is snow-covered. The birds are unable to rear all their young under these strenuous conditions, so they also breed in other months whenever there is an abundance of food. Crossbills have become famous not only because of their beaks but also because of their wanderings. Every few years they turn up in phenomenal numbers outside their normal range. These irruptions are supposed to be due to over-population or food shortage in the breeding areas. Sometimes these movements, because of the numbers of birds involved, attract general attention. As early as 1251 the English chronicler Matthew Paris reported these strange birds which had "flooded" his country in great numbers.

Fig. 16-17. Red crossbill *(Loxia curvirostra)*.

BULLFINCHES (genus *Pyrrhula*) have short and bloated-looking beaks.

Fig. 16-18. Red crossbill in display posture.

Fig. 16-19. Display of the bullfinch.

Fig. 16-20. Hawfinch (*Coccothraustes eoccothraustes*).

The wings and tail are glossy black and the rump is white; males generally have a red breast. Their home is mainly Asia, but one of the six species is also widespread in Europe. This is the BULLFINCH (*Pyrrhula pyrrhula*; L 15 to 19 cm, wt 22 to 30 g).

The bullfinch of Europe lives in woods, parks, gardens and meadowland. The nest, of fine twigs lined with rootlets, is placed in dense bushes. In England and other European areas, there was a great increase in the numbers of these birds between 1945 and 1955. Since they eat the buds of fruit trees, the birds have made themselves unpopular with fruit growers. However, they mainly eat these buds in the winter and spring when seeds, otherwise their main food, are less numerous. It is said that they have on occasion stripped whole orchards of their buds.

Bullfinches keep well in captivity and they breed freely in cages. Females, but not males, will breed with canaries or other finches. The song is modest, but this finch is a good imitator. When one repeatedly plays a tune to young bullfinches, they learn to whistle it. In the last century bullfinches were popular cage birds in Germany and they are often still offered for sale.

The hawfinches form a tribe of birds distinguished by massive beaks with small slanting nostrils. There are nine species, among which are: 1. The HAWFINCH (*Coccothraustes coccothraustes*; 17 cm) of the northern Old World. 2. The EVENING GROSBEAK (*Hesperiphona vespertina*; L 17 to 20 cm) of North America. 3. CHINESE GROSBEAKS (genus *Eophona*) with two species: The MIGRATORY CHINESE GROSBEAK (*Eophona migratoria*; L 18 cm) which is not black on the head and the JAPANESE GROSBEAK (*Eophona personata*; L 23 cm) which is grayish-brown with the head at least partly black and the wings black with a white band (but not white tips); both are found in southeastern Asia. 4. The genus *Mycerobas*. These birds have even larger beaks and live in four species at great heights in central Asia, particularly in the Himalayas. They include the BLACK-AND-YELLOW GROSBEAK (*Mycerobas icterioides*; L about 20 cm, wt up to 100 g). The male is bright yellow with black wings and tail. The female is gray instead of yellow and brownish below. Because of their remarkable ability to crack open seeds as large as those of cherries and olives, the hawfinches are of interest to zoologists and ornithologists. Mountford writes on this theme: "Their skulls have been admired by engineers as an outstanding example of adaptation to particularly high pressures. Every straight or curved line on this skull which stiffens, supports or thickens, speaks of its strength". The pressures required to crack cherry and olive stones have measured with special instruments corresponding to the bird's beak. The pressure required in such an instrument to crack cherry stones was 27.5 to 43.2 kilograms and for olive stones, 45.8 to 68.3 kilograms. As Mountford points out, these astonishing figures could be related to the bird's weight, which is about 55 g.

The hawfinch is mainly found in deciduous forests. In contrast to most other finches, there are a number of fossil remains of this species. Thus in a Pleistocene deposit in Poland even a hawfinch stomach containing stones of the wild cherry *(Prunus aviaumn)* was recovered in 1910. These cherries are still a favorite hawfinch food in that area.

The evening grosbeak occupies the equivalent place of the hawfinch in the northern coniferous woods of North America. Its breeding range is constantly being expanded towards the east and it is also well known because of its irruptions. These take place every few years, when the birds wander to the east and south, appearing in New England and the states along the middle of the Atlantic coast. In recent years these invasions have become more frequent. Evening grosbeaks live mainly on hard tree seeds, but they also readily take to feeding tables to get sunflower seeds. Because of this, it was easy to catch and band many of these birds so that more is known about their movements than about those of any of the other birds which habitually show irruptions.

Fig. 16-21. Hawfinch in threat and in defensive posture.

The Canadian ornithologist D. H. Speirs, of Toronto, who has long studied the behavior of evening grosbeaks, reports of their wanderings: In the winters of 1951-64, over 17,000 evening grosbeaks were banded at Pennsylvania State College. Only forty-eight of these were recovered in subsequent winters at the same place. On the other hand, 450 of the banded birds were found in no less than seventeen states and four Canadian provinces. These returns show how far individuals range and what little tendency they have to return to the same place in different winters. Both these features are in contrast to the behavior of regular migrants. Some evening grosbeaks also stay near their breeding areas as was shown by the recoveries obtained by Mrs. B. Downs. She lives in a breeding area and has often caught the same individual birds in successive summers as well as in winter. Such different findings for this one species are important as they probably also apply to other irruptive finches about which little is known as yet.

The two species of Chinese hawfinches live in mixed and deciduous woods, particularly in mountains. In winter they move to lower levels. Sometimes they raid gardens for peas or beans. But the adaptations shown by all the hawfinches to dealing with large hard seeds is far exceeded by the grosbeaks of the genus *Mycerobas.* The largest relatives of the goldfinches, with phenomenally large beaks are found in this genus. Unfortunately, little is known about them as they live at such high altitudes in the Himalayas. One of them, the black-and-yellow grosbeak, occurs between 2,000 and 4,000 m. Bates and Lowther write about it: "Because of its considerable size, its bright plumage and its massive beak, this hawfinch is the most conspicuous finch in the Himalayas. Its loud, pleasant whistles carry far over the tree-grown slopes and it is often delivered when the members of a swarm move through the tree tops. In addition to seeds, these hawfinches also take the green shoots of pines and berries from the undergrowth."

17 Weavers and Weaverfinches

Weavers and weaverfinches, by H. Wendt

The two closely related families of WEAVERS *(Ploceidae)* and WEAVER-FINCHES *(Estrildidae)* are sometimes regarded as forming one family. Many of the weavers make nests which are truly woven on the warp and woof principle. Other weavers as well as the weaverfinches and the sparrows, which are now included among them, assemble their nest material more loosely into similarly shaped nests in so far as they nest in the open or in cavities which offer sufficient space. Some, the parasitic weaver and the whydahs, are brood parasites.

Birds of both families are sociable; many breed in common in loose aggregations or in definite colonies. In size they generally range from 8 to 24 cm, while some, if one includes the long decorative tail feathers, attain lengths of up to 50 cm, and in one whydah subspecies, of up to 67 cm. They have ten primaries; the outermost in many groups is very much reduced, a process which has apparently taken place independently in the different groups. The beak is small, conical, relatively thick at the base, often quite massive as in the Java sparrow, or delicate as in many other whydahs. Male weavers often have a striking nuptial plumage, but after the breeding season assume a duller plumage; sometimes both sexes have the same plumage. In the whydahs and sparrows, males and females both take part in nest building, incubation and feeding of the young. In the true weavers, the males are often polygamous and generally leave the rearing of the young to the females, although they always take the major share in nest building. They construct the artfully woven structure, leaving only its lining to the females. The true weavers are exclusively Old World and mainly African birds.

Family: weavers, by H. E. Wolters

Weavers *(Ploceidae)* are song birds ranging in size from less than that of a serin (the scaly weaver) to barely the size of a starling (buffalo weavers). The beak is strong, of varying shape in different species, some slim, others short and deep. The tenth primary is almost always shortened, but only in the sparrows and the sociable weaver is it displaced onto the upper surface of the wing as in finches. In the more unmodified species (sparrows, sparrow weavers, sociable weavers and scaly weavers), the plumage is mostly grayish-brown with reddish-brown, black, white or yellow markings; in most of the others, however, at least the male nuptial plumage is very colorful, most often bright yellow or red with black, chestnut brown or greenish markings. In the whydah subfamily, the tail feathers of males in nuptial plumage are prolonged. This involves the four central feathers in whydahs or more or less all of them in many bishops. There are nine tribes: 1. SPARROWS *(Passerinae)*; 2. CUCKOO WEAVERS *(Anomalospizinae)*; 3. WHYDAHS *(Viduinae)*; 4. BISHOPS *(Euplectinae)*; 5. WEAVERS PROPER *(Ploceinae)*; 6. SWAMP WEAVERS *(Amblyospizinae)*; 7. BUFFALO WEAVERS *(Bubalornithinae)*; 8. SPARROW WEAVERS *(Plocepasserinae)*; 9. SCALY WEAVERS *(Sporopipinae)*. Altogether there are about sixty-five genera and 142 to 150 species.

By far, the majority of weavers live in Africa south of the Sahara. Others are found on Madagascar and the adjacent islands and in southern Asia eastwards as far as Java and Bali. The sparrows also occur naturally in temperate areas of Europe and Asia. Two species were introduced in Australia and the New World. Most weavers live socially in the steppes or in the savannahs as inhabitants either of tall grass or of trees and bushes. Some have penetrated into forests, evidently from the savannah. They eat grains, mainly grass seeds, or insects, with the type of diet depending on the tribe and genus and its habitat. Many sparrows breed in holes in trees, earth or rock, which they pad to some degree; all other weavers always build roofed-over, often very artful, nests in tall grass, reeds, bushes or trees. This applies particularly to the bishops and the weavers proper. The sociable weaver shows the highest development in colonial nesting. The less highly evolved species and most of the forest dwellers live in monogamy while among the rest polygyny is frequent. Weavers and weaverfinches probably have a common origin. For example, the rather primitive scaly weavers resemble weaverfinches in a number of features. It is not yet known whether there is also a relationship to the finches and buntings; the similarity in color and pattern of plumage of African weavers and American icterids is striking.

Subfamily: sparrows, by H. E. Wolters

At one time the sparrows (subfamily *Passerinae*) were generally included with the finches with which they share, besides the strong bill, the characteristic of an atrophic outermost primary feather. For some time they have usually been classed as a subfamily of weavers but some ornithologists look on them as forming a family of their own. Decisive features in grouping them with the weavers are a number of characteristics of the skull and the fact that sparrows, when they do not breed in holes, build roofed-over nests, in contrast to the finches. However, sparrows evidently branched off along an evolutionary pathway of their own at an early stage, producing some species which in external appearance come very close to the finches, e.g. the snow

Fig. 17-1. Rock sparrow *(Petronia petronia).*

Fig. 17-2. 1. African yellow-throated sparrow *(Petronia superciliaris);* 2. Yellow-spotted petronia *(Petronia pyrgita);* 3. Yellow-throated sparrow *(Petronia xanthocollis).*

Fig. 17-3. 1. Snow finch *(Montifringilla nivalis);* 2. Tibet snow finch *(Montifringilla adamsi).*

finch and some rock sparrows. Originally, sparrows were doubtless inhabitants of dry warm areas of Africa, western and southern Asia. But they are the only weaver subfamily to have also penetrated the northern temperate zone of the Old World as far as the high mountains and the storm-whipped steppes of central Asia. As a commensal of man, the house sparrow has, furthermore, become established in most parts of the world. Sparrows feed principally on seeds, but they also eat many arthropods and these form the bulk of the food for the young.

The almost uniformly sandy-brown PALE ROCK SPARROW (*Carpospiza brachydactyla;* L 15 cm) is appropriately named. Little is known of its life. It breeds in bushes in the semi-desert dry areas of western Asia. After the breeding season, it wanders as far as northeastern Africa and Arabia.

The ROCK SPARROWS (genus *Petronia*) are related to it. They are marked by a more or less developed yellow spot in the crop area. They walk on the ground rather than hopping like house and tree sparrows. This genus includes: 1. The ROCK SPARROW (*Petronia petronia;* L 14 cm), which has white tips on its tail feathers. It inhabits rocky or stony areas from the Canaries and Madeira across the Mediterranean area as far as Mongolia. 2. The AFRICAN YELLOW-THROATED SPARROW (*Petronia superciliaris* L 15 cm) which has a wide white band over the temples. It is found in Africa from the Cape to Tanzania. With the next two species it forms the subgenus *Gymnornis*. 3. The YELLOW-SPOTTED PETRONIA (*Petronia pyrgita;* L 15 cm) which is found from the Senegal to northeastern Africa and northeastern Tanzania. 4. The YELLOW-THROATED SPARROW (*Petronia xanthocollis;* L 13 cm) which is found in western Asia and India.

As a relic of the fauna of a warmer period, the rock sparrow was once native in Germany too, particularly in ruined castles. In the last century some still nested in southern Baden, Nassau, Franconia and Thuringia, as Günter Niethammer wrote in 1951. But their numbers decreased steadily and the breeding places became deserted. In 1944, the last German rock sparrows gave up their breeding place, the Salzburg on the Saale river in Franconia. An attempted reintroduction in the Rhineland failed. Rock sparrows breed in clefts in rocks and holes in walls or trees, singly or in colonies. Only females incubate the three to six whitish eggs which have reddish-brown or blackish-brown spots. In their movements, the rock sparrows of the *Gymnornis* group resemble finches and buntings rather than true sparrows. They are occasionally seen in the flight cages of European bird fanciers. They inhabit treed areas and breed in tree holes.

The SNOW FINCHES (genus *Montifringilla*) are sparrows which have become adapted to life in high mountains or on the bare steppes of central Asia. There are six species, including: 1. The SNOW FINCH (*Montifringilla nivalis;* L 18 cm) of the high mountain ranges of Eurasia from the Pyrenees to Mongolia. 2. The TIBET SNOW FINCH (*Montifringilla adamsi*) which closely resembles the snow finch but has less white on the wings. It is found from the Himalayas to western China. 3. MANDELL'S SNOW FINCH (*Montifringilla taczanowskii*; L 17 cm), brown above, with a white forehead, eyestripe and rump. The sides of the head are gray and the lower parts are whitish. It occurs from Tibet to the Koko Nor. 4. BLANFORD'S SNOW FINCH (*Montifringilla blanfordii*; L 16 cm), is found in the Himalayas and Tibet to the Koko Nor and Nanshan. 5. *Montifringilla theresae*) which lives in Afghanistan.

Fig. 17-4. Blanford's snow finch *(Montifringilla blanfordii).*

The snow finch, well known to mountaineers of Eurasia who are interested in birds, builds its cup-shaped nest above the tree line in holes or clefts of rocky alpine slopes. However, in sparrow-like manner, it also breeds on alpine huts, hospices and hotels at high elevations. Its call is slightly quacking. In flight it utters a chaffinch-like "tshueb" and it also has a soft rhythmically twittering song. Some smaller species, including Mandell's snow finch, usually build their nests in holes made by the central Asiatic pika, often up to fifty centimeters below the ground at the end of a passage one to three meters long. E. Schäfer describes the display of this snow finch: "Males walked continually about the females in a horizontal posture with heads hanging low, the neck stretched far forward and inflated while the tail was fully fanned out." This ground display usually precedes a display flight which Schäfer describes in Blanford's snowfinch: "Even in mid-July, I could still watch displays. Males would rise with whirring wings to about ten meters then often stay put by hovering and sing their rapidly delivered twittering little song before they came down again with far outspread wings."

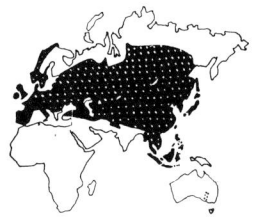

Fig. 17-5. Tree sparrow *(Passer montanus).* Introduction outside Eurasia shown in dots.

Most sparrows are included in the genus, *Passer*, that of the true sparrows. It includes: 1. The TREE SPARROW (*Passer montanus*; L 14 cm) which is found in nearly all of Europe and Asia. 2. The HOUSE SPARROW (*Passer domesticus*; L 15 cm). 3. The RUFOUS SPARROW (*Passer motitensis*; L 15 cm) of southwestern and northeastern Africa and the Cape Verde Islands [not listed in Bannerman, History of the Birds of Cape Verde Islands, translator]. 4. The DEAD SEA SPARROW (*Passer moabiticus*; L 12 cm) of western Asia as far as Afghanistan. Its western subspecies has yellow spots on the side of the head; the eastern one has yellowish lower parts. 5. The PEGU HOUSE SPARROW (*Passer flaveolus*; L 13 cm) is yellow below and occurs from Burma to southern Viet Nam. 6. The CAPE SPARROW (*Passer melanurus*; L 15 cm) which occurs in South Africa. 7. The DESERT SPARROW (*Passer simplex*; L 15 cm) which is sandy-colored and is found in dry river beds in the area of the Sahara, in eastern Iran and Transcaspia. 8. The RUSSET SPARROW (*Passer rutilans*; L 14-15 cm) which has a reddish-brown crown. It occurs from eastern Afghanistan to Japan and Taiwan. 9. The GRAY-HEADED SPARROW (*Passer griseus*; L 15-17 cm) which has a uniformly gray-colored head and is distributed in many subspecies over most of Africa south of the Sahara. 10. The

Fig. 17-6. House sparrow *(Passer domesticus).* 1. Distribution about 1800 (in black); 2. Areas occupied since then (dotted).

Fig. 17-7. Rufous sparrow *(Passer motitensis).*

Fig. 17-8. Russet sparrow (*Passer rutilans*).

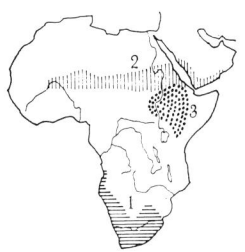

Fig. 17-9. 1. Cape sparrow (*Passer melanurus*); 2. Golden sparrow (*Passer luteus*); 3. Chestnut sparrow (*Passer eminibey*).

Fig. 17-10. House sparrow (*Passer domesticus*) with its subspecies. 1. Typical house sparrows (*Passer domesticus, Passer domesticus tingitanus, Passer domesticus biblicus* and *Passer domesticus niloticus*); 2. Spanish sparrow (*Passer domesticus hispaniolensis*); and 3. a mixed subspecies (*Passer domesticus italiae*).

GOLDEN SPARROW (*Passer luteus*; L 13 cm) which occurs from the southern edge of the Sahara east as far as southwestern Arabia. 11. The CHESTNUT SPARROW (*Passer eminibey*; L 12 cm) which occurs in eastern Africa.

The tree sparrow evidently settled in Europe before the house sparrow. In Europe it occurs mainly at the edge of villages, in gardens, fields, on country roads and on the edge of woodlands, while in central and eastern Asia it takes the place of the house sparrow, inhabiting the central areas of villages and towns as well as rural areas. It eats more arthropods than the house sparrow and often uses nest boxes which have been placed to encourage other hole-breeding birds.

No bird is so well known as the house sparrow. Country people, who persecute it, are very familiar with it, as indicated by its many local names. Originally it probably lived in well-treed steppe areas of western Asia and in part of the Mediterranean area. Here a western form, the SPANISH SPARROW (*Passer domesticus hispaniolensis*), arose and an eastern form (*Passer domesticus domesticus*) developed. In Italy the two met and produced the ITALIAN SPARROW (*Passer domesticus italiae*). This was because the ancestors of the eastern form had, after the Ice Age, linked themselves closely to agricultural man and spread with the westward advance of agriculture. A mingling of the two subspecies occurred, where the Spanish sparrow had already established itself in what became farmland; elsewhere, however, they lived in different habitats and had no connections and still live side by side without interbreeding.

As a follower of European man, the house sparrow has conquered a large part of the whole world in modern times. It is now absent only in a large part of eastern and southern Asia, in western Australia, in the equatorial region and about the poles. It advanced from Russia into Siberia in the course of the 19th Century and reached the mouth of the Amur in 1929. In North America it was introduced in New York in 1850 and 1851 and by 1888, after various other liberations of sparrows in the U.S., it had occupied almost all of the country except for the Pacific states, as well as southern Canada. Since then it has spread into the North American Pacific area as well and also occurs in Mexico, Cuba and in many parts of South America where there were also liberations of sparrows in several places. Sparrows were first introduced in Australia between 1863 and 1872, and now it is only absent from Australia's tropical north and the west. In the 1860's it was introduced in New Zealand and from there, unaided, settled Campbell and Norfolk Islands, which are 560 and 720 kilometers, respectively, from New Zealand. House sparrows of the Indian subspecies (*Passer domesticus indicus*) reached South Africa as a result of an accidental introduction to Durban. Now they can be found almost everywhere in southern and southwestern Africa and Rhodesia and more recently also in Zambia and Malawi. The various North American sparrow populations already show differences in color and size among themselves and from the ancestral European form. This fact is of interest in relation to the time required for the formation of new subspecies.

As a companion of man, the house sparrow in the interior of large cities has to feed largely on garbage but even here it knows how to find many seeds and arthropods. In the country it prefers unripe grain, particularly wheat and oats, and apart from this it feeds on weed seeds, seeds of trees, buds, fruit and small animals including a considerable number of insects and their larvae. In many places city sparrows make long trips into the countryside about the time the cereals begin to ripen.

Fig. 17-11. Displaying male house sparrow.

The nest, built by both sexes but mainly the male, appears disorderly. It is placed in clefts, niches or holes on buildings, in nest boxes, unused swallow nests or even such nests from which the rightful owners have been expelled; also among ivy and other climbing plants. Occasionally nests also stand in the open on old trees as is usual with the Spanish sparrow and in the western Asian range of the house sparrow. Nests are often used as sleeping places and sometimes are built solely for this purpose.

In display the male sparrow approaches the female with drooped wings and erected tail with loud chirping calls. If the female is not ready to mate, it assumes a threat posture which, however, merely intensifies the male's efforts. The noise attracts other males so that such a display often involves several males, sometimes up to a dozen, from which the female then escapes by taking flight. After mating, house sparrows generally form lifelong unions. However, because of many persecutions, widowers and widows are frequent, but they usually find a new partner quickly. In large sparrow populations there are always birds which remain unpaired for lack of nest sites, a prerequisite for pair formation. In central Europe, house sparrows breed in rapid succession, two, three or four times a year, the usual clutch being four to six eggs. However, in Turkistan and Afghanistan, where the house sparrow is a migrant, there is only one annual brood, generally with seven eggs. Both sexes incubate for fourteen days and both feed the young, which are fully fledged in seventeen days. After leaving the nest the young are still fed by the parents for a short time and then they gather with others of their kind into flocks which are often large and noisy at their resting places.

Beside the introduced house sparrow, the Cape sparrow or "Mossie", as it is called there, is found in southern Africa. It has lately started to become injurious to the grape harvest in the wine growing areas of the southwestern Cape Province. The russet sparrow of eastern Asia lives along forest edges, as the tree sparrow does in Europe, and from there it searches for food on adjacent fields. It breeds in tree holes. H. Jahn observed it in Japan and describes its behavior, which differs somewhat from that of other sparrows: "While searching for

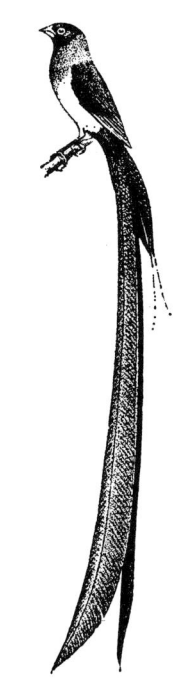

Fig. 17-12. *Steganura togoensis.*

Fig. 17-13.. Distribution of the parasitic weaver (subfamily *Anomalospizinae*).

Subfamily: parasitic weavers, by H. Friedmann

Fig. 17-13a. *Steganura orcuitatis aucupun.*

Subfamily: viduinae, by J. Nicolai

insects in the trees, the birds move calmly, trip about on the branches like chaffinches without many calls...In the breeding areas one generally sees them singly, perched on dry branches of the highest trees at the edge of the forest. The most frequently heard call is a short sparrow-like 'tshe tshe'. A pipit-like 'psee psee psee', always uttered on taking wing, is particularly characteristic." The East African chestnut sparrow has a preference for stands of trees or bushes in papyrus swamps, but in many places it is also found in gardens near settlements. It visits the ripening millet fields together with weavers and weaver finches. For breeding, it often takes over the nests of small sparrow weavers.

The PARASITIC WEAVER *(Anomalospiza imberbis)* is the sole representative of the tribe which has the same name, *(Anomalospizinae)*. It is finch-like and is a brood parasite, mainly on the fan-tailed warblers (Chap. 11) or long-tailed members of the genus *Prinia*, wren warblers. It is found in open savannas from Cameroon, the southern Sudan, and Ethiopia as far as the Transvaal in dry as well as in moist areas.

As the parasitic weaver resembles the serins so much and seeks its food on the ground as do the serins, it has probably often been mistaken for them and therefore little is known about it. Its brood parasitism has evidently developed independently from that of the whydahs which are not particularly closely related to it. The nestling parasitic weaver does not have the striking markings in the mouth shown by the young of the various species of whydah. The specialization of this weaver to grass-dwelling fan-tailed warblers as hosts is interesting as these are species which neither look like it nor are related to it. As far as is known, the eggs of parasitic weavers show no particular adaptation to those of its host birds nor its nestlings. Several times two of its eggs or young have been found in a nest but it is not known in such cases whether the same female laid both eggs in the nest or the eggs came from two different females.

The WHYDAHS (Subfamily *Viduinae*) are also brood parasites, on weaverfinches. Their non-breeding plumage is simple and undecorated, but in the wonderful nuptial plumage males (except among the combassour) have the four central tail feathers prolonged. According to recent researches, they are close relatives of the orange bishop and the bishops; all the characteristics which they share with the weaverfinches are the result of their brood parasitism. There are four genera with fifteen species in the steppes and savannas of tropical Africa south of the Sahara: 1. The WHYDAHS in the restricted sense (genus *Steganura*). Males have decorated feathers which have been particularly developed to produce a sound. There are five species, among them the PARADISE WHYDAH *(Steganura paradisea;* L without the tail feathers of the nuptial plumage, 15 cm; with the tail feathers, 40 cm). It is found in eastern and southern Africa (for further species, see the systematic review).

All whydahs are brood parasites on weaverfinches of the genus *Pytilia*. 2. The INDIGO BIRDS (genus *Hypochera*). The various species differ in color of the feet and bill, color of the flight feathers and a weak or strong violet, blue or green metallic gloss in the male's nuptial plumage. There are six species, among them: the CAMEROONS INDIGO BIRD (*Hypochera chalybeata*; L 11 cm). It has red feet and a white beak (which is, however, salmon pink in one subspecies); depending on the subspecies the plumage is glossy, colored blue, bluish-green or green. It occurs in Africa from the Senegal and Ethiopia to southern Angola and the Transvaal (for other species, see systematic review and illustrations on page 429). All species of this genus parasitize weaverfinches of the genus *Lagonosticta*. 3. Genus *Tetraenura*; two species: The SHAFT-TAILED WHYDAH (*Tetraenura regia*; L without the long tail feathers, 15 cm). The male in nuptial plumage has a black crown, back and tail and is otherwise buffy-brown, with red beak and feet. It breeds in western and central South Africa and also has a small breeding area in Mozambique; FISCHER'S WHYDAH (*Tetraenura fischeri*) has a light-colored crown and occurs from Somalia to Tanzania. These species parasitize, respectively, the violet-eared waxbill and the purple grenadier. 4. Genus *Vidua*; two species: The SHINY BLACK WHYDAH (*Vidua hypocherina*; L without the prolonged tail feathers, 11 cm); males in nuptial plumage are uniformly glossy blue-black. It occurs in eastern Africa from Ethiopia to Uganda and southern Tanzania. The BLACK-CHEEKED WAXBILL (*Estrilda erythronotus*) is suspected to be its brood host. The PIN-TAILED WHYDAH (*Vidua macroura*; L without the prolonged tail feathers, 12.5 cm), is the most widely distributed whydah, found in almost the whole of Africa south of the Sahara. It has a red beak; the male's nuptial plumage is black and white. It is a brood parasite of the ST. HELENA WAXBILL (*Estrilda astrild*), and, where this bird does not occur, the GRAY WAXBILL (*Estrilda troglodytes*).

At the approach of the rainy season, the breeding season of their host birds, a full molt sets in birds of both sexes of the whydah tribe. As a result, the males acquire their wonderful nuptial plumage. Apart from those of the genus *Hypochera*, this plumage is marked by the prolongation of the central four tail feathers. In the pin-tailed whydah, the shiny black and Fischer's whydah, these feathers are of about the same length. In males of the genus *Steganura* the central two tail feathers are very much widened but are only about twice as long as the other tail feathers, while the adjacent two feathers are prolonged to an extraordinary degree and in the Togo whydah are several times as long as the body. Both pairs of decorative feathers are twisted to ninety degrees on the long axis so that their vanes lie close together. When the bird takes wing, the upper edges of the second pair of these feathers rub over the peculiarly grooved vanes of the central pair and this produces a striking rustling sound. This rustling, as well as the display flight, belongs to the male's display.

▷ Whydahs and their hosts: 1. Paradise whydah (*Steganura paradisea*); 2. Sixteen-day-old young (*Steganura paradisea*) (brood parasite); 3. Sixteen-day-old young of the melba finch (host species); 4. Gape of a twelve-day-old paradise whydah; 5. Gape of a twelve-day-old melba finch; 6. Eggs of a paradise whydah; 7. Eggs of the melba finch; 8. Melba finch (*Pytilia melba*); 9. Broad-tailed paradise whydah (*Steganura obtusa*); 10. Red-faced waxbill (*Pytilia afra*); 11. Sixteen-day-old nestling of broad-tailed paradise whydah (brood parasite); 12. Sixteen-day-old nestling of red-faced waxbill (host species); 13. Fischer's whydah (*Tetraenura fischeri*); 14. Purple grenadier (*Uraeginthus iauthinogaster*); 15. Fifteen-day-old nestling of the purple grenadier (host species); 16. Fifteen-day-old nestling of Fischer's whydah, (brood parasite); 17. Gape of a three-day-old purple grenadier; 18. Gape of a three-day-old Fischer's whydah; 19. Black indigo bird (*Hypochera nigerrima*), male; 20. Jameson's fire finch (*Lagonosticta rhodopareia*); 21. Gape of a five-day-old black indigo bird; 22. Gape of a five-day-old Jameson's fire finch; 23. Cameroon's indigo bird (*Hypochera chalylbeata*); continued on page 431.

24. Fire finch (*Lagonosticta senegala*); 25. Gape of a twelve-day-old Cameroon's indigo bird; 26. Gape of a twelve-day-old fire finch; 27. Sixteen-day-old nestling of the Cameroon's indigo bird (brood parasite); 28. Sixteen-day-old nestling of the fire finch (host species).

◁
Weaver males in nuptial plumage: 1. Red-collared whydah (*Coliuspasser ardens*); 2. Red-billed dioch (*Quelea quelea*); 3. Crested malimbe (*Malimbus malimbicus*); 4. Manyar weaver (*Ploceus manyar*); 5. Cape weaver (*Textor capensis*); 6. Little weaver (*Sitagra luteola*); 7. Black swamp weaver (*Amblyospiza albifrons*); 8. Fire-fronted bishop (*Taha diademata*); 9. Grenadier weaver (*Euplectes orix*); 10. Black-headed village weaver (*Textor cucullatus*); 11. Red-headed weaver (*Anaplectes rubriceps*); 12. White-headed buffalo weaver (*Dinemellia dinemelli*); 13. *Hyphanturgus melanogaster*; 14. Scaly weaver (*Sporopipes squamifrons*); 15. Brown-capped weaver (*Phormoplectes insignis*); 16. Parasitic weaver (*Anomalospiza imberbis*).

The most noteworthy fact about the life of the whydahs is undoubtedly their reproductive behavior. All are brood parasites; they do not build nests but lay their eggs into the clutches of other birds. Even until a few years ago it was believed that whydahs parasitized a large number of species of several families, e.g., weavers, sparrows, buntings, finches, weaverfinches and still others. But now it is known that they victimize weaverfinches exclusively and within this family, which is rich in species, they victimize only members of four genera, *Estrilda, Lagonosticta, Uraeginthus,* and *Pytilia.* Each whydah species is specialized to use a single weaverfinch species which adopts and rears the young whydahs.

The weaverfinches, the whydahs' hosts, belong to a bird family whose members have a very accurate conception of how the young of their own species look. Newly hatched weaverfinches have in the gape, i.e. on the palate, tongue and floor of the mouth, complicated patterns and colors which differ from one species to another. They may be developed as horseshoe-shaped lines or as elongated or rounded spots varying in number from three to five, colored black or deep blue. The color of the palate on which these patterns appear can also be very different in different species: ivory, white, yellow, orange or red. Finally, the edges of the beak at the gape which are still thickened at this age can be colored a bright white, yellow, red or cornflower blue and they may appear as simple swellings or a string of wart-like structures. Beyond this, newly hatched weaverfinch nestlings also differ in other features. Their predominant color can be black, brownish or pale flesh-colored; their little bodies can be naked or covered with a dense coat of down. They beg for food with peculiar head movements and are fed by their parents by regurgitation from the crop.

Young whydahs are very closely adapted in all these particulars with respect to the appearance and behavior of their particular hosts. Thus, the young paradise whydah has the same bluish-purple spots on the palate as the young of the MELBA FINCH *(Pytilia melba)* which it parasitizes; it also has the same begging movements and food begging calls. Even its juvenile plumage which grows while it is in the nest is of the same grayish-brown color. The young Fischer's whydah has the same cornflower blue warts at the angles of the beak and an orange palate with the same pattern of three spots as the young of the host bird, the PURPLE GRENADIER *(Uraeginthus ianthinogaster).* This complete correspondence in characters of the young deprives the parent host birds of the ability to detect the young parasites as strangers, to feed them less or pass them over when feeding as they would in the case of the young showing an abnormal gape pattern. Thus young whydahs grow up in the host nest together with the young of their adoptive parents without even one of the host nestlings being sacrificed. The brood parasitism of the whydahs is therefore not coupled with a reduction of reproduction of the host birds as in the case with the cuckoos or the African honeyguides. (All three are shown on Color plate p. 429)

Until a few years ago, the hosts of only three species of whydahs were known, but an unusual ability of male whydahs gave a clue to finding the hosts from among other species. Except for the pin-tailed and the shiny black whydah, all male whydahs imitate in their songs the complete vocal repertoire of their particular host. Thus, the Queen's whydah imitates faithfully all the sounds of the violet-eared waxbill: its song, the phrase indicating excitement, the male's call to attract the female to the nest, the female's greeting phrase, as well as calls indicating anger, contact calls and calls to attract others of the species of both sexes. In the same way, male paradise whydahs imitate the song and calls of the melba finch, always using the quite definite "dialect" of the particular geographical subspecies of this weaverfinch. It was possible to determine the hosts of most whydah species by listening to foreign vocal elements uttered by male whydahs.

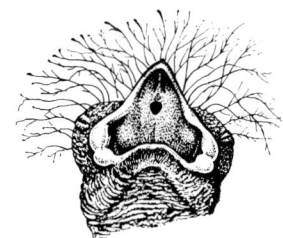

Fig. 17-14. Pattern of the gape of a whydah nestling.

This imitative ability of male whydahs is of decisive importance in regard to their chances of acquiring a mate, because young whydahs, as they grow up in the nest of their host, establish a close and life-long tie to the host species. In the female whydahs this imprinting has effects which influence their later reproductive behavior. The appearance and song of the host father leave permanent impressions in the maturing female whydah. The result is that later she will lay eggs only into nests of weaverfinches of the species to which her host father belonged and she will accept as a mate only a male which sings as her host father used to sing. She will find such a partner only among those male whydahs of her species which were also reared in a nest of the same host species, for only these will deliver such songs. The imprinting of features of the host species and the particular desires this gives rise to in the female whydah ensure that her eggs will be fertilized by a male of the same adaptive type and that the female herself will lay the eggs in the appropriate nests, for only in these will her young be reared successfully. The imitation of the sounds of host birds also performs the important function of preventing hybridization between whydah species. This is particularly important in areas where species of very similar appearance live side by side.

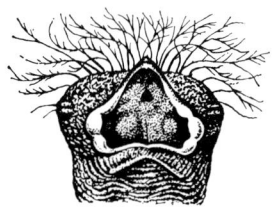

Fig. 17-15. Pattern of the gape of a melba finch nestling.

Fig. 17-16. Adult whydah, female.

Young whydahs as well as their foster siblings are fed exclusively on thin-skinned arthropods during the first few days of life and even later they are still fed mainly on arthropods. From about the eighth day the foster parents mix an increasing amount of half-ripe grass seeds into the predigested paste which they regurgitate into the beaks of the young. Once the young are independent their sole food consists of the seeds of various steppe grasses which they pick off the ground along the sides of trails, on narrow footpaths in the bush or on open areas in grassland, or which they take directly from the seed heads

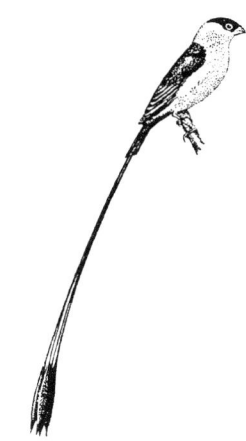

Fig. 17-17. Male Queen's whydah.

Fig. 17-18. Male paradise whydah raises the central tail feathers to an exaggenerated degree in its imposing flight.

Subfamily: whydahs, by H. E. Wolters

Fig. 17-19. The regurgitation display of the male broad-tailed paradise whydah is a ritualized form of feeding behavior.

of plants whose stalks have become broken, for they are unable to climb up the stalks as do many weaverfinches and weavers.

During the breeding season, whydahs live in large or small colony-like populations within the areas of distribution of their host birds. Males occupy large territories during the day. Within these they always fly to particular trees and always into the same dry branches which project beyond the crown. From these they deliver their song, or rise, as the paradise whydah does, high up in the air in a bat-like display flight. Where there is a very high whydah population, the percentage of nests of the host species which contain whydah eggs may be very high. In such an area near Iringa in southern Tanzania only two out of thirty-six melba finch nests did not hold paradise whydah eggs. All the rest had at least one, most of them two or three and some had four and even five whydah eggs, which can be recognized by their greater size. In the same district there were eggs of Fischer's whydah in eleven out of fifteen nests of the purple grenadier.

Even though the main facts about the brood parasitism of whydahs are now largely known, there are still many questions. For example, the host species of some of the whydahs of the genus *Hypochera* and that of the shiny black whydah are still unknown. Thus, there is still much for ornithologists to do before this interesting group of birds is completely investigated.

The BISHOPS (subfamily Euplectinae) are a group of weavers which have developed in parallel with the weavers proper and reached a high degree of differentiation. They have heavy (in some species strikingly heavy) beaks. Males in non-breeding plumage and females are brownish with sparrow-like markings. They have a more markedly reduced outermost primary than do the true weavers. The male's breeding plumage in the genera *Coliuspasser, Euplectes,* and *Taha* is distinguished by velvet-black feathers which seem lacerated by red or yellow markings. They are seed eaters to a greater degree than most of the true weavers. They weave artful nests as do the true weavers, but bishop nests are different in shape, being rounded with an entrance near the top at the side. The less advanced species, those of the genera *Foudia, Queliopsis, Quelea,* and *Brachycope,* live in monogamy; the rest are polygamous. With the exception of some *Foudia* species which have become a forest dweller, they live in open grass and bush country. There are seven genera with twenty-six species in Africa south of the Sahara and on Madagascar and the neighboring islands.

In many species of the genus *Coliuspasser* males in breeding plumage have prolonged tail feathers, the central ones being longest. The breeding plumage is predominantly black, often with yellow or red markings; the lesser and medium wing coverts are often contrastingly colored. There are eight species which include: The RED-COLLARED WHYDAH (*Coliuspasser ardens;* L about 15 cm; males in breeding plumage, depending on the subspecies, 22-36 cm). It occurs in bushy areas and grassland from Senegal to Ethiopia and south to the eastern Cape Province. *Coliuspasser jacksoni*; L 18 cm; the male in breeding plumage, 30 cm. In the breeding plumage it is uniformly black with yellowish-brown wing coverts and wide tail feathers which are bent inwards at the tip. It is found in the highlands of Kenya and northern Tanzania. The LONG-TAILED WHYDAH BIRD (*Coliuspasser progne*; L of males in breeding plumage, 40-60 cm; in non-breeding plumage, 20-24 cm; females 15-17.5 cm). This species has the longest tail of all whydahs. The breeding plumage is black with red lesser and white medium wing coverts. It occurs in several widely separated areas of grassland in Africa, in the eastern half of southern Africa, the highlands of Angola to northern Zambia and the highlands of Kenya. The FAN-TAILED WHYDAH BIRD (*Coliuspasser axillaris*; L 15 to 19 cm) resembles it, but is short-tailed even in its breeding plumage. It is found in moist, tall grass in western Africa and from Ethiopia to Angola and the eastern Cape Province. The YELLOW BISHOP (*Coliuspasser capensis*; L 15-16 cm) has a black breeding plumage with a yellow lower back and lesser and medium wing coverts but no prolonged tail. It is found from Ethiopia to Angola and southern and eastern South Africa, and from the Cameroon highlands as far as northeastern Nigeria.

Fig. 17-20. Red-collared whydah (*Coliuspasser ardens*).

Fig. 17-21. Long-tailed whydah bird (*Coliuspasser progne*).

Male whydahs defend their breeding territories against rivals by assuming an upright imposing posture with fluffed out nape feathers and fanned-out tail; they display to the females in much the same way. They continually fly over their territories in a flight which varies depending on the species, some slow, some fast, some in a curve, others in a straight line. In these flights, the shape of the tail and the plumage patterns are displayed. This attracts the attention of females to territories occupied by males of their own species. The song consists of crackling and hissing sounds. Each male builds the groundwork of several nests in its territory. Skead describes the nest building of the fan-tailed whydah bird: The tips of a number of grass stalks are pulled together and occasionally knotted together. The male then weaves among these stalks the framework of the nest using long, thin stalks. He strengthens it around the entry hole. The female which selects this nest adds an inner layer of fine grass tips, often adding new lining material when it is well into the incubation period.

Fig. 17-22. Yellow bishop (*Coliuspasser capensis*).

The breeding biology of *Coliuspasser jacksoni*, which V. D. van Somerren has studied closely, differs considerably from that of other whydahs. Males in their gray non-breeding plumage move about together with the females in flocks. In February they begin to molt into the black breeding plumage. They now leave the flocks, perch on tufts of grass and start to make "dancing places". Each male has several of these; they consist of a tuft of grass about twenty centimeters high around which the grass is trodden down, the whole having a diameter of 90 to 120 centimeters. It keeps the ring of down-trodden grass around the tuft clean. The bird makes an indentation on each side of the oval tuft of grass which is left standing. The dancing places of different males are usually over four meters apart. When a male approaches his dancing place he erects his tail feathers like a cockerel. If another male comes to the dancing place in this posture, he is immedately attacked and chased away, but if he adopts some other posture, he is left in peace. The male never perches on the grass tuft in the center of his dancing ground, instead keeping to the neighboring grass from where he delivers his short, wooden-sounding song phrases. With the head thrown back, fluffed-out neck feathers, curved tail and wildly beating wings, he jumps into the air to a height of up to sixty centimeters. There are rarely more than six such jumps made in rapid succession, after which the bird walks around the central tuft, displays by bowing and further rounds out the indentation of the central tuft of grass with its breast.

If a female approaches the dancing place, the male stops his jumps, erects his neck feathers and shakes the erect tail feathers, moving them so as to brush the female's face. Meanwhile, the female occupies herself about the tuft of grass and pecks at it; finally mating follows. Females, which apparently build the nests alone, come to the dancing places of the males for mating.

The BISHOPS (genus *Euplectes*) are closely related to the whydahs and are reckoned as part of the same subfamily. Males are always short-tailed. The breeding plumage shows, in addition to black, also red (in the golden-backed weaver, golden yellow) areas of plumage. The plumage of females and of males in the non-breeding season is as in the whydah birds, sparrow-like. In voice, mode of life and reproduction, they are also in many regards like the whydahs; thus their nests are generally among grass or reed stalks. The eggs, however, as far as they are known, are less strongly spotted and have a greenish-blue background color. Of the six species, we mention the following: The GOLDEN-BACKED WEAVER (*Euplectes aureus;* L 14 cm) which is thick-billed; the male in breeding plumage is largely black with an intensely yellow back and rump. Originally found only on the island of Sao Tome, it was introduced in coastal areas of Angola, probably by man. The FIRE-CROWNED BISHOP (*Euplectes hordaeceus;* L 14 cm), distributed over most of tropical Africa. The GRENADIER WEAVER (*Euplectes orix;* L 13.5 cm), is found in reeds, tall grass and cereal fields of southern and eastern Africa. The ORANGE BISHOP (*Euplectes franciscanus;* L 12 cm), black only on the crown and sides of the head, occurs from the Senegal to Ethiopia, Somalia and northwestern Kenya. It is often kept as a cage bird in Europe; the red of its plumage is then generally replaced by orange, which accounts for its common name. Brehm gives an enthusiastic description of the impression orange bishops made on him when "like

Fig. 17-23. 1. Orange bishop (*Euplectes franciscanus*); 2. Grenadier weaver (*Euplectes orix*).

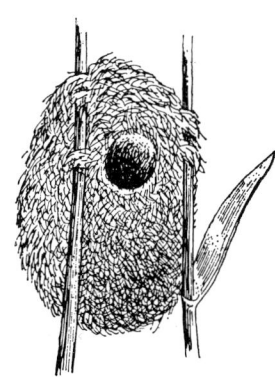

Fig. 17-24. Nest of the grenadier weaver.

bright little flames", they perched on the seed heads and turned to and fro, in a green durra field. (Grenadier weaver, Color plate, p. 430.)

The YELLOW-CROWNED BISHOP *(Taha afra;* L 11.5 cm), widely distributed in the grasslands of Africa in subspecies, is light yellow and black in its breeding plumage. A closely related species, the FIRE-FRONTED BISHOP *(Taha diademata),* lives in eastern Kenya and northeastern Tanzania. The thick-billed *Brachyope anomala;* L 11 cm, looks like a transitional form between the whydah birds and bishops and the diochs and their relatives, which are described next. In inhabits forest clearings along rivers in the Congo area and with its brownish plumage with black sides of the head, black throat, yellow crown and throat, it is also reminiscent of the Indian weavers of the genus *Ploceus.* Its nest, however, resembles that of the bishops but it is not placed in grass or reeds but rather two to six meters above the ground in bushes or trees. (Fire-fronted bishops, Color plate, p. 430.)

The RED-BILLED DIOCH *(Quelea quelea;* L 12 cm) is short-tailed and is distributed in African steppe areas from the Senegal and southern Mauritania to Somalia and southwards to Southwest Africa and central and eastern South Africa. Its habits have been intensively studied recently by several ornithologists. These birds are highly sociable and outside the breeding season occur in swarms which are often enormous. As the breeding season comes on they form large or small breeding colonies from these flocks. The largest breeding colonies may cover an area of several hundred hectares and are estimated to contain up to ten million nests. The nests are usually in trees, particularly in acacias, and sometimes in sedges. They are artfully woven, rounded in shape with an entrance at the side, resembling the nests of bishops and whydah birds. Males offer their not-quite-completed nests to the females, with open beak, raised and at times trembling wings and cocked-up tail. If the female is interested, the male guides her into the nest and accompanies her out again, all the while trembling his wings almost continuously. Pair formation arises from this ceremony and lasts for the rest of the breeding season. Red-billed diochs, in contrast to bishops and whydahs, live in monogamy. The parents share the breeding duties. (Color plate, p. 430.)

Fig. 17-25. Red-billed dioch *(Quelea quelea).*

The young remain in the nest for sixteen days. During the first few days they are fed mainly on arthropods, but later more and more grass seeds are added to their diet. As grass seeds get scarcer, with the progress of the dry season, the birds are forced to take larger grains as well. If the grass seed shortage happens early, red-billed diochs raid the fields of small grains, which at that time have not yet been harvested, where they may take vast amounts of grain. They are therefore attacked by every possible means nowadays, even with flame throwers or by spraying the breeding colonies or roosting places with toxic chemicals. This may one day result in the red-billed dioch sharing the fate of the passenger pigeon (see Vol. VIII) of North America. These birds readily weave their nests in cages; in small cages they weave the nest material between the wires of the cage wall so busily that they were formerly given the name "Le Travailleur" (the worker) in France. The RED-HEADED QUELEA *(Queleopsis erythrops)* and the CARDINAL QUELEA *(Queleopsis cardinalis)* are, like the red-billed dioch, rather inconspicuous African weavers in which, however, the male in breeding plumage does have a red head. They form the link between the red-billed dioch and the Madagascar weavers or fodis.

One species of the Fodis (genus *Foudia*), the REUNION WEAVER (+ *Foudia bruante*) has become extinct and is only known from an 18th Century illustration. There are six living species, originally restricted to Madagascar and neighboring islands, among them: the MADAGASCAR WEAVER *(Foudia madagascariensis;* L 13 cm). Males in breeding plumage are mainly bright red with dark spots on the back, a cheek stripe, with blackish tail and wings with olive-colored feathered edges. The *Foudia eminentissima* is red only on the head, neck, breast and a spot on the rump. It occurs on the Comoro Islands, Aldabra and eastern Madagascar. The RODRIGUES WEAVER *(Foudia flavicans)* resembles it, but red is replaced by yellow and orange-yellow; it occurs only on Rodrigues Island.

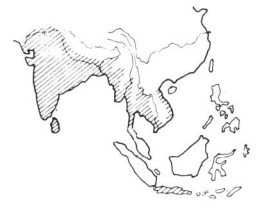

Fig. 17-26. Manyar weaver *(Ploceus manyar).*

The Madagascar weaver originally only lived on Madagascar and was introduced by man on St. Helena. Its present distribution on the Mascarene and Chago Islands, the Comoros, Seychelles and other islands in the Indian Ocean is probably also due to human agency. It lives in open country and forms large flocks after the breeding season. It is regarded as harmful in the rice fields. Like all the weavers of this genus, it lives in monogamy and builds an elongated round nest. The *Foudia emintissima* is much more a forest dweller, which feeds largely on arthropods which it seizes while climbing about like a chickadee. The voice of the Madagascar weaver is not quite so unpleasant as that of most other weavers; the Rodrigues weaver even has a pleasant goldfinch-like song.

Fig. 17-27. Golden weaver *(Plocella hypoxantha).*

Subfamily: true weavers, by H. E. Wolters

The TRUE WEAVERS (subfamily *Ploceinae*) are closely related to the whydahs, bishops, etc., and possibly hardly separable from them. Here those weaver species which not only share many features of appearance, structure and behavior, but which also show the maximal skill in nest building, are joined together. Of the sixty-eight species which can be assigned to fifteen genera, all but five southern Asiatic and two Madagascar species live in Africa south of the Sahara; one also lives in southern Arabia.

It appears that the true weavers were originally inhabitants of the open savanna where numerous species, generally living in polygyny, built large nestling colonies. In the course of evolution, some species penetrated into the forests and then switched to a monogamous and non-communal life. The shape of the nest also changed; while in ancestral species the nest looked like that of a whydah bird, it is retort-shaped but without a long entry hole in the more highly developed steppe weavers, while the forest dwellers construct nests which often have a very long entry tube. Forest dwellers which later moved back to the steppes generally retained their nest form. Nest building proceeds as follows: The male first weaves an upright ring onto branches or between reeds, then it extends it to one side to form the breeding chamber while it weaves the entrance and in some cases also the entry tube onto the other side. The female's duty is merely to line the interior.

True weavers eat, besides grass seeds, far more arthropods than do the whydah birds, bishops and diochs and some even feed mainly on insects. The diet of some species, particularly the forest dwellers, also includes fruit. As in the whydahs, the majority of true weavers show an annual alternation between a colorful breeding plumage and an inconspicuous non-breeding plumage, but the latter is absent in most of the forest dwellers. The major plumage colors are yellow, yellowish-green, chestnut brown and black. Only members of the genera *Malimbus* and *Anaplectes* have red patterns.

The weavers of southeastern Asia (three species of the genus *Ploceus* and both species of the genus *Plocella*) show primitive features in their plumage and nest construction. The following will be mentioned: The BAYA WEAVER (*Ploceus philippinus*; L 13.5 cm), is found from Pakistan, India and Ceylon to Thailand and Sumatra; the MANYAR WEAVER *(Ploceus manyar)* which occurs from Pakistan, India and Ceylon to southwestern China, Java and Bali; and the GOLDEN WEAVER (*Ploceella hypoxantha*; L 13.5 cm) which has a short, thick beak. The male's breeding plumage is golden yellow with black sides of the head and throat, the upper back has black spots and the lower back is brown, the wings and tail are blackish-brown. Females and non-breeding males are rust brown with dark spots on the upper parts. It is found in Burma, Thailand, Cambodia, South Vietnam, Sumatra and Java.

The baya weaver breeds colonially. Its nests, with their long entry tubes, are seen on trees, particularly palms and in the eastern part of its range often on the straw roofs of verandahs; the manyar weaver on the other hand, builds its nests in reeds or tall grass; their nests do not always have an entry tube. The nest of the golden weaver has an entrance at the side but no entry tube.

In the STEPPE WEAVERS (genus *Textor*) as well as in the *Othyphantes baglafecht*; L 15 cm and its relatives, the roof of the nest consists of a double layer; this is evidently for protection from the sun's rays. With twenty-seven species, the steppe weavers constitute the largest genus of true weavers and they show their peculiarities in the most highly developed form. All live in Africa, one species also living in southern Arabia. They build retort-shaped nests with an entry towards the

▷
Weaverfinches: 1. Cut-throat finch (*Amadina fasciata*); 2. Three-colored mannikin (*Lonchura malacca malacca*); 3. Spice finch (*Lonchura punctulata*); 4. Bronze mannikin (*Spermestes cucullatus*); 5. Gouldian finch (*Chloebia gouldiae*); 6. Diamond sparrow (*Stagonopleura guttata*); 7. Chestnut-breasted finch (*Lonchura castaneothrorax*); 8. Pin-tailed nonpareil (*Erythrura prasina*); 9. Long-tailed grass finch (*Poephila acuticauda*); 10. Zebra finch (*Taeniopygia guttata castanotis*); 11. Parrot finch (*Eyrthrura psittacea*); 12. Crimson finch (*Neochmia phaeton*); 13. Painted finch (*Emblema picta*); 14. Java sparrow (*Padda oryzivora*); 15. Sydney waxbill (*Aegintha temporalis*).

▷▷
Weaverfinches: 1. Avadavat (*Amandava amandava*); 2. St. Helena waxbill (*Estrilda astrild*); 3. Orange-cheeked waxbill (*Estrilda melpoda*); 4. Golden-breasted waxbill (*Amandava subflava*); 5. Yellow-bellied waxbill (*Estrilda melanotis*); 6. South African quail finch (*Ortygospiza atricollis*); 7. Violet-eared waxbill (*Uraeginthus granatinus*); 8. Lavender finch (*Estrilda caerulescens*); 9. Red-cheeked waxbill (*Uraeginthus bengalus*); 10. Blue-billed fire finch (*Lagonosticta rubricata*); 11. Masked waxbill (*Lagonosticta larvata*); 12. White-collared oliveback (*Nesocharis ansorgei*); 13. Peter's twin-spot (*Hypargos niveoguttatus*); 14. Green-backed twin-spot (*Mandingoa nitidula*); 15. Black-bellied seed-cracker (*Pyrenestes ostrinus*).

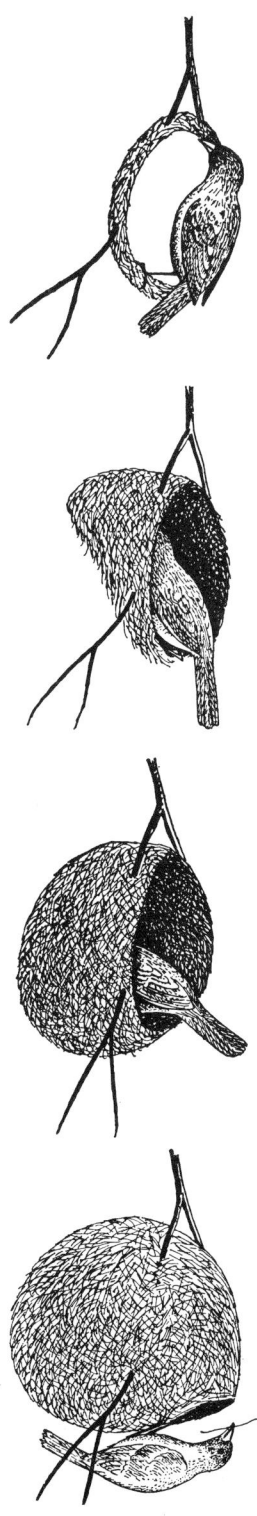

Fig. 17-28. Stages in the formation of the nest of a weaver of the genus *Textor*.

bottom which has, in some cases, a short entry tube. Depending on the species, the nests stand in reeds or are on trees, those on trees often attached to branches which hang down over water. Nests are often near human settlements, even in the middle of villages. These weavers generally live in polygyny and often breed in large colonies, sometimes of more than two hundred nests. After the breeding season, they form large flocks. The male's breeding plumage is usually yellow with greenish, black and reddish-brown areas. The male's non-breeding plumage and that of the female is greenish or brownish. In our zoos these weavers and the orange bishop are the most frequently exhibited weavers. They are less popular as domestic cage birds because of their loud, unpleasant utterances.

Textor xanthops; L 17 cm, has no non-breeding plumage. It lives in reeds and the bushes of shores. Its behavior is somewhat different from the others; it feeds mainly on arthropods and fruit and also eats coffee beans in plantations. The CAPE WEAVER (*Textor capensis*; L 17 cm), of South Africa, is the commonest weaver of the southern and western Cape Province. Besides seeds and arthropods, it also takes the nectar of flowers. The BLACK-FRONTED WEAVER (*Textor velatus*; L 15 cm) is the commonest weaver of the dry areas in western and central South Africa. It breeds in small colonies in reeds or in trees, often in the gardens of farms or at the edge of settlements. RÜPPELL'S WEAVER (*Textor galbula*; L 14 cm) occurs in dry steppeland in northeastern Africa and southwestern Arabia. The BLACK-HEADED WEAVER (*Textor melanocephalus*; L 15 cm) of central Africa is yellow with a black head in the breeding plumage. *Textor jacksoni* is similar but its lower parts in the breeding plumage are reddish-brown; it is found in eastern Africa. The GIANT WEAVER (*Textor grandis*; L 20 to 22 cm) resembles the black-headed weaver, but it is much larger. It is found on the island of Sao Tome. The BLACK-HEADED VILLAGE WEAVER (*Textor cucullatus*; L 17 cm) occurs, in many subspecies which differ considerably in the color of their plumage, over most of tropical Africa and eastern South Africa.

The black-headed village weaver has advanced from the steppes and savannas into the clearings of the jungle. It builds large colonies of nests on palms or other tall trees, often in the middle of villages. As Bannerman says, "A colony in the breeding season makes an unforgetable impression. From November onwards, hundreds of nests are swinging from the tips of the branches. Each nest is suspended from its own twig and artfully woven from strips of palm leaves. The birds collect these strips by biting a small slit into the leaf, holding onto one end of this with the beak and flying off, thus tearing off a strip." In other colonies the nests are constructed from grass stalks. Due to the fact that one-year-old males do not reproduce, there is an excess of females. Males offer their nests to them by hanging beak down and beating with the wings from the entrance of the short entry tubes, a form of behavior which in different forms is also shown by other weavers.

The nine species which form the genus *Malimbus* are predominantly inhabitants of the African primeval forests. Most of them are red and black. This genus includes the only weaver with a crest, the CRESTED MALIMBE (*Malimbus malimbicus*; L 16 cm). The RED-HEADED WEAVER (*Anaplectes rubriceps*), an inhabitant of well-treed savannas and dry forests, is closely related to the malimbes. The peculiar brownish *Thomasophantes, sanctithomae*; L 14 cm) of the island of Sao Tome, the BAR-WINGED WEAVER (*Notiospiza angolensis*; L 14 cm) and the six species of the genus *Phormoplectes*, among them the BROWN-CAPPED WEAVER (*Phormoplectes insignis*) all climb about like chickadees or tree creepers. The FOREST WEAVER (*Symplectes bicolor*; L 15 to 18 cm) has a peculiar voice which differs considerably according to locality. For further genera and species, see the systematic review.

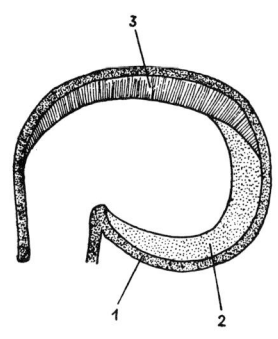

Fig. 17-29. Section of the nest of a weaver of the genus *Textor*. 1. framework; 2. lining; 3. inner roof.

The THICK-BILLED WEAVERS (tribe *Amblyospizinae*) have remained more primitive than the whydahs and the true weavers, but they are related to both. There are two species: The BLACK SWAMP WEAVER or WHITE-FRONTED GROSBEAK (*Amblyospiza albifrons*; L 18 cm) in which the male is blackish-brown to rust-brown with a white forehead and a white area on the base of the secondaries and the female is brown above and paler beneath with dense dark striping. It is widely distributed in Africa in numerous subspecies. The SAO TOME GROSBEAK (*Neospiza concolor*), brown, was only found in Sao Tome and is apparently extinct. It is debateable whether it really belongs to the weavers or finches. The black swamp weaver draws attention by its woodpecker-like undulating flight. It lives in reed beds, papyrus thickets or tall grass where it builds a nest skillfully woven from stalks which has an entrance at the side near the top. (Black swamp weaver, Color plate, p. 430.)

Subfamily: thick-billed weavers, by H. E. Wolters

The BUFFALO WEAVERS (subfamily *Bubalornithinae*) have been regarded by several ornithologists as a distinct family. They are the largest of the weavers, ranging in length from 22 to 24 cm. They are reminiscent of starlings in many ways, but are immediately distinguished from them by their thick beaks. The two sexes are colored similarly; there is no alternation between a breeding and a non-breeding plumage. There are two genera, each with only one species, both confined to Africa. The BLACK BUFFALO WEAVER (*Bubalornis albirostris*) males are mainly black, females blackish-brown. In the breeding season, the male, according to subspecies, has a white or salmon pink beak; the beak may also be peculiarly bloated. It inhabits dry thornbush steppes. The other species, the WHITE-HEADED BUFFALO WEAVER (*Dinemellia dinemelli*) is found in eastern Africa. (Color plate, p. 430.)

Subfamily: buffalo weavers, by H. E. Wolters

One or more male buffalo weavers initially build several closely placed nests in the tree branches; the nests are made from thorny twigs. Then the nests are surrounded with thorny branches and the resulting

Fig. 17-30. 1. Rüppel's weaver (*Textor galbula*); 2. Cape weaver (*Textor capensis*).

structure can often be quite large, containing several incubation chambers which looks like that of the sociable weaver (see below). The females carefully line the egg chambers with shoots. Nests of the white-headed buffalo weaver are often nearby.

Subfamily: sparrow weavers, by H. E. Wolters

The SPARROW WEAVERS (tribe *Plocepasserinae*) are related to the buffalo weavers. They are primitive-looking weavers which are brown, gray, black and white. There are eight species which include the RUFOUS-TAILED WEAVER (*Histurgops ruficauda*; L 21 cm) of eastern Africa; the GRAY-HEADED WEAVER (*Pseudonigrita arnaudi*; L 12 cm) also of eastern Africa; and the MAHALI WEAVER BIRD (*Plocepasser mahali*; L 17 to 18 cm), distinguished by its loud, pleasant song; it is found in dry steppe areas of central South Africa, southwestern Angola and eastern Africa as far as Somalia. In the SOCIABLE WEAVER (*Philetairus socius*; L 14 cm) both sexes are grayish-brown above, with a back which, because of pale feather edges, looks scaly; a cheek stripe and the throat are black, the rest of the lower parts are dull grayish-brown with a few black feathers with pale edges on the flanks. It occurs in arid areas of southern and southwestern Africa.

Fig. 17-31. Crested Malimbe (*Malimbus malimbicus*).

Nests of sparrow weavers are not woven as well as those of the true weavers and whydah birds and bishops. They stand in trees, are retort-shaped and have two entries, (except in the sociable weaver). If the birds use the nest for breeding, they close one entry; if they use it only to sleep in, they leave both open. In many places, the nests of the gray-headed social weaver are regularly placed on acacias, whose galls are inhabited by aggressive ants. The communal nests of the sociable weaver are among the most striking structures made by birds. Building is begun by several birds working together erecting on a strong tree branch, or now sometimes on a telephone pole, a roof of twigs and stalks of the grasses characteristic of the arid areas. In time this grows to a huge mass of material and individual pairs of the birds, which live in monogamy, build their nest chambers into the mass. All that can be seen of the nests are the downward-directed entries. Such communal nests may be used for years and enlarged year after year. Finally the supporting branch breaks under the load and the collective nest tumbles down. N. E. and E. C. Collias saw a colony of these communal nests which was already a century old. The largest nest mass was 4.8 meters long and 3.6 meters wide and had 125 nest entries. Pigmy falcons, small parrots, scaly weavers and red-headed finches often breed in vacant nest chambers.

Fig. 17-32. Black buffalo weaver (*Bubalornis albirostris*).

Subfamily: scaly weavers, by H. E. Wolters

The SCALY WEAVERS (subfamily *Sporopipinae*) are reminiscent of the weaver finches in their external appearance and some aspects of their behavior. According to the recent studies of Kunkel and Ziswiler, they do, however, definitely belong to the weavers, of which they represent a group which has remained primitive. There are only two species: the SPECKLE-FRONTED WEAVER (*Sporopipes frontalis*; L 13 cm), which has a scaly-looking black and white front of the crown and a reddish-brown nape, is otherwise grayish-brown above and has whitish underparts, as well as a black stripe with white dots on each side of the throat. It is found in dry steppe country from Senegal to northern Ethiopia and northern Tanzania. The SCALY WEAVER (*Sporopipes squamifrons*; L 11 cm) inhabits dry areas in South Africa. (Color plate, p. 430.)

Fig. 17-33. 1. Gray-headed weaver (*Pseudonigrita arnaudi*); 2. Sociable weaver (*Philetairus socius*).

Scaly weavers are less tied to water holes than many other birds of the dry areas for they can make do without water for a long time. T. J. Cade has experimented with captive scaly weavers which had free access to water. They consumed only five per cent of their body weight of liquid in twenty-four hours compared to a twenty to forty per cent intake in other small seed-eaters. They could endure sixty-two days without any water at all without showing any signs of impairment whereas most comparable small birds died after a few days without water. As long as the experiment lasted, these weavers were fed only air-dried seeds. Their round, disorderly nests resemble those of weaverfinches and have a projecting entrance at the side.

WEAVERFINCHES (family *Estrildidae*) are related to the weavers. They differ clearly from them not only in external appearance, but also in behavior and in a number of characteristics of the digestive tract. For these reasons, they are now generally regarded as a distinct family which has diverged far from the common ancestral stock which was shared with the weavers. In size they vary from that of a wren to that of a linnet. The beak in those species which are almost exclusively insectivorous is slim as in a warbler; in the eaters of large seeds like the bluebills and the black-bellied seedcracker, it is almost as thick and strong as in the hawfinches. Particularly characteristic are the wart-like projections or swellings of thickened connective tissue shown by the young at the edges of the beak and at the gape. These are a striking white, blue or yellow color, often emphasized by black surroundings. In the Gouldian finch and the parrot finches these structures have been developed into organs which reflect light and thus show up in the semi-darkness of the nest. The gape pattern, dark spots or lines on the palate, the tongue and the floor of the mouth of the nestlings which differs according to the genus and species, is also characteristic of the weaverfinches. In contrast to the colored bulges and warts at the angles of the gape, the patterns in the interior of the mouth are in many cases retained for life. The plumage is sometimes inconspicuous but often very attractively colored; it is never, as in many weavers, striped in a sparrow or bunting-like fashion. As in the whydahs, the outermost primary is generally very much shortened. There are about thirty-five genera with about 125 species.

Family: weaver finches, by H. E. Wolters and K. Immelmann

Fig. 17-34. Communal nest of the sociable weaver.

Most weaverfinches live in grass or bush steppes, savannas and open dry area forests. A few have penetrated deserts and semi-deserts, particularly in South Africa and central Australia. Others resumed their

Fig. 17-35. 1. Speckle-fronted weaver (*Sporopipes frontalis*); 2. Scaly weaver (*Sporopipes squamifrons*).

tribe's earlier position as true forest dwellers, particularly in western and central Africa, southeastern Asia and the Indo-Australian insular area. Recently, several species have become closely linked with man and have moved into fields and gardens, some even into parks of great cities. The red-billed fire finch comes into native huts in its search for food; zebra finches and crimson finches even breed on and in buildings.

The predominant food of weaverfinches is half-ripe and ripe grass seeds. Particularly in the breeding season many also take arthropods. They are attracted by the nuptial swarms of ants and termites at the beginning of the rainy season and they pick the insects up from the ground or catch them like flycatchers in a short fluttering flight. Some Australian weaverfinches have developed a manner of drinking which is evidently unique among passerine birds and occurs outside this order in only a few groups of birds. They suck in the water like pigeons, immersing the beak almost up to its base. This behavior has evolved independently several times in birds of arid areas probably because the birds are exposed to danger at the water holes; by sucking up the water they can reduce the time required to stock up with fluid.

In recent years the display behavior of weaverfinches has been studied with particular intensity. In most species the male has a "display dance" in which it either hops towards the female while singing or, usually also while singing, performs characteristic bows or stretching movements and hops about in front of the female. Many avadavats and related species and some Australian weaverfinches hold a feather or a grass stalk in the beak during this display, apparently as a nest symbol. The female weaverfinch's way of indicating readiness for mating is unique among song birds. She cowers on a branch and trembles her tail, which is held vertically while the wings are kept still. In contrast, other song bird females tremble their wings and keep the tail quite still.

The nests are always roofed over and are, as a rule, almost spherical with a diameter of about ten to twenty centimeters. Many species attach a long entry tube to the nest, but this, in contrast to many weaver nests, never hangs down vertically. Usually both partners participate in nest building; males mainly bring the nest material and the females build with it. Most species use fresh or dry grass stalks and many line the nest cup with feathers or other soft materials. In many cases nests are built outside the breeding season as well and these are used for roosting. In many species a whole group of birds use such sleeping nests together. Nests are usually placed in bushes or low trees. Some species breed on the ground, while others suspend their nests between grass stalks or reeds or breed in tree holes.

The clutch usually consists of four to six, with rare cases of up to nine eggs. Both sexes incubate. In the daytime they relieve one another at approximately equal intervals of about one and a half hours, while at night both sexes in many species sit together in the nest. The male when appearing for relief at the nest often brings as a "present" a bit of grass or feather. The incubation period is twelve to sixteen days. The young receive mainly half-ripe seeds. The parents regurgitate these in small portions from the crop and push the food into the young bird's gape. The nestling with wide open beak grabs the adult's beak about the angles. The begging posture of the young is also unique among song birds. They do not stretch the head and neck towards the parents, but lay the neck flat on the nest floor, turning only the gape upwards. Their beaks are wide open, displaying the characteristic pattern inside the mouth and the head is moved from side to side and turned in a lively manner. The trembling wing movements so characteristic in other food-begging young birds are quite absent. This begging posture is retained after leaving the nest. The fledging period lasts exactly three weeks, which is surprisingly long for such small birds. Even after fledging, the young have not definitely left the nest, for the parents guide them back to it for sleeping and, at first, even for every feeding. They remain dependent for one or two weeks after fledging.

Though weaverfinches in contrast to many other song birds have neither long nor very attractive songs and virtually never become tame, they are at present among the most popular and frequent cage birds. Their deficiencies are balanced in most species by attractive colors and patterns; they are lively, sociable, in most cases peaceable and not demanding in their maintenance. They are particularly suitable for flight cages where a mixed group of different types and colors can be made up. If planting within the cage and feeding are suitable, breeding can confidently be expected in a short period. Density in the flight cage should be such as to allow one cubic meter of space per pair. Higher densities reduce the chances of breeding.

Many species are less suitable for maintenance in smaller indoor cages. Because of their need for company, weaverfinches should never be kept alone; there should be at least a pair of every species. A cage for a pair should be at least seventy centimeters long and forty centimeters high. A cage for a group should be correspondingly larger. As tropical birds, most weaverfinches are very sensitive to cold and must be maintained at an even temperature of eighteen to twenty-three degrees centigrade. The North Australian Gouldian finch only feeds really well at thirty degrees. The basic food for all weaverfinches is a millet mixture; to this are added chickweed, various herbs and wild grasses, minced salad, meal worms, hard-boiled eggs, ant pupae and various other supplements. In winter the lack of green food must be made up from artificially-germinated millet. In large, well-planted flight cages, the birds often build their round nests in dense shrubs, but it is advisable also to hang up ordinary or half-open nest boxes, which they readily accept. In small cages, nest boxes usually provide the only possibilities for nesting. For building material the birds should be provided with fresh and dry grasses and various fibers such as that of the coconut and bast, as well as feathers and other soft materials for nest lining.

If the weaverfinches share a common origin with the weavers, then the two species of the genus *Amadina* are probably close to the appearance of the ancestral weaverfinches. The CUT-THROAT FINCH (*Amadina fasciata;* L 11.5 cm) is widely distributed in dry steppe country from the Senegal to Ethiopia and southwards through eastern Africa as far as the Transvaal. In the RED-HEADED FINCH (*Amadina erythrocephala;* L 14 cm), males have a red head and throat and the rest of the upper parts grayish-brown; the lower parts are a drab red-brown with black and white bars on the tips of all the feathers; the beak is a brownish color. Females have no red on the head and are dull brown below with dark bars. It is found in central and western South Africa as far as the coastal region of Angola.

Fig. 17-36. 1. Cut-throat finch *(Amadina fasciata)*; 2. Red-headed finch *(Amadina erythrocephala)*.

These two weaverfinches belong to the few in their family which have only a poorly developed nest building drive. They rarely build nests standing clear; they prefer to use the old nests of various weavers which they merely reline somewhat, perhaps reducing the size of the entry. In the Kalahari Desert, the red-headed finch mainly chooses the communal nests of the sociable weaver. Because of the latter's own sociability, several pairs of the finches often breed in neighboring nest chambers. As the most marked inhabitants of arid country among the African weaverfinches, both species wander about outside the breeding season in flocks often of thousands in search of water holes. Their song is a ventriloquial purring which can only be heard at a short distance.

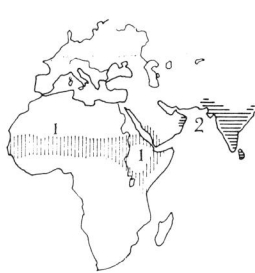

Fig. 17-37. 1. African silver-bill *(Euodice cantans);* 2. Indian silver-bill *(Euodice malabarica)*.

The two species of the genus *Euodice* are also dully colored inhabitants of dry steppe country. The AFRICAN SILVER-BILL (*Euodice cantans;* L 11 cm) is grayish-brown with a black rump. It is found from the Senegal to eastern Africa and southwestern Arabia. The INDIAN SILVER-BILL (*Euodice malabarica;* L 11 cm) has a white rump. It occurs in India and Pakistan. Both species have a silvery-gray beak.

The more attractively colored GRAY-HEADED SILVER-BILL (*Odontospiza caniceps;* L 12 cm) has a gray head, white dots on the front of the head and throat, a brown back and cinnamon-colored lower parts. Its home is eastern Africa. It is related both to the preceding species and to the MANNIKINS *(Spermestes)*, particularly the MAGPIE MANNIKIN (*Spermestes fringilloides;* L 12 cm). This has a black head, brown wings and black and white lower parts with a cinnamon-colored spot on the flanks. The other three species of mannikins are much smaller. They are the BLACK-AND-WHITE MANNIKIN (*Spermestes bicolor;* L 10 cm), the BRONZE MANNIKIN (*Spermestes cucullatus;* L 9 cm), both African, and the BIB FINCH (*Spermestes*

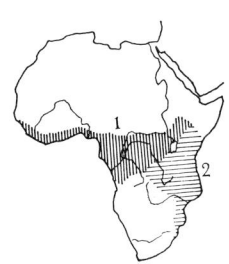

Fig. 17-38. Black-and-white mannikin *(Spermestes bicolor);* 1. Black-backed subspecies; 2. Brown subspecies.

nanus; L 9 cm), brownish with gray sides of the head and a black throat spot, of Madagascar. The magpie mannikin has a great liking for rice grains, and its local distribution in its African home is much influenced by rice cultivation.

Most of the weaverfinches of southeastern Asia and New Guinea belong to the genus *Lonchura.* They have strong black or bluish-gray beaks. The plumage is mainly brown, black and white, in many cases with yellow or reddish-brown edges on the tail feathers. The central tail feathers are pointed in most species (hence the name lancet tails and the scientific name of the genus). There are twenty-eight species ranging from India to Australia. Many are well known domestic cage birds in Europe. Among them are: the SPICE FINCH (*Lonchura punctulata;* L 12 cm) occurs from India to Taiwan and the Lesser Sunda Islands. The SHARP-TAILED MUNIA (*Lonchura striata;* L 11 cm) has the front of the head blackish, is brown above with pale stripes along the shafts of the feathers; the rump has a white bar across it, the upper breast is brown with pale feathered edges and the rest of the lower parts are white with fine brown stripes in many subspecies. It is the ancestral form of the domesticated society finch. It occurs from India and Ceylon to southern China and Sumatra. The SPECTACLED FINCH (*Lonchura spectabilis;* L 10 cm) has a black head and throat, brown back and wings, with a yellow lower rump and yellow edges of the tail feathers. The lower tail coverts are black and the rest of the lower parts are white or brownish. It ranges from New Guinea to New Britain. The THREE-COLORED MANNIKIN (*Lonchura malacca;* L 11 cm) is found from India and Ceylon to the Philippines, Halmahera, Java and Bali. The WHITE-HEADED MANNIKIN (*Lonchura maja;* L 11 cm) is dull chestnut brown, with the middle of the lower parts black and the head and throat white. It occurs in southern Thailand, Malaya, Sumatra, Java and Bali. The YELLOW-TAILED FINCH (*Lonchura flaviprymna;* L 11 cm) has a whitish-gray head and neck, chestnut brown back and wings, an ochre yellow lower rump and central tail feathers and black lower tail coverts. The rest of the lower parts are creamy yellow. It lives in western and central Australia. The CHESTNUT-BREASTED FINCH (*Lonchura castaneothorax;* L 11 cm) frequently hybridizes with the yellow-tailed finch. The PECTORELLA FINCH (*Lonchura pectoralis;* L 12 cm) is a brownish-silver gray above, the wing coverts and inner secondaries having white spots on the tips. There is a rusty-brown stripe above the eye, the sides of the head and throat are black, the breast feathers are white, each feather having a partly concealed black bar, and the abdomen is reddish-gray with black and white spots at the sides. It occurs in western and central Australia and is highly adapted to life on the ground.

Several centuries ago, the Chinese bred the society finch, which became a completely domesticated bird, from the sharp-tailed munia. For about 250 years, breeding of the society finch was carried

Fig. 17-39. 1. Bronze mannikin (*Spermestes cucullatus*); 2. Bib finch (*Spermestes nanus*).

Fig. 17-40. 1. Spice finch (*Lonchura punctulata*); 2. Areas of introductions.

▷

Starlings: 1. Black-necked mynah (*Gracupica nigricollis*); 2. Common mynah, Ceylonese subspecies (*Acridotheres tristis melanosternum*); 3. Yellow-billed ox-pecker (*Buphagus africanus*); 4. Singing starling (*Aplonis cantaroides*); 5. Coleto mynah (*Sarcops calvus*); 6. Dumont's mynah (*Mino dumontii*); 7. Wattled starling (*Creatophora cinerea*); 8. Celebesian starling (*Scissirostrum dubium*); 9. Javan hill mynah (*Gracula religiosa*); 10. Bald-headed wood shrike (*Pityriasis gymnocephala*); 11. Daurian mynah (*Sturnia sturnina*); 12. Pagoda starling (*Temenuchus pagodarum*); 13. Rose-colored starling (*Pastor roseus*); 14. Bali mynah (*Leucopsar rothschildi*).

Fig. 17-41. Three-colored mannikin *(Lonchura malacca)*; 1. Subspecies *(Lonchura malacca malacca)*; 2. Subspecies *(Lonchura malacca ferruginosa)*; 3. Other subspecies.

Fig. 17-42. Pectorella finch *(Lonchura pectoralis)*.

◁
1. Flame minivet *(Pericrocotus flammeus)*; 2. Greater racket-tailed drongo *(Dicrurus paradiseus)*; 3. Papuan mountain drongo *(Chaetorhynchus papuensis)*; 4. Spangled drongo *(Dicrurus hottentottus)*; 5. Ashy minivet *(Pericrocotus divaricatus)*; 6. Cuckoo-shrike *(Campephaga phoenicea)*; 7. Barred cuckoo-shrike *(Coracina lineata)*; 8. *Lalage aurea*; 9. White-winged triller *(Lalage sueurii)*; 10. Maroon oriole *(Oriolus traillii)*; 11. *Coracina azurea*; 12. Golden oriole *(Oriolus oriolus)*; 13. Black-naped oriole *(Oriolus chinensis)*; 14. Yellow figbird *(Sphecotheres flaviventris)*.

out mainly in Japan and, for the past hundred years, also in Europe. The most variously colored types, brown and colored, yellow and colored, and white, were produced by selective breeding. The birds were often used to rear young weaverfinches of other species. Ernst Schäfer describes the behavior of the southern Chinese subspecies of the SHARP-TAILED MUNIA *(Lonchura striata swinhoei)* as follows: "These extraordinarily sociable birds like to keep to the edges of rice fields, weedy gardens, pastures, cemeteries and the copses near temples, where, quite fearless, they often let people approach to five to ten meters from them. Their sociability is such that they seek food in flocks even in the breeding season; they do not, however, mingle with other species. Their large, very bulky nests of grasses and moss are generally placed on pines or spruce one or two meters above the ground."

The spectacled finch feeds largely on grass pollen. Heinroth writes about it: "They fly about the dense grass stalks which are as tall as a man, with a whirring, somewhat clumsy flight. Then they all make for a bush together, for they prefer to rest on some horizontal branch rather than on the vertical stalks. They are found only in the true grasslands with their tall hard alang alang plants. No doubt they wear down their claws, which in captivity reach great lengths, on this sharp-edged plant. A flock of these birds is not exactly shy, but nevertheless cautious and shows far less confidence than the birds of our gardens and parks."

The *Lonchura* weaverfinches without exception are good climbers, moving skillfully even on vertical reed stalks. They often form large flocks and keep close together even in flight. Like flocks of starlings, these finch flocks can perform sudden sharp turns and drop almost vertically into reed beds or stands of grass. In contrast to the majority of weaverfinches, most Lonchura weaverfinches do not spend the night in sleeping nests but, again like starlings, roost in flocks in reed and tall grass, or occasionally in dense shrubbery. When caged they have little opportunity to wear down their claws by climbing and the claws often reach an abnormal length. It is therefore necessary to trim them regularly with scissors or else the birds may get caught in the cage wire and injure themselves.

The songs of birds of this genus are remarkable. In some species the song is literally inaudible; only the beak movements show that the bird is really singing, although often towards the end of the song one can hear a few very high-pitched soft tones. It has not been ascertained whether they sing at sound frequencies beyond the human range, and assuming a different sensitivity in their organs of hearing, can hear one another, if only over short distances; or whether some segments of the song are, in fact, soundless and act on other birds only as visual stimuli. Recent investigations suggest the latter. All other calls of these birds, especially their call of attraction, are in the usual sound frequency range and can be heard over great distances.

Closely related to those are the two species of the genus *Padda*. The JAVA SPARROW (*Padda oryzivora*; L 14 cm) is the better known. The largest of the weaverfinches, it has been carried from its original home in Java and Bali to many areas in southeastern Asia, the Hawaiian Islands, St. Helena, Zanzibar and the coasts of eastern Africa and has become established there. The widely distributed and popular white variety was first bred in captivity in China and later in Japan.

The GOULDIAN FINCH (*Chloebia gouldiae*; L 11 cm, with the thread-like prolonged central tail feathers 15 cm) is probably the most colorful of all weaverfinches. It too is a popular cage bird. It lives near water in open grassy areas, particularly the eucalyptus savannas of northern Australia. Among all the weaverfinches, it is one of the most sociable species. Some have given up breeding in the open, but use holes in trees and in such cases occasionally lay their eggs directly on the debris of the floor of the hole without building a nest. (Color plate, p. 439.)

Fig. 17-43. 1. Gouldian finch *(Chloebia gouldiae)*; 2. Areas visited in the non-breeding season; 3. breeding area.

The Gouldian finch needs more heat and sunshine than do other weaverfinches. When the temperature in the shade is between forty-four and forty-five degrees centigrade, when other weaverfinches are panting with open beaks and begin returning to the interior of trees and shrubs, it evidently feels particularly good and at such times shows the greatest liveliness. At lower temperatures and under overcast skies, on the other hand, it is considerably less active. Like a number of other Australian weaverfinches, it too is distinguished by signs of a certain degree of early maturity. Young males often begin to sing just before leaving the nest and a few days later perform their first display actions. Under favorable conditions, these birds begin to breed when only eight months old. In such cases, their molt is interrupted or continued very slowly. Thus, there may be families in which both the parents and the young are still in the juvenile plumage. The purpose of this early maturation is evidently to ensure that if conditions become favorable, as many individuals as possible shall be capable of breeding so that the population, which becomes reduced during the long dry periods, can recover rapidly.

The PARROT FINCHES (genus *Erythrura*) are almost as colorful as the Gouldian finch. Parrot finches inhabit the Indo-Australian insular world from the Greater Sunda Islands and Philippines to Samoa in ten species. Two species also occur on the Malayan mainland and one has settled northern Australia from New Guinea. Some of the species are: the GREEN-TAILED PARROT FINCH (*Erythrura hyperythra*; L 11 cm) which has a narrow black band followed by a wider blue band on the forehead; the rest of its upper parts are green, and the sides of the head and lower parts are ochre yellow. It occurs in bamboo stands and at the edge of mountain forests in the highlands of Malaya, Java, Borneo, Celebes,

Luzon, Mindao and the Lesser Sunda Islands. The PIN-TAILED NONPAREIL (*Erythrura prasina*; L including the thread-like prolonged central tail feathers about 15 cm) is locally distributed from Laos and northern Thailand to Malaya, Sumatra, Java and Borneo. The BLUE-FACED PARROT FINCH (*Erythrura trichroa*; L 12 cm) has the forehead and sides of the head blue; the lower rump, upper tail coverts and the central tail feathers are red, while the rest of it is green. It occurs in Celebes, the Moluccas, New Guinea, the Cape York Peninsula of northern Australia and many islands of the southwestern Pacific. The RED-HEADED PARROT FINCH (*Erythrura psittacea*; L 12 cm) is found in New Caledonia.

The three species of GRASS FINCHES (genus *Poephila*) are marked by their delicate brown and gray plumage with a wide black stripe at the sides of the body. They are confined to the north and east of Australia. The LONG-TAILED GRASS FINCH (*Poephila acuticauda*; L including the pointed and prolonged tail feathers 17 cm) has a yellow or orange-red beak. Its home is northeastern Australia. The PARSON FINCH (*Poephila cincta*; L 11 cm) has a black beak; it is also found in northeastern Australia. The MASKED GRASS FINCH (*Poephila personata*; L 12 cm) of northern Australia, is brown with a black facial mask and a yellow beak. The different beak colors of the individual species evidently help prevent the formation of mixed pairs.

Fig. 17-44. 1. Long-tailed grass finch *(Poephila acuticauda)*; 2. Parson finch *(Poephila cincta)*.

The ZEBRA FINCH (*Taeniopygia guttata*; L 10 cm) is more widely known that any other weaverfinch and it is bred in many color phases. Those normally seen in captivity are descended from the subspecies *Taeniopygia guttata castanotis,* which lives in most of Australia except certain coastal areas. In the eastern subspecies, *Taeniopygia guttata guttata* of the Lesser Sunda Islands, the dark cross-barring of the gray throat in males is lacking in the middle. Zebra finches inhabit open grasslands and grasslands with isolated trees or bushes scattered about. It also occurs in the spinifex grassland of central Australia and in the settled areas has become an occupant of man-made habitats such as cultivated fields, pastures and gardens. (Color plate, p. 439.)

Fig. 17-45. Zebra finch *(Taeniopygia guttata)* and its subspecies. 1. *(Taeniopygia guttata castanotis,* and 2. *Taeniopygia guttata guttata).*

It adapted to life in the dry areas of Australia to a remarkable degree; apart from the painted finch it is the only species which penetrates as far as the deserts and semi-deserts of the interior. Its most difficult problem is undoubtedly the supply of water. In central Australia, droughts lasting for months or even years are not infrequent. The water holes over large areas may all dry up or the water in them become so concentrated by evaporation that the salt content becomes excessive. Some answers to the question of how zebra finches can survive the droughts were found by the American ornithologists Cade, Tobin and Gould. They were able to show that zebra finches can manage without water for weeks or months, in one case even for as long as 250 days, and that when given free access to water, they drink much less than other species. The astonishing capacity to survive without water is due mainly to a strong reduction of water loss from the body. The kidneys of these birds can hold back almost all the water normally lost in the urine; because of this zebra finch droppings are remarkably dry and very little water is lost in the excreta. Zebra finches can also drink and utilize a fairly concentrated solution of common salt which would be fatal for other small birds.

Another adaptation involves reproduction. In central Australia, rains are not only very rare, but they are also extremely irregular. Rain may occur during any month of the year, but there may also be none for months or years. Only after the rains is there a brief period of plant growth and with this a guarantee of enough food to make the rearing of young possible. Zebra finches therefore do not have regular breeding seasons but breed independently of the season, always after rainfalls. Rain stimulates them at once to display and to nest building. Sometimes after months of drought, they begin to gather nest material during the very first showers. The first clutches can then be found only a few days later. The rains also act as a signal for the start of breeding in a number of other species of birds in the arid areas of central Australia.

Next to the society finch, the zebra finch is today the weaverfinch most frequently kept and bred in captivity. Beside the natural color, there are numerous other artificial color strains; among them are spotted, various brown, gray and also white birds. In England a special zebra finch society has even laid down standards for the evaluation of the varieties and conducts regular shows and competitions. Apart from this, the zebra finch has become a preferred experimental subject for studies of behavior. There have been experiments to determine how zebra finches recognize one another, how they distinguish males and females, adults and young and what role is played in this by the beak color and plumage pattern. It was shown that recognition of these features is not inborn but must be acquired by the young, for if one arranges that zebra finches are reared by another species of weaverfinches (e.g., the society finch) they will later pair with members of the species of their adoptive parents, the characters of which they learned in their youth, and show no interest in members of their own species. They act as if they were society finches and not zebra finches. This preference for another species may be retained for life, for a later relearning is virtually impossible. Only the early impressions and experiences have an effect on their behavior with respect to their choice of mate.

Closely related to the zebra finch, but quite different in color, is BICHENO'S FINCH (*Stitzoptera bichenovii*; L 10 cm). It is white about the eyes, sides of the head and throat, these areas being surrounded by black; there is a black band across the whitish lower parts; the back and wings are brownish, with a white network pattern, on the wings, and the beak is blue-gray. It occurs in northern and eastern Australia. The CHERRY FINCH (*Aidemosyne modesta*; L 11 cm) of eastern Australia is marked by

Fig. 17-46. Bicheno's finch *(Stitzoptera bichenovii)* with its subspecies: 1. *Stitzoptera bichenovii bichenovii*; 2. *Stitzoptera bichenovii annulosa*.

Fig. 17-47. Crimson finch *(Neochmia phaeton)*: 1. Black-bellied subspecies; 2. White-bellied subspecies.

Fig. 17-48. 1. Painted finch *(Emblema picta)*; 2. Diamond sparrow *(Staganopleura guttata)*.

Fig. 17-49. Avadavat *(Amandava amandava)*: 1. Black-bellied subspecies; 2. Yellow-bellied subspecies.

a dark scarlet forehead, with grayish-brown upper parts and brown and white wavy bars on the lower parts. The STAR FINCH *(Bathilda ruficauda;* L 11 cm) lives in northern Australia. It is red on the forehead, sides of the head and upper throat, otherwise, apart from the reddish tail, it is greenish above, and below yellowish along the middle and greenish at the sides. At the sides, and in the lower part of the red sides of the head there are many white spots. The beak is red. The CRIMSON FINCH *(Neochmia phaeton;* L 13 cm) of northern Australia and southern New Guinea occurs in two black-bellied and two white-bellied subspecies and is distinguished by a fairly long, graduated tail. The CRIMSON-BELLIED MOUNTAIN FINCH *(Oreosthrus fuliginosus;* L 12 cm) is dark olive-brown with red marks; it lives at heights between 2800 and 3700 meters in the mountains of New Guinea and little is known of its mode of life.

Further Australian weaverfinches are: The PAINTED FINCH *(Emblema picta;* L 10 to 15 cm), a slender-beaked inhabitant of the dry areas of northwestern and central Australia where it has become adapted in great measure to life on the ground; the DIAMOND SPARROW *(Staganopleura guttata;* L 12 cm) of southeastern Australia. The FIRE-TAILED FINCH *(Staganopleura bella)* is marked by a dark barring of the plumage, as is the following species; it is found in southeastern Australia and Tasmania. The RED-EARED FIRE-TAILED FINCH *(Staganopleura oculata)* is rare, occuring in the eucalyptus forests of southwestern Australia. The SYDNEY WAXBILL *(Aegintha temporalis,* L 12 cm) is found in northern, eastern and southern Australia.

The diamond sparrow is a popular cage bird in Europe. The Sydney waxbill is a prounouced follower of cultivation; it has advanced into the parks and gardens of suburbs and has become one of the best known birds of Australia. Because of its delicacy, it is rarer in the cages of European aviculturalists than are most other Australian weaverfinches.

The three species of the genus *Amandava* are far hardier cage birds. The GREEN AVADAVAT *(Amandava formosa;* L 11 cm) is olive-green above and yellowish below with olive-green, white-banded sides and a red beak. Its home is central India. The AVADAVAT *(Amandava amandava;* L 10 cm) lives in southeastern Asia and the GOLDEN-BREASTED WAXBILL *(Amandava subflava;* L 9 to 10 cm) in Africa. (Color plate, p. 440.)

The avadavat is the only weaverfinch to alternate between a colorful male nuptial plumage and a dull non-breeding plumage which resembles that of the female. It was introduced by man from its southern and southeastern Asiatic home to many tropical islands and also in lower Egypt; this species has even bred in Germany. Its preferred habitats are reeds and tall grass near water, as well as strips of bushes with grass in rice fields. The golden-breasted waxbill inhabits similar areas in Africa.

The QUAIL FINCHES (genus *Ortygospiza*), as indicated by their common name, look like small quail. The LOCUST FINCH (*Ortygospiza locustella*; L 9.5 cm), in which males have red sides of the head and a red throat, is found only from the northeastern Congo to southwestern Angola, Rhodesia and Mozambique. The South African QUAIL FINCH (*Ortygospiza atricollis*; L 10 cm) occurs in many subspecies in African short grass and tufted grass steppes. Both species are very much ground birds which breed on the ground and never perch on branches. "When flushed from the nest, the bird rises rapidly fifty or more feet into the air," as Bannerman writes. "After a few circular flights it drops straight down again. When it feels it is being watched, it flies in circles which gradually get narrower, then it suddenly lands near the nest and walks to its eggs."

Fig. 17-50. South African quail finch *(Ortygospiza atricollis)*: 1. Subspecies without eye-ring; 2. Subspecies with eye-ring.

The TRUE WAXBILLS (genus *Estrilda*), the scientific name of which is also used in the scientific name of the weaverfinch family, can be arranged into three groups. The first consists merely of the subgenus *Neisna* with only one species, the YELLOW-BELLIED WAXBILL (*Estrilda melanotis*; L 9 to 10 cm) of African mountain forests from Ethiopia to the Cape Province; the southern subspecies, which is illustrated on the color plate on page 440, is also known as "black cheeks".

The second group with the two subgenera *Melpoda* includes: the ORANGE-CHEEKED WAXBILL (*Estrilda melpoda*; L 10 cm) an inhabitant of tall grass on shores, in swamps along forest edges and in clearings in western and central Africa. The FAWN-BREASTED WAXBILL (*Estrilda paludicola*; L 10 to 11 cm), which lacks the orange cheeks of the preceding species, and has yellowish or whitish lower parts, a pink spot on the flanks and a brown or gray crown. It lives in similar habitats to the preceding species from Ethiopia to Angola and Tanzania as well as in a small area in southern Nigeria.

Fig. 17-51. Yellow-bellied waxbill *(Estrilda melanotis)*: 1. Subspecies with black side of the head; 2. Subspecies with gray side of the head.

The four species of the subgenus *Estrilda* are marked by a red (in one species of the St. Helena waxbill a black) stripe through the eye; the plumage is otherwise mainly brown or brownish-gray with more or less distinct dark wavy bars. The ST. HELENA WAXBILL (*Estrilda astrild*; L 10.5 to 12.5 cm) has a relatively long wedge-shaped tail. It occurs in Africa from Sierra Leone and Ethiopia to the Cape. The GRAY WAXBILL (*Estrilda troglodytes*; L 10 cm) has a completely red beak, black tail and hardly discernible barring of the plumage; its distribution is more northern than that of the St. Helena waxbill. The Arabian race of this waxbill, *Estrilda astrild rufibarba* of southwestern Arabia, has browner and darker bars, as has the CRIMSON-RUMPED WAXBILL *(Estrilda rhodopyga)* of eastern Africa, which has a partly black bill.

The St. Helena waxbill inhabits reeds and tall grass near water as well as cultivated land and grass-grown clearings in bush or forest. V. C. van Someren writes about it: "If one wants to see hundreds of these birds one must visit a reed bed or a papyrus stand towards evening when the birds are about to drop into one of their preferred roosts. At sunset flocks arrive from all directions. They customarily roost at particular places in the swamps. Nevertheless they fly at first with whirring wings and lively twittering from one place to another. Finally they arrive at the roosting place and settle in rows on reed or papyrus stalks until their weight bends these over so much that they come to lie on adjacent stalks. Then some bird at the outside of the row tries to push his way in among the others, the equilibrium is disturbed, they all fly up, make a circle, come down and perch again. Not all the swarms roost in swamps; some will use a bush as a roosting place for the night."

The third group, the true waxbills, consists of the subgenera *Krimhilda, Brunhilda* and *Glaucestrilda,* with seven species between them. All have a red rump but lack red eye stripes. The beak is gray or only partly red; red feathers, if any, are most intensely colored and wide spread at the sides of the body. This group includes: The BLACK-CROWNED WAXBILL (*Estrilda nonnula;* L 10.5 cm), which has a black crown and a black and red beak, occurs from Fernando Po and the Cameroons to western Kenya. The BLACK-HEADED WAXBILL *(Estrilda atricapilla),* which has less red on a mainly black beak, and is a darker twin species of the preceding, occurs partly in the same area, but more often in forest clearings. The BLACK-CHEEKED WAXBILL (*Estrilda erythronotos;* L 12.5 cm) is long tailed, colored a delicate gray with fine dark bars and a reddish flush in certain parts; the secondaries are broadly barred black and white, the rump is red, the tail, sides of the head, throat, middle of the abdomen and lower tail coverts are black, the flanks are purple and the beak lead gray. It occurs in South Africa and parts of eastern Africa. The *Estrilda charmosyna* has a yellow belly and occurs from Somalia to northeastern Tanzania. The LAVENDER FINCH (*Estrilda caerulescens;* L 11 cm) occurs from Senegal to the areas south of Lake Chad.

One of the most elegant birds of this group and among the weaverfinches as a whole is the black-cheeked waxbill, a characteristic bird of the southwestern and South African acacia steppe, which occurs, widely separated from the area just mentioned, in the interior of northern Tanzania and southwestern Kenya. Besides small seeds and arthropods, it eats many acacia flowers.

The five species of the genus *Uraeginthus* are marked by blue areas in their plumage. In the subgenus *Uraeginthus* there is the RED-CHEEKED WAXBILL or CORDON-BLUE WAXBILL (*Uraeginthus bengalus;* L 12 cm) from the Senegal and Ethiopia to the southern Congo, northern Zambia and Tanzania; the ANGOLA WAXBILL *(Uraeginthus angolensis),* very similar but without a red ear patch. Its range is further south; the BLUE-CRESTED WAXBILL *(Uraeginthus cyanocephalus),* in which males have a blue crown, of southern Somalia and northern Kenya to Tanzania. The subgenus *Granatina* consists of the VIOLET-EARED WAXBILL (*Uraeginthus granatinus;* L 14 cm) of South Africa and the PURPLE GRENADIER *(Uraeginthus ianthinogaster)* of eastern Africa.

The Cordon-bleu is known almost everywhere as a cage bird. The

Fig. 17-52. Violet-eared waxbill *(Uraeginthus granatinus);* Purple grenadier *(Uraeginthus ianthinogaster).*

purple grenadier is one of the most beautiful weaverfinches; it is still in great demand as a cage bird in Europe. The Marquise de Pompadour, the mistress of Louis XV of France, owned some. Purple grenadiers inhabit dry tree and thorn scrub steppes, also occuring in open dry forests.

Male FIRE FINCHES (genus *Lagonosticta*) are mainly red; there are seven species in Africa: the FIRE FINCH (*Lagonosticta senegala*; L 10 cm) of the Senegal, Sudan and Ethiopia, to Natal and along the warm valley of the Orange River almost to its mouth. The BAR-BREASTED FIRE FINCH (*Lagonosticta rufopicta*; L 10 cm) of western Africa to Uganda and western Ethiopia. The BLUE-BILLED FIRE FINCH (*Lagonosticta rubricata*; L 11 cm) is distributed over a large part of Africa in four subspecies. JAMESON'S FIRE FINCH (*Lagonosticta rhodopareia*; L 11 cm) occurs from Ethiopia to the Transvaal and Angola. The MASKED WAXBILL (*Lagonosticta larvata*; L 11 cm) is found from the Senegal to Ethiopia. (Color plate, p. 440.)

The fire finch is the best known of these species for it is a common cage bird. Like most of the other fire finch species, it has minute white dots on its red breast; these dots are also found on the female's grayish-brown plumage. The beak of this small weaverfinch is reddish, as is that of the closely related bar-breasted fire finch, while it is blue gray in the other species listed above. In large portions of its range the fire finch has linked itself closely to man and has become an inhabitant of villages. Bannerman writes: "It visits shops to snatch some of the rice on display there, enters houses to nest in the rafters and sings its song consisting of a few tones, only a few feet away from people towards whom it does not show the least concern." The blue-billed fire finch and Jameson's fire finch deliver loud and varying song phrases which are reminiscent of the songs of the green finch, the woodlark and the tree pipit. The blue-billed fire finch in most areas inhabits dense growths of bushes along forest edges or in moist savannahs while Jameson's fire finch, which resembles it, takes its place in the drier thorn bush steppes.

PETER'S TWIN-SPOT (*Hypargos niveoguttatus*; L 12 to 13 cm) of eastern Africa and the ROSY TWIN-SPOT (*Hypargos margaritatus*; L 12 cm) of southern Mozambique to Zululand are magnificently colored birds. Male rosy twin-spots are pink on the sides of the head, throat and upper breast instead of red as in Peter's twin-spot, and on their black lower parts they have reddish-white instead of pure white drop-shaped spots. The GREEN-BACKED TWIN-SPOT (*Mandingoa nitidula*; L 10 to 11 cm; Color plate p. 440) is a small, colorful weaverfinch of the forest areas of western and central Africa which also occurs in eastern Africa and eastern South Africa. Its soft song is noteworthy because, in addition to whistling tones which fade away in a peculiar manner, it contains a phrase which sounds like the striking of minute bells.

The CRIMSON-WINGS (genus *Cryptospiza*) are dark green and red or gray and red forest birds of the higher mountains of eastern and central

Fig. 17-53. Fire finch (*Lagonosticta senegala*).

Fig. 17-54. Blue-billed fire finch (*Lagonosticta rubricata*).

Fig. 17-55. Jameson's fire finch (*Lagonosticta rhodopareia*).

▷
Above and below: The chestnut-headed weaver (*Textor castaneiceps*) forms noisy aggregations for eating and sleeping and breeding. These flocks are part of the East African scene.

Fig. 17-56. Immature melba finch.

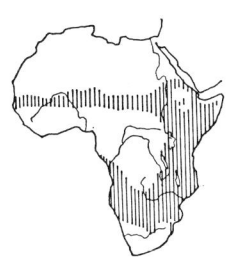

Fig. 17-57. Melba finch *(Pytilia melba).*

◁
Upper photo left: Purple glossy starling *(Lamprotornis purpureus).* Upper photo right: *Lamprotornis corruscus.* Below: Head of the violet-eared waxbill *(Uraeginthus granatinus).* It lives in southern and southwestern Africa and is one of the most beautiful cage birds.

Africa. REICHENOW'S CRIMSON-WING *(Cryptospiza reichenowvii;* L 12 cm) is the best known of them. The BLACK-BELLIED SEEDCRACKER *(Pyrenestes ostrinus;* L 12 to 14 cm) has a phenomenally thick-billed appearance. Depending on the subspecies the width of its black beak, measured at the base of the lower mandible, lies between 10.2 and 21.1 mm. Presumably the most thick-billed birds of this species originally lived in the savannahs which surround the central African jungle area, while those with the weakest beaks lived in forest clearings and swamps in the forests. Evidently the originally separate areas of distribution of the subspecies came to overlap in part later on. This led to a mingling, with the result that one may now come across thick-billed and thin-billed black-bellied seedcrackers in many areas in central Africa. In spite of its massive beak, this bird mainly eats small, though in part hard, seeds (for further species of this genus see the systematic review).

The BLUE BILLS (genus *Spermophaga*) also have quite massive beaks which are, however, more slender and instead of being black are blue and red with a mother-of-pearl-like gloss. They are large forest-dwelling weaverfinches of western and central Africa and of some eastern African highland areas. Their plumage, like that of the black-bellied seedcracker, is black and red, but gray and red in one subspecies; in females the black lower parts are densely spotted with white. The best known species is the BLUE BILL *(Spermophaga haematina;* L 13 to 14 cm) which occurs from Gambia to the Congo.

The AURORA WAXBILL *(Pytilia phoenicoptera;* L 12 cm) the *Pytilia lineata* with a bright red beak, the RED-FACED WAXBILL *(Pytilia afra;* L 12 cm) and the MELBA FINCH *(Pytilia melba;* L 12 to 13 cm) are all closely related species. The melba finch is a secretive inhabitant of African steppes and savannahs and above all of the dry but well-treed and bush-grown thornbush steppes.

Members of the genus *Nesocharis* can climb skillfully like chickadees. A long-tailed species found on Fernando Po and the higher mountains in the Cameroons is called *Nesocharis shelleyi* (L 18 cm). Two larger relatives are the two WHITE-COLLARED OLIVE-BACK *(Nesocharis ansorgei;* L 10 cm) of the central African lake region and the GRAY-HEADED OLIVE-BACK *(Nesocharis capistrata;* L 12 cm) found locally from Gambia to the southwestern Sudan and Uganda.

Both the white-collared olive-back and the *Nesocharis shelleyi* breed in old weaver nests. Mrs. Ruth T. Chapin has observed the white-collared olive-back in the Kiva District (eastern Congo) and she writes about this attractive bird: "One always sees these birds in swamp areas with abundant water or along streams with bush and tree-grown shores. Usually one finds them in pairs or in family units, never in large flocks which are so characteristic of the St. Helena and black-crowned waxbills which inhabit the same area. One is very impressed by the chickadee-like actions of these birds when they cling to the stalks of the composite *Melanthera brownei,* the seeds of which are their principal food. These stalks, which carry the round green seed heads, are thin and weak and when the birds cling to them in order to pull out the seeds, their movements are at times truly acrobatic. One may, for example, see a bird holding onto a stalk with its left foot, as it pecks at the rounded seed head which it holds with its right foot. The male's song is a soft but pleasant trill, preceded by two short tones. I have only heard it a few times. The singing male was sitting still on a branch about thirty meters above the ground. While singing, it raised the head, stretched up the beak and fluffed out the feathers at the side of the body."

The NEGRO FINCHES (genus *Nigrita*) of the African forest areas feed mainly on arthropods but also on plant material including seeds and husks of the oil palm fruit. The four species of this genus differ considerably from other members of the weaverfinch family. We will only name the small CHESTNUT-BREASTED NEGRO FINCH (*Nigrita bicolor*; L 11 to 12 cm) which occurs from Sierra Leone to northern Angola and Uganda and even on the island of Principe and the larger GRAY-HEADED NEGRO FINCH (*Nigrita canicapilla*; L 13 to 14 cm). The brown, gray and black-colored negro finches are forest birds which also visit plantations and gardens. The chestnut-breasted species keep mainly to the undergrowth; the gray-headed species, however, lives mostly in the crowns of trees.

Fig. 17-58. Gray-headed negro finch (*Nigrita canicapilla*).

To conclude the review of this family so rich in species, many of them very colorful, and which is so well liked by people, the ANTPECKER (*Parmoptila woodhousei*; L 11 cm), a relative of the negro finches, will be considered. In males, the forehead and front of the crown are red or at least the feathers of these parts have red tips; the rest of the upper parts are mainly olive-brown with paler stripes and spots while the sides of the head and throat, and in many subspecies the lower parts as well, are reddish-brown. The warbler-like, fine thin beak of this bird indicates that it is an insect eater. It particularly likes small ants. It inhabits the western and central African forest area from Ghana to the eastern Congo in four subspecies which show considerable color differences.

The multi-colored weaverfinches evidently fulfill to an outstanding degree the human desire to care for and tend birds, and year after year they are caught in virtually unimaginable numbers to fill cages in many countries. There is no bird dealer who cannot offer some weaverfinches. One would assume that such maintained catching of the birds in great numbers must have reduced their population in the wild considerably, but so far this is hardly the case. The populations of most weaverfinches have not yet been threatened as a result of their popularity as cage birds. Three species, the Java sparrow, the society finch and the zebra finch, are passerine birds which have just like the canary and the budgerigar become domesticated birds.

18 Starlings, Old World Orioles, and Drongos

The starling family, by H. Bruns

STARLINGS (family Sturnidae) are small- to medium-sized birds varying in length from 17 to 45 cm. The strong beak is usually straight, but in some species is shaped like that of a raven. They have strong legs and feet and walk on the ground, sometimes with a waddle, nodding the head; only a few species hop. The wings are long and pointed or short. The tail is generally short, only rarely long. The plumage of the two sexes is generally alike and often has a silky or metallic gloss. The commonest background color is black, other colors are gray or brown, often with white, less often yellow and rarely with red. Several species have a crest, while others have fleshy wattles or naked areas on the head. The juvenile plumage of many species is dark, with stripes or spots. There is only one molt a year, immediately after the breeding season, which gives rise to the seasonal plumage change. There are two subfamilies: A. TRUE STARLINGS (Sturninae) with five groups of genera: Starlings in the restricted sense (*Sturnus* group), Mynahs, Aplonis and Glossy Starlings. B. TICK BIRDS or OX-PECKERS (Buphaginae) with only one genus and two species. Altogether there are thirty to thirty-two genera and 111 species.

Starlings are usually sociable, even during the breeding season. They are good, swift fliers. Those which live in temperate areas migrate away in part in the winter, sometimes in huge flocks. Many starlings are rather noisy and chatter continuously in flight and when at rest. Though they are tree birds, they often and readily come down to the ground to seek food. Their food consists of arthropods, particularly their larvae, as well as worms and molluscs, and occasionally representatives of other animal groups such as fish, amphibia, lizards, birds and bird eggs. As destroyers of grasshoppers, insect pests and parasites, starlings are regarded as beneficial; on the other hand, they are not liked by fruit and wine growers because of their consumption of cherries, grapes and other fruit.

A characteristic action of many starlings is the insertion of the closed beak into the ground, then opening it so that a space is produced in which they can search for their food. When caged, starlings have no opportunity to do this; they carry it out as a "vacuum activity", opening the beak in the air and touching a perch with it. Observations on starlings reared in indoor cages show that this action is inborn for they push their beaks into the gaps between floor boards. If one holds out a fist to them they will push the beak between the fingers and try to separate them. The vocalizations of starlings are quite varied; there are pleasant whistles but also croaking, cackling, hissing, and chattering sounds.

Insofar as there are detailed observations on the breeding behavior of starlings, individual pairs remain firmly linked and carry out the duties connected with breeding together. Nests are usually in holes in trees, rocks or walls, more rarely on the ground; the nests of some species are like those of weavers. Many species breed in colonies. The eggs are mostly blue-green but in some species are whitish with brown spots. The clutch size varies between two and ten but is most commonly three to five.

Originally starlings were confined to the Old World. Most species are native to Africa and southeastern Asia. They were brought to Australia and America by man, e.g. the common starling was introduced in the United States about 1900 and the common mynah was introduced in Australia, South Africa and on various islands.

The twenty-one species of typical starlings (the *Sturnus* group) include the COMMON STARLING (*Sturnus vulgaris*; L 21.5 to 22 cm; weight 61 to 94 gm). It is black with a bronze-green and purple gloss; in the fall and winter it has bright colored feather tips and spots which are somewhat more marked in females. It has a short tail, a sharp pointed beak and a triangular outline in flight. It is distributed over Eurasia east as far as Lake Baikal in about eleven subspecies. It has been introduced in North America, Australia, New Zealand and the southern Cape Province of South Africa.

After the molt which begins in July and in adults is completed about the end of September but in the young not till October, the feathers have white tips. In the course of the winter, these feather tips get worn off until the bird is a uniform black color. The blackish-brown beak also changes color in the late winter and becomes yellow as the sex glands enlarge. It has been shown experimentally that this color change is due to male sex hormone in both sexes. Female birds produce this hormone in their sex glands in appreciable amounts, as well as the female sex hormone. It has, incidentally, also been shown that the seasonal blackening of the beak in all male and some female house sparrows is due to male sex hormone. At this time, the base of the lower mandible is dark in males, but pale pinkish-white in females. The male has a uniformly dark-brown-colored iris and females can be distinguished by a white demarcated paler ring around the iris. The young are mouse brown with a whitish throat and white stripes below. Abnormally white (albinistic) or yellow (xanthochroistic) starlings have been observed repeatedly.

The SPOTLESS STARLING (*Sturnus unicolor;* L 21.5 cm) is a close relative of the common starling and is regarded as a mere subspecies of it by some ornithologists. It is uniformly black with a purple gloss and shows little spotting even in the fall and winter. It occurs in the Iberian peninsula, Corsica, Sardinia and Sicily and from Morocco to Tunisia.

The common starling now occurs almost everywhere in farming country, in deciduous and mixed woods and gardens, and even in the middle of towns. In the mountains, as it spreads more widely, it has passed the thousand meter line. We are only at the beginning of a detailed study of its distribution.

Starlings are among the fastest-flying European birds: they have been clocked at 20.6 meters per second and 74.2 kilometers per hour. Starlings are distinctive in flight because they do not close the wings after a few wing beats but keep them spread out so as to glide. On warm days one can observe that they do not seek prey only from a perch or on the ground but that they also fly about hawking insects, like swallows, e.g. about church towers or over peaks. On the ground they move in steps, nodding the head.

Starlings are eager to sing and will do so not only from trees, masts and houses, but also from the ground and occasionally even at night or in winter. The song is accompanied by wing flapping and it contains creaking, crunching, squeaking and whistling sounds. Imitations of the sounds of other birds as well as of other noises are often included in the song; such imitations may include the cackling of hens, the crowing of a cock, the call of the golden oriole, the squeaking of a door or the creaking of a weather vane. Females sing less loudly and their songs are shorter. Captive starlings can learn to imitate tunes whistled for them and even spoken words. There are reports of one which delivers the sentence "I'm a wonderful starling". Another exhibited at a cage bird show in Vienna was able to speak seventy words quite clearly.

The short attraction call which sounds like "spreen" is characteristic of the starling and is the basis of its name in the north German dialect. It uses three alarm calls, "spett spett", "vaa" or a noisy "brrrch"; when chasing an opponent, it calls "tshreetshree".

As mentioned above, starlings are almost omnivorous and use animal as well as plant food. Included in the first category are snails, worms, crabs, spiders and particularly many species of insects and their larvae, beetles, diptera, dragonflies, ants and butterflies and exceptionally even small vertebrates such as fish, small birds and lizards. Starlings occasionally act as food parasites; one was observed chasing a least tern until it dropped the fish it was carrying, which the starling then seized. One often hears or reads the statement that only the cuckoo and the golden oriole take hairy caterpillars. This is quite untrue. Starlings as well as many other song birds feed such caterpillars to their young.

Items in the starling's plant diet are mainly fruit and seeds; cherries, elderberries, red currants, gooseberries, raspberries, ripe pears, apples which have started to rot, grapes, mulberries, olives, dates, cereals, peas, lettuce, weed seeds and many more. Depending on place and season, the composition of the diet of starlings can vary to an extraordinary degree. Young starlings are easily reared with fresh ant pupae, worms, a little meat, and a soft food mixture with traces of soil.

Changes in the available food supply may lead to changes in migratory behavior. Usually starlings from the Baltic countries migrate from the end of June onwards to northwestern Germany where they invade cherry orchards in the lower Elbe fruit growing area. However, when for example in 1935 there was an invasion of the oak and willow trees by caterpillars of the leaf-roller moth in eastern Prussia, the starlings did not leave there at the end of June as in other years. The migration takes them in the first place as far as northwestern Germany, is continued in later months to the Netherlands, Great Britain, Belgium and northern France. Other starling populations such as those which breed in central Germany, Poland and Czechoslovakia migrate in the fall over the Alpine passes to southern Europe and northern Africa. Some starlings winter in the breeding area. I have observed such winterings repeatedly on the Norwegian group of islets called Taven, off Trondheimfjord. Starlings also winter in considerable numbers in many parts of Germany but around Hamburg I see them only very rarely in winter. Starlings are also known to be day and also night migrants and as experimental subjects they have contributed to our knowledge of the orientation of birds by the sun's position (Vol. VII, Chapter 1).

Collective flights are a characteristic feature of starling behavior; they start on a small scale during the breeding season but later on involve huge flocks. A gathering of thousands of starlings before sunset performing admirable aerial maneuvers is an impressive sight. All the birds in the flock turn at the same moment as if in obedience to a command and finally, as it gets darker, plunge down into reed beds, willow thickets or a stand of spruce. Such swarms will attack raptors on the wing and force them to retreat. In the U.S., Britain and other countries, and to some extent also in Germany starlings roost in cities. Such starling roosts have been observed for example in Munich on Stachus Square, on the Cathedral at Cologne, in front of the main railway station in Leipzig, in Hamburg and particularly in London and Birmingham.

Female starlings are generally mature when one year old but males do not usually reach maturity until they are two years old. Starlings generally live in monogamy but cases are known of males paired to two females; in which cases the male took part in the rearing of both sets of young. Instances of a female mated to two males have also occurred. Starlings do not maintain proper breeding territories which they defend against others of their kind; they only fend off rivals from the actual nest hole. Immediately after returning from winter quarters in February, starlings look about for nest sites and select a nest hole in March. Even before pair formation, males carry nest materials and flowers such as crocus, primrose, narcissus and others into the nest hole and in part also out again. After pair formation, the nest is built with grass, dry leaves, pieces of bark, fine twigs, feathers, paper and other materials. Starlings readily nest on buildings under roof tiles, behind shutters, in holes in walls, in belfrys and in attics as well as on the masts of high tension power lines, and in all sorts of natural holes as well as nest boxes. Starling nest boxes were made as early as the 17th Century and in times of famine people ate the young starlings.

The clutch consists usually of five or six, more rarely four to nine, light blue or bluish-green eggs which are incubated by both parents for eleven to fourteen days. The young stay in the nest for almost three weeks, leaving it only when they are fully fledged. Winter broods have been observed repeatedly. During incubation the members of the pair relieve one another every twenty to thirty minutes. In many areas of Germany, e.g. Saxony, Thuringia, Brunswick, northern Bavaria and others, starlings raise a second brood. The parent birds remove the feces of the young during the first ten to fifteen days. If the young are fed excessively on caterpillars, their droppings remain liquid, a condition which may cause some of the young to perish in the nest. After the period during which the parents remove the droppings, the young defecate with the rear raised up to the entry hole of the nest and the cleanliness of the nest deteriorates. As long as they are small, the young are brooded by one of the adults between feedings during the day but at night by the female only. As soon as brooding of the young is no longer necessary, the parents spend the night in a collective roost.

Life expectancy of the starling

The average life expectancy of a starling is 1.2 to 1.4 years. An age of three years is rarely exceeded and banded starlings which have lived to nineteen or twenty years are quite exceptional. Starlings are persecuted by man; and they amount to about five per cent of the prey of raptors and owls. During bird catching operations for the market in Italy, 24,000 starlings were caught in nets near Verona in a single night. Up to the end of the last century, the starling was still quite rare in certain places in Germany; then they became progressively more common. This increase is probably due to the nest boxes and the increase of nest sites provided by buildings. Because of its adaptability, the starling is one of the four vertebrates which man has intentionally (starling and house sparrow) or accidentally (house mouse and Norway rat) spread over most of the world.

Repeated attempts to introduce the starling in North America were made between 1870 to 1900. Only the releases in 1900 and 1901 of a hundred starlings in New York's Central Park was successful. The offspring of these hundred starlings reached new areas in which some of them remained and bred. Starlings became breeding birds from Canada to Florida. In 1940 the first starlings appeared beyond the Rocky Mountains, in California. At the time of the introduction, only the good aspects of the starling had been considered; later with their great increase their harmful activities among grain seedlings and in orchards became noticeable. Apart from that, they roost in huge flocks in cities, e.g. in Washington, D.C., where their presence gives rise to complaints about their noise and droppings.

Damage done by starlings is due to increases in its populations, which were caused by man. When starlings make their annual summer invasion in the fruit growing area near Hamburg, they destroy, on the average, ten per cent of the cherry crop. They may be even more destructive in vineyards and in the olive orchards of northern Africa, when they occur in great masses. Attempts to scare starlings away from cities, orchards and vineyards have been made with scarecrows, bird models, various noise makers, nets, traps, flickering lights, by shooting or poisoning them, and with the use of loudspeakers. But only those methods which were based on scientific study, using specific visual stimuli together with reproductions of the inborn warning and alarm calls, were reasonably successful. On the other hand, farmers regard starlings as beneficial because of their consumption of daddy long-legs spider larvae and foresters respect them because they take oak leafrollers and other insects harmful to trees. Finally, we must not overlook the fact that for many people starlings are messengers of spring for which they put up nest boxes.

Damage brought about by starlings

▷
1. Alpine chough *(Pyrrhocorax graculus)*; 2. Chough *(Pyrrhocorax pyrrhocorax)*; 3. Apostle bird *(Struthidea cinerea)*; 4. Magpie lark *(Grallina cyanoleuca)*; 5. White-winged chough *(Corcorax melanorhamphos)*; 6. Saddleback *(Creadion carunculatus)*; 7. Kikako *(Callaeas cinerea)*; 8. Whitebacked magpie *(Gymnorhina hypoleuca)*; 9. Gray butcher bird *(Cracticus torquatus)*; 10. Huia (+ *Heteralocha acutirostris)*; 11. Hume's ground chough *(Pseudopodoces humilis)*; 12. Pied currawong *(Strepera graculina)*; 13. Pander's ground chough *(Podoces panderi)*;

The GRAY STARLING *(Spodiopsar cineraceus;* L 21 cm) is the eastern Asiatic counterpart of the common starling. Its plumage is dark gray, blackish on the head and whitish on the forehead, about the ears and below. It occurs from Transbaikalia to northern China, Korea, Japan and Sakhalin. Closely related to it is the PIED MYNAH *(Sturnopastor contra;* L 22 cm) which is more contrastingly colored and has an area bare of feathers around the eye. It occurs in southern Asia from northern India to Bali. The ROSE-COLORED STARLING *(Pastor roseus;* L 22 cm) also resembles the common starling in its shape and movements. It has a black head with a crest and is also black on the neck, the wings, tail and lower abdomen; the rest of its plumage is a whitish-pink. It occurs in the steppes of southwestern Europe and southwestern Asia, occasionally visiting other parts of Europe and India. (Color plate, p. 449.)

Other true starlings

It often breeds in colonies which may number thousands of pairs, in heaps of stones, holes in the ground, or on buildings. It is not tied to a specific breeding season; rather its breeding is determined both in time and place by the regular mass occurrences of migratory locusts.

Evidently because of the short-lived appearance of the locusts on which it feeds to a great extent, it breeds only once a year. It builds nests very rapidly and produces a rather disorderly structure of twigs and grass. The female incubates the five or six eggs for only eleven or twelve days. The young leave the nest rather early, between the fourteenth and the nineteenth day. The food of this starling, apart from the locusts, includes fruit, e.g. grapes, which it likes just as much as the common starling does.

The young are mainly fed on migratory locusts. When these insects occur in great numbers, large flocks of rose-colored starlings fly out into the steppes to seek food. "At times they form veritable clouds; then again these break up into smaller troops," as Serebrennikov describes it. "They descend onto the locust swarms and walk in a particular direction, picking up the larvae; usually the front rows of starlings move faster than those in the rear, but these do not want to be left behind; they fly up and settle among those in the front of the flock and continue to walk, driven by the general hunting fever. This continual flying up and settling suggests a wave motion and makes a deep impression on the observer."

There are further starlings of the *Sturnus* group in western and southwestern Asia. The PAGODA STARLING (*Temenuchus pagodarum*; L 20 cm) occurs from eastern Afghanistan to India and Ceylon. Like the rose-colored starling, it has a black crest while the rest of its upper parts are mainly gray and the lower parts reddish-buff. The MALABAR MYNAH (*Temenuchus malabaricus*; L 19 cm) is rather similar but has a gray head without a crest. It is found from India to southwestern China, Thailand and Indochina. Three rather small species of eastern Asiatic starlings form the genus *Sturnia*. They are the MANDARIN MYNAH *(Sturnia sinensis)* of southern China and northern Indochina, the VIOLET-BACKED STARLING *(Sturnia philippensis;* L 16.5 cm) of Sakhalin and northern Japan and the STARLET *(Sturnia sturnia)* of southeastern Siberia, northern China and Mongolia. The BLACK-NECKED MYNAH *(Gracupica nigricollis;* L 22 cm) of Java, Bali and Lombok, is white with a gray back and black wings and tail. The BALI MYNAH or ROTHSCHILD'S GRACKLE ($\diamond$ *Leucopsar rothschildi;* L 25 cm) is closely related to the black-necked mynah; apart from its black flight feathers it is white and has a marked crest. It occurs only in a very limited area in one part of Bali and is thus a threatened species. Attempts are in progress to gather up representatives of this species from various zoos for a breeding program.

◁
Birds of paradise:
1. Superb bird of paradise *(Lophorina superba);* 2. Red bird of paradise *(Paradisaea rubra);* 3. Arfak six-wired parotia *(Parotia sefilata);* 4. King bird of paradise *(Cincinnurus regius);* 5. Blue bird of paradise *(Paradisaea rudolphi).*

The black-necked and Bali mynahs lead to the TRUE MYNAHS (genus *Acridotheres*) with six species in southern Asia. They also belong to the *Sturnus* group and have strong, compact bodies, a rounded tail and, in most of them, erect feathers on the forehead. They are not to be confused with the birds of the genus *Gracula* which are also called mynahs in English. The COMMON MYNAH (*Acridotheres tristis;* L 24 cm) is brown and black with a white spot on the wing, a short crest and an orange-yellow area of bare skin about the eyes. Its range is from India to Vietnam and Yunnan; in the last century it has spread through Afghanistan as far as Turkistan and Transcaspia. Other species are: the BANK MYNAH (*Acridotheres ginginianus*; L 21 cm) which is gray with a black head and a red area of bare skin about the eye. It occurs from Afghanistan to India; the CHINESE CRESTED MYNAH (*Acridotheres cristatellus*; L 26 cm) is black with a white area on the wing and large crest. It is the commonest mynah species in Vietnam.

The Chinese crested mynah became established at Vancouver, British Columbia (Canada) in 1900. The number of these birds there now is an estimated twenty thousand. The common mynah favors settlements where it looks for ticks on cattle and grasshoppers which the mynah flushes. In some places, it is kept as a cage bird and it readily learns to repeat words. It has been introduced in South Africa, Malaya, Australia, New Zealand and various islands in the Pacific, Indian and Atlantic Oceans.

The WATTLED STARLING (*Creatophora cinerea*; L 21 cm) is a more distant relative of the common starling which it resembles in its appearance in flight and song. It is pale grayish-brown and black and white on the wings and tail. It is found in eastern Africa. At the beginning of the breeding season, males and occasionally females lose the feathers over the crop, while the wattles on the forehead, crown and throat enlarge. Like the rose-colored starling, the wattled starling follows swarms of locusts and probably because of this has no fixed breeding season. It nests in colonies in thornbushes. The nests are roofed over and are built from twigs and small sticks. Sometimes they are so crowded together that many of the birds can only get through to their nests with difficulty. (Color plate, p. 449.)

As the last of the birds related to the starling, the two species of Mascarene starlings which are now both extinct may be mentioned: the REUNION STARLING (+ *Fregilupus varius*) which had a crest of fluffy feathers and the LEGUA STARLING (+ *Necropsar leguati*). Both were confined to the small area of the Mascarene Islands and became victims of the advance of civilization. The Reunion starling was originally very common on Reunion Island but was easily killed with a stick and was also preyed upon by cats and dogs which had been introduced to the island in the 19th Century. The Legua starling is possibly identical with the RODRIGUEZ STARLING (+ *Necropsar rodericanus*), known only from fossil remains. It was smaller than the Reunion species and had no crest.

Extinct starlings

Among the mynah's six genera and twelve species, those of the genus *Gracula* (L 24 to 37 cm) are best known. These strong birds, which have a black plumage with a metallic gloss, are often kept in captivity in their home countries. The JAVAN HILL MYNAH (*Gracula religiosa*) has yellow lobes of bare skin on the back of the head which differ

Mynahs

in shape in different subspecies; in the CEYLON MYNAH *(Gracula ptilogenys)* there are no such lobes. The Javan hill mynah occurs from Ceylon to Hainan in eleven subspecies, some of which differ considerably in size. These beautiful birds wander through the forests in noisy flocks, producing a pleasant humming sound with their wing beats and uttering loud whistles. They feed mainly on fruit but also take insects and other animal food, particularly in captivity. Their vocal sounds are highly varied and they have astonishing vocal talents. Captives can learn to imitate human speech sometimes better than a parrot can, with quite accurate pronunciation. They become attached to their owners and are quite content in large flight cages.

Related to them is the GOLD-CRESTED GRACKLE *(Ampeliceps coronatus;* L 22 cm) found from Malaysia and Burma to Bengal. It is shiny black above with a yellow crown and base of the wings. The GOLDEN-BREASTED MYNAH *(Mino anais)* and DUMONT´S MYNAH *(Mino dumonti),* both from New Guinea, and the *Basilornis corythaix* of Ceram with a large crest, the *Streptocitta albicollis* of Celebes, the *Streptocitta albertinae* of the Sula Islands and the gray and black COLETO MYNAH *(Sarcops calvus)* which has most of the head bare of feathers and is native to the Philippines, are further relatives. (Dumont's and coleto mynahs, Color plate, p. 449.)

Enodes erythrophris and the Celebesian starling

Enodes erythrophris (L 27 cm) and the CELEBESIAN STARLING *(Scissirostrum dubium)* form a separate group of only two species. *Enodes erythrophris* is slate gray with olive yellow wings, tail and lower abdomen and a bright red stripe above the eye. It occurs in the dark mountain forests of Celebes up to 1500 meters. The Celebesian starling is blackish and slate gray with red feather tips on the rump and flanks. It chisels out holes in dead trees with its unusually strong beak. Sometimes it makes over a hundred in a single tree, so that the bark may show a sieve-like pattern. It too is native of Celebes. (Color plate, p. 449.)

The *Aplonis* starlings

The *Aplonis* starlings are mainly glossy black and occur in twenty-four species from Java and Indochina to the Fiji and Samoan Islands. Two species have become extinct in recent times. The most widely known of these starlings is the SHINING STARLING *(Aplonis metallica;* L 24 cm). Like the weavers, it builds elongated nests more than fifty centimeters long which have an entrance at the side. Often hundreds hang from a single tree. It is found from the Moluccas and New Guinea to northeastern Australia and the Solomon Islands. The rare WHITE-EYED STARLING *(Aplonis bruneicapilla;* L 20 cm) lives on two of the Solomon Islands, Bougainville and Rendova.

Glossy starlings

The glossy starlings, with forty-five almost exclusively African species, form a group of genera, most members of which have a plumage with a marked metallic gloss. The fifteen species of the first subgroup, which includes the MAGPIE STARLING *(Speculipastor bicolor;* L 20 cm) are mainly bluish-black. The magpie starling is bluish-black with white lower parts and a white wing speculum; the female has a gray head. It breeds from southern Ethiopia to Kenya, generally in holes in termite nests. In the WHITE-COLLARED STARLING (*Grafisia torquata*; L 22 cm), the male has a wide white collar around the neck. It occurs from the Cameroons to the northern Congo. The members of the genus *Onychognathus* are marked by reddish-brown or buff-colored flight feathers. The best known among them is the RED-WING STARLING (*Onychognathus morio*; L 27 cm) which occurs from the upper Niger River to northeastern Africa and from there as far as South Africa's Cape Province. It breeds in clefts in rocks or on buildings and feeds on fruit and insects. The PALE-WING STARLING *(Onychognathus nabouroup)* has pale brown flight feathers. It occurs from the central western Cape Province through southwestern Africa to Angola. TRISTRAM'S GRACKLE *(Onychognathus tristramii)* lives in rocky ravines from the Red Sea region as far as southwestern Arabia.

The remaining glossy starlings form another subgroup. Among them is the PIED STARLING (*Spreo bicolor*; L 27 cm), widely distributed in open grassy and cultivated areas of southern Africa. It is brown with a white belly and thus less colorful than its relatives. Like the common starling, it likes to be in the vicinity of cattle. It feeds on insects and fruit and breeds in holes, which it usually digs in the soft earth of river banks or in holes in walls or rocks or under roofs. The VIOLET-BACKED STARLING (*Cinnyricinclus leucogaster*; L 19 cm) is found in the savannahs of Africa, and in southern Arabia. Males are violet above and white below, while females are brown above and have dark stripes on a white background below. Most members of the genus *Lamprospreo* are unusually colorful. The SUPERB GLOSSY STARLING (*Lamprospreo superbus*; L 21 cm) has upper parts which show iridescent black, blue and green colors. The steel blue of the throat and upper breast is separated by a white band from the reddish-brown color of the rest of the lower parts. In this species, as in most of those of the genera *Lamprospreo*, *Cosmopsarus*, and *Lamprotornis*, the wing coverts have velvety black spots on their tips.

The ROYAL STARLING (*Cosmopsarus regius*; L 35 cm) is undoubtedly the most beautiful of all starlings. This slim, long-tailed bird has a metallic green and blue head, neck and upper parts, a violet upper breast and golden yellow lower parts. It inhabits the acacia savannahs of eastern Africa from southern Ethiopia to the area around Mount Kilimanjaro and breeds in holes in trees. One of its favorite foods is termites; in pursuit of them it chisels openings into their nests quite rapidly with its beak and snatches up the insects as they fly out.

The glossy starlings in the restricted sense, i.e. those of the genus *Lamprotornis*, show mainly metallic green, blue or purple colors and, like many related species, a striking yellow-colored iris. The sixteen species of this genus can be divided into a short-tailed and a long-tailed group. The short-tailed group includes the RED-SHOULDERED GLOSSY

STARLING (*Lamprotornis nitens*; L 25 cm) of Gabon and Angola to South Africa; the BLUE-EARED GLOSSY STARLING (*Lamprotornis chalybaeus*; L 22 cm) which occurs from tropical Africa south as far as Botswana and the eastern Transvaal; the SPLENDID GLOSSY STARLING (*Lamprotornis splendidus*; L 25 cm) with purple, violet and blue iridescent lower parts, found in tropical Africa from the Senegal to Ethiopia, western Kenya and Angola; and the PURPLE GLOSSY STARLING *(Lamprotornis purpureus)* which ranges from western Africa to western Kenya and on which not only the iris but also the margin of the eyelids is bright yellow.

The LONG-TAILED GLOSSY STARLING (*Lamprotornis caudatus*; L 45 to 50 cm) is the most familiar species in the long-tailed group of glossy starlings. It shows a shiny dark gold color on the head and the center of the abdomen, with the rest iridescent green, blue and purple. BURCHELL'S GLOSSY STARLING (*Lamprotornis australis*; L 33 cm) has a tail of medium length. It is distributed from southern Angola and southwestern Africa to the Transvaal. The *Lamprotornis* glossy starlings usually nest in holes in trees. One of these species is usually to be seen in European zoos.

Three inconspicuously colored species should probably also be classed with the glossy starlings. They are the MADAGASCAR STARLING (*Hartlaubius auratus*; L 20 cm) of Madagascar, which is mainly brown with a bronze sheen; the SPOTTED-WINGED STARE (*Saroglossa spiloptera*; L 18 cm) which is probably closely related to the preceding Madagascar starling, although it occurs in northern India; and the WHITE-WINGED BABBLING STARLING (*Neocichla gutturalis*; L 20 cm) which is found from southern Angola and Zambia eastwards as far as Tanzania.

Fig. 18-1. Yellow-billed ox-pecker *(Buphagus africanus).*

Subfamily: ox-peckers, by C. W. Benson

The OX-PECKERS or TICK BIRDS (subfamily Buphaginae) differ from the true starlings in their short, compact beaks, short legs with markedly curved claws on the toes, and stiff pointed tail. They inhabit the steppe country of Africa south of the Sahara. Ox-peckers became well known for their habit of perching on domestic and wild mammals such as cattle, zebras, rhinoceros and elephants. They walk about on these animals like woodpeckers on a tree trunk searching the skin for ticks and fly larvae. This activity, while probably not quite painless for the mammal concerned, confers two benefits on it—the removal of ticks, some of which are disease carriers, is a sanitary service; and the birds give warning of the approach of danger. There are two species, the YELLOW-BILLED OX-PECKER *(Buphagus africanus)* and the RED-BILLED OX-PECKER (*Buphagus erythrorhynchus;* L 22.5 cm). Ox-peckers breed in holes in trees, in clefts in rocks and even under the eaves of roofs. The three to five eggs hatch after eleven or twelve days of incubation and the young stay in the nest for twenty-eight or twenty-nine days, rather longer than those of the true starlings.

Fig. 18-2. Red-billed ox-pecker *(Buphagus erythrorhynchus).*

ORIOLES (family Oriolidae) are roughly thrush-sized, tree-dwelling birds, marked by ten primaries and the absence of bristles on the thrush-like beak. The two sexes differ in color; males are conspicuous, yellow and black, red and black or even silvery-white; females are inconspicuous, colored mainly with greenish striping. Their food consists of insects and also fruit and berries. The main area of distribution of orioles extends from India to northern Australia, six in Africa; one species, the black-naped oriole, extends into eastern Asia and another, the golden oriole, into Europe.

The GOLDEN ORIOLE (*Oriolus oriolus;* L 24 cm) is also called Whitsun bird in Germany because of its late arrival near the end of spring. In spite of its striking appearance, few people get to see it, for it stays mainly within the dense foliage of the crowns of trees where only its loud and unmistakable call betrays its presence. The male is bright yellow with black wings, a black and yellow tail and a reddish beak. Females and fledged young, however, are a dull grayish-green with dark stripes on the otherwise paler lower parts. The golden oriole's flute-like whistle, "duedleeoo" is the basis of the bird's scientific name as well as the name "oriole" and German equivalent, "pirol".

Its large area of distribution includes almost all of Europe except the north and the British Isles (in the last decades it has appeared in southeastern England) as well as western, central and southern Asia. In Europe, it inhabits deciduous woods, parklands and large orchards with older trees. It also occurs in open coniferous woods which alternate with areas of heath or deciduous trees. In mountains it is generally not found above 600 meters.

Both partners participate in nest building; nests are well concealed high in the crowns of trees. The birds glue some strips of bast across a horizontal fork using their saliva. These strips will hold the nest. This type of nest construction is characteristic of all the oriole species. Full clutches consist of two to five eggs; the young hatch after fourteen or fifteen days of incubation and require another fourteen or fifteen to reach the fledging stage. Golden orioles leave Germany in August and therefore only have time for a single brood per year. Their winter quarters are in eastern and southern Africa. (Color plate, p. 450.)

The BLACK-NAPED ORIOLE (*Oriolus chinensis;* L 23 cm) is found from Celebes in the south to Amurland in the north in twenty-two subspecies. Like the golden oriole, it has a harsh warning call and soft chattering song. The MAROON ORIOLE (*Oriolus traillii;* L 23 cm) of the area between Vietnam and Taiwan and the Himalayas has part of its plumage blood red. The *Oriolus mellianus* inhabits the mountains of southwestern China; it is silvery-gray with a rust-colored tail and black wings and head. The BLACK-HEADED FOREST ORIOLE (*Oriolus monacha;* L 23 cm) of Ethiopia and Eritrea owes its name to the black color of its head, nape and breast. The MASKED ORIOLE *(Oriolus larvatus)* of eastern and southern Africa and Angola, is rather similar to the black-headed forest oriole. The *Oriolus chlorocephalus* of the mountain forests from

Family: orioles, by H. H. Reinsch

Fig. 18-3. Golden oriole *(Oriolus oriolus)*— distribution in Europe.

Tanzania to Mozambique, has a moss green head, neck and back, blue-gray wings, a yellow collar, and yellow lower parts.

The FIGBIRDS (genus *Sphecotheres*; L 25 cm) are confined to Australia. There are two species, the SOUTHERN FIGBIRD *(Sphecotheres vieilloti;* L 25 cm) and the YELLOW FIGBIRD *(Sphecotheres flaviventris).* In both the ear and eye regions are bare of feathers and are colored. They, like the other orioles, suspend their nests from tree forks.

Family: drongos, by B. E. Smythies

DRONGOS (family Dicruridae) have a strong, laterally-compressed beak which is somewhat hooked at the tip. The bristles at the base of the beak are long and well developed. In length they range from 17 to 37 cm (prolonged tail feathers not included). The tarsi are short but the toes and claws are long. The iris is usually red. The plumage is black, either velvety or glossy, apart from two species in which it varies from dark gray to very pale, almost white gray. Except in the two gray species the plumage has a more or less greenish, bluish or purple gloss. The intensity and distribution of the glossy plumage areas are the best means of distinction among the various species. Some species have beautiful shiny feather tips on the throat and breast, others have glossy lancet-shaped feathers on the side of the neck; many species are crested. A Pleistocene fossil drongo has been found.

Two genera are recognized nowadays: 1. MOUNTAIN DRONGOS *(Chaetorhynchus)* with one species, the PAPUAN MOUNTAIN DRONGO *(Chaetorhynchus papuensis;* L 21 cm) with twelve tail feathers, of the mountain forests of New Guinea; 2. DRONGOS (in the restricted sense) *(Dicrurus)* with ten tail feathers, distributed over most of Africa, southern Asia, Indonesia, parts of Australia and some islands of the western Pacific. There are nineteen species, including the DRONGO *(Dicrurus adsimilis;* L 26 cm) of the African savannahs, the KING-CROW *(Dicrurus macrocercus;* L 30 cm) which occurs from India to Amurland, Vietnam, Taiwan, and sparingly in Java and Bali, the ASHY DRONGO *(Dicrurus leucophaeus)* of southern and eastern Asia; the CROW-BILLED DRONGO *(Dicrurus annectans);* the LESSER RACKET-TAILED DRONGO *(Dicrurus remifer;* L 24 cm); the GREATER RACKET-TAILED DRONGO *(Dicrurus paradiseus;* L 36 cm), all found in eastern Asia; and the SPANGLED DRONGO *(Dicrurus hottentottus),* one of the most variable members of the family, found, in thirty-three subspecies, from the northwestern Himalayas and northern China as far as the eastern Solomon Islands. (Color plate, p. 450.)

Fig. 18-4. Distribution of the mountain drongos (genus *Chaetorhynchus).*

Fig. 18-5. Distribution of the drongos (genus *Dicrurus).*

C. H. Vaurie in 1964 wrote about the general characteristics of drongos: "Drongos are tree-dwelling insect-eaters of medium size which live solitary lives; if the prolonged tail feathers of some species are included they reach a length of about 70 centimeters. The shape of the tail is highly variable; it may be deeply forked or square ended. The outermost tail feathers are sometimes prolonged to an extraordinary degree, twisted or without vanes over certain segments so as to appear spatula-like or like flags. Usually the tail is deeply forked and the outer tail feathers curved outwards. The wings are long, pointed and are so effectively shaped as to ensure, along with the deeply forked tail, great manuverability in flight."

The appearance of the crest in many species may be as variable as that of the tail. Sometimes it consists of a small tuft of feathers behind the base of the beak, while in others it stands out over the culmen of the beak; when more highly developed it arches back over the head and nape. The feathers of the crest are without vanes in part or along their whole length, such feathers showing all transitions between a small tuft about two centimeters long to "hairy" strands which may be twelve centimeters long. Decorative feathers of the trunk, crest and tail often vary within a species from one locality to another. Some species are white on the belly, upper tail coverts, and the cheeks or at the angles of the beak, while one species on the Comoro Islands has reddish-brown wings and tail.

Most drongos utter various sounds and have gay though somewhat metallic calls in which they include pleasant-sounding pure whistling. As a whole they are found in all habitats from open swampy rice fields to the wooded steppes, on forest edges as well as deep in forests both in low country and mountains. Their principal food consists of insects of all shapes and size, including dragonflies, cicadas, grasshoppers and even wasps and hornets. Termites swarming and even more so a grassland fire which scares away insects, attracts all drongos from the vicinity.

It seems that the seven species of Africa and Madagascar are the most primitive. They show little geographical differentiation or development of decorative features. J. P. Chapin found that in the Congo, the drongo eats large insects in the manner of a falcon. It holds the insect down with a foot, dismembers it with the beak and swallows it in pieces of suitable size.

The drongo family is most highly developed in southern Asia, where there are ten species, one of which ranges as far as southeastern Australia and the Solomon Islands, while two other species have penetrated northwards as far as Iran and Manchuria. Three species are confined to Australia, New Guinea and neighboring islands. The kingcrow, in spite of its name, is not related to the crows but resembles them somewhat in color and is as cantankerous and fearless in defense of the nests as they are. One often sees a pair of them chasing a crow; they swoop at it in full strength and surround it. With these falcon-like swoops, they utter a series of angry calls which are very noticeable. The drongos inhabit plains, open wooded country and bushy thickets. It often sits on electric wires, dead trees or other projecting perches from which it keeps a watch for insects on the wing or on the ground. While doing so, it moves the tail to and fro and also opens and closes it like

▷
Birds of Paradise: 1. Trumpet bird (*Phonygammus keraudrenii*); 2. Brown sickle-billed bird of paradise (*Epimachus meyeri*); 3. White-billed sickle-billed bird of paradise (*Drepanornis bruijnii*); 4. Waigeu bird of paradise (*Diphyllodes respublica*); 5. King of Saxony bird of paradise (*Pteridophora alberti*); 6. Magnificent rifle-bird (*Ptiloris magnificus*).

a pair of scissors. It often sits on the backs of cattle to snap up insects flushed by their movements.

The ashy drongo is found in Yunnan above 4,000 meters. Two of its subspecies which breed in eastern Asia occur as migrants in Malaya in the winter. According to Whistler, the only drongo of India "is a marvelous flier which can turn and twist in flight with the utmost speed and skill". The crow-billed drongo is a partial migrant; it passes from Burma to Borneo for the winter. The two most conspicuous drongo species of southern Asia are the lesser racket-tailed drongo and the greater racket-tailed drongo. The latter are forest birds with long wire-like outer tail feathers with a drop-shaped vane at the tip. They rarely occur in the same area as lesser racket-tailed drongo, preferring higher altitudes.

Drongo nests are flat "swings" which look very small compared to the size of the birds. They attach their nests to forks of small branches up to fifteen meters above the ground. Usually three or four, but sometimes two or five eggs are laid. The incubation and fledging periods of all species are still unknown. A few species have been kept for fairly long periods in zoos. The greater racket-tailed drongo with its metallic blue gloss and peculiar vanes at the end of long bare tail feather shafts, is fairly often kept as a cage bird by Indians and Indonesians.

◁
1. Emperor of Germany's bird of paradise *(Paradisaea guilielmi)*; two males in display; 2. Ribbon-tailed astrapia *(Astrapia mayeri)*; 3. Wallace's standard wing *(Semioptera wallacei)*; 4. Long-tailed paradigalla *(Paradigalla carunculata)*.

19 Crows and Related Birds

The crow-like birds and their closer relatives, the birds of paradise and the bower birds, have only recently been combined with a few small families of Papuan, Australian and New Zealand birds to form a group of families. They are probably related by common descent from the ancestors of the babblers, although the connection is hardly recognizable in the crow-like birds. A series of characters of structure and behavior forms the links from one family to another in spite of differences in detail between the families. Among these characters are the strong feet and toes (the wood swallows are an exception), the strong beak in the majority of species, the striking plumage colors, good flying endurance, the loud voice, the resident mode of life and often the urge for a sociable life. Many birds of this group show a strong ability to adapt to their environment and it is therefore believed that the most highly developed species of birds are in this group. Others, however, are so specialized that their relationships were long in dispute and interpreted in different ways from that adopted here. Evidently the component groups have been evolving along different lines since an early period of the earth's history so that common features became indistinct. There are therefore considerable differences within the families which may lead to a further division into subfamilies, particularly in the extensive bird of paradise and crow families. Six families of crow-like birds are recognized: 1. The NEW ZEALAND WATTLEBIRDS (Callaeidae); 2. The AUSTRALIAN MAGPIE LARKS (Grallinidae); 3. The WOOD SWALLOWS (Artamidae); 4. The SONG SHRIKES (Cracticidae); 5. The BIRDS OF PARADISE AND BOWER BIRDS (Paradisaeidae); 6. The CROW-LIKE BIRDS (Corvidae).

Crows, by J. Steinbacher

The NEW ZEALAND WATTLEBIRDS (family Callaeidae) live only in New Zealand. There are three genera, each with one species. They probably developed by adaptations in different directions from a single population which reached the primeval forests of New Zealand between one and sixty-five million years ago. The wattlebirds have a fleshy,

Family: New Zealand wattlebirds, by C. A. Fleming

generally orange-colored wattle on each side in the corners of the beak. They all also share skeletal characteristics, rounded wings and tails, and strong legs and toes with a long claw on the rear toe. Originally they inhabited all of the wooded areas of both islands, but now are found in only a few areas. The HUIA (+ *Heteralocha acutirostris*; L 45 cm) is extinct. It was noteworthy because the beak differed considerably in the two sexes, which cooperated in extracting larvae from rotten wood. The male used its strong beak as a chisel and opened up clefts in the wood in which the female then probed for food with her curved beak. (Color plate, p. 469.)

The SADDLEBACK (⚥ *Creadion carunuculatus*; L 25 cm; Color plate, p. 469) is glossy black with a chestnut-brown saddle on the back. There are two subspecies which live only on the coastal islands separated by the Cook Straits. The KOKAKO (⚥ *Callaeas cinerea*; L 40 cm) is bluish-gray with black cheek stripes. The two subspecies can be distinguished according to the color of the wattle: orange in the south, bright blue in the north. The population of this species has been greatly reduced. As inhabitants of forest interiors, wattlebirds often look for food on the ground, where they hop about in long jumps. They feed on insects, fruit and young leaves. Their flat nests stand up to ten meters above the ground; they are loosely built and are generally roofed over. The clutch consists of two or three (sometimes four) eggs which are pale gray or pale brown with dark spots. (Color plate, p. 469.)

Family: magpie-larks, by D. L. Serventy

The three groups of birds now united in the MAGPIE-LARK family (Grallinidae) were not recognized by ornithologists as being related to one another for a long time. About the only striking feature which is shared by the four species of this family is the fact that they all build nests of mud. The apostle bird and the white-winged chough seem more closely related to one another than to the magpie larks proper (Grallininae), hence two subfamilies are now generally recognized: 1. MAGPIE-LARKS (Grallininae) a genus with two species. 2. MUDNEST CHOUGHS (Corcoracinae) two genera each with one species.

The widely distributed black and white MAGPIE LARK (*Grallina cyanoleuca*; L 29 cm, wt. 126 g) the female of which has a white throat, has been intensively studied by A. Robinson. Because of its nesting habits, it is restricted to areas where there is fresh surface water; this includes the vicinity of clay pits or water reservoirs in the semidesert areas of the interior. Robinson found that these birds form lifelong pairs and use the same territory year after year. Both sexes defend the territory but each only against birds of its own sex. Probably because of this, both sexes sing; they often sing distinctly as a duet. They raise two broods each year and both parents build the nest, incubate and rear the young together. Outside the breeding season, some give up the occupation and defense of territories. Families living in neighboring territories then form a flock of birds which roost together at night. In

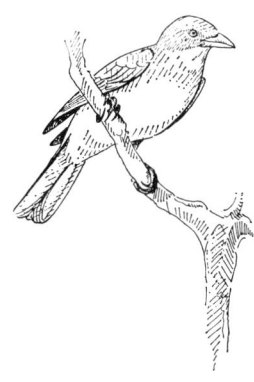

Fig. 19-1. The New Zealand thrush (*Turnagra capensis*) which has become rare. It has been classed by some with the New Zealand wattlebirds.

some places, swarms of several thousand birds may be formed in this way. They feed on insects beside the water and they also take water snails. The TORRENT LARK *(Grallina bruijni)* inhabits New Guinea.

The mudnest choughs have soft, fluffy feathers and they show no color difference in the two sexes. There are two genera, each with one species: 1. The WHITE-WINGED CHOUGH *(Corcorax melanorhamphos;* L 45 cm, wt. about 350 g); it is a dirty black color with white inner vanes on the secondaries. 2. The APOSTLE BIRD *(Struthidea cinerea;* L 32 cm) is gray with brown wings and a black tail. Both live in dry areas and, in the absence of open water, build their mudnests after rainfall or when settlers build artesian wells. (Both shown in Color plate, p. 469.)

Both species differ markedly from the magpie-larks in their communal nesting habits. Several females lay their eggs into a shared nest. John Gould (1814-1881) suggested this to be the case in the chough and in 1869 and 1879 the German zoologist Hermann Lau proved it to hold for both species. In 1944, Michael Sharland showed that this communal nesting proceeds very similarly in both species. White-winged choughs form groups of two to eight birds and the clutch size in the group nests range from two to eight eggs, with an average of four. All members of a group take part in nest building, incubation and the care of the young, but only females high in the rank order lay eggs. However, in no case have more than two young from a single nest been found to survive to sexual maturity.

In the opinion of I. Rowley the advantage of these "breeding groups" is that they enable the bird to seek food more effectively. A local food abundance is less likely to be overlooked by the group and since they tolerate one another, will be utilized by all of them as soon as it is discovered. Communal behavior is also significant in defense against enemies and in the training of the young, which take up to three or four years to reach sexual maturity.

Bre   ing groups

G. S. Chapman has shown that in the apostle bird, breeding groups consist of a top-ranking pair of adults and their descendants from several earlier breeding seasons. In contrast to the chough, neighboring breeding groups may temporarily join into larger flocks and at a common feeding or watering place there may sometimes be several hundred of the birds. These groups are usually strictly resident; only a long drought forces them to wander. Apostle birds may raise two broods in a year, but it is possible that successive clutches of eggs are laid by different females. The white-winged chough is omnivorous, taking a wide variety of insects, frogs and lizards as well as seeds and fruits. Apostle birds take insects, but if these become unavailable at some season they will make do with seeds.

WOOD SWALLOWS or SWALLOW SHRIKES (Artamidae) inhabit Australia, India to southeastern China and the southwestern Pacific area. In almost all the species the plumage shows a mixture of black, gray and

Family: wood swallows by B. E. Smythies

Fig. 19-2. Distribution of the wood swallows (family Artamidae).

white. The two sexes are quite or almost similar. The beak is strong, of medium length, somewhat curved downward and pointed, and the gape is wide. The tail is rather short and not pointed, the wings are long and pointed and almost reach the tip of the tail. There is one genus *Artamus*, with ten species. These include the WHITE-BREASTED SWALLOW SHRIKE (*Artamus leucorhynchus*; L 19 cm) from the Fiji Islands to Borneo and the Philippines; the WHITE-BROWED WOOD SWALLOW (*Artamus superciliosa*; L 20 cm), the largest species with chestnut-colored lower parts, of Australia; the ASHY SWALLOW SHRIKE (*Artamus fuscus*; L 17 cm), reddish-pale-gray, occurs from India to southern China; and the LITTLE WOOD SWALLOW (*Artamus minor*; L 15 cm), dark brown, is found in Australia and breeds in clefts in rocks.

Wood swallows inhabit open forest at low altitudes. They are usually perched on a prominent branch from which they pursue insects in the manner of flycatchers. Very often four or five sit close together, frequently moving the tail from side to side. Some species roost communally. Their flight is extraordinarily graceful; a series of wing beats is often followed by a long glide for which the shape of the wings is quite apt. They are very cantankerous birds and readily attack passing crows and raptors. Their calls are twittering; in part they are harsh.

Nests of wood swallows are built at the end of projecting branches, or sometimes in a cave. Nests may be ten to fifteen meters above the ground, though some Australian species build below three meters in shrubs.

Family: song shrikes by D. L. Serventy

The SONG SHRIKES (family Cracticidae; L 26 to 52 cm) of the Australo-Papuan area are superficially crow-like. Among them, the Australian butcher birds were formerly classed with the shrikes, and the currawongs and Australian magpies with the crows. As early as 1848 John Gould recognized that these three genera belonged together, but he included them in the shrike family. Their plumage is black, gray and white; in some of the butcher birds a reddish-brown phase occasionally occurs. The beak is strong, straight or hook-like (in the butcher birds); there are horny plates on the side of the tarsus divided only in the lower part; the tenth primary is long. There are three genera: 1. AUSTRALIAN BUTCHER BIRDS *(Cracticus)* with six species; 2. CURRAWONGS *(Strepera)* with two species; 3. AUSTRALIAN MAGPIES *(Gymnorhina)* with two or three species.

The smallest species of song shrikes are found among the butcher birds, which inhabit all of Australia and New Guinea. We will name here the GRAY BUTCHER BIRD (*Cracticus torquatus*; L 26 cm), the WHITE-RUMPED BUTCHER BIRD *(Cracticus louisiadensis)* and the BLACK BUTCHER BIRD (*Cracticus quoyi*; L 35 cm). (Gray butcher bird, Color plate, p. 469.)

Butcher birds are cantankerous meat-eaters with a strongly hooked beak. They make food "larders" like the true shrikes with which they were at one time classified. They kill small birds and impale them on twigs in order to eat them later. However, the main function of this habit seems to be to use the end of the twig as a holder to facilitate dividing up the prey. The gray butcher bird must therefore be included among the very few birds (see the woodpecker finch, Chapter 15, and bower birds in this chapter) which use tools. The butcher birds also eat lizards and insects. The black butcher bird feeds on small crabs and insects. Both sexes of butcher birds utter their lively songs at all seasons. Pairs defend permanent territories and males feed the incubating females on the nest.

Like the butcher birds, the CURRAWONGS (genus *Strepera*) also inhabit more or less densely wooded areas. They enrich their diet with fruit and are therefore unpopular in orchards. There are two species, the PIED CURRAWONG (*Strepera graculina*; L 45 to 50 cm) of eastern Australia and Lord Howe Islands with the Tasmanian subspecies (*Strepera graculina fuliginosa*) and the GRAY CURRAWONG (*Strepera versicolor*; L 52 cm) of Tasmania and southern Australia. Currawongs have calls which are to the human ear perhaps the least melodious of all bird sounds. The gray currawong is often called "squeeker" in western Australia from its call. Yet the ornithologist A. H. Chisholm finds the calls of the piping crows to be generally "lively, ringing and not unmusical". In contrast, the songs of butcher birds, Australian magpies, are by general consent almost the best of Australian bird songs. Currawongs take up territories only during the breeding season. Afterwards they form small flocks. The Tasmanian pied currawong regularly undertakes limited migrations after the breeding season.

It is still undecided whether there are one, two or three species of AUSTRALIAN MAGPIES (genus *Gymnorhina*) which range from Australia and Tasmania to southern New Guinea. There are two color groups, in one of which the adult males have a white back, in the other a black back. Accordingly, we separate the BLACK-BACKED MAGPIE (*Gymnorhina tibicen*; L 36 cm) from the WHITE-BACKED MAGPIE (*Gymnorhina hypoleuca*; L 40 cm, wt. 250 g). The white-backed birds occur in southern Australia and the black-backed ones in the north. The two groups presumably evolved this color difference during their long separation. In the west, there is still a gap between the two species, but in southeastern Australia where both forms meet in the mountains of Victoria, which run east and west, a narrow zone of hybridization was formed.

John Gould, who was not in general impressed by the vocal utterances of Australian birds, made an exception for the Australian magpies, writing: "the description of the sounds of this bird exceeds my ability as a writer and I regret that my readers cannot listen to it in its native wilderness as I have been able to do." He had in mind particularly the choral song of these birds, which they deliver in groups. In 1956 A. H. Robinson investigated these songs thoroughly in the southwestern AUSTRALIAN WHITE-BACKED MAGPIE *(Gymnorhina hypoleuca dorsalis)*. He describes the famous dusk and dawn choirs as communal territorial songs. He writes "It is a cheering experience to get up at five o'clock on an ice cold August morning, which corresponds to February in the northern hemisphere, to listen to the song of hundreds of male magpies which within a perimeter of two miles sound like a single mixed choir." As these birds maintain territories throughout the year, one can always hear something of their territorial song.

Australian magpies need trees as breeding sites but they look for food on the ground, almost entirely in open country. They probe the ground for insects, but also eat spiders and small reptiles; they also catch flying insects and occasionally take some plant food. Though these magpies are strictly residents, they do not defend territories as pairs like the butcher birds, but rather lay claims to territories as groups. Robinson studied the southwestern Australian white-backed magpie in a very favorable area in western Australia. The groups there consisted of six to twenty birds and the average size of a group's territory was about forty hectares. All members of a group took part in territory defense. During the breeding season, there was no definite link between individual birds; in matings males did not keep to particular females. A dominant male was the leader of the group. In addition to and apart from the group which remains in a given locality and which is orderly with the leader keeping a sharp watch, there are also groups of young birds which a leader has expelled from a group or which were excluded from their original groups in some other way.

Only the females build nests and incubate. Males do help to feed the young, although most of the feeding is done by the females. The nests are cup-shaped structures made of small sticks and twigs which are fixed in forks of tree branches twelve to fifteen meters above the ground. The eggs lie on a lining of grass, bark or hair; they are bluish or greenish with extensive brown or purple spotting. The clutch usually consists of three or four eggs. Breeding success is low; Robinson found that fifty per cent of the eggs and chicks have a long life expectancy. Wilson calculated an average age of nine years for those birds which had survived the first year. Individuals may live to twenty years.

Birds of paradise, by B. Grzimek & Th. Schultze-Westrum

The birds of paradise and the bower birds (family Paradisaeidae) are quite a heterogenous group. Twenty genera and forty species belong to the BIRD OF PARADISE (Paradisaeinae) subfamily; their relationships to one another have not been definitely established. Presumably they are descendants of crow-like ancestors which were inconspicuous in color and which were probably segregated in the Papuan area since the Tertiary. They are often grouped with their nearest relatives, the bower birds, as "arena birds" (see page 500). The two subfamilies differ from one another mainly in plumage and display behavior. Birds of paradise range in size from that of a starling to that of a raven; L 17 to 120 cm. Their food consists mainly of berries and fruit, as well as insects. Most have relatively massive feet suitable for grasping branches. The males are among the most conspicuously decorated birds; parts of their plumage are reminiscent of silk or velvet, or have a metallic gloss. They almost always have extremely variegated colors and often prolonged decorative plumes of various types. Their song is of only moderate quality; some species are sound imitators. Most display in trees, but a few species display in a dancing area on the ground. Nests are almost always cup-shaped; only the multi-crested bird of paradise builds a roofed-over nest near the ground. Only the males of the more primitive species, which do not have a decorative plumage, take part in raising the young. They are distributed over the tropical forest areas of New Guinea and neighboring islands and the Moluccas and northeastern Australia.

Fig. 19-3. Distribution of the birds of paradise (subfamily Paradisaeinae).

The fantastically beautiful plumage of the birds of paradise has aroused the attention of Occidentals for about five hundred years. Several centuries ago merchants brought the first skins of some of these birds from their New Guinea home to the Orient, to China and the Sunda Islands. The "Victoria," the only surviving ship of Magellan's world circumnavigating convoy, entered the harbor of Seville on September 8th, 1522. Among other curiosities on board were several bird of paradise skins with their unbelievably delicate and colorful feathers. It seems strange that well-preserved skins lacked feet and leg bones. Though the Italian Antonia Pigafetta declared a few years later that these birds in life must have feet which would be about a handbreadth in length, he was not believed, for the legless skins which merchants, mainly Dutch, brought to Europe later on seemed to prove the opposite quite unequivocally.

Thus, all sorts of legends about these wonder birds arose. It was believed that the birds of paradise, as they were named by the Dutchman Jan van Linschoten in about 1590, remained air-borne all the time, never coming down to earth before their death, thus not requiring feet. Pairs were supposed to breed in the air, the egg lying in a groove on the male's back and being covered for incubation by the female's belly. It was also seriously believed that birds of paradise took no food other than dew. As these stories sounded attractively marvelous, they were popular, although serious scientists declared them to be nonsense. Only in 1824 did a ship's apothecary, René Lesson, discover live birds of paradise in their New Guinea home, being the first white man to do so. It was now clear that they were not footless, ethereal creatures living on the dew of the heavens, but real, though very beautiful, birds with strong feet.

The discovery and the early enthusiasm over the beauty of birds of paradise was soon followed by uncontrolled hunting of the birds and a lively trade by middlemen and fashion houses in Paris, London and other cities. Fashion had discovered bird of paradise feathers as decorations for women's hats. During the first five years of German colonial rule over northeastern New Guinea more than fifty thousand skins were exported from there and used for ladies' hats. The native Papuans had been in the habit for centuries, possibly for thousands of years, of using the feathers of certain species as headgear on ceremonial occasions and as a form of money when trading women or goods. This traditional hunting of birds of paradise did not disturb the biological balance. But the depredation of the populations resulting from the bird hunts organized by Europeans, which were accompanied by something like a gold rush fever, and the criminal offenses against native Papuans which accompanied this trade, finally went so far that in 1924, when the value of the feathers had in any case fallen very low, the taking of birds of paradise or possession of them was declared illegal in New Guinea. Since then, only scientific expeditions have been allowed to collect and own birds of paradise.

The natives were not legally prevented from hunting the birds for their own traditional use. It would hardly have been feasible to attempt to restrict them. Thus, the men can still wear feathers for the great dancing feasts as in the past. Since tourists have reached New Guinea, native feather decorations have, in certain areas, become even more opulent than before. It has been calculated that at the annual Mount Hagen show in the central highlands, dancers wear bird of paradise feathers worth no less than about $250,000. This has produced a new danger for the birds which, unfortunately, is not the only one. As the old tribal traditions of the Papuans are melting away under the influence of civilization, old property rights have also become shaky. Formerly male birds of paradise at their display areas in the forests were the property of individual men or of particular families and the birds were not killed until they were fully mature and had reproduced. Also, the local wars which formerly were generally prevalent have ceased, so that now any Papuan can safely go alone to hunt in the forest area of his own group. More Papuans now own modern shotguns.

Quite recently two attempts were made to lift the 1924 ban on bird of paradise hunting. A submission in this regard was made to the Parliament of Australian New Guinea based on the argument that the hunt would increase earnings in the poorest provinces. Presently attempts are underway to abolish the prohibition of the export of bird of paradise feathers from Australia; this would mean that it would once more be profitable to take skins illegally from New Guinea. In addition, some of these birds are threatened by the destruction of their habitats. This would not endanger species like the Raggiana bird of paradise, which is a pronounced follower of cultivation, but it is a serious threat to other species like the blue bird of paradise and the King of Saxony bird of paradise which are specialized to a restricted habitat in central New Guinea's mountain forests which have already been much reduced. Conditions in the formerly Dutch part of New Guinea which are now Indonesian can only be surmised. The new masters of this area are probably not active enough to bother about enforcing the protective regulations.

The MULTI-CRESTED BIRD OF PARADISE (*Cnemophilus macgregorii*; L 23 cm) is regarded as the most primitive of the birds of paradise, and is sometimes even thought to form a distinct subfamily (Cnemophilinae). It is fiery orange-red with blackish lower parts; the female is olive-yellow and yellowish below. The male has four narrow boat-shaped three-centimeter-long feathers on the forehead which can fit into a furrow in the plumage of the crown. These feathers are shorter in females. It occurs in southeastern New Guinea, inhabiting jungles with tall tree trunks, feeding in the main on fruit. It is probably the only bird of paradise which builds a roofed-over and spherical nest near the ground. The female builds the nest, which has an entry at the side. The single young is fed on fruit by regurgitation. The young spits the stones of the fruit out of the nest. In the geographical race of the Kubor Mountains, males have cinnamon-colored upper parts.

Ornithologists are still uncertain about the systematic position of some other primitive species, most of which are not closely related to one another. These are colorful forms but they lack the decorative feathers of the typical birds of paradise. Possibly many of them belong to the beginning of an evolutionary series which leads in the higher forms to the specialized but insignificantly colored bower birds. In this group, the sexes are similar in color and males take part in the care of the brood. It includes the GREEN-BREASTED MANUCODE (*Manucodia chalybata*; L 40 cm) of New Guinea and the related TRUMPET BIRD (*Phonygammus keraudrenii*; L 33 cm) which also occurs in northeastern Australia. The LONG-TAILED PARADIGALLA (*Paradigalla carunculata*; L 37 cm) of the Arfak Mountains of western New Guinea is almost unknown as far as its mode of life is concerned. Perhaps research on this and other primitive species like the YELLOW-BREASTED BIRD OF PARADISE (*Loboparadisaea sericea*; L 17 cm) and LORIA'S BIRD OF PARADISE (*Loria loriae*; L 23 cm) in the pathless moss-and-lichen-grown jungles of New Guinea might lead to a better understanding of the evolution of this group.

In the typical BIRDS OF PARADISE (genus *Paradisaea*) males have on each flank a tuft of fine silky feathers which are directed towards the rear and which can be erected by skin muscles. Their size corresponds to that of a jay, L with the tail feathers ranging from 63 to 100 cm; females are smaller and have inconspicuous colors. The most important species are: 1. The GREATER BIRD OF PARADISE (*Paradisaea apoda*) with eight

The multi-crested bird of paradise

▷
1. Greater bird of paradise *(Paradisaea apoda)*;
2. Arfak astrapia *(Astrapia nigra)*; 3. Magnificent bird of paradise *(Diphyllodes magnificus)*; 4. Twelve-wired bird of paradise *(Seleucides melanoleuca)*.

very different subspecies; among them that of the Aru Islands southwest of New Guinea (*Paradisaea apoda apoda;* L 100 cm) and the RAGGIANA BIRD OF PARADISE *(Paradisaea apoda raggiana)* of southwestern New Guinea. 2. The LESSER BIRD OF PARADISE *(Paradisaea minor)* only slightly smaller. 3. The BLUE BIRD OF PARADISE *(Paradisaea rudolphi;* L 63 cm) of the mountain forests of New Guinea, and 4. the EMPEROR OF GERMANY'S BIRD OF PARADISE *(Paradisaea guilielmi;* L 81 cm; Color plate, p. 480).

It was the feathers of the Aru greater bird of paradise which were most often brought to Europe for use on women's hat. The shafts of the central tail feathers were known as "wires", because these feathers are modified and curved. Since Alfred Russell Wallace published his description of its display in 1869, not much information about this bird has emerged. He writes: "Males danced in certain trees in the forest which have an immense space for the birds to play and exhibit their plumes. On one of these trees, a dozen or twenty full-plumaged male birds assemble together, raise up their wings, stretch out their necks and elevate their exquisite plumes, keeping them in a continual vibration. Meanwhile they fly across from branch to branch in great excitement so that the whole tree is filled with waving plumes in every variety of attitude and motion. During the excitement of display, the wings are raised vertically over the back, the head is bent down and stretched out, and the long plumes are raised up and expanded till they form two magnificent golden fans striped with deep red at the base and fading off into the pale brown tint of the finely divided and softly waving points." Females of this subspecies differ from those of the others by having white underparts.

The habits of the subspecies which live on the mainland of New Guinea are known better, particularly those of the Raggiana bird of paradise. It has partly become a follower of cultivation and displays in casuarina trees at the edges of villages. In addition to mature males, some immature ones which have not yet developed the full plumage also come to the display areas and they also "dance". Individual females are also present as apparently unconcerned onlookers. The display which is much like that of the Aru Island greater bird of paradise is introduced by loud calls of the males before the first rays of the rising sun have reached the branches. Before the morning sun is well up in the sky, the whole overpowering display of flaming tufts of feathers waving to and fro is already over. Sometimes it is repeated in the evening. The raising of the wings described by Wallace is rather different in the Raggiana bird of paradise; the latter shows a rhythmical raising and lowering of the wings while the trunk is horizontal or is slanted somewhat, head down; the wing movements are almost jerks which involve the trunk as well. Evidently, the male becomes immobile in this posture at the peak of its excitement. In 1959, Eugen Schumacher succeeded in filming this display and that of other species of

◁
1. Multi-crested bird of paradise *(Cnemophilus macgregorii);* 2. Black-eared green catbird *(Ailuroedus crassirostris melanocephalus);* 3. Golden bird *(Sericulus aureus);* 4. Golden bower bird *(Prionodura newtoniana);* 5. Tooth-billed bower bird *(Scenopoeetes dentirostris);* 6. Regent bower bird *(Sericulus chrysocephalus);* 7. Striped bower bird *(Amblyornis subalaris);* 8. Spotted bower bird *(Chlamydera maculata).*

birds of paradise. The male's contact call can be heard from afar and is uttered repeatedly; it is one of the characteristic sounds of the Papuan jungle. A young male which the animal trader Fred Shaw Mayer kept in the highlands and which was quite tame, could clearly imitate the words "number one". Females of the Raggiana bird of paradise are buff below with a suggestion of barring.

The plumes of the lesser bird of paradise are bright yellow. The female is dark below. This species occurs in western New Guinea. Possibly it is no more than a subspecies of the greater bird of paradise. The blue bird of paradise is a veritable jewel. Carl Hunstein, the German gold hunter and animal dealer, who drowned in a mountain torrent about the turn of the century, dedicated this bird to the Austrian crown prince Rudolph; hence its scientific name. In display, the male hangs head down from a branch; before that he perches on his display sites, calling. Then carefully and slowly the bird rotates backwards and as soon as it is hanging head down, it shakes itself and this spreads out iridescent blue plumes. By means of a repeated movement at the hip joint, it dances its body and the plumage up and down. With the "dance", it utters a soft, monotonous song. The Emperor of Germany's bird of paradise also hangs head down from a branch during its display.

Fig. 19-4. The display of the Raggiana bird of paradise is a magnificent sight.

Hybrids between different species and even genera are not rare among birds of paradise and were often named as new species when specimens reached the hands of systematists. Only now, when judgement of relationships are based not only are structural features but also on behavior, has a clearer picture emerged of the separation of forms into species and genera. The view which was formerly prevalent, that two species were the more distantly related the more they differed externally and with respect to behavior, certainly does not apply to the birds of paradise.

There is relatively frequent hybridization between the KING BIRD OF PARADISE (*Cincinnurus regius*; L 16 cm) and the magnificent bird of paradise, which belongs to a different genus. E. Thomas Gilliard has suggested that hybrids between species which look so different occur because of the communal display of these birds. Since there is evidently no pair formation, a male which is particularly attractive to females can distribute its genes very effectively. If some of these genes undergo a mutation, e.g. so as to bring about a change in color, this new feature could spread relatively rapidly. The evolution of new characters would then escape from the selection pressures which favor sexual isolation. Unfortunately, our knowledge of the birds of paradise is much too sketchy for a test of the correctness of these views.

The king bird of paradise, somewhat smaller than a starling, is the smallest of the bird of paradise family, but it is one of the most beautiful of these "wonder birds". Wallace gave an enthusiastic description

Fig. 19-5. Lawe's six-wired parotia in display looks like a ballet dancer.

Fig. 19-6. The superb bird of paradise in display.

Fig. 19-7. Wallace's standard wing in full display.

of a male which he received on the Aru Islands more than a hundred years ago. "The greater part of its plumage was of an intense cinnabar red, with a gloss as of spun glass. On the head, the feathers became short and velvety and shaded into rich orange. Beneath, from the breast downwards was pure white, with the softness and gloss of silk, and across the breast a band of deep metallic green separated this color from the red of the throat. Above each eye was a round spot of the same metallic green; the bill was yellow and the feet and legs were a fine cobalt blue."

About the characteristic tufts and spiral feathers, he writes: "Springing from each side of the breast and ordinarily concealed by the wings, were little tufts of grayish feathers about two inches long, each terminated by a band of intense emerald green. These plumes can be raised at the will of the bird and spread out into a pair of elegant fans when the wings are elevated. The two middle feathers of the tail are in the form of slender wires about five inches long, and which diverge in a beautiful curve. About half an inch at the end of this wire is webbed on the outer side and colored a fine metallic green, and being curled spirally inwards forms a pair of elegant glistening buttons hanging five inches below the body and the same distance apart. These two ornaments, the breast feathers and the spiral-tipped tail wires, are altogether unique, not occurring on any other species of bird known to exist upon the earth."

The female, as in all species of this group, is inconspicuous, brownish above, paler below with dark bars. The king bird of paradise lives in dense crowns of trees and though it is not rare throughout the lowlands and hills of New Guinea and adjacent islands, one only occasionally gets to see it and never for more than a few moments. It is the only species which nests in tree holes. (Color plate, p. 470.)

The two species of the genus *Diphyllodes,* the MAGNIFICENT BIRD OF PARADISE *(Diphyllodes magnificus)* and the WAIGEU BIRD OF PARADISE *(Diphyllodes respublica;* L 20 cm) show a veritable palette of bright colors. In the Waigeu bird of paradise, the bare skin of the head in both sexes is divided into blue areas by lines of small black feathers. Apart from this the female is hardly distinguishable from that of the king bird of paradise. The males of these two magnificent species dance in display on a young shoot of a tree. Around it they clean the forest floor of all plants in a circle five to six meters in diameter. They also remove the leaves from overhanging branches so that the light can fully strike their iridescent feathers. (Color plates, p. 479 and p. 491.)

The parotia birds of paradise perform their peculiar dances on the ground in mountain forests or on branches. There are several species, including LAWE'S SIX-WIRED PAROTIA *(Parotia lawesii)* and the ARFAK SIX-WIRED PAROTIA *(Parotia sefilata;* L 31.5 cm) and the QUEEN CAROLA'S PAROTIA *(Parotia carolae).* Males carry six thin feather shafts on the head, each of which has a small web at the tip. In display they erect the body in such a way that an umbrella-like circular formation is produced. The birds dance to and fro and in a circle, looking like tremulous greenish-blue glossy fungi with a feather crown on the head. The parotias are distributed in the higher-lying areas of New Guinea.

The KING OF SAXONY BIRD OF PARADISE (*Pteridophora alberti*; L 21 cm) is restricted to the Papuan highlands. According to the latest researchers, it is possibly not a true bird of paradise at all. Males of this small species bear two feathers, each twenty-five centimeters long, which arise from the head and which are probably the most peculiar feathers in the bird world; in both feathers the bed is only formed on one side of the shaft and it is divided into thirty to forty small enamel-like segments which are colored blue above and brown below. It is no wonder that when ornithologists first unpacked speciments in Dresden, Germany in 1894, they suspected at first that some collector, eager for gain, had inserted the "enamel feathers" into the skin of the head. The female has in place of these long feathers only two insignificant "ear" tufts. (Color plate, p. 479.)

The call of the TWELVE-WIRED BIRD OF PARADISE (*Seleucides melanoleuca*; L 34 cm) is a familiar sound in the coastal lowlands of New Guinea. The male in display erects forward a shield-shaped collar of dark blue-green iridescent feathers so that all that remains visible of the head is the wide open beak with its yellow interior. From each flank a tuft of golden yellow plumes, which are as soft as silk, stand out and over the bird project six wire-like feather shafts on each side which are directed forwards. (Color plate, p. 491.)

The dull black SUPERB BIRD OF PARADISE (*Lophorina superba*; L 24 cm) is roughly the size of a starling. It lives in the mountain forests of New Guinea, and is known to hybridize in the wild with four other species. In display it shows two feather shields, one, a metallic iridescent green-violet arising from the breast feathers and the sides, and on the back a wide semi-circular fan of black feathers which look as if they had been dusted over with metallic green particles. The beak is wide open in display, showing off its greenish-yellow interior. WALLACE'S STANDARD-WING (*Semioptera wallacei*; L 29 cm) of the Moluccas is named after its discoverer, the explorer-scientist Alfred Russell Wallace (1823–1913). It has a similar projection of iridescent green to each side of the breast when erecting its feathers in display, while over the back arise four long-shafted white decorative feathers which are modified wing coverts. (Color plates, p. 470 and p. 480.)

The astrapias are a remarkable group of long-tailed birds of paradise of the highlands of New Guinea. The last species to be described was discovered as recently as 1938. They grow to the greatest size of all birds of paradise and look even larger because of the two central tail feathers which in the RIBBON-TAILED ASTRAPIA (*Astrapia mayeri*; L 110 to

▷
1. Siberian jay (*Perisoreus infaustus*); 2. Azure jay (*Cyanocorax caeruleus*); 3. Unicolored jay (*Aphelocoma unicolor*); 4. Cayenne jay (*Cyanocorax cayanus*); 5. Magpie (*Pica pica*); 6. Green magpie (*Cissa chinensis*); 7. Azure-winged magpie (*Cyanopica cyana*); 8. Ceylon magpie (*Urocissa ornata*); 9. Red-billed blue magpie (*Urocissa erythrorhyncha*); 10. Common jay (*Garrulus glandarius*); 11. Clark's nutcracker (*Nucifraga columbiana*); 12. Nutcracker (*Nucifraga caryocatactes*); 13. Collared jay (*Cyanolyca viridicyana*); 14. Pinyon jay (*Gymnorhinus cyanocephalus*); 15. Blue jay (*Cyanocitta cristata*); 16. Hartlaub's blue jay (*Cissilopha melanocyanea*).

◁
1. Hooded crow *(Corvus corone cornix)*; 2. Fish crow *(Corvus ossifragus)*; 3. Pied crow *(Corvus albus)*; 4. Thick-billed raven *(Corvus crassirostris)*; 5. Raven *(Corvus corax)*; 6. Rook *(Corvus frugilegus)*; 7. Jackdaw *(Corvus monedula)*; 8. Indian house crow *(Corvus splendens)*; 9. Carrion crow *(Corvus corone)*; 10. Common crow *(Corvus brachyrhynchos)*.

120 cm) are almost a meter long. In flight they are dragged behind like a white train with a black tip. These feathers are also long in the PRINCESS STEPHANIE ASTRAPIA *(Astrapia stephaniae;* L 84 cm); they are velvety-black and are a favorite item of head decoration among the Papuans. The ARFAK ASTRAPIA *(Astrapia nigra)* is restricted to the Arfak Mountains in western New Guinea. (Color plate, p. 491.)

Closely related to the astrapias and also very large are the RIFLE BIRDS (genus *Ptiloris*) of which the MAGNIFICENT RIFLE BIRD *(Ptiloris magnificus)* together with two other birds of paradise has settled the northern tip of Australia from southern New Guinea. Although the male is not particularly striking in its color, it still forms a very impressive sight in display. It opens the wings fully, draws the head back and to one side so as to show off its violet iridescent neck and remains in this pose. Its loud, abruptly cut off call is one of the most familiar bird sounds in lowland New Guinea. (Color plate, p. 479.)

The SICKLE-BILLED BIRDS OF PARADISE (genus *Epimachus*) owe their name to their beaks which are up to 7 cm long and curved downward. The BROWN SICKLE-BILLED BIRD OF PARADISE *(Epimachus meyeri;* L 100 cm) has a pair of "false" wings which are normally hidden by the real wings, but which are spread out in display. Together with the BLACK SICKLE-BILLED BIRD OF PARADISE *(Epimachus fastosus)* and the WHITE-BILLED SICKLE-BILLED BIRD OF PARADISE *(Drepanornis bruijnii)*, it belongs to the rarer and less well known birds of paradise. (Color plate, p. 479.)

John Gould, the pioneer of Australian ornithology, first saw the structure made by a bower bird in 1839 in a natural history collection in Sydney. Although a description of the species involved had been published twenty years earlier, knowledge of the bower-like structures it makes had not reached Europe at that time. Gould set out with enthusiasm into the Australian bush to study how bowers were made. In his classical works on the birds of Australia he gave the first objective description of a bower. Some time later, the Italian collector C. Beccari found a bower of gardener bowerbirds in the dark, rainy and moss-grown forests of New Guinea, which looked like a miniature native hut with a well-kept garden around it. The English translation of Beccari's report was published in 1878 together with an illustration of the bower in "The Gardener's Chronicle", a horticultural magazine. But it was not until the middle of this century that A. J. Marshall and E. Th. Gilliard threw some light in the biological significance of bower building. Much about the life of these peculiar birds remains unknown. G. Evelyn Hutchinson has said about their behavior that "in its many-sidedness and uniqueness, it has no parallel in the animal world."

The bower birds, by Th. Schultze-Westrum

The BOWER BIRDS (subfamily Ptilonorhynchinae) are the closest relatives of the birds of paradise. They differ from them only in the structure of the palate, length of the rear toe and plumage. In size they range from 23 to 38 cm. The males of several species have erectile golden yellow, orange or pink crests. In external features the sexes are often indistinguishable. Their food consists of fruit, berries and insects. There are eight genera with eighteen species in New Guinea and its neighboring islands and in parts of Australia.

At least twelve species of bower birds build bowers with bits of twigs, grass stalks and other plant materials and decorate them with shiny or colorful objects. Two species even paint their bowers. From a circular area around the bower, they remove all leaves and other loose objects. The male displays in or in front of the bower to the female and they mate there. Each bower is the product of the efforts of a single male; females take no part in making it and the bower is never used as a nest. The young are raised in cup-shaped nests solely by the female.

According to modern views of avian evolution, the birds of paradise and bower birds form a phylogenetic series in which the bower birds are the more advanced group. Both are arena birds, in which the male prepares and defends a display and mating place which is not related to food requirements or nest building. By the term "arena" is meant an area which is claimed by a particular number of males which are in some sort of relationship to one another. Within the arena, which may be larger than a square kilometer, each male has its private "yard" in which it displays and which it defends against other males. In their arena display birds of paradise show off their plumage, while bower birds mainly show off their bowers. These structures may, therefore, be considered as secondary sex characters of the males which have been carried over into objects in their environment. Since because of this males can "afford" to be inconspicuous in color, they are safer from raptors and other enemies.

A. J. Marshall first pointed out that bower building might be a component of the original nest building behavior transferred to a new function. Males of the true bower birds have nothing to do with nest building or the care of the young. The four-walled bower of the yellow-breasted bower bird does, in fact, suggest a large nest. Gilliard has pointed out that males of certain arena birds step as carefully and quietly into their "yards" as parent birds visiting the nest. Many arena birds are polygynous; in some others the type of sexual relationship is still unknown. It is known for some, however, that certain females repeatedly visit the same male. In view of this loose polygyny, a few males in each generation are enough to ensure the survival of a species.

According to the type of bower, four groups of bower birds are recognized: 1. catbirds; 2. yard bower birds; 3. Maypole bower birds and "gardeners"; and 4. avenue bower birds.

Of the CATBIRDS, those of the genus *Ailuroedus* are evidently the most primitive and they build no bowers. 1. The GREEN CATBIRD (*Ailuroedus*

Above: Young carrion crows *(Corvus corone corone)*. Below: Magpie *(Pica pica)*.

Fig. 19-8. Distribution of the bower birds (subfamily Ptilonorhynchinae).

The catbirds

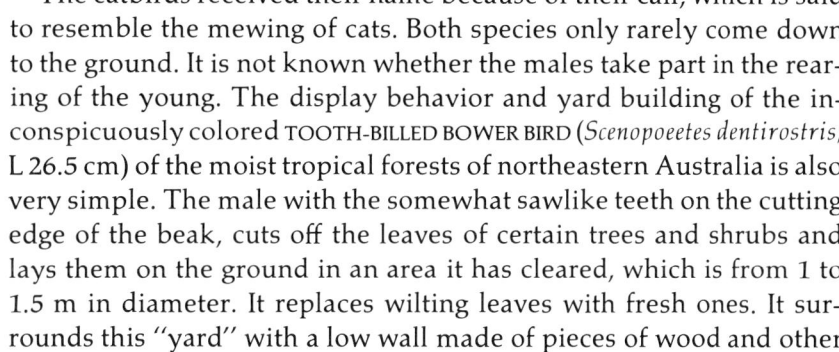
Male greater bird of paradise *(Paradisaea apoda)* in display.

crassirostris; L 30 cm) of eastern and northern Australia and the Aru Islands off New Guinea, is bright emerald green above with a yellowish-green breast with buffy gray spots. 2. The WHITE-EARED CATBIRD *(Ailuroedus buccoides)* has a white ear and throat region and occurs in New Guinea. In both species, the two sexes are similar.

The catbirds received their name because of their call, which is said to resemble the mewing of cats. Both species only rarely come down to the ground. It is not known whether the males take part in the rearing of the young. The display behavior and yard building of the inconspicuously colored TOOTH-BILLED BOWER BIRD *(Scenopoeetes dentirostris;* L 26.5 cm) of the moist tropical forests of northeastern Australia is also very simple. The male with the somewhat sawlike teeth on the cutting edge of the beak, cuts off the leaves of certain trees and shrubs and lays them on the ground in an area it has cleared, which is from 1 to 1.5 m in diameter. It replaces wilting leaves with fresh ones. It surrounds this "yard" with a low wall made of pieces of wood and other objects it finds on the ground. (Color plate, p. 492.)

Two New Guinea forms of the group called gardeners above, behave similarly. They too build a yard or court. They are the ARCHBOLD BOWER BIRD *(Archboldia papuensis;* L 32 cm) which was only discovered in 1938, and SANFORD'S BOWER BIRD *(Archboldia papuensis sanfordi).* Both line their yards with ferns. The outstanding bird photographer Loke Wan Tho was able to photograph a display yard of 1.4 m in diameter in the dense undergrowth of the mountain forests in central New Guinea. It was decorated with little heaps of snail shells and beetle wings.

Fig. 19-9. Display yard of a Maypole bower bird.

The Maypole-building bower birds build more complex yards. They are named after the tower or tent-like structure which the males build around a small vertical tree shoot. All around the shoot they heap small bits of twigs horizontally and intertwine them to a height of 1.2 to 1.8 m. The northern Australian GOLDEN BOWER BIRD *(Prionodura newtoniana;* L 24.5 cm) actually builds towers over several years which may be up to 3 m high. Low down on the tower it builds, again from bits of twigs, a U-shaped connection to a smaller twin tower and just above this connection which lies on the ground it fastens a piece of a branch or root. This is evidently where it perches to show off its golden yellow plumage to the female when one visits the display ground.

Fig. 19-10. The Maypole of MacGregor's bower bird.

Most of the others in this group, the birds of the genus *Amblyornis,* live in the dark mountain forests of New Guinea. Because of the decorative features of their yards, they deserve to be called gardeners. Thomas Gilliard has pointed out that the species which builds the most artful display ground is also the one in which the males are least conspicuous in color. The Maypole of MACGREGOR'S BOWER BIRD *(Amblyornis macgregoriae;* L 25 cm) is still relatively simple. In the middle of a saucer-shaped area very carefully lined with moss and about 1.1 m in diameter, there is a shoot of a tree around which a tower 2 to 4 meters thick of small twigs has been built. The tower too is carefully covered with pieces of moss. During the display the male erects its bright golden to orange-yellow crest.

The male STRIPED BOWER BIRD (*Amblyornis subalaris;* L 24 cm) has a somewhat shorter crest than does the MacGregor's bower bird. It was discovered in 1884 in the Owen Stanley Range in New Guinea. When Goodwin first described its bower, he noted that it was the most beautiful structure ever built by a bird. Within a circular wall there rises from the moss-lined floor a tent-like roof which is about 60 cm high and about 90 cm across and which is attached in the center to a shoot which holds it up like a tent pole. This hut-like structure is open to one side and within it a "Maypole" is built up around the central shoot which divides the opening of the "hut" into two entrances, the passages from which connect behind the Maypole. The center Maypole is made of the dark fibers of the trunk of a tree fern and is richly decorated with flowers, berries, colorful leaves and beetle wings. The male also scatters red fruit and some flowers and berries over the moss-covered floor of his yard. (Color plate, p. 492.)

Fig. 19-11. The display yard and "hut" of a striped bower bird.

In the GARDENER BOWER BIRD (*Amblyornis inornatus;* L 28 cm), the male has no crest and in color is just as inconspicuous as the female. Its bower is more like a man-made structure than those of any other species. Hunters of birds of paradise in the last century in the Arfak region of western New Guinea thought the bowers of these birds were toy huts made by the natives. Beccari wrote enthusiastically about this bird: "Its passion for flowers and 'gardens' is a sign of its good taste and delicate way of life." The carefully cleaned yard is about 1.2 m in diameter and has no wall along the rim. The hut itself is built around a central support and is open on one side. Dillon Ripley, who took the first photos of these bowers, measured the hut as 1 m high and 1.5 m wide at the base. "Flower beds" and little heaps of yellow objects covered the yard, including fungi, berries, flowers, a red cartridge case, and pieces of charcoal. Only in 1964 Heinz Sielmann made the first films of the displays of this and other bower birds.

Fig. 19-12. The bower of the gardener bower bird.

The structures of the AVENUE-BUILDING BOWER BIRDS, those of the genera *Ptilonorhynchus, Sericulus* and *Chlamydera,* are less reminiscent of man-made buildings. Yet they too are built in quite a complex manner. The floor of these bowers is covered with interwoven fibers and fine twigs over an area of 1 to 1.5 m in diameter. In the center there are two walls made of vertical sticks anchored in the floor and interwoven above. Only rarely do the two walls meet to form a roof; they are just sufficiently far apart that the bird can walk through the alley without being hemmed in. The entrance to the alley and its side walls are decorated with berries, little stones, bleached bits of bones, beetle carapaces, mussel shells, bright bits of metal, glass and other objects of human origin, including even car keys.

Fig. 19-13. The alley of the satin bower bird (diagrammatic plan).

"Avenue" building bower birds

The SATIN BOWER BIRD (*Ptilonorhynchus violaceus*; L 32 cm) and the REGENT BOWER BIRD (*Sericulus chrysocephalus*; L 30 cm) paint the walls of their bowers. This painting is particularly interesting in the satin bower bird, for it is one of the few examples of the use of a tool by a bird. The male takes a small piece of bark in its beak and uses it as a paint brush to apply a pigment made from charcoal, fruit juice and its saliva, on the walls of the bower. As the male holds its "brush" in the beak at right angles to the beak and rubs it up and down on the walls, saliva and the pigment flow from the fibrous brush. Males prefer a blue color in their paint and since they are themselves blue, this color is probably attractive to the female. (Regent bower bird, Color plate, p. 492.)

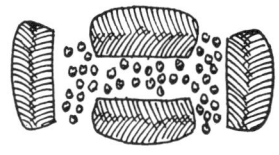

Fig. 19-14. The yellow-breasted bower bird builds four firm walls (diagrammatic plan).

The male of the AUSTRALIAN GREAT BOWER BIRD (*Chlamydera nuchalis*; L 38 cm) shows off its pink crest to the female in display by twisting its neck. Gilliard has observed that the male FAWN-BREASTED BOWER BIRDS (*Chlamydera cerviniventris*) of New Guinea also twist their necks in display, though they have no crest. Thus the crest has been lost in the course of evolution, but not the display behavior that goes with it. In the grasslands of the central highlands of New Guinea Gilliard found an arena in which at least sixteen males of the YELLOW-BREASTED BOWER BIRD (*Chlamydera lauterbachi*) had their yards or bowers. One was built from more than 3,000 little sticks and fibrous stalks which were so densely interwoven that they formed four strong walls. This yard held about 5 kg of pale pebbles. When a female entered the yard, the male began to "dance" excitedly. Then it took a red berry and held it up as if it had been showing a bright tuft of feathers. The berry seemed to function as a substitute for the collar of iridescent feathers still found in other species of this genus, like the great bower bird and the SPOTTED BOWER BIRD (*Chlamydera maculata*; L 28.5 cm; Color plate, p. 492).

Fig. 19-15. Bower of the spotted bower bird (diagrammatic plan).

The yellow-breasted bower bird can be compared with the gardener bower birds. In both groups the inconspicuous males are colored like the females. Their secondary sex characters have thus been lost and were replaced by the artful bowers and their decoration. Various zoos have shown bower birds which have built bowers and shown their displays. At present, however, few arrive in Europe because of the ban on the export of Australian animals.

Family:
Crow-like birds,
by W. Wüst

The CROW-LIKE BIRDS (family Corvidae) undoubtedly belong to the song birds, mainly because of complex structure of their vocal organ, the syrinx. The raven, which weighs about 1.25 kg, is the world's largest song bird. One of the smallest corvids, the *Podoces panderi*, with a L of about 25 cm and a weight of 90 g, is still large as song birds go, being larger than the song thrush for example, and almost as large as a blackbird. The dwarf of the crow-like birds, Hume's ground chough, with a L of 19 cm and a wt of 35 g, lies between the woodlark and skylark in size.

With these comparatively clever, mobile, curious and adaptable birds our description of the world of birds comes to its conclusion. Corvids range from 18 to 70 cm in length. They have ten primaries, the outermost (which in many song birds is more or less atrophied), is longer than the secondaries. There are twelve tail feathers in almost all species, but in *Ptilostomus afer*, there are only ten. The plumage is generally inconspicuous, black, gray or brown, but in a number the plumage and even the beak and feet are beautifully colored. Their distribution is world-wide except for New Zealand (where some have been introduced) and Antarctica. There are seven groups. A. Jays; B. Magpies; C. Ground Jays; D. Nutcrackers; E. Choughs; F. Pipiacs; G. Ravens. There are twenty to thirty-one genera and 101 to 103 species and about 370 subspecies.

Crow-like birds feed on a mixed diet, some of which they obtain by astonishing slyness. Few other birds reach their level of intelligence. For this reason, ethologists have paid particular attention to them since the very beginning of animal psychology. Their relatively large brain alone suggests a high degree of learning ability. All the birds of this family live in lifelong monogamy. In some species, the males incubate and brood the young. They feed and guard the young, give warning of danger and chase off persistent raptors or mammals which could be dangerous to the young. Carrion crows, magpies, common jays and some other species owe their continued existence in the face of merciless human persecution to their own learning ability. The oldest corvid of which we have evidence, a representative of an extinct genus, *Miocorax*, lived seven million years ago in France.

The members of the jay group tend to have prolonged feathers on the crown. The CRESTED JAY (*Platylophus galericulatus*; L 28 cm) has a particularly well developed crest. When erect it stands higher above the crown than the length of the head. Apart from the crest its plumage is dark. It inhabits the tropical jungles from southern Burma to Borneo and Java in three subspecies. The BLACK-CRESTED MAGPIE (*Platysmurus leucopterus*; L 38 cm) of Malaya, Sumatra and Borneo, has a black plumage with a white wing spot and a double crest on the forehead and on the back of the head.

Jays

Jays show a rich development with more than thirty species in America; particularly in the mountainous Pacific region they are often very colorful. Of the species with predominantly blue plumage the genera *Gymnorhinus, Cyanocitta, Aphelocoma, Cyanolyca* and *Cissilopha*, the following will be mentioned: 1. The PINYON JAY (*Gymnorhinus cyanocephalus*; L 25 to 27 cm); it is uniformly steel blue in color and short-tailed, and occurs in the southwestern U.S. 2. The BLUE JAY (*Cyanocitta cristata*; L 29 cm) occurs in the eastern half of the U.S. and a large part of southern Canada. 3. STELLER'S JAY (*Cyanocitta stelleri*; L 32 cm) has a dark plumage with a blue belly, lower back, wing and tail with fine crossbars on the tail. Its range is western, northern and middle

America. 4. The SCRUB JAY (*Aphelocoma coerulescens*; L 30 cm) has, like the next species, no crest; it occurs in western America from the state of Washington in the north to Honduras in the south; one of its sixteen subspecies occurs in Florida. 5. The MEXICAN JAY (*Aphelocoma ultramarina*; L 30 cm) occurs in seven subspecies in southwestern, northern and central America. 6. The UNICOLORED JAY (*Aphelocoma unicolor*; L 30 cm) has five subspecies and occurs in the mountains of Mexico up to about 3,300 m. 7. The COLLARED JAY (*Cyanolyca viridicyana*; L 34 cm) is particularly beautifully bright blue, as are the other six species of this genus. 8. The BLACK-AND-BLUE JAY (*Cissilopha sanblasiana*; L 29 cm) is black on the head, neck, upper back and on its lower parts except for the lower tail coverts which are blue like the rest of the upper parts. It occurs in Central America in four subspecies.

The pinyon jay in shape and behavior is reminiscent of the Old World nutcracker. It lacks the bristles at the base of the beak which are otherwise present in crow-like birds, so that its nostrils are visible. It wanders about in large flocks in search of food. Often it moves on the ground like crows do and catches grasshoppers and other insects there with its pointed beak.

The blue jay with its colorfully patterned plumage and mobile crest is the best known jay of North America. Because of its maneuverability and vocal capacity, it is known as "a clown among the birds of its homeland", though it is at times disliked as a nest robber. It largely occupies the ecological niche in North America which is held by the common jay in the Old World. Like the latter, it "plants" acorns by removing them and pushing them into the ground. Again like its European cousin, it may spoil a hunter's chances by warning the game with its alarm call. No person readily escapes detection by this bright bird and it learns to judge whether an individual is dangerous or harmless. To a greater extent than the common jay, it has become used to settlements and it is a frequent guest at winter feeding tables. Steller's jay greatly resembles the blue jay in its behavior and the variety of its calls. It generally inhabits coniferous forests including those of Douglas firs and the giant redwoods. Its scientific name comes from that of the German naturalist Georg Wilhelm Steller (1709–1746) from Windsheim in central Franconia, who in 1741 was on a Russian expedition and who landed for a few hours in Alaska and on this occasion discovered the bird.

The Central and South American jays of the genus *Cyanocorax* have velvety or brush-like feathers on the crown and a well-rounded tail. There are eleven or twelve species. They include the GREEN JAY (*Cyanocorax yncas*; L 27 to 28 cm, wt 230 g) which is bright green with the back of the head sky blue, deep black cheeks and breast and lemon yellow outer edge of the tail. It is the most colorful of the jays. The CAYENNE JAY (*Cyanocorax cayanus*; wt 170 to 196 g) has only a slight green flush, a black throat and upper breast, deep blue upper parts with a slight violet gloss and a white back of the head and nape. The PLUSH-CRESTED JAY (*Cyanocorax chrysops*; L 35 to 37 cm) and its closest relatives have a crest as long as the head, which consists of fine short black feathers which are always erect. The plush-crested jay is widely distributed in Central America in five subspecies. The birds wander through the forests in troops of about six to eleven birds; they imitate animal sounds and give warning of predators or the approach of hunters.

The WHITE-TIPPED BROWN JAY (*Psilorhinus morio*; L 40 cm) and the MAGPIE JAY (*Calocitta formosa*; L 48 to 49 cm) are closely related as is demonstrated by hybrids of these two species. Both inhabit Central America. The inconspicuous white-tipped brown jay has a loud voice evidently reinforced by the air sac over the furcula. The magpie jay with its long graduated tail, suggests a magpie but it has a pointed crest. Its main color is blue-gray, while the head is black above but white at the sides, as are the lower parts.

The genus *Garrulus* brings us back to the Old World. It includes: 1. The COMMON JAY (*Garrulus glandarius*; L 34 to 35 cm, wt 120 to 192 g); it has an extensive range in the northern Old World; this and the fact that there are a number of isolated populations has led to the formation of thirty-four subspecies. 2. LIDTH'S JAY (*Garrulus lidthi*; L 40 cm) is largely dark blue except for the reddish-brown rump. It occurs in southern Japan. 3. The BLACK-THROATED JAY (*Garrulus lanceolatus*; L 30 cm) is found in the western Himalayas.

Fig. 19-16. Common jay (*Garrulus glandarius*).

The fact that in the German language area alone the common jay has far more than a hundred local names, shows how well known it is. In Europe it is *the* jay. Everyone has seen it on walks in the woods or has been startled by its sudden shriek. It is generally known that it has small feathers with black, blue and white bars on the wings, that it can erect the feathers of the crown to a small crest and meow like a buzzard. It can also imitate other bird sounds and other noises so closely that the most experienced bird watcher is sometimes deceived. During the breeding season this jay is very quiet and this makes it more difficult to discover its nest in a forest or park tree. It is found in the lowlands as well as in mountains, in the Alps up to 1,600 meters. The five or six eggs of a full clutch are, unlike those of other crow-like birds, greenish- or grayish-yellow with very fine brown spots.

The famous surgeon August Bier (1861-1949), in his book "The Soul", has devoted five pages to the jay under the chapter heading "To overlook what is essential and to over-estimate what is non-essential is a widely distributed error of the soul". Our most important deciduous trees, the red beech and two species of oak, owe their present distribution in great part to jays, which plant their seeds and those of other deciduous trees more skillfully than a forester plants them, and Bier writes "Only superficial observers believe that the jay plants its acorns in order to find them again in winter in time of need. Others simply repeat this erroneous view. Where I live, the support of oaks by jays is so exemplary that the birds can hardly have taken back many of the acorns they planted. Also, if they wanted to find them again, they would not plant them singly but close together. Thus an instinct drives the bird to this peculiar behavior, the utility of which we can see at once. There is thus a mutual relationship between the plant and the bird, which is beneficial to both. The plant is widely distributed and the bird continues for at least part of the year to live off its fruit."

The extraordinary significance of the common jay as an agent in spreading certain forest trees is still not fully appreciated, a generation after Bier wrote. While in the USSR the jay is fully protected as being beneficial to forestry, it is still senselessly and quite legally persecuted in Germany. It is just as well that this clever bird cannot be exterminated. While it is true that it does sometimes take eggs or nestlings of small birds, it loses some of its own to the stronger crows which man has largely freed from their natural predators, the goshawks, peregrine falcons and the eagle owl. One or two pairs of jays have nested in the Nymphenburg Park in Munich, for at least a hundred years, yet song birds live there in good numbers. In general, the common jay is resident or at most a short distance wanderer. However, at intervals of a few years, thousands which must have come from further east wander through central Europe.

Fig. 19-17. Siberian jay (*Perisoreus infaustus*).

The JAYS of the genus *Perisoreus* with their short beaks and the lack of a crest look short-headed and almost "cuddly" which befits their far northern homes. 1. The GRAY JAY (*Perisoreus canadensis*; L 28 to 29 cm) is bluish-gray and when adult has much white on the forehead, about the shoulders and lower parts; fully-feathered young are blackish-gray with a white cheek stripe. It occurs in eleven subspecies in the cold coniferous forest areas of northern North America. 2. The SIBERIAN JAY (*Perisoreus infaustus*; L 30 to 31 cm) is very similar, though somewhat more colorful. It occurs in ten subspecies in northern Eurasia.

The further south the gray jay occurs, the more it prefers higher altitudes, while in the north it is found in lowland muskegs. It generally stays near its breeding area even in winter, but occasionally some populations wander. Nests are placed in conifers one to five meters above the ground. The full clutch consists of only three or four eggs, which are incubated by the female. They hatch after sixteen to eighteen days of incubation.

The Siberian jay occasionally straggles into central Europe. The superstituous formerly regarded it as an omen of ill fortune, hence the German name of this harmless bird which means "bad luck jay". In this regard, I prefer the view of the Lapps, who look on it as a sign of good fortune. In northern Europe no one is allowed to harm this bird and it has therefore lost all fear of man. In Finnish Lapland, one took away our soap while we were washing and allowed itself to be photographed at close range when offered slices of sausage. This jay greatly helps to liven up the still, apparently endless forests of northern Europe. Its flight is so silent that often I only noticed when one of them settled only a few meters from me, by the faint sound from the bark of the branches it perched on. It is rarely encountered alone; usually it joins with others of its kind to form a loose flock.

Magpies with their nineteen species developed in the Old World. There are many species, particularly in Asia, most of them marked by colorful plumage and some by deep red or yellow beaks. Among them are the very beautiful genera *Urocissa* and *Cissa* from central, eastern and southern Asia, including Indochina. They include: 1. The CEYLON MAGPIE (*Urocissa ornata;* L 45 cm) which has a bright rust-colored forehead and flight feathers and is otherwise azure blue. It lives in Ceylon. 2. The BLUE MAGPIE (*Urocissa caerulea*) of Taiwan. 3. The YELLOW-BILLED BLUE MAGPIE (*Urocissa flavirostris;* wt 123 to 151 g) which lives in the Himalayas, its range overlapping that of the next species. 4. The RED-BILLED BLUE MAGPIE (*Urocissa erythrorhyncha;* L 60 to 70 cm), of which the tail accounts for 45 cm; wt 225 to 230 g); it is blue above with the head and neck black, and occurs in eastern Asia. 5. The WHITEHEAD'S BLUE MAGPIE (*Urocissa whiteheadi;* L 38 cm) is paler in color; it lives in southeastern Asia. 6. The GREEN MAGPIE (*Cissa chinensis;* L 38 cm) is green above, has brown to red wings, pale green lower parts and red legs and beak. It occurs from the Himalayas to Burma and Borneo. 7. The SHORT-TAILED GREEN MAGPIE (*Cissa thalassina;* L 35 cm) is generally similar in color but has bright yellow lower parts.

The magpies of these two genera are curious and noisy birds which show a preference for forest as well as less densely treed areas. They eat various small animals and fruit. Like the jays, they give warning of a hunter's approach and also follow animal predators, spoiling their chances by their noise. The red-billed blue magpie is a popular cage bird in its native home and also abroad in zoos.

The AZURE-WINGED MAGPIE (*Cyanopica cyana;* L 34 cm) presents a puzzle to zoogeographers, which still remains unsolved, for one subspecies of this well-known bird with its black crown and blue wings and tail (*Cyanopica cyana cooki*) lives in Spain and Portugal, while the eight or nine subspecies live in central and eastern Asia. There is a huge gap between these two areas of distribution. Possibly the two populations were once connected through Eurasia, but there is no present-day evidence of this. This species has no close relatives. On the Iberian peninsula, it favors dense deciduous and coniferous woods as well as tree plantations; in eastern Asia it prefers stands of trees or bushes in cultivated land. Its large nests, built of sticks and twigs, are not roofed over.

Fig. 19-18. Azure-winged magpie (*Cyanopica cyana*).

The Asiatic magpies also include the TREE PIES (*Dendrocitta*) with six species, the RACKET-TAILED MAGPIES (*Crypsirina*) with two species, and the monotypic genus *Temnurus*. The INDIAN TREE PIE (*Dendrocitta vagabunda;* L 42 cm) has much the same figure as the common magpie but has a relatively stronger, more curved beak. The front of its body is dark blue and this changes to a rusty yellow on the back and belly. There is a wide white bar on the wings and the tail. The two sexes incubate the four or five eggs for two weeks. Males which incubate are exceptional among the crow-like birds. Indian tree pies have an undulating flight as they noisily flutter their wings for a spell and then glide for a short distance, spreading the wings and tail. The HOODED RACKET-TAILED TREE PIE (*Crypsirina cucullata;* L 30 cm) which is gray and black, is restricted to northern and central Burma. The RACKET-TAILED TREE PIE (*Temnurus temnurus;* L 30 cm) of southeastern Asia, has peculiarly formed tail feathers.

Fig. 19-19. Magpie *(Pica pica)*.

The only magpie to have reached America is the COMMON MAGPIE (*Pica pica;* L 39 to 57 cm, wt 173 to 245 g). In nineteen species it occupies a large area in the Old and New World; among North American forms is the YELLOW-BILLED MAGPIE *(Pica pica nuttalli)* which may be an independent species. The many folk names given to the magpie show how much it has attracted human attention and imagination. In Europe it is spoken of as a thief, but in magpies living free I have never seen any evidence of this, although I have examined dozens of nests. Magpies tend to follow cultivation, but not without exceptions. On the one hand, they penetrate into the interior of cities but on the other hand they are found far from known habitation in forests and in mountains. In the Alps they nest as high as 1,700 meters and I have seen them in winter 400 meters higher up than that. They are very much resident birds which also maintain the pair bond throughout the year. Particularly in winter they form flocks which also roost as a unit and which may, at the most, exceed five hundred birds.

Though magpies are persecuted in most areas in central Europe, they have been exterminated only locally. The large nests built of twigs and usually roofed over are conspicuous but the birds give them up if they are disturbed. The pair then builds another, and often a third nest. The large average number of eggs, seven in a full clutch, must contribute to making up for losses. Young magpies are often reared by people and among these birds are some which show an urge to seize bright objects and to hide them. Magpies are less often victims of peregrine falcons, goshawks and eagle owls than are jays and crows. Their mode of flight and the fact that they stay close to cover probably enable them to escape their enemies more frequently.

The ground jay group includes three genera which possibly are closely related to one another. They are *Zavattariornis* with one species; *Podoces* with four species; and *Pseudopodoces* with one species. *Zavattariornis* was only discovered in 1937. When the explorer E. Savattari was collecting birds in southern Ethiopia in that year, E. Moltoni, who examined his material, described a new genus and species, STRESEMANN'S BUSH CROW (*Zavattariornis stresemanni*; L 27 cm). It is gray above and partly whitish below with black secondaries and tail feathers and a ring of bare skin around the eye. The fact that this bird was only discovered so recently is the more astonishing in view of the fact that within its range it is locally quite numerous and as conspicuous in its behavior as the common starling. It is sociable and inhabits savannahs with acacia thornbushes, but does not breed in colonies.

The typical ground jays include PANDER'S GROUND CHOUGH (*Podoces panderi*; L 25 cm, wt 90 g, Color plate, p. 469) and HENDERSON'S GROUND CHOUGH (*Podoces hendersoni*; L 33 cm); both live in central Asia. Their plumage colors and walking movements indicate that they are inhabitants of sandy deserts. They build their nests in shrubs, live in pairs and incubate their three to five eggs between March and May. They eat insects, fruit and seeds. Still more adapted to desert life is the smallest of the crow-like birds, HUME'S GROUND CHOUGH (*Pseudopodoces humilis*; L 18 cm, wt 35 g; Color plate, p. 469) of the high country of central Asia. It is buffy-brown above. It hops swiftly over the ground and builds its nest at the end of a long tunnel in earth or among rocks on slopes. As with other hole breeders, its eggs are white. All ground jays are resident. ((Pander's and Hume's ground choughs, Color plate, p. 469.)

Fig. 19-20. Stresemann's bush crow (*Zavattariornis stresemanni*).

The NUTCRACKER group consists of only one genus (*Nucifraga*) with two species. The two are so similar in details of shape, voice, behavior and their requirements from the environment that in spite of differences in pattern and color of the plumage, they are immediately recognized as belonging to the same genus. The NORTH AMERICAN CLARK' NUTCRACKER (*Nucifraga columbiana*; L 30 to 32 cm; Color plate, p. 497) differs only in its gray plumage, white face and black wings from the brown, white-spotted NUTCRACKER (*Nucifraga caryocatactes*; L 32 to 34 cm, wt 200 g; Color plate, p. 497) of Eurasia. Nutcrackers follow coniferous mountain forests and prefer the seeds of conifers to any other food. In hard winters both species move down into the valleys and often feed without any fear of man. Their full egg clutch consists of two to four pale green eggs with fine brown spots. The young hatch after sixteen or seventeen days of incubation.

Fig. 19-21. Nutcracker (*Nucifraga caryocatactes*).

Two subspecies of the nutcrackers can be found in central Europe; the ALPINE NUTCRACKER *(Nucifraga caryocatactes caryocatactes)* and the more slender-billed SIBERIAN NUTCRACKER *(Nucifraga caryocatactes machrorhynchos)*. The Alpine form breeds in coniferous forests of high mountains as well as those of medium elevation and in country which is merely hilly from 300 to 2,000 m; it has been seen as high as 3,850 m. Some of its folk names indicate its preference for the seeds of the pine *(Pinus cembra)* and hazel nuts for food. It often makes stores of these foods and is able to find them again several months later, even when they have been covered by half a meter of snow. In order to fill its storehouse, a nutcracker may fly as far as twelve kilometers. One hundred and thirty-four *Pinus cembra* seeds were found in the crop of one which was brought to the natural history museum of the Swiss town, Freiburg. By means of this habit of storing food, nutcrackers help to spread this pine as well as hazel bushes.

In Scandinavia particular pairs occupy the same territories through the whole year, but in the Alps families wander about in summer and fall. These birds, which are often surprisingly confident towards people, appear in the orchards and hedges of villages during summer and fall. Siberian nutcrackers invade central Europe at irregular intervals, e.g. in 1968 they appeared in the northern German plain. These wanderers seem not to return to their country of origin. On the other hand, there is evidence that individual pairs of the Siberian subspecies remain and breed in the range of the Alpine nutcracker. Thus, a pair of Siberian nutcrackers of the 1954 invasion raised young in 1955 in a stand of conifers in a garden at Gröbenzell near Munich.

The CHOUGHS (genus *Pyrrhocorax*) are two species of Old World crow-like birds which nest in rocks. 1. The CHOUGH (*Pyrrhocorax pyrrhocorax*; L 34 to 44 cm; wt 350 to 450 g; Color plate, p. 469) exists in eight subspecies and apart from living in mountains also occurs in coastal cliffs as in Britain and Ireland. The ALPINE CHOUGH (*Pyrrhocorax graculus*; L 38 to 41 cm; wt 200 to 240 g; Color plate, p. 469) occurs in two subspecies both of which are mountain birds. They are particularly fond of high mountains, e.g. Atlas, Alps, Abruzzi, Altai, occurring in fact, locally in mountains all the way from Spain to China. Choughs are found in the high Atlas at elevations from 2,000 to 2,500 m; in the Himalayas most commonly at about 4,000 m and they even occur up to 6,000 m, although it is not known whether they breed at that altitude. Both species of choughs like to form flocks consisting of only one species, the individuals of which know one another. Choughs may reach at least fifteen years of age. Their range and probably also their total population is greater than that of the Alpine chough. In the Alps, it is, however, by far less common than the Alpine chough, and it is decreasing, possibly facing local extinction. In the Alps, Alpine choughs are the competitors of the chough. In some localities jackdaws also reveal themselves as a stronger species than the chough. Sometimes renovations on a building have deprived choughs of a breeding site. The fact that their population has recently become reduced to isolated pairs as in the Engadin (Switzerland) may have weakened the vital forces in a bird which normally lives in flocks.

In Switzerland choughs begin to lay their three to six eggs in April. The fledging period is thirty-seven to forty days. At the castle of Tarasp (Canton Grisons, Switzerland), where choughs have nested for over a hundred years, the young leave the nest at the latest at the end of June. In the Canary Islands the chough, strangely enough, occurs on only one island, Palma, as a distinct subspecies *(Pyrrhocorax pyrrhocorax barbarus)*. The young are fully fledged there earlier than in Switzerland, but in the Himalayas fledging takes place later than in Switzerland. Breeding pairs remain in their territories even in winter, but it is not certain that this also applies to immatures. Choughs fly with as much endurance and skill as Alpine choughs and more skillfully and more joyfully than other crow-like birds, even during their molt.

In high mountains where people are scarce, as in the Himalayas between 4,000 and 5,000 m, Alpine choughs still show their original habits. In the Alps, they have become largely dependent on man. There they even breed occasionally on or in buildings, but later in the season than choughs. Near huts or restaurants on mountain peaks they mingle with the tourists and eat tidbits offered to them by hand, as well as kitchen refuse. They fly down into the valleys not only at the onset of bad weather in winter but also in summer in order to harvest cherries and in the fall in order to plunder the grapes. All the same, they are generally well liked by tourists and the native populations. Their resounding whistles and trills and their skill in gliding and down-plunging flights are things that probably no mountain wanderer wants to miss. Therefore one can expect that these lively birds will be preserved for the future. Though the Alpine chough is so popular, its biology has not been worked out extensively. Its natural breeding places are almost inaccessible and it is generally impossible to see the nest even if the site can be reached. In the Alps it breeds between 1,320 and 3,000 m.

The PIPIAC *(Ptilostomus afer;* L 42 to 45 cm) is the representative of a group. In shape it resembles a magpie but is almost all black, with the wings and the markedly graduated tail dark brown. In contrast to other crow-like birds, it has only ten tail feathers. It inhabits Africa from the southern edge of the Sahara south to the equator. It is often seen in flocks near palm trees. Flocks also accompany herds of cattle.

I have never been able to see why German laws protect the JACKDAW *(Corvus monedula;* L 33 to 34 cm, wt 193 to 290 g; Color plate, p. 498) while its relatives, crows, magpies and jays are unprotected, because all these birds of the crow tribe eat eggs or the young of other birds with equal rapacity. It seems that a habit for which the others are condemned is not resented in the case of the jackdaw. Perhaps we feel deep down about these "timeless fellows" something of what the great ethologist Konrad Lorenz has described so beautifully: "Twenty-four years have passed since the first jackdaw has been flying about the roofs of Altenberg and since I have lost my heart to these birds with the silvery eyes. And as so often happens with the great loves of our lives, I thought nothing of it as I got to know my first young jackdaw." He writes further about this bird "Tshok": "Once it had become independent I wanted to let it go. I did set it free but not with the expected result, for even nowadays jackdaws are still breeding under our roof. Never have I been so rewarded for an act of mercy towards an animal. Few birds or even higher animals as a whole (the colonial insects are a case of their own) show such a highly developed family and social life as jackdaws. Because of this the young of a very few other species are so touchingly helpless and cling to the person who tends them as attractively as young jackdaws do."

Lorenz's further writings about his jackdaws are among the choicest animal stories in world literature. His observations about the development of the rank order among these birds culminates with the following: "Thus the bride at her betrothal necessarily moves up at once into the rank order of her bridegroom. However, the reverse does not happen. There seems to be a law which cannot be broken which says that no male jackdaw shall marry a female of higher rank. If, for example, the male jackdaw of highest rank chooses a five-month old jackdaw 'girl' as his bride, she immediately becomes the 'First Lady' and she is fully aware of this and without compunction chases other females away from food." Jackdaws have also played a significant role in experiments designed to test the ability of birds to count.

Together with the chough and the Alpine chough, jackdaws belong to the few hole breeders among the crow-like birds. They nest in holes in trees and cliffs as well as in walls, castles, ruins and church towers, even in the interiors of large cities. They are not found above 1,230 meters in the Alps. They therefore do not compete with the choughs and Alpine choughs. They are residents and partial migrants. In winter, central Europe receives an influx of many thousands of jackdaws from the north and east which regularly occur in mixed flocks with rooks. In part, these jackdaws belong to the eastern European-western Siberian subspecies.

The name "crow" is applied to a number of medium-sized crow-like birds. Among them are: 1. The INDIAN HOUSE CROW (*Corvus splendens*; L 45 cm; Color plate, p. 498) of India with five subspecies, one of which *Corvus splendens zugmayeri* was not discovered until 1919 by Laubmann, who described it as a new species. 2. The GRAY CROW *(Corvus tristis)* of New Guinea, which has part of the head bare of feathers; adults are pale gray with a whitish head. 3. The BLACK CROW (*Corvus capensis*; L 42 to 43 cm) which is black with a steel blue gloss; it occurs from the Sudan to South Africa but is only locally numerous. 4. The ROOK (*Corvus frugilegus*; L 48 to 50 cm; wt 360 to 670 g; Color plate, p. 498) which occurs in three subspecies from western Europe to eastern Asia. 5. The COMMON CROW (of America) (*Corvus brachyrhynchos*; L 50 cm; Color plate, p. 498) occurs in North America. 6. The FISH CROW (*Corvus ossifragus*; L 45 cm; Color plate, p. 498) of the North American Atlantic coast. 7. The CARRION CROW (*Corvus corone*: L 47 to 49 cm; wt 460 to 800

g) occurs in six subspecies. They include *Corvus corone corone* (Color plates, pp. 498, 501) which is completely black, of western Europe, the eastern CARRION CROW *(Corvus corone orientalis)* which is also all black and occurs in eastern Asia and the HOODED CROW *(Corvus corone cornix;* Color plate, p. 498) with mainly gray contour feathers, distributed from central Europe to the Urals. 8. The MAORI CROW (+ *Corvus moriorum*) had been extinguished by the natives before the European discovery of New Zealand.

The house crow is probably one of the best known and most tame birds of Indian towns and villages. In the evening they gather noisily into large flocks to roost communally in trees, sometimes in the forest. Males and females alternate in incubation of the four or five eggs. The significance of this crow as a host for the brood parasitic koel is described in Chapter 11 of Volume 8.

The rook looks rather similar to the house crow. In recent years it has become a problem for European conservationists. As it is sociable and breeds in colonies, it is more noticeable and therefore more easily exterminated than the carrion crow for which it is often mistaken. Because of this misidentification, it has already become extinct over large areas. Rooks are occasionally harmful when they feed on fields where wheat or corn is sprouting, but in the main, they feed on cockchafer larvae, wire worms, maggots, caterpillars, snails, worms and mice. Thus they are much more beneficial than harmful to agriculture.

In Germany, rooks have sought protection from human persecution in towns like Königsberg, Stralsund, Schwerin, Greifswald, Lübeck, Flensburg, Hamburg, Berlin, Eisleben, Wiesbaden, Würzburg, Wiesenfels, Leipzig, Munich and Augsburg. There they are surprisingly safer than in the country unless shooting them is permitted, as happened in Augsburg. It has been realized in only a few countries that rooks are more highly deserving of protection, and also in greater need of it, than many other song birds. In towns rooks breed in elms, plantains and pines; elsewhere on spruce and other trees, both coniferous and deciduous. There are often a number of nests in the crown of a single tree. A breeding colony may have from ten up to a few hundred pairs. Nests in deserted rookeries are often used by kestrels which do not build nests of their own and depend on such opportune finds.

Rooks which stay through the winter visit the same feeding areas and use the same roosts for many decades, moving to and fro between these daily, particularly in and around towns. Since rooks can reach an age of at least twenty years, the older birds can transmit their travel experience by example to younger birds. Banding has shown that certain individuals may transfer to another colony up to several hundred kilometers away from their original one. Immatures less than a year old have black feathers up to the root of the beak while in adults the skin of this area is bare of feathers and looks whitish.

In North America, the common crow corresponds to the carrion crow of Eurasia, and everyone knows this clever bird. Hand-reared tame birds perform the same tricks as carrion crows; they often hide scissors and thimbles and are sources both of amusement and annoyance to their owners. In the country they are appreciated as destroyers of harmful insects but their liking for young corn impairs their popularity. Their call is to my mind more reminiscent of the Old World nutcracker than of the carrion crow. The smaller fish crow does not feed solely on fish but is as onmivorous as other crow-like birds. Like Old World crows, it watches for opportunities to rob bird eggs and swiftly makes use of its opportunity to plunder the nests of herons or egrets when photographers scare these away from their nests for even only short periods. Thus man often unintentionally does harm to bird life.

In Europe the carrion crow is *the* crow. Its black forms are confused by the inexperienced with rooks and even with ravens. But the gray subspecies (like the hooded crow) cannot be confused with any other species or even with the black forms of the carrion crow. The two black subspecies, the western and the eastern carrion crows, form the western and eastern wings of the area of distribution of the species as a whole. They are separated over a wide area between the Elbe and the Yenisey Rivers by four subspecies in which the contour feathers are gray. The hooded crow is one of these four; it is quite remarkable from the viewpoint of zoogeography. The zone of hybridization between carrion and hooded crows can vary in depth from twenty-five to a hundred and sixty kilometers. It passes over Schleswig-Holstein, over the island of Fehmarn and Lübeck Bay to the middle and upper reaches of the Elbe and Ultava Rivers, then between Linz and Vienna over the Alps where it curves to the west and passes along the southern edge of the Alps to the Ligurian Sea. In the border area the two subspecies form hybrids of various types which are fertile with either of the original subspecies. New mixed pairs of hooded and carrion crows form every year. Evidently in these cases neither plumage color has great selective value. We do not know what influences the birds to choose one of their own or of the other subspecies for a partner. In the time that attention has been paid to it, there have been no significant shifts of the 2,100 kilometer-long belt of hybridization between these two subspecies. But mixed pairs and even pairs where both partners were hooded crows have also been found far to the west, well within the area of the carrion crow.

Some hooded crows avoid the continental winter and regularly migrate to the west. It is therefore possible every year to see hooded crows in western Germany where the carrion crow predominates. In voice, behavior, structure and the eggs, carrion and hooded crows are indistinguishable. Carrion crows of all subspecies have different calls from those of the rook, which has a low-pitched rough voice.

Carrion crows never breed colonially like rooks even though they may form flocks of more than a hundred birds. In contrast to the sociable rook, old and young carrion crows are rather silent around the nest. The three to six (generally five) eggs are laid between March and April and are incubated solely by the female in seventeen or eighteen days. At first only the male feeds the young but ten days after hatching both parents do so. The fledging period lasts a month. Immature birds do not reach sexual maturity until the end of their second year; then they pair up, forming a lifelong bond. As residents they keep their territories throughout the year and even for the length of their lives. Territories may be fifteen to fifty hectares in area. Hunters resent the fact that carrion crows often take the eggs of pheasants, partridges, ducks, coots, curlews, lapwings and other ground-nesting birds. This happens mainly early in the season when there is not sufficient plant cover to hide nests, or when wires provide elevated perches from which the crows can spot nests.

In tropical and southern Africa, the PIED CROW (*Corvus albus*; L 45 cm; Color plate, p. 498) can readily be seen with little effort. Though scarcely the size of a crow, this black and white bird gives the impression of being powerful and even kites respect it. Hundreds of these birds often inhabit towns and villages where they clear the streets of refuse. In relation to agriculture, it is sometimes beneficial; at other times it damages the harvest of certain crops. Pied crows also like to visit shores and beaches. In the course of irregular wanderings, they show up in areas where they do not breed. They do not like high temperatures, particularly when the air is very dry, but they are also largely absent in the tropical rain forests of the Congo. Nests are placed in trees or in niches in rocks. The eggs are immediately recognizable as crow eggs by their color and spots. Eggs can be found in every month of the year. The pied crow is often a host for the great spotted cuckoo (see Volume 8, Chapter 11).

Raven is the name for the largest crow-like birds but all birds so named are not necessarily closely related. Thus the raven proper is related to the crows while two species, the thick-billed raven and the fan-tailed raven, which were formerly placed in genera of their own, represent another evolutionary line. Among the ravens are: 1. The RAVEN (*Corvus corax*; L 57 to 67 cm; wt about 1,250 g; Color plate, p. 498); it is the largest bird of the crow-like tribe, distributed in eight subspecies over the north of the New and Old World. 2. The BROWN-NECKED RAVEN (*Corvus ruficollis*); it is chocolate-colored, particularly on the neck and nape. It inhabits northern Africa and is sometimes regarded as a subspecies of the raven. 3. The FAN-TAILED RAVEN (*Corvus rhipidurus*; L 45 cm); it is short-tailed and has fan-shaped bristles at the root of the beak. It occurs from the Near East to eastern Africa. 4. The AFRICAN WHITE-NECKED RAVEN (*Corvus albicollis*; L 50 cm) has a white band across the nape and a huge, relatively deep beak. It occurs

Species of ravens

Fig. 19-22. Raven (*Corvus corax*) and in Africa brown-necked raven (*Corvus ruficollis*).

in eastern and southern Africa. 5. The THICK-BILLED RAVEN (*Corvus crassirostris*; L 57 cm; Color plate, p. 498) has a beak like the African white-necked raven; the head and neck are brown, while the rest of the plumage is black except for a white patch at the back of the head and nape. It occurs in Eritrea and Ethiopia.

The raven is the most characteristic of the crow-like birds, although it exceeds all of them in body size as well as in the size of the area of distribution. Among the eight subspecies, not all are absolutely black all over; thus the FAROESE RAVEN *(Corvus corax varius)* has a tendency towards spotting in white. Apart from its greater size, it differs from black crows in the wedge-shaped ending of the tail, the deep beak which is strongly curved along the culmen and above all in its calls. Its deep-pitched rough "kvok" is easily reproduced by the human voice. It is often repeated once or twice and carries surprisingly far. Thus, one often hears a raven before catching sight of it. These large birds glide and circle for long spells and when they are in the mood enjoy looping the loop and tumbling over sideways high up in the air. Hardly any bird can harm a raven in flight; on the contrary, ravens enjoy pestering golden eagles and other raptors by diving at them. On the other hand, one may also see them circling peacefully with griffon vultures.

An uncanny ability enables the raven to find the most hidden animal corpses; often, however, it is forced to leave such prey to eagles or vultures who make use of the raven as their guide. The raven cannot, of course, stand up to these predators on the ground and so it reluctantly yields to them. It does, however, rob eagles of their eggs when they are far from the nest. Often people give the raven this opportunity when they try to watch the eagles from too near the nest or when in attempts to photograph them they keep them away from the nest. On coasts the eggs and young of even the large gulls fall victim to the raven. Ravens eat carrion as well as any living animals they can overcome, as well as a range of plant materials such as fruit and cereals. They feed the young at first mainly on insects, and later on meat of vertebrates.

Breeding behavior

A few years ago Gwinner studied the breeding behavior of ravens in detail. According to him the nest consists of four parts. First, a framework of large sticks is built; between these sticks the pair insert clumps of earth, grass, moss, or stones. Next follows the nest cup of smaller twigs matted together with moss, rags or similar material by means of squeezing movements and tramping with the feet; then the rim of the nest is made from unbranched twigs and finally the nest cup is lined with grass, fibers and other pliable materials. Experienced ravens do not take so long to build as younger ones, which are new to the job. The members of the pair incite one another to building activities. The female generally starts to incubate as early as February when it has laid the next-to-last or the last of the four to six eggs. The female is fed by the male while incubating. The young hatch after eighteen to twenty days of incubation. At hatching the female turns the eggs into the most suitable position for the emerging youngster; she also cleans the nestling and swallows the egg shells and membranes.

Subsequent aspects of breeding behavior also show the high development of the raven. The parents make and adjust the nest lining in accordance with the prevailing temperature so as to provide either warmth or a cooling effect for the young. One might say they make the beds of the young. The adults also keep the naked young and the nest clean; they brood the young, and give them water from their beaks. The female will even partly immerse herself in water and then "bathe" the young by applying her wet belly feathers to them. For about two weeks the parents "eat" the droppings of the nestlings, spitting them from the cloaca as they emerge from the nest. The young leave the nest when they are forty days old, usually in May. Nests are in niches, in rocks or on tall trees, old beech trees or spruce, in the Alps up to 2,400 m, but in the Himalayas evidently up to 5,500 m. The size of the breeding territory of a pair varies between ten and fifty to sixty square kilometers. At one time, however, two occupied nests in Schleswig-Holstein (Germany) were only four hundred meters apart.

While sexually mature ravens remain paired all through the year, the younger ones often gather into flocks of sometimes two hundred birds. In the Alps such flocks may surround mountain hotels at a respectable distance or gather at refuse dumps in the floors of valleys. They take flight when people approach to within a distance much greater than that tolerated by the Alpine chough. There are usually ravens waiting around Alpine peaks for the tourists to descend, when the ravens make do with such remains of food as the much tamer choughs have left.

At one time ravens were distributed continuously over Canada, New England and the midwestern states of the US as well as western and central Europe and northern and central Asia. Now there are many blanks in this gigantic area where it no longer occurs. This is due to the agency of man. In Germany, ravens retreated from their persecutors to Schlewig-Holstein and into the Alps. After it had finally been decided to protect these birds, its much reduced populations recovered somewhat. The raven even spread again into Mecklenburg, western Pomerania, Brandenburg, lower Saxony, in the Alpine foothills and the Black Forest. In northwestern Germany the breeding population fluctuates between a hundred and fifty and three hundred pairs. In the Bavarian Alps the nature of the ground makes it more difficult to assess numbers, but they are probably no higher than those in the north of Germany. Certainly the fact that Bavaria on April 1, 1963 passed an order permitting the shooting of ravens between the first of September and January 31 was an unnecessary offense against bird conservation. Even worse is the placing of poisoned eggs to kill these birds. The ancient Germans regarded the raven as the symbol of wisdom, just as the ancient Greeks venerated the much less intelligent little owl in this role.

The fan-tailed raven likes to circle up on updrafts until it is lost to sight. When walking on the ground, it usually has its beak open. The African white-necked raven needs niches in rocks for its nest; tree nests are much rarer in this species. After the young have fledged these ravens wander about generally in pairs, but sometimes they gather into flocks which may include hundreds of birds. They visit settlements and probably also follow locust swarms. Because of this and because they also catch mice and rats, they are well liked in their African home, although they sometimes steal a baby chick.

This species, as well as the closely-related thick-billed raven, perform wonderful aerial maneuvers. In the latter the wing beats produce a rushing sound.

Systematic Classification

Ornithologists are not yet in agreement about a satisfactory natural system of classification of the song birds (Suborder *Oscines*; see Ch. 5) Depending on their opinions, they recognize from forty-two to fifty-three families of these birds. The arrangement of these in groups can only in a few cases be based on good grounds and with certainty that the group combines all families of common origin. Therefore, such family groups are not given scientific names. In this work, the newest revisions of the Check List of Birds of the World by Peters, Mayr, Greenway and Paynter are followed. For comparison, the sequence adopted by H. von Boetticher (1953) and the grouping used by Wilhelm Meise in "Naturgeschichte der Vögel" by Berndt and Meise (1962) are reproduced below. When one compares the letters which designate groups of families in the table below, it is evident that just as in the matter of the sequence of orders of birds (Vol. VII, page 488 and 489), so in the sequence of song bird families there is yet no general agreement. A great deal of further research is needed to elucidate the true relationships.

Grzimek's Animal Life (1970)

A. Larks (Alaudidae)
B. Swallows (Hirundinidae)
C. Wagtails (Motacillidae), Cuckoo Shrikes (Campephagidae), Bulbuls (Pycnonotidae), Leafbirds (Irenidae)
D. Shrike and Waxwing-like birds with Shrikes (Laniidae), Waxwings (Bombycillidae), Vangas (Vangidae) and Palm Chats (Dulidae)
E. Dippers (Cinclidae), Wrens (Troglodytidae), Mockingbirds (Mimidae) and Hedge Sparrows (Prunellidae)
F. Flycatcher-like Birds (Muscicapidae)
G. Tit-like Birds including Long-tailed Tits (Aegithalidae), Penduline Tits (Remizidae), True Tits (Paridae), Nuthatches (Sittidae), Tree Runners (Climacteridae), Phillipine Creeper (Rhabdornithidae), Spotted Creepers (Salpornithidae) and Tree Creepers (Certhiidae)
H. Sunbirds including Flower-peckers (Dicaeidae) Sunbirds proper (Nectariniidae), Whiteyes (Zosteropidae) and Honey-eaters (Meliphagidae)
I. Bunting-like Birds and Finches: Buntings (Emberizidae), Wood Warblers (Parulidae), Wren Thrushes (Zeledoniidae), Honeycreepers (Drepanididae), Vireos (Vireonidae), Icterids (Icteridae) and Finches (Fringillidae)
K. Weavers (Ploceidae) and Weaverfinches (Estrildidae)
L. Starlings (Sturnidae), Orioles (Oriolidae) and Drongos (Dicruridae)
M. Crow-like Birds with New Zealand Wattlebirds (Callaeidae), Australian Magpie Larks (Grallinidae), Wood Swallows (Artamidae), Song Shrikes (Cracticidae), Birds of Paradise and Bower Birds (Paradisaeidae) and Crow-like Birds proper (Corvidae)

W. Meise (1962)

Birds related to Flycatchers (Muscicapoidae): E & F
Larks (Alaudoidae): A
Tit-like birds and Sunbirds (Paroidae): G & H
Swallows (Hirundinoidae): B
Shrike-like Birds (Lanioidae): D
Crow-like Birds (Corvoidae): M
Oriole-like Birds (Orioloidae): C without wagtails, L without starlings
Weavers and Finches (Fringilloidae): I & K

H. von Boetticher (1953)

Crow-like Birds (Corvidae): M
Shrike-like Birds (Laniidae): D
Insect Eaters (Muscicapidae): E & F without hedge sparrows
Swallows (Hirundinidae): B
Orioles (Oriolidae): C without wagtails and L without starlings
Starlings (Sturnidae): K and only the starlings from L
Tit-like Birds (Paridae): G without the penduline tits
Sunbirds (Promeropidae): H with the penduline tits
Finch-like Birds (Fringillidae): I with the hedge sparrows
Larks (Alaudidae): A

Fossil forms are not listed. Page numbers refer to the main article; numbers in parentheses refer to illustrations or distribution maps of species not mentioned in the text. Species and subspecies without page numbers are not mentioned in the text, nor illustrated or shown on maps. Species and subspecies marked with are endangered; those marked with + are extinct.

Order Coraciiformes

Family Kingfishers (Alcedinidae)	22
Subfamily Water Kingfishers (Alcedininae)	22
Genus *Alcedo*	22
Kingfisher, *A. atthis* (Linné, 1758)	22
Genus *Corythornis*	25
Malachite Kingfisher, *C. cristatus* (Pallas, 1764)	25
Genus *Ispidina*	25
Natal Kingfisher, *I. picta* (Boddaert, 1783)	25
Genus *Myiocepyx*	25
Dwarf Kingfisher, *M. lecontei* (Cassin, 1856)	25
Genus *Ceyx*	25
Celebes Three-toed Kingfisher, *C. fallax* (Schlegel, 1866)	25
Genus *Megaceryle*	26
Belted Kingfisher, *M. alcyon* (Linné, 1758)	26
Ringed Kingfisher, *M. torquata* (Linné, 1766)	27
Giant Kingfisher, *M. maxima* (Pallas, 1769)	27
Pied Kingfisher, *M. lugubris* (Temminck, 1834)	—
Genus *Ceryle*	22
Lesser Pied Kingfisher, *C. rudis* (Linné, 1758)	27
Eastern Lesser Pied Kingfisher, *C. rudis leucomelanura* Reichenbach, 1851	27
Genus Green Kingfishers *(Chloroceryle)*	27
Amazon Kingfisher, *Ch. amazona* (Latham, 1790)	27
Green and Rufous Kingfisher, *Ch. india* (Linné, 1766)	27
Green Kingfisher, *Ch. americana* (Gmelin, 1788)	27
Pygmy Kingfisher, *Ch. aenea* (Pallas, 1764)	27
Subfamily Tree Kingfishers (Daceloninae)	28
Genus *Pelargopsis*	28
Stork-billed Kingfisher, *P. capensis* (Linné, 1766)	28
Genus *Halcyon*	22
White-breasted Kingfisher, *H. smyrnensis* (Linné, 1758)	28
White-collared Kingfisher, *H. chloris* (Boddaert, 1783)	28
Sacred Kingfisher, *H. sancta* Vigors and Horsfield, 1827	28
Senegal Kingfisher, *H. senegalensis* (Linné, 1766)	28
Striped Kingfisher, *H. chelicuti* (Stanley, 1814)	28
Gray-headed Kingfisher, *H. leucocephala* (Müller, 1776)	28
Mangrove Kingfisher, *H. senegaloides* Smith, 1834	28
Genus *Dacelo*	33
Kookaburra, *D. gigas* (Boddaert, 1783)	33
Blue-winged Kookaburra, *D. leachii* Vigors and Horsfield, 1826	33
Genus *Clytoceyx*	34
Earthworm-eating Kingfisher, *C. rex* Sharpe, 1880	34
Genus *Melidora*	34
Hook-billed Kingfisher, *M. macrorrhina* (Lesson, 1827)	34
Genus *Tanysiptera*	34
Caroline Racket-tail, *T. sylvia* Gould, 1850	34
Family Todies (Todidae)	34
Genus *Todus*	34
Cuban Tody, *T. multicolor* Gould, 1837	34
Jamaican Tody, *T. viridis* (Linné, 1758)	34
Narrow-billed Tody, *T. angustirostris* Lafresnaye, 1851	34
Broad-billed Tody, *T. subulatus* Gray, 1847	34
Puerto Rico Tody, *T. mexicanus* Lesson, 1838	34
Family Motmots (Momotidae)	35
Genus *Aspatha*	35
Blue-throated Motmot, *A. gularis* Lafresnaye, 1840	35
Genus *Electron*	36
Broad-billed Motmot, *E. platyrhynchum* (Leadbeater, 1829)	36
Genus *Eumomota*	36
Turquoise-browed Motmot, *E. superciliosa* (Sandbach, 1837)	36
Genus *Baryphthengus*	36
Rufous Motmot, *B. ruficapillus* (Vieillot, 1818)	36
Genus *Momotus*	36
Blue-diademed Motmot, *M. momota* (Linné, 1766)	36
Family Bee-eaters (Meropidae)	39
Genus *Merops*	39
Carmine Bee-eater, *M. nubicus* Gmelin, 1788	40
Southern Carmine Bee-eater, *M. nubicoides* Des Murs and Pucheran, 1846	40
M. malimbicus Shaw, 1806	—
M. leschenaulti Vieillot, 1817	—
Common Bea-eater, *M. apiaster* Linné, 1758	39
Rainbow Bee-eater, *M. ornatus* Latham, 1801	40

Blue-cheeked Bee-eater, *M. superciliosus* Linné, 1766 — 40
Chestnut-headed Bee-eater, *M. viridis* Linné, 1758 — 40
M. orientalis Latham, 1801 — —
Genus *Aerops* — 42
Boehm's Bee-eater, *A. boehmi* (Reichenow, 1882) — 42
White-throated Bee-eater, *A. albicollis* (Vieillot, 1817) — 42
Genus *Dicrocercus* — 41
Green Swallow-tailed Bee-eater, *D. hirundineus* (A. L. Lichtenstein, 1793) — 41
Genus *Melittophagus* — 41
M. revoillii (Oustalet, 1882) — —
M. variegatus (Vieillot, 1817) — 48
M. lafresnayii (Guerin, 1843) — —
Little Bee-eater, *M. pusillus* (Müller, 1776) — 41
M. bullockoides (Smith, 1834) — —
Red-throated Bee-eater, *M. bullocki* (Vieillot, 1817) — 41
Blue-headed Bee-eater, *M. muelleri* (Cassin, 1857) — 41
Black Bee-eater, *M. gularis* (Shaw, 1798) — 41
Genus *Bombylonax* — 42
Black-headed Bee-eater, *B. breweri* (Cassin, 1859) — 42
Genus *Nyctyornis* — 42
Red-bearded Bee-eater, *N. amicuctus* (Temminck, 1824) — 42
Blue-bearded Bee-eater, *N. athertoni* (Jardine and Selby, 1830) — 42
Genus *Meropogon* — 42
Celebes Bee-eater, *M. forsteni* Bonaparte, 1850 — 42

Family Coraciidae — 21
Subfamily Cuckoo Rollers (Leptosomatinae) — 42
Genus *Leptosomus* — 42
Cuckoo Roller, *L. discolor* (Hermann, 1783) — 42

Subfamily Ground Rollers (Brachypteraciinae) — 43
Genus *Brachypteracias* — 43
Short-legged Ground Roller, *B. leptosomus* (Lesson, 1832) — 43
Scaled Ground Roller, *B. squamigera* Lafresnaye, 1838 — 43
Genus *Atelornis* — 43
Blue-tailed Ground Roller, *A. pittoides* (Lafresnaye, 1843) — 43
Crossley's Ground Roller, *A. crossleyi* Sharpe, 1875 — 43
Genus *Uratelornis* — 43
Long-tailed Ground Roller, *U. chimaera* Rothschild, 1895 — 43

Subfamily Tree Rollers (Coraciinae) — 44
Genus *Coracias* — —

Common Roller, *C. garrulus* Linné, 1758 — 44
Lilac-breasted Roller, *C. caudata* Linné, 1766 — 44
Racket-tailed Roller, *C. spatulata* Trimen, 1880 — 44
Rufous-crowned Roller, *C. naevia* Daudin, 1800 — 44
Indian Roller, *C. benghalensis* (Linné, 1758) — 44
Blue-bellied Roller, *C. cyanogaster* Cuvier, 1817 — 44
Senegal Roller, *C. abyssinica* Hermann, 1783 — 44
Genus *Eurystomus* — 45
Dollarbird, *Eu. orientalis* (Linné, 1776) — 45
Broad-billed Roller, *Eu. glaucurus* (Müller, 1776) — 45
Blue-throated Roller, *Eu. gularis* Vieillot, 1819 — 45

Family Hoopoes (Upupidae) — 45
Subfamily Upupinae — 45
Genus *Upupa* — 46
Eurasian Hoopoe, *U. epops* (Linné, 1758) — 46
Subfamily Wood Hoopoes (Phoeniculinae) — 49
Genus *Phoeniculus* — 49
Ph. purpureus (Miller, 1784) — 57
Genus *Rhinopomastus* — 49
Scimitar-bill, *Rh. cyanomelas* (Vieillot, 1819) — 49

Family Hornbills (Bucerotidae) — 49
Genus *Tockus* — 53
Red-billed Hornbill, *T. erythrorhynchus* (Temminck, 1823) — 53
T. erythrorhynchus rufiventris (Sundevall, 1851) — —
T. erythrorhynchus damarensis (Shelley, 1888) — 54
Von Der Decken's Toko, *T. deckeni* (Cabanis, 1869) — 54
T. flavirostris (Rüppell, 1835) — 53
Genus *Ceratogymna* — 54
Black-casqued Hornbill, *C. atrata* (Temminck, 1835) — 54
Genus *Bycanistes* — 54
Trumpeter Hornbill, *B. buccinator* (Temminck, 1824) — 54
Genus *Berenicornis* — 55
White-crested Hornbill, *B. comatus* (Raffles, 1822) — 55
Genus *Aceros* — 55
Rufous-necked Hornbill, *A. nipalensis* (Hodgson, 1829) — 55
Genus *Ptilolaemus* — 55
Tickel's Hornbill, *P. tickelli* (Blyth, 1855) — 55
Genus *Penelopides* — 55
Tarictic Hornbill, *P. panini* (Boddaert, 1783) — 55
Genus *Rhyticeros* — 56
Wreathed Hornbill, *R. undulatus* (Shaw, 1811) — 55
R. undulatus undulatus (Shaw, 1811) — 58
R. undulatus equabilis Sanft, 1960 — 55
Blyth's Hornbill, *R. plicatus* (J. R. Forster, 1781) — 56

R. plicatus plicatus (J. R. Forster, 1781)	50
R. plicatus jungei Mayr, 1937	56
Narcondam Hornbill, R. narcondami Hume, 1873	56
Genus *Buceros*	56
Great Indian Hornbill, B. bicornis Linné, 1758	56
B. bicornis bicornis Linné, 1758	56
B. bicornis homrai Hodgson, 1832	56
Rhinoceros Bird, B. rhinoceros Linné, 1758	56
Rufous Hornbill, B. hydrocorax Linné, 1766	56
Genus *Rhinoplax*	—
Helmeted Hornbill, R. vigil (J. R. Forster, 1781)	—
Genus *Bucorvus*	60
Abyssinian Ground Hornbill, B. abyssinicus (Boddaert, 1783)	60
Ground Hornbill, B. cafer (Schlegel, 1862)	61

Order Piciformes

Suborder Jacamars (Galbuloidea)

Family Jacamars (Galbulidae)	63
Genus *Galbalcyrhynchus*	63
White-eared Jacamar, G. leucotis Des Murs, 1845	63
Genus *Brachygalba*	63
Brown Jacamar, B. lugubris (Swainson, 1838)	63
B. goeringi Sclater and Salvin, 1869	—
B. salmoni Sclater and Salvin, 1879	—
B. albogularis (Spix, 1824)	—
Genus *Jacamaralcyon*	63
Three-toed Jacamar, J. tridactyla (Vieillot, 1817)	63
Genus *Galbula*	63
G. albirostris Latham, 1790	—
Green-tailed Jacamar, G. galbula (Linné, 1766)	63
G. tombacea Spix, 1824	—
G. cyanescens Deville, 1849	—
G. pastazae Taczanowski and Berlepsch, 1885	—
Rufous-tailed Jacamar, G. ruficauda Cuvier, 1817	63
G. leucogastra Vieillot, 1817	—
Paradise Jacamar, G. dea (Linné, 1758)	63
Genus *Jacamerops*	63
Great Jacamar, J. aurea (Müller, 1776)	63
Family Puffbirds (Bucconidae)	64
Genus *Notharchus*	64
White-necked Puffbird, N. macrorhynchus (Gmelin, 1788)	68
Black-and-white Puffbird, N. pectoralis (Gray, 1846)	64
N. ordii (Casin, 1851)	—
N. tectus (Boddaert, 1783)	—
Genus *Bucco*	64
B. macrodactylus (Spix, 1824)	—
B. tamatia Gmelin, 1788	—
B. noanamae Hellmayr, 1909	—
Collared Puffbird, B. capensis Linné, 1766	64
Genus *Nystalus*	64
Barred Puffbird, N. radiatus (Sclater, 1853)	64
N. chacuru (Vieillot, 1816)	—
N. striolatus (Pelzeln, 1856)	—
N. maculatus (Gmelin, 1788)	—
Genus *Hypnelus*	—
H. ruficollis (Wagler, 1829)	—
Genus *Malacoptila*	63
Crescent-chested Puffbird, M. striata (Spix, 1822)	64
M. fusca (Gmelin, 1788)	—
M. fulvogularis (Sclater, 1853)	—
M. rufa (Spix, 1824)	—
White-winged Soft-wing, M. panamensis Lafresnaye, 1847	64
M. mystacalis (Lafresnaye, 1850)	—
Genus *Micromonacha*	—
M. lanceolata (Deville, 1849)	—
Genus *Nonnula*	64
Red-throated Puffbird, N. rubecula (Spix, 1824)	64
N. sclateri Hellmayr, 1907	—
N. brunnea Sclater, 1881	—
N. frontalis (Sclater, 1854)	—
N. ruficapilla (Tschudi, 1844)	—
N. amaurocephala Chapman, 1921	—
Genus *Haploptila*	—
H. castanea (J. Verraeux, 1866)	—
Genus *Monasa*	64
Black Puffbird, M. atra (Boddaert, 1776)	65
M. nigrifrons (Spix, 1824)	—
M. morphoeus (Hahn and Küster, 1823)	—
M. flavirostris Strickland, 1850	—
Genus *Chelidoptera*	65
Swallow Wing, C. tenebrosa (Pallas, 1782)	65

Suborder Picoidea

Family Barbets (Capitonidae)	65
Genus *Semnornis*	66
Toucan Barbet, S. ramphastinus (Jardine, 1855)	66
Prong-billed Barbet, S. frantzii (Sclater, 1864)	66
Genus *Capito*	69
Scarlet-crowned Barbet, C. aureovirens (Cuvier, 1829)	69
Plaintive Barbet, C. niger (Müller, 1776)	69
Genus *Eubucco*	70

Salvin's Barbet, *E. bourcierii* (Lafresnaye, 1845)	70
Genus *Psilopogon*	76
Fire-tufted Barbet, *P. pyrolophus* S. Müller, 1835	76
Genus *Megalaima*	75
Great Hill Barbet, *M. virens* (Boddaert, 1783)	75
Gaudy Barbet, *M. mystacophanos* (Temminck, 1824)	75
Green Barbet, *M. zeylanica* (Gmelin, 1788)	75
Blue-throated Barbet, *M. asiatica* (Latham, 1790)	75
Golden-throated Barbet, *M. franklinii* (Blyth, 1842)	75
Coppersmith, *M. haemacephala* (Müller, 1776)	75
M. haemacephala rosea (Dumont, 1816)	75
Genus *Caloramphus*	76
Brown Barbet, *C. fuliginosus* (Temminck, 1830)	76
Genus *Gymnobucco*	73
Bristle-necked Barbet, *G. peli* Hartlaub, 1857	73
Naked-faced Barbet, *G. calvus* (Lafresnaye, 1841)	73
Genus *Stactolaema*	73
White-eared Barbet, *S. leucotis* (Sundevall, 1850)	73
Genus *Pogoiulus*	73
Yellow-throated Tinker-bird, *P. subsulphureus* (Fraser, 1843)	74
Golden-rumped Tinker-bird, *P. bilineatus* (Sundevall, 1850)	74
Yellow-fronted Tinker-bird, *P. chrysoconus* (Temminck, 1832)	74
Speckled Tinker-bird, *P. scolopaceus* (Bonaparte, 1850)	74
Genus *Tricholaema*	72
Pied Barbet, *T. leucomelan* (Boddaert, 1783)	72
Diadem Barbet, *T. diadematum* (Heuglin, 1861)	97
Genus *Lybius*	70
Bearded Barbet, *L. dubius* (Gmelin, 1788)	70
L. rolleti (Defilippi, 1853)	70
Double-toothed Barbet, *L. bidentatus* (Shaw, 1798)	70
African Collared Barbet, *L. torquatus* (Dumont, 1816)	71
Pied Barbet, *L. leucocephalus* (Defilippi, 1853)	71
Genus *Trachyphonus*	—
Levaillant's Barbet, *T. vaillantii* Ranzani, 1821	74
Red-and-yellow Barbet, *T. erythrocephalus* Cabanis, 1878	74
D'Arnaud's Barbet, *T. darnaudii* Prevost and Des Murs, 1847	74
Yellow-breasted Barbet, *T. margaritatus* (Cretzschmar, 1826)	74

Family Honey Guides (Indicatoridae) 77

Genus *Indicator*	77
Black-throated Honey Guide, *I. indicator* (Sparrman, 1777)	77
Scaly-throated Honey Guide, *I. variegatus* (Lesson, 1830)	77
Lesser Honey Guide, *I. minor* Stephens, 1815	77
Least Honey Guide, *I. exilis* (Cassin, 1856)	77
Malayan Honey Guide, *I. archipelagicus* Temminck, 1832	77
Honey Guide, *I. xanthonotus* Blyth, 1842	77
Genus *Melichneutes*	77
Lyre-tailed Honey Guide, *M. robustus* (Bates, 1909)	77
Genus *Prodotiscus*	77
Wahlberg's Honey Guide, *P. regulus* Sundevall, 1850	77
Genus *Melginomon*	—
Tinker Honey Guide, *M. zenkeri* Reichenow, 1898	—

Family Toucans (Ramphastidae) 79

Genus *Aulacorhynchus*	79
Groove-billed Toucanet, *Au. sulcatus* (Swainson, 1820)	80
Emerald Toucanet, *Au. prasinus* (Gould, 1834)	80
Blue-banded Toucanet, *Au. prasinus caeruleocinctus* d'Orbigny, 1840	107
Genus *Pteroglossus*	80
Collared Aracari, *P. torquatus* (Gmelin, 1788)	80
Fiery-billed Aracari, *P. frantzii* Cabanis, 1861	80
Black-necked Aracari, *P. aracari* (Linné, 1758)	107
P. viridis (Linné, 1766)	—
P. viridis inscriptus Swainson, 1822	—
Red-necked Aracari, *P. bitorquatus* Vigors, 1826	—
Genus *Beauharnaisius*	107
Curl-crested Aracari, *B. beuharnaesii* Wagler, 1832	107
Genus *Selenidera*	80
Spot-billed Toucanet, *S. maculirostris* (Lichtenstein, 1823)	80
Genus *Baillonius*	80
Yellow-breasted Toucan, *B. bailloni* (Vieillot, 1819)	80
Genus *Andigena*	80
Black-beaked Toucan, *A. nigrirostris* (Waterhouse, 1839)	80
Genus *Ramphastos*	85
Yolk Toucan, *R. vitellinus* Lichtenstein, 1823	85
Orange Toucan, *R. vitellinus ariel* Vigors, 1826	98
Red-breasted Toucan, *R. dicolorus* Linné, 1766	85
Rainbow-billed Toucan, *R. sulfuratus* Lesson, 1830	85
Brown-backed Toucan, *R. swainsonii* Gould, 1833	85

Giant Toucan, *R. toco* St. Müller, 1776	85
Cuvier's Toucan, *R. cuvieri* Wagler, 1827	98

Family Woodpeckers (Picidae) 63
Subfamily Wrynecks (Jynginae) 88
  Genus *Jynx* 88
    European Wryneck, *J. torquilla* Linné, 1758 88
    Red-breasted Wryneck, *J. ruficollis* Wagler, 1830 88

Subfamily Piculets (Picumninae) 89
  Genus *Picumnus* 89
    P. cinnamomeus Wagler, 1829 —
    P. rufiventris (Bonaparte, 1837) —
    P. fulvescens Stager, 1961 —
    P. castelnau Malherbe, 1862 —
    P. spilogaster Sundevall, 1866 —
    Scaly-eared Piculet, *P. minutissimus* (Pallas, 1782) 89
    P. minutissimus pallidus Snethlage, 1924 —
    P. minutissimus guttifer Sundevall, 1866 —
    P. minutissimus albosquamatus D'Orbigny, 1840 —
    P. squamulatus Lafresnaye, 1854 —
    P. limae Snethlage, 1924 —
    P. olivaceus Lafresnaye, 1845 —
    P. granadensis Lafresnaye, 1847 —
    P. nebulosus Sundevall, 1866 —
    P. nigropunctatus Zimmer and Phelps, 1950 —
    P. exilis Lichtenstein, 1823 —
    P. borbae Pelzeln, 1870 —
    P. aurifrons Pelzeln, 1870 —
    P. temmincki Lafresnaye, 1845 —
    P. cirrhatus Temminck, 1825 —
    P. dorbygnianus Lafresnaye, 1845 —
    P. sclateri Taczanowski, 1877 —
    P. steindachneri Taczanowski, 1882 —
    P. varzeae Snethlage, 1912 —
    P. pygmaeus Lichtenstein, 1823 —
    P. asterias Sundevall, 1866 (= *P. pygmaeus?*) —
    P. pumilus Cabanis and Heine, 1863 —
  Genus *Nesoctites* 89
    Hispaniola Piculet, *N. micromegas* (Sundevall, 1866) 89
  Genus *Verreauxia* 89
    African Piculet, *V. Africana* (Verreaux and Verreaux, 1855) 89
  Genus *Sasia* 89
    Three-toed Piculet, *S. abnormis* (Temminck, 1825) 89
  Genus *Vivia* —
    V. innominata (Burton), 1836 —

Subfamily True Woodpeckers (Picinae) 89
  Genus *Geocolaptes* 95
    Ground Woodpecker, *G. olivaceus* (Gmelin, 1788) 95
  Genus *Colaptes* 95
    Yellow-shafted Flicher, *C. auratus* (Linné, 1758) 95
    Red-shafted Flicher, *C. auratus cafer* (Gmelin, 1788) 95
    C. auratus chrysocaulosus Gundlach, 1858 96
    C. auratus chrysoides (Malherbe, 1852) —
    C. auratus mexicanoides (Lafresnaye, 1844) 96
    Andean Flicker, *C. rupicola* d'Orbigny, 1840 96
    Campos Flicher, *C. campestris* (Vieillot, 1818) 96
  Genus *Nesoceleus* 96
    Fernandina's Woodpecker, *N. fernandinae* (Vigors, 1827) 96
  Genus *Chrysoptilus* 96
    Green-barred Woodpecker, *Ch. melanochloros* (Gmelin, 1788) 96
  Genus *Piculus* 96
    Golden Olive Woodpecker, *P. rubiginosus* (Swainson, 1820) 96
  Genus *Campethera* 99
    Nubian Woodpecker, *C. nubica* Boddaert, 1783 108
    Spotted Woodpecker, *C. maculosa* (Valenciennes, 1826) 99
    Buff-spotted Woodpecker, *C. nivosa* (Swainson, 1854) 99
  Genus *Celeus* —
    C. loricatus (Reichenbach, 1854) —
  Genus *Micropternus* —
    M. brachyurus (Vieillot, 1818) —
  Genus *Picus* 99
    Green Woodpecker, *P. viridis,* Linné, 1758 99
    Gray Woodpecker, *P. canus* Gmelin, 1788 99
    Crimson-winged Woodpecker, *P. puniceus* Horsfield, 1821 119
  Genus *Dinopium* —
    Golden-backed Woodpecker, *D. benghalense* (Linné, 1758) —
  Genus *Gecinulus* —
    G. grantia (McClelland, 1840) —
  Genus *Meiglyptes* —
    M. tukki (Lesson, 1839) —
  Genus *Muelleripicus* 102
    M. pulverulentus (Temminck, 1826) —
  Genus *Dryocopus* 102
    Black Woodpecker, *D. martius* (Linné, 1758) 102
    Indian Great Black Woodpecker, *D. javensis* (Horsfield, 1821) 102
    Pileated Woodpecker, *D. pileatus* (Linné, 1758) 102
    D. lineatus (Linné, 1766) —
    D. erythrops (Valenciennes, 1826) is probably only a variant of *D. lineatus* —
    D. schulzi (Cabanis, 1883) —
    D. galeatus (Temminck, 1822) —
  Genus *Asyndesmus* —
    A. lewis (G. R. Gray, 1849) —
  Genus *Leuconerpes* probably = *Melanerpes* 119
    L. candidus (Otto, 1796) 119
  Genus *Melanerpes* 104
    Red-headed Woodpecker, *M. erythrocephalus*

(Linné, 1758)	
Acorn Woodpecker, *M. formicivorus* (Swainson, 1827)	104
Genus *Centurus*	104
Gila Woodpecker, *C. uropygialis* Baird, 1854	104
Red-bellied Woodpecker, *C. carolinus* (Linné, 1758)	104
C. aurifrons (Wagler, 1829)	—
C. radiolatus (Wagler, 1827)	—
Golden-naped Woodpecker, *C. chrysauchen* (Salvin, 1870)	104
Yellow-tufted Woodpecker, *C. cruentatus* (Boddaert, 1783)	104
Genus *Syphyrapicus*	106
Yellow-bellied Sapsucker, *S. varius* (Linné, 1766)	106
S. varius nuchalis Baird, 1858	106
S. varius daggetti Grinnell, 1901	106
S. varius ruber (Gmelin, 1788)	106
Williamson's Sapsucker, *S. thyroideus* (Cassin, 1851)	106
Genus *Trichopicus*	—
T. cactorum (d'Orbigny, 1840)	—
Genus *Veniliornis*	—
V. passerinus (Linné, 1766)	—
Genus *Dendropicos*	115
Cardinal Woodpecker, *D. fuscecens* (Vieillot, 1818)	115
Genus *Dendrocopos*	109
Greater Spotted Woodpecker, *D. major* (Linné, 1758)	109
Syrian Woodpecker, *D. syriacus* (Ehrenberg, 1833)	109
Middle Spotted Woodpecker, *D. medius* (Linné, 1758)	109
White-backed Woodpecker, *D. leucotos* (Bechstein, 1803)	109
Lesser Spotted Woodpecker, *D. minor* (Linné, 1758)	109
Hairy Woodpecker, *D. villosus* (Linné, 1766)	110
Downy Woodpecker, *D. pubescens* (Linné, 1766)	110
Red-cockaded Woodpecker, *D. borealis* (Vieillot, 1807)	110
Genus *Picoides*	114
Northern Three-toed Woodpecker, *P. tridactylus* (Linné, 1758)	115
Black-backed Three-toed Woodpecker, *P. arcticus* (Swainson, 1832)	115
Genus *Sapheopipo*	—
S. noguchii (Seebohm, 1887)	—
Genus *Xiphidiopicus*	—
X. percussus (Temminck, 1826)	—
Genus *Polipicus*	—
P. elliotii Cassin, 1863	—
Genus *Thripias*	—
T. namaquus (A. Lichtenstein, 1793)	—
Genus *Hemicircus*	—
H. concretus (Temminck, 1821)	—
Genus *Blythipicus*	115
Maroon Woodpecker, *B. rubiginosus* (Swainson, 1837)	115
B. pyrrhotis (Hodgson, 1837)	—
Genus *Chrysocolaptes*	115
Crimson-backed Woodpecker, *Ch. validus* (Temminck, 1825)	116
Black-backed Woodpecker, *Ch. festivus* (Boddaert, 1783)	116
Sultan Woodpecker, *Ch. lucidus* (Scopoli, 1796)	—
Genus *Phloeoceastes*	117
Red-crested Woodpecker, *Ph. melanoleucos* (Gmelin, 1788)	117
Ph. leucopogon (Valenciennes, 1826)	—
Genus *Campephilus*	116
Ivory-billed Woodpecker, ♀/ + *C. principalis* (Linné, 1758)	116
Imperial Woodpecker, ♀ *C. imperialis* (Gould, 1832)	116
Megallanic Woodpecker, *C. magellanicus* (King, 1828)	116

Order Passerine Birds (Passeriformes)

Suborder Desmodactylae

Family Broadbills (Eurylaimidae)	123
Subfamily Broadbills, i.n.s. (Eurylaiminae)	123
Genus *Smithornis*	123
African Broadbill, *S. capensis* (A. Smith, 1840)	124
Red-sided Broadbill, *S. rufolateralis* Gray, 1864	124
Genus *Pseudocalyptomena*	124
Grauer's Broadbill, *P. graueri* Rotschild, 1909	124
Genus *Corydon*	124
Dusky Broadbill, *C. sumatramus* (Raffles, 1822)	124
Genus *Cymbirhynchus*	124
Black and Red Broadbill, *C. macrorhynchos* (Gmelin, 1788)	124
Genus *Eurylaimus*	120
Banded Broadbill, *Eu. javanicus* Horsfield, 1821	120
Black-and-Yellow Broadbill, *Eu. ochromalus* Raffles, 1822	124
Wattled Broadbill, *Eu. steerii* Sharpe, 1876	124
Genus *Psarisomus*	120
Long-tailed Broadbill, *P. dalhousiae* (Jameson, 1835)	120

Subfamily Green Broadbill (Calyptomeninae)	123	Green Broadbill, *C. viridis* Raffles, 1822	120
Genus *Calyptomena*	123	Hose's Broadbill, *C. hosii* Sharpe, 1892	124

Suborder Clamatores or Tyranni

Superfamily Furnarioidea

Family Woodcreepers (Dendrocolaptidae)	127	Genus *Synallaxis*	131
Genus *Dendrocincla*	128	Red-capped Spinetail, *S. ruficapilla* Vieillot, 1819	131
D. fuliginosa (Vieillot, 1818)	120	White-throated Spinetail, *S. albescens* Temminck, 1823	125
Tawny-winged Woodcreeper, *D. anabatina* Sclater, 1859	128	Genus *Asthenes*	131
Genus *Sittasomus*	120	Creamy-breasted Canastero, *A. dorbignyi* (Reichenbach, 1853)	131
Olivaceous Woodcreeper, *S. griseicapillus* (Vieillot, 1818)	120	Genus *Phacellodomus*	131
Genus *Glyphorhynchus*	120	Rufous-fronted Thornbird, *Ph. rufifrons* (Wied, 1821)	131
Wedge-bill, *G. spirurus* (Vieillot, 1819)	120	Genus *Anumbius*	125
Genus *Xiphocolaptes*	120	*A. annumbi* (Vieillot, 1817)	125
Strong-billed Woodcreeper, *X. promeropirhynchus* (Lesson, 1840)	120	Genus *Philydor*	125
Genus *Dendrocolaptes*	128	Black-capped Foliage Gleaner, *Ph. atricapillus* (Wied, 1821)	125
Barred Woodcreeper, *D. certhia* (Boddaert, 1783)	128	Genus *Xenops*	125
Genus *Xiphorhynchus*	235	*X. rutilans* Temminck, 1821	125
X. lachrymosus (Lawrence, 1862)	–	Genus *Pigarrhichas*	–
Genus *Lepidocolaptes*	128	White-throated Treerunner, *P. albogularis* (King, 1831)	–
Narrow-billed Woodcreeper, *L. angustirostris* (Vieillot, 1818)	120	Genus *Sclerurus*	132
Spot-crowned Woodcreeper, *L. affinis* (Lafresnaye, 1839)	128	Black-tailed Leaf-scraper, *S. caudacutus* (Vieillot, 1816)	132
Genus *Campylorhamphus*	120	Genus *Lochmias*	125
Red-billed Scythebill, *C. trochilirostris* (Lichtenstein, 1820)	120	Sharp-tailed Streamcreeper, *L. nematura* (Lichtenstein, 1823)	125
Brown-billed Scythebill, *C. pusillus* (Sclater, 1860)	–		
		Family Antbirds (Formicariidae)	132
Family Ovenbirds (Furnariidae)	129	Genus *Batara*	132
Genus *Geositta*	129	Giant Antshrike, *B. cinerea* (Vieillot, 1819)	132
Common Miner, *G. cunicularia* (Vieillot, 1816)	129	Genus *Mackenziaena*	132
Genus *Upucerthia*	125	Long-tailed Antshrike, *M. leachii* (Such, 1825)	132
Striated Earthcreeper, *U. serrana* Taczanowski, 1875	125	Genus *Thamnophilus*	132
Genus *Cinclodes*	129	Barred Antshrike, *T. doliatus* (Linné, 1764)	132
Dark-bellied Cinclodes, *C. patagonicus* (Gmelin, 1789)	129	*T. caerulescens* Vieillot, 1816	126
Genus *Furnarius*	129	Genus *Myrmotherula*	132
Rufous Hornero, *F. rufus* (Gmelin, 1788)	129	Pygmy Antwren, *M. brachyura* (Hermann, 1783)	137
Genus *Sylviorthorhynchus*	125	Genus *Formicivora*	126
Des Mur's Spinetail, *S. desmursii* Des Murs, 1847	125	White-fringed Antwren, *F. grisea* (Boddaert, 1783)	126
Genus *Phleocryptes*	125	Genus *Pyriglena*	137
Ph. melanops (Vieillot, 1817)	125	White-shouldered Fire-eye, *P. leucoptera* (Vieillot, 1818)	137
Genus *Leptasthenura*	–	Genus *Gymnocichla*	126
L. aegithaloides (Kittlitz, 1830)	–	Bare-crowned Antbird, *G. nudiceps* (Cassin, 1850)	126
Genus *Schoeniophylax*	125	Genus *Formicarius*	137
White-cheeked Spinetail, *Sch. phryganophila* (Vieillot, 1817)	125	Colma Antthrush, *F. colma* Boddaert, 1783	137

Black-faced Antbird, *F. analis* (d'Orbigny and Lafresnaye, 1837)	126
Genus *Myrmeciza*	126
Chestnut-backed Antbird, *M. exsul* Sclater, 1858	126
Genus *Chamaeza*	137
Short-tailed Antthrush, *Ch. campanisona* (Lichtenstein, 1818)	137
Genus *Pithys*	137
White-faced Antcatcher, *P. albifrons* (Linné, 1766)	137
Genus *Rhegmatorhina*	137
Hairy-crested Antbird, *R. melanosticta* (Sclater and Salvin, 1880)	137
Genus *Hylophylax*	137
Spotted Antbird, *H. naevia*	137
Scale-backed Antbird, *H. poecilonota* (Cabanis, 1847)	126
Genus *Phlegopsis*	137
Black-spotted Bare-eye, *Ph. nigromaculata* (d'Orbigny and Lafresnaye, 1837)	137
Genus *Grallaricula*	126
Slate-crowned Antpitta, *G. nana* (Lafresnaye, 1842)	126
Genus *Grallaria*	137
Undulated Antpitta, *G. squamigera* Prévost and des Murs, 1846	126
Variegated Antpitta, *G. varia* (Boddaert, 1783)	137
Chestnut-crowned Antpitta, *G. ruficapilla* Lafresnaye, 1842	126
Genus *Conopophaga*	139
Silvery-tufted Gnateater, *C. lineata* (Wied, 1831)	139
Genus *Corythopis* (should perhaps be classed with the Tyrants)	140
Delalande's Gnateater, *C. delalandei* (Lesson, 1831)	140
Family Tapaculos (Rhinocryptidae)	140
Genus *Pteroptochos*	140
Moustached Turca, *P. megapodius* Kittlitz, 1830	140
Genus *Scelorchilus*	140
Tapaculo, *S. albicollis* (Kittlitz, 1830)	140
Genus *Rhinocrypta*	140
Gray Gallito, *R. lanceolata* (Geoffroy St. Hilaire, 1832)	140
Genus *Liosceles*	140
Rusty-belted Tapaculo, *L. thoracicus* (Sclater, 1865)	140
Genus *Scytalopus*	140
Mouse-colored Tapaculo, *S. speluncae* (Ménétriés, 1835)	140
Genus *Acropternis*	140
Ocellated Tapaculo, *A. orthonyx* (Lafresnaye, 1843)	140
Genus *Psilorhamphus*	140
Spotted Bamboowren, *P. guttatus* (Ménétriés, 1835)	140

Superfamily Tyrannoidea

Family Pittas (Pittidae)	141
Genus *Pitta*	142
Fairy Pitta, *P. brachyura* (Linné, 1766)	142
Japanese Fairy Pitta, *P. brachyura nympha* (Temminck and Schlegel, 1850)	142
African Pitta, *P. angolensis* Vieillot, 1816	142
P. superba Rothschild and Hartert, 1914	149
Garnet Pitta, *P. granatina* Temminck, 1830	149
Red-breasted Pitta, *P. erythrogaster* Temminck, 1832	142
Hooded Pitta, *P. sordida* (Müller, 1776)	142
Indian Hooded Pitta, *P. sordida cucullata* Hartlaub, 1843	142
Steere's Pitta, *P. steerii* (Sharpe, 1876)	142
Rainbow Pitta, *P. iris* Gould, 1842	142
P. maxima Müller and Schlegel, 1839	149
Noisy Pitta, *P. versicolor* (Swainson, 1825)	142
Giant blue Pitta, *P. caerulea* (Raffles, 1822)	142
P. caerulea willoughbyi Delacour, 1926	149
Banded Pitta, *P. guajana* (Müller, 1776)	142
Elliot's Pitta, *P. ellioti* Oustalet, 1874	149
Eared Pitta, *P. phayrii* (Blyth, 1862)	142
Family Velvet Pittas (Philipittidae)	147
Genus *Philepitta*	147
Velvet Pitta, *Ph. castanea* (Müller, 1776)	147
Genus *Neodrepanis*	147
False Sunbird, *N. coruscans* Sharpe, 1875	147
Family New Zealand Wrens (Xenicidae)	147
Genus *Acanthisitta*	147
Rifleman, *A. chloris* (Sparrman, 1787)	147
Genus *Xenicus*	148
Bush Wren, ◊*X. longipes* (Gmelin, 1789)	147
Rock Wren, *X. gilviventris* Pelzeln, 1867	148
Genus *Traversia*	147
Stephen Island Wren, + *T. lyalli* Rothschild, 1894	148
Family Tyrants (Tyrannidae)	148
Subfamily Water Tyrants (Fluvicolinae)	151
Genus *Agriornis*	151
Kittlitz Ground Tyrant, *A. lividus* (Kittlitz, 1835)	151
Genus *Muscisaxicola*	151
White-fronted Ground Tyrant, *M. albifrons* (Tschudi, 1844)	151
Genus *Sayornis*	151

Eastern Phoebe, *S. phoebe* (Latham, 1790) 151
Black Phoebe, *S. nigricans* (Swainson, 1827) 151
Genus *Colonia* 151
Long-tailed Tyrant, *C. colonus* (Vieillot, 1818) 151
Genus *Fluvicola* 143
Pied Water Tyrant, *F. pica* (Boddaert, 1783) 143
Genus *Arundinicola* 151
White-headed Marsh Tyrant, *A. leucocephala* Linné, 1764 151
Genus *Pyrocephalus* 151
Vermillion Flycatcher, *P. rubinus* (Boddaert, 1783) 151
Genus *Tachuris* 151
Many-colored Tyrant, *T. rubigastra* (Vieillot, 1817) 151
Genus *Machetornis* 151
Cattle Tyrant, *M. rixosa* (Vieillot, 1819) 151

Subfamily Kingbirds (Tyranninae) 151
Genus *Muscivora* 151
Fork-tailed Flycatcher, *M. tyrannus* (Linné, 1766) 151
Scissor-tailed Flycatcher, *M. forficata* (Gmelin, 1788) 151
Genus *Tyrannus* 151
Eastern Kingbird, *T. tyrannus* Linné, 1758 151
Tropical Kingbird, *T. melancholicus* Vieillot, 1819 151
Genus *Legatus* 151
Piratic Flycatcher, *L. leucophaius* (Vieillot, 1818) 151
Genus *Myiodynastes* 151
M. maculatus (Müller, 1776) —
Sulphur-bellied Flycatcher, *M. luteiventris*, Sclater, 1859 151
Genus *Megarynchus* 151
Boat-billed Flycatcher, *M. pitangua* (Linné, 1766) 151
Genus *Myiozetetes* 151
Vermillion-crowned Flycatcher, *M. similis* (Spix, 1825) 151
Genus *Pitangus* 151
Great Kiskadee, *P. sulphuratus* (Linné, 1766) 152

Subfamily Tyrant Flycatchers (Myiarchinae) 152
Genus *Myiarchus* 152
Great-crested Flycatcher, *M. crinitus* (Linné, 1758) 152
Genus *Contopus* 152
Tropical Pewee, *C. cinereus* (Spix, 1825) 152
Genus *Empidonax* 143
Western Flycatcher, *E. difficilis* Baird, 1858 143
Yellow-bellied Flycatcher, *E. flaviventris* (Baird, 1843) 152
Genus *Myiobius* 152
Sulphur-rumped Flycatcher, *M. sulphureipygius* (Sclater, 1857) 152
Genus *Onychorhynchus* 152
Northern Royal Flycatcher, *O. mexicanus* (Sclater, 1857) 152
Genus *Tolmomyias* 152
Yellow-olive Flycatcher, *T. sulphurescens* (Spix, 1825) 152
Genus *Rhynchocyclus* 152
Eye-ringed Flatbill, *R. brevirostris* (Cabanis, 1847) 152

Subfamily Tody Flycatchers (Euscarthminae) 152
Genus *Todirostrum* 152
Common Tody Flycatcher, *T. cinereum* (Linné, 1766) 152
Genus *Oncostoma* 152
Northern Bent-bill, *O. cinereigulare* (Sclater, 1857) 152

Subfamily Torrent Flycatchers, (Serpophaginae) 152
Genus *Serpophaga* 152
Torrent Flycatcher, *S. cinerea* (Tschudi, 1844) 152
Genus *Myiornis* 143
Eared Pygmy Tyrant, *M. auricularis* (Vieillot, 1818) 143
Genus *Perissotriccus* 152
Black-capped Pygmy Tyrant, *P. atricapillus* (Lawrence, 1875) 152

Subfamily Elaenias and related species, (Elaeniinae) 152
Genus *Elaenia* 152
Yellow-bellied Elaenia, *E. flavogaster* (Thunberg, 1822) 152
Lesser Elaenia, *E. chiriquensis* Lawrence, 1867 152
Genus *Tyranniscus* 152
Paltry Tyrannulet, *T. vilissimus* (Sclater and Salvin, 1859) 152
Genus *Pipromorpha* 152
Ochre-bellied Flycatcher, *P. oleaginea* (Lichtenstein, 1823) 152

Family Sharpbills (Oxyruncidae) 157
Genus *Oxyruncus* 157
Crested Sharpbill, *O. cristatus* (Swainson, 1821) 157

Family Manakins (Pipridae) 157
Genus *Pipra* 158
Crimson-hooded Manakin, *P. aureola* (Linné, 1758) 143
Golden-bearded Manakin, *P. erythrocephala* (Linné, 1758) 158
Genus *Teleonema* 143
Wire-tailed Manakin, *T. filicauda* (Spix, 1825) 143
Genus *Machaeropterus* 158
Red-capped Manakin, *M. pyrocephalus* (Sclater, 1852) 158
Genus *Antilophia* 158

Helmeted Manakin, *A. galeata* (Lichtenstein, 1823) ... 158
Genus *Neopelma* ... 158
Wied's Tyrant Manakin, *N. aurifrons* (Wied, 1831) ... 158
Genus *Chiroxiphia* ... 158
Swallow-tailed Manakin, *Ch. caudata* (Shaw and Nodder, 1893) ... 158
Genus *Manacus* ... 158
White-bearded Manakin, *M. manacus* (Linné, 1766) ... 158

Family Cotingas (Cotingidae) ... 160
Genus *Phoenicircus* ... 144
Black-necked red Cotinga, *Ph. nigricollis* (Swainson, 1825) ... 144
Genus *Phibalura* ... 144
Swallow-tailed Cotinga, *Ph. flavirostris* (Vieillot, 1816) ... 144
Genus *Cotinga* ... 161
Purple-breasted Cotinga, *C. cotinga*, (Linné, 1766) ... 144
Banded Cotinga, *C. maculata* (Müller, 1776) ... 161
Lovely Cotinga, *C. amabilis* Gould, 1857 ... 133
Genus *Xipholena* ... 144
Pompadour Cotinga, *X. punicea* (Pallas, 1764) ... 144
Genus *Pipreola* ... 161
Green and Black Fruiteater, *P. riefferii* (Boissonneau, 1840) ... —
Barred Fruiteater, *P. arcuata* (Lafresnaye, 1843) ... 161
Genus *Lipaugus* ... 144
Rose-collared Pina, *L. strephophorus* (Salvin and Godman, 1884) ... 144
Genus *Pachyramphus* ... 161
Green-backed Becard, *P. viridis* (Vieillot, 1816) ... 161
White-winged Becard, *P. polychopterus* (Vieillot, 1818) ... —
Genus *Platypsaris* ... 144
Rose-throated Becard, *P. aglaiae* (Lafresnaye, 1839) ... —
Jamaican Becard, *P. niger* (Gmelin, 1788) ... 144
Genus *Tityra* ... 161
Black-tailed Tityra, *T. cayana* (Linné, 1766) ... 144
Masked Tityra, *T. semifasciata* (Spix, 1825) ... 161
Genus *Pyroderus* ... 161
Red-ruffed Fruitcrow, *P. scutatus* (Shaw, 1792) ... 161
Genus *Cephalopterus* ... 161
Umbrella Bird, *C. ornatus* (Geoffroy St. Hilaire, 1809) ... 161
Long-wattled Umbrella Bird, *C. ornatus penduliger* Sclater, 1859 ... 162
Genus *Perissocephalus* ... 161
Capuchinbird, *P. tricolor* (Müller, 1776) ... 161
Genus *Procnias* ... 161
White Bellbird, *P. alba* (Hermann, 1783) ... 161
Bare-throated Bellbird, *P. nudicollis* (Vieillot, 1817) ... 161
Mossy-throated Bellbird, *P. averano* (Hermann, 1783) ... 161
Genus *Rupicola* ... 161
Cock-of-the-Rock, *R. rupicola* (Linné, 1766) ... 144
Andean Cock-of-the-Rock, *R. peruviana* (Latham, 1790) ... 161

Family Plantcutters (Phytotomidae) ... 163
Genus *Phytotoma* ... 164
Red-breasted Plantcutter, *Ph. rutila* Vieillot, 1818 ... 164
Chilean Plantcutter, *Ph. rara* Molina, 1782 ... 164

Suborder Lyrebirds and Scrub-birds

(Suboscines or Menurae)

Family Lyrebirds (Menuridae) ... 164
Genus *Menura* ... 164
Lyrebird, *M. novaehollandiae* Latham, 1801 ... 164
Prince Albert's Lyrebird, *M. alberti* Bonaparte, 1850 ... 164

Family Scrub-birds (Atrichornithidae) ... 165
Genus *Atrichornis* ... 166
Western Scrub-bird, ♀ *A. clamosus* (Gould, 1844) ... 166
Rufous Scrub-bird, *A. rufescens* (Ramsay, 1866) ... 166

Suborder Songbirds (Oscines)

Family Larks (Alaudidae) ... 171
Genus *Mirafra* ... 172
Red-winged Black Lark, *M. hypermetra* (Reichenow, 1879) ... 179
Clapper Lark, *M. apiata* (Vieillot, 1816) ... 172
Flappet Lark, *M. rufocinnamomea* (Salvadori, 1865) ... 172
M. africana (Smith, 1836) ... —
M. sabota (Smith, 1836) ... —
Genus *Chersomanes* ... —
Ch. albofaciata (Lafresnaye, 1836) ... —
Genus *Certhilauda* ... —
C. curvirostris (Hermann, 1783) ... —
Genus *Calendulauda* ... —
C. albescens (Lafresnaye, 1839) ... —
Genus *Ammomanes* ... 172

Desert Lark, *A. deserti* (Lichtenstein, 1823)	172
Genus *Alaemon*	173
Bifasciated Lark, *A. alaudipes* (Desfontaine, 1789)	173
Genus *Ramphocorys*	174
Thick-billed Lark, *R. clotbey* (Bonaparte, 1850)	174
Genus *Pinarocorys*	—
P. nigricans (Sudevall, 1850)	—
Genus *Eremopterix*	172
E. leucotis (Stanley, 1814)	—
E. verticalis (Smith, 1836)	—
White-fronted Lark, *E. nigriceps* (Gould, 1841)	172
Genus *Aethocorys*	—
Ae. personata (Sharpe, 1895)	—
Genus *Calandrella*	174
Subgenus *Spizocorys*	—
C. (Spizocorys) starki Shelley, 1902	—
C. (Spizocorys) conirostris (Sundevall, 1850)	—
Subgenus *Alaudala*	—
Lesser Short-toed Lark, *C. (Alaudala) rufescens* (Vieillot, 1820)	—
C. (Alaudala) cheleensis (Swinhoe, 1871)	—
Subgenus *Calandrella*	—
Short-toed Lark, *C. (Calandrella) cinerea* (Gmelin, 1789)	—
C. (Calandrella) acutirostris Hume, 1873	—
Genus *Melanocorypha*	174
M. mongolica (Pallas, 1776)	—
White-winged Lark, *M. leucoptera* (Pallas, 1811)	179
M. yeltoniensis (Forster, 1767)	—
M. maxima Blyth, 1867	—
M. bimaculata (Menetries, 1832)	—
Calandra Lark, *M. calandra* (Linné, 1766)	174
Genus *Eremophila*	181
Horned Lark, *E. alpestris* (Linné, 1766)	181
Genus *Pseudalaemon*	—
P. fremantlii (Phillips, 1897)	—
Genus *Chersophilus*	—
Ch. duponti (Vieillot, 1820)	—
Genus *Calendula*	—
C. magnirostris (Stephens, 1826)	—
Genus *Heliocorys*	—
H. modesta Heuglin, 1864	—
Genus *Lullula*	176
Woodlark, *L. arborea* (Linné, 1758)	176
Genus *Galerida*	175
G. malabarica (Scopoli, 1786)	—
Thekla Lark, *G. theklae* Brehm, 1858	179
Crested Lark, *G. cristata* (Linné, 1758)	175
Genus *Alauda*	177
Skylark, *A. arvensis* Linné, 1758	177
Family Swallows (Hirundinidae)	**182**
Subfamily Pseudochelidoninae	**185**
Genus *Pseudochelidon*	185
P. eurystomina Hartlaub, 1861	185
P. siantarae Thonglongya, 1968	185
Subfamily Hirundininae	**185**
Genus *Tachycineta*	186
Tree Swallow, *T. bicolor* (Vieillot, 1808)	186
T. albilinea (Lawrence, 1863)	—
T. albiventer (Boddaert, 1783)	—
T. leucorrhoa (Vieillot, 1817)	—
T. thalassina (Swainson, 1827)	—
Genus *Challichelidon*	—
C. cyaneoviridis (Bryant, 1859)	—
Genus *Kalochelidon*	—
K. euchrysea (Grosse, 1847)	—
Genus *Progne*	186
P. tapera (Linné, 1766)	—
Purple Martin, *P. subis* (Linné, 1758)	—
P. chalybea (Gmelin, 1789)	186
P. modesta Gould 1837	—
Genus *Notiochelidon*	—
N. murina (Cassin, 1853)	—
N. cyanoleuca (Vieillot, 1817)	—
N. pileata (Gould, 1858)	—
N. flavipes (Chapman, 1922)	—
Genus *Atticora*	—
A. fasciata (Gmelin, 1789)	—
A. melanoleuca (Wied, 1820)	—
Genus *Neochelidon*	—
N. tibialis (Cassin, 1853)	—
Genus *Alopochelidon*	—
A. fucata (Temminck, 1822)	—
Genus *Stelgidopteryx*	186
Rough-winged Swallow, *S. ruficollis* (Vieillot, 1817)	186
Genus *Cheramoeca*	—
Ch. leucosternum (Gould, 1841)	—
Genus *Pseudhirundo*	186
African Gray-rumped Swallow, *P. griseopyga* (Sundevall, 1850)	186
Genus *Riparia*	186
R. cincta (Boddaert, 1783)	—
R. conciga (Reichenow, 1887)	—
Bank Swallow, *R. riparia* (Linné, 1758)	186
Genus *Phedina*	185
Muscarene Martin, *Ph. borbonica* (Gmelin, 1789)	185
Congo Martin, *Ph. brazzae* Oustalet, 1866	185
Genus *Ptyonoprogne*	191
Crag Martin, *P. rupestris* (Scopoli, 1769)	191
P. obsoleta (Cabanis, 1850)	—
P. concolor (Sykes, 1832)	—
Genus *Hirundo* (including *Cecropsis*)	187
Barn Swallow, *H. rustica* Linné, 1758	187
American subspecies, *H. rustica erythrogaster* Boddaert, 1783	—
H. lucida Hartlaub, 1858	—
Angola Swallow, *H. angolensis* Bocage, 1868	187

South Sea Swallow, *H. tahitica* Gmelin, 1789	187
White-throated Swallow, *H. albigularis* Strickland, 1849	187
H. aethiopica Blanford, 1869	–
H. smithii Leach, 1818	–
H. atrocaerulea Sundevall, 1850	–
H. nigrita Gray, 1845	–
H. leucosoma Swainson, 1837	–
H. megaensis Benson, 1942	–
H. nigrorufa Bocage, 1877	–
H. dimidiata Sundevall, 1850	–
H. cucullata Boddaert, 1783	–
Striped Swallow, *H. abyssinica* Guérin-Méneville, 1843	187
H. semirufa Sundevall, 1850	–
H. senegalensis Linné, 1766	–
Red-rumped Swallow, *H. daurica* Linné, 1771	187
H. striolata Temminck and Schlegel, 1847	–
Genus *Petrochelidon*	186
P. rufigula (Bocage, 1878)	–
P. preussi (Reichenow, 1898)	–
P. andecola (d'Orbigny and Lafresnaye, 1837)	–
Tree Martin, *P. nigricans* (Vieillot, 1817)	186
P. spilodera (Sundevall, 1850)	–
Cliff Swallow, *P. pyrrhonota* (Vieillot, 1817)	186
P. fulva (Vieillot, 1807)	–
P. fluvicola (Blyth, 1855)	–
P. ariel (Gould, 1843)	–
P. fuliginosa (Chapin, 1925)	–
Genus *Delichon*	191
House Martin, *D. urbica* (Linné, 1758)	191
D. dasypus (Bonaparte, 1851)	–
D. nipalensis Horsfield and Moore, 1854	–
Genus *Psalidoprocne*	186
P. nitens (Cassin, 1857)	–
P. fuliginosa Shelley, 1887	–
White-headed Rough-winged Swallow, *P. albiceps* Sclater, 1864	180
P. pristoptera (Rüppell, 1836)	–
Black Rough-winged Swallow, *P. holomelaena* (Sundevall, 1850)	180
P. obscura (Hartlaub, 1855)	–
Family Wagtails (Motacillidae)	192
Genus *Pipits (Anthus)*	193
A. correndera Vieillot, 1818	–
Meadow Pipit, *A. pratensis* (Linné, 1758)	193
A. gustavi Swinhoe, 1863	–
Water Pipit, *A. spinoletta* (Linné, 1758)	193
Red-throated Pipit, *A. cervinus* (Pallas, 1811)	193
Tree Pipit, *A. trivialis* (Linné, 1758)	193
A. hodgsoni Richmond, 1907	–
Tawny Pipit, *A. campestris* (Linné, 1758)	193
A. vaalensis Shelley, 1900	–
A. similis (Jerdon, 1840)	–
New Zealand Pipit, *A. novaeseelandiae* (Gmelin, 1789)	194
A. novaeseelandiae richardi Vieillot, 1818	–
Genus *Macronyx*	193
Cape Longclaw, *M. capensis* (Linné, 1766)	193
Eastern Cape Longclaw, *M. capensis colleti* Schou, 1908	189
Genus *Dendronanthus*	194
Tree Wagtail, *D. indicus* (Gmelin, 1789)	194
Genus Wagtails *(Motacilla)*	194
Yellow Wagtail, *M. flava* Linné, 1758	194
American Yellow Wagtail, *M. flava tshutschensis* Gmelin, 1789	194
Yellow-headed Wagtail, *M. citreola* Pallas, 1776	194
Gray Wagtail, *M. cinerea* Tunstall, 1771	194
White Wagtail, *M. alba* Linné, 1758	194
Pied Wagtail, *M. alba yarrellii* Gould, 1837	–
Family Cuckoo Shrikes (Campephagidae)	198
Tribe campephagini	198
Genus *Pteropodocys*	198
Ground Cuckoo Shrike, *P. maxima* (Rüppell, 1839)	198
Genus *Coracina*	198
Black-faced Cuckoo Shrike, *C. novaehollandiae* (Gmelin, 1789)	198
Barred Cuckoo Shrike, *C. lineata* (Swainson, 1825)	450
C. azurea (Cassin, 1852)	450
Genus *Campochaera*	198
Orange Cuckoo Shrike, *C. sloetii* (Schlegel, 1866)	198
Genus *Chlamydochaera*	–
Black-breasted Triller, *Ch. jefferyi* Sharpe, 1887	–
Genus *Lalage*	198
Varied Triller, *L. leucomela* (Nigors and Horsfield, 1825)	198
L. nigra (Forster, 1781)	–
Australian Triller, *L. sueurii* (Vieillot, 1818)	450
L. aurea (Temminck, 1827)	450
Genus *Campephaga*	198
Wattled Cuckoo Shrike, *C. lobata* (Temminck, 1824)	198
Black Cuckoo Shrike, *C. flava* (Latham, 1790)	198
Genus *Hemipus*	198
Pygmy Triller, *H. picatus* (Sykes, 1832)	198
Genus *Tephrodornis*	198
Hook-billed Graybird, *T. gularis* Raffles, 1822	198
Tribe Pericrocotini	198
Genus *Pericrocotus*	198
Ashy Minivet, *P. divaricatus* (Raffles, 1822)	198
Flame Minivet, *P. flammeus* (Forster, 1781)	450
Family Bulbuls (Pycnonotidae)	200
Genus *Spizixos*	204
Finch-billed Bulbul, *S. canifrons* Blyth, 1845	204
Taiwan Finch-billed Bulbul, *S. semitorques*	

cinereicapillus Swinhoe, 1871	190
Genus *Pycnonotus*	200
Yellow-crowned Bulbul, *P. zeylanicus* (Gmelin, 1789)	200
Black-and-White Bulbul, *P. melanoleucos* (Eyton, 1839)	200
Black-headed Bulbul, *P. atriceps* (Temminck, 1822)	200
P. leucogenys (J. E. Gray, 1835)	—
Red-whiskered Bulbul, *P. jocosus* (Linné, 1758)	203
Red-vented Bulbul, *P. cafer* (Linné, 1766)	200
African Bulbul, *P. barbatus* (Desfontaines, 1789)	204
P. capensis (Linné, 1766)	—
Nieuwenhuis Bulbul, *P. nieuwenhuisi* (Finsch, 1901)	203
Stripe-throated Bulbul, *P. finlaysoni* Strickland, 1844	204
Blyth's Bulbul, *P. flavescens* Blyth, 1845	203
Yellow-vented Bulbul, *P. goiavier* (Scopoli, 1786)	203
P. gracilirostris (Strickland, 1844)	203
Olive-breasted Mountain Bulbul, *P. tephrolaemus* (Gray, 1862)	203
Genus *Phyllastrephus*	203
Yellow-streaked Bulbul, *Ph. flaviostriatus* (Sharpe, 1876)	203
Genus *Criniger*	184
C. flaveolus (Gould, 1836)	184
Brown White-throated Bulbul, *C. ochraceus* Moore, 1854	200
Genus *Hypsipetes*	200
Black Bulbul, *H. madagascariensis* (Müller, 1776)	203

Family Leafbirds (Irenidae) — 204
Subfamily Leafbirds proper (Chloropseinae) — 204
Genus *Chloropsis*	204
Gold-fronted Leafbird, *Ch. aurifrons* (Temminck, 1829)	204
Orange-bellied Leafbird, *Ch. hardwickei* Jardine and Selby, 1830	204
Genus *Aegithina*	204
Common Iora, *Ae. tiphia* (Linné, 1758)	205

Subfamily Fairy Bluebirds (Ireninae)
Genus *Irena*	206
Blue-backed Fairy Bluebird, *I. puella* (Latham, 1790)	203
Philippine Fairy Bluebird, *I. cyanogaster* Vigors, 1831	206

Family Shrikes (Laniidae) — 207
Subfamily Helmeted Shrikes (Prionopinae) — 207
Genus *Prionops*	207
P. poliolopha Fischer and Reichenow, 1884	208
White Helmeted Shrike, *P. plumata* (Shaw, 1809)	207
P. alberti Schouteden, 1933	208
P. scopifrons (Peters, 1854)	208
P. gabela Rand, 1957	208
P. retzii Wahlberg, 1856	—
P. caniceps (Bonaparte, 1851)	—
Genus *Eurocephalus*	208
White-crowned Shrike, *Eu. anguitimens* Smith, 1836	208
Eu. anguitimens rueppelli Bonaparte, 1853	—

Subfamily Bush Shrikes (Malaconotinae) — 208
Genus *Malaconotus*	209
Gray-headed Bush Shrike, *M. blanchoti* Stephens, 1826	209
M. lagdeni (Sharpe, 1884)	—
M. alius Friedman, 1927	—
M. multicolor (Gray, 1845)	—
M. viridis (Vieillot, 1817)	—
M. kupeensis (Serle, 1915)	—
M. dohertyi (Rothschild, 1901)	—
Genus *Rhodophoneus*	—
R. cruentus (Ehrenberg, 1828)	—
Genus *Laniarius*	208
Crimson-breasted Shrike, *L. atrococcineus* (Burchell, 1822)	208
L. mufumbiri Ogilivie-Grant, 1911	—
Gonolek, *L. barbarus* (Linné, 1766)	195
L. atroflavus Shelley, 1887	—
L. ferrugineus (Gmelin, 1788)	—
L. fuelleborni (Reichenow, 1900)	—
Genus *Lanioturdus*	208
Chat Shrike, *L. torquatus* Waterhouse, 1838	208
Genus *Nilaus*	208
Brubru, *N. afer* (Latham, 1801)	208
Genus *Dryoscopus*	208
Black-backed Puff-back, *D. cubla* (Shaw, 1809)	208
D. pringlii Jackson, 1893	—
Genus *Tchagra*	208
Redwing Shrike, *T. tchagra* (Vieillot, 1861)	208
T. senegala (Linné, 1766)	—
T. australis (Smith, 1836)	208

Subfamily True Shrikes (Laniinae) — 209
Genus *Corvinella*	209
Long-tailed Shrike, *C. corvina* (Shaw, 1809)	213
Genus *Urolestes*	209
Magpie-Shrike, *U. melanoleucus* (Jardine, 1831)	212
Genus *Lanius*	209
L. vittatus Valenciennes, 1826	—
Red-backed Shrike, *L. collurio* Linné, 1758	209
Isabelline Shrike, *L. collurio isabellinus* Ehrenberg, 1833	212
Brown Shrike, *L. cristatus* Linné, 1758	212
Rufous-backed Shrike, *L. schach* Linné, 1758	212
L. collaris Linné, 1766	—

Masked Shrike, *L. nubicus* Lichtenstein, 1823	212
Woodchat Shrike, *L. senator* Linné, 1758	210
L. bucephalus Temminck and Schlegel, 1847	—
Northern Shrike, *L. excubitor* Linné, 1758	195
L. ludovicianus Linné, 1766	—

Subfamily Wood Shrikes (Pityriasinae) 213
  Genus *Pityriasis* 213
    Black-headed Wood Shrike, *P. gymnocephala* (Temminck, 1835) 213

Family Vangas (Vangidae) 213
  Genus *Calicalicus* 214
    Red-tailed Vanga, *C. madagascariensis* (Linné, 1766) 214
  Genus *Schetba* 214
    Rufous Vanga, *Sch. rufa* (Linné, 1766) 214
  Genus *Vanga* 213
    Hook-billed Vanga, *V. curvirostris* (Linné, 1766) 213
  Genus *Xenopirostris* 213
    Pollen's Vanga, *X. polleni* (Schlegel, 1868) 213
  Genus *Falculea* 213
    Sicklebill, *F. palliata* Geoffroy St. Hilaire, 1836 213
  Genus *Tylas* 214
    Gray Kinkimavo, *T. eduardi* Hartlaub, 1862 214
  Genus *Hypositta* 214
    Madagascar Nuthatch, *H. corallirostris* (Newton, 1863) 214
  Genus *Leptopterus* 213
    White-headed Vanga, *L. viridis* (Müller, 1776) 213
    Chabert's Vanga, *L. chabert* (Müller, 1776) 213
    Blue Vanga, *L. madagascarinus* (Linné, 1766) 213
  Genus *Oriolia* 214
    Bernier's Vanga, *O. bernieri* Geoffroy St. Hilaire, 1838 214
  Genus *Euryceros* 213
    Helmet Bird, *Eu. prevostii* Lesson, 1831 213

Family Waxwings (Bombycillidae) 215
Subfamily True Waxwings (Bombycillinae) 215
  Genus *Bombycilla* 215
    Bohemian Waxwing, *B. garrulus* (Linné, 1758) 215
    Japanese Waxwing, *B. japonica* (Siebold, 1824) 215
    Cedar Waxwing, *B. cedrorum* Vieillot, 1808 215

Subfamily Phainopeplas (Ptilogonatinae) 216
  Genus *Ptilogonys* 216
    Long-tailed Silky Flycatcher, *P. caudatus* Cabanis, 1861 216
    Gray Silky Flycatcher, *P. cinereus* Swainson, 1827 216
  Genus *Phainoptila* 216

    Black-and-Yellow Silky Flycatcher, *Ph. melanoxantha* Salvin, 1877 216
  Genus *Phainopepla* 217
    Phainopepla, *Ph. nitens* (Swainson, 1838) 217

Subfamily Hypocoliinae 217
  Genus *Hypocolius* 217
    Gray Hypocolius, *H. ampelinus* Bonaparte, 1851 217

Family Palmchats (Dulidae) 217
  Genus *Dulus* 217
    Palmchat, *D. dominicus* (Linné, 1766) 217

Family Dippers (Cinclidae) 219
  Genus *Cinclus* 219
    Eurasian Dipper, *C. cinclus* (Linné, 1758) 219
    Brown Dipper, *C. pallasii* Temminck, 1820 219
    American Dipper, *C. mexicanus* Swainson, 1827 219
    White-capped Dipper, *C. leucocephalus* Tschudi, 1844 220

Family Wrens (Troglodytidae) 223
  Genus *Campylorhynchus* —
    Cactus Wren, *C. bruneicapillus* (Lafresnaye, 1835) 223
    Band-backed Wren, *C. zonatus* (Lesson, 1832) 224
  Genus *Salpinctes* 224
    Rock Wren, *S. obsoletus* (Say, 1823) 224
    Canyon Wren, *S. mexicanus* (Swainson, 1829) 224
  Genus *Cistothorus* 223
    Short-billed Marsh Wren, *C. platensis* (Latham, 1790) —
    Long-billed Marsh Wren, *C. palustris* (Wilson, 1807) 223
  Genus *Thryomanes* —
    Bewick's Wren, *T. bewickii* (Audubon, 1827) —
  Genus *Thryothorus* 223
    Black-capped Wren, *T. nigricapillus* Sclater, 1860 —
    Carolina Wren, *T. ludovicianus* (Latham, 1790) 223
  Genus *Troglodytes* 223
    Winter Wren, *T. troglodytes* (Linné, 1758) 223
    Icelandic Winter Wren, *T. troglodytes islandicus* Hartert, 1907 223
    House Wren, *T. aedon* Vieillot, 1808 223
    Mountain Wren, *T. solstitialis* Sclater, 1859 226
  Genus *Cyphorinus* 224
    Musician Wren, *C. aradus* (Hermann, 1873) 224

Family Mockingbirds and Thrashers (Mimidae) 227
  Genus *Dumetella* 227
    Catbird, *D. carolinensis* (Linné, 1766) 227
  Genus *Melanotis* 221
    Blue Mockingbird, *M. caerulescens*

(Swainson, 1827)	221	and Horsfield, 1827)	222
Genus *Mimus*	227	Genus *Xiphirhynchus*	235
Mockingbird, *M. polyglottus* (Linné, 1758)	227	Slender-billed Scimitar Babbler, *X. superciliaris* Blyth, 1842	235
Genus *Nesomimus*	227	Genus *Jabouilleia*	—
Galapagos Mockingbird, *N. parvulus* (Gould, 1837)	227	*J. donjoui* (Robinson and Kloss, 1919)	—
Genus *Oreoscoptes*	227	Genus *Rimator*	235
Sage Thrasher, *O. montanus* (Townsend, 1837)	227	Long-billed Wren Babbler, *R. malacoptilus* Blyth, 1847	235
Genus *Toxostoma*	227	Genus *Ptilocichla*	236
Brown Thrasher, *T. rufum* (Linné, 1758)	227	*P. mindanensis* (Blasius, 1890)	—
California Thrasher, *T. redivivum* (Gambel, 1845)	227	Genus *Kenopia*	236
		Striped Wren Babbler, *K. striata* (Blyth, 1842)	236
Genus *Donacobius*	227	Genus *Napothera*	236
Black-capped Mocking Thrush, *D. atricapillus* (Linné, 1766)	227	*N. macrodactyla* (Strickland, 1844)	—
		Genus *Pnoepyga*	236
Family Accentors (Prunellidae)	229	Greater Scaly-breasted Wren Babbler, *P. albiventer* (Hodgson, 1837)	237
Genus *Prunella*	230		
Alpine Accentor, *P. collaris* (Scopoli, 1769)	230	Lesser Scaly-breasted Wren Babbler, *P. pusilla* Hodgson, 1845	237
Rufous-breasted Accentor, *P. strophiata* (Blyth, 1843)	230	Genus *Spelaeornis*	236
Siberian Accentor, *P. montanella* (Pallas, 1776)	230	Barred-wing Wren Babbler, *S. troglodytoides* (Verreaux, 1870)	238
P. ocularis (Radde, 1884)	—	Long-tailed Wren Babbler, *S. chocolatinus* (Godwin-Austen and Walden, 1875)	238
Hedge Sparrow, *P. modularis* (Linné, 1758)	229	Genus *Sphenocichla*	—
		Wedge-billed Wren Babbler, *S. humei* (Mandelli, 1873)	—
Family Flycatcher-like Birds (Muscicapidae)	233		
Subfamily Babbling Thrushes (Timaliinae)	238	**Tribe Timaliini**	238
Tribe Pellorneini	234	Genus *Neomixis*	238
Genus *Pellorneum*	234	Green Jery, *N. viridis* (Sharpe, 1883)	238
Striped Jungle Babbler, *P. ruficeps* Swainson, 1832	235	Genus *Stachyris*	238
Genus *Trichastoma*	235	Red-fronted Babbler, *S. rufifrons* Hume, 1873	—
Blyth's Jungle Babbler, *T. rostratum* Blyth, 1842	235	Golden-headed Babbler, *S. chrysaea* Blyth, 1844	239
Abbott's Babbler, *T. abbotti* (Blyth, 1845)	235	Genus *Dumetia*	239
Black-browed Jungle Babbler, *T. perspicillatum* (Bonaparte, 1850)	235	Rufous-bellied Babbler, *D. hyperythra* (Franklin, 1831)	239
Genus *Ptyrticus*	235	Genus *Rhopocichla*	239
Thrush Babbler, *P. turdinus* Hartlaub, 1883	235	Black-headed Babbler, *R. atriceps* (Jerdon, 1839)	239
Genus *Leonardina*	235	Genus *Macronous*	239
Bagabu Babbler, *L. woodi* (Mearns, 1905)	235	Striped Babbler, *M. gularis* (Horsfield, 1822)	—
Genus *Malacopteron*	235	Fluffy-backed Tit Babbler, *M. ptilosus* Jardine and Selby, 1835	239
M. affine (Blyth, 1842)	222	Genus *Timalia*	239
Greater Red-headed Babbler, *M. magnum* Eyton, 1839	235	Red-capped Babbler, *T. pileata* Horsfield, 1821	239
Tribe Pomatorhinini	235		
Genus *Pomatorhinus*	235	**Tribe Chamaeini**	239
Yellow-billed Scimitar Babbler, *P. erythrogenys* Vigors, 1832	236	Genus *Chrysomma*	240
P. montanus Horsfield, 1821	236	Oriental Golden-eyed Babbler, *Ch. sinense* (Gmelin, 1789)	240
Rufous-necked Scimitar Babbler, *P. ruficollis* Hodgson, 1836	237	Genus *Moupinia*	239
Genus *Pomatostomus*	235	Rufous-crowned Babbler, *M. poecilote* (Verreaux, 1870)	240
Rufous Babbler, *P. isidorei* (Lesson, 1827)	237		
Gray-crowned Babbler, *P. temporalis* (Vigors			

Genus *Chamaea*	239
Wrentit, *Ch. fasciata* (Gambel, 1845)	239
Tribe Turdoidini	234
Genus *Turdoides*	240
Spring Babbler, *T. nipalensis* (Hodgson, 1836)	240
White-throated Babbler, *T. gularis* (Blyth, 1855)	240
T. fulvus (Desfontaines, 1789)	—
Seven Sisters, *T. somervillei* (Sykes, 1832)	240
Ceylonese subspecies, *T. somervillei rufescens* (Blyth, 1847)	222
T. bicolor (Jardine, 1831)	—
Genus *Babax*	243
Chinese Babax, *B. lanceolatus* (Verreaux, 1870)	243
Genus *Garrulax*	243
White-crested Laughing Thrush, *G. leucolophus* (Hardwicke, 1815)	243
Black-gorgetted Laughing Thrush, *G. pectoralis* (Gould, 1836)	243
Necklaced Laughing Thrush, *G. moniliger* (Hodgson, 1836)	243
G. canorus (Linné, 1758)	—
G. erythrocephalus (Vigors, 1832)	—
Genus *Liocichla*	—
Crimson-winged Laughing Thrush, *L. phoenicea* (Gould, 1837)	—
Genus *Leiothrix*	244
Pekin Robin, *L. lutea* (Scopoli, 1786)	244
Silver-eared Mesia, *L. argentauris* (Hodgson, 1837)	244
Genus *Myzornis*	244
Firetail, *M. pyrrhoura* Blyth, 1843	244
Genus *Cutia*	244
Cutia, *C. nipalensis* Hodgson, 1837	244
Genus *Pteruthius*	244
Red-winged Shrike Babbler, *P. flaviscapis* (Vigors, 1931)	244
Genus *Gampsorhynchus*	245
White-headed Babbler, *G. rufulus* Blyth, 1844	245
Genus *Actinodura*	—
A. egertoni Gould, 1836	—
Genus *Siva*	—
S. cyanouroptera Hodgson, 1838	—
S. strigula Hodgson, 1837	—
Genus *Alcippe*	245
A. vinipectus (Hodgson, 1837)	—
Nun Babbler, *A. poioicephala* (Jerdon, 1844)	245
A. nipalensis (Hodgson, 1837)	222
A. chrysotis (Blyth, 1845)	—
Genus *Lioptilus*	—
L. nigricapillus (Vieillot, 1818)	—
Genus *Phyllanthus*	—
Ph. atripennis (Swainson, 1837)	—
Genus *Crocias*	—
C. albonotatus (Lesson, 1832)	—
Genus *Heterophasia*	245
Long-tailed Sibia, *H. picaoides* (Hodgson, 1839)	245
H. capistrata (Vigors, 1831)	—
Genus *Yuhina*	—
Y. nigrimenta Blyth, 1845	—
Y. gularis Hodgson, 1836	—
Tribe Rockfowl (Picahartini)	245
Genus *Picathartes*	245
Gray-necked Rockfowl, *P. gymnocephalus* (Temminck, 1825)	245
White-necked Rockfowl, *P. oreas* Reichenow, 1899	245
Subfamily Parrotbills (Panurinae)	247
Genus *Panurus*	247
Bearded Tit, *P. biarmicus* (Linné, 1758)	247
Genus *Paradoxornis*	247
P. paradoxa (Verreaux, 1870)	247
Vinous-throated Parrotbill, *P. webbiana* (Taczanowski, 1855)	247
Gray-headed Parrotbill, *P. gularis* Gray, 1845	247
Subfamily Rail Babblers (Orthonychinae)	248
Genus *Orthonyx*	222
O. temminckii Ranzani, 1822	222
Genus *Cinclosoma*	248
Spotted Quail Thrush, *C. punctatum* (Shaw, 1794)	248
C. castanorum Gould, 1840	222
Chestnut Quail Thrush, *C. cinnamomeum* Gould, 1846	248
New Guinea Quail Thrush, *C. ajax* (Temminck, 1835)	249
Genus *Ptillorhoa*	—
Blue Scrub Thrush, *P. caerulescens* (Temminck, 1835)	—
Genus *Eupetes*	249
Rail Babbler, *Eu. macrocercus* Temminck, 1831	249
Genus *Melampitta*	249
Lesser Malampitta, *M. lugubris* Schlegel, 1873	249
Genus *Ifrita*	249
Ifrita, *I. kowalki* (De Vis, 1890)	249
Subfamily Old World Warblers (Sylviinae)	249
Genus *Macrosphenus*	250
Genera Tailorbirds (*Orthotomus* and *Phyllergates*)	265
Tailor Bird, *O. sutorius* (Pennant, 1769)	265
O. atrogularis Temminck, 1836	—
Genus *Camaroptera*	250
Green-backed Camaroptera, *C. brachyura* (Vieillot, 1820)	250
Gray-backed Camaroptera, *C. brevicaudata* (Cretzschmar, 1831)	250
Genus *Calamonastes*	—
C. fasciolatus (Smith, 1847)	—
Genus *Sylvietta*	250

S. rufescens (Vieillot, 1817)	250
Genus *Eremomela*	250
E. icteropygialis (Lafresnaye, 1839)	—
Brown-crowned Eremomela, *E. badiceps* (Fraser, 1842)	250
E. atricollis Bocage, 1894	—
Genus *Apalis*	250
Bar-throated Warbler, *A. thoracica* (Shaw and Nodder, 1811)	250
Yellow-chested Apalis, *A. flavida* (Strickland, 1852)	250
Genus *Prinia*	251
P. flavicans (Vieillot, 1820)	—
Tawny-flanked Prinia, *P. subflava* (Gmelin, 1789)	251
P. gracilis (Lichtenstein, 1823)	—
P. flaviventris (Delessert, 1840)	—
Genus *Cisticola*	251
Cisticola, *C. ayresii* Hartlaub, 1863	251
Desert Cisticola, *C. aridula* Witherby, 1900	252
Fan-tailed Warbler, *C. juncidis* (Rafinesque, 1810)	251
Streaked Fan-tailed Warbler, *C. exilis* (Vigors and Horsfield, 1827)	251
C. natalensis (Smith, 1843)	—
C. tinniens (Lichtenstein, 1842)	—
Red-faced Cisticola, *C. erythrops* (Hartlaub, 1857)	252
Singing Cisticola, *C. cantans* (Heuglin, 1896)	252
Genus *Sphenoeacus*	251
Cape Grass Bird, *S. afer* (Gmelin, 1789)	251
Genus *Achaetops*	—
A. pycnopygius (Strickland and Sclater, 1852)	—
Genus *Dromaeocercus*	251
Gray Emu-tail, *D. seebohmi* Sharpe, 1879	251
Genus *Bradypterus*	253
Little Rush Warbler, *B. baboecala* (Vieillot, 1817)	253
Genus *Cettia*	253
Cetti's Warbler, *C. cetti* (Temminck, 1820)	253
Bush Warbler, *C. diphone* (Kittlitz, 1831)	250
Genus *Locustella*	253
Taiga Grasshopper Warbler, *L. fasciolata* (Gray, 1860)	253
River Warbler, *L. fluviatilis* (Wolf, 1810)	253
Savi's Warbler, *L. luscinioides* (Savi, 1824)	253
Streaked Grasshopper Warbler, *L. lanceolata* (Temminck, 1840)	253
Grasshopper Warbler, *L. naevia* (Boddaert, 1783)	253
Pallas' Grasshopper Warbler, *L. certhiola* (Pallas, 1811)	253
Middendorf's Warbler, *L. ochotensis* (Middendorff, 1853)	253
Genus *Acrocephalus*	253
Moustached Warbler, *A. melanopogon* (Temminck, 1823)	252
Sedge Warbler, *A. schoenebaenus* (Linné, 1758)	252
Indian Reed Warbler, *A. agricola* (Jerdon, 1845)	253
Blyth's Reed Warbler, *A. dumetorum* (Blyth, 1849)	253
Marsh Warbler, *A. palustris* (Bechstein, 1798)	253
Reed Warbler, *A. scirpaceus* (Hermann, 1804)	253
Great Reed Warbler, *A. arundinaceus* (Linné, 1758)	253
Egyptian Great Reed Warbler, *A. stentoreus* (Hemprich and Ehrenberg, 1833)	253
Iraq Great Warbler, *A. griseldis* (Hartlaub, 1891)	253
Thick-billed Warbler, *A.* (or *Pragmaticola*) *aedon* (Pallas, 1776) (systematic position uncertain)	253
Genus *Calamocichla*	253
Greater Swamp Warbler, *C. gracilirostris* (Hartlaub, 1964)	253
Genus *Hippolais*	254
Icterine Warbler, *H. icterina* (Vieillot, 1817)	253
Melodious Warbler, *H. polyglotta* (Vieillot, 1817)	253
Olive Tree Warbler, *H. olivetorum* (Strickland, 1837)	253
Upcher's Warbler, *H. languida* (Hemprich and Ehrenberg, 1833)	254
Olivaceous Warbler, *H. pallida* (Hemprich and Ehrenberg, 1833)	254
Babbling Warbler, *H. caligata* (Lichtenstein, 1823)	254
Genus *Sylvia*	258
Blackcap, *S. atricapilla* (Linné, 1758)	258
Garden Warbler, *S. borin* (Boddaert, 1783)	258
Barred Warbler, *S. nisoria* (Bechstein, 1783)	258
Orphean Warbler, *S. hortensis* (Gmelin, 1789)	258
Lesser White Throat, *S. curruca* (Linné, 1758)	258
White Throat, *S. communis* (Latham, 1787)	258
Spectacled Warbler, *S. conspicillata* Temminck, 1820	258
Tristram's Warbler, *S. deserticola* Tristram, 1859	258
Desert Warbler, *S. nana* (Hemprich and Ehrenberg, 1833)	258
Subalpine Warbler, *S. cantillans* (Pallas, 1769)	258
Sardinian Warbler, *S. melanocephala* (Gmelin, 1789)	258
Dartford Warbler, *S. undata* (Boddaert, 1783)	258
Marmora's Warbler, *S. sarda* Temminck, 1820	258
Rüppell's Warbler, *S. rueppelli* Temminck, 1820	258
Genus Leafwarblers (*Phylloscopus*)	263
Radde's Bush Warbler, *Ph. schwarzi* (Radde, 1863)	263
Brown Willow Warbler, *Ph. fuscatus* (Blyth, 1842)	263
Chiff-chaff, *Ph. collybita* (Vieillot, 1817)	263
Willow Warbler, *Ph. trochilus* (Linné, 1758)	263
Wood Warbler, *Ph. sibilatrix* (Bechstein, 1793)	263

Bonelli's Warbler, *Ph. bonelli* (Vieillot, 1819) 263
Arctic Warbler, *Ph. borealis* (Blasius, 1858) 263
Greenish Willow Warbler, *Ph. trochiloides* (Sundevall, 1837) 263
Ph. inornatus (Blyth, 1842) 262
Kinglet Warbler, *Ph. proregulus* (Pallas, 1811) 262
Mountain Leaf Warbler, *Ph. trivirgatus* Strickland, 1849 263
Green Warbler, *Ph. nitidus* Blyth, 1843 263
Genus *Seicercus* 263
Yellow-throated Woodland Warbler, *S. ruficapillus* (Sundevall, 1850) 263
Yellow-eyed Flycatcher Warbler, *S. burkii* (Burton, 1836) 263

Subfamily Wren Warblers (Malurinae) 265
Genus *Aphelocephala* —
A. nigricincta (North, 1895) —
Genus *Acanthiza* 265
Little Thornbill, *A. nana* Vigors and Horsfield, 1827 265
A. pusilla (White, 1790) —
A. chrysorrhoa (Quoy and Gaimard, 1830) 259
Genus *Gerygone* 265
Mangrove Fly-eater, *G. sulphurea* Wallace, 1864 265
G. olivacea (Gould, 1838) —
G. palpebrosa Wallace, 1865 259
Genus *Sericornis* 265
Arfak Buff-faced Sericornis, *S. rufescens* Stresemann, 1921 265
S. frontalis Vigors and Horsfield, 1827 —
Genus *Stipiturus* 265
Southern Emu Wren, *S. malachurus* (Shaw, 1798) 265
Genus *Malurus* 266
Superb Blue Wren, *M. cyaneus* (Latham, 1783) 266
Variegated Wren, *M. lamberti* Vigors and Horsfield, 1827 266
M. melanocephalus (Latham, 1801) —
M. elegans Gould, 1837 —
Genus *Amytornis* 266
A. striatus (Gould, 1839) —
Western Grass Wren, *A. textilis* (Dumont, 1824) 266
A. modestus (North, 1902) —
Genus *Ephtianura* 266
Crimson Chat, *E. tricolor* Gould, 1840 266

Subfamily Kinglets (Regulinae) 267
Genus *Regulus* 267
Goldcrest, *R. regulus* (Linné, 1758) 267
Firecrest, *R. ignicapillus* (Temminck, 1820) 267
Taiwan Goldcrest, *R. goodfellowi* Ogilvie-Grant, 1906 268
Golden-crowned Kinglet, *R. satrapa* Lichtenstein, 1823 268

Ruby-crowned Kinglet, *R. calendula* (Linné, 1766) 268
Genus *Leptopoecile* 268
Stoliczka's Tit Warbler, *L. sophiae* Severtzov, 1872 268
Crested Tit Warbler, *L. elegans* Przewalski, 1887 268

Subfamily Hylias (Hyliinae), systematic position uncertain 268
Genus *Hylia* 268
Green Hylia, *H. prasina* (Cassin, 1855) 268

Subfamily Gnatcatchers (Polioptilinae) 271
Genus *Polioptila* 271
Blue-gray Gnatcatcher, *P. caerulea* (Linné, 1766) 271
Black-tailed Gnatcatcher, *P. melanura* Lawrence, 1857 271
Genus *Ramphocaenus* 271
Long-billed Gnatwren, *R. melanurus* Vieillot, 1819 271
Genus *Microbates* 271
Collared Gnatwren, *M. collaris* (Pelzeln, 1868) 271

Subfamily Flycatchers (Muscicapinae) 272
Genus *Eumyias* 276
Eu. thalassina (Swainson, 1838) —
Indigo Flycatcher, *Eu. indigo* (Horsfield, 1821) 276
Genus *Cyanoptila* 275
Blue Flycatcher, *C. cyanomelana* (Temminck, 1829) 275
Genus *Niltava* 275
Large Niltava, *N. grandis* (Blyth, 1842) 275
Rufous-bellied Niltava, *N. sundara* Hodgson, 1837 275
Genus *Cyornis* 275
C. unicolor Blyth, 1843 —
Blue-throated Flycatcher, *C. rubeculoides* (Vigors, 1831) 275
C. banyumas (Horsfield, 1822) —
C. rufigastra (Raffles, 1822) 275
Genus *Ficedula* 274
Subgenus *Dendrobiastes* 275
Rusty-breasted Blue Flycatcher, *F. (Dendrobiastes) hodgsonii* (Verreaux, 1871) 275
Subgenus *Digenea* 275
Slaty Blue Flycatcher, *F. (Digenea) tricolor* (Hodgson, 1845) 275
Subgenus *Muscicapula* 275
White-browed Flycatcher, *F. (Muscicapula) superciliaris* (Jerdon, 1840) 275
F. (Muscicapula) westermanni (Sharpe, 1888) —
Subgenus *Ficedula* 273
Pied Flycatcher, *F. (Ficedula) hypoleuca* (Pallas, 1764) 273

Half-collared Flycatcher, *F. (Ficedula) semitorquata* (Homeyer, 1885)	274	Chestnut Wattleye, *D. castanea* (Fraser, 1842)	277
Collared Flycatcher, *F. (Ficedula) albicollis* (Temminck, 1815)	274	Genus *Erythrocercus* (Systematic position uncertain)	277
Subgenus *Zanthopygia*	–	Little Yellow Flycatcher, *E. holochlorus* Erlanger, 1901	277
Korean Flycatcher, *F. (Zanthopygia) zanthopygia* (Hay, 1845)	–	Livingstone's Flycatcher, *E. livingstonei* Gray, 1870	277
Narcissus Flycatcher, *F. (Zanthopygia) narcissina* (Temminck, 1835)	–	**Subfamily Flat-billed Flycatchers (Myiagrinae)**	–
Subgenus *Poliomyias*	–	Genus *Myiagra*	278
Black-and-Orange Flycatcher, *F. (Poliomyias) mugimaki* (Temminck, 1835)	–	*M. cyanoleuca* (Vieillot, 1818)	–
Subgenus *Erythrosterna*	–	Micronesian Broad-bill, *M. oceanica* Pucheran, 1853	278
Least Flycatcher, *F. (Erythrosterna) parva* (Bechstein, 1794)	–	Genus *Machaerirhynchus*	277
Subgenus *Siphia*	–	Papuan Flycatcher, *M. flaviventer* Gould, 1851	277
Orange-gorgetted Flycatcher, *F. (Siphia) strophiata* (Hodgson, 1837)	–	Genus *Arses*	278
Genus *Rhinomyias*	276	Frilled Flycatcher, *A. telescophthalmus* (Garnot, 1827)	278
Olive-backed Jungle Flycatcher, *R. olivacea* (Hume, 1877)	276	Genus *Petroica*	278
R. ruficauda (Sharpe, 1877)	–	*P. goodenovii* (Vigors and Horsfield, 1827)	–
R. gularis Sharpe, 1888	–	Scarlet Robin, *P. multicolor* (Gmelin, 1789)	278
Genus *Microeca*	276	Tomtit, *P. macrocephala* (Gmelin, 1789)	278
Lemon-breasted Flycatcher, *M. flavigaster* Gould, 1843	276	*P. rosea* Gould, 1840	–
Genus *Muscicapa*	276	Genus *Eopsaltria*	241
Brown Flycatcher, *M. latirostris* Raffles, 1822	276	*E. austalis* (Shaw, 1790)	241
Siberian Flycatcher, *M. sibirica* Gmelin, 1789	276	Genus *Miro*	278
Spotted Flycatcher, *M. striata* (Pallas, 1764)	260	Southern Yellow Robin, *M. australis* (Sparrman, 1788)	–
M. gambagae (Alexander, 1901)	276	Genus *Chasiempis*	278
Dusky Flycatcher, *M. adusta* (Boie, 1828)	276	Elepaio, *Ch. sandvicensis* (Gmelin, 1789)	278
Genus *Parisoma* (systematic position uncertain)	276	**Subfamily Rhipidurinae**	278
Gray Tit Flycatcher, *P. plumbeum* (Hartlaub, 1858)	276	Genus *Rhipidura*	278
Tit Warbler, *P. subcaeruleum* (Vieillot, 1817)	276	Willie Wagtail, *R. leucophrys* (Latham, 1801)	278
Genus *Newtonia*	–	Rufous Fantail, *R. rufifrons* (Latham, 1801)	278
Mountain Newtonia, *N. brunneicauda* Newton, 1863	276	Genus *Chelidorynx*	278
Genus *Bradornis*	277	Yellow-bellied Fantail Flycatcher, *Ch. hypoxantha* (Blyth, 1843)	278
Mariqua Flycatcher, *B. mariquensis* (A. Smith, 1836)	277	**Subfamily Paradise Flycatchers (Monarchinae)**	281
Pale Flycatcher, *B. pallidus* (V. Müller, 1851)	277	Genus *Monarcha*	281
Genus *Melaenornis*	277	Island Gray-headed Monarch Flycatcher, *M. cinerascens* (Temminck and Laugier, 1827)	281
Black Flycatcher, *M. pammelaina* (Stanley, 1814)	277	Black-faced Flycatcher, *M. melanopis* (Vieillot, 1818)	281
Subfamily Platysteirinae	–	Yap Monarch Flycatcher, *M. godeffroyi* (Hartlaub, 1867)	281
Genus *Batis*	277	Genus *Pomarea*	281
Puff-back Flycatcher, *B. capensis* (Linné, 1766)	277	Marquesas Flycatcher, *P. mendozae* (Hartlaub, 1854)	281
Chin-spot Puff-back Flycatcher, *B. molitor* (Hahn and Küster, 1850)	277	Genus *Hypothymis*	281
Pririt Puff-back Flycatcher, *B. pririt* (Vieillot, 1818)	277	Black-naped Flycatcher, *H. azurea* (Boddaert, 1783)	281
Genus *Platysteira*	277	Genus *Terpsiphone*	282
Black-throated Wattleye, *P. peltata* Sundevall, 1850	277	African Paradise Flycatcher, *T. viridis* (Müller, 1776)	282
Genus *Dyaphorophyia*	277	Indian Paradise Flycatcher, *T. paradisi*	

(Linné, 1758)	282
Japanese Paradise Flycatcher, *T. atrocaudata* (Eyton, 1838)	282
Subfamily Thickheads (Pachycephalinae)	283
Genus *Falcunculus*	283
Eastern Shrike Tit, *F. frontatus* (Latham, 1801)	283
Genus *Pachycephala*	283
Golden Whistler, *P. pectoralis* (Latham, 1801)	283
P. rufiventris (Latham, 1801)	—
Genus *Pitohui*	283
Black-headed Pitohui, *P. dichrous* (Bonaparte, 1850)	284
Subfamily Thrushes (Turdinae)	284
Genus *Brachypteryx*	285
Lesser Shortwing, *B. leucophrys* (Temminck, 1827)	285
Genus *Drymodes*	285
Northern Scrub Robin, *D. superciliaris* Gould, 1850	286
Genus *Chaetops*	286
Rufous Rock Jumper, *Ch. frenatus* (Temminck, 1826)	286
Genus *Erythropygia,* probably = *Cercotrichas*	285
E. leucophrys (Vieillot, 1817)	—
Rufous Warbler, *E. galactotes* (Temminck, 1820)	286
E. paena A. Smith, 1836	—
Genus *Cercotrichas*	286
Black Bush Robin, *C. podobe* (Müller, 1776)	286
Genus *Cichladusa*	—
C. arquata Peters, 1863	—
Genus *Alethe*	287
Fire-crested Aletha, *A. diademata* (Bonaparte, 1851)	287
Genus *Cossypha*	287
Snowy-headed Robin Chat, *C. niveicapilla* (Lafresnaye, 1838)	287
C. dichroa (Gmelin, 1789)	—
Robin Chat, *C. caffra* (Linné, 1771)	279
Genus *Pogonocichla*	—
P. stellata (Vieillot, 1818)	—
Genus *Phoenicurus*	291
Ph. erythrogaster (Güldenstädt, 1775)	—
Ph. auroreus (Pallas, 1776)	—
Redstart, *Ph. phoenicurus* (Linné, 1758)	291
Black Redstart, *Ph. ochruros* (Gmelin, 1758)	291
European Black Redstart, *Ph. ochruros gibraltariensis* (Gmelin, 1789)	—
Ph. erythronotus (Eversmann, 1841)	—
Moussier's Redstart, *Ph. moussieri* (Olphe-Galliard, 1852)	291
Genus *Tarsiger*	279
Orange-flashed Bush Robin, *T. cyanurus* (Pallas, 1773)	279
Genus *Erithacus*	285
Robin, *E. rubecula* (Linné, 1758)	286
Genus *Luscinia*	286
Subgenus *Cyanosylvia*	286
Bluethroat, *L. (Cyanosylvia) svecica* (Linné, 1758)	286
Subgenus *Calliope*	286
Rubythroat, *L. (Calliope) calliope* (Pallas, 1776)	286
L. (Calliope) pectoralis (Gould, 1838)	—
Subgenus *Luscinia*	286
Nightingale, *L. (Luscinia) megarhynchos* (Brehm, 1831)	286
Thrush Nightingale, *L. (Luscinia) luscinia* (Linné, 1758)	286
Subgenus *Icoturus*	—
L. (Icoturus) akahige (Temminck, 1824)	—
Subgenus *Larvivora*	—
L. (Larvivora) pectardens (David and Oustalet, 1877)	—
L. (Larvivora) cyane (Pallas, 1776)	—
Genus *Saxicola*	295
Whinchat, *S. rubetra* (Linné, 1758)	295
S. dacotiae (Meade-Waldo, 1889)	—
Stonechat, *S. torquata* (Linné, 1766)	295
S. caprata (Linné, 1766)	—
Genus *Cercomela*	295
Blackstart, *C. melanura* (Temminck, 1824)	295
Red-tailed Chat, *C. familiaris* (Stephens, 1826)	295
Genus *Oenanthe*	295
Oe. pileata (Gmelin, 1789)	280
Wheatear, *Oe. oenanthe* (Linné, 1758)	295
Greenland Wheatear, *Oe. oenanthe leucorrhoa* (Gmelin, 1789)	297
Oe. pleschanka (Lepechin, 1770)	—
Black-eared Wheatear, *Oe. hispanica* (Linné, 1758)	295
Red-rumped Wheatear, *Oe. moesta* (Lichtenstein, 1823)	295
Desert Wheatear, *Oe. deserti* (Temminck, 1825)	295
Mourning Wheatear, *Oe. lugens* (Lichtenstein, 1823)	295
White-crowned Black Wheatear, *Oe. leucopyga* (Brehm, 1855)	295
Pied Chat, *Oe. picata* (Blyth, 1847)	295
Mountain Chat, *Oe. monticola* Vieillot, 1818	295
Genus *Myrmecocichla*	295
M. arnoti (Tristram, 1869)	—
M. aethiops Cabanis, 1850	295
M. formicivora (Vieillot, 1818)	—
Genus *Thamnolaea*	—
Th. cinnamomeiventris (Lafresnaye, 1836)	—
Genus *Copsychus*	291
C. albospecularis (Eydoux and Gervais, 1836)	—
Magpie Robin, *C. saularis* (Linné, 1758)	291
C. luzoniensis (Kittlitz, 1832)	—
Shama, *C. malabaricus* (Scopoli, 1788)	291
C. pyrropygus (Lesson, 1839)	—
Genus *Chaimarrornis*	291

White-capped Redstart, *Ch. leucocephalus* (Vigors, 1831)	291
Genus *Rhyacornis*	291
Plumbeous Redstart, *R. fuliginosus* (Vigors, 1831)	291
Genus *Enicurus*	285
Leschenault's Forktail, *E. leschenaulti* (Vieillot, 1818)	293
Chestnut-naped Forktail, *E. ruficapillus* Temminck, 1823	293
Little Forktail, *E. scouleri* Vigors, 1832	293
Genus *Saxicoloides*	295
Indian Robin, *S. fulicata* (Linné, 1766)	295
Genus *Monticola*	298
M. cinclorhynchus (Vigors, 1832)	—
M. angolensis Sousa, 1888	—
M. imerinus (Hartlaub, 1860)	—
M. rupestris (Vieillot, 1818)	—
M. explorator (Vieillot, 1818)	—
M. brevipes (Waterhouse, 1838)	—
Rock Thrush, *M. saxatilis* (Linné, 1766)	298
M. rufocinereus (Rüppell, 1837)	—
Blue Rock Thrush, *M. solitarius* (Linné, 1758)	298
M. rufiventris (Jardine and Selby, 1832)	—
Genus *Sialia*	292
S. currucoides (Bechstein, 1798)	—
Eastern Bluebird, *S. sialis* (Linné, 1758)	292
Genus *Myiophoneus*	299
Whistling Thrush, *M. caeruleus* (Scopoli, 1786)	299
Sunda Whistling Thrush, *M. glaucinus* (Temminck, 1823)	299
Genus *Cochoa*	293
Purple Cochoa, *C. purpurea* Hodgson, 1836	293
Green Cochoa, *C. viridis* Hodgson, 1836	293
Malayan Cochoa, *C. azurea* (Temminck, 1824)	293
Genus *Zoothera*	300
Varied Thrush, *Z. naevia* (Gmelin, 1789)	300
Siberian Ground Thrush, *Z. sibirica* (Pallas, 1776)	300
Z. interpres (Temminck, 1826)	—
Dama Thrush, *Z. citrina* (Latham, 1790)	300
Z. gurneyi (Hartlaub, 1864)	—
White's Thrush, *Z. dauma* (Latham, 1790)	300
Genus *Turdus*	303
Ground-Scraper Thrush, *T. litsipsirupa* (Smith, 1836)	304
Mistle Thrush, *T. viscivorus* Linné, 1758	304
T. mupinensis Laubmann, 1920	—
Redwing, *T. iliacus* Linné, 1766	304
T. naumanni Temminck, 1820	—
T. naumanni eunomus Temminck, 1831	—
Red-capped Thrush, *T. ruficollis* Pallas, 1776	304
T. ruficollis atrogularis Jarocki, 1819	304
T. ruficollis ruficollis Pallas, 1776	304
Fieldfare, *T. pilaris* Linné, 1758	304
Blackbird, *T. merula* Linné, 1758	304
T. albocinctus Royle, 1839	—
Ring Ouzel, *T. torquatus* Linné, 1758	304
T. boulboul (Latham, 1790)	—
T. poliocephalus Latham, 1801	—
T. rubrocanus Hodgson, 1846	—
T. obscurus Gmelin, 1789	—
American Robin, *T. migratorius* Linné, 1766	304
Bare-eyed Thrush, *T. nudigenis* Lafresnaye, 1848	289
T. swalesi (Wetmore, 1827)	—
White-necked Thrush, *T. albicollis* Vieillot, 1818	304
Cuban Thrush, *T. plumbeus* Linné, 1758	289
T. serranus Tshudi, 1844	—
T. flavipes Vieillot, 1818	—
Rufous-bellied Thrush, *T. rufiventris* Vieillot, 1818	304
Olive Thrush, *T. olivaceus* Linné, 1766	304
T. libonyana (A. Smith, 1836)	—
T. cardis Temminck, 1831	—
T. hortulorum Sclater, 1863	—
Song Thrush, *T. philomelos* Brehm, 1831	304
Genus *Hylocichla*	303
Wood Thrush, *H. mustelina* (Gmelin, 1789)	221
Hermit Thrush, *H. guttata* (Pallas, 1811)	303
Swainson's Thrush, *H. ustulata* (Nuttall, 1840)	303
Gray-cheeked Thrush, *H. minima* (Lasfresnaye, 1848)	303
H. fuscecens (Stephens, 1817)	—
Genus *Catharus*	303
C. dryas (Gould, 1855)	—
Orange-billed Nightingale-thrush, *C. aurantiirostris* (Hartlaub, 1850)	303
Genus *Myadestes*	285
Townsend's Solitaire, *M. townsendi* (Audubon, 1838)	294
M. elisabeth (Lembeye, 1850)	—
Genus *Phaeornis*	303
Hawaiian Thrush, *Ph. obscurus* (Gmelin, 1789)	303
Oahu subspecies, *Ph. obscurus oahensis* Wilson and Evans, 1899	303
Genus *Neocossyphus*	294
Red-tailed Ant Thrush, *N. rufus* (Fischer and Reichonow, 1884)	294
Genus *Stitzorhina*	—
Rufous Flycatcher, *S. fraseri* (Strickland, 1844)	—

Family Long-tailed Tits (Aegithalidae) 310
Genus *Aegithalos* 311
  Long-tailed Tit, *Ae. caudatus* (Linné, 1758) 311
  Ae. ionschistos (Blyth, 1844) —
  Red-headed Tit, *Ae. concinnus* (Gould, 1855) —
  Ae. fuliginosus (Verreaux, 1870) —
  Ae. leucogenys (Horsfield and Moore, 1854) —
Genus *Psaltria* 311

Java Long-tailed Tit, *P. exilis* Temminck, 1836 ... 311
Genus *Psaltiparus* ... 311
Common Bush Tit, *P. minimus* (Townsend, 1837) ... 311
Black-eared Bush Tit, *P. melanotis* (Hartlaub, 1844) ... 311

Family Penduline Tits (Remizidae) ... 311
Genus *Remiz* ... 312
Penduline Tit, *R. pendulinus* (Linné, 1758) ... 312
Genus *Anthoscopus* ... 313
Mouse-colored Penduline Tit, *A. musculus* (Hartlaub, 1882) ... 308
African Penduline Tit, *A. caroli* (Sharpe, 1871) ... 313
Cape Penduline Tit, *A. minutus* (Shaw and Nodder, 1812) ... 313
Genus *Cephalopyrus* ... 313
Fire-capped Tit, *C. flammiceps* (Burton, 1836) ... 313
Genus *Auriparus*, (systematic position uncertain) ... 313
Verdin, *Au. flaviceps* (Sundevall, 1850) ... 313

Family True Tits (Paridae) ... 314
Genus *Sylviparus* ... 321
Yellow-browed Tit, *S. modestus* Burton, 1836 ... 321
Genus *Melanochlora* ... 321
Sultan Tit, *M. sultana* (Hodgson, 1837) ... 321
Genus *Parus*
Brown-crested Tit, *P. dichrous* Blyth, 1844 ... 318
Crested Tit, *P. cristatus* Linné, 1758 ... 317
Plain Titmouse, *P. inornatus* Gambel, 1845 ... 318
Tufted Titmouse, *P. bicolor* Linné, 1766 ... 318
Bridled Titmouse, *P. wollweberi* (Bonaparte, 1850) ... 318
P. superciliosus (Przewalski, 1876) ... 318
Mountain Chickadee, *P. gambelli* Ridgway, 1886 ... 318
Chestnut-backed Chickadee, *P. rufescens* Townsend, 1837 ... 318
Boreal Chickadee, *P. hudsonicus* Forster, 1772 ... 318
Gray-headed Chickadee, *P. cinctus* Boddaert, 1783 ... 318
Sombre Tit, *P. lugubris* Temminck, 1820 ... 317
Willow Tit, *P. montanus* Conrad, 1827 ... 317
Black-capped Chickadee, *P. atricapillus* Linné, 1766 ... 317
Caroline Chickadee, *P. carolinensis* Audubon, 1834 ... 318
Gray-sided Chickadee, *P. sclateri* Kleinschmidt, 1897 ... 318
Marsh Tit, *P. palustris* Linné, 1758 ... 317
Dusky Tit, *P. funereus* (J. and E. Verreaux, 1855) ... 318
Southern Black Tit, *P. niger* (Vieillot, 1818) ... 318
Black Tit, *P. leucomelas* Rüppell, 1840 ... 318
P. rufiventris Bocage, 1877 ... —
Gray Tit, *P. griseiventris* Reichenow, 1882 ... 318
Cape Gray Tit, *P. afer* Gmelin, 1789 ... 318
Bukhara Great Tit, *P. bokharensis* Lichtenstein, 1823 ... 318
Great Tit, *P. major* Linné, 1758 ... 314
Green-backed Tit, *P. monticolus* Vigors, 1831 ... 318
Blue Tit, *P. caeruleus* Linné, 1758 ... 316
Azure Tit, *P. cyanus* Pallas, 1770 ... 318
Black-spotted Yellow Tit, *P. xanthogenys* Vigors, 1831 ... 318
P. venustulus Swinhoe, 1870 ... 318
P. rubidiventris Blyth, 1847 ... 318
Crested Black Tit, *P. melanolophus* Vigors, 1831 ... 318
Coal Tit, *P. ater* Linné, 1758 ... 316
Varied Tit, *P. varius* Temminck and Schlegel, 1848 ... 318

Family Nuthatches (Sittidae) ... 321
Genus *Sitta* ... 322
Eastern Rock Nuthatch, *S. tephronota* Sharpe, 1872 ... 323
Rock Nuthatch, *S. neumayer* Michahelles, 1830 ... 323
European Nuthatch, *S. europaea* Linné, 1758 ... 321
Chestnut-bellied Nuthatch, *S. europaea castanea* Lesson, 1830 (probably a full species) ... 322
Asiatic White-tailed Nuthatch, *S. himalayensis* Jardine and Selby, 1835 ... 323
Corsican Nuthatch, *S. whiteheadi* Sharpe, 1884 ... 323
Turkish Nuthatch, *S. krueperi* Pelzeln, 1863 ... 323
Yunnan Nuthatch, *S. yünnanensis* Ogilvie-Grant, 1900 ... 323
Korean Nuthatch, *S. villosa* Verreaux, 1865 ... 323
Kansu Nuthatch, *S. villosa bangsi* Stresemann, 1929 ... —
Red-breasted Nuthatch, *S. canadensis* Linné, 1766 ... 323
Giant Nuthatch, *S. magna* Ramsay, 1876 ... 323
Velvet-fronted Nuthatch, *S. frontalis* Swainson, 1820 ... 323
Azure Nuthatch, *S. azurea* Lesson, 1830 ... 323
Beautiful Nuthatch, *S. formosa* Blyth, 1843 ... 323
White-cheeked Nuthatch, *S. leucopsis* Gould, 1850 ... 323
White-breasted Nuthatch, *S. carolinensis* Latham, 1790 ... 323
Pygmy Nuthatch, *S. pygmaea* Vigors, 1839 ... —
Brown-headed Nuthatch, *S. pusilla* Latham, 1790 ... 323

Subfamily Wallcreepers (Tichodrominae) ... 323
Genus *Tichodroma* ... 323
Wall Creeper, *T. muraria* (Linné, 1766) ... 323

Family Australian Tree Creepers (Climacteridae) ... 324
Subfamily Tree Runners (Neosittinae) ... 324

Genus *Neositta*	325
Orange-crowned Sitella, *N. chrysoptera*	
(Latham, 1801)	325
Western subspecies, *N. chrysoptera pileata*	
(Gould, 1838)	319
N. papuensis (Schlegel, 1873)	—
Genus *Daphoenositta*	
D. miranda De Vis, 1897	—

Subfamily Australian Tree Creepers (Climacterinae)   324

Genus *Climacteris*	—
Red-browed Tree Creeper, *C. erythrops* Gould, 1841	—
Brown Tree Creeper, *C. picumnus* Temminck and Laugier, 1824	—
C. leucophaea (Latham, 1801)	—

Family Philippine Creepers (Rhabdornithidae)   325

Genus *Rhabdornis*	325
Striped-headed Creeper, *R. mystacalis* (Temminck, 1825)	325
Plain-headed Creeper, *R. inornatus* Ogilvie-Grant, 1896	325

Family Spotted Creeper (Salpornithidae)   325

Gray Genus *Salpornis*	326
Spotted Creeper, *S. spilonotus* (Franklin, 1831)	326

Family True Tree Creepers (Certhiidae)   326

Genus *Certhia*	326
Short-toed Tree Creeper, *C. brachydactyla* Brehm, 1820	326
Tree Creeper, *C. familiaris* Linné, 1758	326
Himalayan Tree Creeper, *C. himalayana* Vigors, 1832	326
Brown-throated Tree Creeper, *C. discolor* Blyth, 1845	326
Stoliczka's Tree Creeper, *C. nipalensis* Blyth, 1845	326

Family Flower Peckers (Dicaeidae)   329

Genus *Melanocharis*	330
Black Berry Pecker, *M. niger* (Lesson, 1830)	331
Genus *Rhamphocharis*	330
Spotted Berry Pecker, *R. crassirostris* Salvadori, 1876	330
Genus *Prionochilus*	331
Yellow-rumped Flower Pecker, *P. xanthopygius* Salvadori, 1868	331
Genus *Dicaeum*	331
Striped Flower Pecker, *D. agile* (Tickell, 1833)	331
Yellow-bellied Flower Pecker, *D. melanoxanthum* (Blyth, 1843)	329
Orange-breasted Flower Pecker, *D. trigonostigma* (Scopoli, 1786)	330
Tickel's Flower Pecker, *D. erythrorhynchos* (Latham, 1790)	331
Plain Flower Pecker, *D. concolor* Jerdon, 1840	331
Mistletoe Bird, *D. hirundinaceum* (Shaw and Nodder, 1792)	329
Scarlet-throated Black-sided Flower Pecker, *D. celebicum* St. Müller, 1843	—
Bronze-backed Flower Pecker, *D. ignipectus* (Blyth, 1843)	329
Scarlet-backed Flower Pecker, *D. cruentatum* (Linné, 1758)	330
Genus *Oreocharis* (systematic position uncertain)	330
Tit Berry Pecker, *O. arfaki* (Meyer, 1875)	330
Genus *Paramythia* (systematic position uncertain)	330
Crested Berry Pecker, *P. montium* De Vis, 1892	330
Genus *Pardalotus*	330
Spotted Pardolate, *P. punctatus* (Shaw and Nodder, 1792)	330

Family Sunbirds (Nectariniidae)   331

Genus *Anthreptes*	338
Subgenus *Lamprothreptes*	
A. (Lamprothreptes) longuemarei (Lesson, 1831)	—
Subgenus *Chalcopareia*	338
Ruby-cheeked Sunbird, *A. (Chalcopareia) singalensis* (Gmelin, 1789)	338
Subgenus *Anthreptes*	338
Plain-throated Sunbird, *A. (Anthreptes) malacensis* (Scopoli, 1786)	338
Subgenus *Anthodiaeta*	—
Collared Sunbird, *A. (Anthodiaeta) collaris* (Vieillot, 1819)	—
Subgenus *Hedydipna*	—
A. (Hedydipna) platurus (Vieillot, 1819)	—
Genus *Nectarinia*	337
Subgenus *Adelinus*	—
N. (Adelinus) olivacea (Smith, 1840)	—
Subgenus *Chalcomitra*	338
N. (Chalcomitra) amethystina (Shaw, 1811/12)	—
N. (Chalcomitra) senegalensis (Linné, 1766)	—
Subgenus *Hermotimia*	338
Small Sunbird, *N. (Hermotimia) minima* (Sybes, 1832)	338
van Hasselt's Sunbird, *N. (Hermotimia) sperata* (Linné, 1766)	—
Subgenus *Cyrtostomus*	—
N. (Cyrtostomus) jugularis (Linné, 1766)	—
Subgenus *Eucinnyris*	—
N. (Eucinnyris) venusta (Saw and Nodder, 1799)	—
N. (Eucinnyris) venusta albiventris Strickland, 1852	—
N. (Eucinnyris) talatala (A. Smith, 1836)	—
Subgenus *Eremicinnyris*	—
N. (Eremicinnyris) fusca (Vieillot, 1819)	—
Subgenus *Anthobaphes*	—

N. *(Anthobaphes) violacea* (Linné, 1766) —
Royal Sunbird, N. *(Anthobaphes) regia* (Reichenow, 1893) —
Subgenus *Cinnyris* ... 338
N. *(Cinnyris) osea* (Bonaparte, 1856) —
N. *(Cinnyris) coccinigastra* (Latham, 1801) ... 338
N. *(Cinnyris) superba* (Shaw, 1811/12) ... 338
Subgenus *Aidemonia* —
Golden-winged Sunbird, N. *(Aidemonia) reichenowi* (Fischer, 1884) —
Subgenus *Nectarinia* ... 338
Melachite Sunbird, N. *(Nectarinia) famosa* (Linné, 1766) ... 338
Genus *Aethopyga* ... 338
Lovely Sunbird, Ae. *shelleyi* Sharpe, 1876 —
Ae. *gouldiae* (Vigors, 1831) —
Yellow-backed Sunbird, Ae. *siparaja* (Raffles, 1822) ... 338
Genus *Arachnothera* ... 338
Greater Yellow-eared Spider Hunter, A. *chrysogenys* (Temminck, 1826) ... 339

Family White-eyes (Zosteropidae) ... 339
Genus *Zosterops* ... 339
Abyssian White-eye, Z. *abyssinica* Guerin-Meneville, 1843 ... 343
Yellow White-eye, Z. *senegalensis* Bonaparte, 1851 ... 340
Pale White-eye, Z. *pallida* Swainson, 1838 ... 343
Subspecies: Z. *pallida capensis* Sundevall, 1850 ... 340
Annobon White-eye, Z. *griseovirescens* Bocage, 1893 ... 339
Dark Madagascar White-eye, Z. *maderaspatana* (Linné, 1766) ... 339
Z. *mayottensis semiflava* Newton, 1867 —
Chestnut-flanked White-eye, Z. *erythropleura* Swinhoe, 1863 ... 343
Japanese White-eye, Z. *japonica* Temminck and Schlegel, 1847 ... 343
Oriental White-eye, Z. *palpebrosa* (Temminck, 1824) ... 340
Subspecies: Z. *palpebrosa nilgiriensis* Ticehurst, 1927 ... 339
Z. *everetti* Tweeddale, 1878 ... 344
Mountain White-eye, Z. *montana* Bonaparte, 1851 ... 343
Z. *wallacei* Finsch, 1901 ... 340
Z. *citrinella citrinella* Bonaparte, 1850 ... 339
New Guinea Mountain White-eye, Z. *novaeguinae* Salvadori, 1878 ... 339
Gray-breasted Silver-eye, Z. *lateralis* (Latham, 1801) ... 343
Tasmanian subspecies: Z. *lateralis lateralis* (Latham, 1801) ... 343
Southwestern Australian Subspecies: Z. *lateralis gouldi* Bonaparte, 1851 ... 343
Z. *strenua* Gould, 1855 ... 344
Z. *natalis* Lister, 1889 ... 339
Z. *tenuirostris* Gould, 1837 ... 344
Z. *albogularis* Gould, 1837 ... 344
Genus *Chlorocharis* ... 340
Mountain Blackeye, Ch. *emiliae* Sharpe, 1888 ... 340
Genus *Lophozosterops* ... 339
L. *dohertyi* Hartert, 1896 ... 339

Family Honey Eaters (Meliphagidae) ... 344
Subfamily Meliphaginae ... 344
Genus *Lichmera* ... 345
Brown Honey Eater, L. *indistincta* (Vigors and Horsfield, 1827) ... 345
Genus *Myzomela* —
M. *sanguinolenta* (Latham, 1801) —
M. *cardinalis* (Gmelin, 1788) —
Genus *Meliphaga* ... 345
Singing Honey Eater, M. *virescens* (Vieillot, 1817) ... 345
Mangrove Honey Eater, M. *fasciogularis* (Gould, 1854) ... 347
Yellow-faced Honey Eater, M. *chrysops* (Latham, 1801) ... 347
Yellow-plumed Honey Eater, M. *ornata* (Gould, 1838) ... 346
White-gaped Honey Eater, M. *unicolor* (Gould, 1843) ... 346
Genus *Apalopteron* ... 347
Apalopteron, A. *familiaris* (Kittlitz, 1831) ... 347
Genus *Melithreptes* ... 345
White-naped Honey Eater, M. *lunatus* (Vieillot, 1802) ... 320
Genus *Entomyzon* ... 346
Blue-faced Honey Eater, E. *cyanotis* (Latham, 1801) ... 346
Genus *Notiomystis* ... 320
N. *cincta* (DuBus, 1839) ... 320
Genus *Philemon* ... 345
Noisy Friarbird, Ph. *corniculatus* (Latham, 1790) —
Genus *Melidectes* ... 320
Cinnamon-breasted Wattle Bird, M. *torquatus* (Sclater, 1873) ... 320
Genus *Moho* ... 388
Ooaa, M. *braccatus* (Cassin, 1855) ... 388
+ M. *nobilis* (Merrem, 1786) —
Genus *Chaetoptila* —
+ Ch. *angustipluma* (Peal, 1848) —
Genus *Phylidonyris* ... 345
Crescent Honey Eater, Ph. *pyrrhoptera* (Latham, 1804) ... 320
Yellow-winged Honey Eater, Ph. *novaehollandiae* (Latham, 1790) ... 346
Yellow-tufted Honey Eater, Ph. *melanops* (Latham, 1801) ... 345
Ph. *undulata* (Sparrman, 1787) ... 320
Genus *Ramsayornis* ... 346
R. *fasciatus* (Gould, 1843) —

R. *modestus* (Gray, 1858)	—
Genus *Conopophila*	345
Pied Honey Eater, *C. picta* (Gould, 1838)	345
Genus *Xanthomyza*	320
X. phrygia (Shaw, 1794)	320
Genus *Acanthorhynchus*	345
Eastern Spinebill, *A. tenuirostris* (Latham, 1801)	346
Western Spinebill, *A. superciliosus* (Gould, 1801)	345
Genus *Manorina*	333
Bell Miner, *M. melanophrys* (Latham, 1801)	345
M. flavigula (Gould, 1840)	333
Genus *Anthochaera*	345
Little Wattle Bird, *A. chrysoptera* (Latham, 1801)	346
Genus *Prosthemadera*	—
Tui, *P. novaeseelandiae* (Gmelin, 1788)	—
Subfamily Sugarbirds (Promeropinae)	347
Genus *Promerops*	347
Cape Sugarbird, *P. cafer* (Linné, 1758)	347
Family Buntings and Related Birds (Emberizidae)	345
Subfamily Buntings (Emberizinae)	348
Tribe Emberizini	349
Genus True Buntings (*Emberiza*)	349
Corn Bunting, *E. calandra* Linné, 1758	349
Yellow Hammer, *E. citrinella* Linné, 1758	350
Pine Bunting, *E. leucocephala* Gmelin, 1771	349
Rock Bunting, *E. cia* Linné, 1766	341
Siberian Meadow Bunting, *E. cioides* Brandt, 1843	341
Ortolan Bunting, *E. hortulana* Linné, 1758	349
Cirl Bunting, *E. cirlus* Linné, 1766	341
Little Bunting, *E. pusilla* Pallas, 1776	—
Black-headed Bunting, *E. melanocephala* Scopoli, 1769	350
Red-headed Bunting, *E. bruniceps* Brandt, 1841	350
Reed Bunting, *E. schoeniclus* (Linné, 1758)	350
Bullfinch-billed Reed Bunting, *E. schoeniclus pyrrhuloides* Pallas, 1811	—
Genus Striped Buntings (*Fringillaria*)	350
Cinnamon-breasted Rock Bunting, *F. tahapisi* Smith, 1836	350
Pale Rock Bunting, *F. impetuani* (Smith, 1836)	350
Cape Bunting, *F. capensis* (Linné, 1766)	350
House Bunting, *F. striolata* (Lichtenstein, 1823)	350
Genus Crested Buntings (*Melophus*)	350
Crested Bunting, *M. lathami* (Gray, 1831)	350
Genus Snow Buntings (*Plectrophenax*)	353
Snow Bunting, *P. nivalis* (Linné, 1758)	354
McKay's Bunting, *P. hyperboreus* Ridgway, 1884	354
Genus Longspurs (*Calcarius*)	353
Lapland Longspurs, *C. lapponicus* (Linné, 1758)	355
Smith's Longspur, *C. pictus* (Swainson, 1832)	—
Chestnut-collared Longspur, *C. ornatus* (Townsend, 1837)	—
Genus *Rhyncophanes*	—
McCown's Longspur, *R. mccownii* (Lawrence, 1851)	—
Genus *Poospiza*	342
Ringed Warbling Finch, *P. torquata* (Lafresnaye and d'Orbigny, 1837)	342
Cinereous Warbling Finch, *P. cinerea* Bonaparte, 1850	—
Genus Crowned "Sparrows" (*Zonotrichia*)	353
White-throated Sparrow, *Z. albicollis* (Gmelin, 1789)	353
Rufous-collared Sparrow, *Z. capensis* (Müller, 1776)	353
White-crowned Sparrow, *Z. leucophrys* (Forster, 1772)	—
Golden-crowned Sparrow, *Z. atricapilla* (Gmelin, 1789)	—
Harris' Sparrow, *Z. querula* (Nuttall, 1840)	—
Genus Juncos (*Junco*)	353
Slate-colored Junco, *J. hyemalis* (Linné, 1758)	353
Oregon Junco, *J. oreganus* (Townsend, 1837)	353
Genus Savannah Sparrows (*Passerculus*)	353
Savannah Sparrow, *P. sandwichensis* (Gmelin, 1789)	353
Ipswich Sparrow, *P. princeps* Maynard, 1782	353
Genus Song Sparrows (*Melospiza*)	353
Song Sparrow, *M. melodia* (Wilson, 1810)	353
Genus Lark Sparrows (*Chondestes*)	—
Lark Sparrow, *Ch. grammacus* (Say, 1823)	—
Genus *Aimophila*	—
Rufous-crowned Sparrow, *Ai. ruficeps* (Cassin, 1852)	—
Genus Sharp-tailed Sparrows (*Amnospiza*)	354
Seaside Sparrow, *A. maritima* (Wilson, 1811)	354
Genus Chipping Sparrows (*Spizella*)	—
Tree Sparrow, *S. arborea* (Wilson, 1810)	—
Chipping Sparrow, *S. passerina* (Bechstein, 1789)	—
Genus *Arremon*	342
Orange-billed Sparrow, *A. aurantiirostris* Lafresnaye, 1847	342
Genus *Calamospiza*	—
Lark Bunting, *C. melanocorys* Stejneger, 1885	—
Genus Towhees (*Pipilo*)	342
Rufous-sided Towhee, *P. erythrophthalmus* (Linné, 1758)	342
Genus Brush Finches (*Atlapetes*)	354
Chestnut-capped Brush Finch, *A. brunneinucha* (Lafresnaye, 1839)	354
Genus Crested Finches (*Coryphospingus*)	354
Red-crested Finch, *C. cucullatus* (Müller, 1776)	357
Genus *Melanodera*	354
Black-throated Finch, *M. melanodera* (Quoy	

and Gaimard, 1824) — 354
Genus Tristan Buntings *(Nesospiza)* — 354
    Cunha Bunting, *N. acunhae* Cabanis, 1873 — 354
Genus Gough Island Buntings, *(Rowettia)* — 357
    Gough Island Bunting, *R. goughensis* Eagle Clarke, 1904 — 357
Genus *Diuca* — —
    Diuca Finch, *D. diuca* (Molina, 1782) — —

Tribe Grassquits (Tiaridini) — 357
Genus Yellow Finches *(Sicalis)* — 357
    Saffron Finch, *S. flaveola* (Linné, 1766) — 357
Genus Seed Eaters *(Sporophila)* — 357
    Variable Seed Eater, *S. americana* (Gmelin, 1789) — 357
    Ruddy-breasted Seed Eater, *S. minuta* (Linné, 1758) — 342
Genus Black Finches *(Melanospiza)* — 358
    St. Lucia Black Finch, *M. richardsoni* (Cory, 1886) — 358
Genus *Volatinia* — 358
    Blue-black Grassquit, *V. jacarina* (Linné, 1766) — 358
Genus Tiaris Grassquits *(Tiaris)* — 358
    Yellow-faced Grassquit, *T. olivacea* (Linné, 1766) — 358
    Cuban Grassquit, *T. canora* (Gmelin, 1789) — 358

Tribe Darwin's Finches (Geospizini) — 358
Genus Ground Finches *(Geospiza)* — 359
    Large Ground Finch, *G. magnirostris* Gould, 1837 — 359
    Medium Ground Finch, *G. fortis* Gould, 1837 — 363
    Small Ground Finch, *G. fuliginosa* Gould, 1837 — 359
    Sharp-beaked Ground Finch, *G. difficilis* Sharpe, 1888 — 359
    Small Sharp-beaked Ground Finch, *G. difficilis acutirostris* Ridgway, 1894 — 364
    Medium Sharp-beaked Ground Finch, *G. difficilis debilirostris* Ridgway, 1894 — 366
    Large Sharp-beaked Ground Finch, *G. difficilis septentrionalis* Rothschild and Hartert, 1931 — 364
    Cullpepper Sharp-beaked Ground Finch, *G. difficilis nigrescens* Swarth, 1931 — 366
    Cactus Ground Finch, *G. scandens* (Gould, 1837) — 359
    G. conirostris Ridgway, 1890 — —
Genus Vegetarian Tree Finches *(Platyspiza)* — 359
    Vegetarian Tree Finch, *P. crassirostris* (Gould, 1837) — 359
Genus Insectivorous Tree Finches *(Camarhynchus)* — 360
    Small Insectivorous Tree Finch, *C. parvulus* (Gould, 1837) — 360
    Medium Insectivorous Tree Finch, *C. pauper* (Ridgway, 1890) — 365
    Large Insectivorous Tree Finch, *C. psittacula* Gould, 1837 — 360
Genus Woodpecker Finches *(Cactospiza)* — 360
    Woodpecker Finch, *C. pallida* (Sclater and Salvin, 1870) — 360
    Mangrove Finch, *C. heliobates* (Snodgrass and Heller, 1901) — 360
Genus Warbler Finches *(Certhidea)* — 360
    Warbler Finch, *C. olivacea* Gould, 1837 — 360
Genus Cocos Finches *(Pinaroloxias)* — 360
    Cocos Finch, *P. inornata* (Gould, 1843) — 360

Subfamily Cardinals (Cardinalinae) — 368
Genus *Cardinalis* — 366
    Red Cardinal, *C. cardinalis* (Linné, 1758) — 366
Genus Pyrrhuloxias *(Pyrrhuloxia)* — 366
    Pyrrhuloxia, *P. sinuata* (Bonaparte, 1838) — 366
Genus Paroria Cardinals *(Paroaria)* — 367
    Red-crested Cardinal, *P. coronata* (Müller, 1776) — 367
    Red-cowled Cardinal, *P. dominicana* (Linné, 1758) — 367
Genus Yellow Cardinals *(Gubernatrix)* — 367
    Yellow Cardinal, *G. cristata* (Vieillot, 1817) — 367
Genus *Pheucticus* — 367
    Yellow Grosbeak, *Ph. ohrysopeplus* (Vigors, 1832) — —
    Rose-breasted Grosbeak, *Ph. ludovicianus* (Linné, 1766) — 367
    Black-headed Grosbeak, *Ph. melanocephalus* (Swainson, 1827) — 367
Genus *Guiraca* — 367
    Blue Grosbeak, *G. caerulea* (Linné, 1758) — 367
Genus *Saltator* — 352
    Black-cheeked Ant Tanager, *S. atriceps* (Lesson, 1832) — 352
Genus Painted Buntings *(Passerina)* — 367
    Indigo Bunting, *P. cyanea* (Linné, 1766) — 367
    Lazuli Bunting, *P. amoena* (Say, 1823) — 367
    Painted Bunting, *P. ciris* (Linné, 1758) — 367
    Varied Bunting, *P. versicolor* (Bonaparte, 1837) — 367
    Orange-breasted Bunting, *P. leclancherii* Lafresnaye, 1840 — 367

Subfamily Tanagers (Thraupinae) — 370
Tribe True Tanagers (Thraupini) — 370
Genus Chlorophonias *(Chlorophonia)* — 370
    Ch. occipitalis (DuBus, 1847) — —
    Chestnut-breasted Chlorophonia, *Ch. pyrrhophrys* (Sclater, 1851) — 361
Genus Euphonias *(Euphonia)* — 370
    Eu. musica (Gmelin, 1789) — —
Genus *Chlorospingus* — 376
    Glistening Green Tanager, *Ch. phoenicotis* (Bonaparte, 1851) — 335
Genus *Tangara* — 373
    Paradise Tanager, *T. chilensis* (Vigors, 1832) — 373

T. chilensis coelicolor (Sclater, 1851)	—
Seven-colored Tanager, *T. fastuosa* (Lesson, 1831)	—
Green-headed Tanager, *T. seledon* (Müller, 1776)	373
T. chrysophris (Sclater, 1851)	374
Flame-faced Tanager, *T. parzudakii* (Lafresnaye, 1843)	361
Masked Tanager, *T. nigrocincta* (Bonaparte, 1838)	—
Plain-colored Tanager, *T. inornata* (Gould, 1855)	373
Genus *Stephanophorus*	361
Diademed Tanager, *S. diadematus* (Temminck, 1823)	361
Genus *Anisognathus*	335
Scarlet-bellied Mountain Tanager, *A. igniventris* (Lafresnaye and d'Orbigny, 1837)	377
Genus *Compsocoma*	—
Blue-winged Mountain Tanager, *C. flacinucha* (Lafresnaye and d'Orbigny, 1837)	—
Genus *Thraupis*	374
Blue-gray Tanager, *T. episcopus* (Linné, 1766)	374
T. virens (Linné, 1766)	—
Sayaca Tanager, *T. sayaca* (Linné, 1766)	374
Palm Tanager, *T. palmarum* (Wied, 1821)	374
Blue and Yellow Tanager, *T. bonariensis* (Gmelin, 1789)	375
Genus *Ramphocelus*	—
R. bresilius (Linné, 1766)	—
Silver-beaked Tanager, *R. carbo* (Pallas, 1764)	375
Scarlet-rumped Tanager, *R. passerinii* Bonaparte, 1831	375
Yellow-rumped Tanager, *R. passerini* Bonaparte, 1838	—
Genus *Phlogothraupis*	361
Crimson-collared Tanager, *Ph. sanguinolenta* (Lesson, 1831)	361
Genus *Piranga*	—
Summer Tanager, *P. rubra* (Linné, 1758)	375
Scarlet Tanager, *P. olivacea* (Gmelin, 1789)	375
Genus Ant Tanagers (*Habia*)	376
Red-crowned Ant Tanager, *H. rubica* (Vieillot, 1817)	376
Genus *Tachyphonus*	361
Flame-crested Tanager, *T. cristatus* (Linné, 1766)	361
Genus *Eucometis*	376
Gray-headed Tanager, *Eu. penicillata* (Spix, 1825)	376
Genus *Calyptophilus*	370
C. frugiforus (Cory, 1883)	370
Genus *Cholorspingus*	376
Common Bush Tanager, *Ch. ophthalmicus* (DuBus, 1847)	376
Genus *Chlorornis*	377
Grass Green Tanager, *Ch. riefferii* (Boissoneau, 1840)	377
Genus *Rhodinocichla*	370
Rose-breasted Thrush Tanager, *R. rosea* (Lesson, 1832)	370
Genus *Cissopis*	377
Magpie Tanager, *C. leveriana* (Gmelin, 1788)	377
Tribe Swallow Tanagers (Tersinini)	370
Genus *Tersina*	377
Swallow Tanager, *T. viridis* (Illiger, 1811)	377
Tribe Plush-capped Finches (Catamblyrhynchini)	378
Genus Plush-capped Finches (*Catamblyrhynchus*)	378
Plush-capped Finch, *C. diadema* Lafresnaye, 1842	378
Tribe Dacnis (Dacnidini)	378
Genus *Diglossa*	378
Slaty Flower Piercer, *D. baritula* Wagler, 1832	380
Genus *Euneornis*	378
Orangequit, *Eu. campestris* (Linné, 1758)	378
Genus *Cyanerpes*	378
Red-legged Honey Creeper, *C. cyaneus* (Linné, 1766)	378
Blue Honey Creeper, *C. caeruleus* (Linné, 1758)	335
Yellow-footed Shining Honey Creeper, *C. lucicus* (Sclater and Salvin, 1859)	379
Genus *Chlorophanes*	378
Green Honey Creeper, *Ch. spiza* (Linné, 1758)	378
Genus *Dacnis*	378
Black-throated Dacnis, *D. cayana* (Linné, 1766)	379
Family Wood Warblers (Parulidae)	348
Subfamily Wood Warblers (Parulinae)	383
Genus Wood Warblers (*Protonotaria*)	380
Prothonotary Warbler, *P. citrea* (Boddaert, 1783)	383
Genus *Mniotilta*	380
Black and White Warbler, *M. varia* (Linné, 1766)	371
Genus *Vermivora*	380
Golden-winged Warbler, *V. chrysoptera* (Linné, 1766)	384
Blue-winged Warbler, *V. pinus* (Linné, 1766)	383
Genus *Dendroica*	380
Yellow Warbler, *D. petechia* (Linné, 1766)	384
Myrtle Warbler, *D. coronata* (Linné, 1766)	383
Black-throated Blue Warbler, *D. caerulescens* (Gmelin, 1789)	371
Magnolia Warbler, *D. magnolia* (Wilson, 1811)	371
Genus *Seiurus*	380
Northern Water Thrush, *S. aurocapillus* (Linné, 1766)	384

Louisiana Water Thrush, *S. motacilla* (Vieillot, 1808)	383
Genus *Wilsonia*	380
Hooded Warbler, *W. citrina* (Boddaert, 1783)	385
Genus *Setophaga*	380
American Redstart, *S. ruticilla* (Linné, 1758)	—
Genus *Ergaticus*	380
Red Warbler, *E. ruber* (Swainson, 1827)	385
Genus *Basileuterus*	380
Golden-crowned Warbler, *B. culicivorus* (Deppe, 1830)	—
Genus *Myioborus*	380
Collared Redstart, *M. torquatus* (Baird, 1865)	385
Painted Redstart, *M. pictus* (Swainson, 1829)	385
Genus *Peucedramus*	—
Olive Warbler, *P. taeniatus* (DuBus, 1847)	—
Genus *Granatellus*	380
Rose-breasted Chat, *G. pelzelni* Sclater, 1865	385
Genus *Icteria*	380
Yellow-breasted Chat, *I. virens* (Linné, 1758)	383

Subfamily Conebills and Banaquits (Coerebinae) 385
Genus *Conirostrum* 385
Bicolored Conebill, *C. bicolor* (Vieillot, 1808) 385
Genus *Coereba* 386
Bananaquit, *C. flaveola* (Linné, 1758) 386

Family Wren Thrushes (Zeledoniidae) 386
Genus *Zeledonia* 386
Wren Thrush, *Z. coronata* Ridgway, 1889 386

Family Honey Creepers (Drepanididae) 386
Subfamily Parrot Bills (Psittirostrinae) 387
Genus *Viridonia* 387
Amakihi, *V. virens* (Gmelin, 1788) 387
Large amakihi, *V. sagittirostris* Rothschild, 1892 387
Genus Hawaiian Creepers *(Paroreomyzos)* (or *Loxops*) 387
Hawaiian Creeper, *P.* (or *Loxops*) *maculata* (Cabanis, 1851) 387
Genus Akepas *(Loxops)* 387
Akepa, *L. coccinea* (Gmelin, 1789) 387
Genus *Hemignathus* 387
Akialoa, *H. obscurus* (Gmelin, 1788) 387
Nukupuu, *H. lucidus* (Lichtenstein, 1839) 389
Akiapolaau, *H. wilsoni* (Rothschild, 1893) 387
Genus *Pseudonestor* 387
Maui Parrotbill, *P. xanthophrys* Rothschild, 1893 387
Genus *Psittirostra* 387
Ou, *P. psittacea* (Gmelin, 1789) 387
Laysan Finch, *P. cantans* (Wilson, 1890) 387
Nihoan Laysan Finch, *P. cantans ultima* (Bryan, 1917) 390
Greater Koa Finch, + *P. palmeri* (Rothschild, 1892) 388
Lesser Koa Finch, + *P. flaviceps* (Rothschild, 1892) 394
Kona Finch, + *P. kona* (Wilson, 1888) 388

Subfamily Black and Red Honey Creepers (Drepanidinae) 390
Genus *Himatione* 390
Apapane, *H. sanguinea* (Gmelin, 1788) 390
Laysan Apapane, + *H. sanguinolenta freethii* Rothschild, 1892 390
Genus *Palmeria* 372
Crested Honey Creeper, *P. dolei* (Wilson, 1891) 372
Genus *Ciridops* —
Ula-Ai-Hawane, + *C. anna* (Dole, 1879) —
Genus *Vestiaria* 390
Iiwi, *V. coccinea* (Forster, 1780) 390
Genus *Drepanis* 390
Mamo, + *D. pacifica* (Gmelin, 1788) 390
Black Mamo, + *D. funerea* Newton, 1893 390

Family Vireos (Vireonidae) 395
Subfamily Vireos (Vireoninae) 395
Genus *Vireo* 395
Yellow-throated Vireo, *V. flavifrons* Vieillot, 1808 372
Red-eyed Vireo, *V. olivaceus* (Linné, 1766) 395
Warbling Vireo, *V. gilvus* (Vieillot, 1808) 395
Genus *Hylophilus* 395
Gray-headed Greenlet, *H. decurtatus* Bonaparte, 1838 395

Subfamily Shrike Vireos (Vireolaniinae) 348
Genus *Vireolanius* 395
V. melitophrys Bonaparte, 1850 395
Chestnut-sided Shrike Vireo, *V. pulchellus* Sclater and Salvin, 1859 372

Subfamily Pepper Shrikes (Cyclarhinae) 395
Genus *Cyclarhis* 395
Rufous-browed Pepper Shrike, *C. gujanensis* (Gmelin, 1789) 396

Family Icterids (Icteridae) 396
Genus *Psarocolius* 396
Crested Oropendola, *P. decumanus* (Pallas, 1769) 398
Green Oropendola, *P. viridis* (Müller, 1776) 381
Russet-backed Oropendola, *P. angustifrons* (Spix, 1824) 381
Wagler's Oropendola, *P. wagleri* (Gray and Mitchell, 1844) 381
Genus *Gymnostinops* 396
Montezuma Oropendola, *G. montezuma* (Lesson, 1830) 381
Genus *Cacicus* 396
Yellow-rumped Cacique, *C. cela* (Linné, 1758) 398

Genus *Scaphidura*	396
Giant Cowbird, *S. oryzivora* (Gmelin, 1788)	396
Genus Cowbirds *(Molothrus)*	396
Bay-winged Cowbird, *M. badius* (Vieillot, 1819)	398
Shining Cowbird, *M. bonariensis* (Gmelin, 1789)	399
Brown-headed Cowbird, *M. ater* (Boddaert, 1783)	399
Genus Grackels *(Quiscalus)*	399
Common Grackle, *Qu. quiscula* (Linné, 1758)	400
Carib Grackle, *Qu. lugubris* (Swainson, 1838)	400
Genus *Cassidix*	381
Boat-tailed Grackle, *C. mexicanus* (Gmelin, 1788)	381
Genus American Orioles *(Icterus)*	400
Baltimore Oriole, *I. galbula* (Linné, 1758)	400
Bullock's Oriole, *I. bullockii* (Swainson, 1827)	400
Spotted-breasted Oriole, *I. pectoralis* (Wagler, 1829)	381
Yellow-backed Oriole, *I. chrysater* Lesson, 1844	400
Orange-crowned Oriole, *I. auricapillus* Cassin, 1847	400
Black Troupial, *I. icterus* (Linné, 1766)	400
Yellow Oriole, *I. nigrogularis* (Hahn, 1819)	—
Genus American Blackbirds *(Agelaius)*	397
Red-winged Blackbird, *A. phoeniceus* (Linné, 1766)	400
Yellow-hooded Blackbird, *A. icterocephalus* Linné, 1766	401
Yellow-winged Blackbird, *A. thilius* (Molina, 1782)	401
Genus *Pezites*	
P. militaris (Linné, 1771)	—
Genus Meadowlarks *(Sturnella)*	397
Eastern Meadowlark, *S. magna* (Linné, 1758)	401
Western Meadowlark, *S. neglecta* Audubon, 1844	401
Genus *Dolichonyx*	397
Bobolink, *D. oryzivorus* (Linné, 1758)	397

Family Finches (Fringillidae) 402
Subfamily Chaffinches (Fringillinae) 402
Genus *Fringilla* 402
Chaffinch, *F. coelebs* Linné, 1758 402
Canarian Caffinch, *F. teydea* Webb, Berthelot and Moquin-Tandon, 1841 402
Brambling, *F. montifringilla* Linné, 1758 402

Subfamily Other Finches (Carduelinae) 404
Genus *Serinus* 407
Serin, *S. serinus* (Linné, 1766) 407
Canary, *S. canaria* (Linné, 1758) 407
Citril Finch, *S. citrinellus* (Pallas, 1764) 407
White-rumped Seed Eater, *S. leucopygius* (Sundevall, 1850) 407
Yellow-fronted Canary, *S. mozambicus*

(Müller, 1776)	407
S. mozambicus caniceps (d'Orbigny, 1839)	407
Yellow Canary, *S. flaviventris* (Swainson, 1828)	407
Brown-rumped Seed Eater, *S. tristriatus* Rüppell, 1840	382
Genus *Carduelis*	409
Greenfinch, *C. chloris* (Linné, 1758)	409
Siskin, *C. spinus* (Linné, 1758)	409
Pine Siskin, *C. pinus* (Wilson, 1810)	409
Red Siskin, *C. cucullata* (Swainson, 1820)	409
Lesser Goldfinch, *C. psaltria* (Say, 1823)	409
Lawrence's Goldfinch, *C. lawrencei* Cassin, 1852	409
American Goldfinch, *C. tristis* (Linné, 1758)	409
Goldfinch, *C. carduelis* (Linné, 1758)	409
Black-headed forms, *C. carduelis carduelis* (Linné, 1758)	—
Gray-headed forms, *C. carduelis caniceps* Vigors, 1831	—
Genus *Acanthis*	411
Common Redpoll, *A. flammea* (Linné, 1758)	411
Hoary Redpoll, *A. hornemanni* (Holboll, 1843)	—
Twite, *A. flavirostris* (Linné, 1758)	411
Linnet, *A. cannabina* (Linné, 1758)	411
Yemen Linnet, *A. yemenensis* (Ogilvie-Grant, 1913)	—
Warsanglia Linnet, *A. johannis* (Clarke, 1919)	—
Genus *Leucosticte*	412
Gray-crowned Rosy Finch, *L. arctoa* (Pallas, 1811)	412
Genus *Rhodopechys*	412
Trumpeter Bullfinch, *Rh. githaginea* (Lichtenstein, 1823)	412
Genus *Uragus*	
Long-billed Rose Finch, *U. sibiricus* (Pallas, 1773)	—
Japanese subspecies, *U. sibiricus sanguinolentus* (Temminck and Schlegel, 1848)	—
Genus *Carpodacus*	417
Scarlet Grosbeak, *C. erythrinus* (Pallas, 1770)	417
Purple Finch, *C. purpureus* (Gmelin, 1789)	417
Cassin's Finch, *C. cassinii* (Baird, 1854)	417
House Finch, *C. mexicanus* (Müller, 1776)	417
Red-breasted Rose Finch, *C. puniceus* (Blyth, 1845)	417
Genus *Chaunoproctus*	404
Bonin Finch, + *Ch. ferreorostris* (Vigors, 1828-29)	418
Genus *Pinicola*	404
Pine Grosbeak, *P. enucleator* (Linné, 1758)	418
Genus *Loxia*	418
Parrot Crossbill, *L. pytyopsittacus* Borkhausen, 1793	418
Red Crossbill, *L. curvirostra* Linné, 1758	418
White-winged Crossbill, *L. leucoptera* Gmelin, 1789	418

Genus *Pyrrhula*	418
Chinese Bullfinch, *P. erythaca* Blyth, 1962	392
Bullfinch, *P. pyrrhula* (Linné, 1758)	419
Azorean Bullfinch, *P. pyrrhula murina* Godman, 1866	392
Japanese Bullfinch, *P. pyrrhula griseiventris* Lafresnaye, 1841	392
Genus *Cocothraustes*	404
Hawfinch, *C. cocothraustes* (Linné, 1758)	419
Genus *Hesperiphona*	404
Evening Grosbeak, *H. vespertina* Cooper, 1825	419
Genus *Eophonia*	370
Migratory Chinese Grosbeak, *E. migratoria* Hartert, 1903	—
Japanese Grosbeak, *E. personata* Temminck and Schlegel, 1848	—
Genus *Mycerobas*	404
Black and Yellow Grosbeak, *M. icterioides* (Vigors, 1831)	419

Family Weavers and Weaver Finchs (Ploceidae) 421
Subfamily Sparrows (Passerinae) —

Genus *Carpospiza*	423
C. brachydactyla (Bonaparte, 1850)	423
Genus *Petronia*	423
Subgenus *Petronia*	423
Rock Sparrow, *P. (Petronia) petronia* (Linné, 1766)	423
Subgenus *Gymnoris*	423
P. (Gymnoris) dentata (Sundevall, 1850)	—
Yellow-thraoted Sparrow, *P. (Gymnoris) superciliaris* (Blyth, 1845)	423
Yellow-spotted Petronia, *P. (Gymnoris) pyrgita* (Heuglin, 1862)	423
P. (Gymnoris) xanthocollis (Burton, 1838)	423
Genus *Montifringilla*	423
Snow Finch, *M. nivalis* (Linné, 1766)	423
Tibet Snow Finch, *M. adamsi* Adams, 1858	423
Mandell's Snow Finch, *M. taczanowskii* Przewalski, 1876	424
M. davidiana (Verreaux, 1871)	—
M. ruficollis Blanford, 1871	—
Blanford's Snow Finch, *M. blanfordi* Hume, 1876	424
M. theresae Meinterzhagen, 1937	424
Genus *Passer*	424
Russet Sparrow, *P. rutilans* (Gould, 1836)	424
Tree Sparrow, *P. montanus* (Linné, 1758)	424
Desert Sparrow, *P. simplex* (Lichtenstein, 1823)	424
House Sparrow, *P. domesticus* (Linné, 1758)	424
Spanish Sparrow, *P. domesticus hispaniolensis* (Temminck, 1830)	425
Italian Sparrow, *P. domesticus italiae* (Vieillot, 1817)	425
Indian Sparrow, *P. domesticus indicus* Jardine and Selby, 1864	425
Dead Sea Sparrow, *P. moabiticus* Tristram, 1864	424
P. moabiticus moabiticus Tristram, 1864	—
P. moabiticus yatii Sharpe, 1888	—
Pegu House Sparrow, *P. flaveolus* Blyth, 1844	424
Rufous Sparrow, *P. motitensis* A. Smith, 1836	424
Cape Sparrow, *P. melanurus* (Müller, 1776)	424
Gray-headed Sparrow, *P. griseus* (Vieillot, 1817)	424
Golden Sparrow, *P. luteus* (Lichtenstein, 1823)	425
P. luteus luteus (Lichtenstein, 1823)	—
P. luteus euchlorus (Bonaparte, 1850)	—
Chestnut Sparrow, *P. eminibey* (Hartlaub, 1880)	425

Subfamily Parasitic Weaver (Anomalospizinae) 427
Genus *Anomalospiza* 427
Parasitic Weaver, *A. imberbis* (Cabanis, 1868) 427

Subfamily Viduinae 427

Genus *Steganura*	427
Paradise Whydah, *S. paradisaea* (Linné, 1758)	427
S. orientalis (Heuglin, 1871)	—
S. orientalis orientalis (Hueglin, 1871)	—
S. orientalis aucupum Neumann, 1908	—
S. interjecta Grote, 1922	—
Togo Paradise Whydah, *S. togoensis* Grot, 1923	426
S. obtusa Chapin, 1922	429
Genus *Hypochera*	428
Cameroon's Indigo Bird, *H. chalybeata* (Müller, 1776)	428
H. chalybeata amauropteryx Sharpe, 1890	—
H. funerea (De Tarragon, 1847)	—
H. wilsoni Hartert, 1901	—
H. camerunensis Grote, 1922	—
H. nigeriae Alexander, 1908	—
H. nigerrima Sharpe, 1871	429
Genus *Tetraenura*	428
Shaft-tailed Whydah, *T. regia* (Linné, 1766)	428
Fischer's Whyday, *T. fischeri* (Reichenow, 1882)	428
Genus *Vidua*	428
Shiny Black Whydah, *V. hypocherina* J. and E. Verreaux, 1856	428
Pin-tailed Whydah, *V. macroura* (Pallas, 1764)	428

Subfamily Euplectinae 433

Genus *Coliuspasser*	433
Red-collared Whydah, *C. ardens* (Boddaert, 1783)	433
C. jacksoni (Sharpe, 1891)	434
C. hartlaubi (Bocage, 1878)	—
Long-tailed Widow Bird, *C. progne* (Boddaert, 1783)	434
Fan-tailed Widow Bird, *C. axillaris* (A.	

Smith, 1838)	434
C. albonotatus (Cassin, 1848)	—
C. marcourus (Gmelin, 1789)	—
Yellow Bishop, C. capensis (Linné, 1766)	434
Genus Euplectes	435
Golden-backed Weaver, Eu. aureus (Gmelin, 1789)	435
Eu. gierowi Cabanis, 1886	—
Fire-crowned Bishop, Eu. hordeaceus (Linné, 1758)	435
Eu. nigroventris Cassin, 1848	—
Grenadier Weaver, Eu. orix (Linné, 1758)	435
Orange Bishop, Eu. franciscanus (Isert, 1789)	—
Genus Taha	433
Fire-fronted Bishop, T. diademata (Fischer and Richenow, 1878)	436
Yellow-crowned Bishop, T. afra (Gmelin, 1789)	—
T. afra afra (A. Smith, 1836)	436
Genus Brachycope	433
B. anomala (Reichenow, 1887)	436
Genus Quelea	433
Red-billed Dioch, Qu. quelea (Linné, 1758)	436
Genus Queleopsis	433
Red-headed Quelea, Qu. erythrops (Hartlaub, 1848)	437
Qu. cardinalis (Hartlaub, 1880)	436
Genus Foudia	437
Madagascar Weaver, F. madagasariensis (Linné, 1766)	437
Comoro Weaver, F. eminentissima Bonaparte, 1851	437
F. rubra (Gmelin, 1789)	—
Reunion Weaver, +F. bruante (Müller, 1776)	437
F. sechellarum Newton, 1867	—
Rodriguez Weaver, F. flavicans Newton, 1865	437
Subfamily Weavers proper (Ploceinae)	437
Genus Ploceus	438
Baya Weaver, P. philippinus (Linné, 1766)	438
Manyar Weaver, P. manyar (Horsfield, 1821)	438
P. benghalensis (Linné, 1758)	—
Genus Ploceella Oates, 1873	438
P. megarhyncha (Hune, 1869)	—
Golden Weaver, P. hypoxantha (Sparrman, 1788)	438
Genus Othyphantes	438
Baglafecht Weaver, O. baglafecht (Daudin, 1802)	438
O. nigrimentum (Reichenow, 1904)	—
Genus Textor	438
Subgenus Oriolonops	441
Holub's Golden Weaver, T. (Oriolonops) xanthops (Hartlaub, 1862)	441
Subgenus Euploceus	441
Cape Weaver, T. (Euploceus) capensis (Linné, 1766)	441
T. (Euploceus) temporalis (Bocage, 1880)	—
Subgenus Xanthophilus	415
T. (Xanthophilus) subaureus (A. Smith, 1839)	415
Subgenus Microploceus	441
Black-fronted Weaver, T. (Microploceus) velatus (Vieillot, 1819)	441
T. (Microploceus) vitellinus (Lichtenstein, 1823)	—
Rüppell's Weaver, T. (Microploceus) galbula (Rüppell, 1840)	441
Subgenus Hyphantornis	441
Jackson's Weaver, T. (Hyphantornis) jacksoni (Shelley, 1888)	441
Black-headed Weaver, T. (Hyphantornis) melanocephala (Linné, 1758)	441
Giant Weaver, T. (Hyphantornis) grandis (Gray, 1844)	441
Subgenus Textor	441
Black-headed Village Weaver, T. (Textor) cucullatus (Müller, 1776)	441
T. (Textor) cucullatus dilutescens (Clancey, 1831)	—
T. (Textor) cucullatus spilonotus (Vigors, 1831)	—
Subgenus Eremiplectes	
T. (Eremiplectes) rubiginosus (Rüppell, 1840)	—
Subgenus Melanopteryx	
T. (Melanopteryx) nigerrimus (Vieillot, 1819)	—
T. (Melanopteryx) nigerrimus nigerrimus (Vieillot, 1819)	—
T. (Melanopteryx) nigerrimus castaneofuscus (Lesson, 1840)	—
Genus Malimbus	438
M. erythrogaster Reichenow, 1893	—
Crested Malimbe, M. malimbicus (Daudin, 1802)	438
M. nitens (J. E. Gray, 1831)	—
M. coronatus (Sharpe, 1906)	—
M. cassini (Elliot, 1859)	—
M. rubricollis (Swainson, 1838)	—
Genus Anaplectes	438
Red-headed Weaver, A. rubriceps (Sundevall, 1850)	442
Genus Thomasophantes	442
St. Thomas Weaver, T. sanctithomae (Hartlaub, 1848)	442
Genus Notiospiza	442
Bar-winged Weaver, N. angolensis (Bocage, 1878)	442
Genus Phormoplectes	442
Subgenus Phormoplectes	442
Ph. (Phormoplectes) preussi (Reichenow, 1892)	—
Brown-caped Weaver, Ph. (Phormoplectes) insignis (Sharpe, 1891)	442
Subgenus Melanopleucus	
Ph. (Melanopleucus) fuscocastaneus (Bocage, 1880)	—
Genus Symplectes	442
S. olivaceiceps, Reichenow, 1889	—
Forest Weaver, S. bicolor (Vieillot, 1819)	442
Genus Sitagroides	—

S. alienus (Sharpe, 1902)	—
Genus Hyphanturgus	430
H. melanogaster (Shelley, 1887)	430
H. nigricollis (Vieillot, 1805)	—
H. ocularis (A. Smith, 1839)	—
Genus Sitagra	430
S. intermedia (Rüppell, 1845)	—
S. luteola (Lichtenstein, 1823)	430
S. pelzelni Hartlaub, 1887	—
Genus Nelicurvius	—
N. sakalava (Hartlaub, 1861)	—
N. nelicourvi (Scopoli, 1786)	—
Genus Pachyphantes	—
P. superciliosus (Shelley, 1873)	—

Subfamily Amblyospizinae — 442
Genus Amblyospiza — 442
Black Swamp Weaver, A. albifrons (Vigors, 1831) — 442
Genus Neospiza (systematic position uncertain) — 354
Sao Tome Grosbeak, N. concolor (Bocage, 1888) — 442

Subfamily Buffalo Weavers (Bubalornithinae) — 442
Genus Bubalornis — 442
Black Buffalo Weaver, B. albirostris (Vieillot, 1817) — 442
B. albirostris albirostris (Vieillot, 1817) — —
B. albirostris niger A. Smith, 1836 — —
Genus Dinemellia — 442
White-headed Buffalo Weaver, D. dinemelli (Rüppell, 1945) — 442

Subfamily Plocepasserinae — 443
Genus Histurgops — 443
Rufous-tailed Weaver, H. ruficauda Reichenow, 1887 — 443
Genus Plocepasser — 443
Mahali Weaverbird, P. mahali A. Smith, 1836 — 443
P. rufoscapulatus Büttikofer, 1888 — —
Genus Pseudonigrita — 443
Gray-headed Weaver, P. arnaudi (Bonaparte, 1850) — 443
Genus Philetairus — 443
Sociable Weaver, Ph. socius (Latham, 1790) — 443

Subfamily Sporopipinae — 443
Genus Sporopipes — 443
Speckle-fronted Weaver, S. frontalis (Daudin, 1802) — 443
S. squamifrons (A. Smith, 1836) — 444

Family Weaver Finches (Estrildidae) — 444
Genus Amadina — 447
Cut-throat, A. fasciata (Gmelin, 1789) — 447
Red-headed Finch, A. erythrocephala (Linné, 1758) — 447

Genus Euodice — 447
African Silver-bill, Eu. cantans (Gmelin, 1789) — 447
Indian Silver-bill, Eu. malabarica (Linné, 1778) — 447
Genus Odontospiza — 447
Gray-headed Silver-bill, O. caniceps (Reichenow, 1879) — 447
Genus Spermestes — 447
Magpie Mannikin, S. fringilloides (Lafresnaye, 1835) — 447
White Mannikin, S. bicolor (Fraser, 1843) — 447
S. bicolor poensis (Fraser, 1843) — —
S. bicolor nigriceps Cassin, 1852 — —
Bronze Mannikin, S. cucullatus Swainson, 1837 — 447
Bib Finch, S. nanus (Pucherah, 1845) — 447
Genus Lonchura — 448
Subgenus Lonchura — 448
L. (Lonchura) leucogastra (Blyth, 1846) — —
L. (Lonchura) kelaarti (Jerdon, 1863) — —
Spice Finch, L. (Lonchura) punctulata (Linné, 1758) — 448
Sharp-tailed Munia, L. (Lonchura) striata Linné, 1766 — 448
L. (Lonchura) striata striata (Linné, 1766) — —
L. (Lonchura) striata acuticauda (Hodgson, 1836) — —
L. (Lonchura) striata swinhoei (Cabanis, 1882) — 451
L. (Lonchura) leucogastroides (Horsfield and Moore, 1856) — —
L. (Lonchura) molucca (Linné, 1766) — —
Subgenus Munia — 448
Spectacled Finch, L. (Munia) spectabilis (Sclater, 1879) — 448
L. (Munia) hunsteini (Finsch, 1886) — —
L. (Munia) nevermanni Stresemann, 1934 — —
L. (Munia) caniceps (Salvadori, 1876) — —
L. (Munia) grandis (Sharpe, 1882) — —
Three-colored Mannikin, L. (Munia) malacca (Linné, 1766) — 448
Chinese Three-colored Mannikin, L. (Munia) malacca malacca (Linné, 1766) — 439
L. (Munia) malacca atricapilla (Vieillot, 1807) — —
L. (Munia) malacca ferruginosa (Sparrman, 1789) — 451
White-headed Mannikin, L. (Munia) maja (Linné, 1766) — 448
Yellow-tailed Finch, L. (Munia) flaviprymna (Gould, 1845) — 448
Chestnut-breasted Finch, L. (Munia) castaneothorax (Gould, 1837) — 448
L. (Munia) monticola (De Vis, 1897) — —
L. (Munia) teerinki Rand, 1940 — —
L. (Munia) melaena (Sclater, 1880) — —
Subgenus Heteromunia — 448
Pictorella Finch, L. (Heteromunia) pectoralis (Gould, 1841) — 448

Genus *Padda*	452
Java Sparrow, *P. oryzivora* (Linné, 1758)	452
P. fuscata (Vieillot, 1807)	—
Genus *Chloebia*	452
Gouldian Finch, *Ch. gouldiae* (Gould, 1844)	452
Genus *Erythrura*	452
Green-tailed Parrot Finch, *E. hyperthyra* (Reichenbach, 1862)	452
E. viridifacies Hachisuka and Delacour, 1936	—
Pin-tailed Nonpareil, *E. prasina* (Sparrman, 1788)	453
E. tricolor (Vieillot, 1817)	—
E. coloria Ripley and Rabor, 1961	—
Blue-faced Parrot Finch, *E. trichroa* (Kittlitz, 1835)	453
E. papuana Hartert, 1900	—
E. kleinschmidti (Finsch, 1878)	—
E. cyaneovirens (Peale, 1848)	413
Red-headed Parrot Finch, *E. psittacea* (Gmelin, 1789)	453
Genus *Poephila*	453
Long-tailed Grass Finch, *P. acuticauda* (Gould, 1840)	453
Parson Finch, *P. cincta* (Gould, 1837)	453
Masked Grass Finch, *P. personata* Gould, 1842	453
Genus *Taeniopygia*	453
Zebra Finch, *T. guttata* (Vieillot, 1817)	453
Australian Zebra Finch, *T. guttata castanotis* (Gould, 1837)	439
Indoesian Zebra Finch, *T. guttata guttata* (Vieillot, 1817)	453
Genus *Stizoptera*	454
Bicheno's Finch, *S. bichenovii* (Vigors and Horsfield, 1827)	454
Western Bicheno's Finch, *S. bichenovii annulosa* (Gould, 1840)	454
Eastern Bicheno's Finch, *S. bichenovii bichenovii* (Vigors and Horsfield, 1827)	454
Genus *Ademosyne*	—
Cherry Finch, *A. modesta* (Gould, 1837)	—
Genus *Bathilda*	455
Star Finch, *B. ruficauda* (Gould, 1837)	455
Genus *Neochmia*	455
Crimson Finch, *N. phaeton* (Hombron and Jaquinot, 1841)	455
Genus *Oreostruthus*	455
Crimson-bellied Mountain Finch, *O. fuliginosus* (De Vis, 1897)	455
Genus *Emblema*	455
Painted Finch, *E. picta* Gould, 1842	455
Genus *Stagonopleura*	455
Subgenus *Stagonopleura*	455
Diamond Sparrow, *S. (Stagonopleura) guttata* (Shaw, 1796)	455
Subgenus *Zonaeginthus*	455
Red-eared Fire-tailed Finch, *S. (Zonaeginthus) oculata* (Quoy and Gaimard, 1830)	455
Fire-tailed Finch, *S. (Zonaeginthus) bella* (Latham, 1801)	455
Genus *Aegintha*	455
Sidney Waxbill, *Ae. temporalis* (Latham, 1801)	455
Genus *Amandava*	455
Subgenus *Stictospiza*	455
Green Avadavat, *A. (Stictospiza) formosa* (Latham, 1790)	455
Subgenus *Sporaeginthus*	455
Golden-breasted Waxbill, *A. (Sporaeginthus) subflava* (Vieillot, 1819)	455
Subgenus *Amandava*	455
Avadavat, *A. (Amandava) amandava* (Linné, 1758)	455
Genus *Ortygospiza*	455
Subgenus *Paludipasser*	456
Locust Finch, *O. (Paludipasser) locustella* (Neave, 1909)	456
Subgenus *Ortygospiza*	456
South African Quail Finch, *O. (Ortygospiza) atricollis* (Vieillot, 1817)	456
Genus *Nesocharis*	—
Gray-headed Olive-back, *N. capistrata* (Hartlaub, 1801)	—
White-collared Olive-back, *N. ansorgei* (Hartert, 1899)	—
N. shelleyi Alexander, 1903	—
Genus *Estrilda*	456
Subgenus *Neisna*, should perhaps rank as a full genus	456
Yellow-bellied Waxbill, *E. (Neisna) melanotis* (Temminck, 1823)	456
E. (Neisna) melanotis melanotis (Temminck, 1823)	—
E. (Neisna) melanotis bocagei (Shelley, 1903)	—
Subgenus *Melpoda*	456
Orange-cheeked Waxbill, *E. (Melpoda) melpoda* (Vieillot, 1817)	456
Fawn-breasted Waxbill, *E. (Melpoda) paludicola* Heuglin, 1813	456
Subgenus *Estrilda*	456
St. Helena Waxbill, *E. (Estrilda) astrild* (Linné, 1758)	456
Arabian subspecies of St. Helena Waxbill, *E. (Estrilda) astrild rufibarba* (Cabanis, 1851)	—
E. (Estrilda) astrild nigriloris Chapin, 1928	—
Gray Waxbill, *E. (Estrilda) troglodytes* (Lichtenstein, 1823)	456
Crimson-rumped Waxbill, *E. (Estrilda) rhodopyga* Sundevall, 1850	456
Subgenus *Krimhilda*	457
Black, crowned Waxbill, *E. (Krimhilda) nonnula* Hartlaub, 1883	457
Black-headed Waxbill, *E. (Krimhilda) atricapilla* J. and E. Verreaux, 1851	457
Subgenus *Brunhilda*	457
Black-cheeked Waxbill, *E. (Brunhilda) erythronotos* (Vieillot, 1817)	457

    E. (Brunhilda) charmosyna (Reichenow, 1881) — 457
  Subgenus Glaucestrilda — 457
    E. (Glaucestrilda) thomensis Sousa, 1888 — —
    E. (Glaucestrilda) perreini (Vieillot, 1817) — —
    Lavender Finch, E. (Glaucestrilda) caerulescens (Vieillot, 1817) — 457
  Genus Uraeginthus — 457
  Subgenus Uraeginthus — 457
    Red-cheeked Cordon-bleu U. (Uraeginthus) bengalus (Linné, 1766) — 457
    Angola Waxbill, U. (Uraeginthus) angolensis (Linné, 1758) — 457
    Blue-capped Cordon-bleu, U. (Uraeginthus) cyanocephalus (Richmond, 1897) — 457
  Subgenus Granatina (Should possibly be a full genus) — 457
    Violet-eared Waxbill, U. (Granatina) granatinus (Linné, 1766) — 457
    Purple Grenadier, U. (Granatina) ianthionogaster Reichenow, 1879 — 457
  Genus Lagonosticta — 458
    Masked Waxbill, L. larvata (Rüppell, 1840) — 458
    L. larvata nigricollis Heuglin, 1863 — —
    Subspecies of Masked Waxbill, L. larvata vinacea (Hartlaub, 1857) — —
    Jameson's Fire Finch, L. rhodopareia (Heuglin, 1868) — 458
    Blue-billed Fire Finch, L. rubricata (Lichtenstein, 1823) — 458
    L. rara (Antinori, 1864) — —
    L. nitidula Hartlaub, 1886 — —
    Bar-breasted Fire Finch, L. rufopicta (Fraser, 1843) — 458
    Fire Finch, L. senegla (Linné, 1766) — 458
  Genus Hypargos — 458
    Rosy Twin Spot, H. margaritatus (Strickland, 1844) — 458
    Peter's Twin Spot, H. niveoguttatus (Peters, 1868) — 458
  Genus Mandingoa — 458
    Green-backed Twin Spot, M. nitidula (Hartlaub, 1865) — 458
    M. nitidula virginiae (Amadon, 1953) — —
    M. nitidula schlegeli — —
  Genus Cryptospiza — 458
    C. salvadorii Reichenow, 1892 — —
    Reichenow's Crimson-wing, C. reichenovii (Hartlaub, 1874) — 461
    C. jacksoni Sharpe, 1902 — —
    C. shelleyi Sharpe, 1902 — —
  Genus Pyrenestes — 461
    P. minor Shelley, 1894 — —
    Black-bellied Seed-cracker, P. ostrinus (Vieillot, 1805) — 461
    P. sanguineus Swainson, 1837 — —
  Genus Spermophaga — 461
    Blue Bill, S. haematina (Vieillot, 1805) — 461
    S. ruficapilla (Shelley, 1888) — —
    S. poliogenys (Ogilvie-Grant, 1906) — —
  Genus Euschistospiza — —
    Eu. cinercovinacea (Sousa, 1889) — —
    Eu. dybowskii (Oustalet, 1892) — —
  Genus Clytospiza — —
    C. monteiri (Hartlaub, 1860) — —
  Genus Pytilia — 428
    Aurora Waxbill, P. phoenieoptera Swainson, 1837 — 461
    P. lineata Heuglin, 1863 — 461
    P. hypogrammica Sharpe, 1870 — —
    Red-faced Waxbill, P. afra (Gmelin, 1789) — 461
    Melba Finch, P. melba (Linné, 1758) — 461
  Genus Nigrita — 462
    N. fusconata Fraser, 1843 — —
    Gray-headed Negro Finch, N. canicapilla (Strickland, 1841) — 462
    N. luteifrons J. and E. Verreaux, 1851 — —
    Chestnut-breasted Negro Finch, N. bicolor (Hartlaub, 1844) — 462
  Genus Parmoptila — 462
    Antpecker, P. woodhousei Cassin, 1859 — 462

Family Starlings (Sturnidae) — 463
Subfamily True Starlings (Sturninae) — 463
  Genus Creatophora — 472
    Wattled Starling, C. cinerea (Meuschen, 1787) — 472
  Genus Fregilupus — 472
    Reunion Starling, + F. varius (Boddaert, 1783) — 472
  Genus Necropsar — 472
    Legua Starling, + N. leguati Forbes, 1893 — 472
    Rodriguez Starling, + N. rodericanus Sclater, 1879 — 472
  Genus Sturnia — 471
    Violet-backed Starling, S. pilippensis (Forster, 1781) — 471
    Mandarin Mynah, S. sinensis (Gmelin, 1788) — 471
    Starlet, S. sturnina (Pallas, 1776) — 449
  Genus Pastor — 468
    Rose-colored Starling, P. roseus (Linné, 1758) — 468
  Genus Sturnus — 464
    Common Starling, S. vulgaris Linné, 1758 — 464
    Spotless Starling, S. unicolor Temminck, 1820 — 465
  Genus Spodiopsar — 468
    Gray Starling, S. cineraceus (Temminck, 1835) — 468
  Genus Sturnopastor — 468
    Pied Mynah, S. contra (Linné, 1758) — 468
  Genus Temenuchus — 471
    Pagoda Starling, T. pagodarum (Gmelin, 1789) — 471
    T. senex (Bonaparte, 1850) — —
    Malabar Mynah, T. malabaricus Gmelin, 1789 — 471
  Genus Gracupica — 471
    Black-necked Mynah, G. nigricollis (Paykull, 1807) — 471
    G. burmannica (Jerdon, 1862) — —
    G. melanoptera Daudin, 1800 — —
  Genus Leucopsar (probably should be Gracupica) — 471

Bali Mynah, *L. rothschildi* Streseman, 1912	471	Genus *Pholia*	—
Genus *Acridotheres*	471	*P. sharpii* (Jackson, 1898)	—
Common Mynah, *A. tristis* Linné, 1766	471	Genus *Cinnyricinclus*	474
Ceylon subspecies, *A. tristis melanosternus*	449	Violet-backed Starling, *C. leucogaster* (Boddaert, 1783)	474
Bank Mynah, *A. ginginianus* (Latham, 1790)	472	Genus *Cosmopsarus*	474
A. fuscus (Wagler, 1827)	—	Royal Starling, *C. regius* Reichenow, 1879	474
A. javanicus Cabanis, 1850	—	*C. unicolor* Shelly, 1881	—
Crested Mynah, *A. cristatellus* (Linné, 1766)	472	Genus *Lamprospreo*	474
Genus *Gracula*	473	*L. fischeri* (Reichenow, 1884)	—
Javan Hill Mynah, *G. religiosa* Linné, 1758	472	*L. hildebrandti* (Cabanis, 1878)	—
Ceylon Mynah, *G. ptilogenys* Blyth, 1846	473	*L. pulcher* (Müller, 1776)	—
Genus *Ambeliceps*	—	Superb Glossy Starling, *L. superbus* (Rüppell, 1854)	474
Gold-crested Grackle, *A. coronatus* Blyth, 1842)	—	Genus *Lamprotornis*	475
Genus *Mino*	473	*L. purpureiceps* (J. and E. Verreaux, 1851)	—
Golden-breasted Mynah, *M. anais* (Lesson, 1839)	473	*L. corruscus* Nordmann, 1835	460
Dumont's Mynah, *M. dumontii* Lesson, 1827	473	Purple Glossy Starling, *L. purpureus* (Müller, 1776)	—
Genus *Basilornis*	473	*L. chalcurus* Nordmann, 1835	—
B. celebensis Gray, 1861	—	Red-shouldered Glossy Starling, *L. nitens* (Linné, 1766)	475
B. galateus Meyer, 1894	—	Blue-eared Glossy Starling, *L. chalybaeus* Ehrenberg, 1828	475
B. corythaix (Wagler, 1827)	473	*L. chloropterus* Swainson, 1838	—
B. mirandus (Hartert, 1903)	—	*L. acuticaudus* (Bocage, 1870)	—
Genus *Streptocitta*	473	Splendid Glossy Starling, *L. splendidus* (Vieillot, 1822)	475
S. albicollis (Vieillot, 1818)	473	Burchell's Glossy Starling, *L. australis* (A. Smith, 1836)	475
S. albertinae (Schlegel, 1866)	473	*L. mevesii* (Wahlberg, 1856)	—
Genus *Sarcops*	473	*L. purpuropterus* Rüppell, 1845	—
Coleto Mynah, *S. calvus* (Linné, 1766)	473	Long-tailed Glossy Starling, *L. caudatus* (Müller, 1776)	475
Genus *Enodes*	473	Genus *Coccycolius*	—
E. erythrophris (Temminck, 1824)	473	*C. iris* Oustalet, 1879	—
Genus *Scissirostrum*	473	Genus *Spreo*	474
Celebesian Starling, *S. dubium* (Latham, 1802)	473	Pied Starling, *S. bicolor* (Gmelin, 1789)	474
Genus *Aplonis*	473	Genus *Neocichla*	475
+ *A. corvina* (Kittlitz, 1833)	—	White-winged Babbling Starling, *N. guttaralis* (Bocage, 1871)	475
A. tabuensis (Gmelin, 1788)	—	Genus *Saroglossa*	475
A. fusca Gould, 1836	—	Spotted-winged Stare, *S. spiloptera* (Vigors, 1831)	475
A. cantaroides (Gray, 1862)	449	Genus *Hartlaubius*	475
A. panayensis (Scopoli, 1783)	—	Madagascar Starling, *H. auratus* (Müller, 1776)	475
Shining Starling, *A. metallica* (Temminck, 1824)	473		
White-eyed Starling, *A. brunneicapilla* (Danis, 1838)	473	**Subfamily Oxpeckers (Buphaginae)**	475
Genus *Speculipastor*	473	Genus *Buphagus*	475
Magpie Starling, *S. bicolor* Reichenow, 1879	473	Yellow-billed Oxpecker, *B. africanus* Linné, 1766	475
Genus *Grafisia*	474	Red-billed Oxpecker, *B. erythrorhynchus* (Stanley, 1814)	475
White-collared Starling, *G. torquata* (Reichenow, 1900)	474		
Genus *Poeoptera*	—	**Family Orioles (Oriolidae)**	475
P. lugubris Bonaparte, 1854	—	Genus *Oriolus*	476
P. stuhlmanni (Reichenow, 1893)	—	*O. sagittatus* (Latham, 1801)	—
Genus *Onychognathus*	474		
O. walleri (Shelley, 1880)	—		
Pale-wing Starling, *O. nabouroup* (Daudin, 1800)	474		
Red-wing Starling, *O. morio* (Linné, 1766)	474		
Tristram's Grackle, *O. tristramii* (Sclater, 1858)	474		
O. salvadorii (Sharpe, 1891)	—		

O. xanthotonus Horsfield, 1821	—
Golden Oriole, *O. oriolus* (Linné, 1758)	476
Black-naped Oriole, *O. chinensis* Linné, 1766	476
Green-headed Oriole, *O. chlorocephalus* Shelley, 1896	477
O. brachyrhynchus Swainson, 1896	—
African Black-headed Forest Oriole, *O. monacha* (Gmelin, 1789)	476
Masked Oriole, *O. larvatus* Lichtenstein, 1823	476
O. xanthornus (Linné, 1758)	—
Maroon Oriole, *O. traillii* (Vigors, 1832)	476
O. mellianus Stresemann, 1922	476
Genus *Sphecotheres*	477
Yellow Figbird, *S. flaviventris* Gould, 1850	477
Southern Figbird, *S. vieilloti* Vigors and Horsfield, 1827	477
S. viridis Vieillot, 1816	—
Family Drongos (Dicruridae)	477
Genus *Chaetorhynchus*	477
Papua Mountain Drongo, *Ch. papuensis* Meyer, 1874	477
Genus *Dicrurus*	477
D. ludwigii (A. Smith, 1834)	—
Drongo, *D. adsimilis* (Bechstein, 1794)	477
D. forficatus (Linné, 1766)	—
King Crow, *D. macrocercus* (Vieillot, 1804)	477
Ashy Drongo, *D. leucophaeus* Vieillot, 1817	477
Crow-billed Drongo, *D. annectans* (Hodgson, 1836)	477
D. aeneus Vieillot, 1817	—
Lesser Racket-tailed Drongo, *D. remifer* (Temminck, 1823)	477
Spangled Drongo, *D. hottentottus* (Linné, 1766)	477
Greater Racket-tailed Drongo, *D. paradiseus* (Linné, 1766)	477
Family New Zealand Wattlebirds (Callaeidae)	482
Genus *Callaeas*	483
Kokakoo, *C. cinerea* (Gmelin, 1788)	483
Genus *Creadion*	483
Red Wattle Bird, *C. carunculatus* (Gmelin, 1789)	483
Genus + *Heteralocha*	483
Haia, + *H. acutirostris*	483
Family Magpie Larks (Grallinidae)	483
Subfamily Grallininae	483
Genus *Grallina*	484
Magpie Lark, *G. cyanoleuca* (Latham, 1801)	483
Torrent Lark, *G. bruijnii* Salvadori, 1875	484
Subfamily Corcoracinae	483
Genus *Corcorax*	484
White-winged Chough, *C. melanorhamphus* (Vieillot, 1817)	484
Genus *Struthidea*	484
Apostlebird, *S. cinerea* Gould, 1837	484
Family Wood Swallows (Artamidae)	484
Genus *Artamus*	485
Ashy Swallow Shrike, *A. fuscus* Vieillot, 1817	485
White-breasted Swallow Shrike, *A. leucorhynchus* (Linné, 1771)	485
A. personatus (Gould, 1841)	—
White-browed Wood Swallow, *A. superciliosus* (Gould, 1837)	485
A. cinereus Vieillot, 1817	—
A. cyanoptera (Latham, 1801)	—
Little Wood Swallow, *A. minor* Vieillot, 1817	485
Family Song Shrikes (Cracticidae)	485
Genus *Cracticus*	485
C. mentalis Salvadori and d'Albertis	—
Gray Butcher Bird, *C. torquatus* (Latham, 1801)	485
C. cassius (Boddaert, 1783)	—
White-rumped Butcher Bird, *C. louisiadensis* Tristram, 1889	485
C. nigrogularis (Gould, 1837)	—
Black Butcher Bird, *C. quoyi* (Lesson, 1827)	485
Genus *Strepera*	485
Pied Currawong, *S. graculina* (White, 1790)	486
Tasmanian subspecies, *S. graculina fuliginosa* (Gould, 1837)	486
Gray Currawong, *S. versicolor* (Latham, 1801)	486
Genus *Gymnorhina*	485
Black-backed Magpie, *G. tibicen* (Latham, 1801)	486
White-backed Magpie, *G. hypoleuca* (Gould, 1837) (probably only a subspecies of *G. tibicen*)	486
Southeastern Australian subspecies, *G. hypoleuca leuconota* Gould, 1844	—
Tasmanian Subspecies, *G. hypoleuca hypoleuca* (Gould, 1837)	—
Southwestern Australian subspecies, *G. hypoleuca dorsalis* Campbell, 1895	487
Family Birds of Paradise and Bower Birds (Paradisaeidae)	487
Subfamily Birds of Paradise (Paradisaeinae)	487
Genus *Macgregoria*	—
M. pulchra De Vis, 1897	—
Genus *Manucodia*	490
M. atra (Lesson, 1830)	—
Green-breasted Manucode, *M. chalybata* (Pennant, 1781)	490
M. comrii Sclater, 1876	—
Genus *Phonygammus*	490
Trumpet Bird, *Ph. keraudrenii* (Lesson and Garnot, 1826)	490
Genus *Ptiloris*	499
P. paradiseus Swainson, 1825	—

Magnificent Rifle bird, *P. magnifica* (Vieillot, 1819)	499	Genus *Cnemophilus*	490
Genus *Semioptera*	496	Multi-crested Bird of Paradise, *C. macgregorii* De Vis, 1890	490
Wallace's Standard Wing, *S. wallacei* Gould, 1859	496	Genus *Loria*	490
Genus *Seleucidis*	496	Loria's Bird of Paradise, *L. loriae* Salvadori, 1894	490
Twelve-wired Bird of Paradise, *S. melanoleuca* (Daudin, 1800)	496	Genus *Loboparadisaea*	490
Genus *Paradigalla*	490	Yellow-breasted Bird of Paradise, *L. sericea* Rothschild, 1896	490

Long-tailed Paradigalla, *P. carunculata* Lesson, 1835 — 490

Subfamily Bowerbirds (Ptilonorhynchinae) — 499

Genus *Drepanornis*	499	Genus *Ailuroedus*	500
D. albertisi (Sclater, 1873)	—	Green Catbird, *A. crassirostris* (Paykull, 1815)	500
White-bellied Sickle-billed Bird of Paradise, *D. bruijnii* Oustalet, 1880	499	Black-eared Green Catbird, *A. crassirostris melanocephalus*	492
Genus *Epimachus*	499	White-eared Catbird, *A. buccoides* (Temminck, 1835)	503
Brown Sickle-billed Bird of Paradise, *E. meyeri* Finsch, 1885	499	Genus *Scenopoeetes*	503
Black Sickle-billed Bird of Paradise, *E. fastuosus* (Hermann, 1783)	499	Tooth-billed Bower Bird, *S. dentirostris* (Ramsay, 1876)	503
Genus *Astrapia*	496	Genus *Archboldia*	503
Ribbon-tailed Astrapia, *A. mayeri* Stonor, 1939, probably a subspecies of *A. stephaniae*	496	Archbold Bower Bird, *A. papuensis* Rand, 1940	503
Princess Stephanie Astrapia, *A. stephaniae* (Finsch, 1885)	499	Sanford's Bower Bird, *A. papuensis sandfordi* Mayr and Gilliard, 1950	503
Arfak Astrapia, *A. nigra* (Gmelin, 1788)	499	Genus *Prionodura*	503
Genus *Lophorina*	496	Golden Bower Bird, *P. newtoniana* De Vis, 1883	503
Superb Bird of Paradise, *L. superba* (Pennant, 1781)	496	Genus *Amblyornis*	503
Genus *Parotia*	495	Gardener Bower Bird, *A. inornatus* (Schlegel, 1871)	504
Arfak Six-wired Bird of Paradise, *P. sefilata* (Pennant, 1781)	495	Macgregor's Bower Bird, *A. macgregoriae* De Vis, 1890	503
Queen Carola's Parotia, *P. carolae* Meyer, 1894	495	Striped Bower Bird, *A. subalaris* Sharpe, 1884	504
Lawe's Six-wired Bird of Paradise, *P. lawesi* Ramsay, 1885	495	Genus *Sericulus*	504
Genus *Pteridophora*	496	*S. aureus* (Linné, 1758)	492
King of Saxony's Bird of Paradise, *P. alberti* Meyer, 1894	496	Regent Bower Bird, *S. chrysocephalus* (Lewin, 1808)	505
Genus *Cicinnurus*	494	Genus *Ptilonorhynchus*	504
King Bird of Paradise, *C. regius* (Linné, 1758)	494	Satin Bower Bird, *P. violaceus* (Vieillot, 1816)	505
Genus *Diphyllodes*	495	Genus *Chlamydera*	504
Magnificent Bird of Paradise, *D. magnificus* (Pennant, 1781)	495	Spotted Bower Bird, *Ch. maculata* (Gould, 1837)	505
Waigeu Bird of Paradise, *D. respublica* (Bonaparte, 1850)	495	Great Bower Bird, *Ch. nuchalis* (Jardine and Selby, 1830)	505
Genus *Paradisaea*	490	Fawn-breasted Bower Bird, *Ch. cerviniventris* Gould, 1850	505
Greater Bird of Paradise, *P. apoda* Linné, 1758	490	Yellow-breasted Bower Bird, *Ch. lauterbachi* Reichenow, 1897	505
Aru subspecies, *P. apoda apoda* Linné, 1758	493		

Family Crowlike Birds (Corvidae) — 505

Raggiana Bird of Paradise, *P. apoda raggiana* Sclater, 1873	493	Genus *Perisoreus*	509
Lesser Bird of Paradise, *P. minor* Shaw, 1809	493	Gray Jay, *P. canadensis* (Linné, 1766)	509
P. rubra Daudin, 1800	470	Siberian Jay, *P. infaustus* (Linné, 1758)	509
Emperor of Germany's Bird of Paradise, *P. guilielmi* Cabanis, 1888	493	Genus *Platylophus*	506
Blue Bird of Paradise, *P. rudolphi* (Finsch, 1885)	493	Crested Jay, *P. galericulatus* (Cuvier, 1817)	506
		Genus *Platysmurus*	506
		Black-crested Magpie, *P. leucopterus*	

(Temminck, 1824)	506	Hainan Magpie, *U. whiteheadi* Ogilvie-Grant, 1899	510
Genus *Temnurus*	511	Genus *Cissa*	510
Racket-tailed Tree Pie, *T. temnurus* (Temminck, 1825)	511	Green Magpie, *C. chinensis* (Boddaert, 1783)	510
Genus *Crypsirina*	510	Short-tailed Green Magpie, *C. thalassina* (Temminck, 1826)	510
C. temia (Daudin, 1800)	—	Genus *Garrulus*	508
Hooded Racket-tailed, *C. cucullata* Jerdon, 1862	511	Lidth's Jay, *G. lidthi* Bonaparte, 1850	508
Genus *Dendrocitta*	510	Black-throated Jay, *G. lanceolatus* Vigors, 1831	508
D. frontalis Horsfield, 1840	—	Common Jay, *G. glandarius* (Linné, 1758)	508
D. occipitalis (St. Müller, 1835)	—	Genus *Gymnorhinus*	506
D. formosae Swinhoe, 1863	—	Pinyon Jay, *G. cyanocephalus* Wied, 1841	506
Indian Tree Pie, *D. vagabunda* (Latham, 1790)	511	Genus *Nucifraga*	512
Genus *Pica*	511	Clark's Nutcracker, *N. columbiana* (Wilson, 1811)	512
Common Magpie, *P. pica* (Linné, 1758)	511	Nutcracker, *N. caryocatactes* (Linné, 1758)	512
Yellow-billed Magpie, *P. pica nutalli* (Audubon, 1837)	511	Alpine Nutcracker, *N. caryocatactes caryocatactes* (Linné, 1758)	512
Genus *Cyanopica*	510	Siberian Nutcracker, *N. caryocatactes macrorhynchos* (Brehm, 1823)	512
Azure-winged Magpie, *C. cyana* (Pallas, 1776)	510	Genus *Zavattariornis*	511
C. cyana cooki Bonaparte, 1850	510	Streseman's Bush Crow, *Z. stresemanni* Moltoni, 1938	512
Genus *Aphelocoma*	507	Genus *Podoces*	511
Scrub Jay, *A. coerulescens* (Bosc, 1795)	507	Henderson's Ground Chough, *P. hendersoni* Hume, 1871	512
Mexican Jay, *A. ultramarina* (Bonaparte, 1825)	507	*P. biddulphi* Hume, 1874	—
Unicolor Jay, *A. unicolor* (Cassin, 1848)	507	Pander's Ground Chough, *P. panderi* Fischer, 1821	512
Genus *Cyanolyca*	507	Genus *Pseudopodoces*	411
C. cucullata (Ridgway, 1885)	—	Hume's Ground Chough, *P. humilis* (Hume, 1871)	512
C. pulchra (Lawrence, 1876)	—	Genus *Pyrrhocorax*	513
Collared Jay, *C. viridicyana* (Lafresnaye and d'Orbigny, 1838)	507	Chough, *P. phrrhocorax* (Linné, 1758)	513
Genus *Cyanocorax*	507	Canarie's Chough, *P. phrrhocorax barbarus* Vaurie, 1954	513
Azure Jay, *C. caerulens* (Vieillot, 1818)	497	Alpine Chough, *P. graculus* (Linné, 1766)	513
Green Jay, *C. yncas* (Boddaert, 1783)	507	Genus *Ptilostomus*	514
C. violacues Du Bus, 1847	—	Pipiac, *P. afer* (Linné, 1766)	514
Cayenne Jay, *C. cayanus* (Linné, 1766)	507	Genus *Corvus*	514
Plush-crested Jay, *C. chrysops* (Vieillot, 1818)	508	Jackdaw, *C. monedula* Linné, 1758	514
C. mystacalis Geoffroy St. Hilaire	—	West Siberian subspecies, *C. monedula soemmeringii* Fischer, 1811	—
Genus *Calocitta*	508	House Crow, *C. splendens* Vieillot, 1817	515
Magpie Jay, *C. formosa* (Swainson, 1827)	508	*C. splendens zugmayeri* Laubmann, 1913	515
Genus *Psilorhinus*	508	Black Crow, *C. capensis* Lichtenstein, 1823	515.
White-tipped Brown Jay, *P. morio* (Wagler, 1829)	508	Rook, *C. frugilegus* Linné, 1758	515
Genus *Cissilopha*	507	*C. enca* (Horsfield, 1822)	—
C. melanocyanea (Hartlaub, 1844)	497	Gray Crow, *C. tristis* Lesson and Carnot, 1827	—
Black and Blue Jay, *C. sanblasiana* (Lafresnaye, 1842)	507	*C. fuscicapillus* Gray, 1859	—
C. beccheii (Vigors, 1828)	—	*C. macrorhynchus* Wagler, 1827	—
Genus *Cyanocitta*	506	*C. orru* Bonaparte, 1850	—
Blue Jay, *C. cristata* (Linné, 1758)	506	*C. bennetti* North, 1901	—
Steller's Jay, *C. stelleri* (Gmelin, 1788)	506	*C. coronoides* Vigors and Horsfield, 1Q27	—
Genus *Urocissa*	510	*C. mellori* Mathews, 1912	—
Ceylon Magpie, *U. ornata* (Wagler, 1829)	510	Maori Crow, + *C. moriorum* Forbes, 1892	516
Blue Magpie, *U. caerulea* Gould, 1863	510		
Yellow-billed Blue Magpie, *U. flavirostris* (Blyth, 1846)	510		
Red-billed Blue Magpie, *U. erythrorhyncha* (Boddaert, 1783)	510		

C. jamaicensis Gmelin, 1788	—
C. palmarum Württemberg, 1835	—
Fish Crow, *C. ossifragus* Wilson, 1812	515
C. imparatus Peters, 1929	—
C. caurinus Baird, 1858 (possibly a subspecies of *C. brachyrhunchos*)	—
Common Crow, *C. brachyrhynchos* Brehm, 1822	515
Carrion Crow, *C. corone* Linné, 1758	515
Western European Carrion Crow, *C. corone corone* Linné, 1758	516
Eastern Carrion Crow, *C. corone orientalis* Eversmann, 1841	516
Hooded Crow, *C. corone cornix* (Linné, 1758)	516
Southern Hooded Crow, *C. corone sardonius* Kleinschmidt, 1903	—
Siberian Hooded Crow, *C. corone sharpii* Oates, 1889	—
C. torquatus Lesson, 1830	—
Pied Crow, *C. albus* Müller, 1776	518
Brown-necked Raven, *C. ruficollis* Lesson, 1830	518
C. ruficollis edithae Phillips, 1895	—
Raven, *C. corax* Linné, 1758	518
Faroese Raven, *C. corax varius* Brünnich, 1764	519
C. cryptoleucus Couch, 1854	—
Thick-billed Raven, *C. crassirostris* Rüppell, 1836	—
Fan-tailed Raven, *C. rhipidurus* Hartert, 1918	518
African White-necked Raven, *C. albicollis* Latham, 1790	518

On the Zoological Classification and Names

For many years, zoologists and botanists have tried to classify animals and plants into a system which would be a survey of the abundance of forms in fauna and flora. Such a system, of course, may be established under very different aspects. Since Charles Darwin, his predecessors, and his successors have found that all creatures have evolved out of common ancestors, species of animals and plants have been classified according to their natural relationships. Our knowledge about the phylogeny, and thus the relationship of each living being to the other, is augmented every year by new discoveries and insights. Old ideas are replaced with more recent and more appropriate ones. Therefore, the natural classification of the animal kingdom (and the plant kingdom) is subject to changes. Furthermore, the opinions of zoologists, who are working on the classification of animals into the various groups, are anything but uniform. These differences and changes are usually insignificant. The classification of vertebrates into the classes of fish, amphibians, reptiles, birds, and mammals has been fixed for many decades. Only the Cyclostomata were recently separated from the fish and all other classes of vertebrates as the "jawless" Agnatha (comp. Vol. 4).

The animal kingdom has been split into several subkingdoms and these were again divided into further sections, subsections, and so on. The scale of the most important systematic categories follows in a descending rank order:

Kingdom
Subkingdom
Phylum
Subphylum
Class
Subclass
Superorder
Order
Suborder
Infraorder
Family
Subfamily
Tribe
Genus
Subgenus
Species
Subspecies

The scientific names of the animals and their spelling follow the international rules for the zoological nomenclature as agreed upon by the XV International Congress for Zoology and are obligatory for all zoological publications. The name of the genus, which is a Latin or Latinized noun, is singular and capitalized. After the name of the genus follows the name of the species and of the subspecies. The names of the species and subspecies may be nouns or adjectives, and they are spelled in the lower case. The name of a subgenus, which is formed in the same manner as a genus, may be added in brackets following the name of the genus. The names of the tribes, subfamilies, families, and superfamilies are plural capitalized nouns. They are formed from the name of a given genus by adding to the principal word the endings -ini for the tribe, -inae for the subfamily, -idae for the family, and -oidea for the superfamily. The names of the authors who were the first to describe and to name a species, subspecies, or group of animals should be cited with the year of this naming at least once in each scientific publication. The name of the author and year are not enclosed in brackets when the species or subspecies is classified as belonging to the same genus with which the author had originally classified it. They are in brackets when another genus name is used in the present publication. The scientific names of the genus, subgenus, species, and subspecies are supposed to be printed with different letters, usually italics.

ANIMAL DICTIONARY

I. English–German–French–Russian

For scientific names of species see the German-English-French-Russian section of this dictionary or the index.

In most cases names of subspecies are formed by putting an adjective or geographical specification before the name of species. These English names of subspecies will, as a rule, not appear in this part of the zoological dictionary.

ENGLISH NAME	GERMAN NAME	FRENCH NAME	RUSSIAN NAME
Abyssinian barbet	Perlbartvogel	Barbu perlé	
– ground hornbill	Nördlicher Hornrabe	Bucorve d'Abyssinie	Абиссинский рогатый ворон
– fire finch	Rosenamarant	Amaranthe de Jameson	
Accentors	Braunellen	Prunellidés, Accenteurs	Завирушковые, Завирушки
African parson finch	Zwergelsterchen	Spermète naine	
– silver-bill	Silberschnäbelchen	Bec d'argent	
Albert's lyre-bird	Schwarzleierschwanz		Принцев лирохвост
Alpine accentor	Alpenbraunelle	Accenteur alpin	Альпийская завирушка
– chough	Alpendohle	Chocard à bec jaune	Альпийская галка
American blue jay	Blauhäher	Geai bleu d'Amérique	Голубая сойка
– kingbird	Königssatrap	Tyran royal	
– robin	Wanderdrossel	Merle migrateur	Странствующий дрозд
Ant birds	Ameisenvögel	Formicariidés	Птицы-муравьеды
– shrikes	Ameisenwürger	Thamnophiles	
Antpecker	Ameisenpicker	Astrild fourmilier	
Apostle-bird	Gimpelhäher	Glaucope gris	Штрутидея
Aquatic warbler	Seggen-Rohrsänger	Phragmite aquatique	Вертлявая камышевка
Arctic redpoll	Polarbirkenzeisig	Sizerin blanchâtre	Полярная чечетка
Ashy-wood-swallow	Grauschwalbenstar		Серый ласточковый скворец
Aurora waxbill	Aurora-Astrild	Diamant aurore	
Australian bee-eater	Schmuckspint	Guêpier d'Australie	
– butcherbirds	Krähenwürger	Cracticidés	
– laughing jackass	Kookaburra	Martin-chasseur géant	Зимородок-великан
– »wrens«	Südsee-Grasmücken		Малуры
Avadavat	Tigerfink	Bengali rouge	
Azure-winged magpie	Blauelster	Pie bleue	Голубая сорока
Babbling thrushes	Timalien	Timaliinés	Тималии
Baillon's toucan	Goldbrust-Tukan	Toucan de Baillon	
Bald-headed starling	Kahlkopfatzel	Sacrops chauve	
Bali-mynah	Balistar	Martin de Rothschild	Балийский скворец
Baltimore oriole	Baltimore-Trupial		Балтиморский трупиал
Banded aracari	Halsband-Arassari	Aracari à collier	
– chatterer	Halsbandkotinga	Cotinga tityre	Ошейниковая котинга
– cotinga	–	–	– –
Bank mynah	Ufermaina	Martin de rivage	Береговая майна
– swallow (N.A.)	Uferschwalbe	Hirondelle de rivage	– ласточка
Barbets	Bartvögel	Capitonidés	Бородастики
Bar-breasted fire finch	Pünktchenamarant	Sénégali à poitrine barrée	
Barn Swallow (N.A.)	Rauchschwalbe	Hirondelle de cheminée	Деревенская ласточка
Barred warbler	Sperbergrasmücke	Fauvette épervière	Ястребиная славка
Baya weaver	Bajaweber	Tisserin Baya	Бойя
Bearded barbet	Senegalfurchenschnabel	Barbican à poitrine rouge	
– titmouse	Bartmeise	Panure à moustaches	
Bee-eaters	Bienenesser	Méropidés, Guêpiers	Пчелоеды
Bengal pitta	Neunfarbenpitta	Brève du Bengale	Индийская питта
Bellbirds	Glockenvögel	Orapongas	Птицы-звонари
Bib-finch	Zwergelsterchen	Spermète naine	
Bicheno's finch	Ringelastrild	Diamant de Bicheno	
Birds of paradise	Paradies- und Laubenvögel, Eigentliche Paradiesvögel	Paradisiers	Райские и Беседковые птицы, Настоящие райские птицы
Bishop	Tahaweber	Tisserin taha	
Black cuckoo-shrike	Mohrenraupenesser	Echenilleur pourpré	
– redstart	Hausrotschwanz	Rouge-queue noir	Горихвостка-чернушка
– tit	Mohrenmeise	Mésange noire	
– woodpecker	Schwarzspecht	Pic noir	Желна
Black-and-white mannikin	Glanzelsterchen	Spermète bicolore	
Black-backed magpie	Schwarzrücken-Flötevogel	Corbeau flûteur	Черноспинная ворона-свистун
Black-bellied seed-cracker	Purpurastrild	Pyreneste ponceau à ventre noir	
Black-billed magpie	Elster	Pie bavarde	Сорока
– mountain toucan	Schwarzschnabel-Blautukan	Toucan à bec noir	

ANIMAL DICTIONARY

ENGLISH NAME	GERMAN NAME	FRENCH NAME	RUSSIAN NAME
Blackbird	Amsel	Merle noir	Черный дрозд
Blackbirds	Hordenvögel		Болотные трупиалы
Blackcap	Mönchsgrasmücke	Fauvette à tête noire	Черноголовая славка
Black-cheeked waxbill	Elfenastrild	Astrild à moustaches noires	
Black-collared barbet	Halsband-Bartvogel	Barbu à collier noir	
– starling	Schwarzhalsstar	Martin à cou noir	
Black-crowned waxbill	Nonnenastrild	Astrild à cape noire	
Black-eared wheatear	Mittelmeer-Steinschmätzer	Traquet oreillard	Чернопегая каменка
Black-fronted diock	Blutschnabelweber	Travailleur	
– weaver	Maskenweber	Tisserin à front noir	
Black-headed bunting	Kappenammer	Bruant mélanocéphale	Черноголовая овсянка
– waxbill	Kappenastrild	Astrild à tête noire	
– weaver	Textorweber	Tisserin Cap Moor	
Black-naped oriole	Schwarznacken-Pirol	Loriot de Chine	Черноголовая иволга
Black-necked mynah	Schwarzhalsstar	Martin à cou noir	
Black-throated jay	Eichelhäher	Geai des chênes	Сойка
– trush	Schwarzkehldrossel	Grive à gorge noire	Чернозобый дрозд
– waxbill	Larvenamarant	Amaranthe masquée d'Abyssinie	
Black-vented crimson finch	Dunkelroter Amarant	Sénégali à bec bleu	
Blood finch	Sonnenastrild	Phaéton	
Blue rock thrush	Blaumerle	Merle bleu	Синий каменный дрозд
– vanga	Blauvanga	Artamie azurée	
Blue-bellied roller	Blaubauchracke	Rollier à ventre bleu	
Blue-bill	Rotbrust-Samenknacker	Astrild à gros bec bleu	
Blue-billed crimson-breasted weaver	–	– – – –	
– fire finch	Dunkelroter Amarant	Sénégali à bec bleu	
Blue bird of paradise	Blauer Paradiesvogel	Paradisier de Rodophe	
Blue-breasted roller	Blaubauchracke	Rollier à ventre bleu	
– waxbill	Angola-Schmetterlingsfink	Cordon bleu d'Angola	
Blue-eared glossy starling	Grünschwanzglanzstar	Merle métallique à oreilles bleues	Стальной скворец
Blue-faced parrot finch	Dreifarbige Papageiamadine	Diamant tricolore de Kittlitz	
Blue-headed wagtail	Schafstelze	Bergeronette printanière	Желтая трясогузка
– waxbill	Blaukopf-Schmetterlingsfink	Astrild à tête bleue	
Bluethroat	Blaukehlchen	Gorge-bleue à miroir	Варакушка
Blue-throated barbet	Blauwangen-Bartvogel	Barbu à gorge bleue	
Blue-tit, Blue titmouse	Blaumeise	Mésange bleue	Лазоревка
Blue-winged fruit-sucker	Blaubart-Blattvogel	Verdin à barbe bleu	
Blyth's reed warbler	Buschrohrsänger	Rousserolle des buissons	Садовая камышевка
Bobolink	Reisstärling		Рисовый скворец
Bohemian waxwing	Europäischer Seidenschwanz	Jaseur boréal	Обыкновенный свиристель
Bourcier's barbet	Rotbrust-Buntbärtling	Barbu de Bourcier	
Brahminy mynah	Pagodenstar	Martin des pagodes	Браминский скворец
Brambling	Bergfink	Pinson du nord	Юрок
Bristle-nosed barbet	Pel-Bartvogel	Barbu chauve à narines emplumées	
Broad-billed rollers	Breitschnabelracken		Широкоротые сизоворонки
– tody	Breitschnabel-Todi	Todier vert	
Broadbills	Breitrachen	Eurylaimidés	Рогоклювы
Bronze mannikin	Kleinelsterchen	Spermète àcapuchon	
Brown creeper (N.A.)	Waldbaumläufer	Grimpereau des bois	Обыкновенная пищуха
– gardener	Hüttengärtner		Новогвинейская беседковая птица
Buffalo weaver	Büffelweber	Tisserin alecto	Буйволовый ткач
Bulbuls	Haarvögel, Echte Bülbüls		Короткопалые дрозды, Настоящие бульбули
Bullfinch	Dompfaff	Bouvreuil pivoine	Снегирь
Buntings	Ammern, Eigentliche Ammern	Emberizinés, Bruants	Овсянки
Bush-shrikes	Buschwürger		Лесные сорокопуты
Butcherbirds	Würger	Laniidés	Сорокопутовые
Cactus wren	Kaktus-Zaunkönig		Калифорнийский кактусовый крапивник
Canada jay	Kanadischer Unglückshäher		Канадская ронжа
Calandra lark	Kalanderlerche	Alouette calandre	Степной жаворонок
Canary	Kanarienvogel	Serin	Канарейка
Cape red-shouldered glossy starling	Rotschulterglanzstar	Merle à épaulettes rouges	
– sparrow	Kapsperling	Moineau mélanure	Южноафриканский воробей
– weaver	Kapweber	Tisserin à front d'or	
Cardinal	Roter Kardinal		Красный кардинал
Cardinal-grosbeaks	Kardinäle	Cardinaux	Кардиналы
Carolina wren	Karolina-Zaunkönig		Крапивник-пересмешник
Carrion crow	Rabenkrähe	Corneille noire	Обыкновенная черная ворона

ENGLISH NAME	GERMAN NAME	FRENCH NAME	RUSSIAN NAME
Catbird	Katzendrossel	Merle moqueur chat	
Cat-birds	Katzenvögel i. e. S.		Птицы-кошки
Caterpillar birds	Stachelbürzler	Campéphagidés	Личинкоедовые
Cedar waxwing	Zederseidenschwanz		Кедровый свиристель
Cetti's warbler	Seidensänger	Bouscarle de Cetti	Широкохвостая камышевка
Ceylon blue magpie	Blau-Schweifkitta	Pie bleue ornée	
– crackle	Ceylonbeo	Mainate de Ceylan	
– mynah	–	– – –	
Chaffinch	Buchfink	Pinson des arbres	Зяблик
Cherry finch	Zeresfink	Modeste	
Chestnut sparrow	Eminsperling	Moineau doré d'Emin	
Chestnut-bellied nuthatch	Kastanienkleiber	Sittelle à ventre marron	
Chestnut-breasted finch	Braunbrust-Schilffink	Donacole commun	
– negro-finch	Zweifarbenschwärzling	Bengali brun à ventre roux	
Chestnut-eared finch	Zebrafink	Diamant mandarin	
Chikadee	Chickadee-Meise	Mésange à tête noire	
Chinese blue-pie	Rotschnabel-Schweifkitta	Pirolle de la Chine	Красноклювая лазуревая сорока
– blue whistling thrush	Chinesische Pfeifdrossel	Grive siffleuse bleue	Синяя птица
– crested mynah	Haubenmaina	Martin huppé	Хохлатая майна
Chough	Alpenkrähe	Grave à bec rouge	Клушица
Citril finch	Zitronengirlitz	Venturon montagnard	
Clark's nutcracker	Kiefernhäher		Североамериканская ореховка
Coal-tit, Coal-titmouse	Tannenmeise	Petite charbonnière	Черная синица
Cocks of the rock	Klippenvögel	Coqs-de-roche	Скалистые петушки
Coleto mynah	Kahlkopfatzel	Sarcops chauve	
Common African broadbills	Afrikanische Breitrachen	Gobe-mouches	
– bee-eater	Bienenesser	Guêpier méridional	Золотистая щурка
– jacamar	Grünjakamar		Зеленая якамара
– linnet	Bluthänfling	Linotte mélodieuse	Коноплянка
– motmot	Motmot	Momot	
– mynah	Hirtenstar	Martin triste	Саранчовый скворец
– piping crow	Schwarzrücken-Flötenvogel	Corbeau flûteur	Черноспинная ворона-свистун
– roller	Blauracke	Rollier d'Europe	Обыкновенная сизоворонка
– sapsucker	Saftlecker		Дятел-сосун
– starling	Gemeiner Star	Étourneau sansonnet	Обыкновенный скворец
– swallow	Rauchschwalbe	Hirondelle de cheminée	Деревенская ласточка
– waxbill	Graustrild	Bec de coreil	
Congo crested waver	Haubenprachtweber	Malimbe huppé	
Coppersmith	Kupferschmied	Barbu à front rouge	
Corn bunting	Grauammer	Bruant proyer	Просянка
Corsican nuthatch	Korsika-Kleiber	Sittelle corse	
Cotingas	Schmuckvögel, Eigentliche Kotingas	Cotingidés, Cotingas	Котинги
Courol	Kurol	Courol malgache	
Cowbirds	Kuhstärlinge		Коровьи скворцы
Creepers	Baumläuferartige, Baumläufer	Certhiidés, Grimpereaux	Пищуховые, Пищухи
Crested bunting	Haubenammer		Хохлатая овсянка
– lark	Haubenlerche	Cochevis huppé	Хохлатый жаворонок
– malimbe	Haubenprachtweber	Malimbe huppé	
– mynah	Haubenmaina	Martin huppé	Хохлатая майна
– sharpbill	Flammenkopf	Tête de feu	
Crested-tit	Haubenmeise	Mésange huppée	Хохлатая синица
Crimson-bellied mountain finch	Bergamadine	Diamant des montagnes	
Crimson-breasted barbet	Kupferschmied	Barbu à front rouge	
– shrike	Rotbauchwürger	Pie-grièche à ventre cramoisie	
Crimson-crowed weaver	Flammenweber	Euplecte à couronne de feu	
Crimson finch	Sonnenastrild	Phaéton	
Crimson-rumped waxbill	Zügelastrild	Astrild à dos rouge	
Crossbill	Fichtenkreuzschnabel	Bec-croisé des sapins	Клест-еловик
Crossbills	Kreuzschnäbel	Bec-croisés	Клесты
Crows	Rabenvögel	Corbeaux	Вороновые
Cuba finch	Klein-Kubafink	Petit chanteur de Cuba	
Cuckoo-roller	Kurol	Courol malgache	
Cuckoo-shrikes	Stachelbürzler, Raupenesser i. e. S.	Campéphagidés	Личинкоедовые, Гусеницееды
Cuckoo-weaver	Kuckucksweber	Tisserin parasitique	
Currawongs	Würgerkrähen		Кричащие вороны
Cut-throat	Bandfink	Cou coupé	
– whydah	Schildwida	Veuve en feu	
Dark fire finch	Dunkelroter Amarant	Sénégali à bec bleu	
Dartford warbler	Provence-Grasmücke	Fauvette pitchou	Прованская славка

ENGLISH NAME	GERMAN NAME	FRENCH NAME	RUSSIAN NAME
Darwin's finches	Darwinfinken	Géospizinés	
Derby flycatcher (N.A.)	Bentevi	Tyran à ventre jaune	Бентеви
Desert sparrow	Wüstensperling		Пустынный воробей
– wheatear	Wüstensteinschmätzer	Traquet du désert	Пустынная каменка
Dhyal thrush	Dajaldrossel	Merle dyal des Indes	Индийская сорочья славка
Diamond birds	Panthervögel	Pardalots	Пантерные птицы
– sparrow	Diamantfink	Diamant à gouttelettes	
Dinemelli's weaver	Starweber	Dinemelli à tête blanche	
Dipper	Wasseramsel	Cincle plongeur	Обыкновенная оляпка
Dippers	Wasseramseln	Cinclidés	Оляпковые
Dollarbird	Ostroller	Rollier à gros bec	Широкорот
Double-toothed barbet	Doppelzahn-Bartvogel	Barbu à bec denté	
Drongo	Trauerdrongo	Drongo	Траурный дронго
Drongos	Drongos	Drongos	Дронговые, Дронго
Dunnock	Heckenbraunelle	Accenteur mouchet	Лесная завирушка
Dusky Broadbill	Feuerbreitrachen	Eurylaime sombre	
Eastern bluebird	Rotkehl-Hüttensänger	Rouge-gorge bleu d'Amérique	Обыкновенный голубой чекан
– kingbird (N.A.)	Königssatrap	Tyran royal	
Edward's manakin	Weißsäbelpipra	Manakin moine	
Emerald toucanet	Laucharassari	Toucanet vert	
European bee-eater	Bienenesser	Guêpier méridional	Золотистая щурка
– roller	Blauracke	Rollier d'Europe	Обыкновенная сизоворонка
– wryneck	Europäischer Wendehals	Torcol ordinaire	Обыкновенная вертишейка
Evening grosbeak	Abendkernbeißer		Североамериканский дубонос
Exton's barbet	Gelbstirn-Zwergbärtling	Petit barbu à front jaune	
Fairy blue wren	Blauer Staffelschwanz	Roitelet bleu	
– bluebirds	Elfenblauvögel	Oiseaux bleus	
– – and Leafbirds	Blattvögel	Irénidés	
Fan-tailed warbler	Weitraum-Zistensänger	Cisticole des joncs	
Fawn-breasted waxbill	Sumpfastrild	Astrild de marais	
Fieldfare	Wacholderdrossel	Grive litorne	Дрозд-рябинник
Finches	Finken	Frigillidés	Вьюрковые
Fire finch	Gewöhnlicher Amarant	Amaranthe	
Firecrest	Sommergoldhähnchen	Roitelet à triple bandeau	
Fire-crested wren	–	– – – –	
Fire-crowned bishop	Flammenweber	Euplecte à couronne de feu	
– weaver	Feuerweber	Ignocolore	
Fire-tailed finch	Feuerschwanzamadine	Diamant à queue de feu	
Fire-tufted barbet	Rotbüschel-Bartvogel	Barbu à plumules de feu	
Fischer's whydah	Strohwitwe	Veuve de Fischer	
Flicker	Goldspecht	Dolapte doré	Золотой кукушковый дятел
Flower-pecker weaver-bird	Ameisenpicker	Astrild fourmilier	
Flowerpeckers	Mistelesser	Dicaeidés	Цветососы
Forked-tail roller	Grünscheitelracke	Rollier à longue queue	
Flycatchers	Fliegenschnäpperartige	Muscicapidés	Мухоловковые
Gallitos	Strichelstelzer	Gallitos	
Garden warbler	Gartengrasmücke	Fauvette des jardins	Садовая славка
Giant toucan	Riesentukan	Toucan toco	
– weaver	Riesenweber	Tisserin géant	
Gnateaters	Mückenesser	Conopophages	
Goldcrest	Wintergoldhähnchen	Roitelet huppé	
Gold-crested crackle	Kronenatzel	Mainate couronné	
– mynah	–	– –	
– sparrow	Goldsperling	Moineau doré	
Golden-backed weaver	Goldrückenweber	Euplecte à dos doré	
Golden-banded woodpecker	Goldspecht	Colapte doré	Золотой кукушковый дятел
Golden-breasted waxbill	Goldbrüstchen	Bengali zébré	
Golden-crested wren	Wintergoldhähnchen	Roitelet huppé	
Golden oriole	Pirol	Loriot d'Europe	Обыкновенная иволга
Golden-rumped tinker-bird	Goldbürzel-Zwergbärtling	Petit barbu à gorge blanche	
Golden-winged woodpecker	Goldspecht	Colapte doré	Золотой кукушковый дятел
Goldfinch	Stieglitz	Chardonneret élégant	Щегол
Gold-fronted fruit-sucker	Goldstirn-Blattvogel	Verdin à front d'or	
Gouldian finch	Gouldamadine	Diamant de Gould	
Grackles	Bootsschwänze		Челнохвосты
Grag martin	Gewöhnliche Felsenschwalbe	Hirondelle de rochers	Скалистая ласточка
Grasshopper warbler	Feldschwirl	Locustelle tachetée	Обыкновенный сверчок
Gray-backed white-eye	Australischer Brillenvogel		Австралийская белоглазка
Gray-cheeked thrush	Grauwangendrossel		Малый дрозд

ENGLISH NAME	GERMAN NAME	FRENCH NAME	RUSSIAN NAME
Gray silky-flycatcher	Grauseidenschnäpper	Gobe-mouche gris du Mexique	Серый тропический свиристель
Great bower-bird	Flecken-Laubenvogel	Oiseau à bercereaux tacheté	Обыкновенная пятнистая беседковая птица
Great grey shrike	Raubwürger	Pie-grièche grise	Серый сорокопут
– hill barbet	Blaukopf-Bartvogel	Barbu géant	
– hornbill	Doppelhornvogel	Calalo bicorne	
– jacamar	Breitmaul-Glanzvogel		Широкоротая якамара
– nightingale	Sprosser	Rossignol progné	Обыкновенный соловей
– reed warbler	Drosselrohrsänger	Rousserolle turdoïde	Дроздовидная камышевка
– sparrow	Riesensperling	Moineau roux	
– spotted woodpecker	Buntspecht	Pic épeiche	Большой пестрый дятел
– titmouse	Kohlmeise	Mésange charbonnière	Большая синица
Greater bird of paradise	Großer Paradiesvogel	Grand paradisier	– райская птица
– hill mynah	Beo	Mainate religieux	
– niltava	Großniltava	Grand niltava	
– racket-tailed drongo	Flaggendrongo	Drongo à raquettes	Райский дронго
Green Avadavat	Astrild	Bengali vert	
– barbet	Ceylon-Grünbartvogel	Barbu à tête grise du Ceylan	
– cat-bird	Grünkatzenvogel	Oiseau à berceaux vert	Зеленая птица-кошка
– glossy starling	Grünschwanzglanzstar	Merle métallique à oreilles bleues	Стальной скворец
– jay	Peru-Grünhäher		Перуанская голубая ворона
– singing finch	Moçambique-Girlitz	Serin de Moçambique	Мозамбикский вьюрок
– woodpecker	Grünspecht	Pic vert	Зеленый дятел
– woodpeckers	Grünspechte	Pics verts	Зеленые дятлы
Green-backed twin-spot	Grüner Tropfenastrild	Bengali vert pointillé	
Green-billed toucan	Bunttukan	Toucan à bec vert	
Green-breasted pitta	Schwarzkopfpitta	Brève sordide	
Greenfinch	Grünling	Verdier d'Europe	Обыкновенная зеленушка
Green-tailed parrot finch	Bambus-Papageiamadine	Diamant à queue verte	
Grenadier weaver	Oryxweber	Monseigneur	
Grey-headed kingfisher	Graukopfliest	Martin-chasseur à tête grise	
– mynah	Graukopfstar	Martin à tête grise	
– silver-bill	Perlhalsmadine	Spermète à tête grise	
– sparrow	Graukopfsperling	Moineau à tête grise	
– woodpecker	Grauspecht	Pic cendré	Седой дятел
Grey-necked rockfowl	Gelbkopf-Felshüpfer	Picathartes à cou blanc	
Grey-jumper	Gimpelhäher	Glaucope gris	Штрутидея
Grey singing finch	Grau-Edelsänger	Chanteur d'Afrique	Сенегальский вьюрок
– starling	Graustar	Martin gris	Серый скворец
– wagtail	Bergstelze	Bergeronette des ruisseaux	Горная трясогузка
– waxbill	Graustrild	Bec de coreil	
Groove-billed barbet	Senegalfurchenschnabel	Barbican à poitrine rouge	
Grosbeak-weaver	Weißstirnweber	Tisserin à gros bec	
Ground hornbills	Hornraben		Рогатые вороны
– thrush	Bunt-Erddrossel		Пестрый дрозд
Hair-crested drongo	Glanzspitzendrongo	Drongo à crinière	
Hawfinch	Kernbeißer	Gros-bec casse-noyaux	Дубонос
Hedge sparrow	Heckenbraunelle	Accenteur mouchet	Лесная завирушка
Helmet bird	Helmvanga		Шлемоносный сорокопут
Hill robin	Chinesischer Sonnenvogel	Rossignol du Japon	Обыкновенная солнечная птица
Himalayan blue-pie	Rotschnabel-Schweifkitta	Pirolle de la Chine	Красноклювая лазуревая сорока
Hodgson's munia	Spitzschwanz-Bronzemännchen	Domino à longue queue	
Honey creepers	Zuckervögel	Coerébidés	
– eaters	Honigesser	Méliphagidés	Медососовые
– guides	Honiganzeiger	Indicoridés	Кукушки-путеводители
Hooded crow	Nebelkrähe	Corneille mantelée	Серая ворона
– finch	Kleinelsterchen	Spermète à capuchon	
– pitta	Schwarzkopfpitta	Brève sordide	
– racket-tailed tree pie (magpie)	Spatelschwanzelster	Pie bleue de l'Himalaya	
Hoopoe	Wiedehopf	Huppe puput	Обыкновенный удод
Hoopoes	Hopfe	Upupidés	Удодовые
Hornbills	Nashornvögel	Bucérotidés	Птицы-носороги
Horned lark (N. A.)	Ohrenlerche	Alouette hausse-col	Рюм
House martin	Mehlschwalbe	Hirondelle de fenêtre	Городская ласточка
– sparrow	Haussperling	Moineau domestique	Домовый воробей
Huia	Huia	Huia	Туйя
Icterine warbler	Gelbspötter	Hypolaïs ictérine	Зеленая пересмешка
Imperial woodpecker	Kaiserspecht		Американский королевский дятел
Indian pitta	Neunfarbenpitta	Brève du Bengale	Индийская питта
– roller	Bengalenracke	Rollier d'Inde	Бенгальская сизоворонка

ENGLISH NAME	GERMAN NAME	FRENCH NAME	RUSSIAN NAME
– silver-bill	Malabarfasänchen	Bec de plomb	
– white-eye	Ganges-Brillenvogel	Zostérops à lunettes	Гангская белоглазка
Ivory-billed woodpecker	Elfenbeinspecht		Американский белоклювый дятел
Jacamars	Glanzvögel	Galbulidés	Якамары
Jackdaw	Dohle	Choucas des tours	Галка
Jameson's fire finch	Rosenamarant	Amaranthe de Jameson	
Japanese waxwing	Japanischer Seidenschwanz		Восточноазиатский свиристель
– white-eye	– Brillenvogel		Японская белоглазка
Java sparrow	Reisfink	Padda	
Javan hill mynah	Beo	Mainate religieux	
Jucatan motmot	Türkisbrauen-Sägeracke	Motmot à sourcils bleus	
Juncos	Juncos	Juncos	Юнко
King bird of paradise	Königsparadiesvogel	Paradisier royal	Королевская райская птица
– of Saxony's bird of paradise	Albert-Paradiesvogel		Чешуйчатая райская птица
Kingbirds	Satrapen	Tyrans	
Kingfisher	Eisvogel	Martin-pêcheur	Обыкновенный зимородок
Kingfishers	Eisvögel	Alcédinidés	Зимородковые
Kinglets	Goldhähnchen	Roitelets	Корольки
Kiskadee flycatcher	Bentevi	Tyran à ventre jaune	Бентеви
Kitty wren	Zaunkönig	Troglodyte mignon	Обыкновенный крапивник
Kookaburra	Kookaburra	Martin-chausseur	Зимородок-великан
Lanceolated warbler	Strichelschwirl	Locustelle lancéolée	Пятнистый сверчок
Lapland bunting	Spornammer	Bruant lapon	
– longspur (N.A.)	–	– –	
Large niltava	Großniltava	Grand niltava	
Larks	Lerchen	Alouettes	Жаворонковые
Laughing jackass	Kookaburra	Martin-chasseur	Зимородок-великан
– kingfisher	–		
Lavender finch	Schönbürzel	Gris-bleu	
– waxbill	–	–	
Lawe's bird of paradise	Blaunacken-Strahlenparadiesvogel	Sifilet de Lawe	
Leafbirds	Blattvögel	Verdins	
Lesser bird of paradise	Kleiner Paradiesvogel		Малая райская птица
– grey shrike	Schwarzstirnwürger	Pie-grièche à poitrine rose	Чернолобый сорокопут
– short-toed lark	Stummellerche	Alouette pispolette	
– spotted woodpecker	Kleinspecht	Pic épeichette	Малый пестрый дятел
– superb bird of paradise	Kragenhopf	Paradisier superbe	Чудная райская птица
– whitethroat	Klappergrasmücke	Fauvette babillarde	Славка-завирушка
Levaillant's barbet	Schwarzrücken-Bartvogel	Barbu de Levaillant	
Lilac-breasted roller	Grünscheitelracke	Rollier à longue queue	
Lineated barbet	Ceylon-Grünbartvogel	Barbu à tête grise du Ceylan	
Little king	Königsparadiesvogel	Paradisier royal	Королевская райская птица
Locus bird	Lappenstar	Martin caronculé	
Locust-finch	Heuschreckenastrild	Astrild locustelle	
Long tail	Schwarzkehl-Paradieselster		Черная шлемоносная райская птица
– – bird of paradise	–		– – – –
Long-tailed glossy starling	Langschwanzglanzstar	Merle métallique à longue queue	
– grass finch	Spitzschwanzamadine	Diamant à longue queue	
– tit	Schwanzmeise	Mésange à longue queue	Длиннохвостая синица
– tits	Schwanzmeisen		Длиннохвостые синицы
Lyre-birds	Leierschwänze	Menuridés	Птицы-лиры
Madagascar weaver	Madagaskarweber	Foudi rouge	
Magnificent bird of paradise	Prachtparadiesvogel	Paradisier magnifique	Чешуегрудый щитоносец
– riflebird	Prachtreifelvogel	Proméfil magnifique	
Magpie	Elster	Pie bavarde	Сорока
– lark	Drosselstelze	Alouette-pie	Дроздовидная трясогузка
– mannikin	Riesenelsterchen	Spermète pie	
– robin	Dajaldrossel	Merle dyal des Indes	Индийская сорочья славка
Mahali weaver-bird	Mahaliweber	Mahali	
Malabar mynah	Graukopfstar	Martin à tête grise	
Malachite kingfisher	Zwerghaubenfischer	Petit martin-pêcheur huppé	
– sunbird	Malachit-Nektarvogel	Souï-manga malachite	Малахитовая нектарка
Malaysian honey guide	Malayischer Honiganzeiger	Indicateur malais	
Manakins	Schnurrvögel	Pipridés	Манакины
Mandarin mynah	Mandarinstar	Martin de Chine	Южнокитайский малый скворец
Manyar weaver	Manyarweber	Tisserin Manyar	Манья
Marmora's warbler	Sardengrasmücke	Fauvette sarde	Сардинская славка
Marsh-tit, Marsh-titmouse	Nonnenmeise	Mésange nonnette	Черноголовая гаичка
Marsh warbler	Sumpfrohrsänger	Rousserolle verderolle	Болотная камышевка

ENGLISH NAME	GERMAN NAME	FRENCH NAME	RUSSIAN NAME
Masked grass finch	Maskenamadine	Diamant à masque	
– shrike	Maskenwürger	Pie-grièche masquée	Маскированный сорокопут
– waxbill	Larvenamarant	Amaranthe masquée d'Abyssinie	
Meadow pipit	Wiesenpieper	Pipit farlouse	Луговой конек
Melba finch	Buntastrild	Bea-marquet	
Melodious warbler	Orpheusspötter	Hypolaïs polyglotte	Многоголосая камышевка-пеночка
Meyer's sickle bill	Sichelschnabel	Epimagne de Meijer	
Middle spotted woodpecker	Mittelspecht	Pic mar	Средний дятел
Minivets	Mennigvögel	Minivets	
Mistle thrush	Misteldrossel	Grive draine	Дрозд-деряба
Mistletoe bird	Schwalben-Mistelesser		Ласточковый цветосос
Mocking-bird	Spottdrossel	Merle moqueur commun	
Mocking-birds	Spottdrosseln	Mimidés	
Mossie	Kapsperling	Moineau mélanure	Южноафриканский воробей
Motmots	Sägeracken	Momotidés	Момоты
Moustached warbler	Mariskensänger	Lusiniole à moustaches	Тонкоклювая камышевка
Naked-faced barbet	Glatzenbartvogel	Grand barbu chauve	
Naked-throated bellbird	Nacktkehl-Glockenvogel	Oraponga à gorge nue	Бразильская птица-колокольчик
Namaqua masked weaver	Maskenweber	Tisserin à front noir	
Natal kingfisher	Zwergkönigsfischer	Martin-pêcheur de Natal	
New-Zealand wren	Neuseeland-Schlüpfer	Roitelet de la Nouvelle-Zélande	
Nightingale	Nachtigall	Rossignol philomèle	Западный соловей
Noisemakers	Schreivögel	Mésomyodés	Кричащие птицы
Northern shrike (N.A.)	Raubwürger	Pie-grièche grise	Серый сорокопут
– three-toed woodpecker	Dreizehenspecht		Трехпалый дятел
Northwestern shrike (N.A.)	Raubwürger		Серый сорокопут
Nubian carmine bee-eater	Scharlachspint	Guêpier carminé	
Nutcracker	Tannenhäher	Casse-noix moucheté	Ореховка
Nuthatch	Kleiber	Sittelle torchepot	Обыкновенный поползень
Nuthatches	– Eigentliche Kleiber	Sittidés, Sittelles	Поползни, Настоящие поползни
Nutmeg mannikin	Muskatfink	Damier	
Old World warblers	Waldsänger	Parulidés	
Olivaceous warbler	Blaßspötter	Hypolaïs pâle	Большая бормотушка
Olive finch	Groß-Kubafink	Grand chanteur de Cuba	
– weaver	Kapweber	Tisserin à front d'or	
– weaver-finch	Halsbandastrild	Bengali vert d'Ansorge	
Olive-tree warbler	Olivenspötter	Hypolaïs des oliviers	Средиземноморская пересмешка
Orange bishop	Feuerweber	Ignicolore	
Orange-cheeked waxbill	Orangebäckchen	Bengali à joues oranges	
Orange-crested gardener	Rothaubengärtner	Jardinier à huppe orange	
Orange-headed ground thrush	Damadrossel	Grive orangée	
Oregon junco	Oregon-Junco		Орегонский юнко
Oriental broad-billed roller	Ostroller	Rollier à gros bec	Широкорот
– dollar bird	–	– – – –	–
– roller			
Orioles	Stärlinge, Trupiale	Ictéridés	Касиковые, Трупиалы
Oriols	Pirole	Loriots	Иволговые
Ornate umbrella bird	Schirmvogel	Céphaloptère orné	Амазонская зонтичная птица
Orphean warbler	Orpheusgrasmücke	Fauvette orphée	Певчая славка
Ortolan bunting	Ortolan	Bruant ortolan	Садовая овсянка
Ovenbird	Töpfervogel	Fournier roux	Рыжий печник
Ovenbirds	Töpfervögel	Furnariidés	Печники
Oxbirds	Madenhackerstare	Pique-bœufs	Волоклюи
Oxeye-tit	Kohlmeise	Mésange charbonnière	Большая синица
Pagoda starling	Pagodenstar	Martin des pagodes	Браминский скворец
Painted finch	Gemalter Astrild	Emblème peint	
Pallas's grasshopper warbler	Streifenschwirl	Locustelle de Pallas	Певчий сверчок
Palmchat	Palmschmätzer	Oiseau palmiste	
Pampas-flicker	Camposspecht	Colapte des Pampas	
Paradise flycatcher	Paradiesschnäpper	Gobe-mouche paradisier	Индийская длиннохвостая мухоловка
– sparrow	Rotkopf-Amadine	Amadine à tête rouge	
Parasitic weaver	Kuckucksweber	Tisserin parasitique	
Parrot crossbill	Kiefernkreuzschnabel	Parrot crossbill	Клест-сосновик
– finch	Rotköpfige Papageiamadine	Diamant à tête rouge	
Parson finch	Gürtelgrasfink	– à bavette	
Pectoral finch	Weißbrust-Schilffink	Donacole à poitrine blanche	
Pegu (House) sparrow	Gelbbauchsperling	Moineau flavéole	
Pekin nightingale	Chinesischer Sonnenvogel	Rossignol du Japon	Обыкновенная солнечная птица

ENGLISH NAME	GERMAN NAME	FRENCH NAME	RUSSIAN NAME
– robin	Chinesischer Sonnenvogel	Rossignol du Japon	Обыкновенная солнечная птица
Penduline	Beutelmeise	Mésange penduline	Обыкновенный ремез
– pit	–	– –	–
– tits	Beutelmeisen		Ремезовые
Peter's twin-spot	Roter Tropfenastrild	Amarante enflammée	
Phainopepla	Seidenschnäpper i. e. S.	Phainopépla resplendissant	
Philepittas	Lappenpittas	Philepittidés	
Pictorella finch	Weißbrust-Schilffink	Donacole à poitrine blanche	
Pied barbet	Rotstirn-Bartvogel	Barbu pie	
– crow-shrike	Würgerkrähe	Grand calibé	
– flycatcher	Trauerschnäpper	Gobe-mouche noir	Мухоловка-пеструшка
– grallina	Drosselstelze	Alouette-pie	Дроздовидная трясогузка
– kingfisher	Graufischer		Малый пегий зимородок
– mynah	Elsterstar	Martin pie	
– starling	–	–	
– wagtail	Trauerbachstelze	Bergeronette d'Yarrell	Британская белая трясогузка
Pigmy nuthatch	Zwergkleiber		Поползень-крошка
Pileated jay	Kappengrünhäher	Pie akahé	Хохлатая голубая ворона
Pine bunting	Fichtenammer		Белошапочная овсянка
– grosbeak	Hakengimpel	Dur-bec des sapins	Щур
Pin-tailed nonpareil	Lauchgrüne Papageiamadine	Quadricolore	
– whydah	Dominikanerwitwe	Veuve dominicaine	
Piping crows	Würgerkrähen		Кричащие вороны
Pipits	Pieper i. e. S.	Pipits	Коньки
Pittas	Pittas	Pittidés, Brèves	Питты
Plaintive barbet	Trauerbartvogel	Barbu à poitrine orange	
Plantcutters	Pflanzenmäher	Phytotomidés	Косцы
Plum-headed finch	Zeresfink	Modeste	
Preacher birds	Tukane	Rhamphastidés	Туканы
Princess Stephanie's bird of paradise	Prinzessin-Stephanie-Paradiesvogel	Paradisier de la princesse Stéphanie	
Puffbirds	Faulvögel	Bucconidés	Птицы-пуховки
Purple glossy starling	Purpurglanzstar	Merle bronzé pourpre	
Purple grenadier	Veilchenastrild	Grenadin à poitrine bleue	
Purple-bellied waxbill	–		
Queens whydah	Königswitwe	Veuve reine	
Rainbow bird	Schmuckspint	Guêpier d'Australie	
Raven	Kolkrabe	Grand corbeau	Ворон
Red-backed flower-pecker	Scharlach-Mistelesser	Grimpereau à dos rouge	Красноспинный цветосос
– shrike	Neuntöter	Pie-grièche écorcheur	Европейский жулан
– hornbill	Rotschnabeltoko	Calao tock	
Red-bellied waxbill	Wellenastrild	Astrild ondulé	Восковоклювый ткач
Red-billed blue magpie	Rotschnabel-Schweifkitta	Pirolle de la Chine	Красноклювая лазуревая сорока
– hornbill	Rotschnabeltoko	Calao tock	
– ox-pecker	Rotschnabel-Madenhacker	Pique-bœuf à bec rouge	Красноклювый волоклюй
– quelea	Blutschnabelweber	Travailleur	
Red bishop	Oryxweber	Monseigneur	
Redbreast	Rotkehlchen i. e. S.	Rouge-gorge familier	Зарянка
Red-breasted flycatcher	Zwergschnäpper	Gobe-mouche nain	Малая мухоловка
– nuthatch	Kappenkleiber		Черноголовый поползень
– toucan	Bunttukan	Toucan à bec vert	
Red-browed finch	Dornastrild	Astrild à cinq couleurs	
Red-cheeked cordon-bleu	Schmetterlingsfink	Cordon bleu	
– scimitar-babbler	Rotwangen-Säbler	Pomatorhin à joues rouges	
Red-collared whydah	Schildwida	Veuve en feu	
Red-crested finch	Purpur-Kronfink	Pinson couronné	
Red-eared bulbul	Rotohrbülbül	Bulbul à oreillons rouges	
– firetail finch	Rothramadine	Astrild à oreillons rouges	
Red-faced waxbill	Wienerastrild	Pytilie à dos jaune	
Red-fronted finch	Schuppenköpfchen	Sénégali à front pointillé	
Red-headed barbet	Rotbrust-Buntbärtling	Barbu de Bourcier	
– bunting	Braunkopfammer	Bruant à tête rousse	Желчная овсянка
– diock	Rotkopfweber	Quéléa à tête rouge	
– finch	Rotkopf-Amadine	Amadine à tête rouge	
– parrot finch	Rotköpfige Papageiamadine	Diamant à tête rouge	
– quelea	Rotkopfweber	Quéléa à tête rouge	
– weaver	Scharlachweber	Républicain à capuchon écarlate	
– woodpecker	Rotkopfspecht	Pic à tête rouge	Американский красноголовый дятел
Red ovenbird	Töpfervogel	Fournier roux	Рыжий печник
Redpoll	Birkenzeisig	Sizerin flammé	Обыкновенная чечетка

ANIMAL DICTIONARY

ENGLISH NAME	GERMAN NAME	FRENCH NAME	RUSSIAN NAME
Red-rumped swallow	Rötelschwalbe	Hirondelle rousseline	Рыжепоясничная ласточка
Red-sided titmouse	Buntmeise		Японская синица
Redstart	Gartenrotschwanz	Rouge-queue à front blanc	Обыкновенная горихвостка
Red-tailed grass-finch	Binsenastrild	Diamant à queue rousse	
Red-throated pipit	Rotkehlpieper	Pipit à gorge rousse	Краснозобый конек
– thrush	Rotkehldrossel	Grive à cou roux	– дрозд
Red-vented bulbul	Tonkibülbül	Bulbul Indien	
Red-waings	Hordenvögel		Болотные трупиалы
Redwing	Rotdrossel	Grive mauvis	Дрозд-белобровик
Red-wing shrike	Kaptschagra		Чагра
– starling	Rotschwingenstar	Roupenne	
Red-whiskered bulbul	Rotohrbülbül	Bulbul à oreillons rouges	
Reed bunting	Rohrammer	Bruant des roseaux	Камышовая овсянка
– warbler	Teichrohrsänger	Rousserolle effarvatte	Тростниковая камышевка
Regent bower-bird	Samtgoldvogel	Oiseau-régent	
Reichenow's crimson-wing	Reichenows Bergastrild	Bengali vert à face rouge	
Resplendent whydah	Glanzwitwe	Veuve métallique	
Rhinoceros hornbill	Rhinozerosvogel		Малайский калао
Richard's pipit	Spornpieper	Pipit de Richard	
Riflebirds	Reifelvögel		Щитоносные райские птицы
Ring ouzel	Ringdrossel	Merle à plastron	Белозобый дрозд
Robin	Rotkehlchen i. e. S.	Rouge-gorge familier	Зарянка
Robin (N. A.)	Wanderdrossel	Merle migrateur	Странствующий дрозд
Rock bunting	Siebenstreifen-Ammer	Bruant à sept raies	
– flicker	Andenspecht		Андский дятел
– nuthatch	Felsenkleiber	Sittele des rochers	Малый скалистый поползень
– sparrow	Steinsperling	Moineau fou	Каменный воробей
– sparrows	Steinsperlinge		Каменные воробьи
– thrush	Steinrötel	Merle de roches	Пестрый каменный дрозд
Rollers	Racken, Blauracken i. e. S.	Coraciadidés, Rolliers vrais	Сизоворонковые, Сизоворонки
Rook	Saatkrähe	Corbau freux	Грач
Rose-coloured starling	Rosenstar	Martin roselin	Розовый скворец
Rosy finches	Rosenfinken		Горные вьюрки
– pastor	Rosenstar		Розовый скворец
– twin-spot	Perlastrild	Astrild de Verreaux	
Rothschild's crackle	Balistar	Martin de Rothschild	Балийский скворец
Royal antbird	Königs-Ameisenstelzer	Fourmillier	
– ant-thrush	Königs-Ameisenstelzer	–	
– starling	Königsglanzstar	Spréo royal	
Rufous scrub-bird	Kleiner Dickichtschlüpfer	Atrichorne roux	
– sparrow	Riesensperling	Moineau roux	
Rufous-bellied babbler	Rotbauchtimalie	Dumétie à ventre roux	
– niltava	Schwarzkehl-Niltava	Niltava à ventre rouge	
Rufous-fronted fantail	Rotstirn-Fächerschnäpper		Рыжелобая веерохвостка
Rufous-tailed jacamar	Rotschwanzjakamar		Краснохвостая якамара
Rüppell's warbler	Maskengrasmücke	Fauvette masquée	Эгейская славка
Sacred kingfisher	Götzenliest	Halcyon sacré	
Sand martin	Uferschwalbe	Hirondelle de rivage	Береговая ласточка
Sardinian warbler	Samtkopf-Grasmücke	Fauvette mélanocéphale	Средиземноморская славка
Satin bower-bird	Seidenlaubenvogel	Oiseau à berceaux satiné	
Savi's warbler	Rohrschwirl	Locustelle luscinioïde	Соловьиный сверчок
Scaly weaver	Schnurrbärtchen	Tisserin à front pointillé	
Scaly-crowed weaver	–	– – – –	
Scarlet flycatcher	Rubinköpfchen	Tyran rouge écarlate	
– grosbeak	Karmingimpel	Roselin cramoisi	Обыкновенная чечевица
Scarlet-crowned barbet	Trauerbartvogel	Barbu à poitrine orange	
Scrub-birds	Dickichtschlüpfer	Atrichornes	
Sedge warbler	Schilfrohrsänger	Phragmite des joncs	Камышевка-барсучок
Serin	Girlitz	Serin cini	Канареечный вьюрок
Serins	Girlitze	Serins	Канареечные вьюрки
Shama	Schamadrossel	Merle Shama	Малабарская сорочья славка
Sharp-tailed munia	Spitzschwanz-Bronzemännchen	Domino à longue queue	
Shining calornis	Spinnenstar	Stourne métallique	
– starling	–	– –	
Shiny black whydah	Glanzwitwe	Veuve métallique	
Shore lark	Ohrenlerche	Alouette hausse-col	Рюм
Short-toed tree creeper	Gartenbaumläufer	Grimpereau des jardins	Короткопалая пищуха
Shrikes	Würger, Würger i. e. S.	Laniidés, Pie-grièches	Сорокопутовые, Сорокопуты
Siberian accentor	Bergbraunelle		Сибирская завирушка
– jay	Unglückshäher	Geai de Sibérie	Кукша

ENGLISH NAME	GERMAN NAME	FRENCH NAME	RUSSIAN NAME
Silver-eared mesia	Silberohr-Sonnenvogel	Mésia à joues argentée	
Singing birds	Singvögel	Oiseaux chanteurs	Певчие птицы
Siskin	Erlenzeisig	Tarin des aulnes	Чиж
Sitellas	Australkleiber		Австралийские поползни
Six-plumed bird of paradise	Blaunacken-Strahlenparadiesvogel	Sifilet de Lawe	
Sky lark	Feldlerche	Alouette des champs	Полевой жаворонок
Slate-coloured junco	Winter-Junco	Junco ardoisé	Серый юнко
Snow bunting	Schneeammer	Bruant des neiges	
– buntings	Schneeammern	Bruants des neiges	Пуночки
– finch	Schneefink	Niverolle des Alpes	Снежный вьюрок
– finches	Schneefinken	Niverolles	Снежные вьюрки
Sociable weaver	Siedelweber	Tisserin social	Обыкновенный общественный ткач
Social weaver	Siedelweber	Tisserin social	Обыкновенный общественный ткач
Song thrush	Singdrossel	Grive musicienne	Певчий дрозд
South-African quail-finch	Wachtelastrild	Astrild caille à lunettes	
– rock sparrow	Augenbrauensperling	Moineau à sourcils	
Southern masked weaver	Maskenweber	Tisserin à front noir	
Spangled drongo	Glanzspitzendrongo	Drongo à crinière	
Sparrows	Sperlinge, Sperlinge i. e. S.	Passerinés, Moineaux	Воробьи, Настоящие воробьи
Speckle-fronted weaver	Schuppenköpfchen	Sénégali à front pointillé	
Speckled tinker-bird	Flecken-Zwergbärtling	Petit barbu grivelé	
Spectacled finch	Weißbauchnonne	Nonnette à ventre roux	
– warbler	Brillengrasmücke	Fauvette à lunettes	Очковая славка
Spice finch	Muskatfink	Damier	
Spider-hunters	Spinnenjäger		Паукоеды
Splendid glossy starling	Prachtglanzstar	Merle métallique	Великолепный скворец
Spot-billed toucanet	Goldohr-Arassari	Toucanet à bec tâcheté	
Spotless starling	Einfarbstar	étourneau unicolore	Черный скворец
Spotted-backed weaver	Textorweber	Tisserin Cap Moor	
Spotted bower-bird	Flecken-Laubenvogel	Oiseau à bercereaux tacheté	Обыкновенная пятнистая беседковая птица
– flycatcher	Grauschnäpper		Серая мухоловка
– pardalote	Flecken-Panthervogel	Pardalote pointillé	Пятнистая пантерная птица
Spreo starling	Dreifarben-Glanzstar	Spréo superbe	
Standard-wing	Wallace-Paradiesvogel	Paradisier de Wallace	Вымпельная райская птица
Star finch	Binsenastrild	Diamant à queue rousse	
Starling	Gemeiner Star	Étourneau sansonnet	Обыкновенный скворец
Starlings	Stare	Étourneaux	Скворцовые
Steller's jay	Schwarzkopfhäher		Черноголовая голубая сойка
St. Helena waxbill	Wellenastrild	Astrild ondulé	Восококлювый ткач
Stonechat	Schwarzkehlchen	Traquet pâtre	Черноголовый чекан
Stone thrush	Steinrötel	Merle de roches	Пестрый каменный дрозд
St. Thomas weaverbaird	Riesenweber	Tisserin géant	
Subalpine warbler	Weibart-Grasmücke	Fauvette passerinette	Горная славка
Sulphur-breasted toucan	Dottertukan	Toucan à bec caréné	
Sulphury tyrant	Bentavi	Tyran à ventre jaune	Бентеви
Summer tanager	Feuertangare		Красная тангара
Sunbirds	Nektarvögel, Langschwanz-Nektarvögel	Nectariniidés	Нектарковые, Длиннохвостые нектарки
Superb bird of paradise	Kragenhopf	Paradisier superbe	Чудная райская птица
– glossy starling	Dreifarben-Glanzstar	Spréo superbe	
– lyre-bird	Leierschwanz i. e. S.	Oiseaux-lyre	
Swainson's toucan	Braunrückentukan	Toucan tocard	
Swallows	Schwalben, Rauchschwalben	Hirondelles	Ласточковые, Касатки
Swallow-wing	Schwalben-Faulvogel		Ласточкокрылая пуховка
Sydney waxbill	Dornastrild	Astrild à cinq couleurs	
Syrian woodpecker	Blutspecht	Pic syriaque	Сирийский дятел
Tahaweaver	Tahaweber	Tisserin taha	
Tailor bird	Schneidervogel	Fauvette couturière	Славка-портниха
Tanagers	Tangare, Echte Tangare		Тангары, Настоящие тангары
Tasmanian crow-shrike	Weißrücken-Flötenvogel		Белоспинная ворона-свистун
Tawny pipit	Brachpieper	Pipit rousseline	Полевой конек
Thick-billed forest weaver	Purpurastrild	Pyreneste ponceau à ventre noir	
– weaver	Weißstirnweber	Tisserin à gros bec	
Three-coloured mannikin	Schwarzbauchnonne	Jacobin	
– parrot finch	Dreifarbige Papageiamadine	Diamant tricolore de Kittlitz	
Three-toed woodpecker	Dreizehenspecht	Pic tridactyle	Трехпалый дятел

ENGLISH NAME	GERMAN NAME	FRENCH NAME	RUSSIAN NAME
Thrushes	Drosseln	Turdinés	Дроздовые
Tinker-birds	Zwergbartvögel	Petit barbus	
Titmice	Eigentliche Meisen	Paridés	Настоящие синицы
Toco toucan	Riesentukan	Toucan toco	
Todies	Todis	Todidés	Тодиевые
Toucan barbet	Tukan-Bartvogel	Barbu toucan	Туканий бородастик
Toucan-billed barbet	–	– –	– –
Toucans	Tukane	Rhamphastidés	Туканы
Townsend's solitaire	Clarino	Solitaire de Townsend	
Tree creeper	Waldbaumläufer	Grimpereau des bois	Обыкновенная пищуха
– pipit	Baumpieper	Pipit des arbres	Лесной конек
– sparrow	Feldsperling	Friquet	Полевой воробей
Treerunners	Australkleiber		Австралийские поползни
True shrikes	Eigentliche Würger		Настоящие сорокопуты
Turquoise dacnis	Pitpit i. e. S.		Питпит
Twelve-wired bird of paradise	Fadenhopf	Paradisier multifil	Нитчатая райская птица
Twite	Berghänfling	Linotte à bec jaune	Горная чечетка
Two-barred crossbill	Bindenkreuzschnabel	Bec-croisé bifascié	Белокрылый клест
Tyrant flycatchers	Tyrannen	Tyrannidés	Тираны
Umbrella bird	Schirmvogel	Céphaloptère orné	Амазонская зонтичная птица
Vangas	Blauwürger	Vangidés	
Vanga-shrikes	–		
Velvet pitta	Schwarzlappenpitta	Philépitte veloutée	
Velvet-fronted nuthatch	Samtstirnkleiber	Sittelle à front noir	
Vermilion flycatcher	Rubinköpfchen	Tyran rouge écarlate	
Verreaux's twin-spot	Perlastrild	Astrild de Verreaux	
Village weaver	Textorweber	Tisserin Cap Moor	
Violet-backed starling	Amethyst-Glanzstar	Merle violet à ventre blanc	Аметистовый скворец
Violet-eared waxbill	Granatastrild	Grenadin	
Vireos	Vireos	Viréonidés	Листовые сорокопуты
Wagtails	Stelzen	Bergeronettes	Трясогузковые
Wallace's standard-wing	Wallace-Paradiesvogel	Paradisier de Wallace	Вымпельная райская птица
Wall-creeper	Mauerläufer	Tichodrome échelette	Краснокрылый стенолаз
Wall-creepers	Mauerläufer		Стенолазы
Warblers	Grasmücken i. e. S.	Fauvettes	Славкз
Water pipit	Wasserpieper	Pipit spioncelle	Горный конек
Wattle-birds	Neuseeländische Lappenvögel		Лоскутные вороны
Wattled bellbird	Hämmerling		Молотобой
– bird of paradise	Gabelschwanz-Paradiesvogel		Вилохвостая райская птица
– starling	Lappenstar	Martin caronculé	
Waxbills	Eigentliche Astrilde, Prachtfinken	Astrilds, Estrildidés	Кровяно-красные ткачи
Waxwings	Seidenschwänze	Bombycillidés	Свиристелевые
Wheatear	Steinschmätzer	Traquet motteux	Обыкновенная каменка
Weaverbirds	Webervögel i. w. S.	Plocéidés	Ткачиковые
Whinchat	Braunkehlchen	– des prés	Луговой чекан
White-backed woodpecker	Weißrückenspecht	Pic à dos blanc	Белоспинный дятел
White-bellied amethyst-starling	Amethyst-Glanzstar	Merle violet à ventre blanc	Аметистовый скворец
White-breasted kingfisher	Braunliest	Halcyon à poitrine blanche	Красноносый зимородок
White-browed sparrow-weaver	Mahaliweber	Mahali	
– wood swallow	Weißbrauen-Schwalbenstar	Langrayen à sourcils blancs	Белобровый ласточковый скворец
White-capped redstart	Weißkopf-Rotschwanz	Rouge-queue à tête blanche	Белошапочная горихвостка
White-collared flycatcher	Halsbandschnäpper	Gobe-mouche à collier	Мухоловка-белошейка
– olive-back	Halsbandastrild	Bengali vert d'Ansorge	
White-crested laughing jay-trush	Haubenhäherling	Garrulax à huppe blanche	
White-crowned sparrow	Weißkehl-Ammerfink	Pinson à couronne blanche	
White-eyes	Brillenvögel	Zostéropidés	Белоглазковые
White-headed buffalo weaver	Starweber	Dinemelli à tête blanche	
– mannikin	Weißkopfnonne	Nonnette à tête blanche	
White helmet shrike	Brillenwürger		
White-naped honey-eater	Weißnacken-Honigesser	Méliphage à tête noire	Очковый сорокопут
Whitethroat	Dorngrasmücke	Fauvette grisette	Серая славка
White-throated cat-bird	Weißkehl-Katzenvogel		Белозобая птица-кошка
White wagtail	Bachstelze	Bergeronette grise	Белая трясогузка
White-winged crossbill	Bindenkreuzschnabel	Bec-croisé bifascié	Белокрылый клест
– crow (chough)	Bergkrähe	Corcorax à ailes blanches	
Whydahs	Glanz- und Dominikanerwitwen	Veuves	Вдовушки
Willow wren	Fitis	Pouillot fitis	Пеночка-весничка
Wilson's bird of paradise	Blauköpfiger Paradiesvogel	Paradisier républicain	
Winter wren (N.A.)	Zaunkönig	Troglodyte mignon	Обыкновенный крапивник
Wood hewers	Baumsteiger	Dendrocolaptidés	Древолазы

ENGLISH NAME	GERMAN NAME	FRENCH NAME	RUSSIAN NAME
– lark	Heidelerche	Alouette lulu	Лесной жаворонок
– swallows	Schwalbenstare		Ласточковые скворцы
Woodchat shrike	Rotkopfwürger	Pie-grièche à tête rousse	Красноголовый сорокопут
Wood-hoopoes	Baumhopfe	Phoeniculinés	Лесные удоды
Woodpeckers	Spechte	Picidés	Дятловые
Wood-wren	Waldlaubsänger	Pouillot siffleur	Пеночка-трещотка
Wren	Zaunkönig	Troglodyte mignon	Обыкновенный крапивник
Wrens	Zaunkönige, Neuseeland-Schlüpfer	Troglodytidés, Xenicidés	Крапивниковые
Yellow-backed weaver	Samtwida	Gros-bec tacheté du Cap de bonne Espérance	
Yellow-bellied waxbill	Schwarzbäckchen	Joue noire	
Yellow-billed blue-pie	Gelbschnabel-Schweifkitta	Pirolle à bec jaune	Желтоклювая лазуревая сорока
– magpie	Gelbschnabelelster		Калифорнийская сорока
Yellow bishop	Samtwida	Gros-bec tacheté du Cap de bonne Espérance	
Yellow-breasted barbet	Perlbartvogel	Barbu perlé	
Yellow bunting	Goldammer	Bruant jaune	Обыкновенная овсянка
Yellow-eyed babbler	Goldaugentimalie i. e. S.	Timalie aux yeux d'or	
Yellow-fronted canary	Moçambique-Girlitz	Serin de Moçambique	Мозамбикский вьюрок
– tinker-bird	Gelbstirn-Zwergbärtling	Petit barbu à front jaune	
Yellowhammer	Goldammer	Bruant jaune	Обыкновенная овсянка
Yellow-rumped finch	Gelber Schilffink	Donacole à tête grise	
Yellow-shafted flicker	Goldspecht	Colapte doré	Золотой кукушковый дятел
Yellow-tailed finch	Gelber Schilffink	Donacole à tête grise	
Yellow-throated sparrow	Augenbrauensperling	Moineau à sourcils	
– tinker-bird	Gelbkehl-Zwergbärtling	Petit barbu à gorge jaune	
Yellow-vented bulbul	Gelbbauchbülbül	Bulbul à ventre jaune	
Zebra finch	Zebrafink	Diamant mandarin	

II. German–English–French–Russian

Unterartnamen werden meist aus den Artnamen durch Voranstellen von Eigenschaftswörtern oder geographischen Bezeichnungen gebildet. In diesem Teil des Tierwörterbuchs sind so gebildete deutsche Unterartnamen sowie die wissenschaftlichen Unterartnamen in der Regel nicht aufgeführt.

GERMAN NAME	ENGLISH NAME	FRENCH NAME	RUSSIAN NAME
Abendkernbeißer	Evening grosbeak		Североамериканский дубонос
Acanthis		Linottes	Чечетки
– *cannabina*	Common linnet	Linotte mélodieuse	Коноплянка
– *flammea*	Redpoll	Sizerin flammé	Обыкновенная чечетка
– *flavirostris*	Twite	Linotte à bec jaune	Горная чечетка
– *hornemanni*	Arctic redpoll	Sizerin blanchâtre	Полярная чечетка
Acridotheres cristatellus	Chinese crested mynah	Martin huppé	Хохлатая майна
– *ginginianus*	Bank mynah	– de rivage	Береговая майна
– *tristis*	Common mynah	– triste	Саранчовый скворец
Acrocephalus		Rousserolles	Камышевки
– *arundinaceus*	Great reed warbler	Rousserolle turdoïde	Дроздовидная камышевка
– *dumetorum*	Blyth's reed warbler	– des buissons	Садовая камышевка
– *melanopogon*	Moustached warbler	Lusiniole à moustaches	Тонкоклювая камышевка
– *paludicola*	Aquatic warbler	Phragmite aquatique	Вертлявая камышевка
– *palustris*	Marsh warbler	Rousserolle verderolle	Болотная камышевка
– *schoenobaenus*	Sedge warbler	Phragmite des joncs	Камышевка-барсучок
– *scirpaceus*	Reed warbler	Rousserolle effarvatte	Тростниковая камышевка
Aegintha temporalis	Sydney waxbill	Astrild à cinq couleurs	
Aegithalidae	Long-tailed tits		Длиннохвостые синицы
Aegithalos caudatus	Long-tailed tit	Mésange à longue queue	Длиннохвостая синица
Afrikanische Breitrachen	Common African broadbills	Gobe-mouches	
Agelaius	Blackbirds		Болотные трупиалы
Aidemosyne modesta	Cherry finch	Modeste	
Ailuroedes	Cat-birds		Птицы-кошки
– *buccoides*	White-throated cat-bird		Белозобая птица-кошка
– *crassirostris*	Green cat-bird	Oiseau à berceaux vert	Зеленая птица-кошка
Alauda arvensis	Sky lark	Alouette des champs	Полевой жаворонок
Alaudidae	Larks	Alouettes	Жаворонковые
Albert-Paradiesvogel	King of Saxony's bird of paradise		Чешуйчатая райская птица
Alcedinidae	Kingfishers	Alcédinidés	Зимородковые
Alcedo atthis	Kingfisher	Martin-pêcheur	Обыкновенный зимородок

GERMAN NAME	ENGLISH NAME	FRENCH NAME	RUSSIAN NAME
Alpenbraunelle	Alpine accentor	Accenteur alpin	Альпийская завирушка
Alpendohle	– chough	Chocard à bec jaune	– галка
Alpenkrähe	Chough	Crave à bec rouge	Клушица
Amadina erythrocephala	Red-headed finch	Amadine à tête rouge	
– *fasciata*	Cut-throat	Cou coupé	
Amandava amandava	Avadavat	Bengali rouge	
– *formosa*	Green Avadavat	– vert	
– *subflava*	Golden-breasted waxbill	– zébré	
Amblyornis inornatus	Brown gardener		Новогвинейская беседковая птица
– *subalaris*	Orange-crested gardener	Jardinier à huppe orange	
Amblyospiza albifrons	Thick-billed weaver	Tisserin à gros bec	
Amblyospizinae		Amblyospizinés	Толстоклювые ткачи
Ameisenpicker	Antpecker	Astrild fourmilier	
Ameisenvögel	Ant birds	Formicariidés	Птицы-муравьеды
Ameisenwürger	– shrikes	Thamnophiles	
Amethyst-Glanzstar	White-bellied amethyst-starling	Merle violet à ventre blanc	Аметистовый скворец
Ammern	Buntings	Embérizinés	Овсянки
Ampeliceps coronatus	Gold-crested crackle	Mainate couronné	
Amsel	Blackbird	Merle noir	Черный дрозд
Anaplectes rubriceps	Red-headed weaver	Républicain à capuchon écarlate	
Andenspecht	Rock flicker		Андский дятел
Andigena nigrirostris	Black-billed mountain toucan	Toucan à bec noir	
Angola-Schmetterlingsfink	Blue-breasted waxbill	Cordon bleu d'Angola	
Anomalospiza imberbis	Cuckoo-weaver	Tisserin parasitique	
Anthus	Pipits	Pipits	Коньки
– *campestris*	Tawny pipit	Pipit rousseline	Полевой конек
– *cervinus*	Red-throated pipit	– à gorge rousse	Краснозобый конек
– *novaeseelandiae*	Richard's pipit	– de Richard	
– *pratensis*	Meadow pipit	– farlouse	Луговой конек
– *spinoletta*	Water pipit	– spioncelle	Горный конек
– *trivialis*	Tree pipit	– des arbres	Лесной конек
Aplonis metallica	Shining starling	Stourne métallique	
Arachnothera	Spider-hunters		Паукоеды
Artamidae	Wood swallows		Ласточковые скворцы
Artamus fuscus	Ashy-wood-swallow		Серый ласточковый скворец
– *superciliosus*	White-browed wood swallow	Langrayen à sourcils blancs	Белобровый ласточковый скворец
Astrapia nigra	Long tail bird of paradise		Черная шлемоносная райская птица
– *stephaniae*	Princess Stephanie's bird of paradise	Paradisier de la princesse Stéphanie	
Astrild	Green Avadavat	Bengali vert	
Atrichornis	Scrub-birds	Atrichornes	
– *rufescens*	Rufous scrub-bird	Atrichorne roux	
Augenbrauensperling	Yellow-throated sparrow	Moineau à sourcils	
Aulacorhynchus prasinus	Emerald toucanet	Toucanet vert	
Aurora-Astrild	Aurora waxbill	Diamant aurore	
Australischer Brillenvogel	Gray-backed white-eye		Австралийская белоглазка
Australkleiber	Treerunners		Австралийские поползни
Bachstelze	White wagtail	Bergeronette grise	Белая трясогузка
Baillonius bailloni	Baillon's toucan	Toucan de Baillon	
Bajaweber	Baya weaver	Tisserin Baya	Бойя
Balistar	Bali-mynah	Martin de Rothschild	Балийский скворец
Baltimore-Trupial	Baltimore oriole		Балтиморский трупиал
Bambus-Papageiamadine	Green-tailed parrot finch	Diamant à queue verte	
Bandfink	Cut-throat	Cou coupé	
Bartmeise	Bearded titmouse	Panure à moustaches	
Bartvögel	Barbets	Capitonidés	Бородастики
Bathilda ruficauda	Star finch	Diamant à queue rousse	
Baumhopfe	Wood-hoopoes	Moqueurs	Лесные удоды
Baumläufer	Creepers	Grimpereaux	Пищухи
Baumläuferartige	–	Certhiidés	Пищуховые
Baumpieper	Tree pipit	Pipit des arbres	Лесной конек
Baumsteiger	Wood hewers	Dendrocolaptidés	Древолазы
Bengalenracke	Indian roller	Rollier d'Inde	Бенгальская сизоворонка
Bentevi	Kiskadee flycatcher	Tyran à ventre jaune	Бентеви
Beo	Javan hill mynah	Mainate religieux	
Bergamadine	Crimson-bellied mountain finch	Diamant des montagnes	
Bergbraunelle	Siberian accentor		Сибирская завирушка
Bergfink	Brambling	Pinson du nord	Юрок
Berghänfling	Twite	Linotte à bec jaune	Горная чечетка
Bergkrähe	White-winged crow (chough)	Corcorax à ailes blanches	
Berglaubsänger		Pouillot de Bonelli	Светлобрюхая пеночка
Bergstelze	Grey wagtail	Bergeronette des ruisseaux	Горная трясогузка

GERMAN NAME	ENGLISH NAME	FRENCH NAME	RUSSIAN NAME
Beutelmeise	Penduline	Mésange penduline	Обыкновенный ремез
Beutelmeisen	– tits		Ремезовые
Bienenesser	Bee-eaters	Méropidés	Пчелоеды
–	Common bee-eater	Guêpier méridional	Золотистая щурка
– i. e. S.	Bee-eaters	Guêpiers	
Bindenkreuzschnabel	Two-barred crossbill	Bec-croisé bifascié	Белокрылый клест
Binsenastrild	Star finch	Diamant à queue rousse	
Birkenzeisig	Redpoll	Sizerin flammé	Обыкновенная чечетка
Blaßspötter	Olivaceous warbler	Hypolaïs pâle	Большая бормотушка
Blattvögel	Fairy bluebirds and Leafbirds	Irénidés, Verdins	
Blaubart-Blattvogel	Blue-winged fruit-sucker	Verdin à barbe bleu	
Blaubauchracke	Blue-bellied roller	Rollier à ventre bleu	
Blauelster	Azure-winged magpie	Pie bleue	Голубая сорока
Blauer Paradiesvogel	Blue bird of paradise	Paradisier de Rodolphe	
– Staffelschwanz	Fairy blue wren	Roitelet bleu	
Blauhäher	American blue jay	Geai bleu d'Amérique	– сойка
Blaukehlchen	Bluethroat	Gorge-bleue à miroir	Варакушка
Blaukopf-Bartvogel	Great hill barbet	Barbu géant	
– -Schmetterlingsfink	Blue-headed waxbill	Astrild à tête bleue	
Blauköpfiger Paradiesvogel	Wilson's bird of paradise	Paradisier républicain	
Blaumeise	Blue titmouse	Mésange bleue	Лазоревка
Blaumerle	– rock thrush	Merle bleu	Синий каменный дрозд
Blaunacken-Strahlenparadiesvogel	Six-plumed bird of paradise	Silflet de Lawe	
Blauracke	Common roller	Rollier d'Europe	Обыкновенная сизоворонка
Blauracken i. e. S.	Rollers	Rolliers vrais	Сизоворонки
Blau-Schweifkitta	Ceylon blue magpie	Pie bleue ornée	
Blauvanga	Blue vanga	Artamie azurée	
Blauwangen-Bartvogel	Blue-throated barbet	Barbu à gorge bleue	
Blauwürger	Vangas	Vangidés	
Bluthänfling	Common linnet	Linotte mélodieuse	Коноплянка
Blutschnabelweber	Black-fronted diock	Travailleur	
Blutspecht	Syrian woodpecker	Pic syriaque	Сирийский дятел
Bombycilla cedrorum	Cedar waxwing		Кедровый свиристель
Bombycilla garrulus	Bohemian waxwing	Jaseur boréal	Обыкновенный свиристель
– *japonica*	Japanese waxwing		Восточноазиатский свиристель
Bombycillidae	Waxwings	Bombycillidés	Свиристелевые
Bootsschwänze	Grackles		Челнохвосты
Brachpieper	Tawny pipit	Pipit rousseline	Полевой конек
Braunbrust-Schilffink	Chestnut-breasted finch	Donacole commun	
Braunellen	Accentors	Prunellidés, Accenteurs	Завирушковые, Завирушки
Braunkehlchen	Whinchat	Traquet des près	Луговой чекан
Braunkopfammer	Red-headed bunting	Bruant à tête rousse	Желчная овсянка
Braunliest	White-breasted kingfisher	Halcyon à poitrine blanche	Красноносый зимородок
Braunrückentukan	Swainson's toucan	Toucan tocard	
Breitmaul-Glanzvogel	Great jacamar		Широкоротая якамара
Breitrachen	Broadbills	Eurylaimidés	Рогоклювы
Breitschnabelracken	Broad-billed rollers		Широкоротые сизоворонки
Breitschnabel-Todi	– tody	Todier vert	
Brillengrasmücke	Spectacled warbler	Fauvette à lunettes	Очковая славка
Brillenvögel	White-eyes	Zostéropidés	Белоглазковые
– i. e. S.	–		Белоглазки
Brillenwürger	White helmet shrike		Очковый сорокопут
Bubalornis albirostris	Buffalo weaver	Tisserin alecto	Буйволовый ткач
Bubalornithinae		Bubalornithinés	Буйволовые ткачи
Bucconidae	Puffbirds	Bucconidés	Птицы-пуховки
Buceros bicornis	Great hornbill	Calao bicorne	Малайский калао
– *rhinoceros*	Rhinoceros hornbill		
Bucerotidae	Hornbills	Bucérotidés	Птицы-носороги
Buchfink	Chaffinch	Pinson des arbres	Зяблик
Bucorvus	Ground hornbills		Рогатые вороны
– *abyssinicus*	Abyssinian ground hornbill	Bucorve d'Abyssinie	Абиссинский рогатый ворон
Büffelweber	Bubalornithinae, Buffalo weaver	Bubalornithinés, Tisserin alecto	Буйволовые ткачи, Буйволовый ткач
Buntastrild	Melba finch	Beau-marquet	
Bunt-Erddrossel	Ground thrush		Пестрый дрозд
Buntmeise	Red-sided titmouse		Японская синица
Buntspecht	Great spotted woodpecker	Pic épeiche	Большой пестрый дятел
Bunttukan	Green-billed toucan	Toucan à bec vert	
Buphaginae	Oxbirds	Pique-bœufs	Волоклюи
Buphagus erythrorhynchus	Red-billed ox-pecker	Pique-bœuf à bec rouge	Красноклювый волоклюй

GERMAN NAME	ENGLISH NAME	FRENCH NAME	RUSSIAN NAME
Buschrohrsänger	Blyth's reed warbler	Rousserolle des buissons	Садовая камышевка
Buschwürger	Bush-shrikes		Лесные сорокопуты
Calandrella cinerea		Petite alouette à tête rousse	Малый жаворонок
– *rufescens*	Lesser short-toed lark	Alouette pispolette	
Calcarius		Plectrophanes	Подорожники
– *lapponicus*	Lapland bunting	Bruant lapon	
Callaeidae	Wattle-birds		Лоскутные вороны
Campephaga	Cuckoo-shrikes		Гусениееды
– *flava*	Black cuckoo-shrike	Echenilleur pourpré	
Campephagidae	Cuckoo-shrikes	Campéphagidés	Личинкоедовые
Campephilus imperialis	Imperial woodpecker		Американский королевский дятел
– *principalis*	Ivory-billed woodpecker		Американский белоклювый дятел
Camposspecht	Pampas-flicker	Colapte des Pampas	
Campylorhynchus brunneicapillus	Cactus wren		Калифорнийский кактусовый крапивник
Capito aurovirens	Scarlet-crowned barbet	Barbu à poitrine orange	
Capitonidae	Barbets	Capitonidés	Бородастики
Cardinalinae	Cardinal-grosbeaks	Cardinaux	Кардиналы
Cardinalis cardinalis	Cardinal		Красный кардинал
Carduelis		Chardonnerets	Щеглы и чижи
– *carduelis*	Goldfinch	Chardonneret élégant	Щегол
– *chloris*	Greenfinch	Verdier d'Europe	Обыкновенная зеленушка
– *spinus*	Siskin	Tarin des aulnes	Чиж
Carpodacus		Roselins	Чечевицы
– *erythrinus*	Scarlet grosbeak	Roselin cramoisi	Обыкновенная чечевица
Cephalopterus ornatus	Ornate umbrella bird	Céphaloptère orné	Амазонская зонтичная птица
Certhia	Creepers	Grimpereaux	Пищухи
– *brachydactyla*	Short-toed tree creeper	Grimpereau des jardins	Короткопалая пищуха
– *familiaris*	Tree creeper	– – bois	Обыкновенная пищуха
Certhiidae	Creepers	Certhiidés	Пищуховые
Ceryle rudis	Pied kingfisher		Малый пегий зимородок
Cettia		Bouscarles	Широкохвостые камышевки
Cettia cetti	Cetti's warbler	Bouscarle de Cetti	Широкохвостая камышевка
Ceylonbeo	Ceylon mynah	Mainate de Ceylan	
Ceylon-Grünbartvogel	Green barbet	Barbu à tête grise du Ceylan	
Chaimarrornis leucocephalus	White-capped redstart	Rouge-queue à tête blanche	Белошапочная горихвостка
Chelidoptera tenebrosa	Swallow-wing		Ласточкокрылая пуховка
Chickadee-Meise	Chikadee	Mésange à tête noire	
Chinesische Pfeifdrossel	Chinese blue whistling thrush	Grive siffleuse bleue	Синяя птица
Chinesischer Sonnenvogel	Pekin nightingale	Rossignol du Japon	Обыкновенная солнечная птица
Chlamydera maculata	Spotted bower-bird	Oiseau à bercereaux tacheté	Обыкновенная пятнистая беседковая птица
Chloebia gouldiae	Gouldian finch	Diamant de Gould	
Chloropsis	Leafbirds	Verdins	
– *aurifrons*	Gold-fronted fruit-sucker	Verdin à front d'or	
– *hardwickii*	Blue-winged fruit-sucker	– – barbe bleu	
Chrysocolaptes		Pics dorés	Султанские дятлы
Chrysomma sinense	Yellow-eyed babbler	Timalie aux yeux d'or	
Cicinnurus regius	King bird of paradise	Paradisier royal	Королевская райская птица
Cinclidae	Dippers	Cinclidés	Оляпковые
Cinclus cinclus	Dipper	Cincle plongeur	Обыкновенная оляпка
Cinnyricinclus leucogaster	White-bellied amethyst-starling	Merle violet à ventre blanc	Аметистовый скворец
Cisticola		Cisticoles	Травяные певуны
– *juncidis*	Fan-tailed warbler	Cisticole des joncs	
Clamatores	Noisemakers	Mésomyodés	Кричащие птицы
Clarino	Townsend's solitaire	Solitaire de Townsend	
Coccothraustes coccothraustes	Hawfinch	Gros-bec casse-noyaux	Дубонос
Coerebidae	Honey creepers	Coerébidés	
Colaptes auratus	Bolden-banded woodpecker	Colapte doré	Золотой кукушковый дятел
– *campestris*	Pampas-flicker	– des Pampas	
– *rupicola*	Rock flicker		Андский дятел
Coliuspasser ardens	Red-collared whydah	Veuve en feu	
– *capensis*	Yellow bishop	Gros-bec tacheté du Cap de bonne Espérance	
Conopophaga	Gnateaters	Conopophages	
Copsychus malabarius	Shama	Merle Shama	Малабарская сорочья славка

GERMAN NAME	ENGLISH NAME	FRENCH NAME	RUSSIAN NAME
– *saularis*	Dhyal thrush	– dyal des Indes	Индийская сорочья славка
Coracias	Rollers	Rolliers vrais	Сизоворонки
– *benghalensis*	Indian roller	Rollier d'Inde	Бенгальская сизоворонка
– *caudatus*	Lilac-breasted roller	– à longue queue	
– *cyanogaster*	Blue-bellied roller	– à ventre bleu	
– *garrulus*	Common roller	– d'Europe	Обыкновенная сизоворонка
Coraciidae	Rollers	Coraciadidés	Сизоворонковые
Coraciiformes		Coraciadiformes	Ракшеобразные
Corcorax melanoramphos	White-winged crow (chough)	Corcorax à ailes blanches	
Corvidae	Crows	Corbeaux	Вороновые
Corvus albicollis	White-necked raven		Белошеий ворон
– *corax*	Raven	Grand corbeau	Ворон
– *corone cornix*	Hooded crow	Corneille mantelée	Серая ворона
– – *corone*	Carrion crow	– noire	Обыкновенная черная ворона
– *frugilegus*	Rook	Corbeau freux	Грач
– *monedula*	Jackdaw	Choucas des tours	Галка
Corydon sumatranus	Dusky Broadbill	Eurylaime sombre	
Coryphospingus cucullatus	Red-crested finch	Pinson couronné	
Corythornis cristata	Malachite kingfisher	Petit martin-pêcheur huppé	
Cosmopsarus regius	Royal starling	Spréo royal	
Cotinga	Cotingas	Cotingas	Котинги
– *maculata*	Banded cotinga	Cotinga tityre	Ошейниковая котинга
Cotingidae	Cotingas	Cotingidés	Котинги
Cracticidae	Australian butcherbirds	Cracticidés	
Craspedophora	Riflebirds		Щитоносные райские птицы
– *magnifica*	Magnificent riflebird	Proméfil magnifique	Чешуегрудый щитоносец
Creatophora cinerea	Wattled starling	Martin caronculé	
Crypsirina cucullata	Hooded racket-tailed tree pie	Pie bleue de l'Himalaya	
Cryptospiza reichenovii	Reichenow's crimson-wing	Bengali vert à face rouge	
Cyanocitta cristata	American blue jay	Geai bleu d'Amérique	Голубая сойка
– *stelleri*	Steller's jay		Черноголовая голубая сойка
Cyanocorax chryspos	Pileated jay	Pie akahé	Хохлатая голубая ворона
– *yncas*	Green jay		Перуанская голубая ворона
Cynopica cyana	Azure-winged magpie	Pie bleue	Голубая сорока
Dacelo gigas	Australian laughing jackass	Martin-chasseur géant	Зимородок-великан
Dacnis cayana	Turquoise dacnis		Питпит
Dajaldrossel	Dhyal thrush	Merle dyal des Indes	Индийская сорочья славка
Damadrossel	Orange-headed ground thrush	Grive orangée	
Darwinfinken	Darwin's finches	Géospizinés	
Delichon urbica	House martin	Hirondelle de fenêtre	Городская ласточка
Dendrocolaptidae	Wood hewers	Dendrocolapitidés	Древолазы
Dendrocopos leucotos	White-backed woodpecker	Pic à dos blanc	Белоспинный дятел
– *major*	Great spotted woodpecker	– épeiche	Большой пестрый дятел
– *medius*	Middle spotted woodpecker	– mar	Средний дятел
– *minor*	Lesser spotted woodpecker	– épeichette	Малый пестрый дятел
– *syriacus*	Syrian woodpecker	– syriaque	Сирийский дятел
Diamantfink	Diamond sparrow	Diamant à gouttelettes	
Dicaeidae	Flowerpeckers	Dicaeidés	Цветососы
Dicaeum cruentatum	Red-backed flower-pecker	Grimpereau é dos rouge	Красноспинный цветосос
– *hirundinaceum*	Mistletoe bird		Ласточковый цветосос
Dickichtschlüpfer	Scrub-birds	Atrichornes	
Dickschnabelweber		Amblyospizinés	Толстоклювые ткачи
Dicruridae	Drongos	Drongos	Дронговые
Dicrurus adsimilis	Drongo	Drongo	Траурный дронго
– *hottentottus*	Spangled drongo	– à crinière	
– *paradiseus*	Greater racket-tailed drongo	– à raquettes	Райский дронго
Dinemellia dinemelli	Dinemelli's weaver	Dinemelli à tête blanche	
Diphyllodes magnificus	Magnificent bird of paradise	Paradisier magnifique	
– *respublica*	Wilson's bird of paradise	– républicain	
Dohle	Jackdaw	Choucas des tours	Галка
Dolichonyx oryzivorus	Bobolink		Рисовый скворец
Dominikanerwitwe	Pin-tailed whydah	Veuve dominicaine	
Dompfaff	Bullfinch	Bouvreuil pivoine	Снегирь
Doppelhornvogel	Great hornbill	Calao bicorne	
Doppelzahn-Bartvogel	Double-toothed barbet	Barbu à bec denté	
Dornastrild	Sydney waxbill	Astrild à cinq couleurs	
Dorngrasmücke	Whitethroat	Fauvette grisette	Серая славка
Dottertukan	Sulphur-breasted toucan	Toucan à bec caréné	
Dreifarben-Glanzstar	Superb glossy starling	Spréo superbe	
Dreifarbige Papageiamadine	Blue-faced parrot finch	Diamant tricolore de Kittlitz	
Dreizehenspecht	Three-toed woodpecker	Pic tridactyle	Трехпалый дятел

GERMAN NAME	ENGLISH NAME	FRENCH NAME	RUSSIAN NAME
Dreizehenspechte	— woodpeckers	Pics tridactyles	Трехпалые дятлы
Drepanididae		Drépanididés	Нарядные птицы
Drongos	Drongos	Drongos	Дронговые
¬ i. w. S.	—	—	Дронго
Drosseln	Thrushes	Turdinés	Дроздовые
— i. w. S.		Merles	Настоящие дрозды
Drosselrohrsänger	Great reed warbler	Rousserolle turdoïde	Дроздовидная камышевка
Drosselstelze	Magpie lark	Alouette-pie	— трясогузка
Dryocopus martius	Black woodpecker	Pic noir	Желна
Dulus dominicus	Palmchat	Oiseau palmiste	
Dumetella carolinensis	Catbird	Merle moqueur chat	
Dumetia hyperythra	Rufous-bellied babbler	Dumétie à ventre roux	
Dunkelroter Amarant	Blue-billed fire finch	Sénégali à bec bleu	
Echte Bülbüls	Bulbuls		Настоящие бульбули
— Spechte		Picinés	Дятлы
— Tangare	Tanagers		Настоящие тангары
Eichelhäher	Black-throated jay	Geai des chênes	Сойка
Eigentliche Ammern	Buntings	Bruants	Овсянки
— Astrilde	Waxbills	Astrilds	Кровяно-красные ткачи
— Kleiber	Nuthatches		Настоящие поползни
— Kotingas	Cotingas	Cotingas	Котинги
— Meisen	Titmice	Paridés	Настоящие синицы
— Nektarvögel	Sunbirds		Нектарки
— Paradiesvögel	Birds of paradise	Paradisiers	Настоящие райские птицы
— Stare	Starlings		— скворцы
— Stelzen	Wagtails	Bergeronettes	Трясогузки
— Weber		Plocéinés	Древесные ткачи
— Würger	True shrikes		Настоящие сорокопуты
Einfarbstar	Spotless starling	Étourneau unicolore	Черный скворец
Eisvogel	Kingfisher	Martin-pêcheur	Обыкновенный зимородок
Eisvögel	Kingfishers	Alcidénidés	Зимородковые
Elfenastrild	Black-cheeked waxbill	Astrild à moustaches noires	
Elfenbeinspecht	Ivory-billed woodpecker		Американский белоклювый дятел
Elfenblauvögel	Fairy bluebirds	Oiseaux bleus	
Elster	Magpie	Pie bavarde	Сорока
Elsterstar	Pied mynah	Martin pie	
Emberiza	Buntings	Bruants	Овсянки
— *bruniceps*	Red-headed bunting	Bruant à tête rousse	Желчная овсянка
— *calandra*	Corn bunting	— proyer	Просянка
Emberiza citrinella	Yellowhammer	Bruant jaune	Обыкновенная овсянка
— *hortulana*	Ortolan bunting	— ortolan	Садовая овсянка
— *leucocephala*	Pine bunting		Белошапочная овсянка
— *melanocephala*	Black-headed bunting	Bruant mélanocéphale	Черноголовая овсянка
— *scheoniclus*	Reed bunting	— des roseaux	Камышовая овсянка
Emberizinae	Buntings	Emberizinés	Овсянки
Emblema picta	Painted finch	Emblème peint	
Eminsperling	Chestnut sparrow	Moineau doré d'Emin	
Epichmachus meyeri	Meyer's sickle bill	Epimagne de Meijer	
Eremophila alpestris	Shore lark	Alouette hausse-col	
Erithacus		Rouge-gorges	Рюм
— *rubecula*	Robin	Rouge-gorge familier	Зарянки
Erlenzeisig	Siskin	Tarin des aulnes	Зарянка
Erythrura hyperythra	Green-tailed parrot finch	Diamant à queue verte	Чиж
— *prasina*	Pin-tailed nonpareil	Quadricolore	
— *psittacea*	Parrot finch	Diamant à tête rouge	
— *trichroa*	Blue-faced parrot finch	— tricolore de Kittlitz	
Estrilda	Waxbills	Astrilds	Кровяно-красные ткачи
— *astrild*	St. Helena waxbill	Astrild ondulé	Восковклювый ткач
— *atricapilla*	Black-headed waxbill	— à tête noire	
— *caerulescens*	Lavender finch	Gris-bleu	
— *erythronotos*	Black-cheeked waxbill	Astrild à moustaches noires	
— *melanotis*, südliche Unterart	Yellow-bellied waxbill	Joue noire	
— *melpoda*	Orange-cheeked waxbill	Bengali à joues oranges	
— *nonnula*	Black-crowned waxbill	Astrild à cape noire	
— *paludicola*	Fawn-breasted waxbill	— de marais	
— *rhodopyga*	Crimson-rumped waxbill	— à dos rouge	
— *troglodytes*	Grey waxbill	Bec de coreil	
Estrildidae	Waxbills	Estrildidés	
Eubucco bourcierii	Red-headed barbet	Barbu de Bourcier	
Eumomota superciliosa	Jucatan motmot	Motmot à sourcils bleus	
Euodice cantans	African silver-bill	Bec d'argent	
— *malabarica*	Indian silver-bill	— de plomb	
Euplectes aureus	Golden-backed weaver	Euplecte à dos doré	
— *franciscanus*	Orange bishop	Ignicolore	
— *hordeaceus*	Fire-crowned bishop	Euplecte à couronne de feu	

GERMAN NAME	ENGLISH NAME	FRENCH NAME	RUSSIAN NAME
– *orix*	Grenadier weaver	Monseigneur	
Europäischer Seidenschwanz	Bohemian waxwing	Jaseur boréal	Обыкновенный свиристель
– Wendehals	European wryneck	Torcol ordinaire	Обыкновенная вертишейка
Euryceros prevosti	Helmet bird		Шлемоносный сорокопут
Eurylaimidae	Broadbills	Eurylaimidés	Рогоклювы
Eurystomus	Broad-billed rollers		Широкороты сизоворонки
– *orientalis*	Oriental dollar bird	Rollier à gros bec	Широкорот
Fadenhopf	Twelve-wired bird of paradise	Paradisier multifil	Нитчатая райская птица
Faulvögel	Puffbirds	Bucconidés	Птицы-пуховки
Feldlerche	Sky lark	Alouette des champs	Полевой жаворонок
Feldschwirl	Grasshopper warbler	Locustelle tachetée	Обыкновенный сверчок
Feldsperling	Tree sparrow	Friquet	Полевой воробей
Felsenkleiber	Rock nuthatch	Sittelle des rochers	Малый скалистый поползень
Feuerbreitrachen	Dusky broadbill	Eurylaime sombre	
Feuerschwanzamadine	Fire-tailed finch	Diamant à queue de feu	
Feuertangare	Summer tanager		Красная тангара
Feuerweber	Orange bishop	Ignicolore	
Ficedula albicollis	White-collared flycatcher	Gobe-mouche à collier	Мухоловка-белошейка
– *hypolauca*	Pied flycatcher	– noir	Мухоловка-пеструшка
– *parva*	Red-breasted flycatcher	– nain	Малая мухоловка
Fichtenammer	Pine bunting		Белошапочная овсянка
Fichtenkreuzschnabel	Crossbill	Bec-croisé des sapins	Клест-еловик
Finken	Finches	Fringillidés	Вьюрковые
Fitis	Willow wren	Pouillot fitis	Пеночка-весничка
Flaggendrongo	Greater racket-tailed drongo	Drongo à raquettes	Райский дронго
Flammenkopf	Crested sharpbill	Tête de feu	
Flammenweber	Fire-crowned bishop	Euplecte à couronne de feu	
Flecken-Laubenvogel	Spotted bower-bird	Oiseau à bercereaux tacheté	Обыкновенная пятнистая беседковая птица
Flecken-Panthervogel	– pardalote	Pardalote pointillé	Пятнистая пантерная птица
Flecken-Zwergbärtling	Speckled tinker-bird	Petit barbu grivelé	
Fliegenschnäpperartige	Flycatchers	Muscicapidés	Мухоловковые
Formicariidae	Ant birds	Formicariidés	Птицы-муравьеды
Foudia madagascariensis	Madagascar weaver	Foudi rouge	
Fringilla coelebs	Chaffinch	Pinson des arbres	Зяблик
– *montifringilla*	Brambling	– du nord	Юрок
Fringillaria tahapisi	Rock bunting	Bruant à sept raies	
Fringillidae	Finches	Fringillidés	Вьюрковые
Furnariidae	Ovenbirds	Furnariidés	Печники
Furnarius	–		Печники
Furnarius rufus	Ovenbird	Fournier roux	Рыжий печник
Gabelschwanz-Paradiesvogel	Wattled bird of paradise		Вилохвостая райская птица
Galapagosfinken	Darwin's finches	Géospizinés	
Galbula galbula	Common jacamar		Зеленая якамара
– *ruficauda*	Rufous-tailed jacamar		Краснохвостая якамара
Galbulidae	Jacamars	Galbulidés	Якамары
Galerida cristata	Crested lark	Cochevis huppé	Хохлатый жаворонок
Ganges-Brillenvogel	Indian white-eye	Zostérops à lunettes	Гангская белоглазка
Garrulax leucolophus	White-crested laughing jay-thrush	Garrulax à huppe blanche	
Garrulus glandarius	Black-throated jay	Geai des chênes	Сойка
Gartenbaumläufer	Short-toed tree creeper	Grimpereau des jardins	Короткопалая пищуха
Gartengrasmücke	Garden warbler	Fauvette des jardins	Садовая славка
Gartenrotschwanz	Redstart	Rouge-queue à front blanc	Обыкновенная горихвостка
Geierrabe	White-necked raven		Белошеий ворон
Gelbbauchbülbül	Yellow-vented bulbul	Bulbul à ventre jaune	
Gelbbauchsperling	Pegu (House) sparrow	Moineau flavéole	
Gelber Schilffink	Yellow-tailed finch	Donacole à tête grise	
Gelbkehl-Zwergbärtling	Yellow-throated tinker-bird	Petit barbu à gorge jaune	
Gelbkopf-Felshüpfer	Grey-necked rockfowl	Picathartes à cou blanc	
Gelbschnabelelster	Yellow-billed magpie		Калифорнийская сорока
Gelbschnabel-Schweifkitta	– blue-pie	Pirolle à bec jaune	Желтоклювая лазуревая сорока
Gelbspötter	Icterine warbler	Hypolaïs ictérine	Зеленая пересмешка
Gelbstirn-Zwergbärtling	Yellow-fronted tinker-bird	Petit barbu à front jaune	
Gemalter Astrild	Painted finch	Emblème peint	
Gemeiner Star	Starling	Étourneau sansonnet	Обыкновенный скворец
Geospizini	Darwin's finches	Géospizinés	
Gewöhnliche Felsenschwalbe	Grag martin	Hirondelle de rochers	Скалистая ласточка
Gewöhnlicher Amarant	Fire finch	Amaranthe	
Gimpel		Bouvreuils	Снегири
Gimpelhäher	Grey-jumper	Glaucope gris	Штрутидея

ANIMAL DICTIONARY

GERMAN NAME	ENGLISH NAME	FRENCH NAME	RUSSIAN NAME
Girlitz	Serin	Serin cini	Канареечный вьюрок
Girlitze	Serins	Serins	Канареечные вьюрки
Glanzelsterchen	Black-and-white mannikin	Spermète bicolore	
Glanzspitzendrongo	Spangled drongo	Drongo à crinière	
Glanz- und Dominikanerwitwen	Whydahs	Veuves	Вдовушки
Glanzvögel	Jacamars	Galbulidés	Якамары
Glanzwitwe	Shiny black whydah	Veuve métallique	
Glatzenbartvogel	Naked-faced barbet	Grand barbu chauve	
Glockenvögel	Bellbirds	Orapongas	Птицы-звонари
Goldammer	Yellowhammer	Bruant jaune	Обыкновенная овсянка
Goldaugentimalie i. e. S.	Yellow-eyed babbler	Timalie aux yeux d'or	
Goldbrüstchen	Golden-breasted waxbill	Bengali zébré	
Goldbrust-Tukan	Baillon's toucan	Toucan de Baillon	
Goldbürzel-Zwegbärtling	Golden-rumped tinker-bird	Petit barbu à gorge blanche	
Goldhähnchen	Kinglets	Roitelets	Корольки
Goldohr-Arassari	Spot-billed toucanet	Toucanet à bec tâcheté	
Goldrückenweber	Golden-backed weaver	Euplecte à dos doré	
Goldspecht	Golden-banded woodpecker	Colapte doré	Золотой кукушковый дятел
Goldsperling	Golden sparrow	Moineau doré	
Goldstirn-Blattvogel	Gold-fronted fruit-sucker	Verdin à front d'or	
Götzenliest	Sacred kingfisher	Halcyon sacré	
Gouldamadine	Gouldian finch	Diamant de Gould	
Gracula ptilogenys	Ceylon mynah	Mainate de Ceylan	
– religiosa	Javan hill mynah	– religieux	
Gracupica nigricollis	Black-necked mynah	Martin à cou noir	
Grallaria varia	Royal ant-thrush	Fourmillier	
Grallina cyanoleuca	Magpie lark	Alouette-pie	Дроздовидная трясогузка
Granatastrild	Violet-eared waxbill	Grenadin	
Grasmücken i. e. S.	Warblers	Fauvettes	Славки
Grauammer	Corn bunting	Bruant proyer	Просянка
Grauastrild	Grey waxbill	Bec de coreil	
Grau-Edelsänger	– singing finch	Chanteur d'Afrique	Сенегальский вьюрок
Graufischer	Pied kingfisher		Малый пегий зимородок
Graukopfliest	Grey-headed kingfisher	Martin-chasseur à tête grise	
Graukopfsperling	– sparrow	Moineau à tête grise	
Graukopfstar	Malabar mynah	Martin à tête grise	
Grauschnäpper	Spotted flycatcher		Серая мухоловка
Grauschwalbenstar	Ashy-wood-swallow		Серый ласточковый скворец
Grauseidenschnäpper	Gray silky-flycatcher	Gobe-mouche gris du Mexique	– тропический свиристель
Grauspecht	Grey-headed woodpecker	Pic cendré	Седой дятел
Graustar	Grey starling	Martin gris	Серый скворец
Grauwangendrossel	Gray-cheeked thrush		Малый дрозд
Großer Paradiesvogel	Greater bird of paradise	Grand paradisier	Большая райская птица
Groß-Kubafink	Olive finch	– chanteur de Cuba	
Großniltava	Greater niltava	– niltava	
Grüner Tropfenastrild	Green-backed twin-spot	Bengali vert pointillé	
Grünjakamar	Common jacamar		Зеленая якамара
Grünkatzenvogel	Green cat-bird	Oiseau à berceaux vert	Зеленая птица-кошка
Grünling	Greenfinch	Verdier d'Europe	Обыкновенная зеленушка
Grünscheitelracke	Lilac-breasted roller	Rollier à longue queue	
Grünschwanzglanzstar	Green glossy starling	Merle métallique à oreilles bleues	Стальной скворец
Grünspecht	– woodpecker	Pic vert	Зеленый дятел
Grünspechte	– woodpeckers	Pics verts	Зеленые дятлы
Gürtelgrasfink	Parson finch	Diamant à bavette	
Gymnobucco calvus	Naked-faced barbet	Grand barbu chauve	
– peli	Bristle-nosed barbet	Barbu chauve à narines emplumées	
Gymnorhina hypoleuca	Tasmanian crow-shrike		Белоспинная ворона-свистун
– tibicen	Common piping crow	Corbeau flûteur	Черноспинная ворона-свистун
Haarvögel	Bulbuls		Короткопалые дрозды
Hakengimpel	Pine grosbeak	Dur-bec des sapins	Щур
Halcyon leucocephala	Grey-headed kingfisher	Martin-chasseur à tête grise	
– sanctus	Sacred kingfisher	Halcyon sacré	
– smyrnensis	White-breasted kingfisher	– à poitrine blanche	Красноносый зимородок
Halsband-Arassari	Banded aracari	Aracari à collier	
Halsbandastrild	White-collared olive-back	Bengali vert de Ansorge	
Halsband-Bartvogel	Black-collared barbet	Barbu à collier noir	
Halsbandkotinga	Banded cotinga	Cotinga tityre	Ошейниковая котинга
Halsbandschnäpper	White-collared flycatcher	Gobe-mouche à collier	Мухоловка-белошейка
Hämmerling	Wattled bellbird		Молотобой

GERMAN NAME	ENGLISH NAME	FRENCH NAME	RUSSIAN NAME
Hänflinge		Linottes	Чечетки
Haubenammer	Crested bunting		Хохлатая овсянка
Haubenhäherling	White-crested laughing jay-thrush	Garrulax à huppe blanche	
Haubenlerche	Crested lark	Cochevis huppé	Хохлатый жаворонок
Haubenmaina	Chinese crested mynah	Martin huppé	Хохлатая майна
Haubenmeise	Crested-tit	Mésange huppée	Хохлатая синица
Haubenprachtweber	Crested malimbe	Malimbe huppé	
Hausrotschwanz	Black redstart	Rouge-queue noir	Горихвостка-чернушка
Haussperling	House sparrow	Moineau domestique	Домовый воробей
Heckenbraunelle	Dunnock	Accenteur mouchet	Лесная завирушка
Heidelerche	Wood lark	Alouette lulu	Лесной жаворонок
Helmvanga	Helmet bird		Шлемоносный сорокопут
Hesperiphona vespertina	Evening grosbeak		Североамериканский дубонос
Heteraloche acutirostris	Huia	Huia	Туйя
Heuschreckenastrild	Locust-finch	Astrild locustelle	
Hippolais		*Hypolaïs*	Пересмешки-бормотушки
– *icterina*	Icterine warbler	– *ictérine*	Зеленая пересмешка
– *olivetorum*	Olive-tree warbler	– *des oliviers*	Средиземноморская пересмешка
– *pallida*	Olivaceous warbler	– *pâle*	Большая бормотушка
– *polyglotta*	Melodious warbler	– *polyglotte*	Многоголосая камышевка-пеночка
Hirtenstar	Common mynah	Martin triste	Саранчовый скворец
Hirundinidae	Swallows	Hirondelles	Ласточковые
Hirundo	–		Касатки
– *daurica*	Red-rumped swallow	Hirondelle rousseline	Рыжепоясничная ласточка
– *rustica*	Common swallow	– de cheminée	Деревенская ласточка
Honiganzeiger	Honey guides	Indicatoridés	Кукушки-путеводители
Honigesser	– eaters	Méliphagidés	Медососовые
Hopfe	Hoopoes	Upupidés	Удодовые
Hordenvögel	Blackbirds		Болотные трупиалы
Hornraben	Ground hornbills		Рогатые вороны
Huia	Huia	Huia	Туйя
Hüttengärtner	Brown gardener		Новогвинейская беседковая птица
Hylocichla minima	Gray-cheeked thrush		Малый дрозд
Hypargos margaritatus	Rosy twin-spot	Astrild de Verreaux	
– *niveoguttatus*	Peter's twin-spot	Amaranthe enflammée	
Icteridae	Orioles	Ictéridés	Касиковые
Icterus	–		Трупиалы
– *galbula*	Baltimore oriole		Балтиморский трупиал
Indicator archipelagicus	Malaysian honey guide	Indicateur malais	
Indicatoridae	Honey guides	Indicatoridés	Кукушки-путеводители
Irena	Fairy bluebirds	Oiseaux bleus	
Irenidae	Fairy bluebirds and Leafbirds	Irénidés	
Ispidina picta	Natal kingfisher	Martin-pêcheur de Natal	
Jacamerops aurea	Great jacamar		Широкоротая якамара
Japanischer Brillenvogel	Japanese white-eye		Японская белоглазка
– Seidenschwanz	– waxwing		Восточноазиатский свиристель
Junco	Juncos	Juncos	Юнко
– *hyemalis*	Slate-coloured junco	Junco ardoisé	Серый юнко
– *oreganus*	Oregon junco		Орегонский юнко
Junco	Juncos	Juncos	Юнко
Jynginae		Jynginés	Вертишейки
Jynx torquilla	European wryneck	Torcol ordinaire	Обыкновенная вертишейка
Kahlkopfatzel	Coleto mynah	Sarcops chauve	
Kaiserspecht	Imperial woodpecker		Американский королевский дятел
Kaktus-Zaunkönig	Cactus wren		Калифорнийский кактусовый крапивник
Kalanderlerche	Calandra lark	Alouette calandre	Степной жаворонок
Kanadischer Unglückshäher	Canada jay		Канадская ронжа
Kanarienvogel	Canary	Serin	Канарейка
Kappenammer	Balck-headed bunting	Bruant mélanocéphale	Черноголовая овсянка
Kappenastrild	– waxbill	Astrild à tête noire	
Kappengrünhäher	Pileated jay	Pie akahé	Хохлатая голубая ворона
Kappenkleiber	Red-breasted nuthatch		Черноголовый поползень
Kapsperling	Cape sparrow	Moineau mélanure	Южноафриканский воробей
Kaptschagra	Redwing shrike		Чагра
Kapweber	Cape weaver	Tisserin à front d'or	
Kardinäle	Cardinal-grosbeaks	Cardinaux	Кардиналы

ANIMAL DICTIONARY

GERMAN NAME	ENGLISH NAME	FRENCH NAME	RUSSIAN NAME
Karmingimpel	Scarlet grosbeak	Roselin cramoisi	Обыкновенная чечевица
Karolina-Zaunkönig	Carolina wren		Крапивник-пересмешник
Kastanienkleiber	Chestnut-bellied nuthatch	Sittelle à ventre marron	
Katzendrossel	Catbird	Merle moqueur chat	
Katzenvögel i. e. S.	Cat-birds		Птицы-кошки
Kernbeißer	Hawfinch	Gros-bec casse-noyaux	Дубонос
Kiefernhäher	Clark's nutcracker		Североамериканская ореховка
Kiefernkreuzschnabel	Parrot crossbill	Bec-croisé perroquet	Клест-сосновик
Klappergrasmücke	Lesser whitethroat	Fauvette babillarde	Славка-завирушка
Kleiber	Nuthatch, Nuthatches	Sittelle torchepot, Sittelles, Sittidés	Обыкновенный поползень, Поползни
Kleidervögel		Drépanididés	Нарядные птицы
Kleinelsterchen	Bronze mannikin	Spermète à capuchon	
Kleiner Dickichtschlüpfer	Rufous scrub-bird	Atrichorne roux	
– Paradiesvogel	Lesser bird of paradise		Малая райская птица
Klein-Kubafink	Cuba finch	Petit chanteur de Cuba	
Kleinspecht	Lesser spotted woodpecker	Pic épeichette	Малый пестрый дятел
Klippenvögel	Cocks of the rock	Coqs-de-roche	Скалистые петушки
Kohlmeise	Oxeye-tit	Mśange charbonnière	Большая синица
Kolkrabe	Raven	Grand corbeau	Ворон
Königs-Ameisenstelzer	Royal ant-thrush	Fourmillier	
Königsglanzstar	– starling	Spréo royal	
Königsparadiesvogel	King bird of paradise	Paradisier royal	Королевская райская птица
Königssatrap	American kingbird	Tyran royal	
Königswitwe	Queens whydah	Veuve reine	
Kookaburra	Australian laughing jackass	Martin-chasseur géant	Зимородок-великан
Korsika-Kleiber	Corsican nuthatch	Sittelle corse	
Kragenhopf	Lesser superb bird of paradise	Paradisier superbe	Чудная райская птица
Krähenwürger	Australian butcherbirds	Cracticidés	
Kreuzschnäbel	Crossbills	Bec-croisés	Клесты
Kronenatzel	Gold-crested crackle	Mainate couronné	
Kuckucksweber	Cuckoo-weaver	Tisserin parasitique	
Kuhstärlinge	Cowbirds		Коровьи скворцы
Kupferschmied	Coppersmith	Barbu à front rouge	
Kurol	Cuckoo-roller	Courol malgache	
Kurzzehenlerche		Petite alouette à tête rousse	Малый жаворонок
Lagonosticta larvata	Masked waxbill	Amaranthe masquée d'Abyssinie	
– rhodopareia	Jameson's fire finch	Amaranthe de Jameson	
– rubricata	Blue-billed fire finch	Sénégali à bec bleu	
– rufopicta	Bar-breasted fire finch	Sénégali à poitrine barrée	
– senegala	Fire finch	Amaranthe	
Lamprospreo superbus	Superb glossy starling	Spréo superbe	
Lamprotornis caudatus	Long-tailed glossy starling	Merle métallique à longue queue	
– chalybaeus	Green glossy starling	– – à oreilles bleues	Стальной скворец
– nitens	Cape red-shouldered glossy starling	– à épaulettes rouges	
Lamprotornis purpureus	Purple glossy starling	Merle bonzé pourpre	
– splendidus	Splendid glossy starling	– métallique	Великолепный скворец
Langschwanzglanzstar	Long-tailed glossy starling	– – à longue queue	
Langschwanz-Nektarvögel	Sunbirds		Длиннохвостые нектарки
Laniarius atrococcineus	Crimson-breasted shrike	Pie-grièche à ventre cramoisie	
Laniidae	Shrikes	Laniidés	Сорокопутовые
Laniinae	True shrikes		Настоящие сорокопуты
Lanius	Shrikes	Pie-grièches	Сорокопуты
– collurio	Red-backed shrike	Pie-grièche écorcheur	Европейский жулан
– excubitor	Great grey shrike	– grise	Серый сорокопут
– minor	Lesser grey shrike	– à poitrine rose	Чернолобый сорокопут
– nubicus	Masked shrike	– masquée	Маскированный сорокопут
– senator	Woodchat shrike	– à tête rousse	Красноголовый сорокопут
Lappenpittas	Philepittas	Philepittidés	
Lappenstar	Wattled starling	Martin caronculé	
Larvenamarant	Masked waxbill	Amaranthe masquée d'Abyssinie	
Laubsänger i. e. S.		Pouillots	Пеночки
Laucharassari	Emerald toucanet	Toucanet vert	
Lauchgrüne Papageiamadine	Pin-tailed nonpareil	Quadricolore	
Leierschwanz i. e. S.	Superb lyre-bird	Oiseaux-lyre	
Leierschwänze	Lyre-birds	Menuridés	Птицы-лиры
Leiothrix argentauris	Silver-eared mesia	Mésia à joues argentée	
– lutea	Pekin nightingale	Rossignol du Japon	Обыкновенная солнечная птица
Leptopterus madagascarinus	Blue vanga	Artamie azurée	
Leptosomus discolor	Cuckoo-roller	Courol malgache	
Lerchen	Larks	Alouettes	Жаворонковые
Leucopsar rothschildi	Bali-mynah	Martin de Rothschild	Балийский скворец

GERMAN NAME	ENGLISH NAME	FRENCH NAME	RUSSIAN NAME
Leucosticte	Rosy finches		Горные вьюрки
Locustella		Locustelles	Сверчки
– *certhiola*	Pallas's grasshopper warbler	Locustelle de Pallas	Певчий сверчок
– *fluviatilis*		– fluviatile	Речной сверчок
– *lanceolata*	Lanceolated warbler	– lancéolée	Пятнистый сверчок
– *luscinioides*	Savi's warbler	– luscinioïde	Соловьиный сверчок
– *naevia*	Grasshopper warbler	– tachetée	Обыкновенный сверчок
Lonchura castaneothorax	Chestnut-breasted finch	Donacole commun	
– *flaviprymna*	Yellow-tailed finch	– à tête grise	
– *maja*	White-headed mannikin	Nonnette à tête blanche	
– *malacca*	Three-coloured mannikin	Jacobin	
– *pectoralis*	Pictorella finch	Donacole à poitrine blanche	
– *punctulata*	Spice finch	Damier	
– *spectabilis*	Spectacled finch	Nonnette à ventre roux	
– *striata*	Sharp-tailed munia	Domino à longue queue	
Lophorina superba	Lesser superb bird of paradise	Paradisier superbe	Чудная райская птица
Loxia	Crossbills	Bec-croisés	Клесты
– *curvirostra*	Crossbill	Bec-croisé des sapins	Клест-еловик
– *leucoptera*	Two-barred crossbill	– bifascié	Белокрылый клест
– *pytyopsittacus*	Parrot crossbill	– perroquet	Клест-сосновик
Lullula arborea	Wood lark	Alouette lulu	Лесной жаворонок
Luscinia		Rossignols	Соловьи
– *calliope*		Calliope	Соловей-красношейка
– *luscinia*	Great nightingale	Rossignol progné	Обыкновенный соловей
– *megarhynchos*	Nightingale	– philomèle	Западный соловей
– *svecica*	Bluethroat	Gorge-bleue à miroir	Варакушка
Lybius bidentatus	Double-toothed barbet	Barbu à bec denté	
– *dubius*	Bearded barbet	Barbican à poitrine rouge	
– *torquatus*	Black-collared barbet	Barbu à collier noir	
Madenhackerstare	Oxbirds	Pique-bœufs	Волоклюи
Madagaskarweber	Madagascar weaver	Foudi rouge	
Magalaima zeylanica	Green barbet	Barbu à tête grise du Ceylan	
Mahaliweber	Mahali weaver-bird	Mahali	
Malabarfasänchen	Indian silver-bill	Bec de plomb	
Malachit-Nektarvogel	Malachite sunbird	Soui-manga malachite	Малахитовая нектарка
Malaconotinae	Bush-shrikes		Лесные сорокопуты
Malayischer Honiganzeiger	Malaysian honey guide	Indicateur malais	
Malimbus malimbicus	Crested malimbe	Malimbe huppé	
Malurinae	Australian »wrens«		Малуры
Malurus cyaneus	Fairy blue wren	Roitelet bleu	
Manacus manacus	White-bearded manakin	Manakin moine	
Mandarinstar	Mandarin mynah	Martin de Chine	Южнокитайский малый скворец
Mandingoa nitidula	Green-backed twin-spot	Bengali vert pointillé	
Manyarweber	Manyar weaver	Tisserin Manyar	Манья
Mariskensänger	Moustached warbler	Lusiniole à moustaches	Тонкоклювая камышевка
Maskenamadine	Masked grass finch	Diamant à masque	
Maskengrasmücke	Rüppell's warbler	Fauvette masquée	Эгейская славка
Maskenweber	Black-fronted weaver	Tisserin à front noir	
Maskenwürger	Masked shrike	Pie-grièche masquée	Маскированный сорокопут
Mauerläufer	Wall-creeper	Tichodrome échelette	Краснокрылый стенолаз
Megalaima asiatica	Blue-throated barbet	Barbu à gorge bleue	
– *haemacephala*	Coppersmith	– à front rouge	
– *virens*	Great hill barbet	– géant	
Mehlschwalbe	House martin	Hirondelle de fenêtre	Городская ласточка
Melanerpes erythrocephalus	Red-headed woodpecker	Pic à tête rouge	Американский красноголовый дятел
Melanocorypha calandra	Calandra lark	Alouette calandre	Степной жаворонок
Meliphagidae	Honey eaters	Méliphagidés	Медососовые
Melithreptus lunatus	White-naped honey-eater	Méliphage à tête noire	
Melophus lathami	Crested bunting		Хохлатая овсянка
Mennigvögel	Minivets	Minivets	
Menura alberti	Albert's lyre-bird		Принцев лирохвост
– *novaehollandiae*	Superb lyre-bird	Oiseaux-lyre	
Menuridae	Lyre-birds	Menuridés	Птицы-лиры
Meropidae	Bee-eaters	Méropidés	Пчелоеды
Merops		Guêpiers	
– *apiaster*	Common bee-eater	Guêpier méridional	Золотистая щурка
– *nubicus*	Nubian carmine bee-eater	– carminé	
– *ornatus*	Rainbow bird	– d'Australie	
Mimidae	Mocking-birds	Mimidés	
Mimus polyglottus	Mocking bird	Merle moqueur commun	
Misteldrossel	Mistle thrush	Grive draine	Дрозд-деряба
Mistelesser	Flowerpeckers	Dicaeidés	Цветососы
Mittelmeer-Steinschmätzer	Black-eared wheatear	Traquet oreillard	Чернопегая каменка
Mittelspecht	Middle spotted woodpecker	Pic mar	Средний дятел

GERMAN NAME	ENGLISH NAME	FRENCH NAME	RUSSIAN NAME
Moçambique-Girlitz	Green singing finch	Serin de Mozambique	Мозамбикский вьюрок
Mohrenmeise	Black tit	Mésange noire	
Mohrenraupenesser	– cuckoo-shrike	Echenilleur pourpré	
Molothrus	Cowbirds		Коровьи скворцы
Momotidae	Motmots	Momotidés	Момоты
Momotus momota	Common motmot	Motmot	
Mönchsgrasmücke	Blackcap	Fauvette à tête noire	Черноголовая славка
Monticola		Monticoles	Каменные дрозды
– *saxatilis*	Rock thrush	Merle de roches	Пестрый каменный дрозд
– *solitarius*	Blue rock thrush	– bleu	Синий каменный дрозд
Montifringilla	Snow finches	Nivérolles	Снежные вьюрки
– *nivalis*	– finch	Nivérolle des Alpes	Снежный вьюрок
Motacilla	Wagtails	Bergeronettes	Трясогузки
– *alba*	White wagtail	Bergeronette grise	Белая трясогузка
– – *yarrellii*	Pied wagtail	– d'Yarrell	Британская белая трясогузка
– *cinerea*	Grey wagtail	– des ruisseaux	Горная трясогузка
– *citreola*		– citrine	Желтоголовая трясогузка
– *flava*	Blue-headed wagtail	– printanière	Желтая трясогузка
Motacillidae	Wagtails		Трясогузковые
Motmot	Common motmot	Motmot	
Mückenesser	Gnateaters	Conopophages	
Muscicapa striata	Spotted flycatcher		Серая мухоловка
Muscicapidae	Flycatchers	Muscicapidés	Мухоловковые
Muskatfink	Spice finch	Damier	
Myadestes townsendi	Townsend's solitaire	Solaire de Townsend	
Myiophoneus caeruleus	Chinese blue whistling thrush	Grive siffleuse bleue	Синяя птица
Nachtigall	Nightingale	Rossignol philomèle	Западный соловей
Nachtigallen		Rossignols	Соловьи
Nacktkehl-Glockenvogel	Naked-throated bellbird	Oraponga à gorge nue	Бразильская птица-колокольчик
Nashornvögel	Hornbills	Bucérotidés	Птицы-носороги
Nebelkrähe	Hooded crow	Corneille mantelée	Серая ворона
Nectarinia	Sunbirds		Длиннохвостые нектарки, Нектарки
– *famosa*	Malachite sunbird	Soui-manga malachite	Малахитовая нектарка
Nectariniidae, Nektarvögel	Sunbirds	Nectariniidés	Нектарковые
Neochmia phaeton	Crimson finch	Phaéton	
Neosittinae	Treerunners		Австралийские поползни
Nesocharmia ansorgei	White-collared olive-back	Bengali vert d'Ansorge	
Neunfarbenpitta	Indian pitta	Brève du Bengale	Индийская питта
Neuntöter	Red-backed shrike	Pie-grièche écorcheur	Европейский жулан
Neuseeländische Lappenvögel	Wattle-birds		Лоскутные вороны
Neuseeland-Schlüpfer	Wrens, New-Zealand wren	Xenicidés, Roitelet de la Nouvelle-Zélande	
Nigrita bicolor	Chestnut-breasted negro-finch	Bengali brun à ventre roux	
Niltava grandis	Greater niltava	Grand niltava	
– *sundara*	Rufous-bellied niltava	Niltava à ventre rouge	
Nonnenastrild	Black-crowned waxbill	Astrild à cape noire	
Nonnenmeise	Marsh-titmouse	Mésange nonnette	Черноголовая гаичка
Nördlicher Hornrabe	Abyssinian ground hornbill	Bucorve d'Abyssinie	Абиссинский рогатый ворон
Nucifraga caryocatactes	Nutcracker	Casse-noix moucheté	Ореховка
– *columbiana*	Clark's nutcracker		Североамериканская ореховка
Odontospiza caniceps	Grey-headed silver-bill	Spermète à tête grise	
Oenanthe deserti	Desert wheatear	Traquet du désert	Пустынная каменка
– *hispanica*	Black-eared wheatear	– oreillard	Чернопегая каменка
– *oenanthe*	Wheatear	– motteux	Обыкновенная каменка
Ohrenlerche	Shore lark	Alouette hausse-col	Рюм
Olivenspötter	Olive-tree warbler	Hypolaïs des oliviers	Средиземноморская пересмешка
Onychognathus morio	Red-wing starling	Roupenne	
Orangebäckchen	Orange-cheeked waxbill	Bengali à joues oranges	
Oregon-Junco	Oregon junco		Орегонский юнко
Oreostruthus fuliginosus	Crimson-bellied mountain finch	Diamant des montagnes	
Oriolidae	Oriols	Loriots	Иволговые
Oriolus chinensis	Black-naped oriole	Loriot de Chine	Черноголовая иволга
– *oriolus*	Golden oriole	– d'Europe	Обыкновенная иволга
Orpheusgrasmücke	Orphean warbler	Fauvette orphée	Певчая славка
Orpheusspötter	Melodious warbler	Hypolaïs polyglotte	Многоголосая камышевка-пеночка
Orthotomus sutorius	Tailor bird	Fauvette couturière	Славка-портниха
Ortolan	Ortolan bunting	Bruant ortolan	Садовая овсянка
Ortygospiza atricollis	South-African quail-finch	Astrild caille à lunettes	
– *locustella*	Locust-finch	– locustelle	

GERMAN NAME	ENGLISH NAME	FRENCH NAME	RUSSIAN NAME
Oryxweber	Grenadier weaver	Monseigneur	
Oscines	Singing birds	Oiseaux chanteurs	Певчие птицы
Ostroller	Oriental dollar bird	Rollier à gros bec	Широкорот
Oxyruncus cristatus	Crested sharpbill	Tête de feu	
Padda oryzivora	Java sparrow	Padda	
Pagodenstar	Pagoda starling	Martin des pagodes	Браминский скворец
Palmschmätzer	Palmchat	Oiseau palmiste	
Panthervögel	Diamond birds	Pardalots	Пантерные птицы
Panurus biarmicus	Bearded titmouse	Panure à moustaches	
Paradiesschnäpper	Paradise flycatcher	Gobe-mouche paradisier	Индийская длиннохвостая мухоловка
Paradies- und Laubenvögel	Birds of paradise	Paradisiers	Райские и Беседковые птицы
Paradigalla carunculata	Wattled bird of paradise		Вилохвостая райская птица
Paradisaea	Birds of paradise	Paradisiers	Настоящие райские птицы
– *apoda*	Greater bird of paradise	Grand paradisier	Большая райская птица
– *minor*	Lesser bird of paradise		Малая райская птица
– *rudolphi*	Blue bird of paradise	Paradisier de Rodolphe	
Paradisaeidae	Birds of paradise	Paradisiers	Райские и Беседковые птицы
Pardalotus	Diamond birds	Pardalots	Пантерные птицы
– *punctatus*	Spotted pardalote	Pardalote pointillé	Пятнистая пантерная птица
Paridae	Titmice	Paridés	Настоящие синицы
Parmoptila woodhousei	Antpecker	Astrild fourmilier	
Parotia lawesi	Six-plumed bird of paradise	Sifilet de Lawe	
Parulidae	Old World warblers	Parulidés	
Parus ater	Coal-titmouse	Petite charbonnière	Черная синица
– *atricapillus*	Chikadee	Mésange à tête noire	
– *coeruleus*	Blue titmouse	– bleue	Лазоревка
– *cristatus*	Crested-tit	– huppée	Хохлатая синица
– *major*	Oxeye-tit	– charbonnière	Большая синица
– *niger*	Black tit	– noire	
– *palustris*	Marsh-titmouse	– nonnette	Черноголовая гаичка
– *varius*	Red-sided titmouse		Японская синица
Passer	Sparrows	Moineaux	Настоящие воробьи
– *domesticus*	House sparrow	Moineau domestique	Домовый воробей
– *eminibey*	Chestnut sparrow	– doré d'Emin	
– *flaveolus*	Pegu (House) sparrow	– flavéole	
– *griseus*	Grey-headed sparrow	– à tête grise	
– *luteus*	Golden sparrow	– doré	
– *melanurus*	Cape sparrow	– mélanure	Южноафриканский воробей
– *montanus*	Tree sparrow	Friquet	Полевой воробей
– *motitensis*	Great sparrow	Moineau roux	
– *simplex*	Desert sparrow		Пустынный воробей
Passeriformes		Passériformes	Воробьиные
Passerinae	Sparrows	Passerinés	Воробьи
Pastor roseus	Rose-coloured starling	Martin roselin	Розовый скворец
Pel-Bartvogel	Bristle-nosed barbet	Barbu chauve à narines emplumées	
Pericrocotus	Minivets	Minivets	
Perisoreus		Mésangeais	Кукши
Perisoreus canadensis	Canada jay		Канадская ронжа
– *infaustus*	Siberian jay	Geai de Sibérie	Кукша
Perlastrild	Rosy twin-spot	Astrild de Verreaux	
Perlbartvogel	Yellow-breasted barbet	Barbu perlé	
Perlhalsamadine	Grey-headed silver-bill	Spermète à tête grise	
Peru-Grünhäher	Green jay		Перуанская голубая ворона
Petronia	Rock sparrows		Каменные воробьи
– *petronia*	– sparrow	Moineau fou	Каменный воробей
– *superciliaris*	Yellow-throated sparrow	– à sourcils	
Pflanzenmäher	Plantcutters	Phytotomidés	Косцы
Phainopepla nitens	Phainopepla	Phainopépla resplendissant	
Philepitta castanea	Velvet pitta	Philépitte veloutée	
Philepittidae	Philepittas	Philepittidés	
Philetairus socius	Sociable weaver	Tisserin social	Обыкновенный общественный ткач
Phoeniculinae	Wood-hoopoes	Phoeniculinés	Лесные удоды
Phoeniculus		Moqueurs	–
Phoenicurus		Rouge-queues	Горихвостки
– *ochruros*	Black redstart	Rouge-queue noir	Горихвостка-чернушка
– *phoenicurus*	Redstart	– à front blanc	Обыкновенная горихвостка

GERMAN NAME	ENGLISH NAME	FRENCH NAME	RUSSIAN NAME
Phylloscopus		Pouillots	Пеночки
– *bonelli*		Pouillot de Bonelli	Светлобрюхая пеночка
– *collybita*		– véloce	Пеночка-теньковка
– *sibilatrix*	Wood-wren	– siffleur	Пеночка-трещотка
– *trochilus*	Willow wren	– fitis	Пеночка-весничка
Phytotomidae	Plantcutters	Phytotomidés	Косцы
Pica pica	Magpie	Pie bavarde	Сорока
– – *nuttalli*	Yellow-billed magpie		Калифорнийская сорока
Picathartes gymnocephalus	Grey-necked rockfowl	Picathartes à cou blanc	
Picidae	Woodpeckers	Picidés	Дятловые
Piciformes		Piciformes	Дятловидные
Picinae		Picinés	Дятлы
Picoides	Three-toed woodpeckers	Pics tridactyles	Трехпалые дятлы
– *tridactylus*	– woodpecker	Pic tridactyle	Трехпалый дятел
Picumninae		Picuminés	Дятелки
Picus	Green woodpeckers	Pics verts	Зеленые дятлы
– *canus*	Grey-headed woodpecker	Pic cendré	Седой дятел
– *viridis*	Green woodpecker	– vert	Зеленый дятел
Pieper i. e. S.	Pipits	Pipits	Коньки
Pinicola		Dur-becs	Щуры
– *enucleator*	Pine grosbeak	Dur-bec des sapins	Щур
Pipridae	Manakins	Pipridés	Манакины
Piranga rubra	Summer tanager		Красная тангара
Pirol	Golden oriole	Loriot d'Europe	Обыкновенная иволга
Pirole	Oriols	Loriots	Иволговые
Pitangus sulphuratus	Kiskadee flycatcher	Tyran à ventre jaune	Бентеви
Pitpit i. e. S.	Turquoise dacnis		Питпит
Pitta	Pittas	Brèves	Питты
– *brachyura*	Indian pitta	Brève du Bengale	Индийская питта
– *sordida*	Hooded pitta	– sordide	
Pittas, Pittidae	Pittas	Pittidés, Brèves	
Plectrophenax	Snow buntings	Bruants des neiges	Питты
– *nivalis*	– bunting	Bruant des neiges	Пуночки
Ploceidae	Weaverbirds	Plocéidés	Ткачиковые
Ploceinae		Plocéinés	Древесные ткачи
Plocepasser mahali	Mahali weaver-bird	Mahali	
Ploceus manyar	Manyar weaver	Tisserin Manyar	Манья
– *philippinus*	Baya weaver	– Baya	Бойя
Poephila acuticauda	Long-tailed grass finch	Diamant à longue queue	
– *cincta*	Parson finch	– à bavette	
– *personata*	Masked grass finch	– à masque	
Pogoniulus	Tinker-birds	Petits barbus	
– *bilineatus*	Golden-rumped tinker-bird	Petit barbu à gorge blanche	
– *chrysoconus*	Yellow-fronted tinker-bird	– – à front jaune	
– *scolopaceus*	Speckled tinker-bird	– – grivelé	
– *subsulphureus*	Yellow-throated tinker-bird	– –à gorge jaune	
Polarbirkenzeisig	Arctic redpoll	Sizerin blanchâtre	Полярная чечетка
Pomatorhinus erythrogenys	Red-cheeked scimitar-babbler	Pomatorhin à joues rouges	
Prachtfinken	Waxbills	Estrildidés	
Prachtglanzstar	Splendid glossy starling	Merle métallique	Великолепный скворец
Prachtparadiesvogel	Magnificent bird of paradise	Paradisier magnifique	
Prachtreifelvogel	– riflebird	Proméfil magnifique	Чешуегрудый щитоносец
Prinzessin-Stephanie-Paradies-vogel	Princess Stephanie's bird of paradise	Paradisier de la princesse Stéphanie	
Prionops plumata	White helmet shrike		Очковый сорокопут
Procnias	Bellbirds	Orapongas	Птицы-звонари
Procnias nudicollis	Naked-throated bellbird	Oraponga à gorge nue	Бразильская птица-колокольчик
– *tricarunculata*	Wattled bellbird		Молотобой
Provence-Grasmücke	Dartford warbler	Fauvette pitchou	Прованская славка
Prunella	Accenteurs	Accenteurs	Завирушки
– *collaris*	Alpine accentor	Accenteur alpin	Альпийская завирушка
– *modularis*	Dunnock	– mouchet	Лесная завирушка
– *montanella*	Siberian accentor		Сибирская завирушка
Prunellidae	Accenteurs	Prunellidés	Завирушковые
Psilopogon pyrolophus	Fire-tufted barbet	Barbu à plumules de feu	
Pteridophora alberti	King of Saxony's bird of paradise		Чешуйчатая райская птица
Pteroglossus torquatus	Banded aracari	Aracari à collier	
Ptilogonys cinereus	Gray silky-flycatcher	Gobe-mouche gris du Mexique	Серый тропический свиристель
Ptilonorhynchus violaceus	Satin bower-bird	Oiseau à berceaux satiné	
Ptyonoprogne rupestris	Grag martin	Hirondelle de rochers	Скалистая ласточка
Pünktchenamarant	Bar-breasted fire finch	Sénégali à poitrine barrée	
Purpurastrild	Black-bellied seed-cracker	Pyreneste ponceau à ventre noir	
Purpurglanzstar	Purple glossy starling	Merle bronzé pourpre	

GERMAN NAME	ENGLISH NAME	FRENCH NAME	RUSSIAN NAME
Purpur-Kronfink	Red-crested finch	Pinson couronné	
Pycnonotidae	Bulbuls		Короткопалые дрозды
Pycnonotus	—		Настоящие бульбули
— *cafer*	Red-vented bulbul	Bulbul Indien	
— *goiavier*	Yellow-vented bulbul	— à ventre jaune	
— *jocosus*	Red-eared bulbul	— à oreillons rouges	
Pyrenestes ostrinus	Black-bellied seed-cracker	Pyreneste ponceau à ventre noir	
Pyrocephalus rubinus	Vermilion flycatcher	Tyran rouge écarlate	
Pyrrhocorax graculus	Alpine chough	Chocard à bec jaune	Альпийская галка
— *pyrrhocorax*	Chough	Crave à bec rouge	Клушица
Pyrrhula		Bouvreuils	Снегири
— *pyrrhula*	Bullfinch	Bouvreuil pivoine	Снегирь
Pytilia afra	Red-faced waxbill	Pytilie à dos jaune	
— *melba*	Melba finch	Beau-marquet	
— *phoenicoptera*	Aurora waxbill	Diamant aurore	
Quelea quelea	Black-fronted diock	Travailleur	
Queleopsis erythrops	Red-headed quelea	Quéléa à tête rouge	
Quiscalus	Grackles		Челнохвосты
Rabenkrähe	Carrion crow	Corneille noire	Обыкновенная черная ворона
Rabenvögel	Crows	Corbeaux	Вороновые
Racken	Rollers	Coraciadidés	Сизоворонковые
Rackenvögel		Coraciadiformes	Ракшеобразные
Ramphastidae	Toucans	Rhamphastidés	Туканы
Ramphastus dicolorus	Green-billed toucan	Toucan à bec vert	
— *swainsonii*	Swainson's toucan	— tocard	
— *toco*	Toco toucan	— toco	
— *vitellinus*	Sulphur-breasted toucan	— à bec caréné	
Raubwürger	Great grey shrike	Pie-grièche grise	Серый сорокопут
Rauchschwalbe	Common swallow	Hirondelle de cheminée	Деревенская ласточка
Rauchschwalben	Swallows	Hirondelles	Касатки
Raupenesser i. e. S.	Cuckoo-shrikes		Гусеницееды
Regulinae	Kinglets	Roitelets	Корольки
Regulus ignicapillus	Firecrest	Roitelet à triple bandeau	
— *regulus*	Goldcrest	— huppé	
Reichenows Bergastrild	Reichenow's crimson-wing	Bengali vert à face rouge	
Reifelvögel	Riflebirds		Щитоносные райские птицы
Reisfink	Java sparrow	Padda	Рисовый скворец
Reisstärling	Bobolink		Обыкновенный ремез
Remiz pendulinus	Penduline	Mésange penduline	Ремезовые
Remizidae	— tits		
Rhinocrypta	Gallitos	Gallitos	
Rhinozerosvogel	Rhinoceros hornbill		Малайский калао
Rhipidura rufifrons	Rufous-fronted fantail		Рыжелобая веерохвостка
Riesenelsterchen	Magpie mannikin	Spermète pie	
Riesensperling	Great sparrow	Moineau roux	
Riesentukan	Toco toucan	Toucan toco	
Riesenweber	Giant weaver	Tisserin géant	
Ringdrossel	Ring ouzel	Merle à plastron	Белозобый дрозд
Ringelastrild	Bicheno's finch	Diamant de Bicheno	
Riparia riparia	Sand martin	Hirondelle de rivage	Береговая ласточка
Rohrammer	Reed bunting	Bruant des roseaux	Камышовая овсянка
Rohrsänger i. e. S.		Rousserolles	Камышевки
Rohrschwirl	Savi's warbler	Locustelle luscinioïde	Соловьиный сверчок
Rosenamarant	Jameson's fire finch	Amaranthe de Jameson	
Rosenfinken	Rosy finches		Горные вьюрки
Rosenstar	Rose-coloured starling	Martin roselin	Розовый скворец
Rotbauchtimalie	Rufous-bellied babbler	Dumétie à ventre roux	
Rotbauchwürger	Crimson-breasted shrike	Pie-grièche à ventre cramoisie	
Rotbrust-Buntbärtling	Red-headed barbet	Barbu de Bourcier	
— Samenknacker	Blue-bill	Astrild à gros bec bleu	
Rotbüschel-Bartvogel	Fire-tufted barbet	Barbu à plumules de feu	
Rotdrossel	Redwing	Grive mauvis	Дрозд-белобровик
Rötelschwalbe	Red-rumped swallow	Hirondelle rousseline	Рыжепоясничная ласточка
Roter Kardinal	Cardinal		Красный кардинал
— Tropfenastrild	Peter's twin-spot	Amaranthe enflammée	
Rothaubengärtner	Orange-crested gardener	Jardinier à huppe orange	
Rotkehlchen	Robins	Rouge-gorges	Зарянки
— i. e. S.	Robin	Rouge-gorge familier	Зарянка
Rotkehldrossel	Red-throated thrush	Grive à cou roux	Краснозобый дрозд
Rotkehl-Hüttensänger	Eastern bluebird	Rouge-gorge bleu d'Amérique	Обыкновенный голубой чекан
Rotkehlpieper	Red-throated pipit	Pipit à gorge rousse	Краснозобый конек
Rotkopf-Amadine	Red-headed finch	Amadine à tête rouge	
Rotköpfige Papageiamadine	Parrot finch	Diamant à tête rouge	

GERMAN NAME	ENGLISH NAME	FRENCH NAME	RUSSIAN NAME
Rotkopfspecht	Red-headed woodpecker	Pic à tête rouge	Американский красноголовый дятел
Rotkopfweber	– quelea	Quéléa à tête rouge	
Rotkopfwürger	Woodchat shrike	Pie-grièche à tête rousse	Красноголовый сорокопут
Rotohramadine	Red-eared firetail finch	Astrild à oreillons rouges	
Rotohrbülbül	– bulbul	Bulbul à oreillons rouges	
Rotschnabel-Madenhacker	Red-billed ox-pecker	Pique-bœuf à bec rouge	Красноклювый волоклюй
– -Schweifkitta	Chinese blue-pie	Pirolle de la Chine	Красноклювая лазуревая сорока
Rotschnabeltoko	Red-billed hornbill	Calao tock	
Rotschulterglanzstar	Cape red-shouldered glossy starling	Merle à épaulettes rouges	
Rotschwanzjakamar	Rufous-tailed jacamar		Краснохвостая якамара
Rotschwänze		Rouge-queues	Горихвостки
Rotschwingenstar	Red-wing starling	Roupenne	
Rotstirn-Bartvogel	Pied barbet	Barbu pie	
– -Fächerschnäpper	Rufous-fronted fantail		Рыжелобая веерохвостка
Rotwangen-Säbler	Red-cheeked scimitar-babbler	Pomatorhin à joues rouges	
Rubinköpfchen	Vermilion flycatcher	Tyran rouge écarlate	
Rubin-Nachtigall		Calliope	Соловей-красношейка
Rupicola	Cocks of the rock	Coqs-de-roche	Скалистые петушки
Saatkrähe	Rook	Corbeau freux	Грач
Saftlecker	Common sapsucker		Дятел-сосун
Sägeracken	Motmots	Momotidés	Момоты
Samtgoldvogel	Regent bower-bird	Oiseau-régent	
Samtkopf-Grasmücke	Sardinian warbler	Fauvette mélanocéphale	Средиземноморская славка
Samtstirnkleiber	Velvet-fronted nuthatch	Sittelle à front noir	
Samtwida	Yellow bishop	Gros-bec tacheté du Cap de bonne Espérance	
Sarcops calvus	Coleto mynah	Sarcops chauve	
Sardengrasmücke	Marmora's warbler	Fauvette sarde	Сардинская славка
Satrapen	Kingbirds	Tyrans	
Saxicola		Traquets	Чеканы
– *rubetta*	Whinchat	Traquet des près	Луговой чекан
– *torquata*	Stonechat	– pâtre	Черноголовый чекан
Schafstelze	Blue-headed wagtail	Bergeronette printanière	Желтая трясогузка
Schamadrossel	Shama	Merle Shama	Малабарская сорочья славка
Scharlach-Mistelesser	Red-backed flower-pecker	Grimpereau à dos rouge	Красноспинный цветосос
Scharlachspint	Nubian carmine bee-eater	Guêpier carminé	
Scharlachweber	Red-headed weaver	Républicain à capuchon écarlate	
Schildwida	Red-collared whydah	Veuve en feu	
Schilfrohrsänger	Sedge warbler	Phragmite des joncs	Камышевка-барсучок
Schirmvogel	Ornate umbrella bird	Céphaloptère orné	Амазонская зонтичная птица
Schlagschwirl		Locustelle fluviatile	Речной сверчок
Schmätzer i. e. S.		Traquets	Чеканы
Schmetterlingsfink	Red-cheeked cordon-bleu	Cordon bleu	
Schmuckspint	Rainbow bird	Guêpier d'Australie	
Schmuckvögel	Cotingas	Cotingidés	Котинги
Schneeammer	Snow bunting	Bruant des neiges	Пуночки
Schneeammern	– buntings	Bruants des neiges	Снежный вьюрок
Schneefink	– finch	Niverolle des Alpes	Снежные вьюрки
Schneefinken	– finches	Niverolles	Славка-портниха
Schneidervogel	Tailor bird	Fauvette couturière	
Schnurrbärtchen	Scaly weaver	Tisserin à front pointillé	Манакины
Schnurrvögel	Manakins	Piprides	
Schönbürzel	Lavender finch	Gris-bleu	
Schreivögel	Noisemakers	Mésomyodés	Кричащие птицы
Schuppenköpfchen	Speckle-fronted weaver	Sénégali à front pointillé	
Schwalben	Swallows	Hirondelles	Ласточковые
– -Faulvogel	Swallow-wing		Ласточкокрылая пуховка
– -Mistelesser	Mistletoe bird		Ласточковый цветосос
Schwalbenstare	Wood swallows		Ласточковые скворцы
Schwanzmeise	Long-tailed tit	Mésange à longue queue	Длиннохвостая синица
Schwanzmeisen	– tits		Длиннохвостые синицы
Schwarzbäckchen	Yellow-bellied waxbill	Joue noire	
Schwarzbauchnonne	Three-coloured mannikin	Jacobin	
Schwarzhalsstar	Black-necked mynah	Martin à cou noir	
Schwarzkehlchen	Stonechat	Traquet pâtre	Черноголовый чекан
Schwarzkehldrossel	Black-throated thrush	Grive à gorge noire	Чернозобый дрозд
Schwarzkehl-Niltava	Rufous-bellied niltava	Niltava à ventre rouge	
– -Paradieselster	Long tail bird of paradise		Черная шлемоносная райская птица

GERMAN NAME	ENGLISH NAME	FRENCH NAME	RUSSIAN NAME
Schwarzkopfhäher	Steller's jay		Черноголовая голубая сойка
Schwarzkopfpitta	Hooded pitta	Brève sordide	
Schwarzlappenpitta	Velvet pitta	Philépitte veloutée	
Schwarzleierschwanz	Albert's lyre-bird		Принцев лирохвост
Schwarznacken-Pirol	Black-naped oriole	Loriot de Chine	Черноголовая иволга
Schwarzrücken-Bartvogel	Levaillant's barbet	Barbu de Levaillant	
— -Flötenvogel	Common piping crow	Corbeau flûteur	Черноспинная ворона-свистун
Schwarzschnabel-Blautukan	Black-billed mountain toucan	Toucan à bec noir	
Schwarzspecht	Black woodpecker	Pic noir	Желна
Schwarzstirnwürger	Lesser grey shrike	Pie-grièche à poitrine rose	Чернолобый сорокопут
Schwirle		Locustelles	Сверчки
Seidenlaubenvogel	Satin bower-bird	Oiseau à berceaux satiné	
Seidenrohrsänger		Bouscarles	Широкохвостые камышевки
Seidensänger	Cetti's warbler	Bouscarle de Cetti	Широкохвостая камышевка
Seidenschnäpper i. e. S.	Phainopepla	Phainopépla resplendissant	
Seidenschwänze	Waxwings	Bombycillidés	Свиристелевые
Seggen-Rohrsänger	Aquatic warbler	Phragmite aquatique	Вертлявая камышевка
Selenidera maculirostris	Spot-billed toucanet	Toucanet à bec tâcheté	
Seleucidis ignotus	Twelve-wired bird of paradise	Paradisier multifil	Нитчатая райская птица
Semioptera wallacei	Wallace's standard-wing	— de Wallace	Вымпельная райская птица
Semnoris ramphastinus	Toucan barbet	Barbu toucan	Туканий бородастик
Senegalfurchenschnabel	Bearded barbet	Barbican à poitrine rouge	
Sericulus chrysocephalus	Regent bower-bird	Oiseau-régent	
Serinus	Serins	Serins	Канареечные вьюрки
— *canaria*	Canary	Serin	Канарейка
— *citrinellus*	Citril finch	Venturon montagnard	
— *leucopygius*	Grey singing finch	Chanteur d'Afrique	Сенегальский вьюрок
— *mozambicus*	Green singing finch	Serin de Moçambique	Мозамбикский вьюрок
— *serinus*	Serin	— cini	Канареечный вьюрок
Sialia silais	Eastern bluebird	Rouge-gorge bleu d'Amérique	Обыкновенный голубой чекан
Sichelschnabel	Meyer's sickle bill	Epimagne de Meijer	
Siebenstreifen-Ammer	Rock bunting	Bruant à sept raies	
Siedelweber	Sociable weaver	Tisserin social	Обыкновенный общественный ткач
Silberohr-Sonnenvogel	Silver-eared mesia	Mésia à joues argentés	
Silberschnäbelchen	African silver-bill	Bec d'argent	
Singdrossel	Song thrush	Grive musicienne	Певчий дрозд
Singvögel	Singing birds	Oiseaux chanteurs	Певчие птицы
Sitta	Nuthatches	Sittelles	Поползни
— *canadensis*	Red-breasted nuthatch		Черноголовый поползень
— *europaea*	Nuthatch	Sittelle torchepot	Обыкновенный поползень
— — *castanea*	Chestnut-bellied nuthatch	— à ventre marron	
— *frontalis*	Velvet-fronted nuthatch	— à front noir	
— *neumayer*	Rock nuthatch	— des rochers	Малый скалистый поползень
— *pygmaea*	Pigmy nuthatch		Поползень-крошка
— *whiteheadi*	Corsican nuthatch	Sittelle corse	
Sittidae	Nuthatches	Sittidés	Настоящие поползни
Sittinae	—		Поползни
Smithornis	Common African broadbills	Gobe-mouches	
Sommergoldhähnchen	Firecrest	Roitelet à triple bandeau	
Sonnenastrild	Crimson finch	Phaéton	
Spatelschwanzelster	Hooded racket-tailed tree pie (magpie)	Pie bleue de l'Himalaya	
Spechte	Woodpeckers	Picidés	Дятловые
Spechtvögel		Piciformes	Дятловидные
Sperbergrasmücke	Barred warbler	Fauvette épervière	Ястребиная славка
Sperlinge	Sparrows	Passerinés	Воробьи
— i. e. S.	—	Moineaux	Настоящие воробьи
Sperlingsvögel		Passériformes	Воробьиные
Spermestes bicolor	Black-and-white mannikin	Spermète bicolore	
— *cucullatus*	Bronze mannikin	— à capuchon	
— *fringilloides*	Magpie mannikin	— pie	
— *nanus*	Bib-finch	— naine	
Spermophaga haematina	Blue-bill	Astrild à gros bec bleu	
Sphyrapicus varius	Common sapsucker		Дятел-сосун
Spinnenjäger	Spider-hunters		Паукоеды
Spinnenstar	Shining starling	Stourne métallique	
Spitzschwanzamadine	Long-tailed grass finch	Diamant à longue queue	
Spitzschwanz-Bronzemännchen	Sharp-tailed munia	Domino à longue queue	

ANIMAL DICTIONARY

GERMAN NAME	ENGLISH NAME	FRENCH NAME	RUSSIAN NAME
Spodiopsar cineraceus	Grey starling	Martin gris	Серый скворец
Spornammer	Lapland bunting	Bruant lapon	
Spornammern		Plectrophanes	Подорожники
Spornpieper	Richard's pipit	Pipit de Richard	
Sporopipes frontalis	Speckle-fronted weaver	Sénégali à front pointillé	
– *squamifrons*	Scaly weaver	Tisserin à front pointillé	
Spottdrossel	Mocking bird	Merle moqueur commun	
Spottdrosseln	Mocking-birds	Mimidés	
Spötter		Hippolaïs	Пересмешки-бормотушки
Sprosser	Great nightingale	Rossignol progné	Обыкновенный соловей
Stachelbürzler	Cuckoo-shrikes	Campéphagidés	Личинкоедовые
Stagonopleura bella	Fire-tailed finch	Diamant à queue de feu	
– *guttata*	Diamond sparrow	– à gouttelettes	
– *oculata*	Red-eared firetail finch	Astrild à oreillons rouges	
Stare	Starlings	Étourneaux	Скворцовые
Starweber	Dinemelli's weaver	Dinemelli à tête blanche	
Steinrötel	Rock thrushes, Rock thrush	Monticoles, Merle de roches	Каменные дрозды, Пестрый каменный дрозд
Steinschmätzer	Wheatear	Traquet motteux	Обыкновенная каменка
Steinsperling	Rock sparrow	Moineau fou	Каменный воробей
Steinsperlinge	– sparrows		Каменные воробьи
Stelzen	Wagtails		Трясогузковые
Stieglitze und Zeisige		Chardonnerets	Щеглы
Stizoptera bichenovii	Bicheno's finch	Diamant de Bicheno	
Streifenschwirl	Pallas's grasshopper warbler	Locustelle de Pallas	Певчий сверчок
Strepera	Piping crows		Кричащие вороны
– *graculina*	Pied crow-shrike	Grand calibé	
Strichelschwirl	Lanceolated warbler	Locustelle lancéolée	Пятнистый сверчок
Strichelstelzer	Gallitos	Gallitos	
Strohwitwe	Fischer's whydah	Veuve de Fischer	
Struthidea cinerea	Grey-jumper	Glaucope gris	Штрутидеа
Stärlinge	Orioles	Ictéridés	Касиковые
Stieglitz	Goldfinch	Chardonneret élégant	Щегол
Stummellerche	Lesser short-toed lark	Alouette pispolette	
Sturnia sinensis	Mandarin mynah	Martin de Chine	Южнокитайский малый скворец
Sturnidae	Starlings	Étourneaux	Скворцовые
Sturninae	–		Настоящие скворцы
Sturnopastor contra	Pied maynah	Martin pie	Черный скворец
Sturnus unicolor	Spotless starling	Étourneau unicolore	Обыкновенный скворец
– *vulgaris*	Starling	– sansonnet	
Südsee-Grasmücken	Australian »wrens«		Малуры
Sultansspechte		Pics dorés	Султанские дятлы
Sumpfastrild	Fawn-breasted waxbill	Astrild de marais	
Sumpfrohrsänger	Marsh warbler	Rousserolle verderolle	Болотная камышевка
Sylvia	Warblers	Fauvettes	Славки
– *atricapilla*	Blackcap	Fauvette à tête noire	Черноголовая славка
– *borin*	Garden warbler	– des jardins	Садовая славка
– *cantillans*	Subalpine warbler	– passerinette	Горная славка
– *communis*	Whitethroat	– grisette	Серая славка
– *conspicillata*	Spectacled warbler	– à lunettes	Очковая славка
– *curruca*	Lesser whitethroat	– babillarde	Славка-завирушка
– *hortensis*	Orphean warbler	– orphée	Певчая славка
– *melanocephala*	Sardinian warbler	– mélanocéphale	Средиземноморская славка
– *nisoria*	Barred warbler	– épervière	Ястребиная славка
– *rueppelli*	Rüppell's warbler	– masquée	Эгейская славка
– *sarda*	Marmora's warbler	– sarde	Сардинская славка
– *undata*	Dartford warbler	– pitchou	Прованская славка
Taeniopygia guttata	Zebra finch	Diamant mandarin	
Taha afra	Tahaweaver	Tisserin taha	
Tahaweber	–	– –	
Tangare	Tanagers		Тангары
Tannenhäher	Nutcracker	Casse-noix moucheté	Ореховка
Tannenmeise	Coal-titmouse	Petite charbonnière	Черная синица
Tchagra tchagra	Redwing shrike		Чагра
Teichrohrsänger	Reed warbler	Rousserolle effarvatte	Тростниковая камышевка
Temenuchus malabaricus	Malabar mynah	Martin à tête grise	
– *pagodarum*	Pagoda starling	– des pagodes	Браминский скворец
Terpsiphone padarisi	Paradise flycatcher	Gobe-mouche paradisier	Индийская длиннохвостая мухоловка
Tetraenura fischeri	Fischer's whydah	Veuve de Fischer	
– *regia*	Queens whydah	– reine	
Textor capensis	Cape weaver	Tisserin à front d'or	
– *cucullatus*	Black-headed weaver	– Cap Moor	
– *grandis*	Giant weaver	– géant	

GERMAN NAME	ENGLISH NAME	FRENCH NAME	RUSSIAN NAME
– velatus	Black-fronted weaver	– à front noir	
Textorweber	– -headed weaver	– Cap Moor	
Thamnophilus	Ant shrikes	Thamnophiles	
Thraupinae	Tanagers		Тангары
Thraupini	–		Настоящие тангары
Thryothorus ludovivianus	Carolina wren		Крапивник-пересмешник
Tiaris canora	Cuba finch	Petit chanteur de Cuba	
– olivacea	Olive finch	Grand chanteur de Cuba	
Tichodroma muraria	Wall-creeper	Tichodrome échelette	Краснокрылый стенолаз
Tichodrominae	Wallcreepers		Стенолазы
Tigerfink	Avadavat	Bengali rouge	
Timalien	Babbling thrushes	Timaliinés	Тималии
Timaliinae	– –		–
Tockus erythrorhynchus	Red-billed hornbill	Calao tock	
Todidae	Todies	Todidés	Тодиевые
Todis			
Todus subulatus	Broad-billed tody	Todier vert	
Tonkibülbül	Red-vented bulbul	Bulbul Indien	
Töpfervogel	Ovenbird	Fournier roux	Рыжий печник
Töpfervögel	Ovenbirds	Furnariidés	Печники
– i. e. S.			
Trachyphonus margaritatus	Yellow-breasted barbet	Barbu perlé	
– vaillantii	Levaillant's barbet	– de Levaillant	
Trauerbachstelze	Pied wagtail	Bergeronette d'Yarrell	Британская белая трясогузка
Trauerbartvogel	Scarlet-crowned barbet	Barbu à poitrine orange	
Trauerdrongo	Drongo	Drongo	Траурный дронго
Trauerschnäpper	Pied flycatcher	Gobe-mouche noir	Мухоловка-пеструшка
Tricholaema leucomelan	– barbet	Barbu pie	
Troglodytes troglodytes	Wren	Troglodyte mignon	Обыкновенный крапивник
Troglodytidae	Wrens	Troglodytidés	Крапивниковые
Trupiale	Orioles		Трупиалы
Tukan-Bartvogel	Toucan barbet	Barbu toucan	Туканий бородастик
Tukane	Toucans	Rhamphastidés	Туканы
Turdinae	Thrushes	Turdinés	Дроздовые
Turdus		Merles	Настоящие дрозды
– iliacus	Redwing	Grive mauvis	Дрозд-белобровик
– merula	Blackbird	Merle noir	Черный дрозд
– migratorius	American robin	– migrateur	Странствующий дрозд
– philomelos	Song thrush	Grive musicienne	Певчий дрозд
– pilaris	Fieldfare	– litorne	Дрозд-рябинник
– ruficollis atrogularis	Black-throated thrush	– à gorge noire	Чернозобый дрозд
– – ruficollis	Red-throated thrush	– à cou roux	Краснозобый дрозд
– torquatus	Ring ouzel	Merle à plastron	Белозобый дрозд
– viscivorus	Mistle thrush	Grive draine	Дрозд-деряба
Türkisbrauen-Sägeracke	Jucatan motmot	Motmot à sourcils bleus	
Tyrannen	Tyrant flycatchers	Tyrannidés	Тираны
Tyrannidae	– –		–
Tyrannus	Kingbirds	Tyrans	
– tyrannus	American kingbird	Tyran royal	
Ufermaina	Bank mynah	Martin de rivage	Береговая майна
Uferschwalbe	Sand martin	Hirondelle de rivage	– ласточка
Unglückshäher	Siberian jay	Geai de Sibérie	Кукша
Upupa epops	Hoopoe	Huppe puput	Обыкновенный удод
Upupidae	Hoopoes	Upupidés	Удодовые
Upupinae		Upupinés	Удоды
Uraeginthus angolensis	Blue-breasted waxbill	Cordon bleu d'Angola	
– bengalus	Red-cheeked cordon-bleu	Cordon bleu	
– cyanocephalus	Blue-headed waxbill	Astrild à tête bleue	
– granatinus	Violet-eared waxbill	Grenadin	
– ianthinogaster	Purple grenadier	– à poitrine bleue	
Utocissa erythrorhyncha	Chinese blue-pie	Pirolle de la Chine	Красноклювая лазуревая сорока
– flavirostris	Yellow-billed blue-pie	– à bec jaune	Желтоклювая лазуревая сорока
– ornata	Ceylon blue magpie	Pie bleue ornée	
Vangidae	Vangas	Vangidés	
Veilchenastrild	Purple grenadier	Grenadin à poitrine bleue	
Vidua	Whydahs	Veuves	Вдовушки
– hypocherina	Shiny black whydah	Veuve métallique	
– macroura	Pin-tailed whydah	– dominicaine	
Viduinae		Veuves-Combassous	Вдовушковые
Vireonidae	Vireos	Viréonidés	Листовые сорокопуты
Vireos	–		– –
Wacholderdrossel	Fieldfare	Grive litorne	Дрозд-рябинник
Wachtelastrild	South-African quail-finch	Astrild caille à lunettes	

GERMAN NAME	ENGLISH NAME	FRENCH NAME	RUSSIAN NAME
Waldbaumläufer	Tree creeper	Grimpereau des bois	Обыкновенная пищуха
Waldlaubsänger	Wood-wren	Pouillot siffleur	Пеночка-трещотка
Waldsänger	Old World warblers	Parulidés	
Wallace-Paradiesvogel	Wallace's standard-wing	Paradisier de Wallace	Вымпельная райская птица
Wanderdrossel	American robin	Merle migrateur	Странствующий дрозд
Wasseramsel	Dipper	Cincle plongeur	Обыкновенная оляпка
Wasseramseln	Dippers	Cinclidés	Оляпковые
Wasserpieper	Water pipit	Pipit spioncelle	Горный конек
Webervögel i. w. S.	Weaverbirds	Plocéidés	Ткачиковые
Weißbart-Grasmücke	Subalpine warbler	Fauvette passerinette	Горная славка
Weißbauchnonne	Spectacled finch	Nonnette à ventre roux	
Weißbrauen-Schwalbenstar	White-browed wood swallow	Langrayen à sourcils blancs	Белобровый ласточковый скворец
Weißbrust-Schilffink	Pictorella finch		
Weißkehl-Ammerfink	White-crowned sparrow	Donacole à poitrine blanche	
– -Katzenvogel	– -throated cat-bird	Pinson à couronne blanche	Белозобая птица-кошка
Weißkopfnonne	– -headed mannikin	Nonnette à tête blanche	
Weißkopf-Rotschwanz	– -capped redstart	Rouge-queue à tête blanche	Белошапочная горихвостка
Weißnacken-Honigesser	– -naped honey-eater	Méliphage à tête noire	
Weißrücken-Flötenvogel	Tasmanian crow-shrike		Белоспинная ворона-свистун
Weißrückenspecht	White-backed woodpecker	Pic à dos blanc	Белоспинный дятел
Weißsäbelpipra	– -bearded manakin	Manakin moine	
Weißstirnweber	Thick-billed weaver	Tisserin à gros bec	
Weitraum-Zistensänger	Fan-tailed warbler	Cisticole des joncs	
Wellenastrild	St. Helena waxbill	Astrild ondulé	Воскоклювый ткач
Wendehälse		Jynginés	Вертишейки
Wiedehopf	Hoopoe	Huppe puput	Обыкновенный удод
Wiedehopfe		Upupinés	Удоды
Wienerastrild	Red-faced waxbill	Pytilie à dos jaune	
Wiesenpieper	Meadow pipit	Pipit farlouse	Луговой конек
Wintergoldhähnchen	Goldcrest	Roitelet huppé	
Winter-Junco	Slate-coloured junco	Junco ardoisé	Серый юнко
Witwen		Veuves-Combassous	Вдовушковые
Würger	Shrikes	Laniidés	Сорокопутовые
– i. e. S.	–	Pie-grièches	Сорокопуты
Würgerkrähe	Pied crow-shrike	Grand calibé	
Würgerkrähen	Piping crows		Кричащие вороны
Wüstensperling	Desert sparrow		Пустынный воробей
Wüstensteinschmätzer	– wheatear	Traquet du désert	Пустынная каменка
Xenicidae	Wrens	Xenicidés	
Xenicus longipes	New-Zealand wren	Roitelet de la Nouvelle-Zélande	
Zaunkönig	Wren	Troglodyte mignon	Обыкновенный крапивник
Zaunkönige	Wrens	Troglodytidés	Крапивниковые
Zebrafink	Zebra finch	Diamant mandarin	
Zedernseidenschwanz	Cedar waxwing		Кедровый свиристель
Zeisige und Stieglitze		Chardonnerets	Чижи
Zeresfink	Cherry finch	Modeste	
Zilpzalp		Pouillot véloce	Пеночка-теньковка
Zistensänger		Cisticoles	Травяные певуны
Zitronengirlitz	Citril finch	Venturon montagnard	
Zitronenstelze		Bergeronnette citrine	Желтоголовая трясогузка
Zonotrichia albicollis	White-crowned sparrow	Pinson à couronne blanche	
Zoothera citrina	Orange-headed ground thrush	Grive orangée	
– *dauma*	Ground thrush		Пестрый дрозд
Zosteropidae	White-eyes	Zostéropidés	Белоглазковые
Zosterops	–		Белоглазки
– *japonica*	Japanese white-eye		Японская белоглазка
– *lateralis*	Gray-backed white-eye		Австралийская белоглазка
– *palpebrosa*	Indian white-eye	Zostérops à lunettes	Гангская белоглазка
Zuckervögel	Honey creepers	Coerébidés	
Zügelastrild	Crimson-rumped waxbill	Astrild à dos rouge	
Zweifarbenschwärzling	Chestnut-breasted negro-finch	Bengali brun à ventre roux	
Zwergbartvögel	Tinker-birds	Petits barbus	
Zwergelsterchen	Bib-finch	Spermète naine	
Zwerghaubenfischer	Malachite kingfisher	Petit martin-pêcheur huppé	
Zwergkleiber	Pigmy nuthatch		Поползень-крошка
Zwergkönigsfischer	Natal kingfisher	Martin-pêcheur de Natal	
Zwergschnäpper	Red-breasted flycatcher	Gobe-mouche nain	Малая мухоловка
Zwergspechte		Picuminés	Дятелки

III. French–German–English–Russian

Dans la plupart des cas, les noms des sous-espèces sont formés en ajoutant au nom de l'espèce un adjectif ou une désignation géographique. Dans cette partie du dictionnaire zoologique, les noms français des sous-espèces formés de cette manière ne seront en général pas indiqués.

FRENCH NAME	GERMAN NAME	ENGLISH NAME	RUSSIAN NAME
Accenteur alpin	Alpenbraunelle	Alpine accentor	Альпийская завирушка
– mouchet	Heckenbraunelle	Dunnock	Лесная завирушка
Accenteurs	Braunellen	Accentors	Завирушковые, Завирушки
Alcédinidés	Eisvögel	Kingfishers	Зимородковые
Alouette alpestre	Ohrenlerche	Shore lark	Рюм
– calandre	Kalanderlerche	Calandra lark	Степной жаворонок
– commune	Feldlerche	Sky lark	Полевой жаворонок
– de Sibérie	Ohrenlerche	Shore lark	Рюм
– des champs	Feldlerche	Sky lark	Полевой жаворонок
– hausse-col	Ohrenlerche	Shore lark	Рюм
– locustelle	Feldschwirl	Grasshopper warbler	Обыкновенный сверчок
– lulu	Heidelerche	Wood lark	Лесной жаворонок
– pispolette	Stummellerche	Lesser short-toed lark	
Alouette-pie	Drosselstelze	Magpie lark	Дроздовидная трясогузка
Alouettes	Lerchen	Larks	Жаворонковые
Amadine à tête rouge	Rotkopf-Amadine	Red-headed finch	
Amaranthe	Gewöhnlicher Amarant	Fire finch	
– de Jameson	Rosenamarant	Jameson's fire finch	
– enflammée	Roter Tropfenastrild	Peter's twin-spot	
– foncée	Dunkelroter Amarant	Blue-billed fire finch	
– masquée d'Abyssinie	Larvenamarant	Masked waxbill	
– pointillée	Pünktchenamarant	Bar-breasted fire finch	
Amblyospizinés	Dickschnabelweber		Толстоклювые ткачи
Aracari à collier	Halsband-Arassari	Banded aracari	
Ardelle bleue	Blaumeise	Blue titmouse	Лазоревка
Artamie azurée	Blauvanga	– vanga	
Astrild à cape noire	Nonnenastrild	Black-crowned waxbill	
– à cinq couleurs	Dornastrild	Sydney waxbill	
– à dos rouge	Zügelastrild	Crimson-rumped waxbill	
– à flancs rayés	Goldbrüstchen	Golden-breasted waxbill	
– à gros bec bleu	Rotbrust-Samenknacker	Blue-bill	
– à moustaches noires	Elfenastrild	Black-cheeked waxbill	
– à oreillons rouges	Rotohramadine	Red-eared firetail finch	
– à tête bleue	Blaukopf-Schmetterlingsfink	Blue-headed waxbill	
– à tête noire	Kappenastrild	Black-headed waxbill	
– caille à lunettes	Wachtelastrild	South-African quail-finch	
– de marais	Sumpfastrild	Fawn-breasted waxbill	
– de Verreaux	Perlastrild	Rosy twin-spot	
– fourmillier	Ameisenpicker	Antpecker	
– grenadin	Granatastrild	Violet-eared waxbill	
– gris	Grauastrild	Grey waxbill	
– lavande	Schönbürzel	Lavender finch	
– locustelle	Heuschreckenastrild	Locust-finch	
– ondulé	Wellenastrild	St. Helena waxbill	Воскоклювый ткач
– queue de vinaigre	Schönbürzel	Lavender finch	
– rose	Perlastrild	Rosy twin-spot	
Astrilds	Eigentliche Astrilde	Waxbills	Кровяно-красные ткачи
Atrichorne roux	Kleiner Dickichtschlüpfer	Rufous scrub-bird	
Atrichornes	Dickichtschlüpfer	Scrub-birds	
Babillarde épervière	Sperbergrasmücke	Barred warbler	Ястребиная славка
– grisette	Dorngrasmücke	Whitethroat	Серая славка
– ordinaire	Klappergrasmücke	Lesser whitethroat	Славка-завирушка
Barbican à poitrine rouge	Senegalfurchenschnabel	Bearded barbet	
Barbu à bec denté	Doppelzahn-Bartvogel	Double-toothed barbet	
– à collier noir	Halsband-Bartvogel	Black-collared barbet	
– à front rouge	Kupferschmied	Coppersmith	
– à gorge bleue	Blauwangen-Bartvogel	Blue-throated barbet	
– à plumules de feu	Rotbüschel-Bartvogel	Fire-tufted barbet	
– à poitrine cramoisie	Kupferschmied	Coppersmith	
– à poitrine orange	Trauerbartvogel	Scarlet-crowned barbet	
– à tête grise du Ceylan	Ceylon-Grünbartvogel	Green barbet	
– chauve à narines emplumées	Pel-Bartvogel	Bristle-nosed barbet	
– de Bourcier	Rotbrust-Buntbärtling	Red-headed barbet	
– de Levaillant	Schwarzrücken-Bartvogel	Levaillant's barbet	
– géant	Blaukopf-Bartvogel	Great hill barbet	
– perlé	Perlbartvogel	Yellow-breasted barbet	
– pie	Rotstirn-Bartvogel	Pied barbet	

ANIMAL DICTIONARY

FRENCH NAME	GERMAN NAME	ENGLISH NAME	RUSSIAN NAME
– toucan	Tukan-Bartvogel	Toucan barbet	Туканий бородастик
Barbus	Bartvögel	Barbets	Бородастики
Beau-marquet	Buntastrild	Melba finch	
Bec-croisé bifascié	Bindenkreuzschnabel	Two-barred crossbill	Белокрылый клест
– des sapins	Fichtenkreuzschnabel	Crossbill	Клест-еловик
– perroquet	Kiefernkreuzschnabel	Parrot crossbill	Клест-сосновик
Bec-croisés	Kreuzschnäbel	Crossbills	Клесты
Bec d'argent	Silberschnäbelchen	African silver-bill	
– de coreil	Grauastrild	Grey waxbill	
– de plomb	Malabarfasänchen	Indian silver-bill	
Bec-fin à tête noire	Mönchsgrasmücke	Black-cap	Черноголовая славка
– babillard	Klappergrasmücke	Lesser whitethroat	Славка-завирушка
– rossignol	Nachtigall	Nightingale	Западный соловей
– rouge-gorge	Rotkehlchen i. e. S.	Robin	Зарянка
– rouge-queue	Gartenrotschwanz	Redstart	Обыкновенная горихвостка
Bengali à joues oranges	Orangebäckchen	Orange-cheeked waxbill	
– brun à ventre roux	Zweifarbenschwärzling	Chestnut-breasted negro-finch	
– moucheté	Tigerfink	Avadavat	
– rouge	–	–	
– vert	Astrild	Green Avadavat	
– – à face rouge	Reichenows Bergastrild	Reichenow's crimson wing	
– – d'Ansorge	Halsbandastrild	White-collared olive-back	
– – pointillé	Grüner Tropenastrild	Green-backed twin-spot	
– zébré	Goldbrüstchen	Golden-breasted waxbill	
Bergeronnette citrine	Zitronenstelze		Желтоголовая трясогузка
– des ruisseaux	Bergstelze	Grey wagtail	Горная трясогузка
– d'Yarrell	Trauerbachstelze	Pied wagtail	Британская белая трясогузка
– grise	Bachstelze	White wagtail	Белая трясогузка
– jaune	Bergstelze	Grey wagtail	Горная трясогузка
– printanière	Schafstelze	Blue-headed wagtail	Желтая трясогузка
Bergeronettes	Eigentliche Stelzen	Wagtails	Трясогузки
Bombycillidés	Seidenschwänze	Waxwings	Свиристелевые
Bouscarle de Cetti	Seidensänger	Cetti's warbler	Широкохвостая камышевка
Bouscarles	Seidenrohrsänger		Широкохвостые камышевки
Bouvreuil olivert	Groß-Kubafink	Olive finch	
– pivoine	Dompfaff	Bullfinch	Снегирь
– ponceau	–	–	–
Bouvreuils	Gimpel		Снегири
Brève à capuchon	Schwarzkopfpitta	Hooded pitta	
– du Bengale	Neunfarbenpitta	Indian pitta	Индийская питта
– sordide	Schwarzkopfpitta	Hooded pitta	
Brèves	Pittas	Pittas	Питты
Bruant à sept raies	Siebenstreifen-Ammer	Rock bunting	
– à tête rousse	Braunkopfammer	Red-headed bunting	Желчная овсянка
– des neiges	Schneeammer	Snow bunting	
– des roseaux	Rohrammer	Reed bunting	Камышовая овсянка
– jaune	Goldammer	Yellowhammer	Обыкновенная овсянка
– lapon	Spornammer	Lapland bunting	
– mélanocéphale	Kappenammer	Black-headed bunting	Черноголовая овсянка
– ortolan	Ortolan	Ortolan bunting	Садовая овсянка
– proyer	Ammern	Corn bunting	Просянка
Bruants	Grauammer	Buntings	Овсянки
– des neiges	Schneeammern	Snow buntings	Пуночки
Bubalornithinés	Büffelweber		Буйволовые ткачи
Bucconidés	Faulvögel	Puffbirds	Птицы-пуховки
Bucérotidés	Nashornvögel	Hornbills	– носороги
Bucorve d'Abyssinie	Nördlicher Hornrabe	Abyssinian ground hornbill	Абиссинский рогатый ворон
Bulbul à oreillons rouges	Rotohrbülbül	Red-eared bulbul	
Bulbul à ventre jaune	Gelbbauchbülbül	Yellow-vented bulbul	
– Indien	Tonkibülbül	Red-vented bulbul	
Calao bicorne	Doppelhornvogel	Great hornbill	
– tock	Rotschnabeltoko	Red-billed hornbill	
Calaos	Nashornvögel	Hornbills	Птицы-носороги
Calliope	Rubin-Nachtigall		Соловей-красношейка
Campéphagidés	Stachelbürzler	Cuckoo-shrikes	Личинкоедовые
Capitonidés	Bartvögel	Barbets	Бородастики
Capucin tricolore	Schwarzbauchnonne	Three-coloured mannikin	
Cardinaux	Kardinäle	Cardinal-grosbeaks	Кардиналы
Casse-noix moucheté	Tannenhäher	Nutcracker	Ореховка
Céphaloptère orné	Schirmvogel	Ornate umbrella bird	Амазонская зонтичная птица

ANIMAL DICTIONARY

FRENCH NAME	GERMAN NAME	ENGLISH NAME	RUSSIAN NAME
Certhiidés	Baumläuferartige	Creepers	Пищуховые
Chanteur d'Afrique	Grau-Edelsänger	Grey singing finch	Сенегальский вьюрок
Chanteuse des Vignes	Singdrossel	Song thrush	Певчий дрозд
Chardonneret élégant	Stieglitz	Goldfinch	Щегол
Chardonnerets	Stieglitze und Zeisige		Щеглы и чижи
Chat	Katzendrossel	Catbird	
Chocard à bec jaune	Alpendohle	Alpine chough	Альпийская галка
Choucas des tours	Dohle	Jackdaw	Галка
Cincle plongeur	Wasseramsel	Dipper	Обыкновенная оляпка
Cincles, Cinclidés	Wasseramseln	Dippers	Оляпковые
Cisticole des joncs	Weitraum-Zistensänger		
Cisticoles	Zistensänger	Fan-tailed warbler	Травяные певуны
Cochevis huppé	Haubenlerche	Crested lark	Хохлатый жаворонок
Coerébidés	Zuckervögel	Honey creepers	
Colapte des Pampas	Camposspecht	Pampas-flicker	
– doré	Goldspecht	Golden-banded woodpecker	Золотой кукушковый дятел
Conopophages	Mückenesser	Gnateaters	
Coqs-de-roche	Klippenvögel	Cocks of the rock	Скалистые петушки
Coraciadidés	Racken	Rollers	Сизоворонковые
Coraciadiformes	Rackenvögel		Ракшеобразные
Corbeau choucas	Dohle	Jackdaw	Галка
– corneille	Rabenkrähe	Carrion crow	Обыкновенная черная ворона
– flûteur	Schwarzrücken-Flötenvogel	Common piping crow	Черноспинная ворона-свистун
– freux	Saatkrähe	Rook	Грач
– mantelé	Nebelkrähe	Hooded crow	Серая ворона
– noir	Kolkrabe	Raven	Ворон
Corbeaux	Rabenvögel	Crows	Вороновые
Corcorax à ailes blanches	Bergkrähe	White-winged crow (chough)	
Cordon bleu	Schmetterlingsfink	Red-cheeked cordon-bleu	
– – d'Angola	Angola-Schmetterlingsfink	Blue-breasted waxbill	
Corneille mantelée	Nebelkrähe	Hooded crow	Серая ворона
– noire	Rabenkrähe	Carrion crow	Обыкновенная черная ворона
Cotinga tityre	Halsbandkotinga	Banded cotinga	Ошейниковая котинга
Cotingas	Eigentliche Kotingas	Cotingas	Котинги
Cotingidés	Schmuckvögel		–
Cou coupé	Bandfink	Cuckoo-roller	
Courol malgache	Kurol	Cut-throat	
Cracticidés	Krähenwürger	Australian butcherbirds	
Crave à bec rouge	Alpenkrähe	Chough	Клушица
Crécelle	Feldschwirl	Grasshopper warbler	Обыкновенный сверчок
Cul-rouge	Gartenrotschwanz	Spice finch	Обыкновенная горихвостка
Damier	Muskatfink	Redstart	
Dendrocolapitidés	Baumsteiger	Wood hewers	Древолазы
Diamant à bavette	Gürtelgrasfink	Parson finch	
– à goutellettes	Diamantfink	Diamond sparrow	
– à longue queue	Spitzenschwanzamadine	Long-tailed grass finch	
– à masque	Maskanamadine	Masked grass finch	
– à queue de feu	Feuerschwanzamadine	Fire-tailed finch	
– à queue rousse	Binsenastrild	Star finch	
– à queue verte	Bambus-Papageiamadine	Green-tailed parrot finch	
– à tête rouge	Rotköpfige Papageiamadine	Parrot finch	
– aurore	Aurora-Astrild	Aurora waxbill	
– de Bicheno	Ringelastrid	Bicheno's finch	
– de Gould	Gouldamadine	Gouldian finch	
– des montagnes	Bergamadine	Crimson-bellied mountain finch	
– mandarin	Zebrafink	Zebra finch	
– modeste	Zeresfink	Cherry finch	
– tricolore de Kittlitz	Dreifarbige Papageiamadine	Blue-faced parrot finch	
Dicaeidés	Mistelesser	Flowerpeckers	Цветососы
Dicée à dos rouge	Scharlach-Mistelesser	Red-backed flower-pecker	Красноспинный цветосос
Dinemelli à tête blanche	Starweber	Dinemelli's weaver	
Diock à bec rouge	Blutschnabelweber	Black-fronted diock	
Domino à longue queue	Spitzschwanz-Bronzemännchen	Sharp-tailed munia	
Donacole à poitrine blanche	Weißbrust-Schilffink	Pictorella finch	
– à poitrine châtaine	Braunbrust-Schilffink	Chestnut-breastet finch	
– à tête grise	Gelber Schilffink	Yellow-tailed finch	
– commun	Braunbrust-Schilffink	Chestnut-breasted finch	
Drépanididés	Kleidervögel		Нарядные птицы
Drongo	Trauerdrongo	Drongo	Траурный дронго
– à crinière	Glanzspitzendrongo	Spangled drongo	
– à raquettes	Flaggendrongo	Greater racket-tailed drongo	Райский дронго

FRENCH NAME	GERMAN NAME	ENGLISH NAME	RUSSIAN NAME
Drongos	Drongos, Drongos i. w. S.	Drongos	Дронговые, Дронго
Dumétie à ventre roux	Rotbauchtimalie	Rufous-bellied babbler	
Dur-bec des sapins	Hakengimpel	Pine grosbeak	Щур
Dur-becs	–		Щуры
Echenilleur pourpré	Mohrenraupenesser	Black ruckoo-shrike	
Emberizinés	Ammern	Buntings	Овсянки
Emblème peint	Gemalter Astrild	Painted finch	
Epimagne de Meijer	Sichelschnabel	Meyer's sickle bill	
Estrildidés	Prachtfinken	Waxbills	
Étourneau sansonnet	Gemeiner Star	Starling	Обыкновенный скворец
– unicolore	Einfarbstar	Spotless starling	Черный скворец
Étourneaux	Stare	Starlings	Скворцовые
Euplecte à couronne de feu	Flammenweber	Fire-crowned bishop	
– à dos doré	Goldrückenweber	Golden-backed weaver	
Eurylaime sombre	Feuerbreitrachen	Dusky Broadbill	
Eurylaimes, Eurylaimidés	Breitrachen	Broadbills	Рогоклювы
Eurystome d'Asie	Ostroller	Oriental dollar bird	Широкорот
Farlouse des prés	Wiesenpieper	Meadow pipit	Луговой конек
Fauvette à lunettes	Brillengrasmücke	Spectacled warbler	Очковая славка
– à tête noire	Mönchsgrasmücke	Blackcap	Черноголовая славка
– babillarde	Klappergrasmücke	Lesser whitethroat	Славка-завирушка
– couturière	Schneidervogel	Tailor bird	– портниха
– des jardins	Gartengrasmücke	Garden warbler	Садовая славка
– des Saules	Gelbspötter	Icterine warbler	Зеленая пересмешка
– épervière	Sperbergrasmücke	Barred warbler	Ястребиная славка
– grisette	Dorngrasmücke	Whitethroat	Серая славка
– masquée	Maskengrasmücke	Rüppell's warbler	Эгейская славка
– mélanocéphale	Samtkopf-Grasmücke	Sardinian warbler	Средиземноморская славка
– orphée	Orpheusgrasmücke	Orphean warbler	Певчая славка
– passerinette	Weißbart-Grasmücke	Subalpine warbler	Горная славка
– pitchou	Provence-Grasmücke	Dartford warbler	Прованская славка
– sarde	Sardengrasmücke	Marmora's warbler	Сардинская славка
Fauvettes	Grasmücken i. e. S.	Warblers	Славки
Formicariidés	Ameisenvögel	Ant birds	Птицы-муравьеды
Foudi rouge	Madagaskarweber	Madagascar weaver	
Fourmillier	Königs-Ameisenstelzer	Royal ant-thrush	
Fournier roux	Töpfervogel	Ovenbird	Рыжий печник
Fringillidés	Finken	Rinches	Вьюрковые
Friquet	Feldsperling	Tree sparrow	Полевой воробей
Fournariidés	Töpfervögel	Ovenbirds	Печники
Galbulidés	Glanzvögel	Jacamars	Якамары
Gallitos	Strichelstelzer	Gallitos	
Garrulax à huppe blanche	Haubenhäherling	White-crested laughing jay-thrush	
Geai bleu d'Amérique	Blauhäher	American blue jay	Голубая сойка
– – huppé	–	– – –	
– de Sibérie	Unglückshäher	Siberian jay	Кукша
– des chênes	Eichelhäher	Black-throated jay	Сойка
– strié	–	– –	–
Géospizinés	Darwinfinken	Darwin's finches	
Glaucope gris	Gimpelhäher	Grey-jumper	Штрутидеа
Gobe-mouche à collier	Halsbandschnäpper	White-collared flycatcher	Мухоловка-белошейка
– gris du Mexique	Grauseidenschnäpper	Gray silky-flycatcher	Серый тропический свиристель
– nain	Zwergschnäpper	Red-breasted flycatcher	Малая мухоловка
– noir	Trauerschnäpper	Pied flycatcher	Мухоловка-пеструшка
– paradisier	Paradiesschnäpper	Paradise flycatcher	Индийская длиннохвостая мухоловка
Gobe-mouches	Fliegenschnäpperartige, Afrikanische Breitrachen	Flycatchers, Common African broadbills	Мухоловковые
– américains	Tyrannen	Tyrant flycatchers	Тираны
Gorge-bleue à miroir	Blaukehlchen	Bluethroat	Варакушка
Grand barbu chauve	Glatzenbartvogel	Naked-faced barbet	
– calibé	Würgerkrähe	Pied crow-shrike	
– chanteur de Cuba	Groß-Kubafink	Olive finch	
– corbeau	Kolkrabe	Raven	Ворон
Grand niltava	Großniltava	Greater niltava	
– paradisier	Großer Paradiesvogel	– bird of paradise	Большая райская птица
– rossignol	Sprosser	Great nightingale	Обыкновенный соловей
Grande charbonnière	Kohlmeise	Oxeye-tit	Большая синица
Grande nonne	Riesenelsterchen	Magpie mannikin	
Grenadin	Granatastrild	Violet-eared waxbill	
– à poitrine bleue	Veilchenastrild	Purple grenadier	
Grimpereau à dos rouge	Scharlach-Mistelesser	Red-backed flower-pecker	Красноспинный цветосос
– des bois	Waldbaumläufer	Tree creeper	Обыкновенная пищуха

FRENCH NAME	GERMAN NAME	ENGLISH NAME	RUSSIAN NAME
– des jardins	Gartenbaumläufer	Short-toed tree creeper	Короткопалая пищуха
Grimpereaux	Baumsteiger, Baumläufer	Wood hewers, Creepers	Древолазы, Пищухи
Gris-bleu	Schönbürzel	Lavender finch	
Grive à cou roux	Rotkehldrossel	Red-throated thrush	Краснозобый дрозд
– à gorge noire	Schwarzkehldrossel	Black-throated thrush	Чернозобый дрозд
– à pieds noirs	Wacholderdrossel	Fieldfare	Дрозд-рябинник
– champenoise	Rotdrossel	Redwing	– белобровик
– de montagne	Wacholderdrossel	Fieldfare	– рябинник
– de Russie	–		
– de Vigne	Singdrossel	Song thrush	Певчий дрозд
– draine	Misteldrossel	Mistle thrush	Дрозд-деряба
– litorne	Wacholderdrossel	Fieldfare	– рябинник
– mauvis	Rotdrossel	Redwing	– белобровик
– musicienne	Singdrossel	Song thrush	Певчий дрозд
– noire	Amsel	Blackbird	Черный дрозд
– orangée	Damadrossel	Orange-headed ground thrush	
– siffleuse bleue	Chinesische Pfeifdrossel	Chinese blue whistling thrush	Синяя птица
Grives	Drosseln i. e. S.		Настоящие дрозды
Gros-bec casse-noyaux	Kernbeißer	Hawfinch	Дубонос
– ordinaire	–	–	–
– tacheté du Cap de bonne Espérance	Samtwida	Yellow bishop	
Grosse grive	Misteldrossel	Mistle thrush	Дрозд-деряба
Guêpier carminé	Scharlachspint	Nubian carmine bee-eater	
– d'Australie	Schmuckspint	Rainbow bird	
– d'Europe	Bienenesser	Common bee-eater	Золотистая щурка
– méridional		– –	
Guêpiers	–	Bee-eaters	Пчелоеды
Halcyon à poitrine blanche	Braunliest	White-breasted kingfisher	Красноносый зимородок
– sacré	Götzenliest	Sacred kingfisher	
Hirondelle de cheminée	Rauchschwalbe	Common swallow	Деревенская ласточка
– de fenêtre	Mehlschwalbe	House martin	Городская ласточка
– de rivage	Uferschwalbe	Sand martin	Береговая ласточка
– de rochers	Gewöhnliche Felsenschwalbe	Grag martin	Скалистая ласточка
– rousseline	Rötelschwalbe	Red-rumped swallow	Рыжепоясничная ласточка
Hirondelles	Schwalben, Rauchschwalben	Swallows	Ласточковые, Касатки
Hochequeue d'Yarrell	Trauerbachstelze	Pied wagtail	Британская белая трясогузка
– grise	Bachstelze	White wagtail	Белая трясогузка
Huia	Huia	Huia	Туйя
Huppe fasciée	Wiedehopf	Hoopoe	Обыкновенный удод
– puput	–	–	
Hypolaïs	Spötter		Пересмешки-бормотушки
– des oliviers	Olivenspötter	Olive-tree warbler	Средиземноморская пересмешка
– ictérine	Gelbspötter	Icterine warbler	Зеленая пересмешка
– pâle	Blaßspötter	Olivaceous warbler	Большая бормотушка
– polyglotte	Orpheusspötter	Orioles	Многоголосая камышевка-пеночка
Ictéridés	Stärlinge	Melodious warbler	Касиковые
Ignicolore	Feuerweber	Orange bishop	
Indicateur malais	Malayischer Honiganzeiger	Malaysian honey guide	
Indicateurs, Indicatoridés	Honiganzeiger	Honey guides	Кукушки-путеводители
Irénidés	Blattvögel	Fairy bluebirds and Leafbirds	
Jacamars	Glanzvögel	Jacamars	Якамары
Jacobin	Schwarzbauchnonne	Three-coloured mannikin	
Jardinier à huppe orange	Rothaubengärtner	Orange-crested gardener	
Jaseur boréal	Europäischer Seidenschwanz	Bohemian waxwing	Обыкновенный свиристель
– de Bohème	– –	– –	– –
– palmiste	Palmschmätzer	Palmchat	
Joue noire	Schwarzbäckchen	Yellow-bellied waxbill	
– orange	Orangebäckchen	Orange-cheeked waxbill	
Junco ardoisé	Winter-Junco	Slate-coloured junco	Серый юнко
Juncos	Juncos	Juncos	Юнко
Jynginés	Wendehälse		Вертишейки
Langrayen à sourcils blancs	Weißbrauen-Schwalbenstar	White-browed wood swallow	Белобровый ласточковый скворец
Laniidés	Würger	Shrikes	Сорокопутовые
Lavandière	Bachstelze	White wagtail	Белая трясогузка
Linotte à bec jaune	Berghänfling	Twite	Горная чечетка
– de Hornemann	Polarbirkenzeisig	Arctic redpoll	Полярная чечетка
– des montagnes	Berghänfling	Twite	Горная чечетка
– mélodieuse	Bluthänfling	Common linnet	Коноплянка
– montagnarde	Berghänfling	Twite	Горная чечетка
– vulgaire	Bluthänfling	Common linnet	Коноплянка

ANIMAL DICTIONARY 601

FRENCH NAME	GERMAN NAME	ENGLISH NAME	RUSSIAN NAME
Linottes	Hänflinge		Чечетки
Locustelle de Pallas	Streifenschwirl	Pallas's grasshopper warbler	Певчий сверчок
– fluviatile	Schlagschwirl		Речной сверчок
– lancéolée	Strichelschwirl	Lanccolated warbler	Пятнистый сверчок
– luscinioïde	Rohrschwirl	Savi's warbler	Соловьиный сверчок
– tachetée	Feldschwirl	Grasshopper warbler	Обыкновенный сверчок
Locustelles	Feldschirl		Сверчки
Longue-haleine	Schwirle	– –	Обыкновенный сверчок
Loriot de Chine	Schwarznacken-Pirol	Black-naped oriole	Черноголовая иволга
– d'Europe	Pirol	Golden oriole	Обыкновенная иволга
Loriots	Pirole	Oriols	Иволговые
Lusiniole à moustaches	Mariskensänger	Moustached warbler	Тонкоклювая камышевка
Mahali	Mahaliweber	Mahali weaver-bird	
Mainate couronné	Kronenatzel	Gold-crested crackle	
– de Ceylan	Ceylonbeo	Ceylon mynah	
– religieux	Beo	Javan hill mynah	
Malimbe huppé	Haubenprachtweber	Crested malimbe	
Manakin barbu	Weißsäbelpipra	White-bearded manakin	
– moine	–	– –	
Manakins	Schnurrvögel	Manakins	Манакины
Martin à cou noir	Schwarzhalsstar	Black-necked mynah	
– à tête grise	Graukopfstar	Malabar mynah	
– caronculé	Lappenstar	Wattled starling	
– de Chine	Mandarinstar	Mandarin mynah	Южнокитайский малый скворец
– de rivage	Ufermaina	Bank mynah	Береговая майна
– de Rothschild	Balistar	Bali-mynah	Балийский скворец
– des pagodes	Pagodenstar	Pagoda starling	Браминский скворец
– gris	Graustar	Grey starling	Серый скворец
– huppé	Haubenmaina	Chinese crested mynah	Хохлатая майна
– pie	Elsterstar	Pied mynah	
– roselin	Rosenstar	Rose-coloured starling	Розовый скворец
– triste	Hirtenstar	Common mynah	Саранчовый скворец
Martin-chasseur à tête grise	Graukopfliest	Grey-headed kingfisher	
– géant	Kookaburra	Australian laughing jackass	Зимородок-великан
Martin-pêcheur	Eisvogel	Kingfisher	Обыкновенный зимородок
– de Natal	Zwergkönigsfischer	Natal kingfisher	
Mauviette	Rotdrossel	Redwing	Дрозд-белобровик
Méliphage à tête noire	Weißnacken-Honigesser	White-naped honey-eater	
Méliphagidés	Honigesser	Honey eaters	Медососовые
Menuridés	Leierschwänze	Lyre-birds	Птицы-лиры
Merle à bec jaune	Amsel	Blackbird	Черный дрозд
– à épaulettes rouges	Rotschulterglanzstar	Cape red-shouldered glossy starling	
– à plastron	Ringdrossel	Ring ouzel	Белозобый дрозд
– bleu	Blaumerle	Blue rock thrush	Синий каменный дрозд
– bronzé pourpre	Purpurglanzstar	Purple glossy starling	
– de roches	Steinrötel	Rock thrush	Пестрый каменный дрозд
– draine	Misteldrossel	Mistle thrush	Дрозд-деряба
– dyal des Indes	Dajaldrossel	Dhyal thrush	Индийская сорочья славка
– évêque	Amethyst-Glanzstar	White-bellied amethyst-starling	Аметистовый скворец
– grive	Singdrossel	Song thrush	Певчий дрозд
– litorne	Wacholderdrossel	Fieldfare	Дрозд-рябинник
– mauvis	Rotdrossel	Redwing	– белобровик
– métallique	Prachtglanzstar	Splendid glossy starling	Великолепный скворец
– – à longue queue	Langschwanzglanzstar	Long-tailed glossy starling	
– – à oreilles bleues	Grünschwanzglanzstar	Green glossy starling	Стальной скворец
– – royal	Königsglanzstar	Royal starling	
– migrateur	Wanderdrossel	American robin	Странствующий дрозд
– moqueur chat	Katzendrossel	Catbird	
– – commun	Spottdrossel	Mocking bird	
– noir	Amsel	Blackbird	Черный дрозд
– Shama	Schamadrossel	Shama	Малабарская сорочья славка
– violet à ventre blanc	Amethyst-Glanzstar	White-bellied amethyst-starling	Аметистовый скворец
Merles	Drosseln i. e. S.		Настоящие дрозды
Méropidés	Bienenesser	Bee-eaters	Пчелоеды
Mésange à longue queue	Schwanzmeise	Long-tailed tit	Длиннохвостая синица
Mésange à tête noire	Bartmeise	Bearded titmouse	
– à tête bleue	Blaumeise	Blue titmouse	Лазоревка
– à tête grise	Chickadee-Meise	Chikadee	
Mésange bleue	Blaumeise	Blue titmouse	Лазоревка
– cendrée	Nonnenmeise	Marsh-titmouse	Черноголовая гаичка
– charbonnière	Kohlmeise	Oxeye-tit	Большая синица
– des Saules	Chickadee-Meise	Chikadee	
– huppée	Haubenmeise	Crested-tit	Хохлатая синица

FRENCH NAME	GERMAN NAME	ENGLISH NAME	RUSSIAN NAME
– noire	Tannenmeise, Mohrenmeise	Coal-tit mouse, Black tit	Черная синица
– nonnette	Nonnenmeise	Marsh-titmouse	Черноголовая гаичка
– penduline	Beutelmeise	Penduline	Обыкновенный ремез
Mésangeai imitateur	Unglückshäher	Siberian jay	Кукша
Mésangeais	–		Кукши
Mésanges	Eigentliche Meisen	Titmice	Настоящие синицы
Mésia à joues argentée	Silberohr-Sonnenvogel	Silver-eared mesia	
Mésomyodés	Schreivögel	Noisemakers	Кричащие птицы
Meunière	Blaumeise	Blue titmouse	Лазоревка
Mimidés	Spottdrosseln	Mocking-birds	
Minivets	Mennigvögel	Minivets	
Modeste	Zeresfink	Cherry finch	
Moineau à sourcils	Augenbrauensperling	Yellow-throated sparrow	
– à tête grise	Graukopfsperling	Grey-headed sparrow	
– de paradis	Rotkopf-Amadine	Red-headed finch	
– domestique	Haussperling	House sparrow	Домовый воробей
– doré	Goldsperling	Golden sparrow	
– doré d'Emin	Eminsperling	Chestnut sparrow	
– flavéole	Gelbbauchsperling	Pegu (House) sparrow	
– fou	Steinsperling	Rock sparrow	Каменный воробей
– friquet	Feldsperling	Tree sparrow	Полевой воробей
– gris	Graukopfsperling	Grey-headed sparrow	
– mélanure	Kapsperling	Cape sparrow	Южноафриканский воробей
– roux	Riesensperling	Great sparrow	
Moineaux	Sperlinge i. e. S., Sperlinge	Sparrows	Настоящие воробьи, Воробьи
Momot	Motmot	Common motmot	
Momotidés	Sägeracken	Motmots	Момоты
Momots, Momotidés	Oryxweber	Grenadier weaver	
Monseigneur	Steinrötel		Каменные дрозды
Moqueurs	Baumhopf, Spottdrosseln	Wood-hoopoes, Mocking-birds	Лесные удоды
– à sourcils bleus	Türkisbrauen-Sägeracke	Jucatan motmot	
Muscade	Muskatfink	Spice finch	
Muscicapidés	Fliegenschnäpperartige	Flycatchers	Мухоловковые
Nectariniidés	Nektarvögel	Sunbirds	Нектарковые
Niltava à ventre rouge	Schwarzkehl-Niltava	Rufous-bellied niltava	
Niverolle des Alpes	Schneefink	Snow finch	Снежный вьюрок
Niverolles	Schneefinken	Snow finches	Снежные вьюрки
Nonnette à tête blanche	Weißkopfnonne	White-headed mannikin	
– à ventre roux	Weißbauchnonne	Spectacled finch	
– des marais	Nonnenmeise	Marsh-titmouse	Черноголовая гаичка
Oiseau à berceaux satiné	Seidenlaubenvogel	Satin bower-bird	
– à berceaux vert	Grünkatzenvogel	Green cat-bird	Зеленая птица-кошка
– à berceraux tacheté	Flecken-Laubenvogel	Spotted bower-bird	Обыкновенная пятнистая беседковая птица
Oiseau palmiste	Palmschmätzer	Palmchat	
Oiseau-régent	Samtgoldvogel	Regent bower-bird	
Oiseaux bleus	Elfenblauvögel	Fairy bluebirds	
– chanteurs	Singvögel	Singing birds	Певчие птицы
Oiseaux-lyre	Leierschwanz i. e. S.	Superb lyre-bird	
Oiseaux-lyres	Leierschwänze	Lyre-birds	Птицы-лиры
Oiseaux-ombrelles	Schirmvogel	Ornate umbrella bird	Амазонская зонтичная птица
Oraponga à gorge nue	Nacktkehl-Glockenvogel	Naked-throated bellbird	Бразильская птица-колокольчик
Orapongas	Glockenvögel	Bellbirds	Птицы-звонари
Padda, -de riz	Reisfink	Java sparrow	
Panure à moustaches	Bartmeise	Bearded titmouse	
Paradisier de la princesse Stéphanie	Prinzessin-Stephanie-Paradiesvogel	Princess Stephanie's bird of paradise	
– de Rodolphe	Blauer Paradiesvogel	Blue bird of paradise	
– de Wallace	Wallace-Paradiesvogel	Wallace's standard-wing	Вымпельная райская птица
– magnifique	Prachtparadiesvogel	Magnificent bird of paradise	
– multifil	Fadenhopf	Twelve-wired bird of paradise	Нитчатая райская птица
– républicain	Blauköpfiger Paradiesvogel	Wilson's bird of paradise	
– royal	Königsparadiesvogel	King bird of paradise	Королевская райская птица
– superbe	Kragenhopf	Lesser superb bird of paradise	Чудная райская птица
Paradisiers	Paradies- und Laubenvögel, Eigentliche Paradiesvögel	Birds of paradise	Райские и Беседковые птицы, Настоящие райские птицы
Pardalot pointillé	Flecken-Panthervogel	Spotted pardalote	Пятнистая пантерная птица
Pardalots	Panthervögel	Diamond birds	Пантерные птицы

FRENCH NAME	GERMAN NAME	ENGLISH NAME	RUSSIAN NAME
Paresseux	Faulvögel	Puffbirds	Птицы-пуховки
Paridés	Eigentliche Meisen	Titmice	Настоящие синицы
Parulidés	Waldsänger	Old World warblers	
Passériformes	Sperlingsvögel		Воробьиные
Passerinés	Sperlinge	Sparrows	Воробьи
Petit barbu à front jaune	Gelbstirn-Zwergbärtling	Yellow-fronted tinker-bird	
– – à gorge blanche	Goldbürzel-Zwergbärtling	Golden-rumped tinker-bird	
– – à gorge jaune	Gelbkehl-Zwergbärtling	Yellow-throated tinker-bird	
– – grivelé	Flecken-Zwergbärtling	Speckled tinker-bird	
– chanteur de Cuba	Klein-Kubafink	Cuba finch	
– martin-pêcheur huppé	Zwerghaubenfischer	Malachite kingfisher	
Petite alouette à tête rousse	Kurzhaubenlerche		Малый жаворонок
– cendrille bleue	Blaumeise	Blue titmouse	Лазоревка
– charbonnière	Tannenmeise	Coal-titmouse	Черная синица
– grive	Rotdrossel	Redwing	Дрозд-белобровик
Petits barbus	Zwergbartvögel	Tinker-birds	
Pétrocincle bleu	Blaumerle	Blue rock thrush	Синий каменный дрозд
Phaéton	Sonnenastrild	Crimson finch	
Phainopépla resplendissant	Seidenschnäpper i. e. S.	Phainopepla	
Philépitte veloutée	Schwarzlappenpitta	Velvet pitta	
Philépittidés	Lappenpittas	Philepittas	
Phoeniculinés	Baumhopfe	Wood-hoopoes	Лесные удоды
Phragmite aquatique	Seggen-Rohrsänger	Aquatic warbler	Вертлявая камышевка
– des joncs	Schilfrohrsänger	Sedge warbler	Камышевка-барсучок
Phytotomidés	Pflanzenmäher	Plantcutters	Косцы
Pic à dos blanc	Weißrückenspecht	White-backed woodpecker	Белоспинный дятел
– à tête rouge	Rotkopfspecht	Red-headed woodpecker	Американский красноголовый дятел
– cendré	Grauspecht	Grey-headed woodpecker	Седой дятел
– épeiche	Buntspecht	Great spotted woodpecker	Большой пестрый дятел
– épeichette	Kleinspecht	Lesser spotted woodpecker	Малый пестрый дятел
– mar	Mittelspecht	Middle spotted woodpecker	Средний дятел
– noir	Schwarzspecht	Black woodpecker	Желна
– syriaque	Blutspecht	Syrian woodpecker	Сирийский дятел
– tridactyle	Dreizehenspecht	Three-toed woodpecker	трехпалый дятел
– vert	Grünspecht	Green woodpecker	Зеленый дятел
Picathartes à cou blanc	Gelbkopf-Felshüpfer	Grey-necked rockfowl	
Picidés	Spechte	Woodpeckers	Дятловые
Piciformes	Spechtvögel		Дятловидные
Pics, Picinés	Echte Spechte		Дятлы
– dorés	Sultansspechte		Султанские дятлы
– nains	Zwergspechte		Дятелки
– tridactyles	Dreizehenspechte	Three-toed woodpeckers	Трехпалые дятлы
– verts	Grünspechte	Green woodpeckers	Зеленые дятлы
Picuminés	Zwergspechte		Дятелки
Pie akahé	Kappengrünhäher	Pileated jay	Хохлатая голубая ворона
– bavarde	Elster	Magpie	Сорока
– bleue	Blauelster	Azure-winged magpie	Голубая сорока
– – de l'Himalaya	Spatelschwanzelster	Hooded racket-tailed tree pie (magpie)	
– – ornée	Blau-Schweifkitta	Ceylon blue magpie	
Pie-grièche à dos rouge	Neuntöter	Red-backed shrike	Европейский жулан
– à poitrine rose	Schwarzstirnwürger	Lesser grey shrike	Чернолобый сорокопут
– à tête rousse	Rotkopfwürger	Woodchat shrike	Красноголовый сорокопут
– à ventre cramoisie	Rotbauchwürger	Crimson-breasted shrike	
– écorcheur	Neuntöter	Red-backed shrike	Европейский жулан
– grise	Raubwürger	Great grey shrike	Серый сорокопут
– masquée	Maskenwürger	Masked shrike	Маскированный сорокопут
Pies-grièches	Würger	Shrikes	Сорокопутовые
Pinson à couronne blanche	Weißkehl-Ammerfink	White-crowned sparrow	
– couronné	Purpur-Kronfink	Red-crested finch	
– d'Ardennes	Bergfink	Brambling	Юрок
– des arbres	Buchfink	Chaffinch	Зяблик
– du nord	Bergfink	Brambling	Юрок
– ordinaire	Buchfink	Chaffinch	Зяблик
Pipit à gorge rousse	Rotkehlpieper	Red-throated pipit	Краснозобый конек
– de Richard	Spornpieper	Richard's pipit	
– des arbres	Baumpieper	Tree pipit	Лесной конек
– des près	Wiesenpieper	Meadow pipit	Луговой конек
– farlouse	–	– –	– –
– rousseline	Brachpieper	Tawny pipit	Полевой конек
– spioncelle	Wasserpieper	Water pipit	Горный конек
Pipits	Pieper i. e. S.	Pipits	Коньки
Pipridés	Schnurrvögel	Manakins	Манакины
Pique-bœuf à bec rouge	Rotschnabel-Madenhacker	Red-billed ox-pecker	Красноклювый волоклюй
Pique-bœufs	Madenhackerstare	Oxbirds	Волоклюи

FRENCH NAME	GERMAN NAME	ENGLISH NAME	RUSSIAN NAME
Pirolle à bec jaune	Gelbschnabel-Schweifkitta	Yellow-billed blue-pie	Желтоклювая лазуревая сорока
– de la Chine	Rotschnabel-Schweifkitta	Chinese blue-pie	Красноклювая лазуревая сорока
Pittidés	Pittas	Pittas	Питты
Plectrophanes	Spornammern		Подорожники
Plocéidés	Webervögel i. w. S.	Weaverbirds	Ткачиковые
Plocéinés	Eigentliche Weber		Древесные ткачи
Pomatorhin à joues rouges	Rotwangen-Säbler	Red-cheeked scimitar-babbler	
Pouillot de Bonelli	Berglaubsänger		
– fitis	Fitis	Willow wren	Светлобрюхая пеночка
– ictérine	Gelbspötter	Icterine warbler	Пеночка-весничка
– siffleur	Waldlaubsänger	Wood-wren	Зеленая пересмешка
– véloce	Zilpzalp		Пеночка-трещотка
Pouillots	Laubsänger i. e. S.		– теньковка
			Пеночки
Promé́fil magnifique	Prachtreifelvogel	Magnificent riflebird	Чешуегрудный щитоносец
Prunellidés	Braunellen	Accentors	Завирушковые
Pyreneste ponceau à ventre noir	Purpurastrild	Black-bellied seed-cracker	
Pytilie à ailes rouges	Aurora-Astrild	Aurora waxbill	
– à dos jaune	Wienerastrild	Red-faced waxbill	
Quadricolore	Lauchgrüne Papageiamadine	Pin-tailed nonpareil	
Quéléa à bec rouge	Blutschnabelweber	Black-fronted dioch	
– à tête rouge	Rotkopfweber	Red-headed quelea	
Ranzani	Schwarzrücken-Bartvogel	Levaillant's barbet	
Raras	Pflanzenmäher	Plantcutters	Косцы
Rémiz penduline	Beutelmeise	Penduline	Обыкновенный ремез
Républicain à capuchon écarlate	Scharlachweber	Red-headed weaver	
Rhamphastidés	Tukane	Toucans	Туканы
Roitelet	Zaunkönig	Wren	Обыкновенный крапивник
– à triple bandeau	Sommergoldhähnchen	Firecrest	
– bleu	Blauer Staffelschwanz	Fairy blue wren	
– de la Nouvelle-Zélande	Neuseeland-Schlüpfer	New-Zealand wren	
– huppé	Wintergoldhähnchen	Goldcrest	
Roitelets	Goldhähnchen	Kinglets	Корольки
Rollier à gros bec	Ostroller	Oriental dollar bird	Широкорот
– à longue queue	Grünscheitelracke	Lilac-breasted roller	
– à ventre bleu	Blaubauchracke	Blue-bellied roller	
– d'Europe	Blauracke	Common roller	Обыкновенная сизоворонка
– d'Inde	Bengalenracke	Indian roller	Бенгальская сизоворонка
Rolliers	Racken	Rollers	Сизоворонковые
– vrais	Blauracken i. e. S.		Сизоворонки
Roselin cramoisi	Karmingimpel	Scarlet grosbeak	Обыкновенная чечевица
Roselins	–		Чечевицы
Rossignol bâtard	Gelbspötter	Icterine warbler	Зеленая пересмешка
– de muraille	Gartenrotschwanz	Redstart	Обыкновенная горихвостка
– des haies	Nachtigall	Nightingale	Западный соловей
– du Japon	Chinesischer Sonnenvogel	Pekin nightingale	Обыкновенная солнечная птица
– philomèle	Nachtigall	Nightingale	Западный соловей
– progné	Sprosser	Great nightingale	Обыкновенный соловей
Rossignols	Nachtigallen		Соловьи
Rouge-gorge bleu d'Amérique	Rotkehl-Hüttensänger	Eastern bluebird	Обыкновенный голубой чекан
– familier	Rotkehlchen	Robin	Зарянка
Rouge-gorges	Rotkehlchen		Зарянки
Rouge-noir	Rotbrust-Samenknacker	Blue-bill	
Rouge-queue à front blanc	Gartenrotschwanz	Redstart	Обыкновенная горихвостка
– à tête blanche	Weißkopf-Rotschwanz	White-capped redstart	Белошапочная горихвостка
– noir	Hausrotschwanz	Black redstart	Горихвостка-чернушка
Rouge-queues	Rotschwänze		Горихвостки
Roupenne	Rotschwingenstar	Red-wing starling	
Rousserolle des buissons	Buschrohrsänger	Blyth's reed warbler	Садовая камышевка
– des joncs	Schilfrohrsänger	Sedge warbler	Камышевка-барсучок
– effarvatte	Teichrohrsänger	Reed warbler	Тростниковая камышевка
– turdoïde	Drosselrohrsänger	Great reed warbler	Дроздовидная камышевка
– verderolle	Sumpfrohrsänger	Marsh warbler	Болотная камышевка
Rousserolles	Rohrsänger i. e. S.		Камышевки
Rubiette rossignol	Nachtigall	Nightingale	Западный соловей
– rouge-gorge	Rotkehlchen i. e. S.	Robin	Зарянка
Rubin d'Australie	Sonnenastrild	Crimson finch	
Sansonnets	Stare	Starlings	Скворцовые

ANIMAL DICTIONARY

FRENCH NAME	GERMAN NAME	ENGLISH NAME	RUSSIAN NAME
Sarcops chauve	Kahlkopfatzel	Coleto mynah	
Sénégali à bec bleu	Dunkelroter Amarant	Blue-billed fire finch	
– à front pointillé	Schuppenköpfchen	Speckle-fronted weaver	
– à poitrine barrée	Pünktchenamarant	Bar-breasted fire finch	
– rouge	Gewöhnlicher Amarant	Fire finch	
Serin	Kanarienvogel	Canary	Канарейка
– cini	Girlitz	Serin	Канареечный вьюрок
– de Moçambique	Moçambique-Girlitz	Green singing finch	Мозамбикский вьюрок
– méridional	Girlitz	Serin	Канареечный вьюрок
Serins	Girlitze	Serins	Канареечные вьюрки
Sifilet de Lawe	Blaunacken-Strahlenparadiesvogel	Six-plumed bird of paradise	
Sittelle à front noir	Samtstirnkleiber	Velvet-fronted nuthatch	
– à ventre marron	Kastanienkleiber	Chestnut-bellied nuthatch	
– corse	Korsika-Kleiber	Corsican nuthatch	
– des rochers	Felsenkleiber	Rock nuthatch	Малый скалистый поползень
– torchepot	Kleiber	Nuthatch	Обыкновенный поползень
Sittelles, Sittidés	–	Nuthatches	Поползни
Sizerin blanchâtre	Polarbirkenzeisig	Arctic redpoll	Полярная чечетка
– flammé	Birkenzeisig	Redpoll	Обыкновенная чечетка
Solitaire de Townsend	Clarino	Townsend's solitaire	
Souï-manga malachite	Malachit-Nektarvogel	Malachite sunbird	Малахитовая нектарка
Souï-mangas	Nektarvögel	Sunbirds	Нектарковые
Soulcie	Steinsperling	Rock sparrow	Каменный воробей
Spermète à capuchon	Kleinelsterchen	Bronze mannikin	
– à tête grise	Perlhalsamadine	Grey-headed silver-bill	
– bicolore	Glanzelsterchen	Black-and-white mannikin	
– naine	Zwergelsterchen	Bib-finch	
– pie	Riesenelsterchen	Magpie mannikin	
Spréo royal	Königsglanzstar	Royal starling	
– superbe	Dreifarben-Glanzstar	Superb glossy starling	
Stourne métallique	Spinnenstar	Shining starling	
Sucriers	Zuckervögel	Honey creepers	
Tarier ordinaire	Braunkehlchen	Whinchat	Луговой чекан
Tarin des Aulnes	Erlenzeisig	Siskin	Чиж
Tête de feu	Flammenkopf	Crested sharpbill	
– noire	Mönchsgrasmücke	Black-cap	Черноголовая славка
Thamnophiles	Ameisenwürger	Ant shrikes	
Tichodrome échelette	Mauerläufer	Wall-creeper	Краснокрылый стенолаз
Timalie aux yeux d'or	Goldaugentimalie i. e. S.	Yellow-eyed babbler	
Timaliinés	Timalien	Babbling thrushes	Тималии
Tisserin à front d'or	Kapweber	Cape weaver	
– à front noir	Maskenweber	Black-fronted weaver	
– à front pointillé	Schnurrbärtchen	Scaly weaver	
– à gros bec	Weißstirnweber	Thick-billed weaver	
– alecto	Büffelweber	Buffalo weaver	Буйволовый ткач
– Baya	Bajaweber	Baya weaver	Бойя
– Cap Moor	Textorweber	Black-headed weaver	
– des villages	–	– –	
– dinemelli	Starweber	Dinemelli's weaver	
– géant	Riesenweber	Giant weaver	
– Mahali	Mahaliweber	Mahali weaver-bird	
– Manyar	Manyarweber	Manyar weaver	Манья
– parasitique	Kuckucksweber	Cuckoo-weaver	
– social	Siedelweber	Sociable weaver	Обыкновенный общественный ткач
– taha	Tahaweber	Tahaweaver	
Tisserins	Eigentliche Weber		Древесные ткачи
Todidés	Todis	Todies	Тодиевые
Todier vert	Todi	Broad-billed tody	
Todiers	Breitschnabel-Todis	Todies	
Torcol fourmilier	Europäischer Wendehals	European wryneck	Обыкновенная вертишейка
– ordinaire	– –	– –	– –
Torcols	Wendehälse		Вертишейки
Toucan à bec caréné	Dottertukan	Sulphur-breasted toucan	
– à bec noir	Schwarzschnabel-Blautukan	Black-billed mountain toucan	
– à bec vert	Bunttukan	Green-billed toucan	
– de Baillon	Goldbrust-Tukan	Baillon's toucan	
– tocard	Braunrückentukan	Swainson's toucan	
– toco	Riesentukan	Toco toucan	
Toucans	Tukane	Toucans	Туканы
Toucanet à bec tâcheté	Goldohr-Arassari	Spot-billed toucanet	
– émeraude	Laucharassari	Emerald toucanet	
– vert	–	–	
Tourde	Singdrossel	Song thrush	Певчий дрозд

FRENCH NAME	GERMAN NAME	ENGLISH NAME	RUSSIAN NAME
Trachyphone perlé	Perlbartvogel	Yellow-breasted barbet	
Traquet des prés	Braunkehlchen	Whinchat	Луговой чекан
– du désert	Wüstensteinschmätzer	Desert wheatear	Пустынная каменка
Traquet motteux	Steinschmätzer	Wheatear	Обыкновенная каменка
– oreillard	Mittelmeer-Steinschmätzer	Black-eared wheatear	Чернопегая каменка
– pâtre	Schwarzkehlchen	Stonechat	Черноголовый чекан
– tarier	Braunkehlchen	Whinchat	Луговой чекан
Traquets	Schmätzer i. e. S.		Чеканы
Travailleur	Blutschnabelweber	Black-fronted diock	
Troglodyte mignon	Zaunkönig	Wren	Обыкновенный крапивник
Troglodytidés	Zaunkönige	Wrens	Крапивниковые
Turdinés	Drosseln	Thrushes	Дроздовые
Tyran à ventre jaune	Bentevi	Kiskadee flycatcher	Бентеви
– rouge écarlate	Rubinköpfchen	Vermillion flycatcher	
– royal	Königssatrap	American kingbird	
Tyrannidés	Tyrannen	Tyrant flycatchers	Тираны
Tyrans	Satrapen	Kingbirds	
Upupidés	Hopfe	Hoopoes	Удодовые
Upupinés	Wiedehopfe		Удоды
Vangidés	Blauwürger	Vangas	
Vendangeuse	Singdrossel	Song thrush	Певчий дрозд
Venturon montagnard	Zitronengirlitz	Citril finch	
Verdier	Goldammer	Yellowhammer	Обыкновенная овсянка
– d'Europe	Grünling	Greenfinch	– зеленушка
Verdin à barbe bleu	Blaubart-Blattvogel	Blue-winged fruit-sucker	
– à front d'or	Goldstirn-Blattvogel	Gold-fronted fruit-sucker	
Verdins	Blattvögel	Leafbirds	
Veuve à quatre brins	Königswitwe	Queens whydah	
– de Fischer	Strohwitwe	Fischer's whydah	
– dominicaine	Dominikanerwitwe	Pin-tailed whydah	
– en feu	Schildwida	Red-collared whydah	
– métallique	Glanzwitwe	Shiny black whydah	
– Niobe	Schildwida	Red-collared whydah	
– reine	Königswitwe	Queens whydah	
– royale	–	–	
Veuves	Glanz- und Dominikanerwitwen	Whydahs	Вдовушки
Veuves-Combassous	Witwen		Вдовушковые
Viréonidés	Vireos	Vireos	Листовые сорокопуты
Xenicidés	Neuseeland-Schlüpfer	Wrens	
Zostéropidés	Brillenvögel	White-eyes	Белоглазковые
Zostérops à lunettes	Ganges-Brillenvogel	Indian white-eye	Гангская белоглазка

IV. Russian–German–English–French

Названия подвидов отличаются от видовых чаще всего лишь дополнительным прилагательным, главным образом географического характера. Такие русские названия подвидов как правило не включены в данную часть зоологического словаря.

RUSSIAN NAME	GERMAN NAME	ENGLISH NAME	FRENCH NAME
Абиссинский рогатый ворон	Nördlicher Hornrabe	Abyssinian ground hornbill	Bucorve d'Abyssinie
Автсралийская белоглазка	Australischer Brillenvogel	Gray-backed white-eye	
Австралийские поползни	Australkleiber	Treerunners	
Альпийская галка	Alpendohle	Alpine chough	Chocard à bec jaune
Альпиская завирушка	Alpenbraunelle	– accentor	Accenteur alpin
Альпийский вьюрок	Schneefink	Snow finch	Niverolle des Alpes
Амазонская зонтичная птица	Schirmvogel	Ornate umbrella bird	Céphaloptère orné
Американский белоклювый дятел	Elfenbeinspecht	Ivory-billed woodpecker	
– королевский дятел	Kaiserspecht	Imperial woodpecker	
–красноголовый дятел	Rotkopfspecht	Red-headed woodpecker	Pic à tête rouge
Аметистовый скворец	Amethyst-Glanzstar	White-bellied amethyst-starling	Merle violet à ventre blanc
Андский дятел	Andenspecht	Rock flicker	
Балийский скворец	Balistar	Bali-mynah	Martin de Rothschild
Балтиморский трупиал	Baltimore-Trupial	Baltimore oriole	
Белая трясогузка	Bachstelze	White wagtail	Bergeronette grise
Белобровый ласточковый скворец	Weißbrauen-Schwalbenstar	White-browed wood swallow	Langrayen à sourcils blancs
Белоглазки	Brillenvögel i. e. S.	White-eyes	
Белоглазковые	Brillenvögel	–	Zostéropidés

RUSSIAN NAME	GERMAN NAME	ENGLISH NAME	FRENCH NAME
Белозобая птица-кошка	Weißkehl-Katzenvogel	White-throated cat-bird	Merle à plastron
Белозобый дрозд	Ringdrossel	Ring ouzel	
Белокрылый клёст	Bindenkreuzschnabel	Two-barred crossbill	Bec-croisé bifascié
Белоспинная ворона-свистун	Weißrücken-Flötenvogel	Trasmanian crow-shrike	
Белоспинный дятел	Weißrückenspecht	White-backed woodpecker	Pic à dos blanc
Белошапочная горихвостка	Weißkopf-Rotschwanz	White-capped redstart	Rouge-queue à tête blanche
Белошапочная овсянка	Fichtenammer	Pine bunting	
Белошейный ворон	Geierrabe	White-necked raven	
Бенгальская сизоворонка	Bengalenracke	Indian roller	Rollier d'Inde
Бентеви	Bentevi	Kiskadee flycatcher	Tyran à ventre jaune
Береговая ласточка	Uferschwalbe	Sand martin	Hirondelle de rivage
– майна	Ufermaina	Bank mynah	Martin de rivage
Береговушка	Uferschwalbe	Sand martin	Hirondelle de rivage
Бледная пересмешка	Blaßspötter	Olivaceon's warbler	Hypolaïs pâle
Бойя	Bajaweber	Baya weaver	Tisserin Baya
Болотная камышевка	Sumpfrohrsänger	Marsh warbler	Rousserolle verderolle
Болотные трупиалы	Hordenvögel	Blackbirds	
Большая бормотушка	Blaßspötter	Olivaceous warbler	Hypolaïs pâle
– райская птица	Großer Paradiesvogel	Greater bird of paradise	Grand paradisier
– синица	Kohlmeise	Oxeye-tit	Mésange charbonnière
Большой пёстрый дятел	Buntspecht	Great spotted woodpecker	Pic épeiche
Бормотушки	Glanzvögel	Jacamars	Jacamars
Бородастики	Bartvögel	Barbets	Capitonidés
Бородатки	–	–	Barbus
Бразильская птица-колокольчик	Nacktkehl-Glockenvogel	Naked-throated bellbird	Oraponga à gorge nue
Браминский скворец	Pagodenstar	Pagoda starling	Martin des pagodes
Британская белая трясогузка	Trauerbachstelze	Pied wagtail	Bergeronette d'Yarrell
Буйволовые ткачи	Büffelweber		Bubalornithinés
Буйволовый ткач	–	Buffalo weaver	Tisserin alecto
Бульбули	Haarvögel	Bulbuls	
Буроголовая гаичка	Weidenmeise		Mésange à tête noire
Варакушка	Blaukehlchen	Bluethroat	Gorge-bleue à miroir
Вдовушки	Glanz- und Dominikanerwitwen	Whydahs	Veuves
Вдовушковые	Witwen		Veuves-Combassous
Великолепный скворец	Prachtglanzstar	Splendid glossy starling	Merle métallique
Вертишейки	Wendehälse		Jyngines
Вертлявая камышевка	Seggen-Rohrsänger	Aquatic warbler	Phragmite aquatique
Вертлявый дятел	Mittelspecht	Middle spotted woodpecker	Pic mar
Вилохвостая райская птица	Gabelschwanz-Paradiesvogel	Wattled bird of paradise	
Волоклюи	Madenhackerstare	Oxbirds	Pique-bœufs
Воробьи	Sperlinge	Sparrows	Passerines
Воробьиные	Sperlingsvögel		Passériformes
Ворон	Kolkrabe	Raven	Grand corbeau
Вороновые	Rabenvögel	Crows	Corbeaux
Воронок	Mehlschwalbe	House martin	Hirondelle de fenêtre
Восококлювый ткач	Wellenastrild	St. Helena waxbill	Astrild ondulé
Восточноазиатский свиристель	Japanischer Seidenschwanz	Japanese waxwing	
Восточный соловей	Sprosser	Great nightingale	Rossignol progné
Вымпельная райская птица	Wallace-Paradiesvogel	Wallace's standard-wing	Paradisier de Wallace
Вьюрковые	Finken	Finches	Fringillidés
Вьюрос	Bergfink	Brambling	Pinson d'Ardennes
Галка	Dohle	Jackdaw	Choucas des tours
Гангская белоглазка	Ganges-Brillenvogel	Indian white-eye	Zostérops à lunettes
Голубая сойка	Blauhäher	American blue jay	Geai bleu d'Amérique
– сорока	Blauelster	Azure-winged magpie	Pie bleue
Горихвостка-лысушка	Gartenrotschwanz	Redstart	Rouge-queue à front blanc
– -чернушка	Hausrotschwanz	Black redstart	– noir
Горихвостки	Rotschwänze		Rouge-queues
Горная славка	Weißbart-Grasmücke	Subalpine warbler	Fauvette passerinette
– трясогузка	Bergstelze	Grey wagtail	Bergeronette des ruisseaux
– чечётка	Berghänfling	Twite	Linotte à bec jaune
Горные вьюрки	Rosenfinken	Rosy finches	
Горный конёк	Wasserpieper	Water pipit	Pipit spioncelle
Городская ласточка	Mehlschwalbe	House martin	Hirondelle de efnêtre
Горшечники	Töpfervögel i. e. S.	Ovenbirds	
Грач	Saatkrähe	Rook	Corbeau freux
Гусеницееды	Raupenesser i. e. S.	Cuckoo-shrikes	
Деревенская ласточка	Rauchschwalbe	Common swallow	Hirondelle de cheminée
Длиннохвостая синица	Schwanzmeise	Long-tailed tit	Mésange à longue queue

ANIMAL DICTIONARY

RUSSIAN NAME	GERMAN NAME	ENGLISH NAME	FRENCH NAME
Длиннохвостые нектарки	Langschwanz-Nektarvögel	Sunbirds	
– синицы	Schwanzmeisen	Long-tailed tits	
Домовый воробей	Haussperling	House sparrow	Moineau domestique
Древесные ткачи	Eigentliche Weber		Plocéinés
Древолазы	Baumsteiger	Wood hewers	Dendrocolapitidés
Дрозд-белобровик	Rotdrossel	Redwing	Grive mauvis
– -деряба	Misteldrossel	Mistle thrush	– draine
Дрозд-рябинник	Wacholderdrossel	Fieldfare	– litorne
Дроздовидная камышевка	Drosselrohrsänger	Great reed warbler	Rousserolle turdoïde
– трясогузка	Drosselstelze	Magpie lark	Alouette-pie
Дроздовые	Drosseln	Thrushes	Turdinés
Дронго	Drongos i. w. S.	Drongos	Drongos
Дронговые	Drongos	–	
Дубонос	Kernbeißer	Hawfinch	Gros-bec casse-noyaux
Дятел-сосун	Saftlecker	Common sapsucker	
Дятелки	Zwergspechte		Picuminés
Дятловидные	Spechtvögel		Piciformes
Дятловые	Spechte	Woodpeckers	Picidés
Дятлы	Echte Spechte		Picinés
Европейский жулан	Neuntöter	Red-backed shrike	Pie-grièche écorcheur
Жаворонковые	Lerchen	Larks	Alouettes
Желна	Schwarzspecht	Black woodpecker	Pic noir
Желтая трясогузка	Schafstelze	Blue-headed wagtail	Bergeronette printanière
Желтоголовая трясогузка	Zitronenstelze		– citrine
Желтоклювая лазуревая сорока	Gelbschnabel-Schweifkitta	Yellow-billed blue-pie	Pirolle à bec jaune
Желчная овсянка	Braunkopfammer	Red-headed bunting	Bruant à tête rousse
Завирушки	Braunellen	Accentors	Accenteurs
Завирушковые	–	–	Prunellidés
Западный соловей	Nachtigall	Nightingale	Rossignol philomèle
Зарянка	Rotkehlchen i. e. S.	Robin	Rouge-gorge familier
Зарянки	Rotkehlchen		Rouge-gorges
Зеленая пересмешка	Gelbspötter	Icterine warbler	Hypolaïs ictérine
– птица-кошка	Grünkatzenvogel	Green cat-bird	Oiseau à berceaux vert
– якамара	Grünjakamar	Common jacamar	
Зеленые дятлы	Grünspechte	Green woodpeckers	Pics verts
Зеленый дятел	Grünspecht	– woodpecker	Pic vert
Земляной дрозд	Bunt-Erddrossel	Ground thrush	
Земляные птицы носороги	Hornraben	Ground hornbills	
Зимородковые	Eisvögel	Kingfishers	Alcédinidés
Зимородок-великан	Kookaburra	Australian laughing jackass	Martin-chasseur géant
Золотистая щурка	Bienenesser	Common bee-eater	Guêpier méridional
Золотой кукушковый дятел	Goldspecht	Golden-banded woodpecker	Colapte doré
Зяблик	Buchfink	Chaffinch	Pinson des arbres
Иволговые	Pirole	Oriols	Loriots
Индийская длиннохвостая мухоловка	Paradiesschnäpper	Paradise flycatcher	Gobe-mouche paradisier
– питта	Neunfarbenpitta	Indian pitta	Brève du Bengale
– сорочья славка	Dajaldrossel	Dhyal thrush	Merle dyal des Indes
Калифорнийская сорока	Gelbschnabelelster	Yellow-billed magpie	
Калифорнийский кактусовый крапивник	Kaktus-Zaunkönig	Cactus wren	
Каменные воробьи	Steinsperlinge	Rock sparrows	
– дрозды	Steinrötel		Monticoles
Каменный воробей	Steinsperling	Rock sparrow	Moineau fou
Камышевая овсянка	Rohrammer	Reed bunting	Bruant des roseaux
Камышевка-барсучок	Schilfrohrsänger	Sedge warbler	Phragmite des jonos
Камышевки	Rohrsänger i. e. S.		Rousserolles
Канадская ронжа	Kanadischer Unglückshäher	Canada jay	
Канареечные вьюрки	Girlitze	Serins	Serins
Канареечный вьюрок	Girlitz	–	Serin cini
Канарейка	Kanarienvogel	Canary	Serin
Кардиналы	Kardinäle	Cardinal-grosbeaks	Cardinaux
Касатка	Rauchschwalbe	Common swallow	Hirondelle de cheminée
Касатки	Rauchschwalben	Swallows	Hirondelles
Касиковые	Stärlinge	Orioles	Ictéridés
Кедровка	Tannenhäher	Nutcracker	Casse-noix moucheté
Кедровый свиристель	Zederseidenschwanz	Cedar waxwing	
Клест-еловик	Fichtenkreuzschnabel	Crossbill	Bec-croisé des sapins
– -сосновик	Kiefernkreuzschnabel	Parrot crossbill	– perroquet
Клесты	Kreuzschnäbel	Crossbills	Bec-croisés
Клушица	Alpenkrähe	Chough	Crave à bec rouge
Коноплянка	Bluthänfling	Common linnet	Linotte mélodieuse
Коньки	Pieper i. e. S.	Pipits	Pipits
Коровьи скворцы	Kuhstärlinge	Cowbirds	

RUSSIAN NAME	GERMAN NAME	ENGLISH NAME	FRENCH NAME
Королевская райская птица	Königsparadiesvogel	King bird of paradise	Paradisier royal
Корольки	Goldhähnchen	Kinglets	Roitelets
Короткопалая пищуха	Gartenbaumläufer	Short-toed tree creeper	Grimpereau des jardins
Короткопалые дрозды	Haarvögel	Bulbuls	
Косцы	Pflanzenmäher	Plantcutters	Phytotomidés
Котинги	Schmuckvögel	Cotingas	Cotingidés, Cotingas
Крапивниковые	Zaunkönige	Wrens	Troglodytidés
Крапивник-пересмешник	Karolina-Zaunkönig	Carolina wren	
Красная птица	Feuertangare	Summer tanager	
– тангара	–	– –	
Красноголовый сорокопут	Rotkopfwürger	Woodchat shrike	Pie-grièche à tête rousse
Краснозобый дрозд	Rotkehldrossel	Red-throated thrush	Grive à cou roux
– конек	Rotkehlpieper	– pipit	Pipit à gorge rousse
Красноклювая лазуревая сорока	Rotschnabel-Schweifkitta	Chinese blue-pie	Pirolle de la Chine
Красноклювый волоклюй	– Madenhacker	Red-billed ox-pecker	Pique-bœuf à bec rouge
Краснокрылый стенолаз	Mauerläufer	Wall-creeper	Tichodrome échelette
Красноносый зимородок	Braunliest	White-breasted kingfisher	Halcyon à poitrine blanche
Красноспинный цветосос	Scharlach-Mistelesser	Red-backed flower-pecker	Grimpereau à dos rouge
Краснохвостая якамара	Rotschwanzjakamar	Rufous-tailed jacamar	
Красный кардинал	Roter Kardinal	Cardinal	
Кричащие вороны	Würgerkrähen	Piping crows	
– птицы	Schreivögel	Noisemakers	Mésomyodés
Кровяно-красные ткачи	Eigentliche Astrilde	Waxbills	Astrilds
Кукушки-путеводители	Honiganzeiger	Honey guides	Indicatoridés
Кукша	Unglückshäher	Siberian jay	Geai de Sibérie
Кукши	–		Mésangeais
Лазоревка	Blaumeise	Blue titmouse	Mésange bleue
Ласточковые	Schwalben	Swallows	Hirondelles
– скворцы	Schwalbenstare	Wood swallows	
Ласточковый цветосос	Schwalben-Mistelesser	Mistletoe bird	
Ласточкокрылая пуховка	Schwalben-Faulvogel	Swallow-wing	
Ленивки	Faulvögel	Puffbirds	Bucconidés
Лесная завирушка	Heidelerche	Wood lark	Alouette lulu
Леской жаворонок	Heckenbraunelle	Dunnock	Accenteur mouchet
– конек	Baumpieper	Tree pipit	Pipit des arbres
Лесные сорокопуты	Buschwürger	Bush-shrikes	
– удоды	Baumhopfe	Wood-hoopoes	Phoeniculinés, Mogueurs
Листовые сорокопуты	Vireos	Vireos	Viréonidés
Личинкоедовые	Stachelbürzler	Cuckoo-shrikes	Campéphagidés
Лоскутные вороны	Neuseeländische Lappenvögel	Wattle-birds	
Луговой конек	Wiesenpieper	Meadow pipit	Pipit farlouse
– чекан	Braunkehlchen	Whinchat	Traquet des près
Майна	Hirtenstar	Common mynah	Martin triste
Малабарская сорочья славка	Schamdrossel	Shama	Merle Shama
Малайский калао	Rhinozerosvogel	Rhinocercs hornbill	
Малахитовая нектарка	Malachit-Nektarvogel	Malachite sunbird	Souï-manga malachite
Малая мухоловка	Zwergschnäpper	Red-breasted flycatcher	Gobe-mouche nain
– райская птица	Kleiner Paradiesvogel	Lesser bird of paradise	
Малуры	Südsee-Grasmücken	Australian »wrens«	
Малый дрозд	Grauwangendrossel	Gray-cheeked thrush	
– жаворонок	Kurzzehenlerche		Petite alouette à tête rousse
– пегий зимородок	Graufischer	Pied kingfisher	
– пестрый дятел	Kleinspecht	Lesser spotted woodpecker	Pic épeichette
– скалистый поползень	Felsenkleiber	Rock nuthatch	Sittelle des rochers
Манакины	Schnurrvögel	Manakins	Pipridés
Манья	Manyarweber	Manyar weaver	Tisserin Manyar
Маскированный сорокопут	Maskenwürger	Masked shrike	Pie-grièche masquée
Медососовые	Honigesser	Honey eaters	Méliphagidés
Медоуказчика	Honiganzeiger	– guides	Indicateurs
Многоголосая камышевка-пеночка	Orpheusspötter	Melodious warbler	Hypolaïs polyglotte
Мозамбикский вьюрок	Moçambique-Girlitz	Green singing finch	Serin de Moçambique
Молотобой	Hämmerling	Wattled bellbird	
Момоты	Sägeracken	Motmots	Momotidés
Московка	Tannenmeise	Coal-titmouse	Petite charbonnière
Мухоловка-белошейка	Halsbandschnäpper	White-collared flycatcher	Gobe-mouche à collier
– -пеструшка	Trauerschnäpper	Pied flycatcher	– noir
Мухоловковые	Fliegenschnäpperartige	Flycatchers	Muscicapidés
Нарядные птицы	Kleidervögel		Drépanididés
Настоящие бульбули	Echte Bülbüls	Bulbuls	
– воробьи	Sperlinge i. e. S.	Sparrows	Moineaux
– дрозды	Drosseln i. e. S.		Merles

RUSSIAN NAME	GERMAN NAME	ENGLISH NAME	FRENCH NAME
– поползни	Eigentliche Kleiber	Nuthatches	
– райские птицы	– Paradiesvögel	Birds of paradise	Paradisiers
– синицы	– Meisen	Titmice	Paridés
– скворцы	– Stare	Starlings	
– сорокопуты	– Würger	True shrikes	
– тангары	Echte Tangare	Tanagers	
Нектарки	Eigentliche Nektarvögel	Sunbirds	
Нектарковые	Nektarvögel	–	Nectariniidés
Нитчатая райская птица	Fadenhopf	Twelve-wired bird of paradise	Paradisier multifil
Новогвинейская беседковая птица	Hüttengärtner	Brown gardener	
Обыкновенная вертишейка	Europäischer Wendehals	European wryneck	Torcol ordinaire
– горихвостка	Gartenrotschwanz	Redstart	Bouge-queue à front blanc
– зеленушка	Grünling	Greenfinch	Verdier d'Europe
– иволга	Pirol	Golden oriole	Loriot d'Europe
– каменка	Steinschmätzer	Wheatear	Traquet motteux
– овсянка	Goldammer	Yellowhammer	Bruant jaune
– оляпка	Wasseramsel	Dipper	Cincle plongeur
– пищуха	Waldbaumläufer	Tree creeper	Grimpereau des bois
– пятнистая беседковая птица	Flecken-Laubvogel	Spotted bower-bird	Oiseau à bercereaux tacheté
– сизоворонка	Blauracke	Common roller	Rollier d'Europe
– солнечная птица	Chinesischer Sonnenvogel	Pekin nightingale	Rossignol du Japon
– черная ворона	Rabenkrähe	Carrion crow	Corneille noire
– чечевица	Karmingimpel	Scarlet grosbeak	Roselin cramoisi
– чечетка	Birkenzeisig	Redpoll	Sizerin flammé
Обыкновенный голубой чекан	Rotkehl-Hüttensänger	Eastern bluebird	Rouge-gorge bleu d'Amérique
– зимородок	Eisvogel	Kingfisher	Martin-pêcheur
– крапивник	Zaunkönig	Wren	Troglodyte mignon
– общественный ткач	Siedelweber	Sociable weaver	Tisserin social
– поползень	Kleiber	Nuthatch	Sittelle torchepot
– ремез	Beutelmeise	Penduline	Mésange penduline
– сверчок	Feldschwirl	Grasshopper warbler	Locustelle tachetée
– свиристель	Europäischer Seidenschwanz	Bohemian waxwing	Jaseur boréal
– скворец	Gemeiner Star	Starling	Etourneau sansonnet
– соловей	Sprosser	Great nightingale	Rossignol progné
– удод	Wiedehopf	Hoopoe	Huppe puput
Овсянки	Ammern, Eigentliche Ammern	Buntings	Emberizinés, Bruants
Оляпковые	Wasseramseln	Dippers	Cinclidés
Орегонский юнко	Oregon-Junco	Oregon junco	
Ореховка	Tannenhäher	Nutcracker	Casse-noix moucheté
Очковая славка	Brillengrasmücke	Spectacled warbler	Fauvette à lunettes
Очковый сорокопут	Brillenwürger	White helmet shrike	
Ошейниковая котинга	Halsbandkotinga	Banded cotinga	Cotinga tityre
Пантерные птицы	Panthervögel	Diamond birds	Pardalots
Паукоеды	Spinnenjäger	Spider-hunters	
Певчая славка	Orpheusgrasmücke	Orphean warbler	Fauvette orphée
Певчие птицы	Singvögel	Singing birds	Oiseaux chanteurs
Певчий дрозд	Singdrossel	Song thrush	Grive musicienne
– сверчок	Streifenschwirl	Pallas's grasshopper warbler	Locustelle de Pallas
Пеночка-весничка	Fitis	Willow wren	Pouillot fitis
– -теньковка	Zilpzalp		– véloce
– -трещотка	Waldlaubsänger	Wood-wren	– siffleur
Пеночки	Laubsänger i. e. S.		Pouillots
Перуанская голубая ворона	Peru-Grünhäher	Green jay	
Печники	Töpfervögel, – i. e. S.	Ovenbirds	Furnariidés
Пересмешки-бормотушки	Spötter		Hypolaïs
Перцеяды	Tukane	Toucans	Toucans
Пестрый дрозд	Bunt-Erddrossel	Ground thrush	
– каменный дрозд	Steinrötel	Rock thrush	Merle de roches
Питпит	Pitpit i. e. S.	Turquoise dacnis	
Питты	Pittas	Pittas	Brèves
Пищухи	Baumläufer	Creepers	Grimpereaux
Пищуховые	Baumläuferartige	–	Certhiidés
Плоскоклювые	Todis	Todies	Todiers
Подорожники	Spornammern		Plectrophanes
Полевой воробей	Feldsperling	Tree sparrow	Friquet
– жаворонок	Feldlerche	Sky lark	Alouette des champs
– конек	Brachpieper	Tawny pipit	Pipit rousseline
Полярная чечетка	Polarbirkenzeisig	Arctic redpoll	Sizerin blanchâtre
Поползень-крошка	Zwergkleiber	Pigmy nuthatch	
Поползни	Kleiber	Nuthatches	Sittidés, Sittelles
Потатуйка	Wiedehopf	Hoopoe	Huppe puput

RUSSIAN NAME	GERMAN NAME	ENGLISH NAME	FRENCH NAME
Принцев лирохвост	Schwarzleierschwanz	Albert's lyre-bird	
Провансская славка	Provence-Grasmücke	Dartford warbler	Fauvette pitchou
Просянка	Grauammer	Corn bunting	Bruant proyer
Птицы-звонары	Glockenvögel	Bellbirds	Orapongas
− -колокольчики	−	−	−
− -кошки	Katzenvögel i. e. S.	Cat-birds	
− -лиры	Leierschwänze	Lyre-birds	Menuridés
− -муравьеды	Ameisenvögel	Ant birds	Formicariidés
− -носороги	Nashornvögel	Hornbills	Bucérotidés
− -пуховки	Faulvögel	Puffbirds	Bucconidés
Пуночки	Schneeammern	Snow buntings	Bruants des neiges
Пустинная каменка	Wüstensteinschmätzer	Desert wheatear	Traquet du désert
Пустынный воробей	Wüstensperling	Desert sparrow	
Пчелоеды	Bienenesser	Bee-eaters	Méropidés
Пятнистая пантерная птица	Flecken-Panthervogel	Spotted pardalote	Pardalote pointillé
Пятнистый сверчок	Strichelschwirl	Lanceolated warbler	Locustelle lancéolée
Райские и Беседковые птицы	Paradies- und Laubenvögel	Birds of paradise	Paradisiers
Райский дронго	Flaggendrongo	Greater racket-tailed drongo	Drongo à raquettes
Ракшеобразные	Rackenvögel		Coraciadiformes
Ремезовые	Beutelmeisen	Penduline tits	
Речной сверчок	Schlagschwirl		Locustelle fluviatile
Рисовый скворец	Reisstärling	Bobolink	
Рогатые вороны	Hornraben	Ground hornbills	
Рогатый жаворонок	Ohrenlerche	Shore lark	Alouette hausse-col
Роголювы	Breitrachen	Broadbills	Eurylaimidés
Розовый скворец	Rosenstar	Rose-coloured starling	Martin roselin
Ронжа	Unglückshäher	Siberian jay	Mésangeai imitateur
Ронжи	−		Mésangeais
Рыжелобая веерохвостка	Rotstirn-Fächerschnäpper	Rufous-fronted fantail	
Рыжепоясничная ласточка	Rötelschwalbe	Red-rumped swallow	Hirondelle rousseline
Рыжий печник	Töpfervogel	Ovenbird	Fournier roux
Рюм	Ohrenlerche	Shore lark	Alouette hausse-col
Садовая камышевка	Buschrohrsänger	Blyth's reed warbler	Rousserolle des buissons
− овсянка	Ortolan	Ortolan bunting	Bruant ortolan
− славка	Gartengrasmücke	Garden warbler	Fauvette des jardins
Саранчевый скворец	Hirtenstar	Common mynah	Martin triste
Сардинская славка	Sardengrasmücke	Marmora's warbler	Fauvette sarde
Сверчки	Schwirle		Locustelles
Светлобрюхая пеночка	Berglaubsänger		Pouillot de Bonelli
Свиристелевые	Seidenschwänze	Waxwings	Bombycillidés
Североамериканская ореховка	Kieferhäher	Clark's nutcracker	
Североамериканский дубонос	Abendkernbeißer	Evening grosbeak	
Седой дятел	Grauspecht	Grey-headed woodpecker	Pic cendré
Сенегальский вьюрок	Grau-Edelsänger	Grey singing finch	Chanteur d'Afrique
Серая ворона	Nebelkrähe	Hooded crow	Corneille mantelée
− мухоловка	Grauschnäpper	Spotted flycatcher	
− славка	Dorngrasmücke	Whitethroat	Fauvette grisette
Серый ласточковый скворец	Grauschwalbenstar	Ashy-wood-swallow	
− скворец	Graustar	Grey starling	Martin gris
− сорокопут	Raubwürger	Great grey shrike	Pie-grièche grise
− тропический свиристель	Grauseidenschnäpper	Gray silky-flycatcher	Gobe-mouche gris du Mexique
− юнко	Winter-Junco	Slate-coloured junco	Junco ardoisé
Сибирская завирушка	Bergbraunelle	Siberian accentor	
Сизоворонки	Blauracken i. e. S.	Rollers	Rolliers vrais
Сизоворонковые	Racken	−	Coraciadidés
Синий каменный дрозд	Blaumerle	Blue rock thrush	Merle bleu
Синяя птица	Chinesische Pfeifdrossel	Chinese blue whistling thrush	Grive siffleuse bleue
Сирийский дятел	Blutspecht	Syrian woodpecker	Pic syriaque
Скалистая ласточка	Gewöhnliche Felsenschwalbe	Grag martin	Hirondelle de rochers
Скалистые петушки	Klippenvögel	Cocks of the rock	Coqs-de-roche
Скворцовые	Stare	Starlings	Etourneaux
Славка-завирушка	Klappergrasmücke	Lesser whitethroat	Fauvette babillarde
− -мельничек	−	− −	− −
− -портниха	Schneidervogel	Tailor bird	− couturière
Славки	Grasmücken i. e. S.		Fauvettes
Снегири	Gimpel		Bouvreuils
Снегирь	Dompfaff	Bullfinch	Bouvreuil pivoine
Снежные вьюрки	Schneefinken	Snow finches	Niverolles
Снежный вьюрок	Schneefink	− finch	Niverolle des Alpes
Сойка	Eichelhäher	Black-throated jay	Geai des chênes
Соловей-красношейка	Rubin-Nachtigall		Calliope

ANIMAL DICTIONARY

RUSSIAN NAME	GERMAN NAME	ENGLISH NAME	FRENCH NAME
Соловьи	Nachtigallen		Rossignols
Соловьиный сверчок	Rohrschwirl	Savi's warbler	Locustelle luscinioïde
Сорока	Elster	Magpie	Pio bavarde
Сорокопутовые	Würger	Shrikes	Laniidés
Сорокопуты	Würger i. e. S.	—	Pie-grièches
Средиземноморская пересмешка	Olivenspötter	Olive-tree warbler	Hypolaïs des oliviers
– славка	Samtkopf-Grasmücke	Sardinian warbler	Fauvette mélanocéphale
Средний дятел	Mittelspecht	Middle spotted woodpecker	Pic mar
Стальной скворец	Grünschwanzglanzstar	Green glossy starling	Merle métallique à oreilles bleues
Стенолазы	Mauerläufer	Wallcreepers	
Степной жаворонок	Kalanderlerche	Calandra lark	Alouette calandre
Странствующий дрозд	Wanderdrossel	American robin	Merle migrateur
Султанские дятлы	Sultansspechte		Pics dorés
Тангары	Tangare	Tanagers	
Тималии	Timalien	Babbling thrushes	Timaliinés
Тираны	Tyrannen	Tyrant flycatchers	Tyrannidés
Ткачиковые	Webervögel i. w. S.	Weaverbirds	Plocéidés
Тодиевые	Todis	Todies	Todidés
Толстоклювые ткачи	Dickschnabelweber		Amblyospizinés
Тонкоклювая камышевка	Mariskensänger	Moustached warbler	Lusiniole à moustaches
Травяные певуны	Zistensänger		Cisticoles
Траурный дронго	Trauerdrongo	Drongo	Drongo
Трехпалые дятлы	Dreizehenspechte	Three-toed woodpeckers	Pics tridactyles
Трехпалый дятел	Dreizehenspecht	– woodpecker	Pic tridactyle
Тростниковая камышевка	Teichrohrsänger	Reed warbler	Rousserolle effarvatte
Трупиалы	Trupiale	Orioles	
Трясогузки	Eigentliche Stelzen	Wagtails	Bergeronettes
Трясогузковые	Stelzen	—	
Туйя	Huia	Huia	Huia
Туканий бородастик	Tukan-Bartvogel	Toucan barbet	Barbu toucan
Туканы	Tukane	Toucans	Rhamphastidés
Удодовые	Hopfe	Hoopoes	Upupidés
Удоды	Wiedehopfe		Upupinés
Хохлатая голубая ворона	Kappengrünhäher	Pileated jay	Pie akahé
– майна	Haubenmaina	Chinese crested mynah	Martin huppé
– овсянка	Haubenammer	Crested bunting	
– синица	Haubenmeise	Crested-tit	Mésange huppée
Хохлатый жаворонок	Haubenlerche	Crested lark	Cochevis huppé
Цветоклевы, Цветососы	Mistelesser	Flowerpeckers	Dicaeidés
Чагра	Kaptschagra	Redwing shrike	
Чеканы	Schmätzer i. e. S.		Traquets
Челнохвосты	Bootsschwänze	Grackles	
Черная синица	Tannenmeise	Coal-titmouse	Petite charbonnière
– шлемоносная райская птица	Schwarzkehl-Paradieselster	Long tail bird of paradise	
Черноголовая гаичка	Nonnenmeise	Marsh-titmouse	Mésange nonnette
– голубая сойка	Schwarzkopfhäher	Steller's jay	
– иволга	Schwarznacken-Pirol	Black-naped oriole	Loriot de Chine
– овсянка	Kappenammer	Black-headed bunting	Bruant mélanocéphale
– славка	Mönchgrasmücke	Blackcap	Fauvette à tête noire
Черноголовый поползень	Kappenkleiber	Red-breasted nuthatch	
– чекан	Schwarzkehlchen	Stonechat	Traquet pâtre
Чернозобый дрозд	Schwarzkehldrossel	Black-throated thrush	Grive à gorge noire
Чернолобый сорокопут	Schwarzstirnwürger	Lesser grey shrike	Pie-grièche à poitrine rose
Чернопегая каменка	Mittelmeer-Steinschmätzer	Black-eared wheatear	Traquet oreillard
Чернопегий сорокопут	Maskenwürger	Masked shrike	Pie-grièche masquée
Черноспинная ворона-свистун	Schwarzrücken-Flötenvogel	Common piping crow	Corbeau flûteur
Черный дрозд	Amsel	Blackbird	Merle noir
– дятел	Schwarzspecht	Black woodpecker	Pic noir
– скворец	Einfarbstar	Spotless starling	Etourneau unicolore
Чечевицы	Karmingimpel		Roselins
Чечетки	Hänflinge		Linottes
Чешуегрудый щитоносец	Prachtreifelvogel	Magnificent riflebird	Promefil magnifique
Чешуйчатая райская птица	Albert-Paradisvogel	King of Saxony's bird of paradise	
Чиж	Erlenzeisig	Siskin	Tarin des aulnes
Чижи	Zeisige und Stieglitze		Chardonnerets
Чудная райская птица	Kragenhopf	Lesser superb bird of paradise	Paradisier superbe
Широкорот	Ostroller	Oriental dollar bird	Rollier à gros bec
Широкоротая якамара	Breitmaul-Glanzvogel	Great jacamar	
Широкоротые сизоворонки	Breitschnabelracken	Broad-billed rollers	
Широкохвостая камышевка	Seidensänger	Cetti's warbler	Bouscarle de Cetti
Широкохвостка	—	— —	— — —

RUSSIAN NAME	GERMAN NAME	ENGLISH NAME	FRENCH NAME
Широкохвостые камышевки	Seidenrohrsänger		Bouscarles
Шлемоносный сорокопут	Helmvanga	Helmet bird	
Штрутидеа	Gimpelhäher	Grey-jumper	Glaucope gris
Щеврицы	Pieper i. e. S.	Pipits	Pipits
Щеглы	Stieglitze und Zeisige		Chardonnerets
Щегол	Stieglitz	Goldfinch	Chardonneret élégant
Щитоносные райские птицы	Reifelvögel	Riflebirds	
Щур	Hakengimpel	Pine grosbeak	Dur-bec des sapins
Щуры	—		Dur-becs
Эгейская славка	Maskengrasmücke	Rüppell's warbler	Fauvette masquée
Южноафриканский воробей	Kapsperling	Cape sparrow	Moineau mélanure
Южнокитайский малый скворец	Mandarinstar	Mandarin mynah	Martin de Chine
Южный соловей	Nachtigall	Nightingale	Rossignol philomèle
Юла	Heidelerche	Wood lark	Alouette lulu
Юнко	Juncos	Juncos	Juncos
Юрок	Bergfink	Brambling	Pinson du nord
Якамары	Glanzvögel	Jacamars	Galbulidés
Японская белоглазка	Japanischer Brillenvogel	Japanese white-eye	
— синица	Buntmeise	Red-sided titmouse	
Ястребная славка	Sperbergrasmücke	Barred warbler	Fauvette épervière

Conversion Tables of Metric to U.S. and British Systems

Length

U.S. Customary to Metric		Metric to U.S. Customary	
To convert	Multiply by	To convert	Multiply by
in. to mm.	25.4	mm. to in.	0.039
in. to cm.	2.54	cm. to in.	0.394
ft. to m.	0.305	m. to ft.	3.281
yd. to m.	0.914	m. to yd.	1.094
mi. to km.	1.609	km. to mi.	0.621

Area

To convert	Multiply by	To convert	Multiply by
sq. in. to sq. cm.	6.452	sq. cm. to sq. in.	0.155
sq. ft. to sq. mi.	0.093	sq. m. to sq. ft.	10.764
sq. yd. to sq. m.	0.836	sq. m. to sq. yd.	1.196
sq. mi. to ha.	258.999	ha. to sq. mi.	0.004

Volume

To convert	Multiply by	To convert	Multiply by
cu. in. to cc.	16.387	cc. to cu. in.	0.061
cu. ft. to cu. m.	0.028	cu. m. to cu. ft.	35.315
cu. yd. to cu. m.	0.765	cu. m. to cu. yd.	1.308

Capacity (liquid)

To convert	Multiply by	To convert	Multiply by
fl. oz. to liter	0.03	liter to fl. oz.	33.815
qt. to liter	0.946	liter to qt.	1.057
gal. to liter	3.785	liter to gal.	0.264

Mass (weight)

To convert	Multiply by	To convert	Multiply by
oz. avdp. to g.	28.35	g. to oz. avdp.	0.035
lb. avdp. to kg.	0.454	kg. to lb. avdp.	2.205
ton to t.	0.907	t. to ton	1.102
l. t. to t.	1.016	t. to l. t.	0.984

Abbreviations

U.S. Customary

- avdp.—avoirdupois
- ft.—foot, feet
- gal.—gallon(s)
- in.—inch(es)
- lb.—pound(s)
- l. t.—long ton(s)
- mi.—mile(s)
- oz.—ounce(s)
- qt.—quart(s)
- sq.—square
- yd.—yard(s)

Metric

- cc.—cubic centimeter(s)
- cm.—centimeter(s)
- cu.—cubic
- g.—gram(s)
- ha.—hectare(s)
- kg.—kilogram(s)
- m.—meter(s)
- mm.—millimeter(s)
- t.—metric ton(s)

By kind permission of Walker: Mammals of the World
©1968 Johns Hopkins Press, Baltimore, Md., U.S.A.

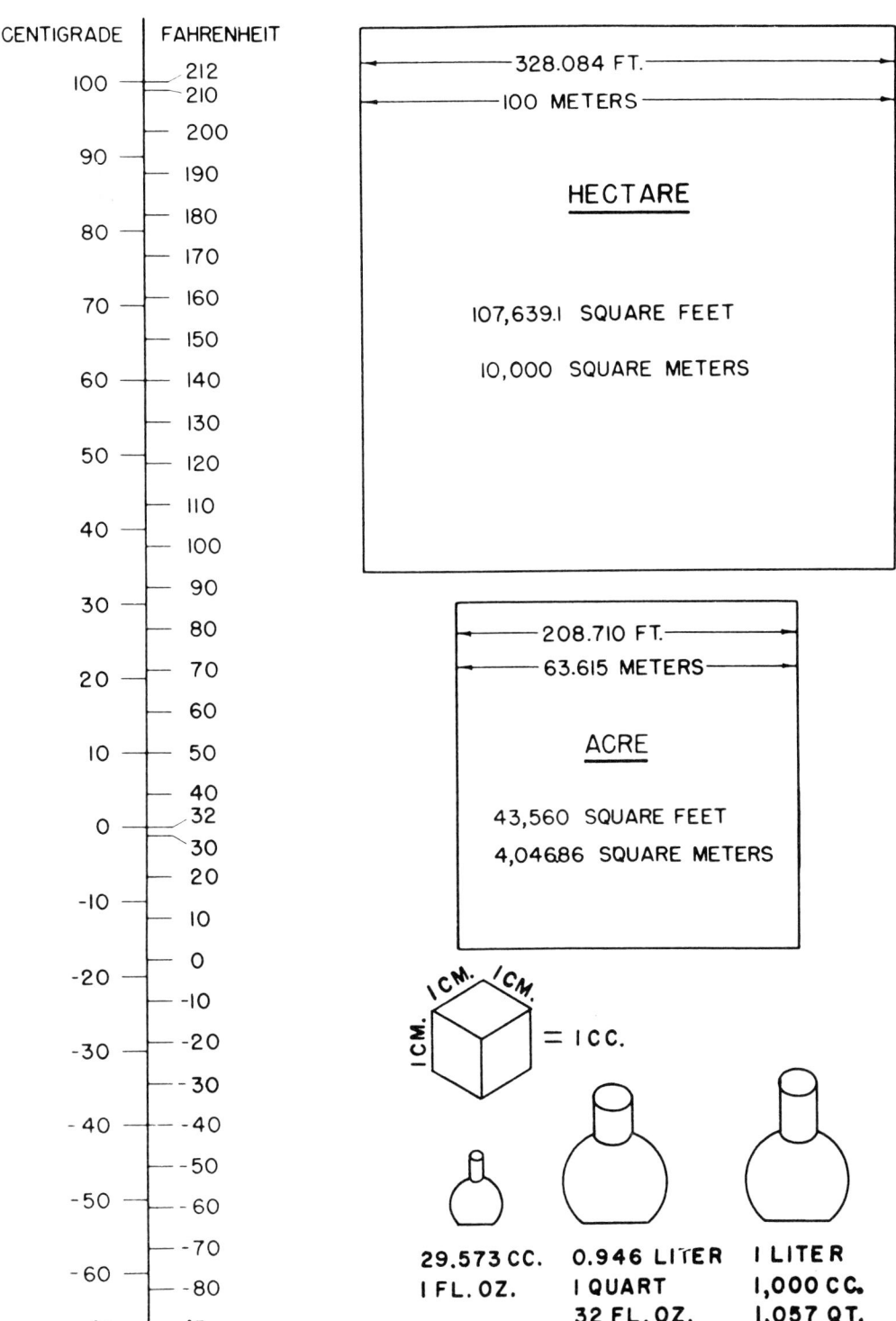

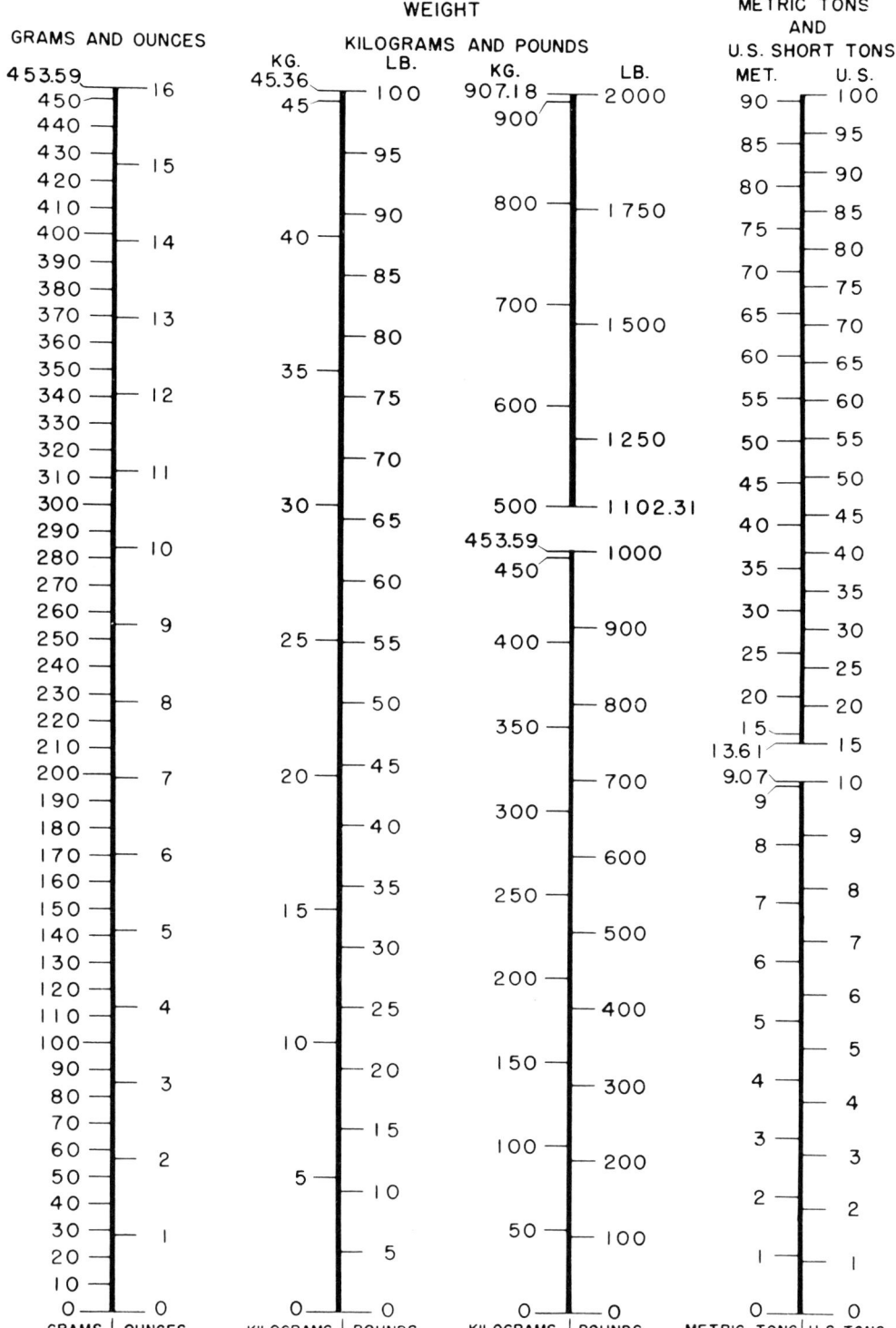

LENGTH: MILLIMETERS AND INCHES

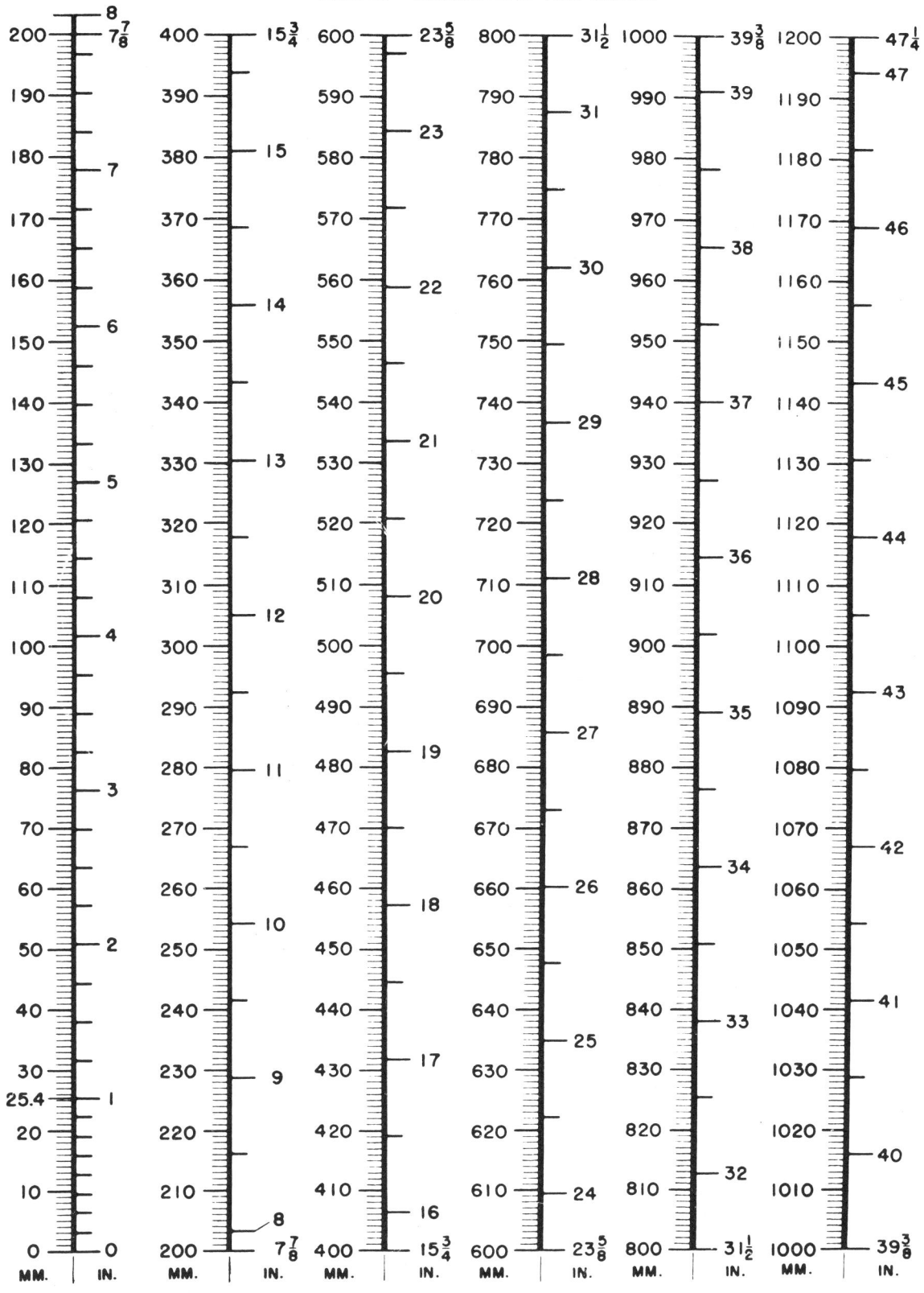

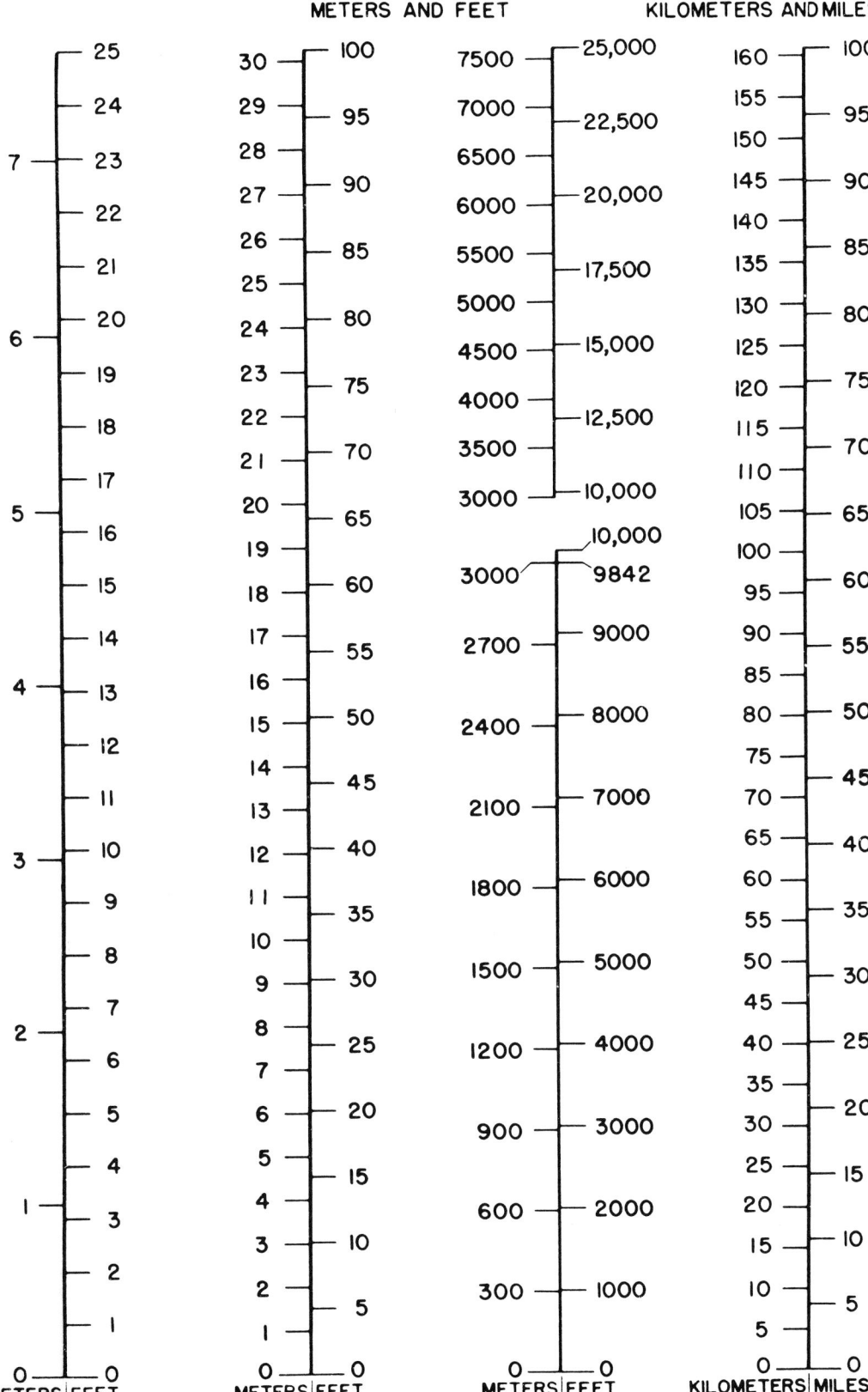

Supplementary Readings

These references of books and articles published in scientific journals deal with animals and topics that are covered in this volume. Some of these were the original sources on which the content of this book is based. These titles are intended as an aid to readers who are interested in additional information and more detailed coverage of the subjects contained in this book.

BOOKS AND MONOGRAPHS

Amadon, D. 1944. A Preliminary Life History Study of the Florida Jay, *Cyanocitta c. coerulescens*. American Museum Novitates. No. 1252.

—. 1950. The Hawaiian Honeycreepers (Aves, Drepanididae). *Bulletin of the American Museum of Natural History* 15.

Armstrong, E. A., 1955. *The Wren*. Collins, London.

Austin, O. L. and N. Kuroda 1953. The Birds of Japan. Their Status and Distributions. *Bulletin of the Museum of Comparative Zoology* (Harvard) No. 109(4).

Bent, A. C. 1939. *Life Histories of North American Woodpeckers*. U.S. National Museum Bulletin 174.

—. 1942. *Life Histories of North American Flycatchers, Larks, Swallows, and Their Allies*. U.S. National Museum Bulletin 179.

—. 1946. *Life Histories of North American Jays, Crows and Titmice*. U.S. National Museum Bulletin 191.

—. 1948. *Life Histories of North American Nuthatches, Wrens, Thrashers, and Their Allies*. U.S. National Museum Bulletin 195.

—. 1949. *Life Histories of North American Thrushes, Kinglets, and Their Allies*. U.S. National Museum Bulletin 196.

—. 1950. *Life Histories of North American Wagtails, Shrikes, Vireos, and Their Allies*. U.S. National Museum Bulletin 197.

—. 1953. *Life Histories of North American Wood Warblers*. U.S. National Museum Bulletin 203.

—. 1958. *Life Histories of North American Blackbirds, Orioles, Tanagers, and Allies*. U.S. National Museum Bulletin 211.

—. 1968. *Life Histories of North American Cardinals, Grosbeaks, Buntings, Towhees, Finches, Sparrows and Allies*. Parts 1, 2, and 3. U.S. National Museum Bulletin 237.

Blake, E. R. 1953. *Birds of Mexico*. University of Chicago Press, Chicago.

Blanchard, B. D. 1941. The White-Crowned Sparrows *(Zonotrichia leucophrys)* of the Pacific Seaboard: Environment and Annual Cycle. *University of California Publications in Zoology*, 46:1-178.

Buxton, J. 1950. *The Redstart*. Collins, London.

Chapin, J. P. 1932-54. *The Birds of the Belgian Congo*. (4 Vols.) Bulletins of the American Museum of Natural History No. 65, 75, 75A and 75B.

Chapman, F. M. 1907. *The Warblers of North America*. D. Appleton and Company, New York (Reprinted by Dover).

Croze, H. 1970. *Searching Image in Carrion Crows*. Paul Parey, Berlin and Hamburg.

de Schauensee, R. M. 1964. *The Birds of Colombia and Adjacent Areas of South and Central America*. Livingston Publishing Co., Narberth, Pa.

—. 1966. *The Species of Birds of South America and Their Distribution*. Livingston Publishing Co., Narberth, Pa.

Eisenmann, E. 1955. *The Species of Middle American Birds*. Transactions of the Linnaean Society of New York.

Friedmann, H. 1929. *The Cowbirds. A Study in the Biology of Social Parasitism*. Charles C. Thomas, Springfield, Illinois.

Gabrielson, I. N. and F. C. Lincoln 1959. *Birds of Alaska*. Stackpole Publishing Co., Harrisburg, Pa.

Gilliard, E. T. 1969. *Birds of Paradise and Bower Birds*. Wedenfeld and Nicolson, London.

Godfrey, W. E. 1966. *The Birds of Canada*. Natural Museum Canadian Bulletin No. 203; Biological Series No. 73.

Gould, J. and A. Rutgers, 1966. *Birds of Europe*. Barnes and Noble, New York.

Griscom, L., A. Sprunt, Jr. and Others 1957. *The Warblers of America*, The Devin-Adair Company, New York.

Hillstead, A. F. C. 1945. *The Blackbird: A Contribution to the Study of a Single Avian Species*. Faber and Faber, London.

Hinde, R. A. 1952. The Behaviour of the Great Tit *(Parus major)* and Some Other Related Species. *Behaviour Supplement* 5. E. J. Brill, Leiden.

Hyde, A. S. 1939. The Life History of Henslow's Sparrow. *Passerherbulus henslowi* (Audubon). University of *Michigan Museum of Zoology Miscellaneous Publications*, No. 41.

Iredale, T. 1950. *Birds of Paradise and Bower Birds*. Georgian House, Melbourne.

Johnston, D. W. 1961. *The Biosystematics of American Crows*. University of Washington Press, Seattle.

Kale, H. W. II. 1965. Ecology and Bioenergetics of the Long-Billed Marsh Wren *Troglodytes palustris griseus* (Brewster) in Georgia Salt Marshes. *Publications of the Nuttall Ornithology Club* No. 5, Cambridge, Massachusetts.

Lack, D. 1943. *The Life of the Robin*. Penguin Books, London.

—. 1947. *Darwin's Finches*. University Press, Cambridge.

Lanyon, W. E. 1957. The Comparative Biology of the Meadowlarks *(Sturnella)* in Wisconsin. *Publications of the Nuttall Ornithology Club* No. 1, Cambridge, Massachusetts.

Lawrence, L. de Lide K. 1967. *A Comparative Life-History of Four Species of Woodpeckers*. American Ornithologists' Union, Ornithology Monograph No. 5.

Linsdale, J. M. 1937. *The Natural History of The Magpies*. Pacific Coast Avifauna No. 25. Cooper Ornithological Club, Berkeley.

Lorenz, K. 1932. Contributions to the Study of the Ethology of Social Corvidae. In: Lorenz, K. 1970 *Studies in Animal and Human Behaviour*, Vol. I. Translated by Robert Martin. Harvard University Press, Cambridge, Massachusetts.

Lunk, W. A. 1962. The Rough-Winged Swallow *Stelgidopteryx ruficollis* (Vieillot): A Study Based on its Breeding Biology in Michigan. *Publications of the Nuttall Ornithology Club* No. 4, Cambridge, Massachusetts.

Mackworth-Praed, C. W. and C. H. B. Grand 1952-55. *Birds of Eastern and Northeast Africa*. (2 Vols.) Longmans Green, New York.

Marler, P. 1956. Behaviour of the Chaffinch *Fringilla Coelebs*. *Behaviour Supplement* No. 5. E. J. Brill, London.

Marshall, A. J. 1954. *Bower-Birds. Their Displays and Breeding Cycles.* Clarendon Press, Oxford.

Mayfield, H. 1960. *The Kirtland's Warbler.* Cranbrook Institute Science Bulletin No. 40, Bloomfield Hills, Michigan.

Miller, A. H. 1931. Systematic Revision and Natural History of the American Shrikes *(Lanius). University of California Publications in Zoology,* 38:11-242.

Mountfort, G. 1957. *The Hawfinch.* Collins, London.

Mumford, R. E. 1964. The Breeding Biology of the Acadian Flycatcher. *University of Michigan Museum of Zoology Miscellaneous Publication,* No. 25.

Nethersole,-Thompson, D. 1966. *The Snow Bunting.* Oliver and Boyd, Edinburgh.

Nice, M. M. 1937. *Studies in the Life History of the Song Sparrow,* I. Transactions of the Linnaean Society of New York. 4:i-vi; 1-247 (Reprinted by Dover).

— . 1943. *Studies in the Life History of the Song Sparrow,* II. Transactions of the Linnaean Society of New York, 6:i-viii; 1-328 (Reprinted by Dover).

Orians, G. H. and G. M. Christman 1964. *A Comparative Study of the Behavior of Red-winged, Tricolored, and Yellow-Headed Blackbirds.* University of California Publications in Zoology, Vol. 84.

Peterson, R. T., G. Mountfort and P. A. D. Hollom 1954. *A Field Guide to the Birds of Britain and Europe.* Houghton-Mifflin Co.; Boston.

Pickwell, G. B. 1931. *The Prairie Horned Lark.* Transactions of the Academy of Science of St. Louis. 27:1-153.

Pough, R. H. 1919. *Audubon Land Bird Guide.* Doubleday & Company, New York.

Prescott, K. W. 1965. Studies in the Life History of the Scarlet Tanager, *Piranga olivacea. New Jersey State Museum Investigations* No. 2.

Roberts, A. 1957. *The Birds of South Africa,* Cape Times Ltd.

Sielmann, H. 1958. *My Year with the Woodpeckers.* Barrie and Rockliff, London.

Skead, C. J. 1967. *The Sunbirds of Southern Africa.* A. A. Balkema, Cape Town.

Skutch, A. F. 1954, 1960. *Life Histories of Central American Birds.* (2 Vols.). Cooper Ornithological Society, Pacific Coast, Avifauna No. 31, 34.

Smith, S. 1950. *The Yellow Wagtail.* Collins, London.

Snow, D. W. 1958. *A Study of Blackbirds.* George Allen and Unwin, London.

Stoner, D. 1939. *Temperature, Growth, and Other Studies on the Eastern Phoebe.* New York State Museum Circular 22.

Summers-Smith, J. D. 1963. *The House Sparrow.* Collins, London.

Tanner, J. T. 1942. *The Ivory-Billed Woodpecker.* Research Report No. 1, National Audobon Society, New York.

Tinbergen, N. 1939. The Behavior of the Snow Bunting in Spring. *Transactions of the Linnaean Society of New York.* 5:1-94.

Voous, K. H. 1960. *Atlas of European Birds.* Nelson, New York.

Wetmore, A. and Others 1964. *Song and Garden Birds of North America.* National Geographic Society, Washington, D.C.

Whistler, H. 1949. *Popular Handbook of Indian Birds.* Gurney & Jackson.

White, H. C. 1953. *The Eastern Belted Kingfisher in the Maritime Provinces.* Fisheries Research Board of Canada Bulletin No. 97.

Witherby, H. E. (ed.) 1938-43. *The Handbook of British Birds.* (5 Vols.) H. F. & G. Witherby, London.

Yeates, G. K. 1934. *The Life of the Rook.* London.

JOURNAL ARTICLES:

Allen, A. A. (1914). The Red-Winged Blackbird: A Study in the Ecology of a Cat-Tail Marsh. *Proceedings of the Linnaean Society of New York.* (24-25): 43-128.

Allen, R. W. and M. M. Nice 1952. A Study of the Breeding Biology of the Purple Martin *(Progne subis). American Midland Naturalist.* 47:606-665.

Amadon, D. 1944. The Genera of Corvidae and Their Relationships. *American Museum Novitates,* 1251:1-21.

Barbour, R. W. 1951. Observations on the Breeding Habits of the Red-Eyed Towhee. *American Midland Naturalist.* 45:672-678.

Barlow, J. C. 1962. Natural History of the Bell Vireo, *Vireo belli,* Audobon. *University of Kansas Publications of the Museum of Natural History.* 12:241-296.

Blanchard, B. D. and M. M. Erickson 1949. The Cycle in the Cambel Sparrow. *University of California Publications in Zoology.* 47:255-318.

Cartwright, B. W., T. M. Shortt, and R. D. Harris 1937. Baird's Sparrow. *Transactions of the Royal Canadian Institute.* 21: 153-197.

Chapman, F. M. 1928. The Nesting Habits of Wagler's Oropendola *(Zarhynchus wagleri)* on Barro Colorado Island. *Bulletin of the American Museum of Natural History.* 58:123-166.

Collias, N. E., and E. C. Collias 1970. The Behaviour of the West African Village Weaverbird. *Ibis.* 112(4):457-480.

Crossin, R. S. 1967. The Breeding Biology of the Tufted Jay. *Proceedings of the Western Foundation for Vertebrate Zoology.* 1(5):265-299.

Eaton, S. W. 1957. A Life History Study of *Seiurus noveboracensis* (with Notes on *Seiurus aurocapillus* and the Species of *Seiurus* Compared). *St. Bonaventure University Science Studies.* 19:7-36.

Erwin, W. G. 1935. Some Nesting Habits of the Brown Thrasher. *Journal of the Tennessee Academy of Science.* 10:170-204.

Graber, J. W. 1961. Distribution, Habitat Requirements, and Life History of the Black-Capped Vireo *(Vireo atricapilla). Ecology Monographs,* 31:313-336.

Hardy, J. W. and R. W. Dickerman 1965. Relationships Between Two Forms of the Red-Winged Blackbird in Mexico. *Living Bird.* 4:107-129.

Hofslund, P. B. 1959. A Life History of the Yellowthroat, *Geothlypis trichas. Proceedings of the Minnesota Academy of Science.* 27:144-174.

Howell, J. C. 1942. Notes on the Nesting Habits of the American Robin *(Turdus migratorius* L.). *American Midland Naturalist.* 28:529-603.

Kessel, B. 1957. A Study of the Breeding Biology of the European Starling *(Sturnus vulgaris* L.) in North America. *American Midland Naturalist,* 58:257-331.

Kuerzi, S. C. 1941. Life History Studies of the Tree Swallow. *Proceedings of the Linnaean Society of New York.* (52-53): 1-52.

Lack, D. 1939. The Behaviour of the Robin. *Proceedings of the Zoological Society of London.* 109:169-219.

Nickell, W. P. 1965. Habitats, Territory, and Nesting of the Catbird. *American Midland Naturalist.* 73:433-478.

Skutch, A. F. Life History of the Broad-Billed Motmot, With Notes on the Rufous Motmot. *Wilson Bulletin.* 83(1):74-94.

Stoner, D. 1920. Nesting Habits of the Hermit Thrush in Northern Michigan. *University of Iowa Studies. Studies in Natural History.* 9:1-21.

Stoner, D. 1936. Studies on the Bank Swallow *Riparia riparia riparia* (Linnaeus), in the Onedia Lake Region. *Roosevelt Wild Life Annals.* 9:126-233.

Verbeek, N. A. M. 1972. Comparison of Displays of the Yellow-Billed

Magpie *(Pica nutalli)* and Other Corvids. *Zeitschrift Für Ornithologie.* 113:297-314.

Woolfenden, G. E. 1956. Comparative Breeding Behavior of *Ammospiza caudacuta* and *A. maritima. University of Kansas Publications of the Museum of Natural History,* 10:45-75.

The following references of books and articles published in scientific journals were listed in the original German volume.

Abs, M. 1963. Vergleichende Untersuchungen an Haubenlerche *(Galerida cristata L.)* und Theklalerche *(Galerida theklae* A. E. Brehm). (Aves Alaudidae, ein Beitrag zur Evolution der Zwillingsarten.) *Bonner Zoologische Beiträge,* Bd. 14, S. 1-128.

Bährmann, U. 1968. *Die Elster.* Neue Brehm-Bücherei, Ziemsen, Wittenberg Lutherstadt.

Bergmann, St. 1952. *Wilde und Paradiesvögel.* Brockhaus, Wiesbaden.

Berndt, R., und W. Meise 1966. *Naturgeschichte der Vögel.* Bd. 3, Franckh, Stuttgart.

Berndt, R., und M. Henβ 1967. Die Kohlmeise, *Parus major,* als Invasionsvogel. *Vogelwarte,* Bd. 24, S. 17-37.

Blume, D. 1962. *Schwarzspecht—Grünspecht—Grauspecht.* Neue Brehm-Bücherei, Ziemsen, Wittenberg Lutherstadt.

—. 1968. *Buntspechte.* Neue Brehm-Bücherei, Ziemsen, Wittenberg Lutherstadt.

Boetticher, H. v. 1952. *Die Widahvögel und Witwen.* Neue Brehm-Bücherei, Ziemsen, Wittenberg Lutherstadt.

—.1959. *Die Pfefferfresser.* Neue Brehm-Bücherei, Ziemsen, Wittenberg Lutherstadt.

Creutz, G. 1966. *Die Wasseramsel.* Neue Brehm-Bücherei, Wittenberg Lutherstadt.

Curio, E. 1959. *Verhaltensstudien am Trauerschnäpper. Zeitschrift für Tierpsychologie,* Beiheft 3, Parey, Hamburg/Berlin.

—. 1964. Zur geographischen Variation des Feinderkennens einiger Darwinfinken (Geospizidae). *Verhandlungen der Deutschen Zoologischen. Gesellschaft Kiel,* S. 446-492.

—.1965. Galapagos—Prüffeld der Evolutionsforschung. *Umschau, Heft 18,* S. 562-567.

— und P. Kramer 1964. Vom Mangrovefinken *(Cactospiza heliobates* Snodgrass und Heller). *Zeitschrift für Tierpsychologie,* Bd. 21, S. 223-234.

Deckert, G. 1955. Beiträge zur Kenntnis der Nestbau-Technik deutscher Sylviiden (Grasmücken). *Journal für Ornithologie,* Bd. 96, S. 186-206.

—. 1965. Nestbau, Jungenaufzucht und postnatale Entwicklung der Kohlmeise, *Parus major. Beiträge zur Vogelkunde,* Bd. 10, S. 212-230.

—.1969. *Der Feldsperling.* Neue Brehm-Bücherei, Ziemsen, Wittenberg Lutherstadt.

Dost, H. 1960. *Die Schamadrossel.* Gefiederte Welt, Bd. 88, Sonderheft, Helène, Pfungstadt.

Eibl-Eibesfeldt, I. 1960. *Galapagos. Die Arche Noah im Pazifik.* München.

—. 1964. *Geospiza fuliginosa* (Fringillidae). Säubern von Meerechsen. *Encyclopaedia Cinematographica Institut für wissenschaftliche Filme, Filmbeiheft* E 576, S. 327-329. Göttingen.

—. 1967. Galapagos—Laboratorium der Natur. *Bild der Wissenschaft,* S. 136-144.

— und H. Sielmann 1962. Beobachtungen am Spechtfinken *Cactospiza pallida* (Sclater und Salvin). *Journal für Ornithologie,* Bd. 103, S. 92-101.

Fiendt, P., und K. Jung 1968. *Bartmeisen (Panurus biarmicus)—Einblicke in ihr verborgenes Leben.* Gerstenberg, Hildesheim.

Franke, H. 1938. *Z-i-i-h—die Beutelmeise.* Deuticke, Wien.

Frisch, O. v. 1966. Beitrag zur Ethologie der Blauracke. *(Coracias garrulus). Zeitschrift für Tierpsychologie,* Bd. 23, S. 44-51.

Gerber, R. 1956. *Die Saatkrähe.* Neue Brehm-Bücherei, Ziemsen, Wittenberg Lutherstadt.

Grzimek, B. 1952. Zum Balzverhalten des westafrikanischen Textor-Webers *Hypophantornis (Ploceus) cucullatus. Zeitschrift für Tierpsychologie,* Bd. 9, S. 289-295.

Gwinner, E. 1964. Untersuchungen über das Ausdrucksund Sozialverhalten des Kolkraben *(Corvus c. corone L.). Zeitschrift für Tierpsychologie,* Bd. 21, S. 657-748.

Hasse, H. 1963. *Die Goldammer.* Neue Brehm-Bücherei, Ziemsen, Wittenberg Lutherstadt.

Hermann, H., und W. Meise 1966. Untersuchungen zur Brutbiologie des Töpfervogels *Furnarius r. rufus* (Gmelin), auf einer argentinischen Hacienda nebst Notizen über Nestbauzeit und Brutzeit. *Abhandlungen und Verhandlungen des Naturwissenschaftlichen Vereins Hamburg,* N.F. Bd. 10, S. 117-152.

Heyder, R. 1953. *Die Amsel.* Neue Brehm-Bücherei, Ziemsen, Wittenberg Lutherstadt.

Hilprecht, A. 1954. *Nachtigall und Sprosser.* Neue Brehm-Bücherei, Ziemsen, Wittenberg Lutherstadt.

Hoesch, W. 1933. Ein Vogelnest mit verschließbarem Eingang: das Nest von *Anthoscopus caroli* (Sharpe). *Ornithologische Monatsberichte,* Bd. 41, S. 1-4.

—. 1937. Über das "Honiganzeigen" von Indicator. *Journal für Ornithologie,* Bd. 85, S. 201-205.

Hubatsch, H. 1966. *Unsere Wiesenschmätzer. Aus dem Leben von Schwarzkehlchen und Braunkehlchen.* Neue Brehm-Bücherei, Ziemsen, Wittenberg Lutherstadt.

Immelmann, K. 1962. Beiträge zu einer vergleichenden Biologie australischer Prachtfinken (Spermestidae). *Zoologisches Jahrbuch (Systematik),* Bd. 90, S. 1 bis 196.

—. 1966. Beiträge zur Biologie und Ethologie australischer Honigfresser, Meliphagidae. *Journal für Ornithologie,* Bd. 102, S. 164-207.

— Steinbacher, J. und H. E. Wolters 1965. *Prachtfinken.* In: Vögel und Käfig und Voliere. Bd. I: Astrilde. Bd. II: Amadinen (im Erscheinen), Aachen.

Impekoven, M. 1962. *Die Jugendentwicklung des Teichrohrsängers (Acrocephalus scirpaceus). Eine Verhaltensstudie.* Revue Suisse Zoologie, Bd. 69, S. 77 bis 191, Societe Suisse de Zoologie, Genf.

Jahn, H. 1939. Zur Biologie des japanischen Paradiesfliegenschnäppers *Tersiphone a. atrocaudata* (Eyton). *Journal für Ornithologie,* Bd. 87, S. 216 bis 223.

— 1952. Zur Oekologie und Biologie der Vögel Japans. *Journal für Ornithologie,* Bd. 90, S. 1-302.

Keve, A. 1969. *Der Eichelhäher.* Neue Brehm-Bücherei, Ziemsen, Wittenberg Lutherstadt.

Kleinschmidt, O. 1929. *Der Formenkreis Parus acredula (Kl.). Berajah, Zoographia infinita,* Gebauer & Schwetschke, Halle.

Koenig, L. 1951. Beitrag zu einem Aktionssystem des Bienenfressers *(Merops paiaster* L.). *Zeitschrift für Tierpsychologie,* Bd. 8, S. 169-210.

Kottek, E. 1965. Die Aufzucht junger Mauerläufer, *Tichodroma muraria* (L). *Beiträge zur Vogelkunde* Bd. 11, S. 48-54.

Küchler, W. 1936. Anatomische Untersuchungen an *Phytotoma rara* Mol. *Journal für Ornithologie,* Bd. 84, S. 352-362.

Kunkel, P. 1962. Bewegungsform, Sozialverhalten, Balz und Nestbau des Gangesbrillenvogels *(Zosterops palpebrosa Temm.). Zeitschrift für Tierpsychologie,* Bd. 19, S. 559-576.

Linsenmair, K. E. 1964. Vogelzwerge des Waldes (Goldhähnchen). *Vogel-Kosmos,* Bd. 1, S. 200-206.

Löhrl, H. 1964. Verhaltensmerkmale der Gattungen *Parus* (Meisen), *Aegithalos* (Schwanzmeisen), *Sitta* (Kleiber), *Tichodroma* (Mauerläufer) und *Certhia* (Baumläufer). *Journal für Ornithologie,* Bd. 105 S. 153-181.

—. 1966. Zur Biologie der Trauermeise *(Parus lugubris). Journal für Ornithologie,* Bd. 107, S. 167-186.

—. 1967. *Die Kleiber Europas.* Neue Brehm-Bücherei, Ziemsen, Wittenberg Lutherstadt.

Lorenz, K. 1931. Beiträge zur Ethologie sozialer Corviden. *Journal für Ornithologie,* Bd. 79, S. 67-127.

Meise, W. 1936. Zur Systematik un Verbreitungsgeschichte des Haus- und Weidensperlings, *Passer domesticus* (L.) und, *P. hispaniolensis* (T.). *Journal für Ornithologie,* Bd. 84, S. 631-671.

—. 1956. Über Nestokkupation durch fremde Vogelarten. *Beiträge zur Vogelkunde,* Bd. 5, S, 117-137.

Melde, M. 1969. *Raben- und Nebelkrähe.* Neue Brehm-Bücherei, Ziemsen. Wittenberg Lutherstadt.

Menzel, H. 1964. *Der Steinschmätzer.* Neue Brehm-Bücherei, Ziemsen, Wittenberg Lutherstadt.

—. 1968. *Der Wendehals.* Neue Brehm-Bücherei, Ziemsen, Wittenberg Lutherstadt.

Münch, H. 1952. *Der Wiedehopf.* Neue Brehm-Bücherei, Ziemsen, Wittenberg Lutherstadt.

Münster, W. 1956. *Der Neuntöter oder Rotrückenwürger.* Neue Brehm-Bücherei, Ziemsen, Wittenberg Lutherstadt.

Nice, M. M. 1933-1934. Zur Naturgeschichte des Singammers. *Journal für Ornithologie,* Bd. 81, S. 552-595 und Bd. 82, S. 1-96.

Nicolai, J. 1956. Zur Biologie und Ethologie des Gimpels *(Pyrrhula pyrrhula* L.). *Zeithschrift für Tierpsychologie,* Bd. 13, S. 93-132.

—. 1964. Der Brutparasitimus der Viduinae (Witwen) als ethologisches Problem. *Zeitschrift für Tierpsychologie,* Bd. 21, S. 129-204.

Niethammer, G. 1937. *Handbuch der deutschen Vogelkunde.* Bd. I, Akademische Verlagsgesellschaft, Leipzig.

—. 1940. Die Schutzanpassung der Lerchen. *Journal für Ornithologie,* Bd. 88, Sonderheft, S. 75-83.

Oehme, H. 1965. Der Flug der Flaggendrongos. *Journal für Ornithologie,* Bd. 106, S. 190-203.

Pätzold, R. 1963. *Die Feldlerche.* Neue Brehm-Bücherei, Ziemsen, Wittenberg Lutherstadt.

Reinsch, A. 1960. Beobachtungen am Nest des Pirols. *Vogelwelt,* Bd. 80, S. 149-156.

Salomonsen, F. 1933. Zur Systematik und Biologie von Promerops. *Ornithologische Monatsberichte,* Bd. 41, S. 37-40.

Sanden-Guja, W. v. 1963. *Der fliegende Edelstein (Eisvogel).* Landbuch-Verlag, Hannover.

Sanft, K. 1960. *Bucerotidae (Nashornvögel).* Das Tierreich, Lieferung 76, de Gruyter, Berlin.

Schäfer, E. 1938. Ornithologische Ergebnisse zweier Forschungsreisen nach Tibet. *Journal für Ornithologie,* 86, Sonderheft.

Schifferli, A. 1961. Einige Beobachtungen am Nest der Bergstelze. *Ornithologischer Beobachter,* Bd. 58, S. 125-133.

Schneider, W. 1960. *Der Star.* Neue Brehm-Bücherei, Ziemsen, Wittenberg Lutherstadt.

Schuster, L. 1941. Zur Brutbiologie der Schafstelze. *Beiträge zur Fortpflanzungsbiologie der Vögel,* Bd. 17, S. 207-210.

Serventy, D. L. 1962. Die Wiederentdeckung von *Atrichornis clamosus* (Gould) (Großer Dickichtschlüpfer) in Westaustralien. *Journal für Ornithologie,* Bd. 103, S. 213/214.

Sick, H. 1954. Zur Biologie des Amazonischen Schirmvogels, *Cephalopterus ornatus. Journal Für Ornithologie,* Bd. 95, S. 233-242.

—. 1957. *Tukani—Unter Tieren und Indianern Zentralbrasiliens.* Parey, Hamburg/Berlin.

—. 1959. Die Balz der Schmuckvögel (Pipridae). *Journal für Ornithologie,* Bd. 100, S. 269-302.

—. 1960. Zur Systematik und Biologie der Bürzelstelzer (Rhinocryptidae) speziell Brasiliens. *Journal für Ornithologie,* Bd. 101, S. 141-174.

—. 1958. und J. Ottow. Vom brasilianischen Kuhvogel, *Molothrus bonariensis,* und seinen Wirten, besonders des Ammerfinken, *Zonotrichia capensis. Bonner Zoologische Beiträge,* Bd. 9, S. 40-62.

Siefke, A. 1962. *Dorn- und Zaungrasmücke.* Neue Brehm-Bücherei, Ziemsen, Wittenberg Lutherstadt.

Sielmann, H. 1958. *Das Jahr mit den Spechten.* Ullstein, Berlin.

Steinfatt, O. 1938. Das Brutleben der Heckenbraunelle, *Prunella m. modularis. Ornithologische Monatsberichte,* Bd. 46, S. 65-76.

—. 1941. Beobachtungen beim Baumpieper. *Journal für Ornithologie,* Bd. 89, S. 393-403.

Stresemann, E. 1953. Lauben und Balz der Laubenvögel (Ptilonorhynchidae). *Vogelwarte,* Bd. 16, S. 148 bis 154.

—. 1954. Die Entdeckungsgeschichte der Paradiesvögel. *Journal für Ornithologie,* Bd. 95, S. 263-291.

—.1920. und H. Sachtleben. Über die europäischen Mattkopfmeisen (Gruppe *Parus atricapillus). Verhandlungen Ornithologischer Gesellschaft Bayern,* Bd. 14, S. 228-269.

Szijj, J. 1958. Beiträge zur Ernährungsbiologie der Blauracke in Ungarn. *Bonner Zoologische Beiträge* Bd. 9, S. 24-39.

Thielcke, G. 1961. Stammesgeschichte und geographische Variation des Gesangs unserer Baumläufer *(Certhia familiaris* L. und *Certhia brachydactyla). Zeitschrift für Tierpsychologie,* Bd. 18, S. 188-204.

—. 1966. Unterschiede im Übernachten von Garten- und Waldbaumläufer *(Certhia brachydactyla* und *Certhia familiaris). Vogelwelt,* Bd. 87, S. 113-117.

Thönen, W. 1962. Stimmengeographische, ökologische und verbreitungsgeschichtliche Studien über die Mönchmeise *(Parus montanus* Conrad). *Ornithologischer Beobachter,* Bd. 59, S. 101-172.

Treuenfels, H. v. 1940. Zur Biologie und Psychologie des Weidenlaubsängers *(Phylloscopus collybita). Journal für Ornithologie,* Bd. 88, S. 509-536.

Vietinghoff-Riesch, A. v. 1955. *Die Rauchschwalbe.* Duncker & Humblot, Berlin.

Wickler, W. 1968. *Mimikry* (Webervögel). Kindlers Universitäts Bibliothek, Kindler, München.

Wittenberg, J. 1968. Freilandbeobachtungen zu Brutbiologie und Verhalten der Rabenkrähe *(Corvus c. corone). Zoologisches Jahrbuch (Systematik),* Bd. 95, S. 15 bis 146.

Wolters, H. E. 1958. Übersicht über die eigentlichen Weber. *Vogelring,* Bd. 27, S. 144-149.

—. 1963. Die Klassifikation der Weberfinken (Estrildae). *Bonner Zoologische Beiträge,* Bd. 8, S. 90-129.

Yamashina, Marquess Y. 1938. Die Lebensweise einiger wenig bekannter Sylviidae (Grasmücken) aus Ostasien. *Journal für Ornithologie,* Bd. 86, S. 497-515.

Zimmer, K. 1943. Der Flug des Nektarvogels *(Cinnyris). Journal für Ornithologie,* Bd. 91, S. 271-287.

List of scientific journals devoted entirely to birds or in which articles on birds frequently appear.

Alauda
American Midland Naturalist
American Museum Novitates
American Naturalist
American Zoologist
Angewandte Ornithologie
Animal Behaviour (Formerly British Journal of Animal Behaviour)
Animal Kingdom
Animals
Annals of the New York Academy of Sciences
Aquila
Ardea
Atlantic Naturalist
Audobon
Audobon Field Notes
The Auk
Aviculture Magazine
Behaviour
Beiträge zur Vogelkunde
Biological Bulletin
Biologisches Zentralblatt
Bird-Banding
Bird Study
British Birds
Bulletin of the American Museum of Natural History
Bulletin of the U.S. National Museum
Canadian Audobon
Canadian Field Naturalist
Canadian Journal of Zoology
The Condor
Ecological Monographs
Ecology
The Emu
Ergebnisse der Biologie
Evolution
Folia Biologica
The Ibis
Japanese Journal of Zoology
Journal of Animal Ecology
Journal of Bombay Natural History Society
Journal of Wildlife Management
The Living Bird
National Wildlife
Natural History
Nature
Naturwissenschaften
L'Oiseau et la Revue Francaise d'Ornithologie
Oiseaux de France
Der Ornothologische Beobachter
Ornithologische Mitteilungen
The Ostrich
Proceedings of the National Academy of Sciences
Proceedings of the Royal Society
Proceeding of the Zoological Society of London
Quarterly Review of Biology
The Ring
Science
Scottish Birds
Sterna
Systematic Zoology
Transactions of the Linnaean Society of New York
University of California Publications in Zoology
Verhandlungen der deutschen Zoologische Gesellschaft
Die Vogelwarte
The Wilson Bulletin
Zeitschrift für Morphologie und Ökologie der Tiere
Zeitschrift für Tierpyschologie
Zoologica
Zoologischer Anzieger
Zoologische Jahrbücher

Picture Credits

Artists: P. Barruel (p. 57, 190, 352, 372). H. Diller (p. 371, 381). K. Grossmann (p. 470, 479, 480, 491, 492). H. Heinzel (p. 280, 342, 382, 392, 416, 430). H. Kacher (p. 429, 439, 440). H. Kühn (p. 179, 189, 221, 391, 449, 469, 497, 498). Dr. N. Kuroda (p. 149). W. Lautz (p. 450). W. Linsenmaier (p. 67). Dr. S. Raethel (p. 351). F. Reimann (p. 231, 232, 259, 289, 290, 307, 341). J. Ritter (p. 37, 38, 58, 68, 97, 98, 107, 108, 119, 120, 125, 126, 143, 144, 150, 180, 195, 196, 222, 260, 279, 308, 319, 320, 361, 362).

Scientific advisors to the artists: Dr. E. Curio (Dr. Raethel). Prof. Dr. H. Dathe (Reimann). Dr. D. Heinemann (Barruel, Dr. Kuroda, Lautz, Linsenmaier). Dr. J. Nicolae (Kacher). Prof. Dr. G. Niethammer (Heinzel). Dr. J. Steinbacher (Diller, Grossmann, Ritter). Prof. Dr. W. Wüst (Kühn). Material for H. Diller was provided by H. Neubüser.

Color photographs: Bracht (p. 84 above and below, 336 above). Collignon (p. 270 below). Czimmek/ CMS-Naturdokumente (p. 83 right). Hosking (p. 169 below, 184 upper left, 270 above, 459 below). Dr. Lachner (p. 30, 32 upper left and upper right, 414 above). Limbrunner (p. 32 lower left, 82 left). Löhr (p. 501 above). Lummer (p. 32 center left above and center right below, 136, 414 below). Müller/Schmida (p. 81 lower right). Dr. Nicolae (p. 413 above from left to right, 460 below). v.d. Nieuwenhuizen (p. 335 center). Okapia (p. 31, 81 lower right, 133 lower left, 201, 269, 333 below). Östmann (p. 170 below). Quedens (p. 501 below). Riese-Barg/Bavaria (p. 183). Schrempp (p. 133 upper left, 184 center left above, center right above, center right below, lower left and lower right, 202, 334 lower left, upper right and lower right, 335 upper left, upper right, center left, center right, lower left and lower right, 413 all nests, 460 above). Sieber Anthony (p. 83 left). Siebrasse (p. 82 right). Siegel/CMS-Naturdokumente (p. 184 upper right). Silvester/Bavaria (p. 29). Smith (p. 134/135, 241). Tilgner (p. 301). Tönges (p. 334 upper left). V-Dia (p. 170 above). Weber-Layer (p. 32 lower right). Wissenbach/Agfa-Gevaert (p. 169 above). Zellmann "Berliner Zoo" (p. 32 center right above, 81 above, middle and below, upper right and lower right). ZFA (p. 32 center left below and lower center, 333 above, 336 center and below, 415). Zingel (p. 242, 302).

Line drawings: Blume (Fig. 2-7, p. 88; Fig. 2-8, p. 89; Fig. 2-9 and 2-10, p. 90; Fig. 2-11, 2-12 and 2-13, p. 91; middle and lower Fig. 2-15 and 2-16, p. 92; Fig. 2-17 and 2-18, p. 93; Fig. 2-25, p. 99; Fig. 2-28, p. 100; Fig. 2-29, p. 102; upper Fig. 2-32, p. 103; Fig. 2-34, p. 104; Fig. 2-36 and 2-37, p. 105; Fig. 2-40, p. 110; Fig. 2-42, p. 111; Fig. 2-43, p. 112; Fig. 2-49, p. 114; Fig. 2-55, p. 116). Dr. Kuroda (Fig. 4-18 upper and 4-19, p. 146). Dr. Raethel (Figs. 15-17, 15-58 and 15-19, p. 363; Figs. 15-20, 15-21, 15-22, 15-23 and 15-24, p. 364; Figs. 15-25, 15-26, 15-27, 15-28, and 15-29, p. 365). Dr. Smythies (Fig. 11-3, p. 247). From Berndt/Meise, Natural History of Birds, with the permission of Franck'sche Verlagslandlung in Stuttgart (Fig. 4-33, p. 155). Diller (Fig. 1-11, p. 36; Fig. 2-17, p. 93; Fig. 2-33, p. 103; Fig. 4-5, p. 128; Fig. 4-9, p. 137; Fig. 4-12, p. 138; Fig. 4-15, p. 140; Fig. 4-26, p. 151; Fig. 4-27, p. 152; Fig. 4-28 and 4-29, p. 153; Fig. 4-30 and upper and middle 4-31, p. 154; Fig. 4-32, p. 155; Fig. 4-34, p. 156; Fig. 4-34, p. 158; Fig. 4-38, p. 159; Fig. 4-39, p. 160; Fig. 4-48, p. 164; Fig. 8-5, p. 194; Fig. 8-10, p. 199; Fig. 9-5, p. 211; Fig. 10-1, p. 219; Fig. 11-7, p. 248; Fig. 11-8 and 11-9, p. 249; Fig. 11-10 and 11-11, p. 250; upper Fig. 11-13, p. 251; middle Fig. 11-29, p. 256; lower Fig. 11-33, p. 257; middle Fig. 11-39, p. 263; Fig. 11-48, p. 267; Fig. 12-4 and upper 12-5, p. 276; Fig. 12-22, p. 287; upper and middle Fig. 12-25, p. 292; Fig. 12-26, p. 293; Fig. 12-33, p. 296; Fig. 12-34, p. 297; Fig. 12-37, p. 303; Fig. 12-45 and 12-46, p. 306; upper Fig. 15-15, p. 359; middle, upper and lower Fig. 15-53 and 15-55, p. 388; Fig. 15-57, lower p. 389; Fig. 15-59, p. 390; lower Fig. 15-63, p. 394; Fig. 16-2, p. 403; Fig. 16-4, p. 404; Fig. 16-10, p. 410; Fig. 16-18, p. 418; Fig. 16-19 and 16-20, p. 419; Fig. 16-21, p. 420; Fig. 19-1, p. 483). Jungreuthmayer (Fig. 17-34, p. 444). Kacher (Fig. 17-12, p. 426; Fig. 17-14, p. 432; Fig. 17-18, p. 433). Kühn and Steffel (Fig. 1-10, p. 35; Fig. 2-4 and 2-5, p. 87; Fig. 2-19 and 2-20, p. 94; Fig. 2-22 and 2-23, p. 95; Fig. 4-2, p. 124; Fig. 4-7, p. 129; Fig. 4-8, p. 130; middle Fig. 4-10, p. 137; lower Fig. 4-13, p. 138; middle Fig. 4-20, p. 146; lower Fig. 4-36, p. 158; Fig. 4-41, p. 161; Fig. 4-42 and 4-43, p. 162; Fig. 4-44, 4-45 and 4-46, p. 165; Fig. 4-54, p. 166; lower Fig. 8-8, p. 198; Fig. 9-9, p. 217; Fig. 10-3, p. 227; middle Fig. 11-14, p. 251; Fig. 11-16 and 11-17, p. 252; Fig. 11-18, p. 253; lower Fig. 11-30 and upper and middle 11-31, p. 256; Fig. 11-32, p. 257; upper Fig. 11-38, p. 262; Fig. 11-45, p. 265; Fig. 11-51, p. 271; lower Fig. 12-6, 12-7 and 12-8, p. 277; Fig. 12-9, p. 278; Fig. 12-11 and 12-12, p. 282; Fig. 12-14, p. 284; lower Fig. 12-27, p. 293; Fig. 13-3 and 13-4, p. 312; Fig. 13-5 and 13-6, p. 313; Fig. 13-7, p. 314; Fig. 13-9 and 13-10, p. 315; Fig. 13-11 and 13-12, p. 316; Fig. 13-14, p. 322; Fig. 15-9 and 15-10, p. 355; Fig. 15-11 and 15-12, p. 356; Fig. 15-13, p. 358; Fig. 15-15, p. 359; Fig. 15-16, p. 360; Fig. 15-31, 15-33, p. 367; Fig. 15-39, p. 376; Fig. 15-43, p. 379; Fig. 15-46, p. 384; Fig. 15-50, p. 386; Fig. 15-53, 15-54, 15-55 and 15-56, p. 388; Fig. 15-57 and 15-58, p. 389; Fig. 15-59, p. 390; Fig. 15-60 and 15-61, p. 393; Fig. 15-62 and 15-63, p. 394; Fig. 15-66, p. 395; Fig. 15-69, and 15-71, p. 399; Fig. 15-72, p. 401; upper Fig. 17-11, p. 426; upper, middle upper and middle Fig. 17-15 and 17-16, p. 432; Fig. 17-17, p. 433; Fig. 17-24, p. 435; Fig. 17-29, p. 442; Fig. 17-56, p. 461; Fig. 19-4, p. 494; Fig. 19-5, 19-6 and 19-7, p. 495; Fig. 19-9 and 19-10, p. 503; Fig. 19-11, 19-12 and 19-13, p. 504; Fig. 19-14 and 19-15, p. 505). Kühn (Distribution maps).

Index

Abbott's babbler *(Trichastoma abbotti)*, **235**
Abyssinian ground hornbill *(Bucorvus abyssinicus)*, 31*, 58*, 59m, 60f
Abyssinian white-eye *(Zosterops abyssinica)*, **343**
Acanthis (Linnets), 404, **411**f
— *cannabina* (Common linnet), 391*, 404, **411**, 411m
— *flammea* (Common redpoll), 391*, 405f, **411**, 411m
— *flavirostris* (Twite), 391*, 404, **411**, 411m
— *hornemanni* (Arctic redpoll), **411**, 411m
— *johannis* (Warsanglia linnet), **411**
— *yemenensis* (Yemen linnet, **411**
— *chrysorrhoa* (Yellow-tailed thornbill), 259*
— *nana* (Little thornbill), **265**
Acanthorynchus supercilious Western spinebill, 345f
— *tenuirostris* (Eastern spinebill), 346
Accentors (Prunellidae), 219, **229**f, 229m
Aceros nipalensis (Rufous-necked hornbill), **55**
Acorn woodpecker *(Melanerpes formicivorus)*, 104f, 105*
Acridotheres (True mynahs), 471f
— *cristatellus* (Chinese crested mynah), **472**
— *ginginianus* (Bank mynah), **472**
— *tristis* (Common mynah), 471f
— *tristis melanosternum*, 449*
Acrocephalus
— *aedon* (Thick-billed warbler), **253**, 255
— *agricola* (Indian reed warbler or paddy field warbler), **253**, 255
— *arundinaceus* (Great reed warbler), 231*, 242*, 247*, **253**, 255
— — *arundiceus*, 245m
— — *orientalis* (Eastern great warbler), 245m
— *baeticatus* (African reed warbler), 252f
— *dumetorum* (Blyth's reed warbler), **253**, 255f
— *griseldis* (Iraq great reed warbler), **253**, 255
— *melanopogon* (Moustached warbler), 231*, **252**, 254f, 256*
— *paludicola* (Aquatic warbler), 231*, **252**, 253m, 254, 256*
— *palustris* (Marsh warbler), 232*, **253**, 254m, 255f
— *schoenobaenus* (Sedge warbler), 231*, **252**, 253m, 254, 256*
— *scirpaceus* (Reed warbler), 232*, **253**f, 253m, 255, 257*
— *stentoreus* (Egyptian reed warbler), **253**
Acromydae, 121
Acropternis
— *orthonyx* (Ocellated tapaculo), 140
Aegintha temporalis (Sydney waxbill), 439*, **455**
Aegithalidae (Long-tailed tits), 310f
Aegithalos caudatus (Long-tailed tit), 308*, **311**, 311m
— *concinnus* (Red-headed tit), 308*, **311**, 311m
Aegithina (Ioras), 204f
— *tiphia* (Common iora), 190*, **205**

Aerops albicollis (White-throated bee-eater), 42
— *boehmi* (Boehm's bee-eater), 42
Aethopyga, **338**
— *siparaja* (Yellow-backed sunbird), **338**
African broadtails *(Smithornis)*, 123, 127
African gray tit *(Parus griseiventris)*, 318
African gray-rumped swallow *(Pseudhirundo griseopyga)*, 186
African paradise flycatcher *(Terpsiphone viridis)*, 282f
African penduline tit *(Anthoscopus caroli)*, 313, 314*
African piculet *(Verreauxia africana)*, 89
African pitta *(Pitta angolensis)*, **142**
African reed warbler *(Acrocephalus baeticatus)*, 252f
African river martin *(Pseudochelidon eurystomina)*, 185
African rough-winged swallows *(Psalidoprocne)*, 186
African silver-bill *(Eudice cantans)*, **447**, 447m
African tit flycatchers *(Parisoma)*, 276
African white-necked raven *(Corvus albicollis)*, **518**, 521
African yellow-throated sparrow *(Petronia superciliaris)*, **423**, 423m
Agelaius (American blackbirds), 397
— *icterocephalus* (Yellow-headed blackbird), 381*, **401**
— *phoeniceus* (Red-winged blackbird), **400**f
— *thilius* (Yellow-winged blackbird) **401**
Agriornis lividus (Great shrike-tyrant), 151
Aidemosyne modesta (Cherry finch), 454f
Ailuroedus (Catbirds), 500f
— *buccoides* (White-eared catbird), **503**
— *crassirostris* (Green catbird), 500f
— — *melanocephalus* (Black-eared green catbird), 492*
Aitlinger, 267
Ajax scrub robin *(Cinclosoma ajax)*, **249**
Akepa (*Loxops coccinea*), 372*, **387**, 388*, 389
Akiapolaau *(Hemignathus wilsoni)*, 387f
Akiola *(Hemignathus obscurus)*, 372*, **387**, 388*
Alaemon alaudipes (Bifasciated lark), **173**, 173m, 179*
Alauda arvensis (Skylark), 169*, 177ff, 177m, 179*
Alaudidae (Larks), 171ff
Alcedinidae (Kingfishers), 21, **22**f
Alcedininae (Water kingfishers), 22
Aledo atthis (Eurasian kingfisher), 22ff, 26m, 37*
Alcippe (Nun babblers), 245
— *nipalensis* (Nepal babbler), 222*
— *poioicephala* (Nun babbler), **245**
Alethe
— *diadema* (Fire-crest aletha), 287ff
Allen, A. A., 410, 412
Alpine accentor *(Prunella collaris)*, 221*, **230**
Alpine chough *(Pyrrhocorax graculus)*, 469*, **513**f

Alpine nutcracker *(Nucifraga caryocatactes caryocatactes)*, 512f
Amadina
— *erythrocephala* (Red-headed finch), **447**, 447m
— *fasciata* (Cut-throat finch), 439*, **447**, 447m
Amakihi *(Viridonia virens* or *Loxops virens)*, **387**, 388*, 389
Amandava
— *amandava* (Avadavat), 440*, **455**, 455m
— *formosa* (Green avadavat), **455**
— *subflava* (Golden-breasted waxbill), 440*, **455**
Amazon kingfisher *(Chloroceryle amazona)*, 26m, **27**, 37*
Amazonian umbrellabird *(Cephalopterus ornatus)*, 144*, **161**f, 162*
Amblyornis
— *inornatus* (Gardener bower bird), **504**, 504*
— *macgregoriae* (MacGregor's bower bird), 503f, 503*
— *subalaris* (Striped bower bird), 492*, 503*, **504**, 504*
Amblyospiza albifrons (Black swamp weaver or white-fronted grosbeak), 430*, **442**
Amblyospizinae (Swamp weavers), 396, 422, **442**
American blackbirds *(Agelaius)*, 397
American dipper *(Cinclus mexicanus)*, 219
American goldfinch *(Carduelis tristis)*, 391*, **409**f
American orioles *(Icterus)*, 397, 400
American redstarts *(Setophaga)*, 380
American robin *(Turdus migratorius)*, 290*, **304**, 309
Ammomanes deserti (Desert lark), 172f, 173m
Amnospiza (Sharp-tailed sparrows), 354
— *maritima* (Seaside sparrows), 342*, **354**, 357
Ampeliceps coronatus (Gold-crested crackle), **473**
Amytornis (Grass wrens), 266
— *textilis* (Western grass wren), **266**
Anaplectes, **438**
— *rubriceps* (Red-headed weaver), 430*, **442**
Andean cock-of-the-rock *(Rupicola peruviana)*, 133*, **161**
Andean flicker *(Colaptes rupicola)*, 96
Andigena (Mountain toucans), 80
— *laminirostris* (Plate-billed mountain toucan), 81*
— *nigrirostris* (Black-billed toucan), 80, 107*
Angola red-billed shrike *(Prionops gabela)*, 208m
Angola swallow *(Hirundo angolensis)*, 187
Angola waxbill *(Uraeginthus angolensis)*, **457**
Anisognathus
— *flavinucha* (Blue-winged mountain tanager), 335*, 361*
— *igniventris* (Scarlet-bellied mountain tanager), 377
Anomalospiza imberbis (Parasitic weaver), **427**, 430*
Anomalospizinae (Cuckoo weavers), 422, **427**, 427m
Ant catchers *(Pithys)*, 137

Ant thrushes *(Neocossyphus)*, 294
Antbirds (Formicariidae), 127, 132f, 137m
Antbirds *(Hylophylax)*, 137ff
Anteater chat *(Myrmecocichla aethiops)*, **295**, 297
Anthochaera (Wattlebirds), 345
— *chrysoptera* (Little wattlebird), 346
Anthoscopus caroli (African penduline tit), 313, 314*
— *minutus* (Cape penduline tit), 313
— *musculus* (Mouse-colored penduline tit), 308*
Anthreptes, **338**
— *malacensis* (Plain-throated sunbird), **338**
— *singalensis* (Ruby-cheeked sunbird), **338**
Anthus (Pipits), 193f
— *campestris* (Tawny pipit), 189*, **193**, 197m
— *cervinus* (Red-throated pipit), 189*, **193**, 193m
— *novaeseelandiae* (New Zealand pipit), 189*, **194**
— *pratensis* (Meadow pipit), 189*, **193**, 193m
— *spinoletta* (Water pipit), 184*, 189*, **193**, 193m
— *trivialis* (Tree pipit), 189*, **193**f, 194*, 194m, 197*
Antilophia galeata (Helmeted manakin), **158**
Antpecker *(Parmoptila woodhousei)*, **462**
Antpipits *(Corythopis)*, 140, 152f
Antpittas *(Grallaria)*, **137**, **139**
Antshrikes *(Thamnophilus)*, 132
Ant-tanagers *(Habia)*, 376
Antthrushes *(Formicarius)*, **137**, **139**
Antwrens *(Myrmotherula)*, 132f
Anumbius
— *annumbi* (Firewood-gatherer), 125*
Apalis (Apalis), 250
— *flavida* (Yellow-chested apalis), 247m, **250**f
— *thoracica* (Yellow bar-throated apalis), 248*, **250**
Apalopteron (*Apalopteron familiaris*), 347
Apalopteron familiaris (Apalopteron), 347
Apapane *(Himatione sanguinea)*, 372*, 388, **390**, 390*, 393
Aphelocoma
— *coerulescens* (Scrub jay), **507**
— *ultramarina* (Mexican jay), **507**
— *unicolor* (Unicolored jay), 497*, **507**
Aplonis brunneicapilla (White-eyed starling), **473**
— *cantaroides* (Singing starling), 449*
— *metallica* (Shining starling), **473**
Apostle bird *(Struthidea cinerea)*, 469*, **484**
Aquatic warbler *(Acrocephalus paludicola)*, 231*, **252**, 253m, 254, 256*
Aracaris (*Pteroglossus*), 80f
Arachnothera (Spider hunters), 332f, **338**f
— *chrysogenys* (Lesser yellow-eared spider hunter), 320*, **339**
Archboldia papuensis (Archbold bower bird), **503**
— — *sanfordi* (Sanford's bower bird), **503**

Heavy type indicates the main entry, an asterisk* indicates an illustration, and m indicates a distribution map.

626 INDEX

Archbold's bower bird *(Archboldia papuensis)*, **50**
Arctic warbler *(Phylloscopus borealis)*, 249f, **263**, 263m
Arena birds, 500
Arfak astrapia *(Astrapia nigra)*, 491*, **499**
Arfak six-wired parotia *(Parotia sefilata)*, 470*, **495**
Arfale buff-faced scrub wren *(Sericornis rufescens)*, **265**
Ariel toucan *(Ramphastos vitellinus aiel)*, 98*
Arremon
— *aurantirostris* (Orange-billed sparrow), **342***
Arses telescophthalmus (Frilled flycatcher), **278**
Artamidae (Wood swallows or swallow shrikes), 482, 484f, 485m
Artamus
— *fuscus* (Ashy swallow shrike), **485**
— *leucorhynchus* (White-breasted swallow shrike), **485**
— *minor* (Little wood swallow), **485**
— *superciliosus* (White-browed wood swallow), 196*, **485**
Aru Island greater bird of paradise *(Paradisaea apoda apoda)*, 493
Arundinicola leucocephala (Whiteheaded marsh-tyrant), 143*, **151**
Aschenbrenner, 264
Ashy drongo *(Dicrurus leucophaeus)*, **477**, 481
Ashy minivet *(Pericrocotus roseus divaricatus)*, 198, 450*
Ashy swallow shrike *(Armatus fuscus)*, **485**
Asiatic paradise flycatcher *(Terpsiphone paradisi)*, 260*, **282f**, 282*
Asitys (Philepittidae), 127, 127m, **147**
Aspatha gularis (Blue-throated motmot), **35**, 48*
Asthenes (Twelve-feathered spine-tails), 131
— *dorbygnyi* (Creamy-breasted canastero), **131**, 137f
Astrapia (Astrapias), 496f
— *mayeri* (Ribbon-tailed astrapia), 480*, **496f**
— *nigra* (Arfak astrapia), 491*, **499**
— *stephaniae* (Princess Stephanie astrapia), **499**
Astrapias *(Astrapia)*, 496f
Atelornis crossleyi (Crossley's ground roller), **43**, 57*
— *pittoides* (Blue-tailed ground roller), **43**
Atlapetes (Brush finches), 354
— *brunneinucha* (Chestnut-capped brush finch), **354**
Atrichornis (Scrub-birds)
— *clamosus* (Western scrub bird), **166**, 166m, 166*
— *rufescens* (Rufous scrub-bird), 150*, **166**, 166m
Atrichornithidae (Scrub birds), 165f
Attwell, G. D., 115
Aulacorhynchus (Green toucanets), 79f
— *caeruleocinctus* (Blue-banded toucanet), 107*
— *prasinus* (Emerald toucanet), **80**, 107*
— *sulcatus* (Groove-billed toucanet), **80**
Auriparus flaviceps (Verdin), 308*, **313f**
Aurora waxbill *(Pytilia phoen-*

ieoptera), **461**
Australian butcher birds *(Cracticus)*, 485
Australian chats *(Ephthianura)*, 266
Australian great bower bird *(Chlamydera nuchalis)*, **505**
Australian magpie-larks (Grallinidae), 482, **483f**
Australian magpies *(Gymnorhina)*, 485ff
Australian wrens or wren warblers (Malurinae), 233, **265f**, 265m
Avadavat *(Amandava amandava)*, 440*, **455**, 455m
Azores bullfinch *(Pyrrhula pyrrhula murina)*, **342***
Azure jay *(Cyanocorax caeruleus)*, 497*
Azure nuthatch *(Sitta azurea)*, **323**
Azure tit *(Parus cyanus)*, 307*, **318f**
Azure-winged magpie *(Cyanopica cyana)*, 497*, **510**, 510m

Babax lanceolatus (Chinese babax), **243**
Babbling thrushes or babblers (Timaliinae), 233f, 234m
Babbling warbler *(Hippolais caligata)*, **254**, 258
Baillonius bailloni (Saffron toucan), 80, 107*
Baker, 76
Bald-headed wood shrike *(Pityriasis gymnocephala)*, **213**, 213m, 449*
Baldwin, Paul H., 393f
Bali mynah *(Leucopsar rothschildi)*, 449*, **471**
Baltimore oriole *(Icterus galbula)*, 381*, **400**
Bamboowren *(Psilorhamphus guttatus)*, 140
Bananaquit *(Coereba flaveda)*, 371*, **386**, 386*
Bananaquits (Coerebinae), 380, **385f**, 386m
Band-backed wren *(Campylorhynchus zonatus)*, **224**
Banded aracari *(Pteroglossus acracari)*, 107*
Banded broadbill *(Eurylaimus javanicus)*, **120***
Banded cotinga *(Cotinga maculata)*, 161f
Banded woodpeckers, 96, 104ff
Bank mynah *(Acridotheres ginginianus)*, **472**
Bank swallow or sand martin *(Riparia riparia)*, 180*, **186**, 186m
Bannerman, D., 407f, 412, 441, 456, 458
Barbets (Capitonidae), 63, **65ff**
Bar-breasted fire finch *(Logonosticta rufopicta)*, **458**
Bare-crowned antbird *(Gymnocichla nudiceps)*, 126*
Bare-eyed thrush *(Turdus nudigenis)*, 289*
Bare-eyes *(Phlepogsis)*, 137
Bare-throated bellbird *(Procnias nudicollis)*, 136*, 144*, **161**, 163*
Barn swallow *(Hirundo rustica)*, 169*, 170*, 180*, **187ff**, 187m
Barred antshrike *(Thamnophilus doliatus)*, 126*, **132**, 138*
Barred cuckoo-shrike *(Coracina lineata)*, 450*
Barred fruiteater *(Pipreola arcuata)*, 161
Barred puffbird *(Nystalus radiatus)*, **64**, 68*
Barred warbler *(Sylvia nisoria)*, 232*, **258**

Barred woodcreeper *(Dendrocolaptes certhia)*, **128***
Barred-winged wren babbler *(Spelaeornis troglodytoides)*, **238**
Bartels, H., 56
Bar-winged weaver *(Notiospiza angolensis)*, 442
Baryphthengus ruficapillus (Rufous motmot), 32*, **36**
Basileuterus, 380
Basilornis celebensis
— *corythaix*, **473**
Batara
— *cinerea* (Giant antshrike), **132**
Bates, Henry Walter, 224, 420
Bathilda ruficauda (Star finch), **455**
Batis (Puff-backs), 277
— *capensis* (Puff-back flycatcher), 260*, **277**, 277*
— *molitor* (Chin-spot puff-back flycatcher), 277
— *pririt* (Pritit puff-back flycatcher), 277
Baya weaver *(Ploceus philippinus)*, **438**
Bay-winged cowbird *(Molothrus badius)*, **398**, 398m, 399*
Bearded barbet *(Lybius dubius)*, 70f
Bearded or mossy-throated bellbird *(Procnias averano)*, 144*, **161**, 163*
Bearded tit *(Panurus biarmicus)*, 222*, **247f**
Beauharnaisius
— *beauharnaesii* (Curl-crested aracari), 107*
Beautiful nuthatch *(Sitta formosa)*, **323**
Becards *(Pachyramphus)*, 161f
Beccari, C., 499, 504
Beecher, W. J., 378
Bee-eaters (Meropidae), 21, **36f**
— *(merops)*, 39f
Bell miner *(Manorina melanophrys)*, 345
Bellbirds *(Procnias)*, 161ff
Belted kingfisher *(Megaceryle alcyon)*, 26f, 26m, 37*
Bent, 226
Berenicoris comatus (White-crested hornbill), **55**
Berndt, Rudolf, 146
Bernier's vanga *(Oriolia bernieri)*, **214**
Berry peckers *(Melanocharis)*, 330
Bib finch *(Spermestes nanus)*, **447f**, 448m
Bicheno's finch *(Stizoptera bichenovii)*, **454**, 454m
Bicolored conebill *(Conirostrum bicolor)*, **385**
Bier, August, 508
Bifasciated lark *(Alaemon alaudipes)*, **173**, 173m, 179*
Billeb, 364
Birds of paradise (Paradisaeinae), 487ff, 488m
Birds of paradise and bower birds (Paradisaeidae), 482, **487f**
Bishops *(Euplectes)*, 433, **435f**
Bishops (Euplectinae), 422, **433f**
Black bee-eater *(Melittophagus gularis)*, **41**
Black berry pecker *(Melanocharis nigra)*, **331**
Black buffalo weaver *(Bubalornis albirostris)*, **442f**, 443m
Black bulbul *(Hypsipetes madagascariensis)*, 203
Black bush robin *(Cercotrichas podobe)*, **286**
Black butcher bird *(Cracticus quoyi)*, **485**

Black crow *(Corvus capensis)*, **515**
Black cuckoo-shrike *(Campephaga flava)*, 198f
Black flycatcher *(Melaenornis pammelaina)*, **277**
Black indigo bird *(Hypochera nigerrima)*, 429*
Black mamo *(Drepanis funerea)*, **390**, 394*
Black nunbird *(Monasa atra)*, **65**
Black phoebe *(Sayornis nigricans)*, 151, 153
Black redstart *(Phoenicurus ochruros)*, 291f, 292m, 293*
Black rough-winged swallow *(Psalidoprocne holomelaena)*, 180*
Black sickle-billed bird of paradise *(Epimachus fastuosus)*, **499**
Black swamp weaver or whitefronted grosbeak *(Amblyospiza albifrons)*, 430*, **442**
Black tit *(Parus leucomelas)*, **318**
Black woodpecker *(Dryocopus martius)*, 93*, 94*, **102f**, 102m, 103*, 108*
Black woodpecker *(Dryocopus)*, 102ff
Black-and-blue jay *(Cissilopha sanblasiana)*, **507**
Black-and-orange flycatcher *(Ficedula mugimaki)*, **275**
Black-and-red broadbill *(Cymbirhynchus macrorhynchos)*, 124
Black-and-red honeycreeper (Orepanidinae), 387, **390**
Black-and-white mannikin *(Spermeste bicolor)*, **447**, 447m
Black-and-white warbler *(Mniotilta varia)*, 371*, 383f, 384*
Black-and-yellow broadbill *(Eurylaimus ochromalus)*, 124
Black-and-yellow grosbeak *(Mycerobas icterioides)*, **419f**
Black-and-yellow silky flycatcher *(Phainoptila melanoxantha)*, 216f
Black-backed magpie *(Gymnorhina tibican)*, **486**
Black-backed puff back *(Dryoscopus cubla)*, 195*, **208**
Black-backed three-toed woodpecker *(Picoides arcticus)*, **115**
Black-backed woodpecker *(Chrysocolaptes festivus)*, **116**, 116m
Black-bellied seed-cracker *(Pyrenestes ostrinus)*, 440*, **461**
Black-billed toucan *(Andigena nigrirostris)*, 80, 107*
Blackbird *(Turdus merula)*, 290*, 304, 305m, 306f
Blackbirds (American) see American blackbirds
Black-breasted barbet *(Lybius rolleti)*, **70**
Black-breasted puff bird *(Notharchus pectoralis)*, **64**
Black-breasted triller *(Clamydochaera jefferyi)*, 198*
Black-browed jungle babbler *(Trichastoma perspicillatum)*, **235**
Blackcap *(Sylvia atricapilla)*, 232*, 258, 261, 413*
Black-capped chickadee *(Parus atricapillus)*, **317**
Black-capped foliage-gleamer *(Philydor atricapillus)*, 125*
Black-capped mocking thrush *(Donacobius atricapillus)*, 221*, **227**
Black-capped pygmy tyrant *(Perissotriccus atricapillus)*, 152
Black-capped wren *(Thryothorus nigricapillus)*, **224f**
Black-casqued hornbill *(Ceratogymna atrata)*, **54f**, 54m, 58*

INDEX 627

Black-casqued hornbills *(Cerato-gymna),* 54
Black-cheeked ant tanager *(Saltator atriceps),* 352*
Black-cheeked waxbill *(Estrilda erythronotos),* **428, 457**
Black-crested magpie *(Platysmurus leucopterus),* **506**
Black-crowned waxbill *(Estrilda nonnula),* **457**
Black-eared bush tit *(Psaltriparus melanotis),* **311**
Black-eared green catbird *(Ailuroedus crassirostris melanocephalus),* 492*
Black-eared wheatear *(Oenanthe hispanica),* **295,** 295m, 297f
Black-faced antthrush *(Formicarius analis),* **126**
Black-faced cuckoo-shrike *(Coracina novaehollandiae)* 198f
Black-faced flycatcher *(Monarcha melanopsis),* 281f, 282*
Black-fronted weaver *(Textor velatus),* **441**
Black-gorgeted laughing thrush *(Garrulax pectoralis),* **243**
Black-headed babbler *(Rhopocichla atriceps),* **239**
Black-headed bee-eater *(Bombylonax breweri),* **42**
Black-headed bulbul *(Pycnonotus atriceps),* **200**
Black-headed bunting *(Emberiza melanocephala),* 341*, **350,** 350m, **353**
Black-headed forest oriole *(Oriolus monacha),* **476**
Black-headed goldfinch *(Carduelis carduelis carduelis),* **409**
Black-headed grosbeak *(Pheucticus melanocephalus),* **367**
Black-headed pitohui *(Pitohui dichrous),* 260*, **284**
Black-headed starling *(Temenuchus pagodarum),* 449*, **471**
Black-headed village weaver *(Textor cucullatus),* 430*, **441**f
Black-headed waxbill *(Estrilda atricapilla),* **457**
Black-headed weaver *(Textor melanocephalus),* **441**
Black-naped flycatcher *(Hypothymis azurea),* 260*, **281**
Black-naped green woodpecker *(Picus canus),* 84*, 94*, **99**ff, 102*, **119**
Black-naped oriole *(Oriolus chinensis),* 450*, **476**
Black-necked mynah *(Gracupica nigricollis),* 449*, **471**
Black-necked red cotinga *(Phoenicircus nigricollis),* 144*
Black-spotted barbet *(Capito niger),* **69**f
Black-spotted yellow tit *(Parus xanthogenys),* **318**
Blackstart *(Cercomela melanura),* 295f
Black-tailed gnatcatcher *(Polioptila melanura),* **271**
Black-tailed leaf-scraper *(Sclerurus caudacutus),* **132**
Black-tailed tityra *(Tityra cayana),* 144*
Black-throated blue warbler *(Dendroica caerulescens),* 371*
Black-throated finch *(Melanodera melanodera),* **354, 357**
Black-throated honeyguide *(Indicator indicator),* 67*, **77**f
Black-throated jay *(Garrulus lanceolatus),* **508**

Black-throated warbler *(Gerygone palpebrosa),* 259*
Black-throated wattle-eyed flycatcher *(Platysteira peltata),* **277**
Blanford's snow finch *(Montifringilla blanfordi),* **424,** 424m
Blue bird of paradise *(Paradisaea rudolphi),* 470*, **493**
Blue dacnis *(Dacnis cayana),* 362*, **379**
Blue flycatcher *(Cyanoptila cyanomelana),* 260*, **275**
Blue grosbeak *(Guiraca caerulea),* 352*, **367, 369**
Blue jay *(Cyanocitta cristata),* 497*, **506**f
Blue magpie *(Urocissa caerulea),* **510**
Blue mockingbird *(Melanotis caerulescens)* 221*
Blue rock thrush *(Monticola solitarius),* 289*, **298**f, 298m
Blue tit *(Parus caeruleus),* 307*, **316**
Blue vanga *(Leptopterus madagascarinus),* 196*, **213**f
Blue vangas *(Leptopterus),* **213**
Blue whistling thrush *(Myiophoneus caeruleus),* 280*, **299**
Blue-and-yellow tanager *(Thraupis bonariensis),* 335*, **375**
Blue-backed fairy bluebird *(Irena puella),* 184*, 190*, **206**
Blue-banded toucanet *(Aulacorhynchus caeruleocinctus),* 107*
Blue-bellied roller *(Coracias cyanogaster),* **44,** 44m
Blue-bill *(Spermophaga haematina),* **461**
Blue-billed fire finch *(Lagonosticta rubricata),* 440*, **458,** 458m
Bluebirds *(Sialia),* **292**
Blue-black grassquit *(Volantinia jacarina),* 342*, **358**
Blue-cheeked barbet *(Megalaima asiatica),* **75,** 97*
Blue-cheeked bee-eater *(Merops superciliosus),* **40,** 40m
Blue-crested waxbill *(Uraeginthus cyanocephalus),* **457**
Blue motmot *(Momotus momota),* 32*, **36,** 36*, 48*
Blue-eared glossy starling *(Lamprotornis chalybaeus),* **475**
Blue-faced honey eater *(Entomyzon cyanotis),* **346**
Blue-faced parrot finch *(Erythura trichroa),* **453**
Blue-gray gnatcatcher *(Polioptila caerulea),* 259*, **271,** 271m
Blue-gray tanager *(Thraupis episcopus),* 361*, **374**
Blue-headed bee-eater *(Melittophagus muelleri),* **41**
Blue-headed bee-eater *(Nyctyornis athertoni),* **42**
Blue-naped pitta *(Pitta nipalensis),* 149*
Blue-tailed ground roller *(Ateloenis pittoides),* **43**
Blue-tailed pitta *(Pitta guajana),* 133*, **142,** 149*
Bluethroat *(Luscinia svecica),* 279*, **286**f, 286m
Blue-throated flycatcher *(Cyornis rubeculoides),* 260*, **275**
Blue-throated motmot *(Aspatha gularis),* **35,** 48*
Blue-throated roller *(Eurystomus gularis),* **45,** 45m
Blue-winged kookaburra *(Dacelo teachi),* **33**
Blue-winged mountain tanager *(Anisognathus flavinucha),* 335*, 361*

Blue-winged pitta *(Pitta brachyura),* **142,** 145m, **145**
Blue-winged warbler *(Vermivora pinus),* 383*, **384**
Blythipicus *(Maroon woodpeckers),* **115**
— *rubiginosus* (Maroon woodpecker), 115f
Blyth's bulbul *(Pycnonotus flavescens),* **203**
Blyth's hornbill *(Rhyticeros plicatus),* **56**
Blyth's jungle babbler *(Trichastoma rostratum),* **235**
Blyth's reed warbler *(Acrocephalus dumetorum),* **253,** 255f
Boat-billed flycatcher *(Megarynchus pitangua),* **151**
Boat-tailed grackle *(Cassidix mexicanus),* 381*
Bobolink *(Dolichonyx oryzivorus),* 381*, **397, 401**
Boehm's bee-eater *(Aerops boehmi),* **42**
Boem, E. M., **26**
Bohemian waxwing *(Bombycilla garrulus),* 196*, **215**f, 215m
Bombycilla cedrorum (Cedar waxwing), **215,** 215m
— *garrulus* (Bohemian waxwing) 196*, **215**f, 215m
— *japonica* (Japanese waxwing), **215,** 215m
Bombycillidae (Waxwings), **207, 215**f
Bombylonax breweri (Black-headed bee-eater), **42**
Bonelli's warbler *(Phylloscopus bonelli),* 259*, **263**f, 264f
Bonin finch *(Chaunoproctus ferreorostris),* **404, 418**
Bower birds *(Ptilonorhynchinae),* **499**f, 500m
Bowman, 363ff
Brachycope, **433**
— *anomala,* **436**
Brachygalba
— *lugubris* (Brown jacamar), **63**
Brachypteracias leptosomus (Short-legged ground roller), **43,** 57*
— *squamigera* (Scaled ground roller), **43**
Brachypteraciinae (Ground rollers), **42,** 43f
Brachypteryx (Shortwings), **285**
— *leucophrys* (Lesser shortwing), **285**
Bradornis mariquensis (Mariqua flycatcher), **277**
— *pallidus* (Pale flycatcher), **277**
Bradypterus (Rush warblers), **253**
— *baboecala* (Little rush warbler), **253**
Brambling *(Fringilla montifringilla),* 336*, 392*, **402**f, 402m
Brazilian tanager *(Ramphocelus carbo bresilius),* 335*
Brehm, Alfred Edmund, **51, 403, 412, 435**
Bridled titmouse *(Parus wollweberi),* **318**
Bright-green leaf warbler *(Phylloscopus nitidus),* **263**
Bristle-nosed barbet *(Gymnobucco peli),* **73,** 97*
Broad-billed motmot (Electron platyrhynchum), 35*, **36**
Broad-billed roller *(Eurystomus glaucurus),* **45,** 45m
Broad-billed rollers *(Eurystomus),* **45**
Broad-billed tody *(Todus subulatus),* **34,** 48*
Broadbills (Desmodactylae), **118,** 123f

Broadbills (Eurylaimidae), **123,** 123m
Broad-tailed paradise whydah *(Steganura obtusa),* 429*, 433*
Broadtails (African) (see African broadtails)
Bronze mannikin *(Spermestes cucullatus),* 439*, **447,** 448m
Bronze-backed flower pecker *(Dicaeum ignipectus),* **329**
Brown barbet *(Caloramphus fuliginosus),* 76f
Brown dipper *(Cinclus pallasii),* **219**
Brown flycatcher *(Muscicapa latirostris),* **276**
Brown honey eater *(Lichmera indistincta),* **345**
Brown jacamar *(Brachygalba lugubris),* **63**
Brown shrike *(Lanius cristatus),* 195*, **212**
Brown sickle-billed bird of paradise *(Epimachus meyeri),* 479*, **499**
Brown thrasher *(Toxostoma rufum),* 221*
Brown tree creeper *(Climacteris picumnus),* **325**
Brown willow warbler *(Phylloscopus fuscatus),* **263**
Brown-capped weaver *(Phormoplectes insignis),* 430*, **442**
Brown-crested tit *(Parus dichrous),* **318**
Brown-crowned eremomela *(Eremomela badiceps),* 249*, **250**
Brown-headed chickadee *(Parus hudsonicus),* **318**
Brown-headed cowbird *(Molothrus ater),* **399,** 399m, 399*
Brown-headed nuthatch *(Sitta pusseilla),* **323**
Brown-necked raven *(Corvus ruficolis),* **518,** 518m
Brown-throated tree creeper *(Certhia discolor),* **326**
Brown-white-throated bulbul *(Criniger ochraceus),* **200**
Brubru *(Nilaus afer),* **208**
Brush finches *(Atlapetes),* **354**
Bubalornis albirostris (Black buffalo weaver), **442**f, 443m
Bubalornithinae (Buffalo weavers), **422, 442**
Bucco capensis (Collared puffbird), **64,** 68*
Bucconidae (Puffbirds), **62, 64**f
Buceros bicornis (Great Indian hornbill), 47*, **56, 59**
— — *bicornis* (Lesser great Indian hornbill), **56,** 56m
— *homrai* (Homrai), **56,** 56m
— *hydrocorax* (Philippine hornbill), **56**
— *rhinoceros* (Rhinoceros hornbill), **56**
Bucerotidae (Hornbills) **21, 49**m, **49**ff
Bucorvus (Ground hornbills), 60f
— *abyssinicus* (Abyssinian ground hornbill), 31*, 58*, **59,** 60f
— *cafer* (Cape ground hornbill), 59m, **61**
Buffalo weavers (Bubalornithinae), **422, 442**
Bukharah great tit *(Parus bokharensis),* **318**
Bulbuls (Pycnonotidae), **192, 200**f
Bulbubs proper *(Pycnonotus),* **200**
Bullfinch *(Pyrrhula pyrrhula),* 336*, **419,** 419*
Bullfinch-billed reed bunting *(Em-*

beriza schoeniclus pyrrhyloides), 350
Bullfinches *(Pyrrhula)*, 404, 406, **418f**
Bullock's oriole *(Icterus bullocki)*, **400**
Buntings, i.n.s. (Emberizini), 350f
Buphaginae (Tick birds or oxpeckers), 463, **475**f
Buphagus africanus (Yellow-billed ox-pecker), **449***, **475**, 475m
— *erythrorhynchus* (Red-billed ox-pecker), **475**, 475m
Burchell's glossy starling *(Lamprotornis australis)*, **475**
Bush and rock wrens *(Xenicus)*, 147f
Bush creepers *(Macrosphenus)*, 250
Bush shrikes *(Mala conotinae)*, 207, 208f
Bush tanagers *(Chlorospingus)*, 376
Bush warbler *(Cettia diphone)*, **250**
Bush wren *(Xenicus longipes)*, 147f, 147*, **150***
Butcher birds (Australian) (see Australian butcher birds)
Bycanistes (Trumpeter hornbills), 54
— *buccinator* (Trumpeter hornbill), **54**f, 54m, 58*

Cacicus, 396
— *cela* (Yellow-rumped cacique), 381*, **398**
Cactospiza (Woodpecker finches)
— *heliobates* (Mangrove finch), 360, 364
— *pallida* (Woodpecker finch), 333*, 351*, **360**, 363ff, 365*
Cactus ground finch *(Geospiza scandens)*, 351*, **359**f, 363f, 364*, 367*
Cactus wren *(Campylorhynchus brunneicapillus)*, 190*, **223**
Cade, T. J., 444, 453
Calamocichla (Swamp warblers), 253
— *gracilirostris* (Greater swamp warbler), **253**, 254m
Calandra lark *(Melanocorypha calandra)*, **174**, 174m, 179*
Calandrella acutirostris
— *brachydactyla* (Short-toed lark), 174f, 174m, 179*
— *rufescens* (Lesser short-toed lark), **174**
Calcarius (Longspurs), 353
— *lapponicus* (Lapland longspur), 342*, **355**, 355m, 355*
Calicalicus madagascariensis Red-tailed vanga), 214
California thrasher *(Toxostoma redivivum)*, **227**f, 227*
Callaeas cinerea (Kokako), 469*, **483**
Callaeidae (New Zealand wattlebirds), **482**f
Calocitta formosa (Magpie jay), **508**
Caloramphus fuliginosus (Brown barbet), 76f
Calyptomena, 123
— *hosii* (Hose's broadbill), 124
— *viridis* (Green broadbill), 120*, 124*, **127**, 133*
Calyptomeninae (Green broadbills), 123
Calyptophilus
— *frugivorus* (Chat tanager), 370
Camaroptera (Camaropteras), 250
— *brachyura* (Green-backed camaroptera), **250**
— *brevicaudata* (Gray-backed camaroptera), 248m, **250**, 250*
Camarhynchus (Insectivorous tree finches), **360**, 364
— *parvulus* (Small insectivorous tree finch), 351*, **360**, 365*

— *pauper* (Medium insectivorous tree finch), 365*
— *psittacula* (Large insectivorous tree finch), **360**, 360*, 364*
Cameroons indigo bird *(Hypochera chalybeata)*, **428**, 429*
Campephaga (Cuckoo-shrikes proper), 198
— *flava* (Black cuckoo-shrike), 198f
— *lobata* (Wattled cuckoo-shrike), **198**f
— *phoenicea* (Cuckoo-shrike), 450*
Campephagidae (Cuckoo-shrikes), 192, **198**f
Campephagini, 198
Campephilus, 116
— *imperialis* (Imperial woodpecker), **116**
— *magellanicus* (Magellanic woodpecker), **116**
— *principalis* (Ivory-billed woodpecker), 108*, **116**, 116m, 116*
Campethera
— *maculosa* (Golden-backed woodpecker), **99**
— *nivosa* (Termite woodpeckers), 99m, **99**
— *nubica nubica* (Spotted woodpecker), 108*
Campochaera sloetii (Orange cuckoo-shrike), 198
Campos flicker (Colaptes campestris), **96**
Campylorhamphus
— *trochilirostris* (Red-billed scythebill), 120*, 128*
Campylorhynchus brunneicapillus (Cactus wren), 190*, **223**
— *zonatus* (Band-backed wren), **224**
Canary *(Serinus canaria)*, **407**
Canary Island Chaffinch *(Fringilla teydea)*, **402**f
Canyon wren *(Salpinctes mexicanus)*, 190*, **224**
Cape broadbill *(Smithornis capensis)*, 120*, **124**
Cape bunting *(Fringillaria capensis)*, 350
Cape grass bird *(Sphenoeacus afer)*, **251**
Cape ground hornbill *(Bucorvus cafer)*, 59m, **61**
Cape long-claw *(Macronyx capensis)*, **193**
Cape pendulne tit *(Anthoscopus minutus)*, **313**
Cape sparrow *(Passer melanurus)*, 416*, **424**, 425m, 426
Cape sugarbird *(Promerops cafer)*, 320*, **344**, **347**
Cape weaver *(Textor capensis)*, 430*, **441**, 442m
Capito
— *aurovirens* (Scarlet-crowned barbet), **69**
— *niger* (Black-spotted barbet), **69**f
Capitonidae (Barbets), 63, **65**ff
Capuchin bird *(Perissocephalus tricolor)*, 161f
Cardinal *(Cardinalis cardinalis)*, 334*, 352*, **366**ff, 368m
Cardinal quelea *(Queleopsis cardinalis)*, **436**
Cardinal woodpecker *(Dendropicus fuscescens)*, **115**, 115m
Cardinalinae (Cardinals), 368*
Cardinalis, (Cardinals proper)
— *cardinalis* (Cardinal), 334*, 352*, **366**ff, 368m
Carduelinae, 402, **404**, 404m
Carduelis (Goldfinches and siskins), 404, **409**
— *carduelis* (Goldfinch), 334*, 391*,

404*, 405, **409**ff, 409m
— — *caniceps* (Gray-headed goldfinch), 409
— — *carduelis* (Black-headed goldfinch), 409
— *chloris* (Greenfinch), 391*, 404, **409**, 409m, 410*, 413*
— *cucullata* (Red siskin), 391*, **409**f
— *lawrencei* (Lawrence's goldfinch), 409
— *pinus* (Pine siskin), 406, **409**f
— *psaltria* (Lesser goldfinch), **409**
— *spinus* (Siskin), 391*, 404f, **409**, 409m
— *tristis* (American goldfinch), 391*, **409**f
Carib grackle (South-American) (See South American carib grackle)
Carmine bee-eater *(Merops nubicus)*, 40, 48*
Carolina chickadee *(Parus carolinensis)*, **318**
Carolina wren *(Thryothorus ludovicanus)*, 190*, **223**, 225f
Caroline racket-tail *(Tanysiptera sylvia)*, 34, 34m
Carpodacus (Scarlet grosbeaks), 404, **417**
— *cassinii* (Cassin's finch), **417**
— *erythrinus* (Scarlet grosbeak), 391*, **417**, 417m
— *mexicanus* (House finch), 391*, 405, **417**f
— *puniceus*, **417**
— *purpureus* (Purple finch), **417**
Carpospiza brachydactyla (Pale rock sparrow), **423**
Carrion crow *(Corvus corone)*, **515**ff
Cassidix
— *mexicanus* (Boat-tailed grackle), 381*
Cassin's finch *(Carpodacus cassinii)*, **417**
Catamblyrhynchini (Plush-capped finches), 370, **378**, 378m
Catamblyrhynchus
— *diadema* (Plush-capped finch), 362*, **378**
Catbird *(Dumetella carolinensis)*, 221*, **227**f
Catbirds *(Ailuroedes)*, **500**f
Cathurus (New World nightingale-like thrushes), **303**
— *aurantiirostris* (Orange-billed nightingale thrush), 221*, **303**
Cattle tyrant *(Machetornis rixosa)*, 143*, **151**, 153
Cayenne jay *(Cyanocorax cayanus)*, 497*, **507**f
Cayley, 146
Cedar waxwing *(Bombycilla cedrorum)*, **215**, 215m
Celebes roller *(Coracias temminckii)*, 44m
Celebes three-toed kingfisher *(Ceyx fallax)*, **25**
Celebesian starling *(Scissirostrum dubium)*, 449*, **473**
Centurus, 104m
— *carolinus* (Red-bellied woodpecker), 95*, **104**ff, 105*
— *chrysauchen* (Golden-naped woodpecker), **104**
— *cruentatus* (Yellow-tufted woodpecker), **104**
— *uropygialis* (Gila woodpecker), 93*, **104**, 106
Cephalopterus ornatus (Amazonian umbrellabird), 144*, **161**f
— — *penduliger*, 162*
Cephalopyrus flammiceps (Fire-capped tit), **313**

Ceram hornbill *(Rhyticeros plicatus plicatus)*, 50
Ceratogymna (Black-casqued hornbills), 54
— *atrata* (Black-casqued hornbill), **54**f, 54m, 58*
Cercomela (Blackstarts), 295
— *familiaris* (Red-tailed chat), **295**f
— *melanura* (Blackstart), **295**f
Cercotrichas podobe (Black-bush robin), **286**
Cerophagy, 78
Certhia (Creepers)
— *brachydactyla* (Short-toed tree creeper), 308*, **326**ff, 326m
— *discolor* (Brown-throated tree creeper), **326**
— *familiaris* (Tree creeper), 308*, **326**, 326m, 328
— *himalayana* (Himalayan tree creeper), **326**
— *nipalensis* (Stoliczka's tree creeper), **326**f
Certhidea
— *olivacea* (Warbler finch), 351*, **360**, 364, 365*
Certhiidae (True tree creepers), **326**ff
Cercyle, 22
— *lugubris* (Pied kingfisher), 27
— *rudis* (Lesser pied kingfisher), 26m, **27**, 38*
— — *leucomelanura*, 27
Cettia
— *cetti* (Cetti's warbler), 231*, **253**f, 255m
— *diphone* (Bush warbler), **250**
Cetti's warbler *(Cettia cetti)*, 231*, **253**f, 255m
Ceylon jungle babbler *(Turdoides somervillei rufescens)*, 222*
Ceylon magpie *(Urocissa ornata)*, 497*, **510**
Ceylon mynah *(Gracula ptilogenys)*, **473**
Ceyx (Three-toed kingfishers), 25
— *erithacus* (Indian forest kingfisher), 37*
— *fallax* (Celebes three-toed kingfisher), **25**
Chabert vanga *(Leptopterus chabert)*, **213**
Chaetops frenatus (Rufous rock jumper), **286**
Chaetorhynchus (Mountain drongos), 477, 477m
— *papuensis* (Papuan Mountain drongo), 450*, **477**
Chaffinch *(Fringilla coelebs)*, 392*, **402**f, 402m, 403*
Chaffinches (Fringillinae), 402
Chaimarrornis leucocephalus (White-capped redstart), 279*, **291**
Chalcomitra, 338
Chalcoparia, 338
Chamaea (North American wren tits), 239
— *fasciata* (Wren tit), **239**
Chamaeini (Wren tits), 234, **239**
Chamaeza
— *campanisona* (Short-tailed antthrush), 137
Channel-billed toucan *(Ramphastus vitellinus)*, 81*, **85**
Chapin, James P., 71, 124, 478
Chapin, Ruth T., 461
Chapman, G. S., 228, 484
Chasiempis sandwicensis (Elepaio), **278**
Chat shrike *(Laniotardus torquatus)*, **208**
Chat tanager *(Calyptrophorus fru-*

givorus), 370
Chats (Australian), (See Australian chats)
Chaunoproctus (Bonin finches), 404
— *ferreorostris* (Bonin finch), 404, **418**
Chelidoptera
— *tenebrosa* (Swallow-wing), 65, 68*
Chelidorynx hypoxantha (Yellow-bellied fantail flycatcher), **278**
Cherry finch *(Aidemosyne modesta)*, **454**f
Chestnut quail thrush *(Cinclosoma castanorum)*, 222*
Chestnut sparrow *(Passer eminibey)*, 416*, **425**, 425m, 427
Chestnut wattle-eyed flycatcher *(Dyaphorophyia castanea)*, **277**, 277*
Chestnut-backed antbird *(Myrmeciza exsul)*, 126*
Chestnut-backed chickadee *(Parus rufescens)*, 318
Chestnut-bellied nuthatch *(Sitta europaea castanea)*, 319*, **322**, 322m
Chestnut-breasted chlorophonia *(Chlorophonia pyrrhophrys)*, 361*
Chestnut-breasted finch *(Lonchura castaneothorax)*, 439*, **448**
Chestnut-breasted negro finch *(Nigrita bicolor)*, 462
Chestnut-capped brush finch *(Atlapetes brunneinucha)*, 354
Chestnut-crowned antpitta *(Gralleria ruficapilla)*, 126*
Chestnut-flanked white-eye *(Zosterops erythropleura)*, **343**
Chestnut-fronted shrike *(Prionops scopifrons)*, 208m
Chestnut-headed bee-eater *(Merops viridis)*, 30*, **40**
Chestnut-headed weaver *(Textor castaneiceps)*, 459*
Chestnut-naped forktail *(Enicurus ruficapillus)*, **293**
Chestnut-sided shrike vireo *(Vireolanius melitophrys)*, **395**f
Chiffchaff *(Phylloscopus collybita)*, 259*, **263**f, 263m
Chilean plantcutter *(Phytotoma rara)*, **164**, 164*
Chinese babax *(Babax lanceolatus)*, **243**
Chinese bullfinch *(Pyrrhula erythaca)*, **392**
Chinese crested mynah *(Acridotheres cristatellus)*, **472**
Chinese grosbeaks *(Eophona)*, 404, **419**
Chin-spot puff-back flycatcher *(Batis molitor)*, **277**
Chipping sparrow *(Spizella passerina)*, **353**f, 357
Chiroxiphia (Chorus manakins), 158
— *caudata* (Swallow-tailed manakin), 143*, **158**ff, 160*
Chisholm, A. H., 486
Chlamydera, 504
— *cerviniventris* (Fawn-breasted bower bird), **505**
— *lauterbachi* (Yellow-breasted bower bird), **505**, 505*
— *maculata* (Spotted bower bird), 492*, **505**, 505*
— *nuchalis* (Australian great bower bird), **504**
Chloebia gouldiae (Gouldian finch), 439*, **452**, 452m
Chloroceryle (Green kingfishers), 27
— *aenea* (Pygmy kingfisher), 27
— *amazona* (Amazon kingfisher), 26m, **27**, 37*
— *americana* (Green kingfisher), **27**
— *india* (Green-and-rufous kingfisher), 27m, **27**
Chlorocharis emiliae (Mountain blackeye), **340**
Chlorophanes
— *spiza* (Green honey creeper), 362*, **378**f
Chlorophonia (Chlorophonias), 370, **373**
— *pyrrhophrys* (Chestnut-breasted chlorophonia), 361*
Chloropseinae (Leafbirds), 204f
Chloropsis (Leafbirds), 204f
— *aurifrons* (Gold-fronted leafbird), 184*, 190*, **204**
— *hardwickei* (Orange-bellied leafbird), 204
Chlorornis
— *riefferi* (Grass-green tanager), 376*, **377**
Chlorospingus (Bush tanagers), 376
— *ophthalmicus* (Common bush tanager), 376f
— *phoenicotis* (Glistening-green tanager), **335**
Chorus manakins *(Chiroxiphia)*, 158
Chotoy spinetail *(Schoeniophylax phryganophila)*, 125*
Chough *(Pyrrhocorax pyrrhocorax)*, 469*, **513**
Choughs *(Pyrrhocorax)*, 513
Chrysocolaptes (Golden-backed woodpeckers), 115
— *festivus* (Black-backed woodpecker), **116**, 116m
— *validus* (Crimson-backed woodpecker), **116**, 119*
Chrysomma
— *sinense* (Oriental golden-eyed babbler), **240**
Chrysoptilus melanochloros (Green-barred woodpecker), **96**
Cicinnurus regius (King bird of paradise), 470*, **494**f
Cinclidae (Dippers), 219ff
Cinclodes
— *patagonicus* (Dark-bellied shake-tail), 125*, **129**
Cinclosoma (Quail thrushes), 248f
— *ajax* (Ajax scrub robin), **249**
— *castanotum* (Chestnut quail-thrush), 222*
— *cinnamomeum* (Cinnamon quail-thrush), 248f
— *punctatum* (Spotted quail thrush), **248**
Cinclosomatinae (Australian quail thrushes), 233
Cinclus (Dippers)
— *cinclus* (Eurasian dipper), 190*, **219**f, 219*
— *leucocephalus* (White-capped dipper), **220**
— *mexicanus* (American dipper), **219**
— *pallasii* (Brown dipper), **219**
Cinnamon quail thrush *(Cinclosoma cinnamomeum)*, 248f
Cinnamon-breasted rock bunting *(Fringillaria tahapisi)*, 350
Cinnamon-breasted wattle bird *(Melidectes torquatus)*, **320**
Cinnyricinclus leucogaster (Violet-backed starling), **474**
Cinnyris, 338
Ciridops
— *anna* (Ulaai-hawane), **390**, 393, 393*
Cirl bunting *(Emberiza cirlus)*, 341*
Cissa
— *chinensis* (Green magpie), 497*, **510**
— *thalassina* (Short-tailed green magpie), **510**
Cissilopha
— *melanocyaena* (Hartlaub's blue jay), 497*
— *sanblasiana* (Black-and-blue jay), **507**
Cissopis
— *leveriana* (Magpie tanager), 361*, **377**
Cisticola (Fan-tailed warblers), 251f
— *aridula* (Desert cisticola), **252**, 252m, 252*
— *ayersii* (Wing-snapping or Ayer's cloud warbler) 251f
— *cantans* (Singing cisticola), **252**
— *erythrops* (Red-faced cisticola), **252**, 253*
— *exilis* (Streaked fan-tailed warbler), **251**f, 251*
— *juncidis* (Fan-tailed warbler), 231*, 250m, **251**, 251*, 252*
Cistothorus palustris (Long-billed marsh-wren), 190*, **223**, 225
Citril finch *(Serinus citrinellus)*, **407**, 407m
Clamatores (Noisemakers), 118, 123, 127f
Clamydochaera jefferyi (Black-breasted triller), 198f
Clapper lark *(Mirafra apiata)*, **172**
Clancey, 72
Clark's nutcracker *(Nucifraga columbiana)*, 497*, **512**
Cliff swallow *(Petrochelidon pyrrhonota)*, 170*, 180*, **186**
Climacteridae (Tree runner-like birds), 324
Climacterinae (Treecreepers), 310, **324**f
Climacteris picumnus (Brown tree creeper), **325**
Clytoceyx rex (Earthworm-eating kingfisher), **34**, 38*
Cnemophilinae, 490
Cnemophilus macgregorii (Multi-crested bird of paradise), **490**, 492*
Coal tit *(Parus ater)*, 307*, **316**f
Coccothraustes (Hawfinches), 404
— *coccothraustes* (Hawfinch), 392*, 404, **419**f, 419m, 420*
Cochoa (Cochoas), 285, **293**
— *azurea* (Malayan cochoa), 293f
— *purpurea* (Purple cochoa), 280*, **293**
— *viridis* (Green cochoa), 293f
Cocks of the rock *(Rupicola)*, 161ff
Cocos finch *(Pinaroloxias inornata)*, 351*, 359m, **360**, 364, 365*
Coereba
— *flaveola* (Bananaquit), 371*, **386**, 386*
Coerebidae (Honey creepers), 348
Coerebinae (Bananaquits), 380, **385**f, 386m
Colaptes auratus (Yellow-shafted flicker), 95, 96m, 99*, 108*
— — *cafer* (Red-shafted flicker), 95, 96m, 108*
— — *chrysocaulosus*, 96m
— — *mexicanoides*, 96m
— *campestris* (Campos flicker), 96
— *rupicola* (Andean flicker), 96
Coleto mynah (Sarcops calvus), 449*, **473**
Coliuspasser, **433**
— *ardens* (Red-collared whydah), 430*, **433**f, 434m
— *axillaris* (Fan-tailed whydah bird), **434**
— *capensis* (Yellow bishop), **434**, 434m
— *jacksoni*, **434**
— *progne* (Long-tailed widow bird), **434**, 434m
Collared aracari *(Pteroglossus torquatus)*, 80, 81*
Collared barbet *(Lybius torquatus)*, 71f, 97*
Collared flycatcher *(Ficedula albicollis)*, 260*, **274**f, 274m, 276*
Collared gnatwren *(Microbates collaris)*, **271**
Collared jay *(Cyanolyca viridicyana)*, 497*, **507**
Collared puffbird *(Bucco capensis)*, **64**, 68*
Collared redstart *(Myioborus torquatus)*, 371*, **385**
Collias, E. C., 443
— N.E., 443
Colma antthrush *(Formicarius colma)*, 137
Colonia colonus (Long-tailed tyrant), 151, 151*
Common bush tanager *(Chlorspingus ophthalmicus)*, 376f
Common bush tit *(Psaltriparus minimus)*, **311**
Common crow *(Corvus brachyrhynchos)*, 498*, **515**ff
Common grackle *(Quiscalus quiscalus)*, 381*, **400**
Common iora *(Aegithina tiphia)*, 190*, **205**
Common jay *(Garrulus glandarius)*, 497*, **508**, 508m
Common linnet *(Acanthis cannabina)*, 391*, 404, **411**, 411m
Common miner *(Geositta cunicularia)*, **129**
Common mynah *(Acridotheres tristis)*, **471**f
Common or European bee-eater *(Merops apiaster)*, 32*, **39**ff, 40m, 48*
Common redpoll *(Acanthis flammea)*, 39*, **405**f, **411**, 411m
Common roller *(Coracias garrulus)*, 32*, **44**f, 44m, 57*
Common starling *(Sturnus vulgaris)*, **464**ff
Common tody flycatcher *(Todirostrum cinereum)*, 143*, **152**
Conebills *(Conirostrum)*, 385
Conirostrum (Conebills), 385
— *bicolor* (Bicolored conebill), **385**
Conopophaga (Gnateaters), 139f
— *lineata* (Rufous gnateater), **139**
Conopophila picta (Painted honey eater), **345**
Conrad, K., 101
Contopus
— *cinereus* (Tropical pewee), **152**
Coppersmith *(Megalaima haemacephala)*, 75f
Copsychus (Magpie-robins), 285, **291**
— *malabaricus* (Shama), 279*, **291**
— *saularis* (Magpie robin), **291**
Coracias (Rollers)
— *abyssinica* (Senegal roller), 44m
— *benghalensis* (Indian roller), 44f, 44m
— *caudata* (Lilac-breasted roller), 44, 44m
— *cyanogaster* (Blue-bellied roller), 44, 44m
— *garrulus* (Common roller), 32*, **44**f, 44m, 57*
— *naevia* (Rufous-crowned roller), 44, 44m
— *spatulata* (Racket-tailed roller),

44f, 44m
— *temminckii* (Celebes roller), 44m
Coraciidae (Rollers), 21, 42f
Coraciiformes, 21ff
Coraciinae (True rollers), 42, **44f**
Coracina
— *azurea*, 450*
— *lineata* (Barred cuckoo-shrike), 450*
— *novaehollandiae* (Black-faced cuckoo-shrike), 198f
Corcoracinae (Mudnest choughs), 483f
Corcorax melanoramphus (White-winged chough), 469*, **484**
Corn bunting (*Emberiza calandra*), 341*, **349**f, 349m
Corsican nuthatch (*Sitta whiteheadi*), 319*, **323**
Corvidae (Crow-like birds), 482, **505**ff
Corvinella (Yellow-billed shrikes), 209
— *corvina* (Long-tailed shrike), 213
Corvus
— *albicollis* (African white-necked raven), **518**, 521
— *albus* (Pied crow), 498*, **518**
— *brachyrhynchos* (Common crow), 498*, **515**ff
— *capensis* (Black crow), **515**
— *corax* (Raven), 498*, **518**ff, 518m
— — *varius* (Faroese raven), 519
— *corone* (Carrion crow), **515**ff
— — *cornix* (Hooded crow), 498*, **516**f
— — *corone*, 498*, 501*, **516**
— — *orientalis* (Eastern carrion crow), 516
— *crassirostris* (Thick-billed raven), 498*, **519**, 521
— *frugilegus* (Rook), 498*, **515**f
— *monedula* (Jackdaw), 498*, **514**f
— *moriorum* (Maori crow), 516
— *ossifragus* (Fish crow), 498*, **515**, 517
— *rhipidurus* (Fan-tailed raven), **518**, 521
— *ruficollis* (Brown-necked raven), **518**, 518m
— *splendens* (Indian house crow), 498*, **515**f
— — *zugmayeri*, 515
— *tristis* (Gray crow), **515**
Corydon sumatranus (Dusky broadbill), 124
Coryphospingus (Crested finches), 354
— *cucullatus* (Red-crested finch), **357**
Corythopis (Antpipits), 140, 152f
— *delalandi* (Southern antpipit), 140
Corythornis cristata (Malachite kingfisher), 25f, 25m
Cosmopsarus
— *regius*, (Royal starling), **474**
Cossypha (Robin chats or forest robins), 287
— *caffra* (Robin chat), 279*
— *niveicapilla* (Snowy-headed robin chat), **287**f
Cotinga (True cotingas), 161
— *amabilis* (Lovely cotinga), 133*
— *cotinga* (Purple-breasted cotinga), 144*
— *maculata* (Banded cotinga), **161**f
Cotingas (Cotingidae), 127, **160**ff, 161m
Cotingidae (Cotingas), 127, **160**ff, 161m
Cowbirds (*Molothrus*), 396
Cracticidae (Song shrikes), 213, 482, **485**f

Cracticus (Australian butcher birds), 485
— *louisiadensis* (White-rumped butcher bird), **485**
— *quoyi* (Black butcher bird), **485**
— *torquatus* (Gray butcher bird), 469*, **485**f
Crag martin (*Ptyonoprogne rupestris*), 180*, **191**
Creadion carunculatus (Saddleback), 469*, **483**
Creamy-breasted canastero (*Asthenes dorbignyi*), **131**, 137*
Creatophora cinerea (Wattled starling), 449*, **472**
Crescent honey eater (*Phylidonyris pyrrhoptera*), 320*
Crescent-chested puffbird (*Malacoptila striata*), **64**, 68*
Crested berry pecker (*Paramythia montium*), **330**f
Crested black tit (*Parus melanolophus*), 318
Crested bunting (*Melophus lathami*), 350
Crested finches (*Coryphospingus*), 354
Crested gallito (*Rhinocrypta lanceolata*), 140f, 140*
Crested honey creeper (*Palmeria dolei*), 372*
Crested jay (*Platylophus galericulatus*), 506
Crested lark (*Garlerida cristata*), 175f, 175m, 179*
Crested malimbe (*Malimbus malimbicus*), 430*, **442**, 443m
Crested oropendola (*Psarocolius decumanus*), **398**
Crested sharpbill (*Oxyruncus cristatus*), 150*, **157**, 158*
Crested tit (*Parus cristatus*), 307*, **317**
Crimson chat (*Ephthianura tricolor*), **266**
Crimson finch (*Neochmia phaeton*), 439*, **455**, 455m
Crimson-backed woodpecker (*Chrysocolaptes validus*), **116**, 119*
Crimson-bellied mountain finch (*Oreostruthus fuliginosus*), **455**
Crimson-breasted shrike (*Laniarius atrococcineus*), **208**f, 208m
Crimson-collared tanager (*Phlogothraupis sanguinolenta*), 361*
Crimson-crested woodpecker (*Phloeoceastes melanoleucos*), **117**
Crimson-hooded manakin (*Pipra aureola*), **143**
Crimson-rumped waxbill (*Estrilda rhodopyga*), **456**
Crimson-winged woodpecker (*Picus puniceus*), 119*
Crimson-wings (*Cryptospiza*), 458f
Criniger
— *flaveolus* (White-throated bulbul), 184*
— *ochraceus* (Brown-white-throated bulbul), **200**
Crombecs (*Sylvietta*), 250
Crossbills (*Loxia*), 404, 406, **418**
Crossley's ground roller (*Atelornis crossleyi*), **43**, 57*
Crow-billed drongo (*Dicrurus annectans*), **477**, 481
Crow-like birds (Corvidae), 482, **505**ff
Crowned "sparrows" (*Zonotrichia*), 353
Crypsirina (Racket-tailed magpies), 510
— *cucullata* (Hooded racket-tailed tree pie), **511**

Cryptospiza (Crimson wings), 458f
— *reichenovii* (Reichenov's crimson-wing), **461**
Cuban grassquit (*Tiaris canora*), **358**
Cuban thrush (*Turdus plumbeus*), 289*
Cuban tody (*Todus multicolor*), **34**
Cuckoo-roller (*Leptosomus discolor*), **42**f, 57*
Cuckoo-rollers (Leptosomatinae), 42f
Cuckoo shrike (*Campephaga phoenicea*), 450*
Cuckoo-shrikes (Campephagidae), 192, **198**f
Cuckoo-shrikes proper (*Campephaga*), 198
Cuckoo weavers (Anomalospizinae), 427, **427**, 427m
Culpepper sharp-beaked ground finch (*Geospiza difficilis nigrescens*), 366m
Cunha bunting (*Neospiza acunhae*), 354
Curl-crested aracari (*Beauharnaisius beauharnaesii*), 107*
Currawongs (*Strepera*), 486f
Cutia (*Cutia nipalensis*), **244**
Cutia nipalensis (Cutia), **244**
Cut-throat finch (*Amadina fasciata*), 439*, **447**, 447m
Cuvier's toucan (*Ramphastos cuvieri*), 98*
Cyanocompsa cyana (Ultramarine grosbeak), 352*
Cyanerpes (Honey creepers)
— *caeruleus* (Purple honey creeper), 335*
— *cyaneus* (Red-legged honey creeper), 335*, 362*, **378**f
— *lucidus* (Shining honey creeper), **379**
Cyanocitta
— *cristata* (Blue jay), 497*, **506**f
— *stelleri* (Steller's jay), **506**f
Cyanocorax
— *caeruleus* (Azure jay), 497*
— *cayanus* (Cayenne jay), 497*, **507**f
— *chrysops* (Plush-crested jay), **508**
— *yncas* (Green jay), **507**
Cyanolyca
— *viridicyana* (Collared jay), 497*, **507**
Cyanopica cyana (Azure-winged magpie), 497*, **510**, 510m
— — *cooki*, 510
Cyanoptila cyanomelana (Blue flycatcher), 260*, **275**
Cyclarhinae (Pepper shrikes), 395
Cyclarhis (Pepper shrikes), 395, 395m
— *gujanensis* (Rufous-browed pepper shrike), 372*, **396**
Cymbirhynchus macrorhynchos (Black-and-red broadbill), 124
Cyornis
— *rubeculoides* (Blue-throated flycatcher), 260*, **275**
— *rufigastra*, 275f
Cyphorinus aradus (Musician wren), **224**
Cyrtostomus, 338

Dacelo (Tree kingfishers), 33f
— *gigas* (Kookaburra or laughing jackass), 33f, 38*
— *leachii* (Blue-winged kookaburra), 33
Daceloninae (Tree kingfishers), 22, **28**f
Dacnidini (Honey creepers), 370, **378**, 379m

Dacnis, 378f, 379m
— *cayana* (Blue dacnis), 362*, **379**
Dark-bellied shake-tail (*Cinclodes patagonicus*), 125*, **129**
Dark-capped thrush (*Turdus ruticollis atrogularis*), **304**
D'Arnaud's barbet (*Trachyphonus darnaudii*), 74f
Dartford warbler (*Sylvia undata*), **258**, 262, 262m
Darwin, Charles, 141, 358, 411
Darwin's finches (Geospizini), 349, **358**ff
Davrian mynah (*Sturnia sturnina*), 449*
Davison, 146, 235
Dead Sea sparrow (*Passer moabiticus*), 416*, **424**
Deignan, 199
Delacour, Jean, 204, 234
Delichon dasypus
— *urbica* (House martin), 180*, **191**, 191m
Delius, J. D., 177f
Dendrocincla, 128
— *anabatina* (Tawny-winged woodcreeper), **128**
— *fuliginosa* (Plain brown woodcreeper), 120*
Dendrocitta (Tree pies), 510
— *vagabunda* (Indian tree pie), **511**
Dendrocolaptes certhia (Barred woodcreeper), 128*
Dendrocolaptidae (Woodcreepers), **127**f, 128m
Dendrocopos borealis (Red-cockaded woodpecker), **110**, 114, 114m
— *leucotos* (White-backed woodpecker), **109**, 113, 113m
— *major* (Greater spotted woodpecker), 83*, 84*, 92*, 94*, **109**ff, 110m, 111*, 112*, 119*
— *major major*, 112
— *medius* (Middle spotted woodpecker), **109**, 113, 113m, 119*
— *minor* (Lesser spotted woodpecker), 94*, **109**, 113f, 113m, 119*
— *pubescens* (Downy woodpecker), **110**, 114
— *syriacus* (Syrian woodpecker), **109**, 112f, 113m, 119*
— *villosus* (Hairy woodpeckers), **110**, 114, 114*, 114m
Dendroica, 380
— *caerulescens* (Black-throated blue warbler), 371*
— *coronata* (Myrtle warbler), **383**
— *magnolia* (Magnolia warbler), 371*, **384**
— *petechia* (Yellow warbler), **384**
Dendronanthus
— *indicus* (Tree wagtail), 189*, **194**, 198*
Dendropicos fuscescens (Cardinal woodpecker), **115**, 115m
Des Mur's wiretail (*Sylviorthorhynchus desmursii*), 125*
Desert cisticola (*Cisticola aridula*), **252**, 252m, 252*
Desert lark (*Ammomanes deserti*), 172f, 173m
Desert sparrow (*Passer simplex*), 416*, **424**
Desert warbler (*Sylvia nana*), **258**
Desert wheatear (*Oenanthe deserti*), **295**, 298
Desmodactylae (Broadbills), 118, 123f
Diacromyodae, 121
Diadem barbet (*Tricholaema diadematum*), 97*
Diademed tanager (*Stephanophorus*

diadematus), 361*
Diamond birds *(Pardalotus)*, 330
Diamond sparrow *(Stagonopleura gutata)*, 439*, **455**, 455m
Dicaeidae (Flower peckers or mistletoe birds), 329f, 330m
Dicaeum
— *agile* (Striped flower pecker), **331**
— *concolor* (Plain flower pecker), **331**
— *cruentatum* (Scarlet-backed flower pecker), **330**, 372*
— *erythrorhynchos* (Tickell's flower pecker), **331**
— *hirundinaceum* (Mistletoe bird), **329**, 372*
— *ignipectus* (Bronze-backed flower pecker), **329**
— *melanoxanthum* (Yellow-bellied flower pecker), **329**
— *trigonostigma* (Orange-breasted flower pecker), **330**
Dicrocercus hirundineus (Green swallow-tailed bee-eater), **41**
Dicruridae (Drongos), 477ff
Dicrurus (Drongos, i.n.s.), **477**, 477m
— *adsimilis* (Drongo), **477**
— *annectans* (Crow-billed drongo), **477**, 481
— *hottentottus* (Spangled drongo), 450*, **477**
— *leucophaeus* (Ashy drongo), **477**, 481
— *macrocercus* (King-crow), **477**f
— *remifer* (Lesser racket-tailed drongo), **477**, 481
— *paradiseus* (Greater racket-tailed drongo), 450*, **477**, 481
Diglossa (Flower piercers), 378ff
Diglossa baritula (Slaty flower piercer), 362*, 379*, **380**
Dinemellia dinemelli (White-headed buffalo weaver), 430*, **442**f
Diphyllodes
— *magnificus* (Magnificent bird of paradise), 491*, **495**
— *respublica* (Waigeu bird of paradise), 479*, **495**
Dipper (American) (See American dipper)
Dipper (Eurasian) (See Eurasian dipper)
Dippers (Cinclidae), 219ff
Dolichonyx
— *oryzivorus* (Bobolink), 381*, **397**, 402
Dollar bird *(Eurystomus orientalis)*, **45**, 45m, 57*
Donacobius atricapillus (Black-capped mocking thrush), 221*, 227
Double-toothed barbet *(Lybius bidentatus)*, **70**f
Downs, B., 420
Downy woodpecker *(Dendrocopos pubescens)*, **110**, 114
Drepanididae (Honey creepers), 348, 386ff, 387f
Drepanidinae (Black-and-red honey creeper), 387, **390**
Drepanis funerea (Black mamo), **390**, 394*
— *pacifica* (Mamo), 372*, **390**, 394, 394*
Drepanornis
— *bruijnii* (White-billed sickle-billed bird of paradise), 479*, **499**
Dromaeocercus (Emu-tails), 251
— *seebohmi* (Gray emu-tail), **251**
Drongo *(Dicrurus adsimilis)*, **477**
Drongos (Dicruridae), 477ff

Drongos, i.n.s. *(Dicrurus)*, **477**, 477m
Drymodes (Scrub robins), 285f
— *superciliaris* (Northern scrub robin), **286**
Dryocopus (Black woodpeckers), 102ff
— *javensis* (Indian great black woodpecker), **102**
— *martius* (Black woodpecker), 93*, **94***, 102f, 102m, 103*, 108*
— *pileatus* (Pileated woodpecker), 90*, 102m, **102**, 104*, 104
Dryoscopus cubla (Black-backed puffback), 195*, **208**
Dulidae (Palm chats), 207, **217**f
Dulus dominicus (Palm chat), 196*, **217**f
Dumetella carolinensis (Catbird), 221*, **227**ff
Dumetia hyperythra (Rufous-bellied babbler), **239**
Dumont's mynah *(Mino dumontii)*, 449*, **473**
Dusky broadbill *(Corydon sumatranus)*, 124
Dusky flycatcher *(Muscicapa adusta)*, **276**
Dusky tit *(Parus funereus)*, **318**
Dwarf kingfisher *(Myioceryx lecontei)*, 71*
Dyaphorophyia castanea (Chestnut wattle-eyed flycatcher), **277**, 277*

Eared pygmy tyrant *(Myiornis auricularis)*, 143*
Earthworm-eating kingfisher *(Clytoceyx rex)*, **34**, 38*
Eastern bluebird *(Sialia sialis)*, 221*, 270*, **292**f
Eastern carrion crow *(Corvus corone orientalis)*, **516**
Eastern great warbler *(Acrocephalus arundinaceus orientalis)*, 254m
Eastern kingbird *(Tyrannus tyrannus)*, 143*, **151**, 153, 155*, 156
Eastern meadowlark *(Sturnella magna)*, 401*
Eastern phoebe *(Sayornis phoebe)*, 151, 153*, **155**
Eastern rock nuthatch *(Sitta tephronota)*, 322m, **323**
Eastern shrike tit *(Falcunculus frontatus)*, 283f, 284*
Eastern spinebill *(Acanthorhynchus tenuirostris)*, 346
Egyptian reed warbler *(Acrocephalus stentoreus)*, **253**
Eibl-Eibesfeldt, I., 365
Eisentraut, Martin, 73f
Elaenia (Elaenia flycatchers), 152
— *chiriquensis* (Lesser elaenia), **152**f
— *flavogaster* (Yellow-bellied elaenia), 143*, **152**
Elaenia flycatchers *(Elaenia)*, 152
Elaeniinae, 152
Electron platyrhynchum (Broad-billed motmot), **35***, **36**
Elepaiu (Chasiempis sandvicensis), 278
Eleuterodactylae, 121
Elliot's pitta *(Pitta ellioti)*, 149*
Emberiza (Typical buntings), 349f
— *bruniceps* (Red-headed bunting), **350**, 353
— *calandra* (Corn bunting), 341*, 349f, 349m
— *cia* (Rock bunting), 341*
— *cioides* (Siberian meadow bunting), 341*
— *cirlus* (Cirl bunting), 341*

— *citrinella* (Yellowhammer), 341*, 349m, **350**f
— *hortulana* (Ortolan bunting), 341*, **349**f, 349m, 353
— *leucocephalos* (Pine bunting), **349**, 349m
— *melanocephala* (Black-headed bunting), 341*, **350**, 350m, 353
— *schoeniclus* (Reed bunting), 341*, **350**, 350m, 353
— — *pyrrhuloides* (Bullfinch-billed reed bunting), **350**
Emberizidae (Buntings), 348
Emberizinae, 348f, 348m
Emberizini (Buntings, i.n.s.), 349
Emblema picta (Painted finch), 439*, **455**, 455m
Emerald toucanet *(Aulacorhynchus prasinus)*, **80**, 107*
Emperor of Germany's bird of paradise *(Paradisaea guilielmi)*, 480*, **493**f
Empidonax
— *difficilis* (Western flycatcher), 143*
— *flaviventris* (Yellow-bellied flycatcher), 152f
Emu-tails *(Dromaeocercus)*, 251
Emu-wrens *(Stipiturus)*, 265
Enicurus (Forktails), 285, 293
— *leschenaulti* (Leschenault's forktail), 280*, **293**, 293*
— *ruficapillus* (Chestnut-naped forktail), **293**
— *scouleri* (Little forktail), **293**
Enodes erythrophris, 473
Entomyzon cyanotis (Blue-faced honey eater), 346
Eophona (Chinese grosbeaks), 404, 419
— *migratoria* (Migratory Chinese grosbeak), **419**f
— *migratoria* (Migratory Chinese grosbeak), **419**f
— — *personata* (Japanese grosbeak), 392*, **419**f
Eopsaltria
— *australis* (Southern yellow robin), 241*
Ephthianura (Australian chats), 266
— *tricolor* (Crimson chat), **266**
Epimachus (Sickle-billed birds of paradise), **499**
— *fastuosus* (Black sickle-billed bird of paradise), **499**
— *meyeri* (Brown sickle-billed bird of paradise), 479*, **499**
Eremomela (Eremomelas), 250
— *badiceps* (Brown-crowned eremomela), 249*, **250**
Eremophila alpestris (Horned lark), 179*, **181**, 181m
Eremopterix (Finch larks), 172
— *nigriceps* (White-fronted lark), **172**, 172m
Ergaticus, 380
— *ruber* (Red warbler), 371*, **385**
Erithacus (Robins), 285f
— *rubecula* (Eurasian robin), 269*, 279*, 285m, **286**f, 287*, 413*
Erythrocercus holochlorus (Little yellow flycatcher), **277**
— *livingstonei* (Livingstone's flycatcher), **277**
Erythropygia (Forktails proper), 285j
— *galactotes* (Rufous warbler), 285m, **286**
Erythrura (Parrot finches), 452f
— *cyaneovirens*, 413*
— *hyperythra* (Green-tailed parrot finch), **452**
— *prasina* (Pintailed nonpareil), 439*, **453**

— *psittacea* (Red-headed parrot finch), 439*, **453**
— *trichroa* (Blue-faced parrot finch), **453**
Estrilda (True waxbills), 456f
— *astrild* (St. Helena waxbill), 428, 440*, **456**
— *rufibarba*, 456
— *atricapilla* (Black-headed waxbill), **457**
— *caerulescens* (Lavender finch), 440*, **457**
— *charmosyna*, 457
— *erythronotos* (Black-cheeked waxbill), 428, **457**
— *melanotis* (Yellow-bellied waxbill), 440*, **456**, 456m
— *melpoda* (Orange-cheeked waxbill), 440*, **456**
— *nonnula* (Black-crowned waxbill), **457**
— *paludicola* (Fawn-breasted waxbill), **456**
— *rhodopyga* (Crimson-rumped waxbill), **435**
— *troglodytes* (Gray waxbill), 428, **456**
Estrildidae (Weaverfinches), 421, **444**ff
Eubucco
— *bourcierii* (Red-headed barbet), **70**
Euchloris riefferii (Green-and-black fruiteater), 144*
Eucometis
— *penicillata* (Gray-headed tanager), **376**
Euodice
— *cantans* (African silver-bill), **447**, 447m
— *malabarica* (Indian silver-bill), **447**, 447m
Eumomota superciliosa (Turquoise-browed motmot), **36**, 48*
Eumyias
— *indigo* (Indigo flycatcher), **276**
Euneornis campestris (Orangequit), 378
Eupetes macrocercus (Rail babbler), 222*, **249**
Euphonia (Euphonias), 370, 373
Euplectes (Bishops), 433, **435**f
— *aureus* (Golden-backed weaver), **435**
— *franciscanus* (Orange bishop), **435**, 435m
— *hordeaceus* (Fire-crowned bishop), **435**
— *orix* (Grenadier weaver), 430*, **435**, 435*, 435m
Euplectinae (Bishops), 422, **433**f
Eurasian dipper *(Cinclus cinclus)*, 190*, **219**ff, 219*
Eurasian hoopoe *(Upupa epops)*, 29*, **46**f, 57*
Eurasian kingfisher *(Alcedo atthis)*, **22**ff, 26m, 37*
Eurasian robin *(Erithacus rubecula)*, 269*, 279*, 285m, **286**f, 287*, 413*
Eurasian wryneck *(Jynx torquilla)*, 82*, **88**f, 88m, 88*, 89*, 108*
Eurocephalus
— *anquitimens* (White-crowned shrike), 196*, **208**
European nuthatch *(Sitta europaea)*, 301*, 319*, **321**f, 322*, 322m
Euryceros prevosti (Helmet bird), 196*, **213**f
Eurylaimidae (Broadbills), 123, 123m
Eurylaiminae (Typical broadbills), 123

Eurylaimus
— *javanicus* (Banded broadbill), 120*
— *ochromalus* (Black-and-yellow broadbill), 124
— *steerii* (Philippine broadbill), 124
Eurystomus (Broad-billed rollers), 45
— *glaucurus* (Broad-billed roller), 45, 45m
— *gularis* (Blue-throated roller), 45, 45m
— *orientalis* (Dollar bird), 45, 45m, 57*
Euscarthminae (Tody flycatchers), 152
Evening grosbeak (*Hesperiphona vespertina*), 392*, 406, 419f
Eye-ringed flatbill (*Rhynchocyclus brevirostris*), 152, 157

Fairy bluebirds (Ireninae), 204
Falculea palliata (Sicklebill or falculea), 213f
Falcunculus frontatus (Eastern shrike-tit), 283f, 284*
False sunbird (*Neodrepanis coruscans*), 147, 150*
Fan-tailed raven (*Corvus rhipidurus*), 518, 521
Fan-tailed warbler (*Cisticola juncidis*), 231*, 250m, 251, 251*, 252*
Fan-tailed warblers (*Cisticola*), 251f
Fan-tailed whydah bird (*Coliuspasser axillaris*), 434
Fantails (Rhipidurinae), 272, 274m, 278f
Faroese raven (*Corvus corax varius*), 519
Faust, Ingrid, 70f, 248
Faust, Richard, 245
Fawn-breasted bower bird (*Chlamydera cerviniventris*), 505
Fawn-breasted waxbill (*Estrilda paludicola*), 456
Feindt, P., 113
Fernandina's woodpecker (*Nesoceleus fernandinae*), 96
Ficedula
— *albicollis* (Collared flycatcher), 260*, 274f, 271m, 276*
— *hodgsonii* (Rusty-breasted blue flycatcher), 275
— *hypoleuca* (Pied flycatcher), 260*, 273f, 274m, 276*
— *mugimaki* (Black-and-orange flycatcher), 275
— *narcissina* (Narcissus flycatcher), 260*, 275
— *parva* (Least flycatcher), 260*, 274m, 275
— *semitorquata* (Half-collared flycatcher), 274, 274m
— *strophiata* (Orange-gorgeted flycatcher), 275
— *superciliaris* (White-breasted flycatcher), 275
— *tricolor* (Slaty blue flycatcher), 275
— *zanthopygia* (Korean flycatcher), 275
Fieldfare (*Turdus pilaris*), 290*, 304f, 304m
Fiery-billed aracari (*Pteroglosus frantzii*), 80, 85
Fiery-capped manakin (*Machaeropterus pyrocephalus*), 158f
Figbirds (Sphecotheres), 477
Finch-billed (*Spizixos canifrons*), 204
Finches (Fringilidae), 348, 402ff
Finch larks (*Eremopterix*), 172
Fire finch (*Lagonosticta senegala*), 413*, 429*, 458, 458m
Fire finches (*Lagonosticta*), 428, 458
Fire-capped tit (*Cephalopyrus flammiceps*), 313
Firecrest (*Regulus ignicapillus*), 259*, 267f, 267m, 267*
Fire-crest aletha (*Alethe diademata*), 287ff
Fire-crowned bishop (*Euplectes hordeaceus*), 435
Fire-eye (*Pyriglena leucoptera*), 137
Fire-fronted bishop (*Taha diademata*), 430*, 436
Firetail (*Myzornis pyrrhoura*), 244
Fire-tailed finch (*Stagonopleura bella*), 455
Fire-tufted barbet (*Psilopogon pyrolophus*), 76f
Firewood-gatherer (*Anumbius annumbi*), 125*
Fish crow (*Corvus ossifragus*), 498*, 515, 517
Fisher's whydah (*Tetraenura fisheri*), 428, 429*
Flame minivet (*Pericrocotus flammeus*), 450*
Flame-crested tanager (*Tachyphonus cristatus*), 361*
Flame-faced tanager (*Tangara parzudakii*), 361*
Flappet lark (*Mirafra rufocinnamomea*), 172, 172m
Flat-billed flycatchers, 273, 277f
Flower peckers or mistletoe birds (Dicaeidae), 329f, 330m
Flower piercers (*Diglossa*), 378ff
Fluffy-backed tit babbler (*Macronous ptilosus*), 239
Fluvicola
— *pica* (Pied water tyrant), 143*
Fluvicolinae, 151
Flycatcher warblers (*Seicercus*), 263
Flycatcher-like birds (Muscicapidae), 233ff, 272ff
Flycatchers proper (Muscicapinae), 272ff, 274m
Fly-eaters (*Gerygone*), 265
Fodis (*Foudia*), 433, 437
Forest weaver (*Symplectes bicolor*), 442
Fork-tailed flycatcher (*Muscivora tyrannus*), 143*, 151
Forktails (*Enicurus*), 285, 293
Forktails proper (*Erythropygia*), 285f
Formicariidae (Antbirds), 127, 132f, 137m
Formicarius (Ant thrushes), 137, 139
— *analis* (Black-faced ant thrush), 126*
— *colma* (Colma ant thrush), 137
Formicivora
— *grisea* (White-fringed ant wren), 126*
Foudia (Fodis), 433, 437
— *bruante* (Reunion weaver), 437
— *eminentissima*, 437
— *flavicans* (Rodrigues weaver), 437
— *madagascariensis* (Madagascar weaver), 437
Franz, J., 113
Fregilupus varius (Reunion starling), 472
Friedmann, 398
Frieling, Heinrich, 275
Frilled flycatcher (*Arses telescophthalmus*), 278
Fringillaria (Striped buntings), 350f
— *capensis* (Cape bunting), 350
— *impetuani* (Pale rock bunting), 350
— *striolata* (House bunting), 350
— *tahapisi* (Cinnamon-breasted rock bunting), 350

Fringilla
— *coelebs* (Chaffinch), 392*, 402f, 402m, 403*
— *montifringilla* (Brambling), 336*, 392*, 402f, 402m
— *teydea* (Canary Island chaffinch), 402f
Fringillidae (Finches), 348, 402ff
Fringillinae (Chaffinches), 402
Frisch, Otto von, 44
Fruit eaters (*Pipreola*), 161
Furnariidae (Ovenbirds), 127, 129, 129m
Furnarius (Ovenbirds), 129
— *rufus* (Rufous hornero or ovenbird), 125*, 129ff, 129*, 130*

Galapagos mockingbird (*Nesomimus parvulus*), 221*, 227
Galbalcyrhynchus leucotis (White-eared jacamar), 63
Galbula (Jacamars)
— *dea* (Paradise jacamar), 63, 68*
— *galbula* (Green-tailed jacamar), 63, 68*
— *ruficauda* (Rufous-tailed jacamar), 63
Galbuloidea (Jacamars), 62
Galbulidae (Jacamars), 62, 63f
Galerida cristata (Crested lark), 175f, 175m, 179*
— *theclae* (Thecla lark), 179*
Gampsorhynchus rufulus (White-headed babbler), 245
Gaping of young, 121
Garden warbler (*Sylvia borin*), 232*, 258, 258m, 261
Gardener bower bird (*Amblyornis inornatus*), 504, 504*
Garnet or red-headed pitta (*Pitta granatina*), 149*
Garrulax (Laughing thrushes), 243
— *leucolophus* (White-crested laughing thrush), 222*, 243
— *moniliger* (Necklaced laughing thrush), 243
— *pectoralis* (Black-gorgeted laughing thrush), 243
Garrulus
— *glandarius* (Common jay), 497*, 508, 508m
— *lanceolatus* (Black-throated jay), 508
— *lidthi* (Lidth's jay), 508
Gaudy barbet (*Megalaima mystacophanos*), 75
Geiseloceros robustus, 50
Geocolaptes olivaceus (Ground woodpecker), 91*, 94m, 95, 108*
Geositta (Miners), 129
— *cunicularia* (Common miner), 129
Geospiza (Ground finches), 359ff, 359*, 363*
— *difficilis* (Sharp-beaked ground finch), 359, 363f, 364*, 367*
— — *acutirostris* (Small sharp-beaked ground finch), 364m, 366m
— — *debilirostris* (Medium sharp-beaked ground finch), 366, 366m, 367*
— *difficilis*, 366m
— *nigrescens* (Culpepper sharp-beaked ground finch), 366m
— *septentrionalis* (Large sharp-beaked ground finch), 364, 364m
— *fortis* (Medium ground finch), 363*
— *fuliginosa* (Small ground finch), 359, 363f, 363*
— *magnirostris* (Large ground finch), 351*, 359, 363*
— *scandens* (Cactus ground finch), 351*, 359f, 363f, 364*, 367*
Geospizini (Darwin's finches), 349, 358ff
Gerygone (Fly-eaters), 265
— *palpebrosa* (Black-throated warbler), 259*
— *sulphurea* (Mangrove fly-eater), 265f
Giant antshrike (*Batara cinerea*), 132
Giant cowbird (*Scaphidura oryzivora*), 381*, 396, 398
Giant kingfisher (*Megaceryle maxima*), 26m, 27
Giant nuthatch (*Sitta magna*), 323
Giant weaver (*Textor grandis*), 441
Gifford, 365
Gila woodpecker (*Centurus uropygialis*), 93*, 104, 106
Gilbert, John, 166
Gilliard, E. Thomas, 86, 124, 200, 237, 239, 408, 494, 499f, 505
Glistening-green tanager (*Chlorospingus phoeniceus*), 335*
Glyphorhynchus
— *spirurus* (Wedge-billed woodcreeper), 120*
Gnatcatchers (Polioptilinae), 233, 271
Gnat eaters (*Conopophaga*), 139f
Godeffroy, Jean Cesar, 282
Goldcrest (*Regulus regulus*), 259*, 267f, 267m, 267*
Gold-crested crackle (*Ampeliceps coronatus*), 473
Golden bird (*Sericulus aurens*), 492*
Golden bower bird (*Prionodura newtoniana*), 492*, 503
Golden oriole (*Oriolus oriolus*), 413*, 450*, 476, 476m
Golden sparrow (*Passer luteus*), 416*, 425, 425m
Golden weaver (*Ploceella hypoxantha*), 437m, 438
Golden weaver (*Textor subaurens*), 415*
Golden whistler (*Pachycephala pectoralis*), 260*, 283
Golden-backed weaver (*Euplectes aureus*), 435
Golden-backed woodpecker (*Campethera maculosa*), 99
Golden-backed woodpeckers (*Chrysocolaptes*), 115
Golden-breasted mynah (*Mino anais*), 473
Golden-breasted waxbill (*Amandava subflava*), 440*, 455
Golden-crowned kinglet (*Regulus satrapa*), 268, 268m
Golden-headed babbler (*Stachyris chrysaea*), 239
Golden-naped woodpecker (*Centurus chrysauchen*), 104
Golden-olive woodpecker (*Piculus rubiginosus*), 96
Golden-rumped tinker-bird (*Pogoniulus bilineatus*), 74, 97*
Golden-throated barbet (*Megalaima franklinii*), 75
Golden-winged sunbird (*Nectarina reichenowi*), 332
Golden-winged warbler (*Vermivora chrysoptera*), 371*, 384, 384m
Goldfinch (*Carduelis carduelis*), 334*, 391*, 404*, 405*, 409ff, 409m
Goldfinch (American) (See American goldfinch)
Goldfinches and siskins (*Carduelis*), 404, 409
Gold-fronted finch (*Serinus pu-*

sillus), 382*
Gold-fronted leafbird (*Chloropsis aurifrons*), 184*, 190*, **204**
Gonolek (*Laniarius barbarus*), 195*
Goodwin, Derek, 76, 504
Gouch Island bunting (*Rowettia goughensis*), 357
Gould, John, 166, 344f, 453, 484ff, 499
Gouldian finch (*Chloebia gouldiae*), 439*, **452**, 452m
Grackles (*Quiscalus*), 396f, **399f**
Gracula
— *ptilogenys* (Ceylon mynah), **473**
— *religiosa* (Javan hill mynah), 449*, **472f**
Gracupica burmannica
— *nigricollis* (Black-necked mynah), 449*, **471**
Grafisia torquata (White-collared starling), **468**
Grallaria (Antpittas), 137, 139
— *ruficapilla* (Chestnut-crowned antpitta), 126*
— *squamigera* (Undulated antpitta), 126*
— *varia* (Royal antpitta), 137, 138*
Grallaricula
— *nana* (Slate-crowned antpitta), 126*
Grallina bruijni (Torrent lark), **484**
— *cyanoleuca* (Magpie lark), 469*, **483**
Grallinidae (Australian magpielarks), 482, **483f**
Grallininae (Magpie-larks), 483
Granatellus, 380
— *pelzelni* (Rose-breasted chat), 371*, **385**, 385m
Grant, C. H. W., 124
Grass finches (*Poephila*), 453
Grass warblers, 250, **251f**
Grass wrens (*Amytornis*), 266
Grass-green tanager (*Chlorornis riefferii*), 376*, **377**
Grasshopper warbler (*Locustella naevia*), 231*, 247*, **253**, 255m, 256
Gray broadbill (*Pseudocalyptomena graueri*), 124
Gray butcher bird (*Cracticus torquatus*), 469*, **485f**
Gray crow (*Corvus tristis*), **515**
Gray currawong (*Strepera versicolor*), **486**
Gray emu-tail (*Dromaeocercus seebohmi*), **251**
Gray hypocolius (*Hypocolius ampelinus*), **217**, 217m, 218m
Gray jay (*Perisoreus canadensis*), **509**
Gray kinkimavo (*Tylas eduardi*), **214**
Gray silky flycatcher (*Ptilogonys cinereus*), **216f**
Gray spotted creeper (*Salpornis spilonotus*), 319*, **326**, 326m
Gray starling (*Spodiopsar, cineraceus*), **468**
Gray tit (*Parus afer*), 308*, **318**, 318m
Gray tit (African) (See African gray tit)
Gray tit flycatcher (*Parisoma plumbeum*), **276**
Gray wagtail (*Motacilla cinerea*), 183*, 184*, 189*, **194**, 200m
Gray waxbill (*Estrilda troglodytes*), **428**, 460
Gray-backed camaroptera (*Camaroptera brevicaudata*), 248*, **250**, 250*
Gray-breasted silver eye (*Zosterops lateralis*), 340, 340m, **343**

Gray-cheeked thrush (*Hylocichla minima*), **303**
Gray-crested helmet shrike (*Prionops poliolopha*), 208m
Gray-crowned babbler (*Pomatostomus temporalis*), **222***
Gray-crowned rosy finch (*Leucosticte arctoa*), 392*, **412**
Gray-headed bush shrike (*Malaconotus blanchoti*), **209**
Gray-headed chickadee (See Siberian tit)
Gray-headed goldfinch (*Carduelis carduelis caniceps*), **409**
Gray-headed greenlet (*Hylophilus decurtatus*), 372*, **395**
Gray-headed kingfisher (*Halcyon leucocephala*), 27m, **28**, 33, 37*
Gray-headed negro finch (*Nigrita canicapilla*), **462**, 462m
Gray-headed olive-back (*Nesocharis capistrata*), **461**
Gray-headed parrotbill (*Paradoxornis gularis*), **247**
Gray-headed silver-bill (*Odontospiza*), **447**
Gray-headed sparrow (*Passer griseus*), **424**
Gray-headed tanager (*Eucometis penicillata*), **376**
Gray-headed weaver (*Pseudonigrita arnaudi*), **443**, 444m
Gray-necked rockfowl (*Picathartes oreas*), 222*, **245ff**
Gray-rumped swallow (African) (See African gray-rumped swallow)
Gray-sided chickadee (*Parus sclateri*), 318
Great ant shrike (*Taraba major*), 126*
Great bower bird (Australian) (See Australian great bower bird)
Great blue pitta (*Pitta caerulea*), 142, 146, 149*
Great crested flycatcher (*Myiarchus crinitus*), 152
Great hill barbet (*Megalaima virens*), **75**, 97*
Great Indian hornbill (*Buceros bicornis*), 47*, **56**, 59
Great jacamar (*Jacamerops aurea*), **63**, 68*
Great kiskadee flycatcher (*Pitangus sulphuratus*), 143*, **152**
Great reed warbler (*Acrocephalus arundinaceus*), 231*, 242*, 247*, **253**, 255
Great shrike tyrant (*Agriornis lividus*), 151
Great tit (*Parus major*), 307*, **314ff**, 315*, 316*
Greater bird of paradise (*Paradisaea apoda*), **490f**, 491*, 502*
Greater koa finch (*Psittirostra palmeri*), 388, **389***, 390
Greater raquet-tailed drongo (*Dicrurus paradiseus*), 450*, **477**, 481
Greater red-headed babbler (*Malacopteron magnum*), **235**
Greater scaly-breasted wren babbler (*Pnoepyga albiventer*), **237**
Greater spotted woodpecker (*Dendrocopos major*), 83*, 84*, 92*, 94*, **109ff**, 110m, 111*, 112*, 119*
Greater swamp warbler (*Calamocichla gracilirostris*), **253**, 254m
Green avadavat (*Amandava formosa*), **455**
Green barbet (*Magalaima zeylanica*), **75f**
Green broadbill (*Calyptomena*

viridis), 120*, 124*, 127
Green catbird (*Ailuroedes crassirostris*), **500f**
Green cochoa (*Cochoa viridis*), **293f**
Green honey creeper (*Chlorophanes spiza*), 362*, **378f**
Green hylia (*Hylia prasina*), **268f**, 268m, 271*
Green jay (*Cyanocorax yncas*), **507**
Green jery (*Neomixis viridis*), **238**
Green kingfisher (*Chloroceryle americana*), 27
Green kingfishers (*Chloroceryle*), 27
Green magpie (*Cissa chinensis*), 497*, **510**
Green oropendola (*Psarocolius viridis*), 381*
Green shrike vireo (*Vireolanius pulchellus*), 372*
Green swallow-tailed bee-eater (*Dicrocercus hirundineus*), **41**
Green toucanets (*Aulacorhynchus*), 79f
Green wood hoopoe (*Phoeniculus purpureus*), 57*
Green woodpecker (*Picus viridis*), 81*, 92*, **99f**, 100m, 100*, 119*
Green woodpeckers (*Picus*), 99f
Green-and-black fruiteater (*Euchloris riefferii*), 144*
Green-and-rufous kingfisher (*Chloroceryle inda*), 27m, **27**
Green-backed becard (*Pachyramphus viridis*), **161**, 161*
Green-backed camaroptera (*Camaroptera brachyura*), **250**
Green-backed tit (*Parus monticolus*), **318**
Green-backed twin-spot (*Mandingoa nitidula*), 440*, **458**
Green-barred woodpecker (*Chrysoptilus melanochloros*), **96**
Green-breasted manucode (*Manucodia chalybata*), **490**
Greenfinch (*Carduelis chloris*), 391*, **404**, 409, 409m, 410*, 413*
Green-headed tanager (*Tangara seledon*), **373**
Greenlets (*Hylophilus*), 395
Greenish warbler (*Phylloscopus trochiloides*), 259*, **263**, 263m
Green-tailed jacamar (*Galbula galbula*), **63**, 68*
Green-tailed parrot finch (*Erythura hyperythra*), **452**
Greenway, 167
Grenadier weaver (*Euplectes orix*), 430*, **435**, 435*, 435m
Groove-billed toucanet (*Aulacorhynchus sulcatus*), **80**
Grosbeak finch or kona finch (*Psittirostra kona*), 388
Grosbeaks (*Pheucticus*), 367f
Ground cuckoo-shrike (*Pteropodocys maxima*), **198f**
Ground finches (*Geospiza*), **359ff**, 359*, 363*
Ground hornbills (*Bucorvus*), 60f
Ground rollers (Brachypteraciinae), 42, **43f**
Ground thrushes (*Zoothera*), 300
Ground woodpeckers (*Geocolaptes olivaceus*), 91*, 94m, **95**, 108*
Ground-scraper thrush (*Turdus litsipsirupa*), **304**
Gubernatrix
— *cristata* (Yellow cardinal), 334*, 342*, **367**, 369m
Guianan cock-of-the-rock (*Rupicola rupicola*), 144*, 162
Guichard, 175
Guiraca
— *caerulea* (Blue grosbeak), 352*,

367, 369
Gurial (*Pelargopsis capensis gurial*), 38*
Gwinner, 264, 519
Gymnobucco
— *calvus* (Naked-faced barbet), 73
— *peli* (Bristle-nosed barbet), **73**, 97*
Gymnocichla
— *nudiceps* (Bare-crowned ant bird), 126*
Gymnorhina (Australian magpies), **485ff**
— *hypoleuca* (White-backed magpie), 469*, **486f**
— — *dorsalis*, 487
— *tibicen* (Black-backed magpie), 486
Gymnorhinus cyanocephalus (Pinyon jay), 497*, **506f**
Gymnoris, 423
Gymnostinops, 396
— *montezuma* (Montezuma oropendola), 381*

Habia (Ant-tanagers), 376
— *rubica* (Red-crowned anttanager), 361*, **376**
Hachisuka, 146
Hairy woodpecker (*Dendrocopos villosus*), 110, 114, 114*, 114m
Halcyon, 22, 28f
— *chelicuti* (Striped kingfisher), **28**
— *chloris* (White-collared kingfisher), **28**, 37*
— *leucocephala* (Gray-headed kingfisher), 27m, **28**, 33, 37*
— — *pallidiventris*, 33
— *sancta* (Sacred kingfisher), 27m, **28**, 33
— *senegalensis* (Senegal kingfisher), **28**
— *senegaloides* (Mangrove kingfisher), 28
— *smyrnensis* (White-breasted kingfisher), **28f**, 32*, 38*
— *torotoro*, 37*
— *winchelli*, 32*
Halfbills (*Hemignathus*), 387
Half-collared flycatcher (*Ficedula semitorquata*), **274**, 274m
Harrisson, Barbara, 331
— Tom, 331
Hartert, Ernst, 233
Hartlaubius auratus (Madagascar starling), **475**
Hartlaub's blue jay (*Cissilopha melanocyanea*), 497*
Hawaiian creeper (*Paroreomyza* or *Loxops maculata*), 387, 388*
Hawaiian thrush (*Phaeornis obscurus oahensis*), **303**, 303*
Hawfinch (*Coccothraustes coccothraustes*), 392*, 404, **419f**, 419m, 420*
Hedge sparrow or dunnock (*Prunella modularis*), 221*, **229f**, 413*
Heinrich, Gerd, 55, 145, 236ff, 243
Heinroth, Oskar, 24, 51, 94, 210, 257, 451
Helmet bird (*Euryceros prevosti*), 196*, **213f**
Helmeted manakin (*Antilophia galeata*), 158
Helmeted shrikes (Prionopinae), **207f**
Hemignathus (Halfbills), 387
— *lucidus* (Nukupau), 389*
— *obscurus* (Akiola), 372*, **387**, 388*
— *wilsoni* (Akiapolaau), **387f**
Hemipus
— *picatus* (Pygmy triller or pied

shrike), 198f
Henderson's ground chough *(Podoces hendersoni)*, **512**
Hermann, Helfried, 130f
Hermit thrush *(Hylocichla guttata)*, 221*, **303**
Hermotimia, 338
Hesperiphona, 404
— *vespertina* (Evening grosbeak), 392*, 406, **419**f
Heteralocha acutirostris (Huia), 469*, **483**
Heterophasia (Sibias), 245
— *picaoides* (Long-tailed sibia), **245**
Heyn, 24
Hillemacher, H., 117
Himalayan tree creeper *(Certhia himalayana)*, **326**
Himatione
— *sanguinea* (Apapane), 372*, 388, **390**, 390*, **393**
— — *freethii* (Laysan apapane), **390**, 393
Hippolais, 254, 257
— *caligata* (Babbling warbler), **254**, 258
— *icterina* (Iceterine warbler), 232*, **253**, 255m, 257
— *languida* (Upcher's warbler), **254**
— *olivetorum* (Olive-tree warbler), **253**f, 257
— *pallida* (Olivaceous warbler), **254**, 255m, 257
— *polyglotta* (Melodious warbler), **253**, 255m, 257, 257*
Hirundinidae (Swallows), 182ff
Hirundininae (True swallows), 182, **185**f
Hirundo (Swallows)
— *abyssinica* (Striped swallow), **187**
— *albigularis* (White-throated swallow), **187**
— *angolensis* (Angola swallow), **187**
— *daurica* (Red-rumped swallow), 180*, **187**, 187m
— *rustica* (Barn swallow), 169*, 170*, 180*, **187**ff, 187m
— *tahitica* (South Sea swallow), **187**
Hispaniola piculet *(Nesoctites micromegas)*, **89**
Histurgops ruficauda (Rufous-tailed weaver), **443**
Hoary redpoll *(Acanthis horemanni)*, **411**, 411m
Hoesch, Walter, 49, 72f
Homrai *(Buceros bicornis homrai)*, 56, 56m
Honey creepers (Dacnidini), 370, **378**, 379m
Honey creepers (Drepanididae), 348, **386**ff, 387m
Honey eaters (Meliphagidae), **344**ff
Honey guides (Indicatoridae), 63, **77**ff, 77m
Hooded crow *(Corvus corone cornix)*, 498*, **516**f
Hooded pitta *(Pitta sordida)*, **142**, 145f, 149*
Hooded racket-tailed tree pie *(Crypsirina cucullata)*, **511**
Hooded warbler *(Wilsonia citrina)*, 371*, **385**, 385m
Hoogerwerf, 146
Hook-billed graybird *(Tephrodornis gularis)*, **198**
Hook-billed kingfisher *(Melidora macrorrhina)*, **34**
Hook-billed vanga *(Vanga curvirostris)*, 196*, **213**f
Hoopoe (Eurasian) (See Eurasian hoopoe)
Hoopoes (Upupidae), 21, **45**f

Hoopoes proper (Upupinae), 45
Hornbills (Bucerotidae), 21, 49m, **49**ff
Horned lark *(Eremophila alpestris)*, 179*, **181**, 181m
Hose's broadbill *(Calyptomena hosii)*, 124
House bunting *(Fringilaria striolata)*, 350
House crow (Indian) (See Indian house crow)
House finch *(Carpodacus mexicanus)*, 391*, 405, **417**f
House martin *(Delichon urbica)*, 180*, **191**, 191m
House sparrow *(Passer domesticus)*, 416*, **424**ff, 42ém, 425m, 426*
House wren *(Troglodytes aedon)*, **223**, 225f
Howard, 250
Howell, Tho Ro, 106
Hudson, W. H., 141
Huia *(Heteralocha acutirostris)*, 469*, **483**
Humboldt, Alexander von, 86
Hume's ground chough *(Pseudopodoces humilis)*, 469*, **512**
Hume's tree babbler *(Stachyris rufifrons)*, 222*
Hunstein, Carl, 494
Hutchinson, G. Evelyn, 499
Hylia
— *prasina* (Green hylia), **268**f, 268m, 271*
Hyliinae (Hylias), 268
Hylocichla (Wood thrushes), 303
— *guttata* (Hermit thrush), 221*, **303**
— *minima* (Gray-cheeked thrush), **303**
— *mustelina* (Wood thrush), 221*
— *ustulata* (Swainson's thrush), **303**
Hylophilus (Greenlets), 395
Hylophilus decurtatus (Gray-headed greenlet), 372*, **395**
Hylophylax (Antbirds), 137ff
— *naevia* (Spotted antbird), 137
— *poecilonata* (Scale-backed antbird), **126***
Hypargos niveoguttatus (Peter's twin-spot), 440*, **458**
— *margaritatus* (Rosy twin-spot), **458**
Hyphanturgus
— *melanogaster*, 430*
Hypochera (Indigo birds), **428**, 433
— *chalybeata* (Cameroon's indigo bird), **428**, 429*
— *nigerrima* (Black indigo bird), 429*
Hypocoliinae, 217
Hypocolius ampelinus (Gray hypocolius), **217**, 217*, 218m
Hypositta corallirostris, (Madagascar nuthatch), **214**, 319*
Hypothymis
— *azurea* (Black-naped flycatcher), 260*, **281**
Hypsipetes, 200
— *madagascariensis* (Black bulbul), **203**

Icteria (Yellow-breasted chats), 380
— *virens* (Yellow-breasted chat), **383**
Icteridae (Icterids or New World Orioles), 348, **396**f, 396m
Icterids or American orioles (Icteridae), 348, **396**f, 396m
Icterine warbler *(Hippolais icterina)*, 232*, **253**, 255m, 257
Icterus (American orioles), **397**, **400**
— *auricapillus* (Orange-crowned oriole), **400**
— *bullockii* (Bullock's oriole), **400**
— *chrysater* (Yellow-backed oriole), **400**
— *galbula* (Baltimore oriole), 381*, **400**
— *icterus* (Troupial), 381*, **400**
— *pectoralis* (Spotted-breasted oriole), 381*
Ifrita (*Ifrita kowaldi*), **249**
Ifrita kowaldi (Ifrita), **249**
Iiwi *(Vestiaria coccinea)*, 372*, 388, 390, 393*, **393**f
Immelmann, K., 346
Imperial woodpecker *(Campephilus imperialis)*, **116**
Indian black-headed pitta *(Pitta sordida cuculata)*, **142**
Indian forest kingfisher *(Ceyx erithacus)*, 37*
Indian great black woodpecker *(Dryocopus javensis)*, **102**
Indian house crow *(Corvus splendens)*, 498*, **515**f
Indian reed warbler or paddy field warbler *(Acrocephalus agricola)*, **253**, 255
Indian robin *(Saxicoloides fulicata)*, **295**f, 298
Indian roller *(Coracias benghalensis)*, **44**f, 44m
Indian silver-bill *(Euodice malabarica)*, **447**, 447m
Indian sparrow *(Passer domesticus indicus)*, **425**
Indian tree pie *(Dendrocitta vagabunda)*, **511**
Indicator, 77, 79
— *archipelagus* (Malayan honey guide), **77**
— *exilis* (Least honey guide), **77**
— *indicator* (Black-throated honey guide), 67*, **77**f
— *minor* (Lesser honey guide), **77**, 97*
— *variegatus* (Scaly-throated honey guide), 77f
— *xanthonotus*, 77
Indicatoridae (Honey guides), 63, **77**ff, 77m
Indigo birds *(Hypochera)*, **428**, 433
Indigo bunting *(Passerina cyanea)*, 352*, **367**, 369f
Indigo flycatcher *(Eumyias indigo)*, **276**
Insectivorous tree finches *(Camarhynchus)*, **360**, 364
Ioras *(Aegithina)*, 204f
Ipswich sparrow *(Passerculus princeps)*, **353**, 356
Iraq great reed warbler *(Acrocephalus griseldis)*, **253**, 255
Irena (Fairy bluebirds)
— *cyanogaster* (Philippine fairy bluebird), **206**
— *puella* (Blue-backed fairy bluebird), 184*, 190*, **206**
Irenidae, 192, **204**
Ireninae (Fairy bluebirds), 204
Iridoprogne
— *bicolor* (Tree swallow), 180*, **186**
Isabelline red-backed shrike *(Lanius collurio isabellinus)*, 195*, **212**
Island gray-headed monarch flycatcher *(Monarcha cinerascens)*, **281**f
Ispidina picta (Natal kingfisher), **25**f, 25m
Italian sparrow *(Passer domesticus italiae)*, **425**, 425m
Ivory-billed woodpecker *(Campephilus principalis)*, 108*, **116**, 116m, 116*

Jacamaralcyon, 63
— *tridactyla* (Three-toed jacamar), 63, **68***
Jacamars (Galbulidae), 62, **63**f
Jacamars (Galbuloidea), 62
Jacamerops aurea (Great jacamar), 63, **68***
Jackdaw *(Corvus monedula)*, 498*, **514**f
Jahn, H., 282, 426
Jamaican becard *(Platypsaris niger)*, 144*
Jamaican tody *(Todus viridis)*, **34**
Jameson's fire finch *(Lagonosticta rhodopareia)*, 429*, **458**, 458m
Japanese bullfinch *(Pyrrhula pyrrhula griseiventris)*, 392*
Japanese fairy pitta *(Pitta brachyura nympha)*, **142**, 145, 146*
Japanese grosbeak *(Eophona personata)*, 392*, **419**f
Japanese paradise flycatcher *(Terpsiphone atrocaudata)*, 260*, **282**f
Japanese waxwing *(Bombycilla japonica)*, **215**, 215m
Japanese white eye *(Zosterops japonica)*, **343**
Java long-tailed tit *(Psaltaria exilis)*, **311**, 311m
Java sparrow *(Padda oryzivora)*, 439*, **452**
Javan hill mynah *(Gracula religiosa)*, 449*, **472**f
Junco (Juncos), 353
— *hyemalis* (Slate-colored junco), **353**, 356
— *oreganus* (Oregon junco), 342*, **353**, 356, 356*
Juncos *(Junco)*, 353
Jungle babblers (Pellorneini), **234**f
Jynginae (Wrynecks), 87, **88**f
Jynx ruficollis (Red-breasted wryneck), **88**, 88m
— *torquilla* (Eurasian wryneck), 82*, **88**f, 88m, 88*, **89***, 108*

Kansu nuthatch *(Sitta bangsi)*, **323**
Keel-billed toucan *(Ramphastus sulfuratus)*, 85, **98***
Kenopia striata (Striped wren babbler), **236**
Kokako *(Callaeas cinerea)*, 469*, **483**
Kilham, L., 94, 104f
King bird of paradise *(Cicinnurus regius)*, 470*, **494**f
King of Saxony's bird of paradise *(Pteridophora alberti)*, 479*, **496**
King-crow *(Dicrurus macrocercus)*, **477***
Kingfisher (Eurasian) (See Eurasian kingfisher)
Kingfishers (Alcedinidae), 21, **22**f
Kinglet warbler *(Phylloscopus proregulus)*, **262**
Kinglets (Regulinae), 233, **267**f
Kiskadee flycatchers *(Pitangus)*, 151f
Kiyosu, 145
Kluyver, 255
Kneitz, G., 95
Koenig, Lilly, 40
— Otto, 248
Kookaburra or laughing jackass *(Dacelo gigas)*, **33**f, **38***
Korean flycatcher *(Ficedula zanthopygia)*, **275**
Korean nuthatch *(Sitta villosa)*, **323**
Kramer, Peter, 363, 365f

Lack, David, 287, 365
Lagonosticta (Fire finches), **428**, **458**

INDEX 635

— *larvata* (Masked waxbill), 440*, **458**
— *rhodopareia* (Jameson's fire finch), 429*, **458**, 458m
— *rubricata* (Blue-billed fire finch), 440*, **458**, 458m
— *rufopicta* (Bar-breasted fire finch), **458**
— *senegala* (Fire finch), 413*, 429*, **458**, 458m
Lalage (Trillers), 198
— *aurea*, 450*
— *leucomela* (Varied triller), 198
— *sueurii* (White-winged triller), 450*
Lamprospreo
— *superbus* (Superb glossy starling), **474**
Lamprotornis
— *australis* (Burchell's glossy starling), **475**
— *caudatus* (Long-tailed glossy starling), **475**
— *chalybaeus* (Blue-eared glossy starling), **475**
— *corruscus*, 460*
— *nitens* (Red-shouldered glossy starling), **475**
— *purpureus* (Purple glossy starling), 460*, **475**
— *splendidus* (Splendid glossy starling), **475**
Laniarius
— *atrococcineus* (Crimson-breasted shrike), 208f, 208m
— *barbarus* (Gonolek), 195*
Laniidae (Shrikes), 207ff
Laniinae (True shrikes), 207, **209**f
Lanioturdus torquatus (Chat shrike), **208**
Lanius (Shrikes proper), 209ff
— *collurio* (Red-backed shrike), 209ff, 209m, 211*
— — *collurio*, 195*
— — *isabellinus* (Isabelline red-backed shrike), 195*, **212**
— *cristatus* (Brown shrike), 195*, **212**
— *excubitor* (Northern shrike), 195*, 201*, 202*, **211**f
— *minor* (Lesser gray shrike), 195*, **211**f
— *nubicus* (Masked shrike), 195*, **212**
— *schach* (Rufous-backed shrike), 195*, **212**
— *senator* (Woodchat shrike), 195*, **210**
Lapland longspur (*Calcarius lapponicus*), 342*, **355**, 355m, 355*
Lapwing (*Vanellus vanellus*), 413*
Large amakihi (*Viridonia* or *loxops sagittirostris*), 387
Large ground finch (*Geospiza magnirostris*), 351*, **359**, 363*
Large insectivorous tree finch (*Camarhynchus psittacula*), **360**, 360*, 364*
Large niltava (*Niltava grandis*), **275**
Large sharp-beaked ground finch (*Geospiza difficilis septentrionalis*), 364, 366m
Larks (Alaudidae), 171ff
Lau, Hermann, 484
Laubmann, 515
Laughing thrushes (*Garrulax*), 243
Lavender finch (*Estrilda caerulsecens*), 440*, **457**
Lawe's six-wired parotia (*Parotia lawesi*), **495**
Lawrence's goldfinch (*Carduelis lawrencei*), **409**
Laysan apapane (*Himatione sanguinea freethii*), **390**, 393
Laysan finch (*Psittirostra cantans*), 387f, 389*, 389f
Lazuli bunting (*Passerina amoena*), 367
Leafbirds (Chloropseinae), 204
Leafbirds (*Chloropsis*), 204f
Leaf-scrapers (*Sclerurus*), 132
Leaf warblers, 250, **262**f
Least flycatcher (*Ficedula parva*), 260*, 274m, **275**
Least honey guide (*Indicator exilis*), **77**
Legatus leucophaius (Piratical flycatcher), **151**, 155*, 156
Legua starling (*Necropsar leguati*), 472
Leiothrix
— *argentauris* (Silver-eared mesia), 222*, **244**
— *lutea* (Pekin robin), 222*, **244**
Lemon-breasted flycatcher (*Microeca flavigaster*), **276**
Leonardina woodi, **235**
Lepidocolaptes, 128
— *affinis* (Spot-crowned woodcreeper), **128**
— *angustirostris* (Narrow-billed woodcreeper), 120*
Leptopoecile
— *elegans*, 268
— *sophiae*, 268
Leptoptilus (Blue vangas), 213
— *chabert* (Chabert vanga), **213**
— *madagascarinus* (Blue vanga), 196*, **213**f
— *viridis* (White-headed vanga), **213**
Leptosomatinae (Cuckoo-rollers), 42f
Leptosomus discolor (Cuckoo-roller), **42**f, 71*f
Leschenault's forktail (*Enicurus leschenaulti*), 280*, **293**, 293*
Lesser bird of paradise (*Paradisaea minor*), **493**f
Lesser elaenia (*Elaenia chiriquensis*), 152f
Lesser goldfinch (*Carduelis psaltria*), **409**
Lesser gray shrike (*Lanius minor*), 195*, **211**f
Lesser great Indian hornbill (*Buceros bicornis bicornis*), 56, 56m
Lesser honey guide (*Indicator minor*), **77**, 97*
Lesser koa finch (*Psittirostra flaviceps*), 394
Lesser melampitta (*Melampitta lugubris*), **249**
Lesser pied kingfisher (*Ceryle rudis*), 26m, **27**, 38*
Lesser racket-tailed drongo (*Dicrurus remifer*), **477**, 481
Lesser scaly-breasted wren babbler (*Pnoepyga pusilla*), 237f
Lesser short-toed lark (*Calandrella rufescens*), **174**
Lesser shortwing (*Brachypteryx leucophrys*), **285**
Lesser spotted woodpecker (*Dendrocopos minor*), 94*, **109**, 113f, 113m, 119*
Lesser white-throat (*Sylvia curruca*), 232*, **258**, 262
Lesser yellow-eared spider hunter (*Arachnothera chrysogenys*), 320*, **339**
Leuconerpes
— *candidus* (White woodpecker), 119*
Leucopsar rothschildi (Bali mynah), 449*, **471**
Leucosticte (Rosy finches), 404, **412**
— *arctoa* (Gray-crowned rosy finch), 392*, **412**
Levaillant's barbet (*Trachyphonus vaillantii*), **74**
Lichmera indistincta (Brown honey eater), 345
Lidth's jay (*Garrulus lidthi*), **508**
Lilac-breasted roller (*Coracias caudata*), **44**f, 44m, 57*
Lindsdale, 410
Linnaeas, Carl von, 403
Linnets (*Acanthis*), 404, **411**f
Linschoten, Jan von, 488
Liucichla
— *phoenicea*, 243
Liosceles thoracicus (Rusty-belted tapaculo), 140
Lipaugus
— *strephophorus* (Rose-collared piha), 144*
Little bee-eater (*Melittophagus pusillus*), **41**
Little forktail (*Enicurus scouleri*), **293**
Little rush warbler (*Bradypterus baboecala*), **253**
Little thornbill (*Acanthiza nana*), **265**
Little wattlebird (*Anthochaera chrysoptera*), 346
Little weaver (*Sitagra luteola*), 430*
Little wood swallow (*Artamus minor*), **485**
Little yellow flycatcher (*Erythrocercus holochlorus*), **277**
Livingstone's flycatcher (*Erythrocercus livingstonei*), **277**
Loboparadisaea sericea (Yellow-breasted bird of paradise), **490**
Lochmias
— *nematura* (Sharp-tailed streamcreeper), **125**
Locustella
— *certhiola* (Pallas' grasshopper warbler), 231*, **253**, 256
— *fasciolata* (Taiga grasshopper warbler), **253**, 256
— *fluviatilis* (River warbler), **253**, 257
— *lanceolata* (Streaked or Temminck's grasshopper warbler), **253**, 256
— *luscinioides* (Savi's warbler), 231*, **253**, 256f
— *naevia* (Grasshopper warbler), 231*, 247*, **253**, 255m, 256
— *ochotensis* (Middendorf's warbler), **253**
Locust finch (*Ortygospiza locustella*), **456**
Loke Wan Tho, 503
Lonchura, 448ff
— *castaneothorax* (Chestnut-breasted finch), 439*, **448**
— *flaviprymna* (Yellow-tailed finch), **448**
— *maja* (White-headed mannikin), **448**
— *malacca* (Three-colored mannikin), **448**, 451m
— — *ferruginosa*, 451m
— — *malacca*, 439*, 451m
— *pectoralis* (Pectorella finch), **448**, 451m
— *punctulata* (Spice finch), 439*, **448**, 448m
— *spectabilis* (Spectacled finch), **448**, 451
— *striata* (Sharp-tailed munia), **448**
— — *swinhoei*, 451
Long-billed crombec (*Sylvietta rufescens*), 248m, 249*, **250**f
Long-billed gnatwren (*Ramphocaenus melanurus*), **271**
Long-billed marsh wren (*Cistothorus palustris*), 190*, **223**, 225
Long-billed wren babbler (*Rimator malacoptilus*), 235f
Long-spurred pipits (*Macronyx*), 193
Longspurs (*Calcarius*), 353
Long-tailed broadbill (*Psarisomus dalhousiae*), 120*, 124
Long-tailed glossy starling (*Lamprotornis caudatus*), **475**
Long-tailed grass finch (*Poephila acuticauda*), 439*, **453**, 453m
Long-tailed ground roller (*Uratelornis chimaera*), **43**f
Long-tailed meadowlark (*Sturnella loyca*), 381*
Long-tailed paradigalla (*Paradigalla carunculata*), 480*, **490**
Long-tailed shrike (*Corvinella corvina*), 213
Long-tailed sibia (*Heterophasia picaoides*), **245**
Long-tailed silky flycatcher (*Ptilogonys caudatus*), **216**f
Long-tailed sunbirds (*Nectarinia*), 337f
Long-tailed tit (*Aegithalos caudatus*), 308*, **311**, 311m
Long-tailed tyrant (*Colonia colonus*), **151**, 151*
Long-tailed widow bird (*Coliuspasser progne*), **434**, 434m
Long-tailed wren babbler (*Spelaeornis chocolatinus*), **238**
Löns, Hermann, 327
Lophorina superba (Superb bird of paradise), 470*, 495*, **496**
Lophozosterops, 339
— *dohertyi*, 339f
Lorenz, Konrad, 514f
Loria loriae (Loria's bird of paradise), **490**
Loria's bird of paradise (*Loria loriae*), **490**
Louisiana waterthrush (*Seiurus motacilla*), 383
Lovely cotinga (*Cotinga amabilis*), 133*
Lowland eupetes (*Ptilorrhoa caerulescens*), **249**
Lowther, 420
Loxia (Crossbills), 404, 406, **418**
— *curvirostra* (Red crossbill), **418**, 418m, 418*
— *leucoptera* (White-winged crossbill), **418**
— *pytyopsittacus* (Parrot crossbill), **418**, 418m
Loxops
— *coccinea* (Akepa), 372*, 387*, 388*, 389
Lucanus, von, 88
Lullula arborea (Woodlark), **176**, 176m, 179*
Luscinia (Nightingales), 286f
— *calliope* (Ruby throat), 279*, **286**, 286m
— *luscinia* (Thrush nightingale), 279*, **286**, 286m
— *megarhynchos* (Nightingale), 279*, 285m, **286**, 288
— *svecica* (Blue throat), 279*, **286**ff, 286m
Lybius
— *bidentatus* (Double-toothed barbet), **70**f
— *dubius* (Bearded barbet), **70**f
— *leucocephalus* (White-headed barbet), **71**f
— *rolleti* (Black-breasted barbet), **70**
— *torquatus* (Collared barbet), **71**f,

97*
— *vielloti* (Viellot's barbet), 75
Lyrebirds *(Menura)*, 164f, 164*, 165*
Lyrebirds (Menuridae), 164f
Lyre-tailed honey guide *(Melichneutes robustus)*, 77

MacGregor's bower bird *(Amblyornis macgregoriae)*, 503f, 503*
Machaerirhynchus flaviventer (Papuan flycatcher), 277, 277*
Machaeropterus
— *pyrocephalus* (Fiery-capped manakin), 158f
Machetornis rixosa (Cattle tyrant), 143*, 151, 153
Mackenziaena (Parana shrikes), 132
— *leachii* (Spotted parana shrike), 132
Mackrodt, P., 246
Mackworth-Praed, C. W., 124
Macronous (True tit babblers), 239
— *ptilosus* (Fluffy-backed tit babbler), 239
Macronyx (Long-claws), 193
— *capensis* (Cape long-claw), 193
— — *colletti*, 189*
Macrosphenus (Bush creepers), 250
— *flavicans* (Yellow longbill), 250f, 250*
Madagascar jerys *(Neomixis)*, 238
Madagascar nuthatch *(Hypositta corailirostris)*, 214
Madagascar starling *(Hartlaubius auratus)*, 475
Madagascar weaver *(Foudia madagascariensis)*, 437
Magalaima zeylanica (Green barbet), 75f
Magellanic woodpecker *(Campephilus magellanicus)*, 116
Magnificent bird of paradise *(Diphyllodes magnificus)*, 491*, 495
Magnificent rifle bird *(Pitloris magnifica)*, 479*, 499
Magnolia warbler *(Dendroica magnolia)*, 371*, 384
Magpie *(Pica pica)*, 497*, 501*, 511, 511m
Magpie jay *(Calocitta formosa)*, 508
Magpie lark *(Grallina cyanoleuca)*, 469*, 483
Magpie-larks (Grallininae), 483
Magpie mannikin *(Spermestes fringilloides)*, 447f
Magpie robin *(Copsychus savlaris)*, 291
Magpie robins *(Copsychus)*, 285, 291
Magpie shrike *(Urolestes melanoleucus)*, 195*, 212f
Magpie shrikes *(Urolestes)*, 209
Magpie starling *(Speculipastor bicolor)*, 473f
Magpie tanager *(Cissopis leveriana)*, 361*, 377
Magpie-larks (Australian) (See Australian magpie-larks)
Magpies (Australian) (See Australian magpies)
Mahali weaver bird *(Plocepasser mahali)*, 443
Malabar mynah *(Temenuchus malabaricus)*, 471
Malachite kingfisher *(Corythornis cristata)*, 25f, 25m
Malachite sunbird *(Nectarinia famosa)*, 338
Malaconotinae (Bush shrikes), 207, 208f
Malaconotus, 209m
— *blanchoti* (Gray-headed bush shrike), 209

Malacopteron, 235
— *affine* (Sooty-headed babbler), 222*
— *magnum* (Greater red-headed babbler), 235
Malacoptila, 63
— *panamensis* (White-whiskered puffbird), 64
— *striata* (Crescent-chested puffbird), 64, 68*
Malayan cochoa *(Cochoa azurea)*, 293f
Malayan honey guide *(Indicator archipelagus)*, 77
Malimbus, 438
— *malimbicus* (Crested malimbe), 430*, 442, 443m
Malurinae (Australian wrens or wren warblers), 233, 265, 265m
Malurus
— *cyaneus* (Superb blue wren), 266
— *lamberti* (Variegated wren), 259*, 266
Mamo (Drepanis pacifica), 372*, 390, 394, 394*
Manacus (Manakins), 158
— *manacus* (White-bearded manakin), 143*, 158ff
Manakins (Pipridae), 127, 157f, 158m
Mandarian mynah *(Sturnia sinensis)*, 471
Mandell's snow finch *(Montifringilla taczanowskii)*, 424
Mandingoa nitidula (Green-backed twin-spot), 440*, 458
Mangrove finch *(Cactospiza heliobates)*, 360, 364
Mangrove fly-eater *(Gerygone sulphurea)*, 265f
Mangrove honey eater *(Meliphaga fasciogularis)*, 347
Mangrove kingfisher *(Halcyon senegaloides)*, 28
Mannikins *(Spermestes)*, 447f
Manorina flavigula (Yellow-throated miner), 333*
— *melanophrys* (Bell miner), 345
Manucodia
— *chalybata* (Green-breasted manucode), 490
Many-colored rush tyrant *(Tachuris rubrigastra)*, 151
Manyar weaver *(Ploceus manyar)*, 430*, 437m, 438
Maori crow *(Corvus moriorum)*, 516
Marchant, S., 217
Mariqua flycatcher *(Bradornis mariquensis)*, 277
Marmora's warbler *(Sylvia sarda)*, 258, 262m
Maroon oriole *(Oriolus traillii)*, 450*, 476
Maroon woodpecker *(Blythipicus rubiginosus)*, 115f
Maroon woodpeckers *(Blythipicus)*, 115
Marquesas flycatcher *(Pomarea mendozae)*, 281
Marsh tit *(Parus palustris)*, 302*, 307*, 317
Marsh warbler *(Acrocephalus palustris)*, 232*, 253, 254m, 255f
Marshall, A. J., 499f
Martini, E., 106
Mascarene martin *(Phedina borbonica)*, 185
Masked grass finch *(Poephila personata)*, 453
Masked oriole *(Oriolus larvatus)*, 476f
Masked shrike *(Lanius nubicus)*, 195*, 212

Masked tanager *(Tangara nigrocinta)*, 361*, 374
Masked tityra *(Tityra semifasciata)*, 161
Masked waxbill *(Lagonosticta larvata)*, 440*, 458
Maversberger, 174
Mavi parrotbill *(Pseudonestor xanthophrys)*, 387, 389
Mayer, Fred Shaw, 494
Mayr, E., 167, 325
Meadow pipit *(Anthus pratensis)*, 189*, 193, 193m
Meadowlarks *(Sturnella)*, 397, 401
Meadows, 216
Medium ground finch *(Geospiza fortis)*, 363*
Medium insectivorous tree finch *(Camarhynchus pauper)*, 365*
Medium sharp-beaked ground finch *(Geospiza difficilis debilirostris)*, 366, 366m, 367*
Mees, G. F., 345
Megaceryle alcyon (Belted kingfisher), 26f, 26m, 37*
— *maxima* (Giant kingfisher), 26m, 27
— *torquata* (Ringed kingfisher), 27
Megalaima
— *asiatica* (Blue-cheeked barbet), 75, 97*
— *franklinii* (Golden-throated barbet), 75
— *haemacephala* (Coppersmith), 75f
— — *rosea*, 75
— *mystacophanos* (Gaudy barbet), 75
— *virens* (Great Hill barbet), 75, 97*
— *zeylanica* (Green barbet), 75f
Megarhynchus pitangua (Boat-billed flycatcher), 151
Meise, Wilhelm, 146, 156f, 160, 162
Melaenornis pammelaina (Black flycatcher), 277
Melampitta (Melampittas), 249
— *lugubris* (Lesser melampitta), 249
Melanerpes, 104m
Melanerpes erythrocephalus (Red-headed woodpecker), 104f, 119*
— *formicivorus* (Acorn woodpecker), 104f, 105*
— *thyroideus*, 106
Melanocharis (Berry peckers), 330
— *nigra* (Blackberry pecker), 331
Melanochlora sultanea (Sultan tit), 308*, 321
Melanocorypha bimaculata
— *calandra* (Calandra lark), 174, 174m, 179*
— *leucoptera* (White-winged lark), 179*
Melanodera (Melanodera finches), 354
— *melanodera* (Black-throated finch), 354, 357
Melanodera finches *(Melanodera)*, 354
Melanospiza
— *richardsoni* (Stolucia black finch), 358
Melanotis
— *caerulescens* (Blue mockingbird), 221*
Melba finch *(Pytilia melba)*, 413*, 429*, 431, 432*, 461, 461*, 461m
Melichneutes
— *robustus* (Lyre-tailed honey guide), 77
Melidectes
— *torquatus* (Cinnamon-breasted wattle bird), 320*
Melidora macrorrhina (Hook-billed kingfisher), 34

Melignomon zenkeri (Tinker honey guide), 77f
Meliphaga chrysops (Yellow-faced honey eater), 347
— *fasciogularis* (Mangrove honey eater), 347
— *ornata* (Yellow-plumed honey eater), 346
— *nunicolor* (White-gaped honey eater), 346
— *virescens* (Singing honey eater), 345f
Meliphagidae (Honey eaters), 344ff
Meliphaginae, 344
Melithreptes, 345
— *lunatus* (White-naped honeyeater), 320*, 347
Melittophagus, 41f
— *bulocki* (Red-throated bee-eater), 41, 48*
— *gularis* (Black bee-eater), 41
— *muelleri* (Blue-headed bee-eater), 41
— *pusillus* (Little bee-eater), 41
— *variegatus*, 48*
Melodious warbler *(Hippolais polyglotta)*, 253, 255m, 257, 257*
Melophus (Crested buntings)
— *lathami* (Crested bunting), 350
Melospiza
— *melodia* (Song sparrow), 342*, 353, 356f, 356*
Menura (Lyrebirds), 164f, 164*, 165*
— *alberti* (Prince Albert's lyrebird), 164f, 164m
— *novaehollandiae* (Superb lyrebird 134, 135*, 164f, 164m
Menuridae (Lyrebirds), 164f
Menzel, H., 89
Meropidae (Bee-eaters), 21, 36f
Meropogon forsteni, 42
Merops (Bee-eaters), 39f
— *apiaster* (Common or European bee-eater), 32*, 39ff, 40m, 48*
— *nubicoides* (Southern carmine bee-eater), 40
— *nubicus* (Carmine bee-eater), 40, 48*
— *ornatus* (Rainbow bee-eater), 40
— *superciliosus* (Blue-cheeked bee-eater), 40, 40m
— *viridis* (Chestnut-headed bee-eater), 30*, 40
Mexican jay *(Aphelocoma ultramarina)*, 507
Microbates collaris (Collared gnatwren), 271
Microeca flavigaster (Lemon-breasted flycatcher), 276
Micronesian broad-bill *(Myiagra oceanica)*, 278
Middendorf's warbler *(Locustella ochotensis)*, 253
Middle spotted woodpecker *(Dendrocopos medius)*, 109, 113, 113m, 119*
Migratory Chinese grosbeak *(Eophona migratoria)*, 419f
Miller, 229
Millikan, 365
Mimidae (Mockingbirds and thrashers), 219, 227f, 227m
Mimus (Mockingbirds)
— *polyglottos* (Mockingbird), 221*, 227ff
Miners *(Geositta)*, 129
Minivets *(Pericrocotus)*, 198f
Mino anais (Golden-breasted mynah), 473
— *dumontii* (Dumont's mynah), 449*, 473

Miocorax, 506
Mirafra
— *apiata* (Clapper lark), **172**
— *hypermetra* (Red-winged bush lark), **179***
— *rufocinnamomea* (Flappet lark), **172**, 172m
Miro, 278
Mistle thrush *(Turdus viscivorus)*, 290*, **304f**, 304m
Mistletoe bird *(Dicaeum hirundinaceum)*, **329**
Mniotilta, 380
— *varia* (Black-and-white warbler), 371*, **383f**, 384*
Mockingbird *(Mimus polyglottos)*, 221*, **227ff**
Mockingbirds and thrashers (Mimidae), 219, **227f**, 227m
Moho
— *braccatus* (Ooaa), 320*, **388**
Molothrus (Cowbirds), 396
— *ater* (Brown-headed cowbird), **399**, 399m, 399*
— *badius* (Bay-winged cowbird), **398**, 398m, 399*
— *bonariensis* (Shining cowbird), 381*, **399**
Moltoni, E., 511
Momotidae (Motmots), 21, **35f**
Momotus momota (Blue-crowned motmot), 32*, **36**, 36*, 48*
Monarch flycatchers *(Monarcha)*, 281
Monarcha (Monarch flycatchers), 281
— *cinerascens* (Island gray-headed monarch flycatchers), **281f**
— *godeffroyi* (Yap monarch flycatchers), **281f**
— *melanopis* (Black-faced flycatcher), **281f**, 282*
Monarchinae (Paradise flycatchers), 272, **281f**, 281m
Monasa (Nunbirds), 64f
— *atra* (Black nunbird), **65**
Montezuma oropendola *(Gymnostinops montezuma)*, 381*
Monticola (Rock thrushes), 285, **298f**
— *saxatilis* (Rock thrush), 289*, **298f**, 298m
— *solitarius* (Blue rock thrush), 289*, 298*, **298f**, 298m
Montifringilla (Snow finches), 423
— *adamsi* (Tibet snow finch), **423f**, 423m
— *blanfordi* (Blanford's snow finch), **424**, 424m
— *nivalis* (Snow finch), 416*, **423**, 423m
— *taczanowskii* (Mandell's snow finch), **424**
— *theresae*, 416*, **424**
Moreau, R. E., 73
Motacilla (True wagtails), 194
— *alba* (White wagtail), 184*, 189*, **194**, 197f, 203m, 205*
— — *yarrellii* (Pied wagtail), 194
— *cinerea* (Gray wagtail), 183*, 184*, 189*, **194**, 200m
— *citreola* (Yellow-headed wagtail), 194, 200m
— *flava* (Yellow wagtail), 184*, 189*, **194f**, 199m, 199*
— — *tschutschensis*, 194
Motacillidae (Wagtails), 192ff
Motmots (Momotidae), 21, **35f**
Mountain blackeye *(Chlorocharis emiliae)*, **340**
Mountain chickadee *(Parus gambeli)*, **318**
Mountain drongos *(Chaetorhynchus)*, 477, 477m
Mountain leaf warbler *(Phylloscopus trivirgatus)*, **263**
Mountain newtonia *(Newtonia brunneicauda)*, **276**
Mountain toucans *(Andigena)*, **80**
Mountain wheatear *(Oenanthe monticola)*, **295**
Mountain white-eye *(Zosterops montana)*, 343, **372***
Mountain wren *(Troglodytes solstitialis)*, **226**
Mountford, 419
Moupinia, 239
— *poecilote* (Rufous-crowned babbler), 239, **240**
Mourning wheatear *(Oenanthe lugens)*, **297**
Mouse-colored penduline tit *(Anthoscopus musculus)*, 308*
Mouse-colored tapaculo *(Scytalopus speluncae)*, 140
Moussier's redstart *(Phoenicurus moussieri)*, **291**
Moustached turca *(Pteroptochos megapodius)*, 140
Moustached warbler *(Acrocephalus melanopogon)*, 231*, **252**, 254f, 256*
Mudnest choughs (Corcoracinae), **483f**
Mudnest choughs (Corcoracinae), **483f**
Muelleripicus (Slaty woodpeckers), 102
Multi-crested bird of paradise *(Cnemophilus macgregorii)*, **490**, 492*
Muscicapa
— *adusta* (Dusky flycatcher), **276**
— *gambagae*, **276**
— *latirostris* (Brown flycatcher), **276**
— *sibirica* (Siberian flycatcher), **276**
— *striata* (Spotted flycatcher), 260*, **276**
Muscicapidae (Flycatcher-like birds), 233ff, 272ff
Muscicapinae (Flycatchers proper), **272ff**, 274m
Muscicapoidea (Primitive insectivores), 219ff
Muscisaxicola albifrons (White-fronted groundtyrant), 151, 153
Muscivora
— *forficata* (Scissor-tailed flycatcher), 151, 153
— *tyrannus* (Fork-tailed flycatcher), 143*, **151**
Musician wren *(Cyphorhinus aradus)*, **224**
Myadestes (Solitaires), 285, 294
— *townsendi* (Townsend's solitaire), 289*, **294**
Mycerobas, 404
— *icterioides* (Black-and-yellow grosbeak), **419f**
Myiagra oceania (Micronesian broad-bill), **278**
Myiarchinae, 152
Myiarchus, 152, 155
— *crinitus* (Great crested flycatcher), **152**
Myiobius sulphureipygius (Sulphur-rumped flycatcher), 152, 154*
Myioborus, 380
— *pictus* (Painted redstart), 371*, **385**
— *torquatus* (Collared redstart), 371*, **385**
Myioceryx lecontei (Dwarf kingfisher), **25f**
Myiodynastes, 151, 155f

— *luteiventris* (Sulphur-bellied flycatcher), 151, 154
Myiophoneus (Whistling thrush group), 299
— *caeruleus* (Blue whistling thrush), 280*, **299**
— *glaucinus* (Sunda whistling thrush), **299**
Myiornis auricularis (Eared pygmy tyrant), 143*
Myiozetetes (Small kiskadee flycatchers), 151
— *similis* (Vermillion-crowned flycatcher), 151, 153*, **155f**
Myrmeciza exsul (Chestnut-backed ant bird), 126*
Myrmecocichla (Anteater chats), 295
— *aethiops* (Anteater chat), **295**, 297
Myrmotherula (Antwrens), **132f**
— *brachyura* (Short-tailed antwren), 126*, **137**
Myrtle warbler *(Dendroica coronata)*, **383**
Myzornis pyrrhoura (Firetail), **244**
McArthur, Robert, 383
McKay's bunting *(Plectrophenax hyperboreus)*, **354**

Naked-faced barbet *(Gymnobucco calvus)*, 73
Napal babbler *(Alcippe nipalensis)*, 222*
Napothera, 236
Narcondam hornbill *(Rhyticeros narcondami)*, **56**
Narcissus flycatcher *(Ficedula narcissina)*, 260*, **275**
Narrow-billed tody *(Todus angustirostris)*, **34**
Narrow-billed woodcreeper *(Lepidocolaptes angustirostris)*, 120*
Natal kingfisher *(Ispindina picta)*, **25f**, 25m
Natal sugarbird *(Promerops gurneyi)*, 347
Necklaced laughing thrush *(Garrulax moniliger)*, **243**
Necropsar leguati (Legua starling), **472**
— *rodericanus* (Rodriguez starling), **472**
Nectarina reichenowi (Golden-winged sunbird), **332**
Nectarinia (Long-tailed sunbirds), **337f**
— *amethystina*, **338**
— *coccinigaster*, **338**
— *famosa* (Malachite sunbird), **338**
— *minima* (Small sunbird), **338**
— *olivacea*, **338**
— *superba*, **338**
Nectariniidae (Sunbirds), 331ff
Negro finches *(Nigrita)*, 462
Neochmia phaeton (Crimson finch), 439*, **455**, 455m
Neocichla gutturalis (White-winged babbling starling), **475**
Neocossyphus (Ant thrushes), 294
— *rufus* (Red-tailed ant thrush), **294**
Neodrepanis coruscans (False sunbird), **147**, 150*
Neomixis (Madagascar jerys), 238
— *viridis* (Green jery), **238**
Neopelma, 158f
— *aurifrons* (Wied's tyrant manakin), **158**
Neositta
— *chrysoptera* (Orange-crowned sitella), **325**
— *pileata*, 319*
Neosittinae (Tree runners), **324f**
Neospiza (Tristan buntings), 354
— *acunhae* (Cunha bunting), **354**

Neospiza concolor (Sao Tome grosbeak), **442**
Nesoceleus fernandinae (Fernandina's woodpecker), **96**
Nesocharis
— *ansorgei* (White-collared oliveback), **461f**
— *capistrata* (Gray-headed oliveback), **461**
— *shelleyi*, 461
Nesoctites micromegas (Hispaniola piculet), **89**
Nesomimus parvulus (Galapagos mockingbird), 221*, **227**
New Guinea mountain white-eye *(Zosterops novaeguineae)*, 339
New World nightingale-like thrushes *(Cathurus)*, 303
New Zealand pipit *(Anthus novaeseelandiae)*, 189*, **194**
New Zealand rock wren *(Xenicus gilviventris)*, 147*, **148**
New Zealand wattlebirds (Callaeidae), **482f**
New Zealand wrens (Xenicidae), 127, 127m, **147f**
Newton, A., 340
Newtonia brunneicauda (Mountain newtonia), **276**
Nice, Margaret Morse, 356
Niethammer, Günther, 72, 432
Nieuwenhuis bulbul *(Pycnonotus nieuwenhuisi)*, 203
Night bee-eaters *(Nyctyornis)*, 42
Nightingale *(Luscinia megarhynchos)*, 279*, 285m, **286**, 288
Nigrita (Negro finches), 462
— *bicolor* (Chestnut-breasted negro finch), 462
— *canicapilla* (Gray-headed negro finch), 462, 462m
Nihoan Laysan finch *(Psittirostra cantans ultima)*, 390
Nilaus afer (Brubru), **208**
Niltava (Niltavas), 275
— *grandis* (Large niltava), **275**
— *sundara* (Rufous-bellied niltava), 260*, **275**
Noisemakers (Clamatores), 118, 123, **127f**
Noisy pitta *(Pitta versicolor)*, **142**, 146
Nonnula rubecula (Red-throated nunlet), 64*, 68*
North American wren tits *(Chamaea)*, 239
Northern bentbill *(Oncostoma cinereignathus)*, **152**, 155
Northern royal flycatcher *(Onychorhinus mexicanus)*, 143*, **152**, 154, 156*
Northern scrub robin *(Drymodes superciliaris)*, **286**
Northern shrike *(Lanius excubitor)*, 195*, 201*, 202*, **211f**
Northern three-toed woodpecker *(Picioides tridactylus)*, 91*, **115**, 119*
Notharchus, 64f
— *macrorhynchus* (White-necked puffbird), 68*, **68**
— *pectoralis* (Black-breasted puffbird), **64**
Notiomystis cincta, 320*
Notiospiza angolensis (Bar-winged weaver), **442**
Nucifraga
— *caryocatactes* (Nutcracker), 497*, **512**, 512m
— — *caryocatactes* (Alpine nutcracker), **512f**
— — *macrorhynchos* (Siberian nutcracker), **512f**
— *columbiana* (Clark's nutcracker),

497*, **512**
Nukupau *(Hemignathus lucidus)*, 389*
Nun babbler *(Alcippe poioicephala)*, 245
Nun babblers *(Alcippe)*, 245
Nunbirds *(Monasa)*, 64f
Nutcracker*(Nucifraga caryocatactes)*, 497*, **512**, 512m
Nuthatch (European) (See European nuthatch)
Nuthatches proper (Sittinae), 310, **321**
Nyctyornis (Night bee-eaters), 42
— *amicuctus* (Red-bearded bee-eater), **42**, 48*
— *athertoni* (Blue-headed bee-eater), **42**
Nystalus chacura
— *radiatus* (Barred puffbird), **64**, 68*

Ocellated tapaculo *(Acropternis orthonyx)*, 140
Odontospiza caniceps, (Gray-headed silver-bill), **447**
Oenanthe (Wheatears), 295, 297
— *deserti* (Desert wheatear), **295**, 298
— *hispanica* (Black-eared wheatear), **295**, 295m, 297f
— *leucopyga* (White-crowned black wheatear), **295**, 297f
— *lugens* (Mourning wheatear), **295**
— *moesta* (Red-rumped wheatear), **295**, 298
— *monticola* (Mountain wheatear), **295**
— *oenanthe* (Wheatear), 270*, 280*, **295**, 295m, 297, 297*
— — *leucorrhoa*, 297f
— *picata* (Pied chat), **295**, 297
— *pileata* (Pied chat), 280*
Old World warblers (Sylviinae), 233, **249**f
Olivaceous warbler *(Hippolais pallida)*, **254**, 255m, 257f
Olivaceous woodcreeper *(Sittasomus griseicapillus)*, 120*
Olive thrush *(Turdus olivaceus)*, **304**
Olive-backed jungle flycatcher *(Rhinomyias olivacea)*, **276**
Olive-bellied flycatcher *(Pipromorpha oleaginea)*, 152
Olive-breasted mountain greenbul *(Pycnonotus tephrolaemus)*, 203
Olive-tree warbler *(Hippolais olivetorum)*, **253**f, 257
Oncostoma cinereigulare (Northern bentbill), **152**, 155
Onychognathus
— *morio* (Red-winged starling), **474**
— *nabouroup* (Pale-winged starling), **474**
— *tristramii* (Tristram's grackle), **474**
Onychorhynchus (Royal flycatchers), 152
— *mexicanus* (Northern royal flycatcher), 143*, **152**, 154, 156*
Ooaa *(Moho braccatus)*, 320*, **388**
Orange bishop *(Euplectes franciscanus)*, **435**, 435m
Orange cuckoo-shrike *(Campochaera sloetii)*, 198
Orange-bellied leafbird *(Chloropsis hardwickei)*, 204
Orange-billed nightingale thrush *(Catharus aurantiirostris)*, 221*, **303**
Orange-billed sparrow *(Arremon aurantiirostris)*, 342*

Orange-breasted bunting *(Passerina leclancherii)*, 367
Orange-breasted flower pecker *(Dicaeum trigonostigma)*, **330**
Orange-cheeked waxbill *(Estrilda melpoda)*, 440*, **456**
Orange-crowned oriole *(Icterus auricapillus)*, **400**
Orange-crowned sitella *(Neositta chrysoptera)*, **325**
Orange-flashed bush robin *(Tarsiger cyanurus)*, 279*
Orange-gorgeted flycatcher *(Ficedula strophiata)*, **275**
Orange-headed ground thrush *(Zoothera citrina)*, 221*, 289*, **300**
Orangequit *(Euneornis campestris)*, 378
Oregon junco*(Junco oreganus)*, 342*, **353**, 356, 356*
Oreocharis, 330
— *arfaki* (Tit berry pecker), **330**, 372*
Oreoscoptes montanus (Sage thrasher), **227**f
Oreosthruthus fuliginosus (Crimson-bellied mountain finch), **455**
Oriental golden-eyed babbler *(Chrysomma sinense)*, **240**
Oriental white-eye *(Zosterops palpebrosa)*, **340**, 343
Orioles (Oriolidae), 475f
Orioles(American) (See American orioles)
Oriolia bernieri (Bernier's vanga), **214**
Oriolidae (Orioles), 475f
Oriolus brachyrhynchus
— *chinensis* (Black-naped oriole), 450*, **476**
— *chlorocephalus*, **477**
— *larvatus* (Masked oriole), 476f
— *mellianus*, **476**
— *monacha* (Black-headed forest oriole), **476**
— *oriolus* (Golden oriole), 413*, 450*, **476**, 476m
— *traillii* (Maroon oriole), 450*, **476**
Orphean warbler *(Sylvia hortensis)*, **258**, 261
Orr, 360f
Orthonychinae (Rail-babblers), 248
Orthonyx temminckii (Southern chowchilla), 222*
Orthotomus atrogularis
— *sutorius* (Tailor bird), 259*, **265**, 265*
Ortolan bunting *(Emberiza hortulana)*, 341*, **349**f, 349m, 353
Ortygospiza (Quail finches), 455f
— *atricollis* (South African quail finch), 440*, **456**, 456m
— *locustella* (Locust finch), **456**
Oscines (Songbirds), 121, **167**f
Othyphantes baglafecht, 438
Ou *(Psittirostra psittacea)*, 387, 389*, **389**, 394
Ovenbird *(Seiurus aurocapillus)*, 371*, **384**f, 385m
Ovenbirds (Furnariidae), 127, **129**, 129m
Ovenbirds (Furnarius), 129
Oxyruncidae (Sharpbills), 127, **157**, 157m
Oxyruncus cristatus (Crested sharpbill), 150*, **157**, 158*

Pachycephala (Whistlers), 283f
— *pectoralis* (Golden whistler), 260*, **283**
Pachycephalinae (Thickheads), 272, **283**, 283m
Pachyramphus (Becards), 161f
— *dorsalis*, 144*
— *viridis* (Green-backed becard), 161, 161*
Padda oryzivora (Java sparrow), 439*, **452**
Paddy field warbler (See Indian reed warbler)
Paget-Wilkes, A. H., 75
Painted bunting *(Passerina ciris)*, 334*, 352*, **367**
Painted finch *(Emblema picta)*, 439*, **455**, 455m
Painted redstart *(Myioborus pictus)*, 371*, **385**
Pale flycatcher *(Bradornis pallidus)*, **277**
Pale flycatchers, 273, **276**f
Pale rock bunting *(Fringillaria impetuani)*, 350
Pale rock sparrow *(Carposiza brachydactyla)*, **423**
Pale white-eye *(Zosterops pallida)*, 343
Pale-breasted spine-tail *(Synalaxis albescens)*, 125*
Pale-wing starling *(Onychognathus nabouroup)*, **474**
Pallas' grasshopper warbler *(Locustella certhiola)*, 231*, **253**, 256
Palm chat *(Dulus dominicus)*, 196*, **217**f
Palm chats (Dulidae), 207, **217**f
Palm tanager *(Thraupis palmarum)*, **374**
Palmeria dolei (Crested honey creeper), 372*
Paltry tyrannulet *(Tyranniscus vilissimus)*, 152f, 154*
Pander's ground chough *(Podoces panderi)*, 469*, 505, **512**
Panurinae (Parrot bills), 233, **247**f
Panurus (Bearded tits), 247
— *biarmicus* (Bearded tit), 222*, **247**f
Papuan flycatcher *(Machaerirhynchus flaviventer)*, **277**, 277*
Papuan mountain drongo *(Chaetorhunchus papuensis)*, 450*, **477**
Paradigalla carunculata (Long-tailed paradigalla), 480*, **490**
Paradisaea (Birds of paradise)
— *apoda* (Greater bird of paradise), **490**f, 491*, 502*
— — *apoda* (Aru Island greater bird of paradise), 493
— — *raggiana* (Raggiana bird of paradise), 489, **493**f, 494*
— *guilielmi* (Emperor of Germany's bird of paradise), 480*, **493**f
— *minor* (Lesser bird of paradise), **493**f
— *rubra* (Red bird of paradise), 470*
— *rudolphi* (Blue bird of paradise), 470*, **493**
Paradisaeidae (Birds of paradise and bower birds), 482, **487**ff
Paradisaeinae (Birds of paradise), **487**f, 488m
Paradise flycatcher (African) (See African paradise flycatcher)
Paradise flycatcher (Asiatic) (See Asiatic paradise flycatcher)
Paradise flycatchers (Monarchinae), 272, **281**, 281m
Paradise jacamar *(Galbula dea)*, **63**, 68*
Paradise tanager *(Tangara chilensis)*, 361*, **373**
Paradise whydah *(Stegenura paradisea)*, **427**, 429*, 432f, 433*

Paradoxornis(Parrotbills,i.n.s.), 247
— *gularis* (Gray-headed parrotbill), 247
— *paradoxa*, 247
— *webbiana* (Vinous-throated parrotbill), 247
Paramythia, 330
— *montium* (Crested berry pecker), **330**f
Parana shrikes *(Mackenziaena)*, 132
Parasitic weaver *(Anomalospiza imberbis)*, **427**, 430*
Pardalotus (Diamond birds), 330
— *punctatus* (Spotted pardalote), **330**, 372*
Paridae (True tits), 310, **314**f
Paris, Matthew, 418
Parisoma (African tit flycatchers), 276
— *plumbeum* (Gray tit flycatcher), **276**
— *subcaeruleum* (Tit babbler), **276**
Parmoptila woodhousei (Antpecker), **462**
Paroaria coronata (Red-crested cardinal), 342*, **367**, 369m
— *dominicana* (Red-cowled cardinal), **367**, 369m
Paroreomyza or *loxops*
— *maculata* (Hawaiian creeper), **387**, 388*
Parotia
— *carolae* (Queen Carola's parotia), **495**f
— *lawesi* (Lawe's six-wired parotia), **495**
— *sefilata* (Arfak six-wired parotia), 470*, **495**
Parrot crossbill *(Loxia pytyopsittacus)*, **418**, 418m
Parrot finches *(Erythrura)*, 452f
Parrotbills (Psittirostrinae), 387f
Parrotbills (Panurinae), 233, **247**f
Parrotbills, i.n.s., *(Paradoxornis)*, 247
Parson finch *(Poephila cincta)*, **453**, 453m
Parulidae (Wood warblers), 348, 380f, 383m
Parulinae (Woodwarblers proper), 380, **383**f
Parus
— *afer* (Gray tit), 308*, 318, 318m
— *ater* (Coal tit), 307*, 316f
— *atricapilla* (Black-capped chickadee), 317
— *bicolor* (Tufted titmouse), 308*, **318**
— *bokharensis* (Bukharah great tit), **318**
— *caeruleus* (Blue tit), 307*, **316**
— *carolinensis* (Carolina chickadee), **318**
— *cinctus* (Siberian tit or gray-headed chickadee), 307*, 317, **318**
— *cristatus* (Crested tit), 307*, **317**
— *cyanus* (Azure tit), 307*, **318**f
— *dichrous* (Brown-crested tit), **318**
— *funereus* (Dusky tit), **318**
— *gambeli* (Mountain chickadee), **318**
— *griseiventris*(African gray tit), **318**
— *hudsonicus* (Brown-headed chickadee), **318**
— *inornatus* (Plain titmouse), **318**
— *leucomelas* (Black tit), **318**
— *lugubris* (Sombre tit), 317f
— *major* (Great tit), 307*, **314**ff, 315*, 316*
— *melanolophus* (Crested black tit), **318**
— *montanus* (Willow tit), 307*, **317**

— *monticolus* (Green-backed tit), 318
— *niger* (Southern black tit), 307*, 318, 318m
— *palustris* (Marsh tit), 302*, 307*, 317
— *rubidiventris* (Rufous-bellied crested tit), 318
— *rufescens* (Chestnut-backed chickadee), 318
— *sclateri* (Gray-sided chickadee), 318
— *superciliosus*, 318
— *varius* (Varied tit), 308*, **318**
— *venustulus*, 318
— *woolweberi* (Bridled titmouse) 318
— *xanthogenys* (Black-spotted yellow tit), 318
Passer (Sparrows),
— *domesticus* (House sparrow), 416*, **424**ff, 424m, 425m, 426*
— — *biblicus*, 425m
— — *domesticus*, 425
— — *hispaniolensis* (Spanish sparrow), 416*, **425**f, 425m
— — *indicus* (Indian sparrow), **425**
— — *italiae* (Italian sparrow), 425, 425m
— — *niloticus*, 425m
— — *tingitanus*, 425m
— *eminibey* (Chestnut sparrow), 416*, **424**, 425m, 427
— *flaveolus* (Pegu house sparrow), **424**
— *griseus* (Gray-headed sparrow), **424**
— *luteus* (Golden sparrow), 416*, **425**, 425m
— *melanurus* (Cape sparrow), 416*, **424**, 425m, 426
— *moabiticus* (Dead Sea sparrow), 416*, **424**
— *montanus* (Tree sparrow), 413*, **424**f, 424m
— *motitensis* (Rufous sparrow), **424**, 424m
— *rutilans* (Russet sparrow), **424**, 425m, 426
— *simplex* (Desert sparrow), 416*, **424**
Passerculus (Savannah sparrows)
— *princeps* (Ipswich sparrow), **353**, 356
— *sandwichensis* (Savannah sparrow), **353**, 356
Passeriformes (Passerine birds), 118ff
Passerina (Painted buntings), 367
— *amoena* (Lazuli bunting), 367
— *ciris* (Painted bunting), 334*, 352*, **367**
— *cyanea* (Indigo bunting), 352*, **367**, 369f
— *leclancherii* (Orange-breasted bunting), **367**
— *versicolor* (Varied bunting), **367**
Passerinae (Sparrows), 422f
Pastor roseus (Rose-colored starling), 449*, **468**f
Pax, 209
Paynter, 167
Pectorella finch *(Lonchura pectoralis)*, **448**, 451m
Pegu house sparrow *(Passer flaveolus)*, **424**
Peitznieier, 210
Pekin robin *(Leiothrix lutea)*, 222*, **244**
Pelargopsis (Stork-billed kingfisher), 28
— *capensis* (Stork-billed kingfisher), **28**

— — *gurial* (Gurial), 38*
Pellorneini (Jungle babblers), 234f
Pellorneum, 234
— *ruficeps* (Striped jungle babbler), 222*, **235**
Penduline tit (African) (See African penduline tit)
Penduline tit *(Remiz pendulinus)*, 308*, **312**f, 312m, 312*, 313*
Penduline tits (Remizidae), 310, **311**f
Penelopides panini (Tarictic hornbill), 55
Peppershrikes (Cyclarhinae, Cyclarhis), 395, 395m
Pericrocotini, 198
Pericrocotus (Minivets), 198f
— *flammeus* (Flame minivet), 450*
— *roseus divaricatus* (Ashy minivet), 198, 450*
Perisoreus
— *canadensis* (Gray jay), **509**
— *infaustus* (Siberian jay), 497*, **509**f, 509m
Perissocephalus tricolor (Capuchin bird), 161f
Perissotriccus (Pygmy tyrant), 152
— *atricapillus* (Black-capped pygmy tyrant), 152
Peters, 167
Peter's twin-spot *(Hypargos niveoguttatus)*, 440*, **458**
Petrochelidon andecola
— *nigricans* (Tree martin), **186**
— *pyrrhonota* (Cliff swallow), 170*, 180*, **186**
Petroica, 278
— *macrocephala* (Tomtit), **278**
— *multicolor* (Scarlet robin), 278
Petronia (Rock sparrows), 423
— *petronia* (Rock sparrow), **423**, 423m
— *pyrgita* (Yellow-spotted sparrow), **423**, 423m
— *superciliaris* (African sparrow), **423**, 423m
— *xanthocollis* (Yellow-throated sparrow), 416*, **423**
Phacellodomus (Thornbirds i.n.s.), 131
— *rufifrons* (Rufous-fronted thornbird), 125*, **131**
Phaeornis (Hawaiian thrushes), 303
— *obscurus oahensis* (Hawaiian thrush), **303**, 303*
Phainopepla *(Phainopepla nitens)*, 196*, **217**
Phainopepla nitens (Phainopepla), 196*, **217**
Phainopeplas (Ptilogonatinae), 216, 217m
Phainoptila melanoxantha (Black-and-yellow silky flycatcher), 216f
Phedina
— *borbonica* (Mascarene martin), 185
— *brazzae*, 185
Pheuctics (Grosbeaks), 367f
— *chrysopeplus* (Yellow grosbeak), 352*
— *ludiovicianus* (Rose-breasted grosbeak), **367**, 369, 369m
— *melanocephalus* (Black-headed grosbeak), **367**
Phibalura flavirostris (Swallow-tailed cotinga), 144*
Philemon (Leatherheads), 345
Philepitta (Asitys)
— *castanea* (Velvet asity), 146*, **147**, 150*
Philepittidae (Asitys), 127, 127m, **147**

Philetairus socius (Sociable weaver), **443**, 444m, 444*
Philippine broadbill *(Eurylaimus steerii)*, 124
Philippine creepers (Rhabdornithidae), 310, **325**f
Philippine fairy bluebird *(Irena cyanogaster)*, 206
Philippine hornbill *(Buceros hydrocorax)*, 56
Philydor atricapillus (Black-capped foliage-gleaner), 125*
Phlegopsis (Bare-eyes), 137
— *nigromaculata* (Spectacled antbird), 126*, **137**
Phleocryptes melanops (Wren-like rushbird), 125*
Phloeoceastes (Sharp-crested woodpeckers), 117m, 117
— *melanoleucos* (Crimson-crested woodpecker), **117**
Phlogothraupis sanguinolenta (Crimson-collared tanager), 361*
Phoenicircus-nigricollis (Black-necked red cotinga), 144*
Phoeniculinae (Wood-hoopoes), 45, **49**
Phoeniculus, 49
— *purpureus* (Green wood hoopoe), 57*
Phoenicurus (Redstart group), 285, 291
— *moussieri* (Moussier's redstart), 291
— *ochruros* (Black redstart), **291**f, 292m, 293*
— — *aterrimus*, 279*
— *phoenicurus* (Redstart), 279*, **291**f, 292m, 292*
Phonygammus keraudrenii (Trumpet bird), 479*, **490**
Phormoplectes insignis (Brown-capped weaver), 430*, **442**
Phylidonyris
— *melanops* (Yellow-tufted honey eater), 345
— *novaehollandiae* (Yellow-winged honey eater), 346
— *pyrrhoptera* (Crescent honey eater), 320*
— *undulata*, 320*
Phyllastrephus flavostriatus (Yellow-streaked greenbul), 190*, **203**
Phylloscopus
— *bonelli* (Bonelli's warbler), 259*, **263**f, 264
— *borealis* (Arctic warbler), 249f, **263**, 263m
— *collybita* (Chiff chaff), 259*, **263**f, 263m
— *fuscatus* (Brown willow warbler), 263
— *inornatus* (Yellow-browed warbler), 259*, **262**f
— *nitidus* (Bright-green leaf warbler), 263
— *proregulus* (Kinglet warbler), 262
— *schwarzi* (Radde's bush warbler), 263
— *sibilatri* (Wood warbler), 259*, **263**f, 263*
— *trivirgatus* (Mountain leaf warbler), 263
— *trochiloides* (Greenish warbler), 259*, **263**, 263m
— *trochilus* (Willow warbler), 259*, **263**f, 264m, 413*
Phytotoma
— *rara* (Chilean plantcutter), **164**, 164*
— *rutila* (Red-breasted plant-

cutter), 150*, **164**
Phytotomidae (Plantcutters), 127, **163**f, 163m
Pica pica (Magpie), 497*, 501*, **511**, 511m
— — *nuttalli* (Yellow-billed magpie), **511**
Picathartes, 245, 245m
— *gymnocephalus* (White-necked rockfowl), 222*, **245**f
— *oreas* (Gray-necked rockfowl), 222*, **245**f
Picathartini (Rockfowl or bald crows), 234, **245**f
Picidae (Woodpeckers), 63, 87ff
Piciformes (Woodpeckers), 62ff
Picinae (True woodpeckers), 87, **89**ff, 90*, 91*
Picoidea (Woodpeckers), 62
Picoides (Three-toed woodpeckers), 114f, 115m
— *arcticus* (Black-backed three-toed woodpecker), 115
— *tridactylus* (Northern three-toed woodpecker), 91*, **115**, 119*
Piculet (African) (See African piculet)
Piculets (Picumninae), 87, **89**
Piculus rubiginosus (Golden-olive woodpecker), 96
Picumninae (Piculets), 87, **89**
Picumnus, 89
— *minutissimus* (Scaly-eared piculet), 89, 108*
Picus (Green woodpeckers), 99f
— *canus* (Black-naped green woodpecker), 84*, 94*, **99**ff, 102*, 119*
— *puniceus* (Crimson-winged woodpecker), 119*
— *viridis* (Green woodpecker), 82*, 92*, **99**f, 100m, 100*, 119*
— *williamsonii*, 106
Pied barbet *(Tricholaema leucomelan)*, 72f
Pied chat *(Oenanthe picata)*, 295, 297
Pied Chat *(Oenanthe pileata)*, 280*
Pied crow *(Corvus albus)*, 498*, **518**
Pied currawong *(Strepera graculina)*, 469*, **486**
Pied fantails *(Rhipidura)*, 278
Pied flycatcher *(Ficedula hypoleuca)*, 260*, **273**f, 274m, 276*
Pied kingfisher *(Ceryle lugubris)*, 27
Pied mynah *(Sturnopastor contra)*, 468
Pied starling *(Spreo bicolor)*, **474**
Pied wagtail *(Motacilla alba yarrellii)*, 194
Pied water tyrant *(Fluvicola pica)*, 143*
Pileated woodpecker *(Dryocopus pileatus)*, 90*, **102**, 102m, 104*, 104
Pigafetta, Antonio, 488
Pinaroloxias inornata (Cocos finch), 351*, 359m, **360**, 364, 365*
Pine bunting *(Emberiza leucocephala)*, 349, 349m
Pine grosbeak *(Pinicola enucleator)*, **418**, 418m
Pine siskin *(Carduelis pinus)*, 406, **409**f
Pinicola (Pine grosbeaks), 404
— *enucleator*, **418**, 418m
Pin tailed nonpareil *(Erythrura prasina)*, 439*, **453**
Pine-tailed whydah *(Vidua macroura)*, **428**, 432
Pinyon jay *(Gymnorhinus cyanocephalus)*, 497*, **506**f
Pipiac *(Ptilostomus afer)*, **514**
Pipilo erythrophthalmus (Rufous-sided towhee), 342*

640 INDEX

Pipit (New Zealand) (See New Zealand pipit)
Pipits, i.n.s. *(Anthus)*, 193f
Pipra (Pipras), 158f
— *aureola* (Crimson-hooded manakin), 143*
— *erythrocephala* (Red-headed manakin or virapuru), 158, 159*
Pipra flycatchers *(Pipromorpha)*, 152, 154, 156
Pipras (Pipra), 158f
Pipreola (Fruit-eaters), 161
— *arcuata* (Barred fruit-eater), 161
Pipridae (Manakins), 127, 157f, 158f
Pipromorpha (Pipra flycatchers), 152, 154, 156
— *oleaginea* (Olive-bellied flycatcher), 152
Piranga
— *olivacea* (Scarlet tanager), 375
— *rubra* (Summer tanager), 361*, 375f
Piratical flycatcher *(Legatus leucophaius)*, 151, 155*, 156
Pitangus (Kiskadee flycatchers), 151f
— *sulphuratus* (Great kiskadee flycatcher), 143*, 152
Pithys (Ant catchers), 137
— *albifrons* (White-faced ant catcher), 126*, 137
Pitohui, 283f
— *dichrous* (Black-headed pitohui), 260*, 284
Pitta (Pittas), 142ff
— *angolensis* (African pitta), 142
— *brachyura* (Blue-winged pitta), 142, 145m, 145
— — *nympha* (Japanese fairy pitta), 142, 145, 146*
— *caerula* (Great blue pitta), 142, 146, 149*
— — *willoughbyi*, 149*
— *ellioti* (Elliot's pitta), 149*
— *erythrogaster* (Red-breasted pitta), 142
— *granatina* (Garnet or red-headed pitta), 149*
— *guajana* (Blue-tailed pitta), 133*, 142, 149*
— *iris* (Rainbow pitta), 142
— *maxima*, 149*
— *nipalensis* (Blue-naped pitta), 149*
— *phayrii* (Sickle-tailed pitta), 142
— *sordida* (Hooded pitta), 142, 145f, 149*
— — *cucullata* (Indian black-headed pitta), 142
— *steerii* (Steer's pitta), 142, 146, 149*
— *superba*, 149*
— *versicolor* (Noisy pitta), 142, 146
Pitta (African) (See African pitta)
Pittas *(Pitta)*, 142ff
Pittidae (Pittas), 127, 141f, 145m
Pityriasinae (Wood shrikes), 207, 213
Pityriasis gymnocephala (Bald-headed wood shrike), 213, 213m, 449*
Plain brown woodcreeper *(Dendrocincla fuliginosa)*, 120*
Plain flower pecker *(Dicaeum concolor)*, 331
Plain titmouse *(Parus inornatus)*, 318
Plain-colored tanager *(Tangara inornata)*, 373f
Plain-headed creeper *(Rhabdornis inornatus)*, 319*, 325
Plain-throated sunbird *(Anthreptes malacensis)*, 338
Plantcutters (Phytotomidae), 127, 163f, 163m
Plate-billed mountain toucan *(Andigena laminirostris)*, 81*
Platylophus galericulatus (Crested jay), 506
Platypsaris niger (Jamaican becard), 144*
Platysmurus leucopterus (Black-crested magpie), 506
Platyspiza crassirostris (Vegetarian tree finch), 351*, 359, 364, 364*
Platysteira (Wattle-eyed flycatchers), 277
— *peltata* (Black-throated wattle-eyed flycatcher), 277
Plectrophenax (Snow buntings), 353
— *hyperboreus* (McKay's bunting), 354
— *nivalis* (Snow bunting), 342*, 354f, 354m
Ploceella hypoxantha (Golden weaver), 437m, 438
Ploceidae (Weaverbirds), 421ff
Ploceinae (True weavers), 414*, 422, 437f
Plocepasser mahali (Mahali weaver bird), 443
Plocepasserinae (Sparrow weavers), 422, 443
Ploceus
— *philippinus* (Baya weaver), 438
— *manyar* (Manyar weaver), 430*, 437m, 438
Plush-capped finch *(Catamblyrhynchus diadema)*, 362*, 378
Plush-capped finches (Catamblyrhynchini), 370, 378, 378m
Plush-crested jay *(Cyanocorax chrysops)*, 508
Pnoepyga, 236
— *albiventer* (Greater scaly-breasted wren babbler), 237
— *pusilla* (Lesser scaly-breasted wren babbler), 237f
Podoces, 511
— *hendersoni* (Henderson's ground chough), 512
— *panderi* (Pander's ground chough), 469*, 505, 512
Poephila (Grass finches), 453
— *acuticauda* (Long-tailed grass finch), 439*, 453, 453m
— *cincta* (Parson finch), 453, 453m
— *personata* (Masked grass finch), 453
Pogoiulus (Tinker-birds), 73f
— *bilineatus* (Golden-rumped tinker-bird), 74, 97*
— *chrysoconus* (Yellow-fronted tinker-bird), 74
— *scolopaceus* (Speckled tinker-bird), 74
— *subsulphureus* (Yellow-throated tinker-bird), 74
Polioptila caerulea (Blue-gray gnatcatcher), 259*, 271, 271m
— *melanura* (Black-tailed gnatcatcher), 271
Polioptilinae (Gnatcatchers), 233, 271
Pollen's vanga *(Xenopirostris polleni)*, 196*, 213f
Pomarea mendozae (Marguesas flycatcher), 281
Pomatorhinini (Scimitar and wren babblers), 234, 235f
Pomatorhinus, 235
— *erythrogenys* (Rusty-cheeked scimitar babbler), 222*, 236
— *montanus* (Yellow-billed scimitar babbler), 236
— *ruficollis* (Rufous-necked scimitar babbler), 237
Pomatostomus, 235
— *isidorei* (Rufous babbler), 237
— *temporalis* (Gray-crowned babbler), 222*
Pompadour cotinga *(Xipholena punicea)*, 144*
Poospiza torquata (Ringed warbling finch), 342*
Porter, Sidney, 216
Potter, 50
Prapamaticola aedon (See *acrocephalus aedon*)
Primary songbirds or lyre-birds relatives (Suboscines), 118, 123, 164ff
Primitive insectivores (Muscicapoidea), 219ff
Prince Albert's lyre-bird *(Menura alberti)*, 164f, 164m
Princess Stephanie's astrapia *(Astrapia stephaniae)*, 499
Prinia (Wren warblers), 251f, 427
— *subflava* (Tawny-flanked prinia), 251
Prionochilus xanthopygius (Yellow-rumped flower pecker), 331
Prionodura newtoniana (Golden bower bird), 492*, 503
Prionopinae (Helmeted shrikes), 207f
Prionops
— *alberti* (Yellow-crested helmet shrike), 208m
— *gabela* (Angola red-billed shrike), 208m
— *plumata* (White helmeted shrike), 196*, 207f
— *poliolopha* (Gray-crested helmet shrike), 208m
— *scopifrons* (Chestnut-fronted shrike), 208m
Pritit puff-back flycatcher *(Batis pririt)*, 277
Procnias (Bellbirds), 161ff
— *alba* (White bellbird), 161, 162*, 163
— *averano* (Bearded or mossy-throated bellbird), 144*, 161, 163*
— *nudicollis* (Bare-throated bellbird), 136*, 144*, 161, 163*
— *tricarunculata* (Three-wattled bellbird), 144*, 161, 163, 163*
Prodoticus regulus (Wahlberg's honey guide), 77
Progne chalybea subis (Purple martin), 180*, 186
Promeropinae (Sugarbirds), 344, 347
Promerops cafer (Cape sugarbird), 320*, 344, 347
— *gurneyi* (Natal sugarbird), 347
Prong-billed barbet *(Semnornis frantzii)*, 66f
Prong-billed barbets *(Semnornis)*, 66f
Protonotaria, 380
— *citrea* (Protonotary warbler), 371*, 383
Protonotary warbler *(Protonotaria citrea)*, 371*, 383
Porzesky, O. P. M., 74
Prunella (Accentors)
— *collaris* (Alpine accentor), 221*, 230
— *modularis* (Hedge sparrow), 221*, 229f, 413*
— *montanella* (Siberian accentor), 230
— *strophiata* (Rufous-breasted accentor), 230
Prunellidae (Accentors), 219, 229f, 229m
Psalidoprocne (African rough-winged swallows), 186
— *albiceps* (White-headed rough-winged swallow), 180*
— *holomelaena* (Black rough-winged swallow), 180*
Psaltaria exilis (Java long-tailed tit), 311, 311m
Psaltriparus melanotis (Black-eared bush tit), 311
— *minimus* (Common bush tit), 311
Psarisomus dalhousiae (Long-tailed broadbill), 120*, 124
Psarocolius, 396
— *angustifrons* (Russett-backed oropendola), 381*, 398
— *decumanus* (Crested oropendola), 398
— *viridis* (Green oropendola), 381*
— *wagleri* (Wagler's oropendola), 381*
Pseudhirundo griseopyga (African gray-rumped swallow), 186
Pseudocalyptomena graueri (Gray broadbill), 124
— *siantarae* (White-eyed river martin), 185
— *siantarae* (White-eyed river martin), 185
Pseudochelidoninae, 182, 185
Pseudonestor xanthophrys (Maui parrotbill), 387, 389
Pseudonigrita arnaudi (Gray-headed weaver), 443, 444m
Pseudopodoces, 411
— *humilis* (Hume's ground chough), 469*, 512
Psilopogon pyrolophus (Fire-tufted barbet), 76f
Psilorhamphus guttatus (Bamboowren), 140
Psilorhinus morio (White tipped brown jay), 508
Psittirostra (Parrotbills)
— *cantans* (Laysan finch), 387f, 389*, 389f
— — *ultima* (Nihoan Laysan finch), 390
— *flaviceps* (Lesser Koa finch), 394
— *kona* (Grosbeak finch or Kona finch), 388
— *palmeri* (Greater Koa finch), 388, 389*, 390
— *psittacea* (Ou), 387, 389*, 389, 394
Psittirostrinae (Parrotbills), 387f
Pteridophora alberti (King of Saxony's bird of paradise), 479*, 496
Pteroglossus (Aracaris), 80f
— *aracari* (Banded aracari), 107*
— *frantzii* (Fiery-billed aracari), 80, 85
— *torquatus* (Collared aracari), 80, 81*
Pteropodocys maxima (Ground cuckoo-shrike), 198f
Pteroptochos megapodius (Moustached turca), 140
Pteruthius (Shrike babblers), 244
— *flaviscapius* (Red-winged shrike babbler), 244f
Ptilochichla, 236
Ptilogonatinae (Phainopeplas), 216, 217m
Ptilogonys caudatus (Long-tailed silky flycatcher), 216f
— *cinereus* (Gray silky flycatcher), 216f
Ptilolaemus tickelli (Tickell's hornbill), 55
Ptilonorhynchinae (Bower birds),

499f, 500m
Ptilonorhynchus, 504
— violaceus (Satin bower bird), 504*, **505**
Ptiloris (Rifle birds), 499
— magnificus (Magnificent riflebird), 479*, **499**
Ptilorrhoa, 249
— caerulescens (Lowland eupetes), **249**
Ptilostomus afer (Pipiae), **514**
Ptyonoprogne rupestris (Crag martin), 180*, **191**
Ptyrticus turdinus (Thrush babbler), **235**
Puerto Rican tody (Todus mexicanus), **34**
Puff-back flycatcher (Batis capensis), 260*, **277**, 277*
Puff-back flycatchers and relatives, 273, **277**
Puff-backs (Batis), 277
Puffbirds (Bucconidae), 62, **64**f
Purple cochoa (Cocha purpurea), 280*, **293**
Purple finch (Carpodacus purpureus), **417**
Purple glossy starling (Lamprotornis purpureus), 460*, **475**
Purple grenadier (Uraeginthus ianthinogaster), 429*, 431, **457**f, 457m
Purple honey creeper (Cyanerpes caeruleus), 335*
Purple martin (Progne subis), 180*, **186**
Purple-breasted cotinga (Cotinga cotinga), 144*
Pycnonotidae (Bulbuls), 192, **200**f
Pycnonotus (Bulbuls proper), 200
— atriceps (Black-headed bulbul), 200
— barbatus (White-vented bulbul), 190*, **204**
— cafer (Red-vented bulbul), **200**
— finlaysoni (Stripe-throated bulbul), 200f
— flavescens (Blyth's bulbul), 203
— goiavier (Yellow-vented bulbul), **203**
— gracilirostris, 203
— jocosus (Red-whiskered bulbul), 184*, 190*, **203**
— melanoleucos, 200
— nieuwenhuisi (Nieuwenhuis bulbul), 203
— tephrolaemus (Olive-breasted mountain greenbul), 203
— zeylanicus (Yellow-crowned bulbul), 200, **203**
Pygarrhichas albogularis (White-throated tree-runner), 131f, 137*
Pygmy kingfisher (Chloroceryle aenea), **27**
Pygmy nuthatch (Sitta pygmaea), 319*, **323**
Pygmy triller or pied shrike (Hemipus picatus), 198f
Pygmy tyrant (Perissotriccus), 152
Pynnönen, 111, 113
Pyrenestes ostrinus (Black-bellied seed-cracker), 440*, **461**
Pyriglena leucoptera (Fire-eye), 137
Pyrocephalus rubinus (Vermillion flycatcher), 143*, **151**, 155
Pyroderus scutatus (Red-ruffed fruitcrow), 161f
Pyrrhocorax (Choughs), 513
— graculus (Alpine chough), 469*, **513**f
— pyrrhocorax (Chough), 469*, **513**
— — barbarus, 514

Pyrrhula (Bullfinches), 404, 406, **418**f
— erythaca (Chinese bullfinch), 392*
— pyrrhula (Bullfinch), 336*, **419**, 419*
— — griseiventris (Japanese bullfinch), 392*
— — murina (Azores bullfinch), 392*
Pyrrhuloxia (Pyrrhuloxias sinuata), 352*, **366**, 368f
Pyrrhuloxias sinuata (Pyrrhuloxia), 352*, **366**, 368f
Pytilia, 428
— afra (Red-faced waxbill), 429*, **461**
— lineata, **461**
— melba (Melba finch), 413*, 429*, 431, 432*, **461**, 461*, 461m
— phoenicoptera (Aurora waxbill), **461**

Quail finch (South African) (See South African quail finch)
Quail finches (Ortygospiza), 455f
Queen Carola's parotia (Parotia carolae), **495**f
Queens whydah (Tetraenura regia), 433*
Quelea, 433
— quelea (Red-billed dioch), 430*, **436**f, 436m
Queleopsis, 433
— cardinalis (Cardinal quelea), **436**
— erythrops (Red-headed quelea), **437**
Quiscalus (Grackles), 396f, 399f
— lugubris (South American carib grackle), **400**
— quiscula (Common grackle), 381*, **400**

Racket-tailed magpies (Crypsirina), 510
Racket-tailed roller (Coracias spatulata), 44f, 44m
Racket-tailed tree pie (Temnurus temnurus), **511**
Radde's bush warbler (Phylloscopus schwarzi), **263**
Raethel, Heinz-Sigurd, 216
Raggiana bird of paradise (Paradisaea apoda raggiana), 489, **493**f, 494*
Rail babbler (Eupetes macrocercus), 222*, **249**
Rail babblers (Orthonychinae), **248**
Rainbow bee-eater (Merops ornatus), **40**
Rainbow pitta (Pitta iris), 142
Ramphastidae (Toucans), 63, **79**ff, 79m
Ramphastos, 85f, 87*
— cuvieri (Cuvier's toucan), 98*
— dicolorus (Red-breasted toucan), 85, 98*
— sulfuratus (Keel-billed toucan), 85, 98*
— swainsonii (Swainson's toucan), 85f
— toco (Toco toucan), 81*, **85**, 98*
— vitellinus (Channel-billed toucan), 81*, **85**
— — ariel (Ariel toucan), 98*
Ramphocaenus melanurus (Long-billed gnatwren), **271**
Ramphocelus
— carbo (Silver-beaked tanager), 361*, **375**
— — bresilius (Brazilian tanager), 335*

— icteronotus (Yellow-rumped tanager), **375**
— passerinii (Scarlet-rumped tanager), 361*, **375**
Ramphocorys clotbey (Thick-billed lark), **174**, 174m, 179*
Ramsayornis, 346
Ranger, 340
Raven (Covus corax), 498*, **518**ff, 518m
Red bird of paradise (Paradisaea rubra), 470*
Red crossbill (Loxia curvirostra), **418**, 418m, 418*
Red siskin (Carduelis cucullata), 391*, **409**f
Red warbler (Ergaticus ruber), 371*, **385**
Red-and-yellow barbet (Trachyphonus erythrocephalus), 74, 97*
Red-backed shrike (Lanius collurio), **209**ff, 209m, 211*
Red-bearded bee-eater (Nyctornis amictus), **42**, 48*
Red-bellied sapsucker (Sphyrapicus vanius ruber), **106**, 106m, 109m
Red-bellied woodpecker (Centurus carolinus), 95*, **104**ff, 105*
Red-billed blue magpie (Urocissa erythrorhyncha), 497*, **511**
Red-billed dioch (Quelea quelea), 430*, **436**f, 436m
Red-billed hornbill (Tockus erhthrorhynchus), 32*, **53**f, 54m, 58*
Red-billed ox-pecker (Buphagus erythrorhynchus), **475**, 475m
Red-billed scythebill (Campylorhamphus trochilirostris), 120*, **128***
Red-breasted nuthatch (Sitta canadensis), 319*, **323**, 323m
Red-breasted pitta (Pitta erythrogaster), **142**
Red-breasted plantcutter (Phytotoma rutila), 150*, **164**
Red-breasted toucan (Ramphastus dicolorus), **85**, 98*
Red-breasted wryneck (Jynx ruficollis), **88**, 88m
Red-capped babbler (Timalia pileata), 222*, **239**
Red-capped spine-tail (Synallaxis ruficapilla), **131**
Red-capped thrush (Turdus ruficollis), **304**, 305m
Red-cheeked waxbill or Cordonbleu (Uraeginthus bengalus), 440*, **457**
Red-cockaded woodpecker (Dendrocopos borealis), **110**, 114, 114m
Red-collared whydah (Coliuspasser ardens), 430*, **433**f, 434m
Red-cowled cardinal (Paroaria dominicana), **367**, 369m
Red-crested cardinal (Paroaria coronata), 342*, **367**, 369m
Red-crested finch (Coryphospingus cucullatus), **357**
Red-crowned ant-tanager (Habia rubica), 361*, **376**
Red-eared fire-tailed finch (Stagonopleura oculata), **455**
Red-eyed vireo (Vireo olivaceus), 372*, **395**, 395*
Red-faced cisticola (Cisticola erythrops), **252**, 253*
Red-faced waxbill (Pytilia afra), 429*, **461**
Red-flanked broadbill (Smithornis rufolateralis), **124**
Red-headed barbet (Eubucco bourcierii), **70**

Red-headed bunting (Emberiza bruniceps), **350**, 353
Red-headed finch (Amadina erythrocephala), **447**, 447m
Red-headed manakin or uirapuru (Pipra erythrocephala), **158**, 159*
Red-headed parrot finch (Erythrura psittacea), 439*, **453**
Red-headed quelea (Queleopsis erythrops), **437**
Red-headed tit (Aegithalos concinnus), 308*, **311**, 311m
Red-headed weaver (Anaplectes rubriceps), 430*, **442**
Red-headed woodpecker (Melanerpes erythrocephalus), **104**f, 119*
Red-legged honey creeper (Cyanerpes cyaneus), 335*, 362*, **378**f
Red-ruffed fruitcrow (Pyroderus scutatus), 161f
Red-rumped swallow (Hirundo daurica), 180*, **187**, 187m
Red-rumped wheatear (Oenanthe moesta), **295**
Red-shafted flicker (Colaptes aurufus cafer), **95**, 96m, 108*
Red-shouldered glossy starling (Lamprotornis nitens), **475**
Redstart (Phoenicurus phoenicurus), 279*, **291**f, 292m, 292*
Redstart group (Phoenicurus), 285, **291**
Redstarts (American) (See American redstarts)
Red-tailed ant thrush (Neocossyphus rufus), **294**
Red-tailed chat (Cercomela familiaris), **295**f
Red-tailed vanga (Calicalicus madagascariensis), **214**
Red-throated bee-eater (Melittophagus bulocki), **41**, 48*
Red-throated nunlet (Nonnula rubecula), 64, **68***
Red-throated pitit (Anthus cervinus), 189*, **193**, 193m
Red-throated thrush (Turdus ruficollis ruficollis), **304**
Red-vented bulbul (Pyenonotus cafer), **200**
Red-whiskered bulbul (Pycnonotus jocosus), 184*, 190*, **203**
Redwing (Turdus iliacus), 290*, **304**, 304m
Redwing shrike (Tchagra tchagra), 195*, **208**
Red-wing starling (Onychognathus morio), **474**
Red-winged blackbird (Agelaius phoeniceus), **400**f
Red-winged bush lark (Mirafra hypermetra), 179*
Red-winged shrike babbler (Pteruthius flaviscapis), 244f
Reed bunting (Emberiza schoeniclus), 341*, **350**, 350m, 353
Reed warbler (Acrocephalus scirpaceus), 232*, **253**f, 253m, 255, 257*
Reed warbler (African) (See African reed warbler)
Reed warblers, 250, **252**f
Reetz, Horst, 25
Regent bower bird (Sericulus chrysocephalus), 492*, **505**
Regulinae (Kinglets), 233, **267**f
Regulus
— calendula (Ruby-crowned kinglet), 259*, **268**, 268m
— goodfellowi, 268
— ignicapillus (Firecrest), 259*, **267**f,

267m, 267*
— *regulus* (Goldcrest), 259*, 267f, 267m, 267*
— *satrapa* (Golden-crowned kinglet), 268, 268m
Reichenow's crimson-wing *(Cryptospiza reichenovii)*, 461
Remiz pendulinus (Penduline tit), 308*, 312f, 312m, 312*, 313*
Remizidae (Penduline tits), 310, 311f
Reunion starling *(Fregiluous varius)*, 472
Reunion weaver *(Foudia bruante)*, 437
Rhabdornis
— *inornatus* (Plain-headed creeper), 325
— *mystacalis* (Striped-headed creeper), 319*, 325
Rhabdornithidae (Philippine creepers), 310, 325f
Rhamphocharis (Spotted berry peckers), 330
— *crassirostris* (Spotted berry pecker), 330
Rhegmatorhina (Rhegmatorhinas), 137
— *melanosticta* (Spotted rhegmatorhina), 137
Rhinoceros hornbill *(Buceros rhinoceros)*, 52, 56
Rhinocrypta lanceolata (Crested gallito), 140f, 140*
Rhinocryptidae (Tapaculos), 127, 140f, 140m
Rhinomyias olivacea (Olive-backed jungle flycatcher), 276
Rhinoplax vigil (Helmeted hornbill), 58*, 59f, 59m
Rhinopomastus (Scimitar-billed hoopoes), 49
— *cyanomelas* (Scimitar-bill hoopoe), 49
— *minor*, 57*
Rhipidura (Pied fantails), 278
— *leucophrys* (Willie wagtail), 260*, 278f
— *rufifrons* (rufous-fronted fantail), 278f, 278*
Rhipidurinae (Fantails), 272, 274m, 278f
Rhodinocichla rosea (Rose-breasted thrush-tanager), 362*, 370
Rhodopechys (Trumpeter bullfinches), 404, 412
— *githaginea* (Trumpeter bullfinch), 412
Rhopocichla atriceps (Black-headed babbler), 239
Rhyacornis fuliginosus, 291
Rhynchocyclus brevirostris (Eye-ringed flatbill), 152
Rhyticeros
— *narcondami* (Narcondam hornbill), 56
— *plicatus* (Blyth's hornbill), 56
— — *jungei*, 56
— — *plicatus* (Ceram hornbill), 50
— *undulatus* (Wreathed hornbill), 55f
— — *equabilis*, 55, 55m
— — *narcondami*, 55m
— — *undulatus*, 58m
Ribbon-tailed astrapia *(Astrapia mayeri)*, 480*, 496f
Richmondena (See Cardinalis)
Rifle birds *(Ptiloris)*, 499
Rifleman *(Acanthisitta chloris)*, 147, 147*, 150*
Rimator malacoptilus (Long-billed wren babbler), 235f
Ring ouzel *(Turdus torquatus)*, 290*, 304, 305m, 309
Ringed kingfisher *(Megaceryle torquata)*, 27
Ringed warbling finch *(Poospiza torquata)*, 342*
Riparia riparia (Bank swallow or sand martin), 180*, 186, 186m
Ripley, Dillon, 243, 504
River martin (African) (See African river martin)
River tyrannulet *(Serpophaga cinerea)*, 152, 152*
River warbler *(Locustella fluviatilis)*, 253, 257
Robin (American) (See American robin)
Robin (Eurasian) (See Eurasian robin)
Robin chat *(Cossypha caffra)*, 279*
Robin chats or forest robins *(Cossypha)*, 287
Robins *(Erithacus)*, 285f
Robinson, A. H., 483, 486f
Rock bunting *(Emberiza cia)*, 341*
Rock nuthatch *(Sitta neumayer)*, 319*, 322m, 323
Rock sparrow *(Petronia petronia)*, 423, 423m
Rock thrush *(Monticola saxatilis)*, 289*, 298f, 298m
Rock thrushes *(Monticola)*, 285, 298f
Rock wren *(Salpinctes obsoletus)*, 224
Rock wren (New Zealand) (See New Zealand rock wren)
Rock fowl or bald crows *(Picathartini)*, 234, 245f
Rodrigues weaver *(Goudia flavicans)*, 437
Rodriguez starling *(Necropsar rodericanus)*, 472
Rollers (Coraciidae), 21, 42f
Rook *(Corvus frugilens)*, 498*, 515f
Rose-breasted chat *(Granatellus pelzelni)*, 371*, 385, 385m
Rose-breasted grosbeak *(Pheucticus ludovicianus)*, 367, 369, 369m
Rose-breasted thrush-tanager *(Rhodinocichla rosea)*, 362*, 370
Rose-collared piha *(Lipaugus strephophorus)*, 144*
Rose-colored starling *(Pastor roseus)*, 449*, 468f
Rosy finches *(Leucosticte)*, 404, 412
Rosy twin-spot *(Hypargos margaritatus)*, 458
Rothschild, Walter, 148
Rough-winged swallow *(Stelgidopteryx ruficollis)*, 180*, 186
Rough-winged swallows (African) (See African rough-winged swallows)
Rowan, W., 356
Rowettia goughensis (Gouch Island bunting), 357
Rowley, I., 484
Royal antpitta *(Grallaria varia)*, 137, 138*
Royal flycatchers *(Onychorhynchus)*, 152
Royal starling *(Cosmopsarus regius)*, 474
Ruby throat *(Luscinia calliope)*, 279*, 286, 286m
Ruby-cheeked sunbird *(Anthreptes singalensis)*, 338
Ruby-crowned kinglet *(Regulus calendula)*, 259*, 268, 268m
Ruddy-breasted seed eater *(Sporophila minuta)*, 342*
Rueppell's warbler *(Sylvia rueppelli)*, 258, 262, 262m
Rufous babbler *(Pomatostomus isidorei)*, 237
Rufous flycatcher *(Stizorhina frazeri)*, 294
Rufous gnateater *(Conopophaga lineata)*, 139
Rufous hornero or ovenbird *(Furnarius rufus)*, 125*, 129ff, 129*, 130*
Rufous motmot *(Baryphthengus ruficapillus)*, 32*, 36
Rufous rock jumper *(Chaetops frenatus)*, 286
Rufous scrub-bird *(Atrichornis rufescens)*, 150*, 166, 166m
Rufous sparrow *(Passer motitensis)*, 424, 424m
Rufous warbler *(Erythropygia galactotes)*, 285m, 286
Rufous-backed shrike *(Lanius schach)*, 195*, 212
Rufous-bellied babbler *(Dumetia hyperythra)*, 239
Rufous-bellied crested tit *(Parus rubidiventris)*, 318
Rufous-bellied niltava *(Niltava sundara)*, 260*, 275
Rufous-bellied thrush *(Turdus rufiventris)*, 304
Rufous-breasted accentor *(Prunella strophiata)*, 230
Rufous-browed pepper shrike *(Cyclarhis gujanensis)*, 372*, 396
Rufous-collared sparrow *(Zonotrichia capensis)*, 353, 355
Rufous-crowned babbler *(Moupinia poecilote)*, 239, 240
Rufous-crowned roller *(Coracias noevia)*, 44, 44m
Rufous-ronted fantail *(Rhipidura rufifrons)*, 278f, 278*
Rufous-fronted thornbird *(Phacellodomus rufifrons)*, 125*, 131
Rufous-necked hornbill *(Aceros nipalensis)*, 55
Rufous-necked scimitar babbler *(Pomatorhinus ruficollis)*, 237
Rufous-tailed jacamar *(Galbula ruficauda)*, 63
Rufous-tailed weaver *(Histurgops ruficauda)*, 440
Rupicola (Cocks-of-the-rock), 161ff
— *peruviana* (Andean cock-of-the-rock), 133*, 161
— *rupicola* (Guianan cock-of-the-rock), 144*, 162
Rüppell's weaver *(Textor galbula)*, 441, 442m
Rush warblers *(Bradypterus)*, 253
Russet sparrow *(Passer rutilans)*, 424, 425m, 426
Russet-backed oropendola *(Psarocolius angustifrons)*, 381*, 398
Rusty-belted tapaculo *(Liosceles thoracicus)*, 140
Rusty-breasted blue flycatcher *(Ficedula hodgsonii)*, 275
Rusty-cheeked scimitar-babbler *(Pomatorhinus erythrogenys)*, 222*, 236
Sacred kingfisher *(Halcyon sancta)*, 27m, 28, 33
Saddleback *(Creadion carunculatus)*, 469*, 471
Saffron finch *(Sicalis flaveola)*, 342*, 357
Saffron toucan *(Baillonius bailloni)*, 80, 107*
Sage thrasher *(Oreoscopes montanus)*, 227f
St. Helena waxbill *(Estrilda astrild)*, 428, 440*, 456
Salomonsen, F., 331, 345
Salpinctes mexicanus (Canyon wren), 190*, 224
— *obsoletus* (Rock wren), 224
Salpornis spilonotus (Gray Spotted creeper), 319*, 326, 326m
Salpornithidae (Spotted creepers), 310, 325f
Saltator atriceps (Black-cheeked ant tanager), 352*
Sanford's bower bird *(Archboldia papuensis sanfordi)*, 503
St. Lucia black finch *(Melanospiza richardsoni)*, 442
Sap Tome grosbeak *(Neospiza concolor)*, 442
Sarcops calvus (Coleto mynah), 449*, 473
Sardinian warbler *(Sylvia melanocephala)*, 258, 262, 262m
Sargent, O. H., 345
Saroglossa spiloptera (Spotted-winged stare), 475
Sasia abnormis (Three-toed piculet), 89
Satin bower-bird *(Ptilonorhynchus violaceus)*, 504*, 505
Saver, 261
Savannah sparrow *(Passerculus sandwichensis)*, 353, 356
Savattari, E., 511
Savi's warbler *(Locustella luscinioides)*, 231*, 253, 256f
Saxicola, 285, 295
— *rubetra* (Whinchat), 295f, 295m, 296*
— *torquata* (Stone chat), 280*, 295f, 295m, 296*
Saxicoloides fulicata (Indian robin), 295f, 298
Sayaca tanager *(Thraupis sayaca)*, 374
Sayornis nigricans (Black phoebe), 151, 153
— *phoebe* (Eastern phoebe), 151, 153*, 155
Scale-backed antbird *(Hylophylax poecilonata)*, 126*
Scaled ground roller *(Brachypteracias squamigera)*, 43
Scaly weaver *(Sporopipes squamifrons)*, 430*, 444, 445m
Scaly-eared piculet *(Picumnus minutissimus)*, 89, 108*
Scaly-throated honey guide *(Indicator variegatus)*, 77f
Scaphidura oryzivora (Giant cowbird), 381*, 396, 398
Scarlet grosbeak *(Carpodacus erythrinus)*, 391*, 417, 417m
Scarlet robin *(Petroica multicolor)*, 278
Scarlet tanager *(Piranga olivacea)*, 375
Scarlet-backed flower pecker *(Dicaeum cruentatum)*, 330
Scarlet-bellied mountain tanager *(Anisognathus igniventris)*, 377
Scarlet-crowned barbet *(Capito aurovirens)*, 69
Scarlet-rumped tanager *(Ramphocelus passerinii)*, 361*, 375
Scelorchilus albicollis (White-throated tapaculo), 140
Scenopoeetes dentirostris (Tooth-billed bower bird), 492*, 503
Schäfer, Ernst, 424, 451
Scherpner, Christopher, 24
Schetba rufa, 214
Schlegel, 211
Schneider, Gustav, 60
Schoeniophylax phryanophila (Chotoy spine-tail), 125*
Schomburgk, Robert, 163

Schumacher, Eugen, 493
Schuster, Ludwig, 197
Scimitar and wren babblers (Pomatorhinini), 234, **235**f
Scimitar-billed hoopoe *(Rhinopomastus cyanomelas)*, **49**
Scissirostrum dubium (Celebesian starling), 449*, **473**
Scissor-tailed flycatcher *(Muscivora forficata)*, 151, 153
Sclerurus (Leaf-scrapers), 132
— *caudacutus* (Black-tailed leafscraper), **132**
Scrub jay *(Aphelocoma coerulescens)*, **507**
Scrub robins *(Drymodes)*, 285f
Scrub wrens *(Sericornis)*, 265
Scrub-birds (Atrichornithidae), 165f
Seytalopus, 140f
— *speluncae* (Mouse-colored tapaculo), 140
Seaside sparrow *(Amnospiza maritima)*, 342*, **354**, 357
Sedge warbler *(Acrocephalus schoenobaenus)*, 231*, **252**, 253m, 254, 256*
Seed eaters *(Sporophila)*, 357f
Seicercus (Flycatcher warblers), 263
— *burkii* (Yellow-eyed flycatcher warbler), **263**
— *ruficapillus* (Yellow-throated woodland warbler), 263
Seiurus (Waterthrushes), 380
— *aurocapillus* (Ovenbird), 371*, **384**f, 385m
— *motacilla* (Louisiana waterthrush), **383**
Selenidera (Short-billed toucans), 80
— *maculirostris* (Spot-billed toucanet), 80, **107***
Seleucides melanoleuca (Twelvewired bird of paradise) 491*, **496**
Semioptera wallacei (Wallace's standard wing), 480*, 495*, **496**
Semnornis (Prong-billed barbets), 66f
— *frantzii* (Prong-billed barbet), **66**
— *rhamphastinus* (Toucan barbet), 66, 97*
Senegal kingfisher *(Halcyon senegalensis)*, **28**
Senegal roller *(Coracias abyssinica)*, 44m
Serebrennikov, 471
Sericornis (Scrubwrens), 265
— *rufescens* (Arfale buff-faced scrub wren), 265
Sericulus, 504
— *aureus* (Golden bird), 492*
— *chrysocephalus* (Regent bower bird), 492*, **505**
Serin (Serinus serinus), 382*, **407**, 407m
Serine finches *(Serinus)*, 404, **407**
Serinus (Serine finches), 404, **407**
— *canaria* (canary), **407**f
— *citinellus* (citril finch), **407**, 407m
— *flaviventris* (Yellow canary), **407**, 409
— *leucopygius* (White-rumped seed eaters), 407f
— *mozambicus* (Yellow-fronted canary), 382*, **407**f
— — *caniceps*, 407
— *pusillus* (Gold-fronted finch), 382*
— *serinus* (serin), 382*, **407**, 407m
— *tristriatus*, 382*
Serpophaga cinerea (River tyrannulet), 152, 152*

Serpophaginae (Tyrannulets and pygmy tyrants), 152f
Setophaga (American redstarts), 380
— *picta* (See *Myioborus pictus*)
Seven sisters *(Turdoides somervillei)*, 240
Shaft-tailed whydah *(Tetraenura regia)*, **428**
Shama *(Copsychus malabaricus)*, 279*, **291**
Sharland, Michael, 484
Sharp-beaked ground finch *(Geospiza difficilis)*, **359**, 363f, 364*, 367*
Sharpbills (Oxyruncidae), 127, **157**, 157m
Sharp-crested woodpeckers *(Phloeoceastes)*, 117m, 117
Sharp-tailed munia *(Lonchura striata)*, **448**
Sharp-tailed sparrows *(Amnospiza)*, **354**
Sharp-tailed stream-creeper *(Lochmias nematura)*, 125*
Shining cowbird *(Molothrus bonariensis)*, 381*, **399**
Shining honey creeper *(Cyanerpes lucidus)*, **379**
Shining starling *(Aplonis metallica)*, **473**
Shiny black whyday *(Vidua hypocherina)*, **428**, 432f
Short-billed toucans *(Selenidera)*, 80
Short-legged ground roller *(Brachypteracias leptosomus)*, 43, 57*
Short-tailed antthrush *(Chamaeza campanisona)*, 137
Short-tailed ant wren *(Myrmotherula brachyura)*, 127*, **137**
Short-tailed green magpie *(Cissa thalassina)*, **510**
Short-toed lark *(Calandrella brachydactyla)*, 174f, 174m, 179*
Short-toed tree creeper *(Certhia brachydactyla)*, 308*, **326**ff, 326m
Shortwings *(Brachypteryx)*, 285
Shrike babblers *(Pteruthius)*, 244
Shrike vireos (Vireolaniinae, *Virelanius)*, 348, 395, 395m
Shrikes (Laniidae), 207ff
Shrikes proper (Lanius), 209ff
Sialia (Bluebirds), 292
— *sialis* (Eastern bluebird), 221*, 270*, **292**f
Siberian accentor *(Prunella montanella)*, 230
Siberian flycatcher *(Muscicapa sibirica)*, **276**
Siberian ground thrush *(Zoothera sibirica)*, 300
Siberian jay *(Perisoreus infaustus)*, 497*, **509**f, 509m
Siberian meadow bunting *(Emberiza cioides)*, 341*
Siberian nutcracker *(Nucifraga caryacatactes macrorhynchos)*, 512f
Siberian tit or gray-headed chickadee *(Parus cinctus)*, 307*, 317, **318**
Sibias (Heterophasia), 245
Sicalis (Yellow finches), 357
— *flaveola* (Saffron finch), 342*, **357**
Sicklebill or falculea *(Falculea palliata)*, 213f
Sickle-billed birds of paradise *(Epimachus)*, 499
Sickle-tailed pitta *(Pitta phayrii)*, 142
Sielmann, Heinz, 91f, 365, 504
Sigmodus retzii (Retz's red-billed shrike), 196*
Siivonen, 216
Silver-eared mesia *(Leiothrix argen-*

tauris), 222*, **244**
Silver-beaked tanager *(Ramphocelus carbo)*, 361*, **375**
Silver-bill (African) (See African Silver-bill)
Silver-throated tanager *(Tangara chrysophrys)*, **374**
Singing cisticola *(Cisticola cantans)*, **252**
Singing honey eater *(Meliphaga virescens)*, 345f
Singing starling *(Aplonis cantoroides)*, 449*
Siskin *(Carduelis spinus)*, 391*, 404f, **409**, 409m
Sitagra luteola (Little weaver), 430*
Sitta (Nuthatches)
— *azurea* (Azure nuthatch), 323
— *bangsi* (Kansu nuthatch), 323
— *canadensis* (Red-breasted nuthatch), 319*, **323**, 323m
— *carolinensis* (White-breasted nuthatch), 319*, **323**, 323m
— *europaea* (European nuthatch), 301*, 319*, **321**f, 322*, 322m
— — *castanea* (Chestnut nuthatch), 319*, **322**, 322m
— *formosa* (Beautiful nuthatch), 323
— *frontalis* (Velvet-fronted nuthatch), 319*, **323**
— *himalayensis* (White-tailed nuthatch), 323
— *krueperi* (Turkish nuthatch), 319*, **323**
— *leucopsis* (White-cheeked nuthatch), 323
— *magna* (Giant nuthatch), 323
— *neumayer* (Rock-nuthatch), 319*, 322m, **323**
— *pusilla* (Brown-headed nuthatch), 319*, **323**
— *tephronota* (Eastern rock nuthatch), 322m, **323**
— *villosa* (Korean nuthatch), 323
— *whiteheadi* (Corsican nuthatch), 319*, **323**
— *yünnanensis* (Yunnan nuthatch), 323
Sittasomus griseicapillus (Olivaceous woodcreeper), 120*
Sittidae (True nuthatches), 321f
Sittinae (Nuthatches proper), 310, 321
Skead, 340
Skutch, Alexander, 63, 65f, 70, 105, 156, 224, 370, 385, 395
Skylark *(Alauda arvensis)*, 169*, **177**ff, 177m, 179*
Slate-colored junco *(Junco hyemalis)*, 353, 356
Slaty blue flycatcher *(Ficedula tricolor)*, 275
Slaty flower piercer *(Diglossa biritula)*, 362*, 379*, **380**
Slender-billed scimitar babbler *(Xiphirhynchus superciliari)*, 235ff
Small ground finch *(Geospiza fuliginosa)*, **359**, 363f, 363*
Small insectivorous tree finch *(Camarhynchus parvulus)*, 351*, **360**, 365*
Small kiskadee flycatchers *(Myiozetetes)*, 151
Small sharp-beaked ground finch *(Geospiza difficilis acutirostris)*, 364, 366m
Small sunbird *(Nectarinia minima)*, **338**
Smith, H. C., 235
Smith, Stuart, 197
Smithornis (African broadbills), 123, 127
— *capensis* (Cape broadbill), 120*,

124
— *rufolateralis* (Red-flanked broadbill), 124
Snethlage, E. H., 396
Snow bunting *(Plectrophenax nivalis)*, 342*, **354**f, 354m
Snow buntings *(Plectrophenax)*, 353
Snow finch *(Montifringilla nivalis)*, 423
Snow finches *(Montifringilla)*, 423
Snowy-headed robin chat *(Cossypha niveicapilla)*, **287**f
Sociable weaver *(Philetairus socius)*, **443**, 444m, 444*
Society finch, 448f
Solitaires *(Myadestes)*, 285, 294
Sombre tit *(Parus lugubris)*, **317**f
Song babblers, i.n.s. *(Turdiodes)*, 240f
Song shrikes (Cracticidae), 213, 482, **485**f
Song sparrow *(Melospiza melodia)*, 342*, **353**, 356f, 356*
Song thrush *(Turdus philomelos)*, 290*, **304**f, 304m, 306*, 413*
Songbirds (Oscines), 121, **167**f
Sooty-headed babbler *(Malacopteron affine)*, 222*
South African quail finch *(Ortygospiza atricollis)*, 440*, **456**, 456m
South American carib grackle *(Quiscalus lugubris)*, **400**
South Sea swallow *(Hirundo tahitica)*, 187
South ant pipit *(Corythopis delalandi)*, 140
Southern black tit *(Parus niger)*, 307*, **318**, 318m
Southern carmine bee-eater *(Merops nubicoides)*, **40**
Southern chowchilla (Orthonyx temminckii), 222*
Southern emu-wren *(Stipiturus malachurus)*, 259*, **265**f
Southern figbird *(Sphecotheres vieilloti)*, **477**
Southern yellow robin *(Eopsalria australis)*, 241*
Spanish sparrow *(Passer domesticus hispaniolensis)*, 416*, **425**f, 425m
Spangled drongo *(Dicrurus hottentottus)*, 450*, **477**
Sparrow weavers (Plocepasserinae), 422, **443**
Sparrows (Passerinae), 422f
Speckled tinker-bird *(Pogoniulus scolopaceus)*, **74**
Speckled-fronted weaver *(Sporopipes frontalis)*, **443**f, 445m
Spectacled antbird *(Phlegopsis nigromaculata)*, 126*, **137**
Spectacled finch *(Lonchura spectabilis)*, **448**, 451
Spectacled warbler *(Sylvia conspicillata)*, **258**, 221m, 262
Speculipastor bicolor (Magpie starling), 473f
Speirs, D. H., 420
Spelaeornis, 236
— *chocolatinus* (Long-tailed wren babbler), **238**
— *troglodytoides* (Barred-wing wren babbler), **238**
Spermestes (Mannikins), 447f
— *bicolor* (Black-and-white mannikin), **447**, 447m
— *cucullatus* (Bronze mannikin), 439*, **447**, 448m
— *fringilloides* (Magpie mannikin), 447f
— *nanus* (Bib finch), **447**f, 448m
Spermophaga haematina (Blue-bill), **461**

Sphecotheres (Figbirds), 477
— *flaviventris* (Yellow figbird), 450*, **477**
— *faeilloti* (Southern figbird), **477**
Sphenocichla humei (Wedge-billed wren), 236
Sphenoeacus afer (Cape grass bird), **251**
Sphyrapicus thyroideus (Williamson's sapsucker), **106**
— *varius* (Yellow-bellied sapsucker), 92*, 94*, **106**, 110*, 119*
— — *daggetti*, 106, 106m, 109m
— — *nuchalis*, 106, 106m, 109m
— — *ruber* (Red-bellied sapsucker, **106**, 106m, 109m
— — *varius*, 106, 106m, 109m
Spice finch (*Lonchura punctulata*), 439*, **448**, 448m
Spider-hunters (*Arachnothera*), 332f, **338**f
Spine-tails (*Synallaxis*), 131
Spinus (See *carduelis*)
Spiny babbler (*Turdoides nipalensis*), **240**f
Spizixos canifrons (Finch-billed bulbul), 204
— *semitorques cinereicapillus*, 190*
Spizella (Chipping sparrows)
— *arborea* (Tree sparrow), 342*
— *passerina* (Chipping sparrow), 353f, 357
Splendid glossy starling (*Lamprotornis splendidus*), **475**
Spodiopsar cineraceus (Gray starling), **468**
Sporophila (Seed eaters), 357f
— *americana* (Variable seed eater), 357f
— *minuta* (Ruddy-breasted seed eater), 342*
Sporopipes frontalis (Speckle-fronted weaver), **443**f, 445m
— *squamifrons* (Scaly weaver), 430*, **444**, 445m
Sporopipinae (Scaly weavers), 422, **443**f
Spot-billed toucanet (*Selenidera maculirostris*), 80, 107*
Spot-crowned woodcreeper (*Lepidocolaptes affinis*), **128**
Spotless starling (*Sturnus unicolor*), **465**
Spotted antbird (*Hylophylax naevia*), 137
Spotted berry pecker (*Rhamphocharis crassirostris*), 330
Spotted bower bird (*Chlamydera maculata*), 492*, **505**, 505*
Spotted creepers (Salpornithidae), 310, **325**f
Spotted flycatcher (*Muscicapa striata*), 260*, 274m
Spotted Parana shrike (*Mackenziaena leachii*), 132
Spotted pardalote (*Pardalotus punctatus*), **330**
Spotted quail thrush (*Cinclosoma punctatum*), **248**
Spotted rhegmatorhina (*Rhegmatorhina melanosticta*), 137
Spotted woodpecker (*Campethera nubica nubica*), 108*
Spotted woodpeckers, 109f
Spotted-breasted oriole (*Icterus pectoralis*), 381*
Spotted winged stare (*Saroglossa spiloptera*), **475**
Spreo bicolor (Pied starling), **474**
Spring, L. W., 91
Stachyris (Tree babblers), 238f
— *chrysaea* (Golden-headed babbler), **239**

— *rufi* (Hume's tree babbler), 222*
Stactolaema leucotis (White-eared barbet), **73**
Stagonopleura
— *bella* (Fire-tailed finch), **455**
— *guttata* (Diamond sparrow), 439*, **455**, 455m
— *oculata* (Red-eared fire-tailed finch), **455**
Star finch (*Bathilda ruficauda*), **455**
Starlings (Sturnidae), 463ff
Steer's pitta (*Pitta steerii*), **142**, 146, 149*
Steganura (Whydahs, i.n.s.), 427f, 432*, 431ff
— *obtusa* (Broad-tailed paradise whydah), 429*, 433*
— *orcuitatis*, 427*
— *paradisea* (Paradise whydah), **427**, 429*, 432f, 433*
— *togoensis*, 426*
Steinbacher, Georg, 186, 408
Steinfatt, Otto, 88, 110
Stelgidopteryx ruficollis (Rough-winged swallow), 180*, **186**
Steller, Georg Wilhelm, 507
Steller's jay (*Cyanocitta stelleri*), **506**f
Stephanophorus diadematus (Diademed tanager), 361*
Stephen Island wren (*Traversia lyalli*), **148**, 148*
Stephen Island wrens (*Traversia*), 147f
Steppe weavers (*Textor*), 438ff, 441*, 442*
Stipiturus (Emu-wrens), 265
— *malachurus* (Southern emu-wren), 259*, **265**f
Stizoptera bichenovii (Biceno's finch), **454**, 454m
— — *annulosa*, 454m
— — *bichenovii*, 454m
Stizorhina fraseri (Rufous flycatcher), **294**
Stoliczka's tree creeper (*Certhia nipalensis*), **326**f
Stonechat (*Saxicola torquata*), 280*, 295f, 295m, 296*
Stonor, C. A., 55
Stork-billed kingfisher (*Pelargopsis capensis*), **28**
Stork-billed kingfishers (*Pelargopsis*), 28
Streaked fan-tailed warbler (*Cisticola exilis*), 251f, 251*
Streaked or Temminck's grasshopper warbler (*Locustella lanceolata*), **253**, 256
Streaked xenops (*Xenops rutilans*), 125*
Strepera (Currawongs), 485f
— *graculina* (Pied currawong), 469*, **486**
— — *fuliginosa*, 486
— — *versicolor* (Gray currawong), **486**
Streptocitta albertinae, **473**
— *albicollis*, **473**
Stresemann's bush crow (*Zavattariornis stresemanni*), **512**, 512m
Striated earthcreeper (*Upucerthis serrana*), 125*
Striped bower bird (*Amblyornis subalaris*), 492*, **503**, **504**, 504*
Striped buntings (*Fringillaria*), 350f
Striped flower pecker (*Dicaeum agile*), **331**
Striped jungle babbler (*Pellorneum ruficeps*), 222*, **235**
Striped kingfisher (*Halcyon chelicuti*), **28**
Striped swallow (*Hirundo abyssinica*), **187**

Striped wren babbler (*Kenopia striata*), **236**
Striped-headed creeper (*Rhabdornis mystacalis*), 319*, **325**
Stripe-throated bulbul (*Pycnonotus finlaysoni*), 200f
Strong-billed woodpecker (*Xiphocolaptes promeropirhynchus*), 120*
Struthidea cinerea (Apostle bird), 469*, **484**
Sturnella (Meadowlarks), 397, 401
— *loyca* (Long-tailed meadowlark), 381*
— *magna* (Eastern meadowlark), 401*
— *neglecta* (Western meadowlark), 381*, **401**
Sturnia
— *philippensis* (Violet-backed starling), **471**
— *sinensis* (Mandarin mynah), **471**
— *sturnia* (Starlet), 471
— *sturnina* (Daurian mynah), 449*
Sturnidae (Starlings), 463ff
Sturninae (True starlings), 463
Sturnopastor contra (Pied mynah), **468**
Sturnus unicolor (Spotless starling), **465**
— *vulgaris* (Common starling), **464**ff
Subalpine warbler (*Sylvia cantillans*), **258**
Suboscines (Primary songbirds or lyre-bird relatives), 118, 123, 164ff
Sudhaus, W., 176
Sugarbirds (Promeropinae), 344, **347**
Sulphur-bellied flycatcher (*Myiodynastes luteiventris*), 151, 154
Sulphur-rumped flycatcher (*Myiobius sulphureipygius*), 152, 154*
Sultan tit (*Melanochlora sultanea*), 308*, **321**
Summer tanager (*Piranga rubra*), 361*, **375**f
Sunda whistling thrush (*Myiophoneus glaucinus*), **299**
Superb bird of paradise (*Lophorina superba*), 470*, 495*, **496**
Superb blue wren (*Malurus cyaneus*), 266
Superb glossy starling (*Lamprospreo superbus*), **474**
Superb lyrebird (*Menura novaehollandiae*), 134/135*, **164**f, 164m
Superb warblers, 266
Svärdson, G., 409
Swainson's thrush (*Hylocichla ustulata*), **303**
Swainson's toucan (*Ramphastos swainsonii*), 85f
Swallow tanager (*Tersina viridis*), 362*, **377**f
Swallow tanagers (Tersinini), 370, 377m
Swallows (Hirundinidae), 182ff
Swallow-tailed cotinga (*Phibalura flavirostris*), 144*
Swallow-tailed manakin (*Chiroxiphia caudata*), 143*, **158**ff, 160*
Swallow-wing (*Chelidaptera tenebrosa*), **65**, 68*
Swamp warblers (*Calamocichla*), 253
Swamp weavers (Amblyospizinae), 396, 422, **442**
Sydney waxbill (*Aegintha temporalis*), 439*, **455**
Sylvia
— *atricapilla* (Blackcap), 232*, **258**, 261, 413*

— *borin* (Garden warbler), 232*, **258**, 258m, 261
— *cantillans* (Subalpine warbler), **258**
— *communis* (White-throat), 232*, **258**, 261f, 262*
— *conspicillata* (Spectacled warbler), **258**, 261m, 262
— *curruca* (Lesser), 232*, **258**, 262
— *deserticola* (Tristram's warbler), **258**
— *hortensis* (Orphean warbler), **258**, 261
— *melanocephala* (Sardinian warbler), **258**, 262, 262m
— *nana* (Desert warbler), **258**
— *nisoria* (Barred warbler), 232*, **258**
— *rueppelli* (Rueppel's warbler), **258**, 262, 262m
— *sarda* (Marmora's warbler), **258**, 262m
— *undata* (Dartford warbler), **258**, 262, 262m
Sylvietta (Crombecs), 250
— *rufescens* (Long-billed crombec), 248m, 249*, **250**f
Sylviinae (Old World warblers), 233, **249**f
Sylviorthorhynchus desmursii (Des Mur's wiretail), 125*
Sylviparus modestus (Yellow-browed tit), **321**
Symplectes bicolor (Forest weaver), **442**
Synallaxis (Spine-tails), 131
— *albescens* (Pale-breasted spinetail), 125*
— *ruficapilla* (Red-capped spinetail), **131**
Syndactylism, 21
Syrian woodpecker (*Dendrocopos syriacus*), **109**, 112f, 112m, 119*
Szlivka, L., 112

Tachuris rubrigastra (Many-colored rush tyrant), 151
Tachycineta (See *Iridoprogne*)
Tachycineta albilinea bicolor (Tree swallow), 180*, **186**
Tachyphonus cristatus (Flame-crested tanager), 361*
Taeniopygis guttata (Zebra finch), 453f, 453m
— — *castanotis*, 439*, 453, 453m
— — *guttata*, 453, 453m
Taha, 433
— *afra* (Yellow-crowned bishop), **436**
— *diademata* (Fire-fronted bishop), 430*, **436**
Taiga grasshopper warbler (*Locustella fasciolata*), **253**, 256
Tailor bird (*Orthotomus sutornis*), 259*, **265**, 265*
Tailor birds, 250, **265**
Tanager fastuosa (Seven-colored tanager), 335*
Tanagra (See *Euphonia*)
Tangara
— *chilensis* (Paradise tanager), 361*, **373**
— *chrysophrys* (Silver-throated tanager), **374**
— *inornata* (Plain-colored tanager), 65, 68*
— *nigrocincta* (Masked tanager), 361*, **374**
— *parzudakii* (Flame-faced tanager), 361*
— *seledon* (Green-headed tanager), **373**
Tanner, J. T., 116

INDEX 645

Tanysiptera sylvia (Caroline racket-tail), 34, 34m
Tapaculos (Rhinocryptidae), 127, **140f**, 140m
Taraba major (Great ant shrike), 126*
Tarictic hornbill *(Penelopides panini),* 55
Tarsiger cyanurus (Orange-flashed bush robin), 279*
Tasmanian gray-breasted silver eye *(Zosterops lateralis lateralis),* 343
Tawny pipit *(Anthus campestris),* 189*, **193**, 197m
Tawny-flanked prinia *(Prinia subflava),* 251
Tawny-winged woodcreeper *(Dendrocincla anabatina),* **128**
Tchagra australis tchagra (Redwing shrike), 195*, **208**
Teleonema filicauda (Wire-tailed manakin), 143*
Temenuchus malabaricus (Malabar mynah), **471**
— *pagodarum* (Black-headed starling), 449*, **471**
Temnurus temnurus (Racket-tailed tree pie), **511**
Tephrodornis, 198
— *gularis* (Hook-billed graybird), 198
Termite woodpecker *(Campethera nivosa),* **99**, 99m
Terpsiphone
— *atrocaudata* (Japanese paradise flycatcher), 260, **282**f
— *paradisi* (Asiatic paradise flycatcher), 260, **282**f, 282*
— *viridis* (African paradise flycatcher), **282**f
Tersina viridis (Swallow tanager), 362*, **377**f
Tersinini (Swallow tanagers), 370, 377m
Tetraenura, 428
— *fischeri* (Fischer's whydah), **428**, 429*
— *regia* (Shagt-tailed whydah), **428**
Textor (Steppe weavers), **438**ff, 441*, 442*
— *badius* (Cinnamon weaver), 414*
— *capensis* (Cape weaver), 430*, **441**, 442m
— *castaneiceps* (Chestnut-headed weaver), 459*
— *cucullatus* (Black-headed village weaver), 430*, **441**f
— *galbula* (Rüppell's weaver), **441**, 442m
— *grandis* (Giant weaver), **441**
— *jacksoni*, **441**
— *melanocephalus* (Black-headed weaver), **441**
— *subaureus* (Golden weaver), 415*
— *velatus* (Black-fronted weaver), **441**
— *xanthops*, **441**
Thamnophilus (Ant shrikes), 132
— *caerulescens* (Variable ant shrike), 126*
— *doliatus* (Barred ant shrike), 126*, **132**, 138*
Theil, August, 25
Thecla lark *(Galerida theclae),* 179*
Thick-billed lark *(Ramphocorys clotbey),* **174**, 174m, 179*
Thick-billed raven *(Corvus crassirostris),* 498*, **519**, 521
Thick-billed warbler *(Acrocephalus aedon),* **253**, 255
Thickheads (Pachycephalinae), 272, **283**f, 283m

Thielke, 328
Thomasophantes sanctithomae, 442
Thompson, D. Nethersole, 355
Thornbills or thornbill warblers *(Acanthiza),* 265
Thornbirds, i.n.s. *(Phacellodomus),* 131
Thorpe, 407
Thrashers *(Toxostoma),* 227
Thraupinae, 370ff, 373m
Thraupini (True tanagers), **370**f
Thraupis
— *bonariensis* (Blue-and-yellow tanager), 335*, **375**
— *episcopus* (Blue-gray tanager), 361*, **374**
— *palmarum* (Palm tanager), **374**
— *sayaca* (Sayaca tanager), **374**
Three-colored mannikin *(Lonchura malacca),* **448**, 451m
Three-toed jacamar *(Jacamaralcyon tridactyla),* **63**, 68*
Three-toed kingfishers *(Ceyx),* **26**
Three-toed piculet *(Sasia abnormis),* **89**
Three-toed woodpeckers *(Picoides),* 114f, 115m
Three-wattled bellbird *(Procnias tricarunculata),* 144*, **161**, 163, 163*
Thrush babbler *(Ptyrticus turdinus),* **235**
Thrush nightingale *(Luscinia luscinia),* 279*, **286**, 286m
Thrushes (Turdinae), 272, **284**f, 284m
Thryothorus ludovicianus (Carolina wren), 190*, **223**, 225f
— *nigricapillus* (Black-capped wren), **224**f
Tiardini, 357
Tiaris (Tiaris grassquits), **358**
— *canora* (Cuban grassquit), **358**
— *olivacea* (Yellow-faced grassquit), 342*, **358**, 358*
Tibet snow finch *(Montifringilla adamsi),* **423**f, 423m
Ticehurst, C. B., 243
Tichodroma muraria (Wall creeper), 319*, **323**f
Tichodrominae (Wall creepers), 323
Tick birds or ox-peckers (Buphaginae), 463, **475**f
Tickell, 59
Tickell's flower pecker *(Dicaeum erythrorhynchos),* **331**
Tickell's hornbill *(Ptilolaemius tickelli),* 55
Timalia pileata (Red-capped babbler), 222*, **239**
Timalinae (Babbling thrushes or babblers), **233**f, 234m
Timaliini (Tit babblers), 234, **238**f
Tinker honey guide *(Melignomon zenkeri),* 77f
Tinker-birds *(Pogoniulus),* 73f
Tit babbler *(Parisoma subcaeruleum),* **276**
Tit babblers (Timaliini), 234, **238**f
Tit berry pecker *(Oreocharis arfaki),* **330**
Tit flycatchers (African) (See African tit flycatchers)
Tityra, 161f
— *cayana* (Black-tailed tityra), 144*
— *semifasciata* (Masked tityra), 161
Tobin, 453
Tockus (Tokos), 53f
— *deckeni* (Von der Decken's hornbill), 32*, **54**, 54m, 58*
— *erythrorhynchus* (Red-billed hornbill), 32*, **53**f, 54m, 58*

— — *damarensis*, 54m
— — *erythrorhynchus*, 54m
— — *rufirostris*, 54m
— *flavirostris* (Yellow-billed hornbill), **53**
— *jacksoni*, 54
Tocotoucan *(Rhamphastus toco),* 81*, **85**, 98*
Todidae (Todies), 21, **34**f
Todies (Todidae), 21, **34**f
Todirostrum cinereum (Common tody flycatcher), 143*, **152**
Todus angustirostris (Narrow-billed tody), **34**
— *mexicanus* (Puerto Rican tody), **34**
— *multicolor* (Cuban tody), **34**
— *subulatus* (Broad-billed tody), **34**, 48*
— *viridis* (Jamaican tody), **34**
Tody flycatchers (Euscarthminae), 152
Tokos *(Tockus),* 53f
Tolmomyias sulphurescens (Yellow-olive flycatcher), **152**, 154*, 156f
Tomtit *(Petroica macrocephala),* **278**
Tooth-billed bower bird *(Scenopoetes dentirostris* 492*, **503**
Torrent lark *(Grallina bruijni),* **484**
Toucan barbet *(Semnornis ramphastinus),* **66**, 97*
Toucans (Rhamphastidae), 63, **79**ff, 79m
Townsend's solitare *(Myadestes townsendii),* 289*, **294**
Toxostoma (Thrashers), 227
— *redivivum* (California thrasher), **227**f, 227*
— *rufum* (Brown thrasher), 221*
Trochyphonus
— *darnaudii* (D'Arnaud's barbet), **74**f
— *erythrocephalus* (Red-and-yellow barbet), **74**, 97*
— *margaritatus* (Yellow-breasted barbet), **74**
— *vaillantii* (Levaillant's barbet), **74**
Traversia (Stephen Island wrens), 147f
— *lyalli* (Stephen Island wren), **148**, 148*
Tree babblers *(Stachyris),* **238**f
Tree creeper *(Certhia familiaris),* 308*, **326**, 326m, 328
Tree creepers (Climacterinae), 310, **324**f
Tree kingfishers (Daceloninae), 22, **28**f
— — *(Dacelo),* 33f
Tree martin *(Petrochelidon nigricans),* 186
Tree pie (Indian) (See Indian tree pie)
Tree pies *(Dendrocitta),* **510**
Tree pipit *(Anthus trivialis),* 189*, **193**f, 194*, 194m, 197*
Tree runners (Neasittinae), 324f
Tree sparrow *(Passer montanus),* 413*, **424**f, 424m
Tree sparrow *(Spizella arborea),* 342*
Tree swallow *(Tachycineta* or *Iridoprogne bicolor),* 180*, **186**
Tree wagtail *(Dendronanthus indicus),* 189*, **194**, 198*
Tretzel, Erwin, 176
Trichastoma, 235
— *abbotti* (Abbott's babbler), **235**
— *perspecillatum* (Black-browed jungle babbler), **235**
— *rostratum* (Blyth's jungle babbler), **235**
Tricholaema
— *diadematum* (Diadem barbet), 97*

— *leucomelan* (Pied barbet), 72f
Trillers *(Lalage),* 198
Tristram's grackle *(Onychognathus tristramii),* 474
Tristram's warbler *(Sylvia deserticola),* **258**
Troglodytes aedon (House wren), **223**, 225f
— *solstitialis* (Mountain wren), **226**
— *troglodytes* (Winter wren), 190*, **223**, 225f
— — *islandicus*, 223
Troglodytidae (Wrens), 219, **223**ff
Tropical kingbird *(Tyrannus melancholicus),* 151
Tropical pewee *(Contopus cinereus),* 152
Troupial *(Icterus icterus),* 381*, **400**
True contingas *(Cotinga),* 161
True mynahs *(Acridotheres),* **471**f
True nuthatches (Sittidae), 321f
True rollers (Coraciinae), 42, **44**f
True shrikes (Laniinae), 207, **209**f
True starlings (Sturninae), 463
True swallows (Hirundininae), 182, **185**f
True tanagers (Thraupini), **370**f
True tit babblers *(Macronous),* 239
True tit (Paridae), 310, **314**f
True tree creepers (Certhiidae), 326ff
True wagtails *(Motacilla),* 194
True waxbills *(Estrilda),* **456**f
True weavers (Ploceinae), 414*, 422, **437**f
True woodpeckers (Picinae), 87, **89**ff, 90*, 91*
Trumpet bird *(Phonygammus keraudrennii),* 479*, **490**
Trumpeter bullfinch *(Rhodopechys githagineus),* **412**
Trumpeter hornbill *(Bycanistes buccinator),* 54f, 54m, 58*
Tufted titmouse *(Parus bicolor),* 308*, **318**
Turdinae (Thrushes), 272, **284**f, 284m
Turdoides (Song babblers, i.n.s.), 240f
— *gularis* (White-throated babbler), 240
— *nipalensis* (Spiny babbler), **240**f
— *somervillei* (Seven sisters), 240
— — *rufescens* (Ceylon jungle babbler), 240
Turdoidini (Song babblers), 234
Turdus (True thrushes), 248, **303**f
— *albicollis* (White-necked thrush), **304**
— *iliacus* (Redwing), 290*, **304**, 304m
— *litsipsirupa* (Ground-scraper thrush), **304**
— *merula* (Blackbird), 290*, **304**, 305m, 306f
— *migratorius* (American robin), 290*, **304**, 309
— *nudigenis* (Bare-eyed thrush), 289*
— *olivaceus* (Olive thrush), **304**
— *philomelos* (Song thrush), 290*, **304**f, 304m, 306*, 413*
— *pilaris* (Fieldfare), 290*, **304**f, 304m
— *plumbeus* (Cuban thrush), 289*
— *ruficollis* (Red-capped thrush), **304**, 305m
— — *atrogularis* (Dark-capped thrush), **304**
— — *ruficollis* (Red-throated thrush), **304**
— *rufiventris* (Rufous-bellied thrush), **304**

— *torquatus* (Ring ouzel), 290*, **304**, 305m, 309
— *viscivorus* (Mistle thrush), 290*, **304f**, 304m
Turkish nuthatch *(Sitta krueperi)*, 319*, **323**
Turnagra capensis, 483*
Turquoise-browed motmot *(Eumomota superciliosa)*, **36**, **48***
Twelve-feathered spine-tails (Asthenes), 131
Twelve-wired bird of paradise *(Seleucides melanoleuca)*, 491*, **496**
Twite *(Acanthis flavirostris)*, 391*, 404, **411**, 411m
Tylas eduardi (Gray kinkimavo), **214**
Typical buntings *(Emberiza)*, 349f
Typical flycatchers, 273
Typical warblers, 250
Tyrant flycatchers (Tyrannidae), 127, **148f**, 151m
Tyrannidae (Tyrant flycatchers), 127, **148f**, 151m
Tyranninae, 151f
Tyranniscus (Tyrannulets), 152
— *vilissimus* (Paltry tyrannulet), 152f, **154***
Tyrannoidea (Tyrant flycatchers), 127
Tyrannulets *(Tyranniscus)*, 152
Tyrannulets and pygmy tyrants (Serpophaginae), 152f
Tyrannus
— *melancholicus* (Tropical kingbird), 151
— *tyrannus* (Eastern kingbird), **143***, **151**, 153, 155*, 156
Tyrant flycatchers (Tyrannidae), 127, **148f**, 151m
Tyrant flycatchers (Tyrannoidea), 127

Ula-ai-hawane *(Ciridops anna)*, **390**, 393, 393*
Ultramarine grosbeak *(Cyanacompsa cyana)*, 352*
Undulated antpitta *(Grallaria squamigera)*, 126*
Unicolored jay *(Aphelocoma unicolor)*, 497*, **507**
Upcher's warbler *(Hippolais languida)*, **254**
Upucerthia serrana (Striated earthcreeper), 125*
Upupa epops (Eurasian hoopoe), 29*, **46f**, **57***
Upupidae (Hoopoes), 21, **45f**
Upupinae (Hoopoes proper), 45
Uraeginthus
— *angolensis* (Angola waxbill), **457**
— *bengalus* (Red-cheeked waxbill or Cordon-bleu), 440*, **457**
— *cyanocephalus* (Blue-crested waxbill), **457**
— *granatinus* (Violet-eared waxbill), 440*, **457**, 457m, 460*
— *ianthinogaster* (Purple grenadier), 429*, **431**, **457f**, 457m
Uratelornis chimaera (Long-tailed ground roller), **43f**
Urocissa caerulea (Blue magpie), **510**
— *erythrorhyncha* (Red-billed blue magpie), **510**
— *flavirostris* (Yellow-billed blue magpie), **510**
— *ornata* (Ceylon magpie), 497*, **510**
— *whiteheadi* (Whitehead's blue magpie), **510**
Urolestes (Magpie shrikes), 209
— *melanoleucus* (Magpie shrike), 195*, **212f**

Van Someren, V. D., 45, 72ff, 204, 434, 456
Vanellus vanellus (Lapwing), 413*
Vanga curvirostris (Hook-billed vanga), 196*, **213f**
Vangas (Vangidae), 207, **213f**
Vangidae (Vangas), 207, **213f**
Variable ant shrike *(Thamnophilus caerulescens)*, 126*
Variable seed eater *(Sporophila americana)*, 357f
Varied bunting *(Passerina versicolor)*, **367**
Varied thrush *(Zoothera naevia)*, **300**, 303
Varied tit *(Parus varius)*, 308*, **318**
Varied triller *(Lalage leucomela)*, 198
Variegated wren *(Malurus lamberti)*, 259*, **266**
Vaurie, C. H., 447
Vegetarian tree finch *(Platyspiza crassirostris)*, 351*, **359**, 364, 364*
Velvet asity *(Philepitta castanea)*, 146*, **147**, **150***
Velvet-fronted nuthatch *(Sitta frontalis)*, 319*, **323**
Verdin *(Auriparus flaviceps)*, 308*, **313f**
Verheyen, 212
Vermillion flycatcher *(Pyrocephalus rubinus)*, **143***, **151**, 155
Vermillion-crowned flycatcher *(Myiozetetes similis)*, **151**, 153*, 155f
Vermivora, 380
— *chrysoptera* (Golden-winged warbler), 371*, **384**, 384m
— *pinus* (Blue-winged warbler), **383f**, 384m
Verreauxia africana (African piculet), **89**
Vestiaria coccinea (Iiwi), 372*, 388, **390**, 393*, 393f
Vidua
— *hypocherina* (Shiny black whydah), **428**, 432f
— *macroura* (Pin-tailed whydah), **428**, 432
Viduinae (Whydahs), 422, **427**, 431ff, 432*
Viellot's barbet *(Lybius vieilloti)*, 75
Vietinghoff-Riesch, Arnold Baron, 187
Vineous-throated parrotbill *(Paradoxornis webbiana)*, **247**
Violet-backed starling *(Cinnuricinclus leucogaster)*, **474**
Violet-backed starling *(Sturnia philippensis)*, **471**
Violet-eared waxbill *(Uraeginthus granatinus)*, 440*, **457**, 457m, 460*
Vireo (Vireos, i.n.s.), 395
— *flavifrons* (Yellow-throated vireo), 372*
— *gilvus* (Warbling vireo), 395
— *olivaceus* (Red-eyed vireo), 372*, **395**, 395*
Vireolaniinae (Shrike vireos), 348, 395
Vireolanius (Shrike vireos), 395, 395m
— *melitophrys* (Chestnut-sided shrike vireo), 395f
— *pulchellus* (Green shrike vireo), 372*
Vireonidae (Vireos), 395
Vireoninae (Vireos), 395f
Vireos (Vireoninae), 395f
Vireos, i.n.s. (*Vireo*), 395
Viridonia or *Loxops*, 387
— *sagittirostris* (Large amakihi), 387
— *virens* (Amakihi), **387**, 388*, 389

Volatinia jacarina (Blue-black grassquit), 342*, **358**
Vonder Decken's hornbill *(Tockus deckeni)*, 32*, **54**, 54m, 58*

Wagler's oropendola *(Psarocolius wagleri)*, 381*
Wagtails (Motacillidae), 192ff
Wahlberg's honey guide *(Prodotiscus regulus)*, 77
Waigeu bird of paradise *(Diphyllodes respublica)*, 479*, **495**
Wall creeper *(Tichodroma muraria)*, 319*, **323f**
Wall creepers (Tichodrominae), 323
Wallace, Alfred Russell, 493f, 496
Wallace's standard wing *(Semioptera wallacei)*, 480*, 495*, **496**
Warbler finch *(Certhidea olivacea)*, 351*, **360**, 364, 365*
Warbling vireo *(Vireo gilvus)*, 395
Warga, 216
Warsanglia linnet *(Acanthis johannis)*, **411**
Water kingfishers (Alcedininae), 22
Water pipit *(Anthus spinoletta)*, 184*, **189***, **193**, 193m
Wattlebirds *(Anthochaera)*, 345
Wattlebirds (New Zealand) (See New Zealand wattlebirds)
Wattled cuckoo-shrike *(Campephaga lobata)*, **198f**
Wattled starling *(Creatophora cinerea)*, 449*, **472**
Wattle-eyed flycatchers *(Platysteira)*, 277
Waxwings (Bombycillidae), 207, **215f**
Weaverbirds (Ploceidae), 421ff
Weaverfinches (Estrildidae), 421, **444ff**
Wedge-billed woodcreeper *(Glyphorhynchus spirurus)*, 127*
Wedge-billed wren *(Sphenocichla humei)*, 236
Western flycatcher *(Empidonax difficilis)*, 143*
Western grass wren *(Amytornis textilis)*, 266
Western meadowlark *(Sturnella neglecta)*, 381*, **401**
Western scrub bird *(Atrichornis clamosus)*, **166**, 166m, 166*
Western spinebill *(Acanthorhynchus superciliosus)*, 345f
Wheatear *(Oenanthe oenanthe)*, 270*, 280*, **295**, 295m, 297, 297*
Wheatears *(Oenanthe)*, 295, 297
Whinchat *(Saxicola rubetra)*, 295f, 295m, 296*
Whistler, H., 236, 244, 299, 481
White bellbird *(Procnias alba)*, **161**, 162*, 163
White helmeted shrike *(Prionops plumata)*, 196*, **207f**
White wagtail *(Motacilla alba)*, 184*, **189***, **194**, 197f, 203m, 205*
White woodpecker *(Leuconerpes candidus)*, 119*
White-backed magpie *(Gymnorhina hyleuca)*, 469*, **486f**
White-backed woodpecker *(Dendrocopos leucotos)*, **109**, 113, 113m
White-bearded manakin *(Manacus manacus)*, 143*, **158f**
White-billed sickle-billed bird of paradise *(Drepanornis bruijnii)*, 479*, **499**
White-breasted kingfisher *(Halcyon smyrnensis)*, 28f, 32*, 38*

White-breasted nuthatch *(Sitta carolinensis)*, 319*, **323**, 323m
White-breasted swallow shrike *(Artamus leucorhynchus)*, **485**
White-browed flycatcher *(Ficedula superciliaris)*, **275**
White-browed wood swallow *(Artamus superciliosus)*, 196*, **485**
White-capped dipper *(Cinclus leucocephalus)*, **220**
White-capped redstart *(Chaimarrornis leucocephalus)*, 279*, **291**
White-cheeked nuthatch *(Sitta leucopsis)*, 323
White-collared kingfisher *(Halcyon chloris)*, **28**, 37*
White-collared olive-back *(Nesocharis ansorgei)*, **461f**
White-collared starling *(Grafisia torquata)*, **474**
White-crested hornbill *(Berenicornis comatus)*, **55**
White-crested laughing thrush *(Garrulax leucolophus)*, 222*, **243**
White-crowned back wheatear *(Oenanthe leucopyga)*, **295**, 297f
White-crowned shrike *(Eurocephalus auguitimens)*, 196*, **208**
White-eared barbet *(Stactolaema leucotis)*, **73**
White-eared catbird *(Ailuroedes buccoides)*, **503**
White-eared jacamar *(Galbalcyrhynchus leucotis)*, **63**
White-eyed river martin *(Pseudochelidon siantarae)*, **185**
White-eyed starling *(Aplonis brunneicapilla)*, **473**
White-eyes (Zosteropidae), 339f
White-eyes proper *(Zosterops)*, 339ff
White-faced ant catcher *(Pithys albifrons)*, 126*, **137**
White-fronted ground tyrant *(Muscisaxicola albifrons)*, 151, 153
White-fronted lark *(Eremopterix nigriceps)*, **172**, 172m
White-gaped honey eater *(Meliphaga unicolor)*, 346
Whitehead, J., 146, 199, 326
White-headed babbler *(Gampsorhynchus rufulus)*, **245**
White-headed barbet *(Lybius leucocephalus)*, **71f**
White-headed buffalo weaver *(Dinemellia dinemelli)*, 430*, **442f**
White-headed mannikin *(Lonchura maja)*, 448
White-headed marshtyrant *(Arundinicola leucocephala)*, 143*, **151**
White-headed rough-winged swallow *(Psalidoprocne albiceps)*, 180*
White-headed vanga *(Leptopterus viridis)*, **213**
Whitehead's blue magpie *(Urocissa whiteheadi)*, **510**
White-naped honey eater *(Melithreptes lunatus)*, 320*, 347
White-necked puffbird *(Notharchus macrorhynchus)*, 68*
White-necked raven (African) (See African white-necked raven)
White-necked rockfowl *(Picathartes gymnocephalus)*, 222*, **245f**
White-necked thrush *(Turdus albicollis)*, **304**
White-rumped butcher bird *(Cracticus lousiadensis)*, **485**
White-rumped seedeater *(Serinus leucopygius)*, **407f**

INDEX

White's thrush or golden mountain thrush (*Zoothera dauma*), 300
White-tailed nuthatch (*Sitta himalayensis*), 323
White-throat (*Sylvia communis*), 232*, **258**, 261f, **262***
White-throated babbler (*Turdoides gularis*), **240**
White-throated bee-eater (*Aerops albicollis*), **42**
White-throated bulbul (*Criniger flaveolus*), 184*
White-throated sparrow (*Zonotrichia albicollis*), 342*, **353**, 355, 355*
White-throated swallow (*Hirundo albigularis*), **187**
White-throated tapacula (*Scelorchilus albicollis*), **140**
White-throated tree runner (*Pygarrhichas albogularis*), 131f, **137***
White-tipped brown jay (*Psilorhinus morio*), **508**
White-vented bulbul (*Pycnonotus barbatus*), 190*, **204**
White-whiskered puffbird (*Malacoptila panamensis*), **64**
White-winged babbling starling (*Neocichla gutturalis*), **475**
White-winged crossbil (*Loxia leucoptera*), **418**
White-winged chough (*Corcorax melanoramphos*), 469*, **484**
White-winged lark (*Melanocorypha leucoptera*), 179*
White-winged thriller (*Lalage sueurii*), 450*
Whydahs (Viduinae), 422, **427**, 431ff, **432***
Whydahs, i.n.s. (Steganura), 427f, 431ff, **432***
Wickler, Wolfgang, 75, 172
Wied's tyrant manakin (*Neopelma aurifrons*), **158**
Williamson's sapsucker (*Sphyrapicus thyroideus*), **106**
Willie wagtail (*Rhipidura leucophrys*), 260*, **278f**
Willow tit (*Parus montanus*), 307*, **317**
Willow warbler (*Phylloscopus trochilus*), 259*, **263**, 264m, 413*
Wilson, 487
Wilsonia, 380
— *citrina* (Hooded warbler), 371*, **385**, 385m
Wing-snapping or Ayer's cloud warbler (*Cisticola ayresii*), 251f
Winter wren (*Troglodytes troglodytes*), 190*, **223**, 225f
Wire-tailed manakin (*Teleonema filicauda*), 143*
Wolley, John, 215
Wood shrikes (Pityriasinae), 207, **213**
Wood swallows or swallow shrikes (Artamidae), 482, **484f**, 485m
Wood thrush (*Hylocichla mustelina*), 221*
Wood thrushes (*Hylocichla*), 303
Wood warbler (*Phylloscopus sibilatrix*), 259*, **263f**, 263*
Wood warblers (Parulidae), 348, 380f, **383f**
Wood warblers proper (Parulinae), 380, **383f**
Woodchat shrike (*Lanius senator*), 195*, **210**
Woodcreepers (Dendrocolaptidae), 127f, 128m
Wood-hoopoes (Phoeniculinae), 45, **49**

Woodlark (*Lullula arborea*), **176**, 176m, 179*
Woodpecker finch (*Cactospiza pallida*), 333*, 351*, **360**, 363ff, 365*
Woodpeckers (Picidae), 63, 87ff
Woodpeckers (Piciformes), 62ff
Wreathed hornbill (*Rhyticeros undulatus*), 55f
Wren thrush (*Zeledonia coronata*), 289*, **386**
Wren thrushes (Zeledoniidae), 348, **386**, 387m
Wren tit (*Chamaea fasciata*), **239**
Wren tits (*Chamaeini*), 234, **239**
Wren tits (North American) (See North American wren tits)
Wren warblers (Prinia), 251f
Wren-like rushbird (*Phleocryptes melanops*), 125*
Wrens (Troglodytidae), 219, **223f**
Wrens (Australian) (See Australian wrens)
Wrens (New Zealand) (See New Zealand wrens)
Wryneck (Eurasian) (See Eurasian wryneck)
Wrynecks (Jynginae), 87, **88f**

Xanthomyza phruygia, 320*
Xenicidae (New Zealand wrens), 127, 127m, **147f**
Xenicus (Bush and rock wrens), 147f
— *gilviventris* (New Zealand rock wren), 147f, **147***
— *longipes* (Bush wren), 147f, 147*, 150*
Xenopirostris polleni (Pollen's vanga), 196*, **213f**
Xenops rutilans (Streaked xenops), 125*
Xiphirhynchus superciliaris (Slender-billed scimitar babbler), 235f
Xiphocalaptes promeropirhynchus (Strong-billed woodcreeper), 120*
Xipholena punicea (Pompadour cotinga), 144*

Yap monarch flycatcher (*Monarcha godeffroyi*), 281f
Yellow bar-throated apalis (*Apalis thoracica*), 248*, **250**
Yellow bishop (*Coliuspasser capensis*), **434**, 434m
Yellow canary (*Serinus flaviventris*), **407**, 409
Yellow cardinal (*Gubernatrix cristata*), 334*, 342*, **367**, 369m
Yellow figbird (*Sphecotheres flaviventris*), 450*, **477**
Yellow finches (*Sicalis*), 357
Yellow grosbeak (*Pheucticus chrysopeplus*), 352*
Yellow longbill (*Macrosphenus flavicans*), **250f**, 250*
Yellow wagtail (*Motacilla flava*), 184*, 189*, **194f**, 199m, 199*
Yellow warbler (*Dendroica petechia*), **384**
Yellow white-eye (*Zosterops senegalensis*), 340
Yellow-backed oriole (*Icterus chrysater*), **400**
Yellow-backed sunbird (*Aethopyga siparaja*), **338**
Yellow-bellied elaenia (*Elaenia flavogaster*), 143*, **152**
Yellow-bellied fantail flycatcher (*Chelidorynx hypoxantha*), **278**

Yellow-bellied flowerpecker (*Dicaeum melanoxanthum*), **329**
Yellow-bellied flycatcher (*Empidonax flaviventris*), **152f**
Yellow-bellied sapsucker (*Sphyrapicus varius*), 92*, 94*, **106**, 110*, 119*
Yellow-bellied waxbill (*Estrilda melanotis*), 440*, **456**, 456m
Yellow-billed blue magpie (*Urocissa flavirostris*), **510**
Yellow-billed hornbill (*Tockus flavirostris*), **53**
Yellow-billed magpie (*Pica pica nuttalli*), 511
Yellow-billed ox-pecker (*Buphagus africanus*), 449*, **475**, 475m
Yellow-billed scimitar babbler (*Pomatorhinus montanus*), **236**
Yellow-billed shrikes (*Corvinella*), **209**
Yellow-breasted barbet (*Trachyphonus margaritatus*), **74**
Yellow-breasted bird of paradise (*Loboparadisaea sericea*), **490**
Yellow-breasted bower bird (*Chlamydera lauterbachi*), **505**, 505*
Yellow-breasted chat (*Icteria virens*), **383**
Yellow-breasted tit (*Sylviparus modestus*), **321**
Yellow-browed warbler (*Phylloscopus inornatus*), 259*, **262f**
Yellow-chested apalis (*Apalis flavida*), 247m, **250f**
Yellow-crested helmet shrike (*Prionops alberti*), 208m
Yellow-crowned bishop (*Taha afra*), **436**
Yellow-crowned bulbul (*Pycnonotus zeylanicus*), 200, 203
Yellow-eyed flycatcher warbler (*Seicercus burkii*), **263**
Yellow-faced grassquit (*Tiaris olivacea*), 342*, **358**, 358*
Yellow-faced honey eater (*Meliphaga chrysops*), 347
Yellow-fronted canary (*Serinus mozambicus*), 382*, **407f**
Yellow-fronted tinker-bird (*Pogoniulus chrysoconus*), **74**
Yellowhammer (*Emberiza citrinella*), 341*, 349m, **350f**
Yellow-headed blackbird (*Agelaius icterocephalus*), 381*, **401**
Yellow-headed wagtail (*Motacilla citreola*), 194, 200m
Yellow-olive flycatcher (*Tolmomyias sulphurescens*), 152, 154*, 156f
Yellow-plumed honey eater (*Meliphaga ornata*), 346
Yellow-rumped cacique (*Cacicus cela*), 381*, **398**
Yellow-rumped flower pecker (*Prionochilus xanthopygius*), **331**
Yellow-rumped tanager (*Ramphocelus icteronotus*), **375**
Yellow-shafted flicker (*Colaptes auratus*), 95, 96m, 99*, 108*
Yellow-spotted sparrow (*Petronia pyrgita*), **423**, 423m
Yellow-streaked greenbul (*Phyllastrephus flavostriatus*), 190*, **203**
Yellow-tailed finch (*Lonchura flaviprymna*), **448**
Yellow-tailed thornbill (*Acanthiza chrysorrhoa*), 259*
Yellow-throated miner (*Manorina flavigula*), 333*
Yellow-throated sparrow (*Petronia xanthocollis*), 416*, **423**
Yellow-throated sparrow (African)

(See African yellow-throated sparrow)
Yellow-throated tinker-bird (*Pogoniulus subsulphureus*), **74**
Yellow-throated vireo (*Vireo flavifrons*), 372*
Yellow-throated woodland warbler (*Seicercus ruficapillus*), **263**
Yellow-tufted honey eater (*Phylidonyris melanops*), **345**
Yellow-tufted woodpecker (*Centurus cruentatus*), **104**
Yellow-vented bulbul (*Pycnonotus goiavier*), **203**
Yellow-winged blackbird (*Agelaius thilius*), **401**
Yellow-winged honey eater (*Phylidonyris novaehollandiae*), **346**
Yemen linnet (*Acanthis yemenensis*), 44, **411**
Yunnan nuthatch (*Sitta yunnanensis*), **323**

Zavattarjornis, 511
— *stresemanni* (Stresemann's bush crow), **512**, 512m
Zebra finch (*Taeniopygia guttata*), **453f**, 453m
Zeledonia coronata (Wren thrush), 289*, **386**
Zeledoniidae (Wren thrushes), 348, **386**, 387m
Zimmer, J. T., 145
Zonotrichia (Crowned "sparrows"), 353
— *albicollis* (White-throated sparrow), 342*, **353**, 355, 355*
— *capensis* (Rufous-collared sparrow), **353**, 355
Zoothera (Ground thrushes), **300**
— *citrina* (Orange-headed ground thrush), 221*, 289*, **300**
— *dauma* (White's thrush or golden mountain thrush), **300**
— *naevia* (Varied thrush), **300**, 303
— *sibirica* (Siberian ground thrush), 300
Zosteropidae (White-eyes), **339f**
Zosterops (White-eyes proper), 339ff
— *abyssinica* (Abyssinian white-eye), 343
— *albogularis*, 344
— *citrinella citrinella*, **339**
— *erythropleura* (Chestnut-flanked white-eye), 343
— *everetti everetti*, 344
— *griseovirescens*, 339
— *japonica* (Japanese white-eye), 343
— *lateralis* (Gray-breasted silver eye), 340m, 340, **343**
— — *gouldi*, 343
— — *lateralis* (Tasmanian gray-breasted silver eye), 343
— *maderaspatana*, 339
— *montana* (Mountain white-eye), 343, 372*
— *natalis*, 339
— *novaeguineae* (New Guinea mountain white-eye), 339
— *pallida* (Pale white-eye), 343
— — *capensis*, 340
— *palpebrosa* (Oriental white-eye), **340**, 343
— — *nilgiriensis*, 339
— *semiflava*, 344
— *senegalensis* (Yellow white-eye), 340
— *strenua*, 344
— *tenuirostris*, 344
— *wallacei*, 340

Abbreviations and Symbols

C, °C	Celsius, degrees centigrade	WL	wing length
C.S.I.R.O.	Commonwealth Scientific and Industrial Res. Org. (Australia)	BL	body length
f	following (page)	♂	male
ff	following (pages)	♂♂	males
L	total length (from tip of bill to end of tail)	♀	female
		♀♀	females
I.R.S.A.C.	Institute for Scientific Res. in Central Africa, Congo	♂♀	pair
		+	extinct
I.U.C.N.	Intern. Union for Conserv. of Nature and Natural Resources	▷	following (opposite page) color plate
		▷▷	Color plate or double color plate on the page following the next
BH	body height		
TL	tail length	▷▷▷	Third color plate or double color plate (etc.)
i.n.s.	in a narrower sense	◊	Endangered species and subspecies